Adaptive Blind Signal and Image Processing

Adaptive Blind Signal and Image Processing

Learning Algorithms and Applications

Andrzej CICHOCKI
Riken Brain Science Institute, JAPAN
and Warsaw University of Technology, POLAND

Shun-ichi AMARI
Riken Brain Science Institute, JAPAN

JOHN WILEY & SONS, LTD

Email (for orders and customer service enquiries): cs-books@wiley.co.uk
Visit our Home Page www.wileyeurope.com or www.wiley.com

Reprinted with corrections April 2003

Other Wiley Editorial Offices

John Wiley & Sons Inc., 111 River Street, Hoboken, NJ 07030, USA

Jossey-Bass, 989 Market Street, San Francisco, CA 94103-1741, USA

Wiley-VCH Verlag GmbH, Boschstr. 12, D-69469 Weinheim, Germany

John Wiley & Sons Australia Ltd, 33 Park Road, Milton, Queensland 4064, Australia

John Wiley & Sons (Asia) Pte Ltd, 2 Clementi Loop #02-01, Jin Xing Distripark, Singapore 129809

John Wiley & Sons (Canada) Ltd, 22 Worcester Road, Etobicoke, Ontario M9W 1L1

Wiley also publishes its books in a variety of electronic formats. Some content that appears in print may not be available in electronic books.

British Library Cataloguing in Publication Data

A catalogue record for this book is available from the British Library

ISBN 0 471 60791 6

Produced from PostScript files supplied by the author.
Printed and bound in Great Britain by Biddles Ltd, Guildford and King's Lynn
This book is printed on acid-free paper responsibly manufactured from sustainable forestry in which at least two trees are planted for each one used for paper production.

Contents

List of Figures

List of Tables

Preface

Signal Processing has always played a critical role in science, technology and development of new systems such as computer tomography, (PET, fMRI, EEG/MEG, optical recordings), wireless communications, digital cameras, HDTV, etc. As demand for high quality and reliability in recording and visualization systems increases, signal processing has an even more important role to play.

Blind Signal Processing (BSP) is now one of the hottest and emerging areas in Signal Processing with solid theoretical foundations and many potential applications. In fact, BSP has become a very important topic of research and development in many areas, especially biomedical engineering, medical imaging, speech enhancement, remote sensing, communication systems, exploration seismology, geophysics, econometrics, data mining, etc. The blind signal processing techniques principally do not use any training data and do not assume *a priori* knowledge about parameters of convolutive, filtering and mixing systems. BSP includes three major areas: Blind Signal Separation and Extraction, Independent Component Analysis (ICA), and Multichannel Blind Deconvolution and Equalization which are the main subjects of the book. Recent research in these areas is a fascinating blend of heuristic concepts and ideas and rigorous theories and experiments.

Researchers from various fields are interested in different, usually very diverse aspects of BSP. For example, neuroscientists and biologists are interested in the development of biologically plausible neural network models with unsupervised learning. On the other hand, they need reliable methods and techniques which will be able to extract or separate useful information from superimposed biomedical source signals corrupted by huge noise and interferences, for example, by using non-invasive recordings of human brain activities, (e.g., by using EEG or MEG) in order to understand the ability of the brain to sense, recognize, store and recall patterns as well as crucial elements of learning: association, abstraction and generalization. A second group of researchers: engineers and computer scientists are fundamentally interested in possibly simple models which can be implemented in hardware in available VLSI technology and in the computational approach, where the aim is to develop flexible and efficient algorithms for specific practical engineering and scientific applications. The third group of researchers: mathematicians and physicists, have an interest in the development of fundamental theory to understand mechanisms, properties and abilities of developed algorithms and in their generalizations to more complex and sophisticated models. The interactions among the groups make real progress in this very interdisciplinary research devoted to BSP and each group benefits from the others.

The theory built up around BSP present so extensive and applications are so numerous that we are, of course, not able to cover all of them. Our selection and treatment of materials reflects our background and our own research interest and results in this area during the last 10 years. We prefer to complement other books on the subject of BSP rather than to compete with them. The book provides wide coverage of adaptive blind signal processing techniques and algorithms both from the theoretical and practical point of view. The main objective is to derive and present efficient and simple adaptive algorithms that work well in practice for real-world data. In fact, most of the algorithms discussed in the book have been implemented in MATLAB and extensively tested. We attempt to present concepts, models and algorithms in possibly general or flexible forms to stimulate the reader to be creative in visualizing new approaches and adopt methods or algorithms for his/her specific applications.

The book is partly a textbook and partly a monograph. It is a textbook because it gives a detailed introduction to BSP basic models and algorithms. It is simultaneously a monograph because it presents several new results and ideas and further developments and explanation of existing algorithms which are brought together and published in the book for the first time. Furthermore, the research results previously scattered in many scientific journals and conference papers worldwide, are methodically collected and presented in the book in a unified form. As a result of its twofold character the book is likely to be of interest to graduate and postgraduate students, engineers and scientists working in the field of biomedical engineering, communications, electronics, computer science, finance, economics, optimization, geophysics, and neural networks. Furthermore, the book may also be of interest to researchers working in different areas of science, because a number of results and concepts have been included which may be advantageous for their further research. One can read this book through sequentially but it is not necessary since each chapter is essentially self-contained, with as few cross references as possible. So, browsing is encouraged.

Acknowledgments

The authors would like to express their appreciation and gratitude to a number of researchers who helped in a variety of ways, directly and also indirectly, in development of this book.

First of all, we would like to express our sincere gratitude to Professor Masao Ito - Director of Brain Science Institute Riken, Japan for creating a great scientific environment for multidisciplinary research and promotion of international collaborations.

Although part of this book is derived from the research activities of the two authors over the past 10 years on this subject, many influential results and well known approaches are developed in collaboration with our colleagues and researchers from the Brain Science Institute Riken and several universities worldwide. Many of them have made important and crucial contributions. Special thanks and gratitude go to Liqing Zhang from the Laboratory for Advanced Brain Signal Processing BSI Riken, Japan; Sergio A. Cruces-Alvarez from E.S. Ingenieros, University of Seville, Spain; Seungjin Choi from Computer Science and Engineering at Pohang University of Science and Technology (POSTECH); and Scott Douglas from Southern Methodist University, USA.

Some parts of this book are based on close cooperation with these and other of our colleagues. Chapters 9-11 are partially based on joint works with Liqing Zhang and they include his important contributions. Chapters 7 and 8 are influenced by joint works with Sergio A. Cruces-Alvarez and Scott Douglas. Chapter 5 is partially based on joint works with Ruck Thawonmas, Allan Barros, Seungjin Choi and Pando Georgiev. Chapters 4 and 6 are partially based on joint works with Seungjin Choi and Adel Belouchrani. Section 2.6 is devoted to the total least squares problem and is based partially on joint work with John Mathews.

We would like also to warmly thank many of our former and actual collaborators: Seungjin Choi, Sergio Cruces, Wlodzimierz Kasprzak, Liqing Zhang, Scott Douglas, Tetsuya Hoya, Ruck Thawonmas, Allan Barros, Jianting Cao, Yuanqing Li, Tomasz Rutkowski, Reda Gharieb, John Mathews, Adel Belouchrani, Pando Georgiev, Ryszard Szupiluk, Irek Sabala, Leszek Moszczynski, Krzysztof Siwek, Juha Karhunen, Ricardo Vigario, Mark Girolami, Noboru Murata, Shiro Ikeda, Gen Hori, Wakako Hashimoto, Toshinao Akuzawa, Andrew Back, Sergyi Vorobyov, Ting-Ping Chen and Rolf Unbehauen, whose contributions were instrumental in the development of many of the ideas presented here.

Over various phases of writing this book, several people have kindly agreed to read and comment on parts or all of the text. For the insightful comments and suggestions we are very grateful to Jonathon Chambers, Farid Hamzei-Sichani, Tariq Durrani, Chong-Yung Chi, Joab Winkler, Tetsuya Hoya, Wlodzimierz Kasprzak, Danilo Mandic, Yuanqing Li, Liqing Zhang, Pando Georgiev, Wakako Hashimoto, Fernando De la Torre, Allan Barros, Jagath C. Rajapakse, Andrew W. Berger, Seungjin Choi, Sergio Cruces, Jim Stone, Stanley Stansell, Carl Leichner, Khurram Waheed, and Gordon Morison.

Those whose works have had strong impact in our book, and are reflected in the text include Yujiro Inoue, Ruey-wen Liu, Sergio A. Cruces-Alvarez, Lang Tong, Scott Douglas Jean-Francois Cardoso, Yingboo Hua, Zhi Ding, Chong-Yung Chi, Jitendra K. Tugnait, Erkki Oja, Juha Karhunen, Aapo Hyvarinen, Jonathon Chambers and Noboru Murata.

Finally, we must acknowledge the help and understanding of our families during the past two years while we carried out this project.

A. CICHOCKI AND S. AMARI

October 2002, Tokyo, Japan

1

Introduction to Blind Signal Processing: Problems and Applications

The fundamental problem of communication is that of reproducing at one point either exactly or approximately a message selected at another point.

—(*Claude Shannon*, 1948)

In this book, we describe various approaches, methods and techniques to blind and semi-blind signal processing, especially principal and independent component analysis, blind source separation, blind source extraction, multichannel blind deconvolution and equalization of source signals when the measured sensor signals are contaminated by additive noise. Emphasis is placed on an information-theoretical unifying approach, adaptive filtering models and the development of simple and efficient associated on-line adaptive nonlinear learning algorithms.

We derive, review and extend the existing adaptive algorithms for blind and semi-blind signal processing with a particular focus on robust algorithms with equivariant properties in order to considerably reduce the bias caused by measurement noise, interferences and other parasitic effects. Moreover, novel adaptive systems and associated learning algorithms are presented for estimation of source signals and reduction of influence of noise. We discuss the optimal choice of nonlinear activation functions for various signals and noise distributions, e.g., Gaussian, Laplacian and uniformly-distributed noise assuming a generalized Gaussian distribution and other models. Extensive computer simulations have confirmed the usefulness and superior performance of the developed algorithms. Some of the research results presented in this book are new and are presented here for the first time.

1.1 PROBLEM FORMULATIONS – AN OVERVIEW

1.1.1 Generalized Blind Signal Processing Problem

A fairly general blind signal processing (BSP) problem can be formulated as follows. We observe records of sensor signals $\mathbf{x}(t) = [x_1(t), x_2(t), \ldots, x_m(t)]^T$ from a MIMO (multiple-input/multiple-output) nonlinear dynamical system[1]. The objective is to find an inverse system, termed a reconstruction system, neural network or an adaptive inverse system, if it exists and is stable, in order to estimate the primary source signals $\mathbf{s}(t) = [s_1(t), s_2(t), \ldots, s_n(t)]^T$. This estimation is performed on the basis of the output signals $\mathbf{y}(t) = [y_1(t), y_2(t), \ldots, y_n(t)]^T$ and sensor signals as well as some *a priori* knowledge of the mixing system. Preferably, the inverse system should be adaptive in such a way that it has some tracking capability in nonstationary environments (see Fig.1.1). Instead of estimating the source signals directly, it is sometimes more convenient to identify an unknown mixing and filtering dynamical system first (e.g., when the inverse system does not exist or the number of observations is less than the number of source signals) and then estimate source signals implicitly by exploiting some *a priori* information about the system and applying a suitable optimization procedure.

In many cases, source signals are simultaneously linearly filtered and mixed. The aim is to process these observations in such a way that the original source signals are extracted by the adaptive system. The problems of separating and estimating the original source waveforms from the sensor array, without knowing the transmission channel characteristics and the sources can be expressed briefly as a number of related problems: Independent Components Analysis (ICA), Blind Source Separation (BSS), Blind Signal Extraction (BSE) or Multichannel Blind Deconvolution (MBD) [26].

Roughly speaking, they can be formulated as the problems of separating or estimating the waveforms of the original sources from an array of sensors or transducers without knowing the characteristics of the transmission channels.

There appears to be something magical about blind signal processing; we are estimating the original source signals without knowing the parameters of mixing and/or filtering processes. It is difficult to imagine that one can estimate this at all. In fact, without some *a priori* knowledge, it is not possible to *uniquely* estimate the original source signals. However, one can usually estimate them up to certain indeterminacies. In mathematical terms these indeterminacies and ambiguities can be expressed as arbitrary scaling, permutation and delay of estimated source signals. These indeterminacies preserve, however, the waveforms of original sources. Although these indeterminacies seem to be rather severe limitations, in a great number of applications these limitations are not essential, since the most relevant information about the source signals is contained in the waveforms of the source signals and not in their amplitudes or order in which they are arranged in the output of the system. For some dynamical models, however, there is no guarantee that the estimated or extracted signals have exactly the same waveforms as the source signals, and then the requirements

[1] Single-input single-output (SISO) or single-input/multiple-output (SIMO) are special cases.

(a)

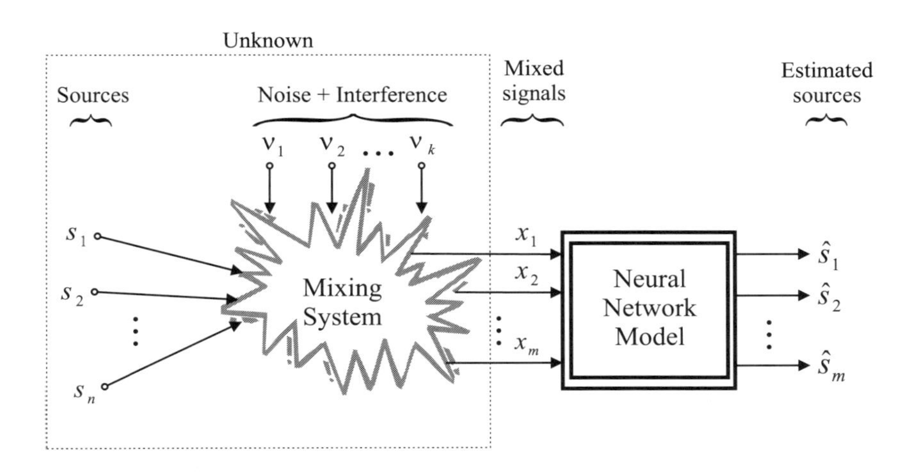

(b)

Fig. 1.1 Block diagrams illustrating blind signal processing or blind identification problem: (a) General schema, (b) nonlinear model with additive noise.

must be sometimes further relaxed to the extent that the extracted waveforms are distorted (filtered or convolved) versions of the primary source signals [169, 1271] (see Fig.1.1).

We would like to emphasize the essential difference between the standard inverse identification problem and the blind or semi-blind signal processing task. In a basic linear identification or inverse system problem we have access to the input (source) signals (see Fig.1.2 (a)). Our objective is to estimate a delayed (or more generally smoothed or filtered) version of the inverse system of a linear dynamical system (plant) by minimizing the mean square error between the delayed (or model-reference) source signals and the output signals.

(a)

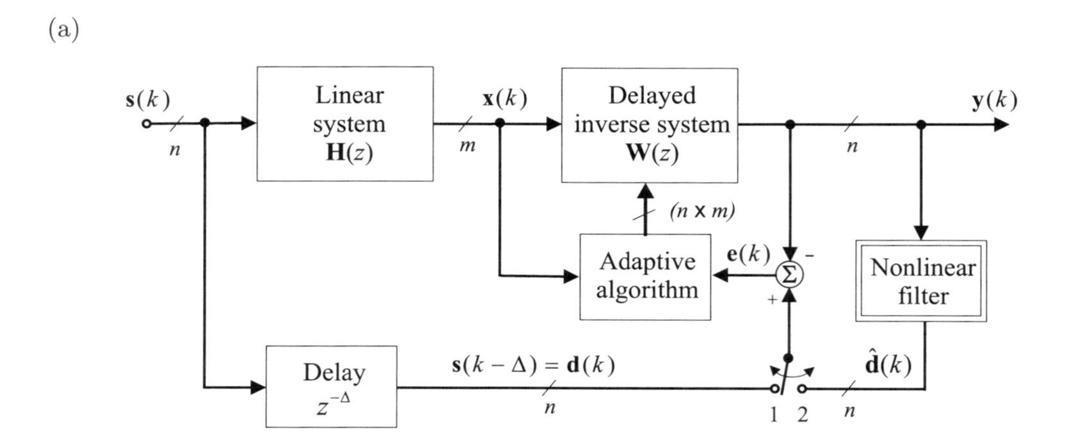

(b)

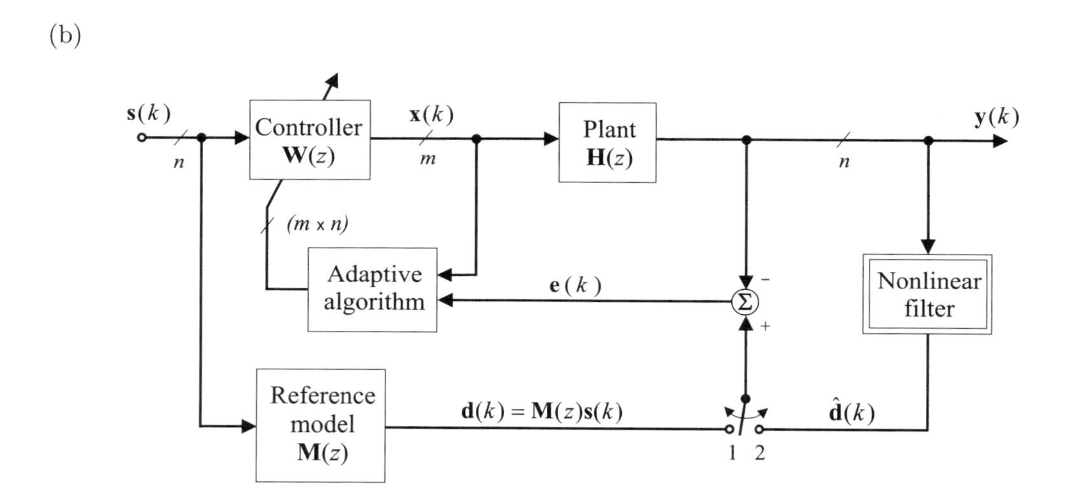

Fig. 1.2 (a) Conceptual model of inverse system problem. (b) Model-reference for adaptive inverse control. For the switch in position 1, the system performs a standard adaptive inverse by minimizing the norm of error vector $\mathbf{e}$, for switch in position 2, the system estimates errors blindly.

In BSP problems we do not have access to source signals (which are usually assumed to be statistically independent), so we attempt, for example, to design an appropriate nonlinear filter that estimates desired signals as illustrated in the case of an inverse system in Fig.1.2 (a). Similarly, in the basic adaptive inverse control problem [1280], we attempt to estimate a form of adaptive controller whose transfer function is the inverse (in some sense) of that of the plant itself. The objective of such an adaptive system is to make the plant to directly follow the input signals (commands). A vector of error signals defined as the difference between the plant outputs and the reference inputs are used by an adaptive learning algorithm to adjust parameters of the linear controller. Usually, it is desirable that the plant outputs do not track the input source (command) signals themselves but rather track a delayed or smoothed (filtered) version of the input signals represented in Fig.1.2 (b) by transfer function $\mathbf{M}(z)$. It should be noted that in the general case the global system consisting of the cascade of the controller and the plant after convergence should model a dynamical response of the reference model $\mathbf{M}(z)$ (see Fig.1.2 (b)) [1280].

1.1.2 Instantaneous Blind Source Separation and Independent Component Analysis

In blind signal processing problems, the mixing and filtering processes of the unknown input sources $s_j(k)$ $(j = 1, 2, ..., n)$ may have different mathematical or physical models, depending on specific applications.

In the simplest case, m mixed signals $x_i(k)$ $(i = 1, 2, \dots, m)$ are linear combinations of n (typically $m \geq n$) unknown mutually statistically independent, zero-mean source signals $s_j(k)$, and are noise-contaminated (see Fig.1.3). This can be written as

$$x_i(k) = \sum_{j=1}^{n} h_{ij}\, s_j(k) + \nu_i(k), \qquad (i = 1, 2, ..., m) \tag{1.1}$$

or in the matrix notation

$$\mathbf{x}(k) = \mathbf{H}\, \mathbf{s}(k) + \boldsymbol{\nu}(k), \tag{1.2}$$

where $\mathbf{x}(k) = [x_1(k), x_2(k), \dots, x_m(k)]^T$ is a vector of sensor signals, $\mathbf{s}(k) = [s_1(k), s_2(k), \dots, s_n(k)]^T$ is a vector of sources, $\boldsymbol{\nu}(k) = [\nu_1(k), \nu_2(k), \dots, \nu_m(k)]^T$ is a vector of additive noise, and $\mathbf{H}$ is an unknown full rank $m \times n$ mixing matrix. In other words, it is assumed that the signals received by an array of sensors (e.g., microphones, antennas, transducers) are weighted sums (linear mixtures) of primary sources. These sources are typically time-varying, zero-mean, mutually statistically independent and totally unknown as is the case of arrays of sensors for communications or speech signals.

In general, it is assumed that the number of source signals n is unknown unless stated otherwise. It is assumed that only the sensor vector $\mathbf{x}(k)$ is available and it is necessary to design a feed-forward or recurrent neural network and an associated adaptive learning algorithm that enables estimation of sources, identification of the mixing matrix $\mathbf{H}$ and/or separating matrix $\mathbf{W}$ with good tracking abilities (see Fig.1.3).

The above problems are often referred to as BSS (blind source separation) and/or ICA (independent component analysis): the BSS of a random vector $\mathbf{x} = [x_1, x_2, \dots, x_m]^T$ is obtained by finding an $n \times m$, full rank, linear transformation (separating) matrix $\mathbf{W}$ such that the output signal vector $\mathbf{y} = [y_1, y_2, \dots, y_n]^T$, defined by $\mathbf{y} = \mathbf{W}\,\mathbf{x}$, contains components that are as independent as possible, as measured by an information-theoretic cost function such as the Kullback-Leibler divergence or other criteria like sparseness, smoothness or linear predictability. In other words, it is required to adapt the weights w_{ij} of the $n \times m$ matrix $\mathbf{W}$ of the linear system $\mathbf{y}(k) = \mathbf{W}\,\mathbf{x}(k)$ (often referred to as a single-layer feed-forward neural network) to combine the observations $x_i(k)$ to generate estimates of the

(a)

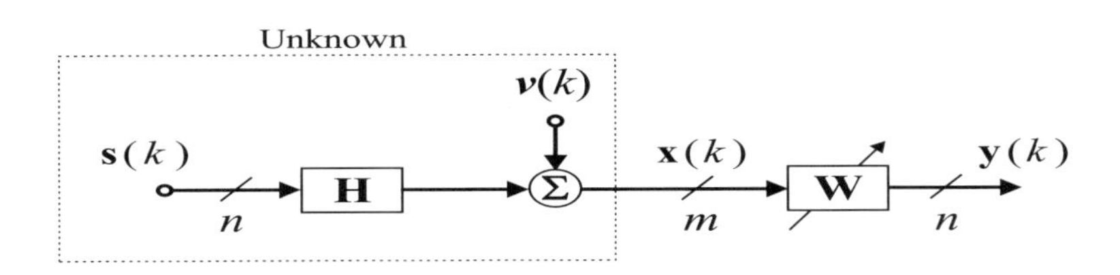

(b)

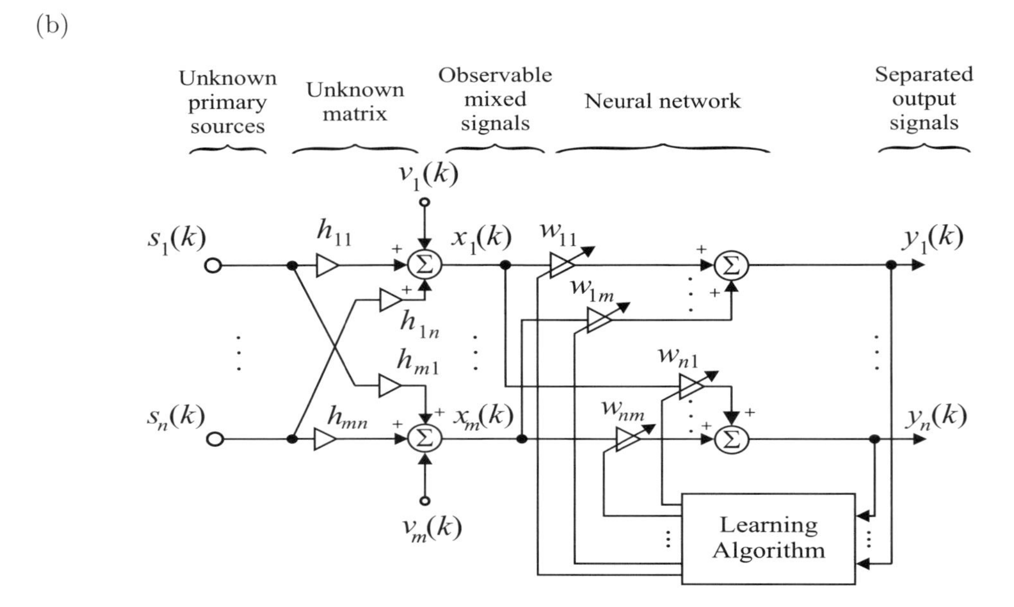

Fig. 1.3 Block diagram illustrating the basic linear instantaneous blind source separation (BSS) problem: (a) General block diagram represented by vectors and matrices, (b) detailed architecture. In general, the number of sensors can be larger, equal to or less than the number of sources. The number of sources is unknown and can change in time [258, 269].

source signals

$$\hat{s}_j(k) = y_j(k) = \sum_{i=1}^{m} w_{ji}\, x_i(k), \qquad (j = 1, 2, \ldots, n). \tag{1.3}$$

The optimal weights correspond to the statistical independence of the output signals $y_j(k)$ (see Fig.1.3).

Remark 1.1 *In this book, unless otherwise mentioned, we assume that the source signals (and consequently output signals) are zero-mean. A non zero-mean source can be modeled by zero-mean source with an additional constant source. This constant source can be usually detected but its amplitude cannot be recovered without some a priori knowledge.*

There are several definitions of ICA. In this book, depending on the problem, we use different definitions given below.

Definition 1.1 (Temporal ICA) *The ICA of a noisy random vector* $\mathbf{x}(k) \in \mathbb{R}^m$ *is obtained by finding an* $n \times m$, *(with* $m \geq n$*), full rank separating matrix* $\mathbf{W}$ *such that the output signal vector* $\mathbf{y}(k) = [y_1(k), y_2(k), \ldots, y_n(k)]^T$ *defined by*

$$\mathbf{y}(k) = \mathbf{W}\,\mathbf{x}(k), \tag{1.4}$$

contains the estimated source components $\mathbf{s}(k) \in \mathbb{R}^n$ *that are as independent as possible, evaluated by an information-theoretic cost function such as minima of Kullback-Leibler divergence.*

Definition 1.2 *For a random noisy vector* $\mathbf{x}(k)$ *defined by*

$$\mathbf{x}(k) = \mathbf{H}\,\mathbf{s}(k) + \boldsymbol{\nu}(k), \tag{1.5}$$

where $\mathbf{H}$ *is an* $(m \times n)$ *mixing matrix,* $\mathbf{s}(k) = [s_1(k), s_2(k), \ldots, s_n(k)]^T$ *is a source vector of statistically independent signals, and* $\boldsymbol{\nu}(k) = [\nu_1(k), \nu_2(k), \ldots, \nu_m(k)]^T$ *is a vector of uncorrelated noise terms, ICA is obtained by estimating both the mixing matrix* $\mathbf{H}$ *and the independent components* $\mathbf{s}(k) = [s_1(k), s_2(k), \ldots, s_n(k)]^T$.

Definition 1.3 *The ICA task is formulated as estimation of all the source signals and their numbers and/or identification of a mixing matrix* $\widehat{\mathbf{H}}$ *or its pseudo-inverse separating matrix* $\mathbf{W} = \widehat{\mathbf{H}}^+$ *assuming only the statistical independence of the primary sources and linear independence of columns of* $\mathbf{H}$.

The mixing (ICA) model can be represented in a batch form as

$$\mathbf{X} = \mathbf{H}\,\mathbf{S}, \tag{1.6}$$

where $\mathbf{X} = [\mathbf{x}(1), \mathbf{x}(2), \ldots, \mathbf{x}(N)]^T \in \mathbb{R}^{m \times N}$ and $\mathbf{S} = [\mathbf{s}(1), \mathbf{s}(2), \ldots, \mathbf{s}(N)]^T \in \mathbb{R}^{n \times N}$. In many applications, especially where the number of ICs is large and they have sparse (or other specific) distributions, it is more convenient to use the following equivalent form:

$$\mathbf{X}^T = \mathbf{S}^T\,\mathbf{H}^T. \tag{1.7}$$

By taking the transpose, we simply interchange the roles of the mixing matrix $\mathbf{H} = [\mathbf{h}_1, \mathbf{h}_2, \ldots, \mathbf{h}_n]$ and the ICs $\mathbf{S} = [\mathbf{s}(1), \mathbf{s}(2), \ldots, \mathbf{s}(N)]^T$, thus the vectors of the matrix $\mathbf{H}^T$ can be considered as independent components and the matrix $\mathbf{S}^T$ as the mixing matrix and vice-versa. In the standard temporal ICA model, it is usually assumed that ICs $\mathbf{s}(k)$ are time signals and the mixing matrix $\mathbf{H}$ is a fixed matrix without imposing any constraints on its elements. In the spatio-temporal ICA, the distinction between ICs and the mixing matrix is completely abolished [1106, 589]. In other words, the same or similar assumptions are made on the ICs and the mixing matrix. In contrast to the conventional ICA the spatio-temporal ICA maximizes the degree of independence over time and space.

Definition 1.4 (Spatio-temporal ICA) *The spatio-temporal ICA of random matrix* $\mathbf{X}^T = \mathbf{S}^T\mathbf{H}^T$ *is obtained by estimating both the unknown matrices* $\mathbf{S}$ *and* $\mathbf{H}$ *in such a way that* *rows of* $\mathbf{S}$ *and columns of* $\mathbf{H}$ *be as independent as possible and both* $\mathbf{S}$ *and* $\mathbf{H}$ *consist of the same or very similar statistical properties (e.g., the Laplacian distribution or sparse representation).*

The real-world sensor data often build up complex nonlinear structures, so applying ICA to global data may lead to poor results. Instead, applying ICA to all available data, we can preprocess this data by grouping it into clusters or sub-bands with specific features and then apply ICA individually to each cluster or sub-band separately. The preprocessing stage of suitable grouping or clustering of data is responsible for an overall coarse nonlinear representation of the data, while the linear ICA models of individual clusters are used for describing local features of the data.

Definition 1.5 (Local ICA) *In local ICA the available sensor data are suitably preprocessed, by grouping them into clusters in space, or in the time, frequency or in the time-frequency domain, and then applying linear ICA to each cluster locally. More generally, an optimal local ICA can be implemented as the result of mutual interaction of two processes: A suitable clustering process and the application of the ICA process to each cluster.*

A globally linear model, as implied by conventional ICA, may be insufficient to represent multivariate data in many situations. A combination of several local ICA's can provide a suitable approach in such cases. An important question is then how to find an appropriate partitioning of the data space together with a proper choice of the local numbers of independent components (IC's).

Despite the success of using standard ICA in many applications, the basic assumptions of ICA may not hold hence some caution should be taken when using standard ICA to analyze real world problems, especially in biomedical signal processing. In fact, by definition, the standard ICA algorithms are not able to estimate statistically dependent original sources, that is, when the independence assumption is violated. A natural extension and generalization of ICA is multiresolution subband decomposition ICA (MSD-ICA) which relaxes considerably the assumption regarding mutual independence of primarily sources. The key idea in this approach is the assumption that the wide-band source signals are dependent, however some narrow band subcomponents are independent In other words, we assume that each unknown source can be modeled or represented as a sum of narrow-band sub-signals (sub-components):

$$s_i(k) = s_{i1}(k) + s_{i2}(k) + \cdots + s_{iK}(k). \tag{1.8}$$

The basic concept of MSD-ICA is to divide the sensor signal spectra into their subspectra or subbands, and then to treat those subspectra individually for the purpose at hand. The subband signals can be ranked and processed independently. Let us assume that only a certain set of sub-components is independent. Provided that for some of the frequency subbands (at least one) all sub-components, say $\{s_{ip}(k)\}_{i=1}^{n}$, are mutually independent or temporally decorrelated, then we can easily estimate the mixing or separating system (under condition that these subbands can be identified by some *a priori* knowledge or detected by some self-adaptive process) by simply applying any standard ICA algorithm, however not for all available raw sensor data but only for suitably preprocessed (band pass filtered) sensor signals. Such explanation can be summarized as follows.

Definition 1.6 (Multiresolution Subband Decomposition ICA) *The MSD-ICA can be formulated as a task of estimation of the mixing matrix* $\mathbf{H}$ *on the basis of suitable multiresolution subband decomposition of sensors signals and by applying a classical ICA (instead for raw sensor data) for one or several preselected subbands for which source sub-components are independent.*

In one of the most simplest cases, source signals can be modeled or decomposed into their low- and high- frequency sub-components:

$$s_i(k) = s_{iL}(k) + s_{iH}(k) \qquad (i = 1, 2, \ldots, n). \tag{1.9}$$

In practice, the high-frequency sub-components $s_{iH}(k)$ are often found to be mutually independent. In such a case in order to separate the original sources $s_i(k)$, we can use a High Pass Filter (HPF) to extract high frequency sub-components and then apply any standard ICA algorithm to such preprocessed sensor (observed) signals. In the preprocessing stage, more sophisticated methods, such as block transforms, multirate subband filter bank or wavelet transforms, can be applied.

In many blind signal separation problems, one may want to estimate only one or several desired components with particular statistical features or properties, but discard the rest of the uninteresting sources and noises. For such problems, we can define Blind Signal Extraction (BSE) (see Chapter 5 for more detail and algorithms).

Definition 1.7 (Blind Signal Extraction) *BSE is formulated as a problem of estimation of one source or a selected number of the sources (smaller than* n*) with particular desired properties or characteristics, sequentially one by one or "a one shot" estimation of a specific group of sources. Equivalently the problem is formulated as an identification of the corresponding vector(s)* $\widehat{\mathbf{h}}_j$ *of the mixing matrix* $\widehat{\mathbf{H}}$ *and/or their pseudo-inverses* $\mathbf{w}_j$ *which are rows of the separating matrix* $\mathbf{W} = \widehat{\mathbf{H}}^+$*, assuming only the statistical independence of its primary sources and linear independence of columns of* $\mathbf{H}$.

Remark 1.2 *It is worth emphasizing that in the literature, BSS/BSE and ICA terms are often confused or interchanged, although they refer to the same or similar models and are solved with the similar algorithms under the assumption that the primary sources are mutually independent. However, in the general case, especially for real-world problems, the objective for ICA and BSS are somewhat different. In fact, the objective of BSS is to estimate the original source signals even if they are not completely mutually statistically independent, while the objective of ICA is to determine a transformation which assures that the output signals are as independent as possible. It should be noted that ICA methods use higher-order statistics (HOS) in many cases, while BSS methods are apt to use only second order statistics (SOS). The second order methods assume that sources have some temporal structure, while the higher order methods assume their mutual independence. Thus, the second statistics methods, generally do not perform independent component analysis. Another difference is that the higher-order statistics methods can not be applied to Gaussian signals*

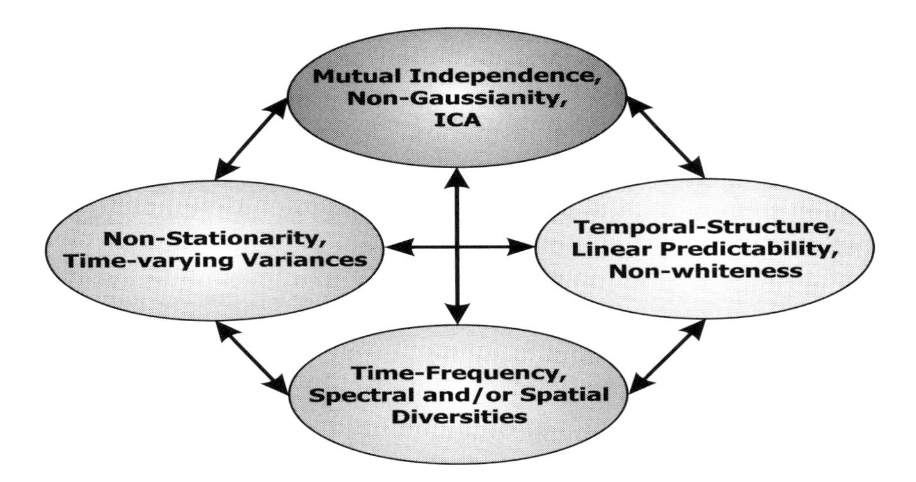

Fig. 1.4 Basic approaches for blind source separation with some *a priori* knowledge.

while second order methods do not have such constraints. In fact, BSS methods do not really replace ICA and vice versa, since each approach is based on different criteria, assumptions and often different objectives.

Although many different source separation algorithms are available, their principles can be summarized by the following four approaches (see Fig.1.4):

- The most popular approach exploits as the cost function some measure of signals independence, non-Gaussianity or sparseness. When original sources are assumed to be statistically independent without a temporal structure, the higher-order statistics (HOS) are essential (implicitly or explicitly) to solve the BSS problem. In such a case, the method does not allow more than one Gaussian source (see Chapters 5 and 6 for more detail).

- If sources have temporal structures, then each source has non-vanishing temporal correlation, and less restrictive conditions than statistical independence can be used, namely, second-order statistics (SOS) are sufficient to estimate the mixing matrix and sources. Along this line, several methods have been developed [1159, 1155, 850, 84]. Note that these SOS methods do not allow the separation of sources with identical power spectra shapes or i.i.d. (independent and identically distributed) sources (see Chapter 4).

- The third approach exploits nonstationarity (NS) properties and second order statistics (SOS). Mainly, we are interested in the second-order nonstationarity in the sense that source variances vary in time. The nonstationarity was first taken into account by Matsuoka *et al.* [828] and it was shown that a simple decorrelation technique is able to perform the BSS task. In contrast to other approaches, the nonstationarity information based methods allow the separation of colored Gaussian sources with identical power spectra shapes. However, they do not allow the separation of sources with identical nonstationarity properties. There are some recent works on nonstationary source separation [217, 218, 966] (see Chapters 4, 6 and 8).

- The fourth approach exploits the various diversities[2] of signals, typically, time, frequency, (spectral or "time coherence") and/or time-frequency diversities, or more generally, joint space-time-frequency (STF) diversity.

Remark 1.3 *In fact, the concept of space-time-frequency diversities are widely used in wireless communications systems. Signals can be separated easily if they do not overlap in either the time-, the frequency- or the time-frequency domain (see Fig.1.5 and Fig.1.6). When signals do not overlap in the time-domain then one signal stops (is silent) before another one begins. Such signals are easily separated when a receiver is accessible only while the signal of interest is sent. This multiple access method is called TDMA (Time Division Multiple Access). If two or more signals do not overlap in the frequency domain, then they can be separated with bandpass filters as is illustrated in Fig.1.5. The method based on this principle is called FDMA (Frequency Division Multiple Access). Both TDMA and FDMA are used in many modern digital communication systems [472]. Of course, if the source power spectra overlap, the spectral diversity is not sufficient to extract sources, therefore, we need to exploit other kinds of diversity. If the source signals have different time-frequency diversity and time-frequency signatures of the sources do not (completely) overlap then still they can be extracted from one (or more) sensor signal by masking individual source signals or interference in the time-frequency domain and then synthesized from time-frequency domain as illustrated in Fig.1.6. However, in such cases some a priori information about source signals is necessary. Therefore, separation is not completely blind but only semi-blind.*

More sophisticated or advanced approaches use combinations or integration of all the above mentioned approaches: HOS, SOS, NS and STF (Space-Time-Frequency) diversity, in order to separate or extract sources with various statistical properties and to reduce the influence of noise and undesirable interferences. Methods that exploit either the temporal structure of sources (mainly second-order correlations) and/or the nonstationarity of sources, lead to the second-order BSS methods. In contrast to BSS methods based on HOS, all the second-order statistics based methods do not have to infer the probability distributions of sources or nonlinear activation functions.

1.1.3 Independent Component Analysis for Noisy Data

As the estimation of a separating (unmixing) matrix $\mathbf{W}$ and a mixing matrix $\widehat{\mathbf{H}}$ in the presence of noise is rather difficult; the majority of past research efforts have been devoted to only the noiseless case, where $\boldsymbol{\nu}(k) = \mathbf{0}$. One of the objectives of this book is to present promising novel approaches and associated algorithms that are more robust with respect to noise and/or that can reduce the noise in the estimated output vector $\mathbf{y}(k)$. Usually, it is

[2]By diversities we mean usually different characteristics or features of the signals.

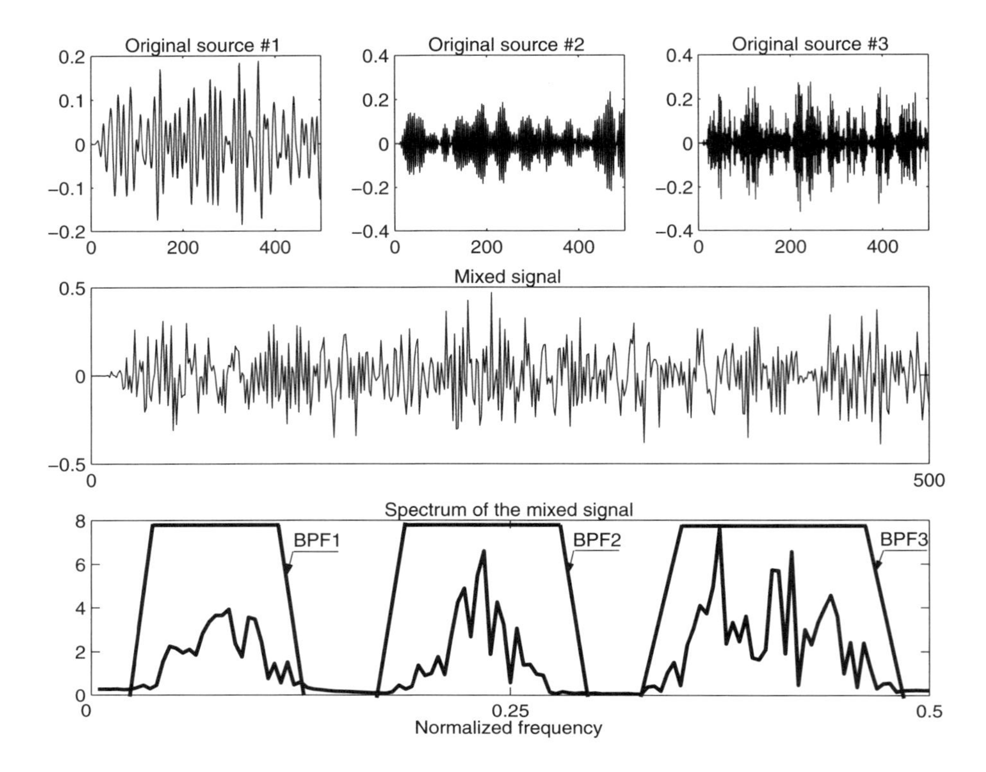

Fig. 1.5 Illustration of exploiting spectral diversity in BSS. Three unknown sources and their available mixture in time domain (horizontal axis represents time scaled in milliseconds)- top and middle plots; corresponding mixed signal spectrum is shown the bottom plot. The sources are extracted by passing the mixed signal through three bandpass filters (BPF) with suitable frequency characteristics depicted in the bottom figure.

assumed that the source signals and additive noise components are statistically independent.

In some models described in this book, it is assumed that sources of additive noise are incorporated as though they were unknown source signals. In other words, the effect of incident noise fields impinging on several sensors may be considered to be equivalent to additional sources, and thus are subject to the same separation process as the desired signals. Of course, there may be more than one noise source. However, for the separation of noise sources, at most one noise source may have a Gaussian distribution, and all other sources must have non-Gaussian distributions. It may well be that one is not interested in separation of the noise sources.

In general, the problem of noise cancellation is difficult and even impossible to treat because we have $(m+n)$ unknown source signals (n sources and m noise signals, see Fig.1.8). Various signal processing methods have been developed for noise cancelling [1230, 1231,

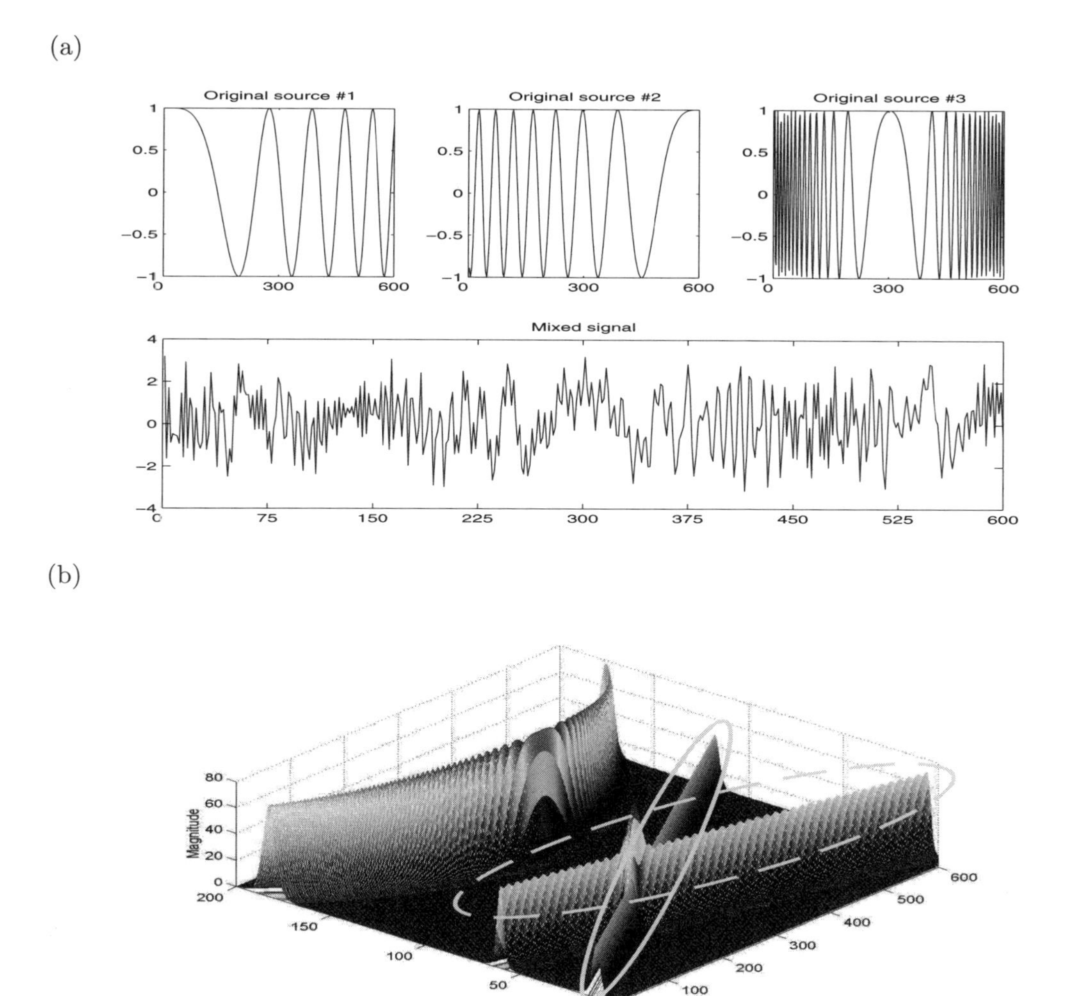

Fig. 1.6 Illustration of exploiting time-frequency diversity in BSS. (a) Original unknown source signals and available mixed signal (horizontal axis represents time scaled in milliseconds). (b) Time-frequency representation of the mixed signal. Due to non-overlapping time-frequency signatures of the sources by masking and synthesis (inverse transform), we are able to extract the desired sources.

1228] and with some modifications they can be applied to noise cancellation in BSS. In many practical situations, we can measure or model the environmental noise. Such noise is termed reference noise (denoted by ν_R in Fig.1.7). For example, in the acoustic "cocktail party" problem, we can measure or record the environmental noise by using an isolated microphone. In a similar way, noise in biomedical applications can be measured by appro-

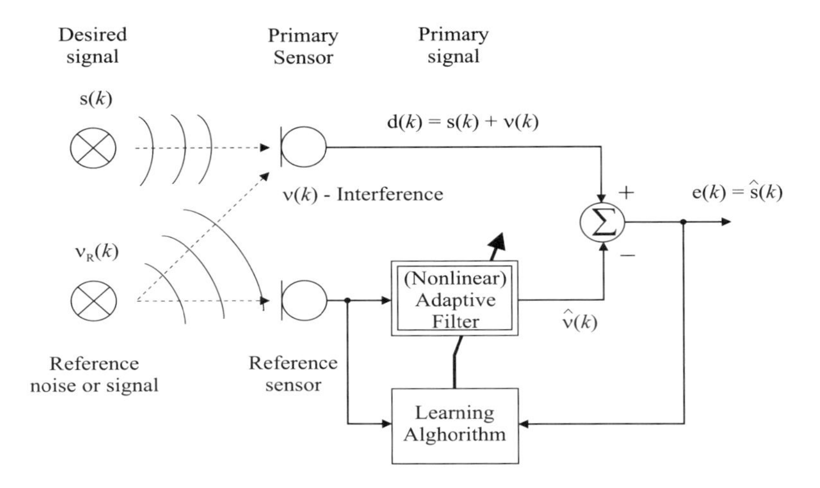

Fig. 1.7 Standard model for noise cancellation in a single channel using a nonlinear adaptive filter or neural network.

priately placed auxiliary sensors (or electrodes). The noise $\nu_R(k)$ may influence each sensor in some unknown manner due to environmental effects; hence, such effects as delays, reverberations, echo, nonlinear distortions etc. may occur. It may be assumed that the reference noise is processed by some unknown dynamical system before reaching the sensors. In a simple case, a convolutive model of noise is assumed where the reference noise is processed by some FIR filters (see Fig.1.8). In this case, two learning processes are performed simultaneously: An un-supervised learning procedure performing blind separation and a supervised learning algorithm performing noise reduction [261]. This approach has been successfully applied to the elimination of noise under the assumption that the reference noise is available [261, 665].

In a traditional linear Finite Impulse Response (FIR) adaptive noise cancellation filter, the noise is estimated as a weighted sum of delayed samples of the reference interference. However, the linear adaptive noise cancellation systems mentioned above may not achieve an acceptable level of cancellation of noise in many real world situations when interference signals are related to the measured reference signals in a complex dynamic and nonlinear way.

In many applications, especially in biomedical signal processing, the sensor signals are corrupted by various interference and noise sources. Efficient interference and noise cancellation usually require nonlinear adaptive processing of the observed signals. In this book, we describe various neural network models and associated on-line adaptive learning algorithms for noise and interference cancellation. In particular, we propose to use the Hyper Radial Basis Function Network (HRBFN) with all of its parameters being fully adaptive. Moreover, we examine Amari-Hopfield recurrent neural networks [254]. We study the problem from the perspective of optimal signal estimation and nonlinear adaptive systems. Our mathematical analysis and computer simulations demonstrate that such neural networks

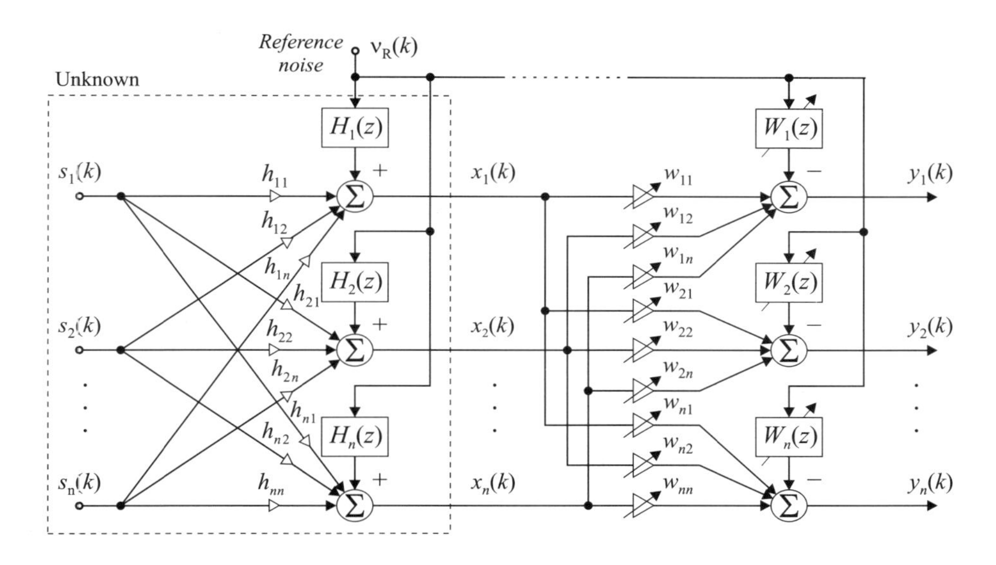

Fig. 1.8 Illustration of noise cancellation and blind separation - deconvolution problem (for $m = n$).

can be quite effective and useful in removing interference and noise. In particular, it will be shown that the Amari-Hopfield recurrent neural network (see Chapter 8) can be more effective than feed-forward networks for certain noise distributions, where the data exhibit a long memory structure (temporal correlation).

1.1.4 Multichannel Blind Deconvolution and Separation

A single channel convolution and deconvolution process is illustrated in Fig.1.9.

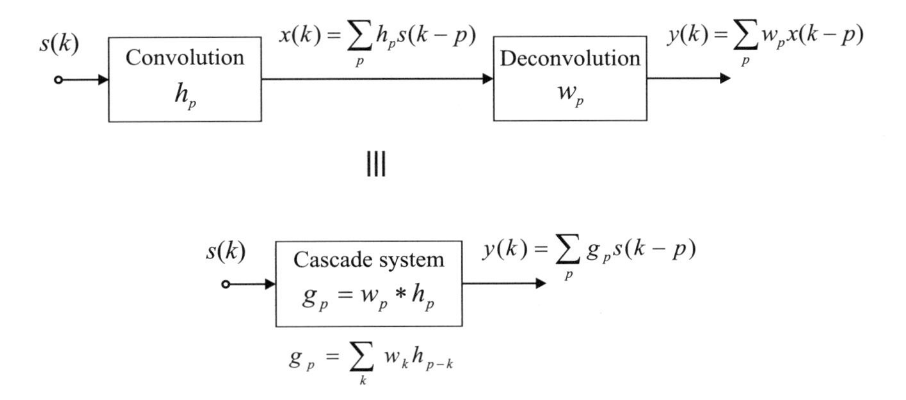

Fig. 1.9 Diagram illustrating the single channel convolution and inverse deconvolution process.

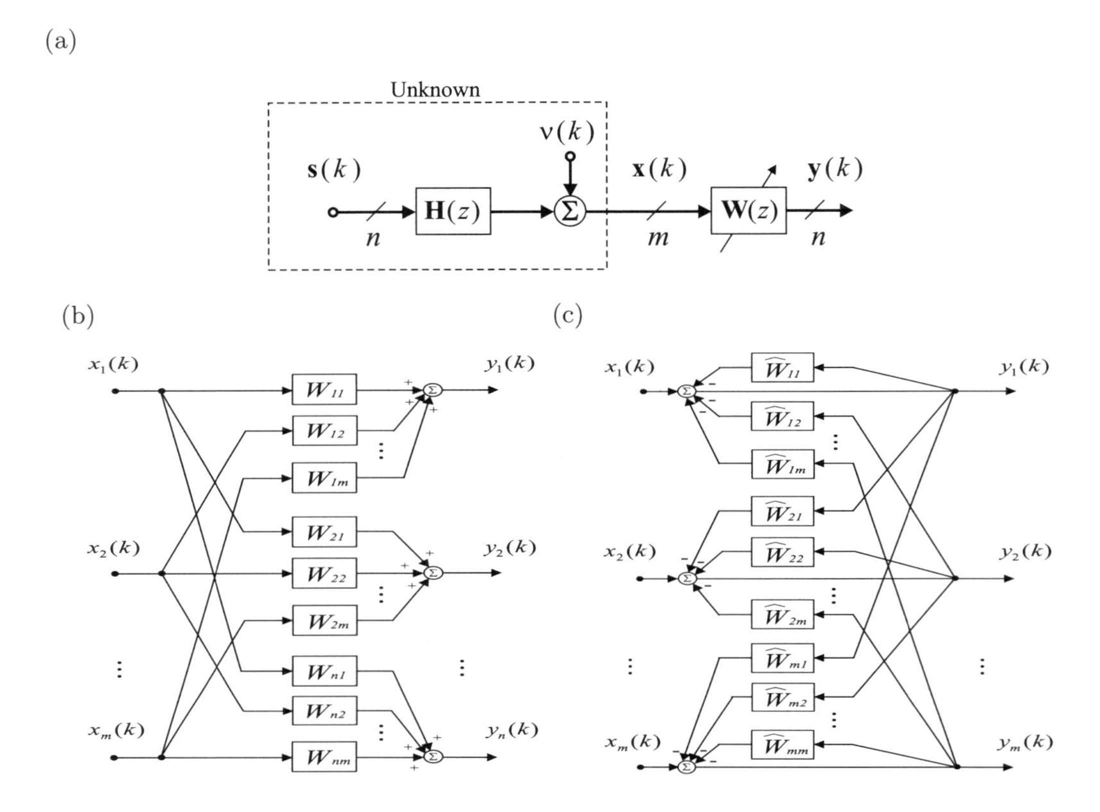

Fig. 1.10 Diagram illustrating standard multichannel blind deconvolution problem (MBD): (a) Functional block diagram of the feed-forward model, (b) architecture of the feed-forward neural network (each synaptic weight $W_{ij}(z,k)$ is an FIR or stable IIR filter, (c) architecture of the fully connected recurrent neural network.

A multichannel blind deconvolution problem can be considered as a natural extension or generalization of the instantaneous blind separation problem (see Fig.1.10). In the multidimensional blind deconvolution problem, an m-dimensional vector of received discrete-time signals $\mathbf{x}(k) = [x_1(k), x_2(k), \ldots, x_m(k)]^T$ at time k is assumed to be produced from an n-dimensional vector of source signals $\mathbf{s}(k) = [s_1(k), s_2(k), \ldots, s_n(k)]^T$, $m \geq n$, by using a stable mixture model [34, 26, 247, 606]

$$\mathbf{x}(k) = \sum_{p=-\infty}^{\infty} \mathbf{H}_p \, \mathbf{s}(k-p) = \mathbf{H}_k * \mathbf{s}(k), \qquad with \sum_{p=-\infty}^{\infty} \|\mathbf{H}_p\| < \infty, \tag{1.10}$$

where $*$ denotes the convolution operator and $\mathbf{H}_p$ is an $(m \times n)$ matrix of mixing coefficients at time-lag p.

Define

$$\mathbf{H}(z) = \sum_{p=-\infty}^{\infty} \mathbf{H}_p \, z^{-p} \tag{1.11}$$

where z^{-1} denotes the unit time-delay (backward shift) operator (i.e. $z^{-p}[s_i(k)] = s_i(k-p)$). It should be noted that if z is replaced with the complex variable $\tilde{z} = \exp(-\sigma + j\omega T)$, then $\mathbf{H}(\tilde{z})$ is the $\mathcal{Z}$-transform of $\{\mathbf{H}_p\}$, i.e., it is the system matrix transfer function [371, 472]. Using (1.11), (1.10) may be rewritten as

$$\mathbf{x}(k) = [\mathbf{H}(z)] \, \mathbf{s}(k). \tag{1.12}$$

The goal of multichannel deconvolution is to calculate the possibly scaled and time-delayed (or filtered) versions of the source signals from the received signals by using approximate knowledge of the source signal distributions and statistics. Typically, every source signal $s_i(k)$ is an i.i.d. (independent and identically-distributed) sequence that is independent of all the other source sequences.

In order to recover the source signals, we can use the neural network models depicted in Fig.1.3 (b) and Fig.1.10 but the synaptic weights should be generalized to filters (e.g., FIR or IIR) as is illustrated in Fig.1.11. In this book, many such extensions and generalizations are described.

Let us consider briefly one example of such a generalization: A standard multichannel blind deconvolution where each weight [33, 210, 606]

$$W_{ji}(z, k) = \sum_{p=0}^{M} w_{jip}(k) \, z^{-p} \tag{1.13}$$

is described by a multichannel finite-duration impulse response (FIR) adaptive filter at discrete-time k [606, 651].

We will consider a stable feed-forward model that estimates the source signals directly by using a truncated version of a doubly-infinite multichannel equalizer of the form [606] (see Fig.1.11 (a))

$$y_j(k) = \sum_{i=1}^{m} \sum_{p=-\infty}^{\infty} w_{jip} \, x_i(k-p), \qquad (j = 1, 2, \ldots, n) \tag{1.14}$$

or in the compact matrix form as

$$\mathbf{y}(k) = \sum_{p=-\infty}^{\infty} \mathbf{W}_p(k) \, \mathbf{x}(k-p) = \mathbf{W}_p(k) * \mathbf{x}(k) = [\mathbf{W}(z, k)] \, \mathbf{x}(k), \tag{1.15}$$

where $\mathbf{y}(k) = [y_1(k), y_2(k), \ldots, y_n(k)]^T$ is an n-dimensional vector of outputs and $\mathbf{W}(k) = \{\mathbf{W}_p(k), \ -\infty \leq p \leq \infty\}$ is a sequence of $n \times m$ coefficient matrices used at time k, and the matrix transfer function is given by

$$\mathbf{W}(z, k) = \sum_{p=-\infty}^{\infty} \mathbf{W}_p(k) \, z^{-p}. \tag{1.16}$$

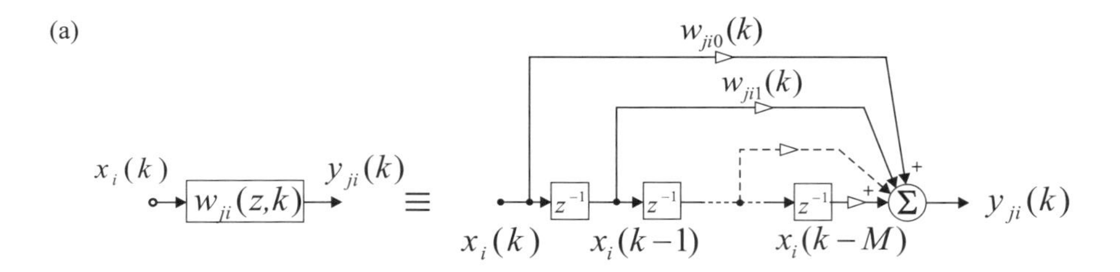

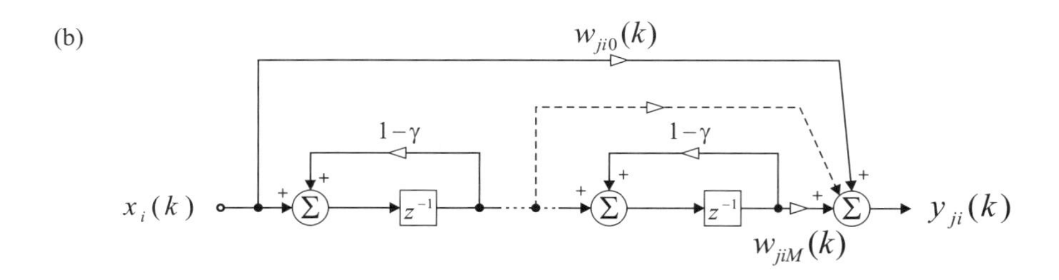

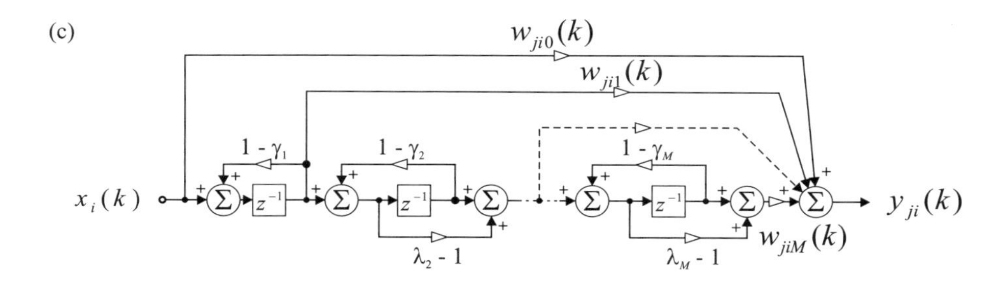

Fig. 1.11 Basic models of synaptic weights for the feed-forward adaptive system (neural network) shown in Fig.1.3 : (a) Basic FIR filter model, (b) Gamma filter model, (c) Laguerre filter model.

The goal of adaptive blind deconvolution or equalization is then to adjust $\mathbf{W}(z,k)$ such that the global system be described as

$$\lim_{k\to\infty} \mathbf{G}(z,k) = \mathbf{W}(z,k)\,\mathbf{H}(z) = \mathbf{P}\,\mathbf{D}(z), \tag{1.17}$$

where $\mathbf{P}$ is an $n \times n$ permutation matrix, $\mathbf{D}(z)$ is an $n \times n$ diagonal matrix whose (i,i)-th entry is $c_i z^{-\Delta_i}$, c_i is a non-zero scalar factor, and Δ_i is an integer delay. We assume that both $\mathbf{H}(z)$ and $\mathbf{W}(z,k)$ are stable with non-zero eigenvalues on the unit circle $|z| = 1$. In addition, the derivatives of quantities with respect to $\mathbf{W}(z,k)$ can be understood as a series of matrices indexed by the lag p of $\mathbf{W}_p(k)$ [33, 34, 606].

Fig.1.11 (b) and (c) show alternative neural network models with the weights in the form of stable constrained infinite impulse response (IIR) filters. In these models, the weights

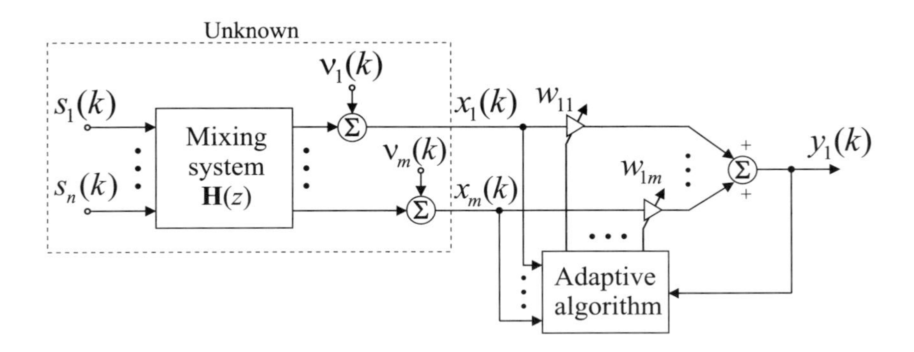

Fig. 1.12 Block diagram illustrating the sequential blind extraction of sources or independent components. Synaptic weights w_{ij} can be time-variable coefficients or adaptive filters (see Fig.1.11).

W_{ji} are generalized to real- or complex-valued Gamma [990, 989] or Laguerre filters (see Fig.1.11 (b) and (c)) or other structures such as state-space models (see Fig.1.13) which may have some useful properties [26, 1352, 1368]. In all these models, it is assumed that only the sensor vector $\mathbf{x}(k)$ is available and it is necessary to design a feed-forward or recurrent neural network and an associated adaptive learning algorithm that enables estimation of the source signals.

1.1.5 Blind Extraction of Signals

There are two main approaches to solve the problem of blind separation and deconvolution. The first approach, which was mentioned briefly in previous sections, is to simultaneously separate all sources. In the second one, we extract sources sequentially in a blind fashion, one by one, rather than separating them all simultaneously. In many applications, a large number of sensors (electrodes, microphones or transducers) are available but only a very few source signals are subjects of interest. For example, in the EEG or MEG devices, we observe typically more than 64 sensor signals, but only a few source signals are interesting; the rest can be considered as interfering noise. In another example, the cocktail party problem, it is usually essential to extract the voices of specific persons rather than separate all the source signals available from a large array of microphones. For such applications it is essential to develop reliable, robust and effective learning algorithms which enable us to extract only a small number of source signals that are potentially interesting and contain useful information (see Fig.1.12). This problem is the subject of Chapter 5. The blind signal extraction approach may have several advantages over simultaneous blind separation/deconvolution, such as.

- Signals can be extracted in a *specified order* according to the statistical features of the source signals, e.g., in the order determined by absolute values of generalized normalized kurtosis. Blind extraction of sources can be considered as a generalization of PCA (principal components analysis), where decorrelated output signals are extracted according to the decreasing order of their variances.

- Only "interesting" signals need to be extracted. For example, if the source signals are mixed with a large number of Gaussian noise terms, we may extract only specific signals which possess some desired statistical properties.
- The available learning algorithms for BSE are purely local and biologically plausible. In fact, the learning algorithms derived below can be considered as extensions or modifications of the Hebbian/anti-Hebbian learning rule. Typically, they are simpler than those of instantaneous blind source separation.

In summary, blind signal extraction is a useful approach when our objective is to extract several source signals with specific statistical properties from a large number of mixtures. Extraction of a single source is closely related to the problem of blind deconvolution [606, 609, 1068, 1079]. In blind signal extraction (BSE), our objective is to extract the source signals sequentially, i.e. one by one, rather than to separate all of them simultaneously. This procedure is called the sequential blind signal extraction in contrast to the simultaneous blind signal separation (BSS). Sequential blind signal extraction can be performed by using a cascade neural network similar to the one used for the extraction of principal components. However, in contrast to PCA, the optimization criteria for BSE are different. A single processing unit (artificial neuron) is used in the first step to extract one source signal with specified statistical properties. In the next step, a deflation technique can be used to eliminate the already extracted signals from the mixtures.

1.1.6 Generalized Multichannel Blind Deconvolution – State Space Models

In the general case, linear dynamical mixing and demixing systems can be described by state-space models. In fact, any stable mixing dynamical system can be described as (see Fig.1.13)

$$\overline{\boldsymbol{\xi}}(k+1) = \overline{\mathbf{A}}\,\overline{\boldsymbol{\xi}}(k) + \overline{\mathbf{B}}\,\mathbf{s}(k) + \overline{\mathbf{N}}\,\boldsymbol{\nu}_P(k), \tag{1.18}$$

$$\mathbf{x}(k) = \overline{\mathbf{C}}\,\overline{\boldsymbol{\xi}}(k) + \overline{\mathbf{D}}\,\mathbf{s}(k) + \boldsymbol{\nu}(k), \tag{1.19}$$

where $\boldsymbol{\xi} \in \mathbb{R}^r$ is the state vector of the system, $\mathbf{s}(k) \in \mathbb{R}^n$ is a vector of unknown input signals (assumed to be zero-mean, non-Gaussian independent and identically distributed (i.i.d.) and mutually (spatially) independent), $\mathbf{x}(k)$ is an available vector of sensor signals, $\boldsymbol{\nu}_P(k)$ is the vector of process noise, $\boldsymbol{\nu}(k)$ is the vector of output noise, and the state matrices have dimensions: $\overline{\mathbf{A}} \in \mathbb{R}^{r\times r}$ is a state matrix, $\overline{\mathbf{B}} \in \mathbb{R}^{r\times n}$ an input mixing matrix, $\overline{\mathbf{C}} \in \mathbb{R}^{m\times r}$ an output mixing matrix, $\overline{\mathbf{D}} \in \mathbb{R}^{m\times n}$ an input-output mixing matrix and $\overline{\mathbf{N}} \in \mathbb{R}^{r\times p}$ is a noise matrix. The transfer function is an $m \times n$ matrix of the form

$$\mathbf{H}(z) = \overline{\mathbf{C}}\,(z\,\mathbf{I} - \overline{\mathbf{A}})^{-1}\,\overline{\mathbf{B}} + \overline{\mathbf{D}}, \tag{1.20}$$

where z^{-1} is a delay operator (i.e., $z^{-1}\,x(k) = x(k-1)$).

Analogously, we can assume that the demixing model is another linear state-space system described as (see Fig.1.13)

$$\boldsymbol{\xi}(k+1) = \mathbf{A}\,\boldsymbol{\xi}(k) + \mathbf{B}\,\mathbf{x}(k) + \mathbf{L}\,\boldsymbol{\nu}_R(k), \tag{1.21}$$

$$\mathbf{y}(k) = \mathbf{C}\,\boldsymbol{\xi}(k) + \mathbf{D}\,\mathbf{x}(k), \tag{1.22}$$

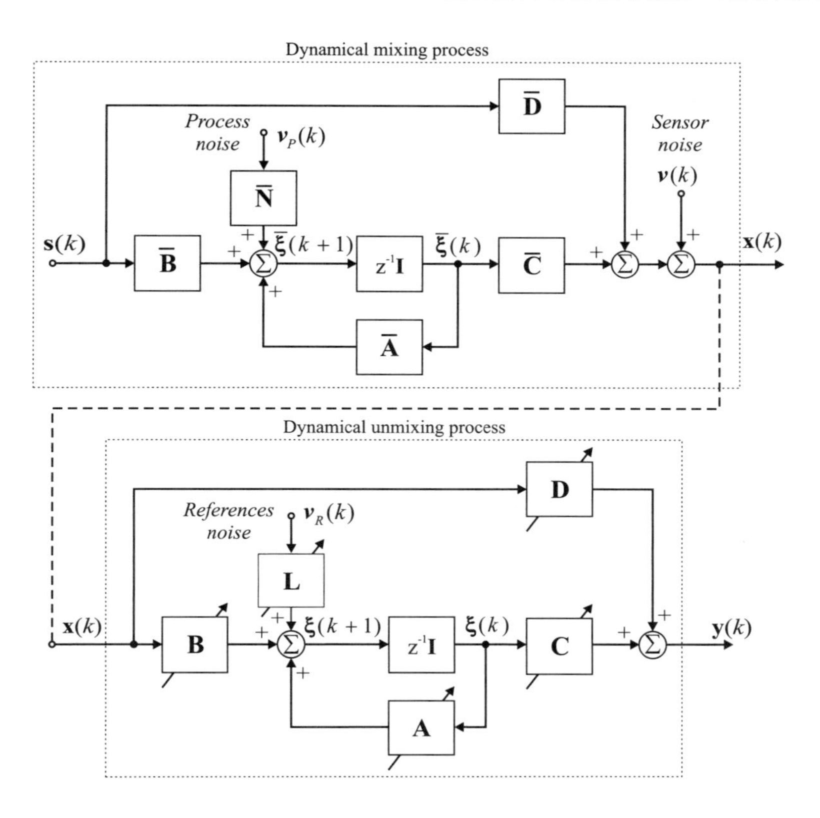

Fig. 1.13 Conceptual state-space model illustrating general linear state-space mixing and self-adaptive demixing model for Dynamic ICA (DICA). The objective of learning algorithms is estimation of a set of matrices $\{\mathbf{A}, \mathbf{B}, \mathbf{C}, \mathbf{D}, \mathbf{L}\}$ [281, 283, 284, 1352, 1353, 1354].

where the unknown state-space matrices, respectively have the dimension: $\mathbf{A} \in \mathbb{R}^{M \times M}$, $\mathbf{B} \in \mathbb{R}^{M \times m}$, $\mathbf{C} \in \mathbb{R}^{n \times M}$, $\mathbf{D} \in \mathbb{R}^{n \times m}$, $\mathbf{L} \in \mathbb{R}^{M \times m}$, with $M \geq r$ (i.e., the order of the demixing system should be at least the same or larger than the order of the mixing system).

It is easy to see that the linear state-space model is an extension of the instantaneous blind source separation model. In the special case when the matrices $\overline{\mathbf{A}}, \overline{\mathbf{B}}, \overline{\mathbf{C}}$ in the mixing model and $\mathbf{A}, \mathbf{B}, \mathbf{C}$ in the demixing model are null matrices, the problem is simplified to the standard ICA problem. In general, the matrices $\boldsymbol{\Theta} = [\mathbf{A}, \mathbf{B}, \mathbf{C}, \mathbf{D}, \mathbf{L}]$ are parameters to be determined in a learning process on the basis of knowledge of the sequence $\mathbf{x}(k)$ and some *a priori* knowledge about the system. The transfer function of the demixing model is $\mathbf{W}(z) = \mathbf{C}\,(z\,\mathbf{I} - \mathbf{A})^{-1}\,\mathbf{B} + \mathbf{D}$. We formulate the dynamical blind separation problem as a task to recover original source signals from the observations $\mathbf{x}(k)$ without *a priori* knowledge of the source signals or the state-space matrices $[\overline{\mathbf{A}}, \overline{\mathbf{B}}, \overline{\mathbf{C}}, \overline{\mathbf{D}}]$, by assuming, for example, that the sources are mutually independent, zero-mean signals. Other assumptions such as smoothness or linear predictability of sources can also be used. We also usually assume that the output signals $\mathbf{y}(k) = [y_1(k), y_2(k), \ldots, y_n(k)]^T$ will recover the source signals for the noiseless case in the following sense

$$\mathbf{y}(k) = [\mathbf{W}(z)\ \mathbf{H}(z)]\ \mathbf{s}(k) = [\mathbf{D}(z)]\ \mathbf{P}\ \mathbf{s}(k), \tag{1.23}$$

where $\mathbf{P}$ is an $n \times n$ generalized permutation matrix which consists of n nonzero elements and only one nonzero element in each column and $\mathbf{D}(z) = \text{diag}\{D_{11}(z), D_{22}, \ldots, D_{nn}(z)\}$ is a diagonal matrix with transfer functions $D_{ii}(z)$ of shaping filters. In some applications, such as equalization problems, it is required that $D_{ii}(z) = \lambda_i z^{-\tau_i}$, where λ_i is a nonzero constant scaling factor and τ_i is any positive integer delay (i.e., constant scaling factors and/or pure delays are only acceptable).

A question arising here is whether matrices $[\mathbf{A}, \mathbf{B}, \mathbf{C}, \mathbf{D}]$ exist for the demixing model shown in Fig.1.13 such that the transfer function $\mathbf{W}(z)$ satisfies (1.23). The answer is affirmative [1349, 1368, 1369]. It will be shown later that if there is a filter $\mathbf{W}_*(z)$, which is the inverse of $\mathbf{H}(z)$ in the sense of (1.23), then for the given specific matrices $[\mathbf{A}, \mathbf{B}]$, there are matrices $[\mathbf{C}, \mathbf{D}]$, such that the transfer matrix $\mathbf{W}(z)$ satisfies equation (1.23).

Remark 1.4 *It should be noted that in general case, we can assume that* $\mathbf{D}$ *is an* $m \times m$ *square matrix, i.e., the number of outputs of the system is equal to the number of sensors, although in practice the number of sources can be less than the number of sensors* ($m \geq n$). *Such a model is justified by two facts. First of all, the number of sources is generally unknown and may change over time. Secondly, in practice we have additive noise signals that can be considered as auxiliary unknown sources; therefore, it is also reasonable to extract these noise signals. In the ideal noiseless case, the redundant* $(m-n)$ *output signals* y_j *should decay to zero during adaptive learning process and then only* n *outputs will correspond to the recovered sources.*

1.1.7 Nonlinear State Space Models – Semi-Blind Signal Processing

The above linear state-space demixing and filtering model is relatively easy to generalize into a flexible nonlinear model as (see Fig.1.14)

$$\boldsymbol{\xi}(k) = \mathbf{f}[\underline{\mathbf{x}}(k), \underline{\boldsymbol{\xi}}(k)], \tag{1.24}$$

$$\mathbf{y}(k) = \mathbf{C}(k)\ \boldsymbol{\xi}(k) + \mathbf{D}(k)\ \mathbf{x}(k), \tag{1.25}$$

where $\boldsymbol{\xi}(k) = [\xi_1(k), \xi_1(k), \ldots, \xi_M(k)]^T$ is the state vector, $\mathbf{x}(k) = [x_1(k), x_2(k), \ldots, x_m(k)]^T$ is an available vector of sensor signals, $\mathbf{f}[\underline{x}(k), \underline{\boldsymbol{\xi}}(k)]$ is an M-dimensional vector of nonlinear functions (with $\underline{\boldsymbol{x}}(k) = [\mathbf{x}^T(k), \mathbf{x}^T(k), \ldots, \mathbf{x}^T(k - L_x)]^T$ and $\underline{\boldsymbol{\xi}}(k) = [\boldsymbol{\xi}^T(k), \boldsymbol{\xi}^T(k - 1), \ldots, \boldsymbol{\xi}^T(k - L_x)]^T$), $\mathbf{y}(k) = [y_1(k), y_2(k), \ldots, y_n(k)]^T$ is the vector of output signals, and $\mathbf{C} \in \mathbb{R}^{n \times M}$ and $\mathbf{D} \in \mathbb{R}^{n \times m}$ are output matrices. It should be noted that equation (1.24) describes the nonlinear autoregressive moving average (NARMA) model while the output model (1.25) is linear. Our objective will be to estimate the output matrices $\mathbf{C}$ and $\mathbf{D}$, as well as to identify the NARMA model by using a neural network on the basis of sensor signals $\mathbf{x}(k)$ and source (desired) signals $\mathbf{s}(k)$ (which are available for short-time windows).

In order to solve this challenging and difficult problem, we attempt to apply a semi-blind approach, i.e., we combine both supervised and un-supervised learning algorithms. Such

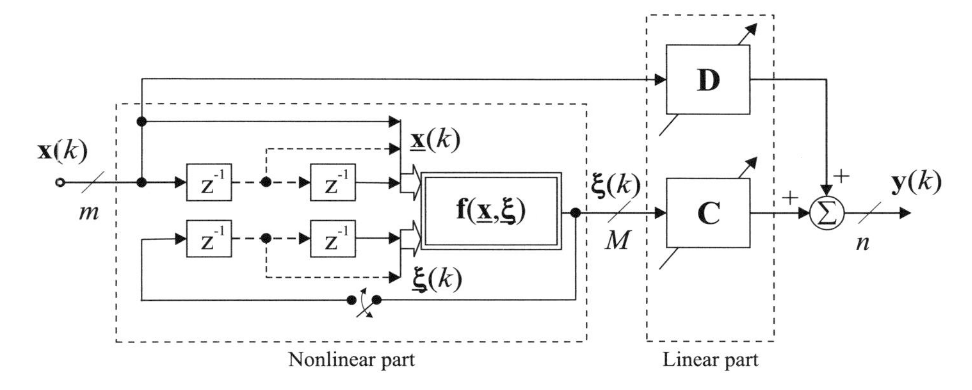

Fig. 1.14 Block diagram of a simplified nonlinear demixing NARMA model. For the switch in the open position we have a feed-forward MA model and for the switch closed we have a recurrent ARMA model.

an approach is justified in many practical applications. For example, for MEG or EEG, we can use a phantom of the human head with known artificial source excitations located in specific places inside of the phantom. For the cocktail party problem, in some case is possible to record for short-time windows original test speech sources. These short-time window training sources enable us to determine, on the basis of a supervised algorithm, a suitable nonlinear demixing model and associated nonlinear basis functions of the neural network and their parameters.

However, we assume that the mixing system is a slowly time-varying system for which some parameters fluctuate slightly over time, mainly due to the change in localization of source signals in space. Furthermore, we assume that training sources are available only for short-time slots. During the time windows in which the training signals are not available, we can apply an unsupervised learning algorithm which performs a fine adjustment of the output matrices **C** and **D** (by keeping the nonlinear model fixed). In this way, we will be able to estimate continuously in time the source signals. An exemplary implementation of the nonlinear state-space model using the radial basis function (RBF) neural network is shown in Figure 1.15 (see Chapter 12 for detail).

1.1.8 Why State Space Demixing Models?

There are several essential reasons why the state-space models provide a useful and powerful approach in blind signal processing:

- The mixing and filtering processes of unknown input sources $s_j(k)$, $(j = 1, 2, ..., n)$ may have different mathematical or physical models, depending on specific applications. The state-space demixing model is a flexible and universal linear model which describes a wide class of stable dynamical systems including standard multichannel deconvolution models with finite impulse response (FIR) filters, Gamma filters or

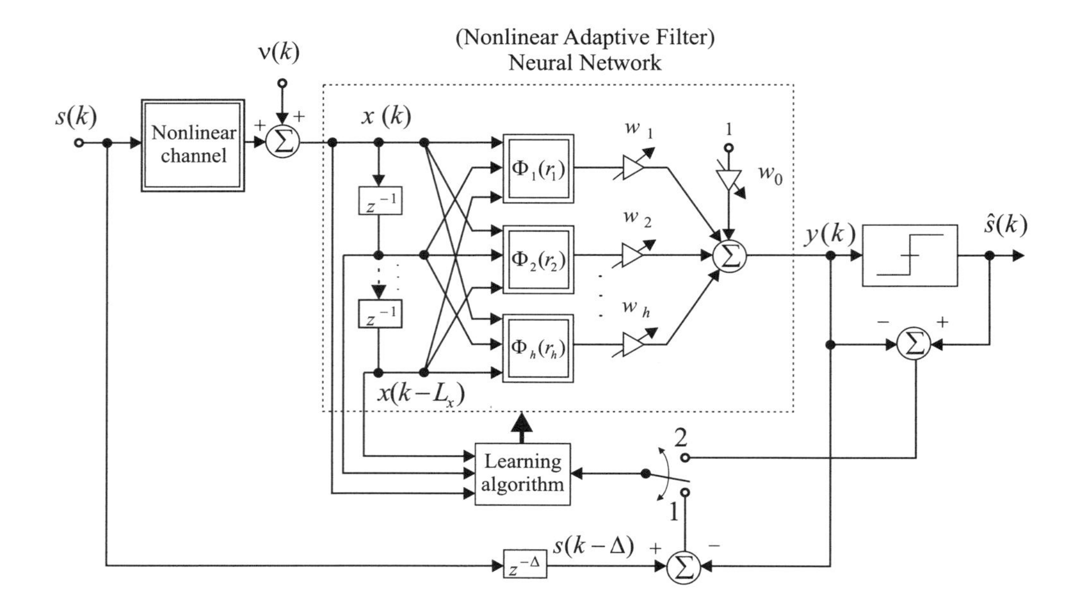

Fig. 1.15 Simplified model of RBF neural network applied for nonlinear semi-blind single channel equalization of binary sources; if the switch is in position 1, we have supervised learning, and unsupervised learning if it is in position 2.

more general models: AR (autoregressive), MA (moving average) and ARMA (autoregressive moving average) models as special cases.

- Moreover, such a dynamical demixing model enables us to generate many canonical realizations of the same dynamical system by using equivalent transformations.

- It is easy to note that the linear state-space model is an extension of the instantaneous mixture blind source separation model.

- State-space models have two subsystems: A linear, memoryless output layer and a dynamical linear or nonlinear recurrent network, which can be identified or updated using different approaches [283, 284, 285, 1358].

1.2 POTENTIAL APPLICATIONS OF BLIND AND SEMI-BLIND SIGNAL PROCESSING

The problems of independent component analysis (ICA), blind separation and multichannel deconvolution of source signals have received wide attention in various fields such as biomedical signal analysis and processing (EEG, MEG, ECG), geophysical data processing, data mining, speech enhancement, image recognition and wireless communications

[26, 34, 448, 1090]. In such applications a number of observations of sensor signals or data that are filtered superpositions of separate signals from different independent sources are available, and the objective is to process the observations in such a way that the outputs correspond to the separate primary source signals.

Acoustic applications are considered in situations where signals, from several microphones in a sound field produced by several speakers (the so-called cocktail-party problem) or from several acoustic transducers in an underwater sound field produced by engine noises of several ships (sonar problem) need to be processed. Radio and wireless communication examples include the observations corresponding to outputs of antenna array elements in response to several transmitters, and the observations may also include the effects of the mutual couplings of the elements. Other radio communication examples include the use of polarization multiplexing in microwave links. The maintenance of the orthogonality of the polarization cannot be perfect and there is still interference between the separate transmissions. Radar examples include the superposition of signals from different target modulating mechanisms as observed by multiple receivers whose elements are sensitive to different polarizations.

Let us consider some exemplary promising biomedical applications in more detail.

1.2.1 Biomedical Signal Processing

A great challenge in biomedical engineering is to non-invasively asses the physiological changes occurring in different internal organs of the human body (Figure 1.16 (a)). These variations can be modeled and measured often as biomedical source signals that indicate the function or malfunction of various physiological systems. To extract the relevant information for diagnosis and therapy, expert knowledge in medicine and engineering is also required.

Biomedical source signals are usually weak, nonstationary signals and distorted by noise and interference. Moreover, they are usually mutually superimposed. Besides classical signal analysis tools (such as adaptive supervised filtering, parametric or non-parametric spectral estimation, time-frequency analysis, and higher-order statistics), intelligent blind signal processing techniques (IBSP) can be used for preprocessing, noise and artifact reduction, enhancement, detection and estimation of biomedical signals by taking into account their spatio-temporal correlation and mutual statistical dependence.

One successful and promising application domain of blind signal processing includes those biomedical signals acquired with multi-electrode devices: Electrocardiography (ECG), electromyography (EMG), electroencephalography (EEG) and magnetoencephalography (MEG).

Exemplary applications in biomedical problems include the following:

- Fetal electrocardiogram (ECG) extraction, i.e., removing/filtering maternal electrocardiogram signals and noise from fetal electrocardiogram signals.

- Enhancement of low-level ECG components.

- Separation of transplanted heart signals from residual original heart signals.

- Separation of heart sounds from gastrointestinal acoustic phenomena (bowel-sounds). Bowel sounds can be measured in a non-invasive way by using microphones or accelerometers positioned on the skin.

- Reduction or blind separation of heart sounds from lung sounds using multichannel blind deconvolution.

- Cancellation of artifacts and noise from electroencephalographic and magnetoencephalographic recordings.

- Enhancement of evoked potentials (EP) and categorization of detected brain signals. (The brain potentials evoked by sensory stimulations such as visual, acoustic or somatosensory are generally called evoked potentials).

- Detection and estimation of sleep-spindles. (Sleep-spindles are specific phenomena of electroencephalograms (EEG) appearing during sleep; they are characterized by a group of oscillations in the range 11.5-15 Hz).

- Decomposition of brain sources as independent components and then localizing them in time and space.

Let us consider in more detail, some exemplary promising biomedical applications.

1.2.2 Blind Separation of Electrocardiographic Signals of Fetus and Mother

The mechanical action of the heart muscles is stimulated by electrical depolarization and repolarization signals. These quasi-periodical signals project potential differences to the skin level which can be measured and visualized as functions of time using electrocardiogram (ECG). As for adults, it would also be possible to measure the electrical activity of a fetal heart [720, 722]. The characteristics of a fetal electrocardiogram (FECG) can be very useful for determining if a fetus is developing or being delivered properly. These characteristics include an elevated heart rate that indicates fetal stress, cardiac arrythmia and ST segment depression which may indicate acidosis.

It is a non-trivial task to obtain an accurate and reliable FECG in a non-invasive fashion by using several electrodes. Problems develop due to the facts that the electrocardiogram (ECG) also contains a maternal electrocardiogram (MECG) which can be from one-half to one-thousandth the magnitude of the MECG. Moreover, the FECG will occasionally overlap the MECG and make it normally impossible to detect. Along with the MECG, extensive electromyographic (EMG) noise also interferes with the FECG and it can completely mask the FECG. The separation of fetal and maternal electrocardiograms from skin electrodes located on a pregnant woman's body may be modeled as a Blind Signal Processing problem (see Figure 1.16). The recordings pick up a mixture of FECG, MECG contributions, and other interferences, such as maternal electromyogram (MEMG), power supply interference, thermal noise from the electrodes and other electronic equipment. In fact, BSP techniques can be successfully applied to efficiently solve this problem and the first results are very promising [224, 226, 879]. Ordinary filtering and signal processing techniques have great difficulties with this problem [1280].

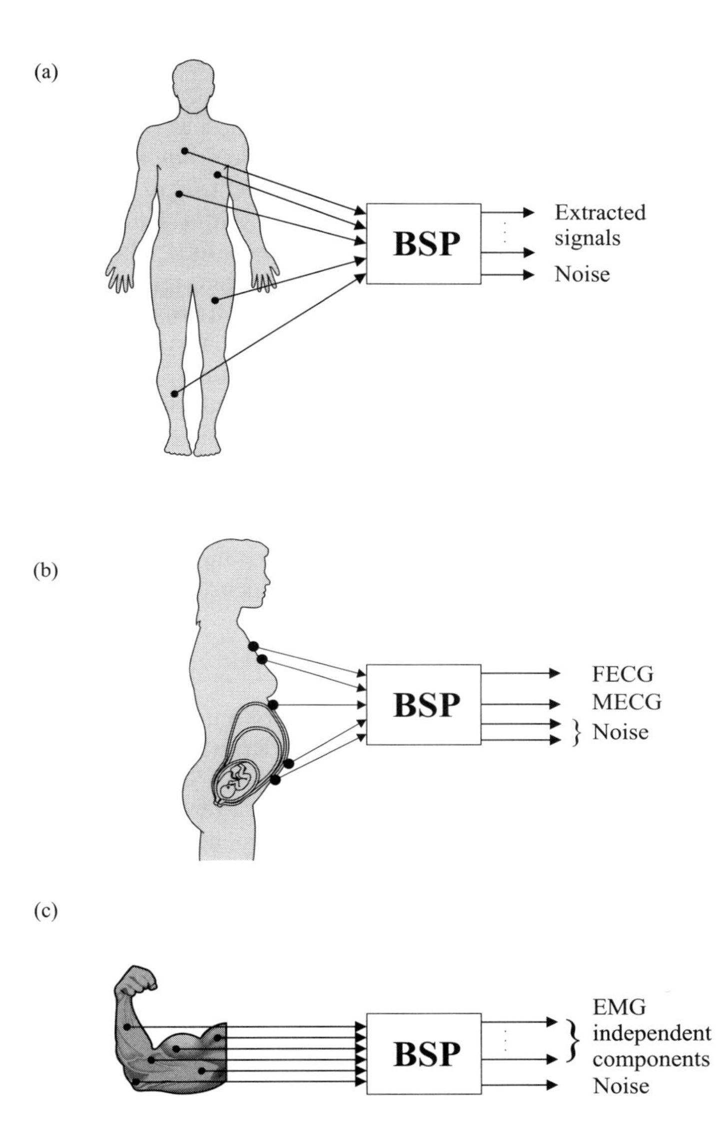

Fig. 1.16 Exemplary biomedical applications of blind signal processing: (a) A multi-recording monitoring system for blind enhancement of sources, cancellation of noise, elimination of artifacts and detection of evoked potentials, (b) blind separation of the fetal electrocardiogram (FECG) and maternal electrocardiogram (MECG) from skin electrode signals recorded from a pregnant woman, (c) blind enhancement and independent components analysis of multichannel electromyographic (EMG) signals.

1.2.3 Enhancement and Decomposition of EMG Signals

The movement and positioning of limbs are controlled by electrical signals travelling back and forth between the central nervous system and the muscles. Electromyography is a technique of recording of the electrical signals in the muscle (muscle action potentials). Electromyographic (EMG) signals recorded by a multi-electrode system provide important information about the brain motor system and the diagnosis of neuromuscular disorders that affect the brain, spinal cord, nerves or muscles. EMG signals, which are recorded simultaneously by several electrodes at low and moderate force levels can be composed of motor unit action potentials (MUAPs) generated by different motor units. The motor unit is the smallest functional unit of the muscle that can be voluntarily activated: It consists of a group of muscle fibers all innervated by the same motor neuron. In other words, MUAP consists of the spatial and temporal summation of all single fiber potentials innervated by the same motor neuron. The MUAP waveforms give information about the structural organization of the motor units [1380].

Blind signal processing techniques can be used for the enhancement of EMG signals. A more challenging problem is to apply BSS for decomposition of EMG signals into independent components and MUAPs. Such blind or semi-blind processing may be able to cluster MUAPs into groups of similar waveforms and provide important information about the brain motor system thus facilitating the assessment of neuromuscular pathology.

1.2.4 EEG and MEG Data Processing

Applications of BSP show special promise in the areas of non-invasive human brain imaging techniques to delineate the neural processes that underlie human cognition and sensoro-motor functions.

To understand human neurophysiology, we rely on several types of non-invasive neuroimaging techniques. These techniques include electroencephalography (EEG), magnetoencephalography (MEG), anatomical magnetic resonance imaging (MRI) and functional MRI (fMRI). While each of these techniques is useful, there is no single technique that provides both the spatial and temporal resolution necessary to make inferences about the intracranial brain sources of activity.

Very recently, several research groups have demonstrated that the techniques and methods of blind source separation (BSS) are related to those currently used in electromagnetic source localization (ESL) [836]. This framework provides a methodology by which several different types of information can be combined to aid in making inferences about a problem. Neural activity in the cerebral cortex generates small electric currents which create potential differences on the surface of the scalp (detected by EEG) as well as very small magnetic fields which can be detected using SQUIDs (SuperConducting QUantum Interference Devices). The greatest benefit of MEG is that it provides information that is complementary to EEG. In addition, the magnetic fields (unlike the electric currents) are not distorted by the intervening biological mass. Under certain circumstances, this allows precise localization of the neural currents responsible for the measured magnetic field.

Here, we give a very brief introduction to EEG and MEG [1244, 1245]. When a region of neural tissue (consisting of about 100,000 neurons) is synchronously active, detectable

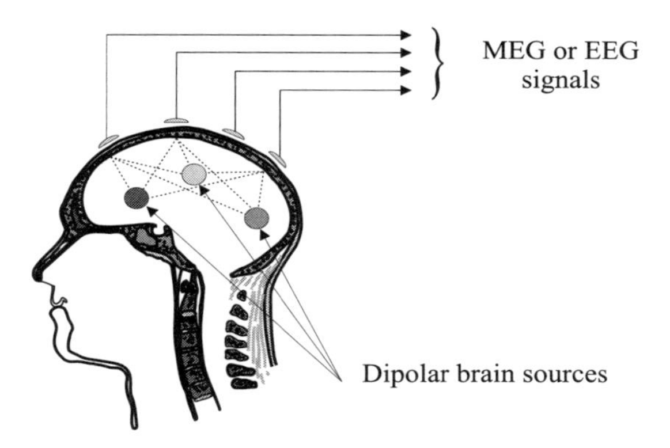

Fig. 1.17 Non-invasive multi-electrode recording of brain activity using EEG or MEG.

extracellular electric currents and magnetic fields are generated. These regions of activity can be modeled as "current dipoles" because they generate a dipolar electric current field in the surrounding volume of the head. These extracellular currents flow throughout the volume of the head and create potential differences on the surface of the head that can be detected with surface electrodes in a procedure called electroencephalography (EEG). One can also place super-conducting coils above the head and detect the magnetic fields generated by the activity in a procedure called magnetoencephalography (MEG).

If one knows the positions and orientations of the sources in the brain, one can calculate the patterns of electric potentials or magnetic fields on the surface of the head. This is called the forward problem. If otherwise one has only the patterns of electric potential or magnetic fields, then one needs to calculate the locations and orientations of the sources. This is called the inverse problem. Inverse problems are notoriously more difficult to solve than forward problems. In this case, given only the electric potentials and magnetic fields on the surface, there is no unique solution to the problem. The only hope is that there is some additional information available that can be used to constrain the infinite set of possible solutions to a single unique solution. This is where intelligent blind signal processing will be used.

The idea is that one must use all the available information to solve the problem. We will demonstrate this by focusing on an inverse problem, where we have information delivered from one or several devices, say EEG and/or MEG.

In Figure 1.17, we depicted three neural sources, represented in this case by equivalent current dipoles, in the cortical gray matter of the brain. The electrodes on the surface of the head detect the potential differences due to the extracellular currents generated by these active sources. The arrows merely demonstrate that each electrode detects some of the current flow from each neural source. The currents do not flow directly from the sources to the electrodes, but instead they flow throughout the volume of the entire head.

Determining active regions of the brain, given EEG/MEG measurements on the scalp is an important problem. An accurate and reliable solution to such a problem can give information about higher brain functions and patient-specific cortical activity. However, estimating the location and distribution of electric current sources within the brain from EEG/MEG recording is an ill-posed problem, because there is no unique solution and the solution does not depend continuously on data. The ill-posedness of the problem and distortion of sensor signals by large noise sources makes finding a correct solution a challenging analytic and computational problem.

The ICA approach and blind signal extraction methods are promising techniques for the extraction of useful signals from the EEG/MEG recorded raw data. The EEG/MEG data can be first decomposed into useful signal and noise subspaces using standard techniques like local and robust PCA, SVD and nonlinear adaptive filtering. Next, we apply ICA algorithms to decompose the observed signals (signal subspace) into independent components. The ICA approach enables us to project each independent component (independent "brain source") onto an activation map at the skull level. For each activation map, we can apply an EEG/MEG source localization procedure, looking only for a single dipole (or 2 dipole) per map. By localizing multiple dipoles independently, we can dramatically reduce the complexity of the computation and increase the likelihood of efficiently converging to the correct and reliable solution.

Figure 1.18 illustrates an example of a promising application of blind source separation and independent component analysis (ICA) algorithms for localization of the brain source signals activated after the auditory and somatosensory stimuli were applied simultaneously. In the MEG experiments performed in collaboration with the Helsinki University of Technology, Finland, the stimulus presented to the subject was produced with a sub-woofer, and the acoustic energy was transmitted to the shielded-room via a plastic tube with a balloon on the end [258]. The subject had his hands in contact with the balloon and sensed the vibration. In addition, the subject listened to the sound produced by the sub-woofer that provided auditory stimulation. Using ICA, we successfully extracted auditory and somatosensory evoked fields (AEF and SEF, respectively) and localized the corresponding brain sources [258] (see Figure 1.18).

1.2.5 Application of ICA/BSS for Noise and Interference Cancellation in Multi-sensory Biomedical Signals

The nervous systems of humans and animals must encode and process sensory information within the context of noise and interference, and the signals which are encoded (the images, sounds, etc.) have very specific statistical properties. One of the challenging tasks is how to reliably detect, enhance and localize very weak, nonstationary brain source signals corrupted by noise (e.g., evoked and event related potentials EP/ERP) by using EEG/MEG data.

Independent Component Analysis (ICA) and related methods like Adaptive Factor Analysis (AFA) are promising approaches for elimination of artifacts and noise from EEG/MEG data [253, 625]. In fact, for these applications, ICA/BSS techniques have been successfully applied to remove artifacts and noise including background brain activity, electrical activity of the heart, eye-blink and other muscle activity, and environmental noise efficiently.

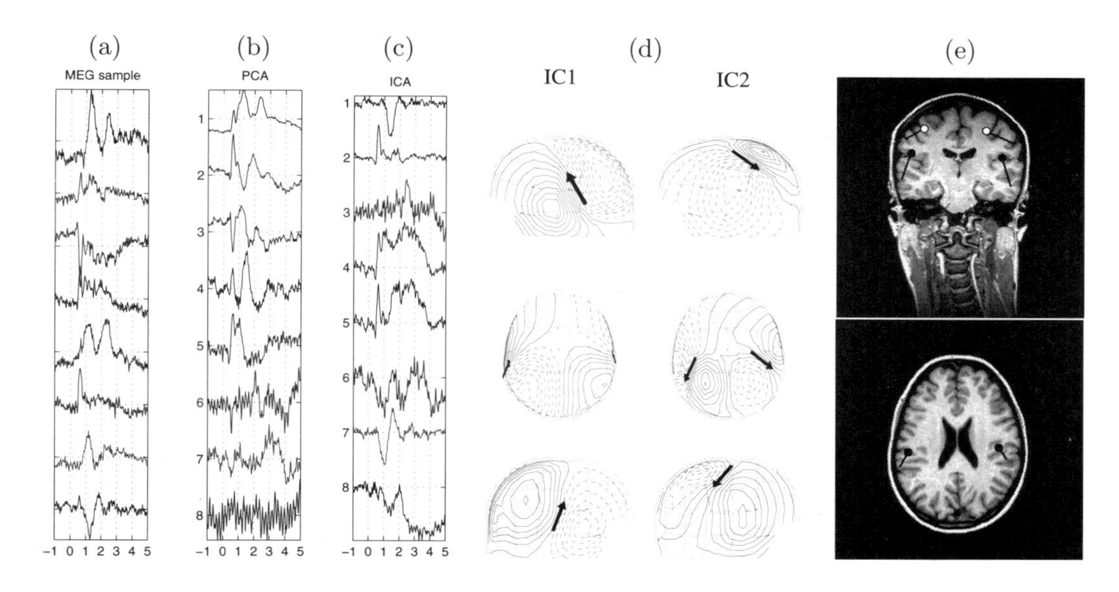

Fig. 1.18 (a) A subset of 122-MEG channels. (b) Principal and (c) independent components of the data. (d) Field patterns corresponding to the first two independent components. In (e) the superposition of the localizations of the dipole originating IC1 (black circles, corresponding to the auditory cortex activation) and IC2 (white circles, corresponding to the SI cortex activation) onto magnetic resonance images (MRI) of the subject. The bars illustrate the orientation of the net source current. Results were obtained in collaboration with researchers from Helsinki University of Technology, Finland [258].

However, most of the methods require manual detection, classification of interference components and the estimation of the cross-correlation between independent components and the reference signals corresponding to specific artifacts [589, 807, 808, 1244, 1245].

One of the important problem is how to automatically detect, extract and eliminate noise and artifacts. Another related problem is how to classify independent "brain sources" and artifacts. The automatic on-line elimination of artifacts and other interference sources is especially important for extended recordings, e.g., EEG/MEG recording during sleep.

Evoked potentials (EPs) of the brain are meaningful for clinical diagnosis and they are important factors in understanding higher order mechanisms in the brain. The EPs are usually embedded within the ongoing EEG/MEG with a signal to noise ratio (SNR) less than 0 dB, making them very difficult to extract by using only a single trial. The traditional method of EPs extraction uses ensemble averaging to improve the SNR. This often requires hundreds or even thousands of trials to obtain a usable waveform. Therefore, it is important to develop novel techniques that can rapidly improve the SNR and reduce the number of trials required to a minimum. Traditional signal processing techniques, such as Wiener filtering, adaptive noise cancellation, latency-corrected averaging [546] and invertible wavelet transform filtering, have recently been proposed for SNR improvements and ensemble reduction. However, these methods require *a priori* knowledge pertaining to the nature of the signal [530, 1138]. Since EP signals are known to be nonstationary, sparse and changing their characteristics from trial to trial, it is essential to develop novel algorithms for enhancement of single trial EEG/MEG noisy data.

The formulation of the problem can be given in the following form: Denote by $\mathbf{x}(k) = [x_1(k), x_2(k), ..., x_m(k)]^T$ the observed m-dimensional vector of noisy signals that must be "cleaned" from the noise and interference. Here we have two types of noise. The first is so called "inner" noise generated by some primary sources that cannot be observed directly but contained in the observations. They are mixtures of useful signals and random noise signals or other undesirable sources. The second type of noise is the sensor additive noise (observation errors) at the output of the measurement system. This noise is not directly measurable, either. Formally, we can write that an observed m-dimensional vector of sensor signals $\mathbf{x}(k)$ is a mixture of source signals plus observation errors

$$\mathbf{x}(k) = \mathbf{H}\,\mathbf{s}(k) + \boldsymbol{\nu}(k), \tag{1.26}$$

where $k = 0, 1, 2, ...$ is a discrete-time index; $\mathbf{H}$ is a full rank $(m \times n)$ mixing matrix; $\mathbf{s}(k) = [s_1(k), s_2(k), ..., s_n(k)]^T$ is an n-dimensional vector of sources containing useful signals and $\boldsymbol{\nu}(k)$ is an m-dimensional vector of additive white noise. We also assume that some useful sources are not necessarily statistically independent. Therefore, we cannot achieve perfect separation of primary sources by using any ICA procedure. However, our purpose here is not the separation of the sources but the removal of independent or uncorrelated noisy sources.

Let us emphasize that the problem consists of cancellation of the noise sources and reduction of observation errors based only on information about observed vector $\mathbf{x}(k)$.

A conceptual model for elimination of noise and other undesirable components from multi-sensory data is depicted in Figure 1.19. Firstly, ICA is performed using any robust (with respect to Gaussian noise) algorithm [20], [26], [254, 255], [850] by a linear transformation of sensory data as $\mathbf{y}(k) = \mathbf{W}\mathbf{x}(k)$, where the vector $\mathbf{y}(k)$ represents independent components. However, robust ICA methods allow us only to obtain an unbiased estimate of the unmixing matrix $\mathbf{W}$. Furthermore, due to memoryless structure such methods by definition, cannot remove the additive noise. Noise removal can be performed using optional nonlinear adaptive filtering and nonlinear noise shaping (see Figure 1.20). In the next stage, we classify independent signals $\hat{y}_j(k)$ and then remove noise and undesirable components by switching corresponding switches "off".

The projection of interesting or useful independent components (e.g., independent activation maps) $\tilde{y}_j(k)$ back onto the sensors (electrodes) can be done by the transformation $\hat{\mathbf{x}}(k) = \mathbf{W}^+\tilde{\mathbf{y}}(k)$, where $\mathbf{W}^+$ is the pseudo-inverse of the unmixing matrix $\mathbf{W}$. In the typical case, where the number of independent components is equal to the number of sensors, we have $\mathbf{W}^+ = \mathbf{W}^{-1}$.

The standard adaptive noise and interference cancellation systems may be subdivided into the following classes [546, 548]:

1. *Noise cancellation* (see Figure 1.20). This term is normally referred to the case, when we have both the primary signal $y_j(k) = \hat{y}_j(k) + n_j(k)$ contaminated with noise and reference noise $n_j(k)$, which is correlated with the noise $n_j(k)$ but is independent of the primary signal $\hat{y}_j(k)$. By feeding the reference signal to the linear adaptive filter

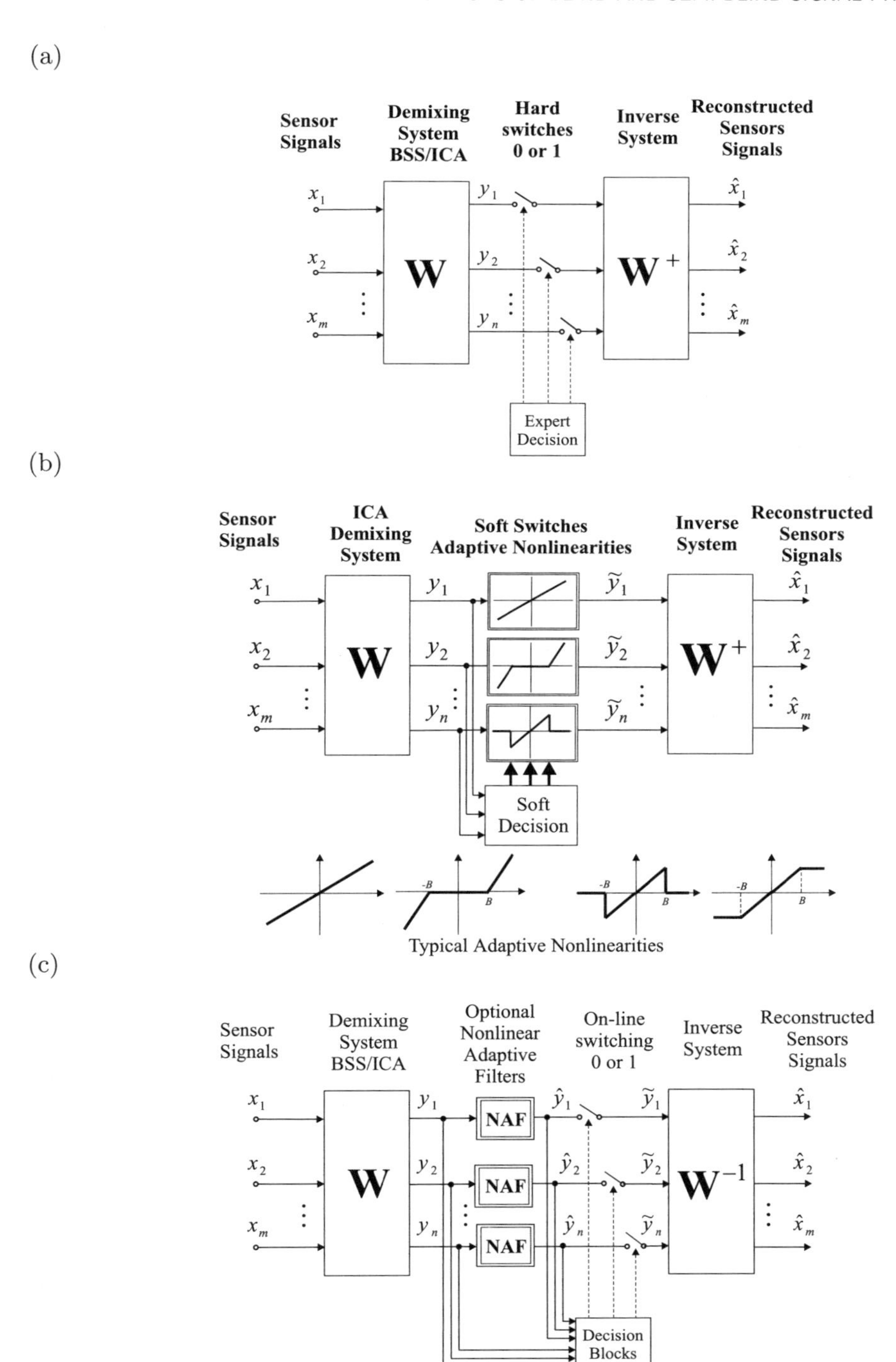

Fig. 1.19 Conceptual models for removing undesirable components like noise and artifacts and enhancing multi-sensory (e.g., EEG/MEG) data: (a) Using expert decision and hard switches, (b) using soft switches (adaptive nonlinearities in time, frequency or time-frequency domain), (c) using nonlinear adaptive filters and hard switches [280, 1248].

we are able to estimate or reconstruct the noise, then subtract it from the primary signal and thereby enhance the signal to noise ratio.

2. *Deconvolution-reverberation and echo cancelling.* This kind of interference cancelling is often referred to as echo cancelling, because it enables the removal of reverberations and echo from a single observed signal. A delayed version of the primary input signal is fed to the linear adaptive filter thus enabling the filter to reconstruct and remove reverberation from the primary signal. The deconvolver may also be used to cancel periodic interference components in the primary input such as power line interference, etc. The adaptive filter is able to extrapolate the periodic interference and subtract this component from the undelayed primary input (see Figure 1.20). This approach normally provides superior performance compared to standard notch or comb filtering techniques.

3. *Line enhancement.* In this case the objective is to estimate or extract a periodic or quasi periodic signal buried in noise. The adaptive filter receives the same input as the deconvolver, however, instead of subtracting the extrapolated periodic signal from the input, it outputs directly the enhanced signal (see Fig.1.20).

4. *Adaptive bandpass filtering.* Often we may take advantage of some *a priori* knowledge regarding the bandwidth of the signal we wish to denoise. By bandpass filtering of the signal, we eliminate a part of the frequency range where the useful signal is weak and the noise is comparatively strong, thus enhancing the overall signal to noise ratio.

In a traditional linear Finite Impulse Response (FIR) adaptive noise cancellation filter, the noise is estimated as a weighted sum of the delayed samples of reference interference. However, for many real world problems (when interference signals are related to the measured reference signals in a complex dynamic and nonlinear way) the linear adaptive noise cancellation systems mentioned above may not achieve acceptable levels of noise cancellation. Optimum interference and noise cancellation usually requires nonlinear adaptive processing of the recorded and measured on-line signals [258, 261].

A common technique for noise reduction is to split the signal in two or more bands. The high-pass bands are subjected to a threshold nonlinearity that suppresses low amplitude values while retaining high amplitude values (see Fig.1.20) [549, 545]. In addition to denoising and artifacts removal, ICA/BSS techniques can be used to decompose EEG/MEG data into separate components, each representing a physiologically distinct process or brain source. The main idea here is to apply localization and imaging methods to each of these components in turn. The decomposition is usually based on the underlying assumption of statistical independence between the activation of different cell assemblies involved. An alternative criterion for decomposition is temporal predictability or smoothness of components. These approaches lead to interesting and exciting new ways of investigating and analyzing brain data and developing new hypotheses how the neural assemblies communicate and process information. This is actually a very extensive and potentially promising research area, however these approaches still remain to be validated at least experimentally.

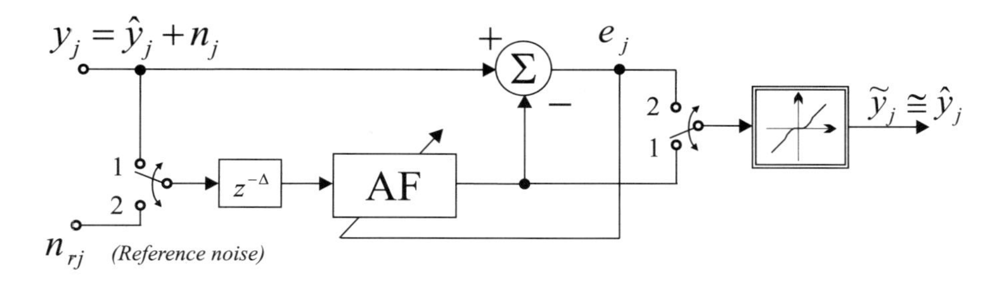

Fig. 1.20 Adaptive filter configured for line enhancement (switches in position 1) and for standard noise cancellation (switches in position 2).

1.2.6 Cocktail Party Problem

The "cocktail party" problem can be described as the ability to focus one's listening attention on a single talker among a cacophony of conversations and background noise. This problem has long been recognized as an interesting and challenging problem. Also known as the "cocktail party effect" or more technically, "multichannel blind deconvolution", the problem of separating a set of mixtures of convolved (filtered) signals, detected by an array of microphones, into their original source signals is performed extremely well by the human brain. Over the years attempts have been made to capture this function by using assemblies of abstracted neurons or adaptive processing units.

Humans are able to concentrate on listening to one voice in the midst of other conversations and noise, but not all the mechanisms for this process are completely understood. This specialized listening ability may be because of characteristics of the human speech production system, auditory system, or high-level perceptual and language processing.

In the EEG/MEG brain source separation algorithms, we make the fundamental assumption that the recorded signals form an instantaneous mixture, meaning that all of the signals are time-aligned so that they enter the sensors simultaneously without any delay.

Consider now an application to speech separation in which the sounds are recorded in a typical room using an array of microphones (see Fig.1.21). Each microphone will receive a direct copy of the sound source (at some propagation delay based on the location of both the sources and the microphone) as well as several reflected and modified (attenuated and delayed) copies of the sound sources (as the sound waves bounce off the walls and objects in the room).

The distortions of the recorded signals are dependent upon the reverberation and absorption characteristics of the room, as well as the objects within the room, and can be modeled as an impulse response in a linear system. The impulse response provides a model of all the possible paths that the sound sources take to arrive at the microphones.

To find a specific original sound source that was recorded with the microphones in a conference room, we must cancel out, or deconvolve, the rooms impulse response to the original sound source. Since we have no prior knowledge of what this impulse response

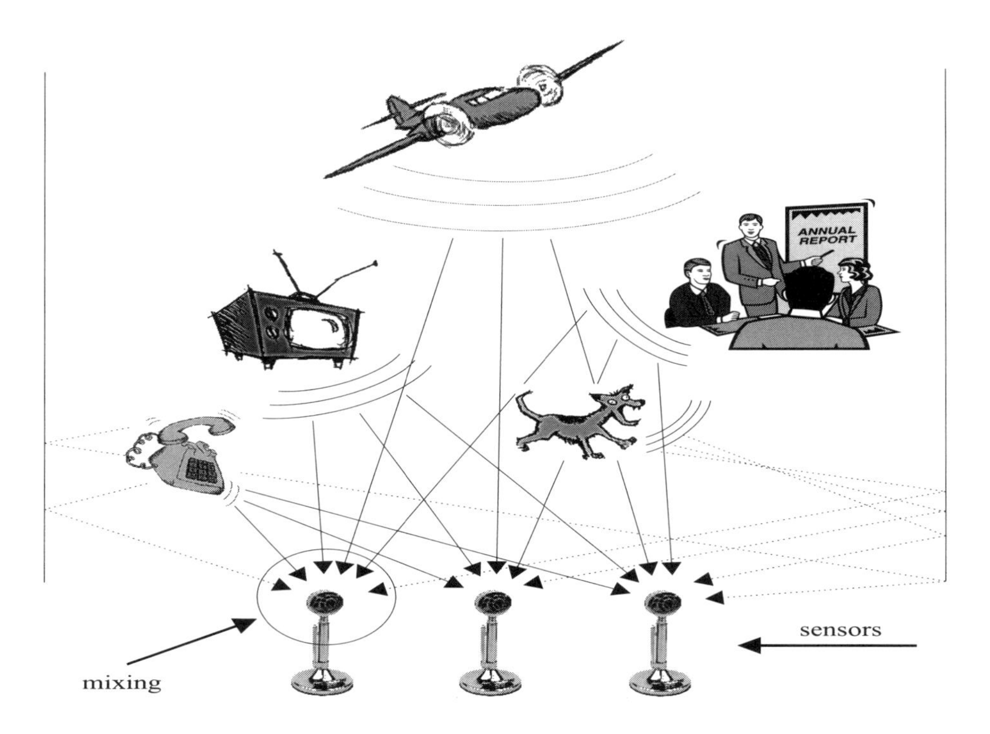

Fig. 1.21 Illustration of the "cocktail party" problem and speech enhancement.

of the room is, we call this process the multichannel blind deconvolution or cocktail party problem.

In the "cocktail party problem" our objective is to design intelligent adaptive systems and associated learning algorithms that have similar abilities to humans to focus attention on one conversation among the many that would be occurring concurrently in a hypothetical cocktail party.

1.2.7 Digital Communication Systems

Blind and semi-blind signal processing models and algorithms also arise in a wide variety of digital communications applications, for example, digital radio with diversity, dually polarized radio channels, high speed digital subscriber lines, multi-track digital magnetic recording, multiuser/multi-access communications systems, multi-sensor sonar/radar systems, to mention just a few. BSP algorithms are promising tools for a unified and optimal design of MIMO equalizers/filters/combiners for suppression of intersymbol interference (ISI), cochannel and adjacent channel interference (CCI and ACI) and multi-access interference (MAI). The state-of-the-art in this area incorporates complete knowledge of the MIMO transfer functions which is unrealistic for practical communication systems. The operating environment may consist of dispersive media involving multipath propagation

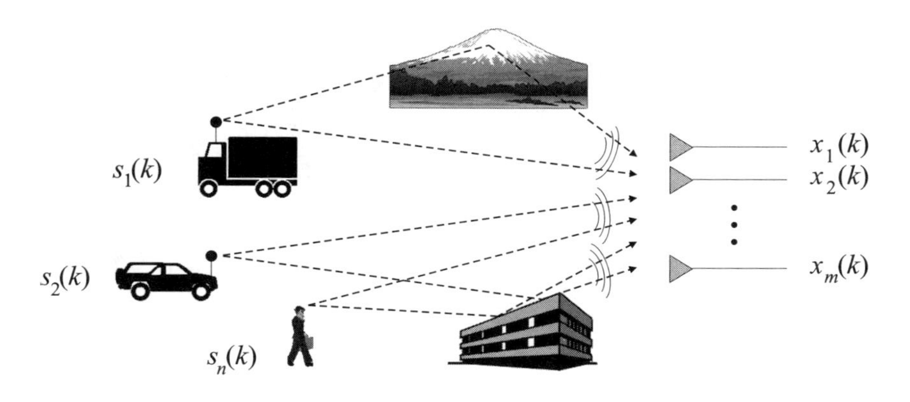

Fig. 1.22 Wireless communication scenario.

and frequency-selective fading, the characteristics of which are unknown at the receiver. The blind signal processing methods may result in more effective and computationally efficient algorithms for a broad class of digital communication systems such as high-speed digital subscriber lines, multi-track digital magnetic recording and multiuser wireless communications [991, 1086, 1087, 1187, 1194].

In Fig.1.22, we have an illustration of multiple signal propagation in a wireless communication scenario; a number of users broadcast digitally modulated signals $s_1, s_2, \ldots, s_n$ towards a base station in a multi-path propagation environment. In other words, via multiple paths digital signals are received at an antenna array from many users. The transmitted signals interact with various objects in the physical region before reaching the antenna array or the base station. Each path follows a different direction, with some unknown propagation delay and attenuation. This phenomenon of receiving a superposition of many time-varying delayed signals is called multi-path fading.

Moreover, in some cellular networks, there is another additional source of distortion, so called co-channel interference. This interference may be caused by multiple users that share the same frequency and time slot. The level of interference depends on the propagation environment, mobile location and mobile transmission power. Each transmitted signal is susceptible to multiple interference, multi-user interference and additive noise. In addition, the channel may be time-varying due to user mobility. Advanced blind signal processing algorithms are required to extract desired signals from the interference noise. An even more challenging signal processing problem is the blind joint space-time separation and equalization of transmitted signals, i.e. to estimate source signals and their channels in the presence of other co-channel signals and noise without the use of a training set.

1.2.7.1 *Why Blind?*

Blind signal processing techniques are promising because they require neither prior knowledge of the array response geometry nor any training signals in order to equalize the channels. Moreover, they are usually robust under severe multi-path fading en-

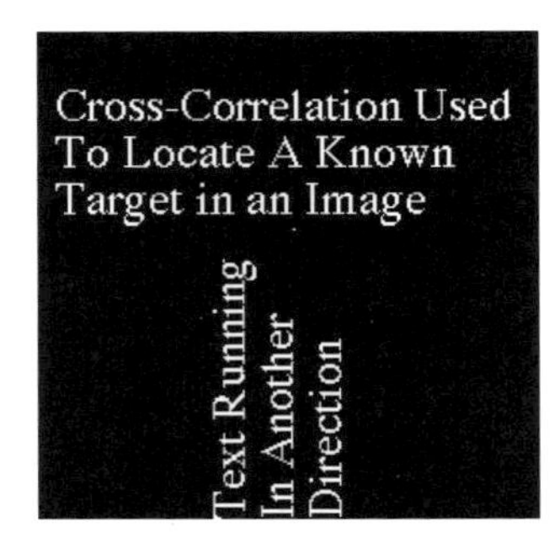

Fig. 1.23 Blind extraction of binary image from superposition of several images [757].

vironments. In situations where prior spatial knowledge or a set of short training sequences is available, the prior information can be incorporated in the semi-blind techniques applied.

There are several reasons to apply blind signal processing techniques [1086, 1133, 1134, 1135], such as

- Training examples for interference are often not available.
- In rapid time-varying channels, training may not be efficient.
- Capacity of the system can be increased by eliminating or reducing training sets.
- Multi-path fading during the training period may lead to poor source or channel estimations.
- Training in distributed systems requires synchronization and/or sending a training set each time a new link is to be set up. This may not be feasible in a multi-user scenario.

1.2.8 Image Restoration and Understanding

Image restoration involves the removal or minimization of degradation (blur, clutter, noise, interference etc.) in an image using *a priori* knowledge about the degradation phenomena. Blind restoration is the process of estimating both the true image and the blur from the degraded image characteristics, using only partial information about degradation sources and the imaging system.

Scientists and engineers are actively seeking to overcome the degradation of image quality caused by optical recording devices, atmospheric turbulence and other image degradation processes.

In many applications, it is necessary to extract or enhance the target image form an image corrupted or superimposed by other images. This is illustrated in Figure 1.23. In some applications, it is necessary to extract or separate all superimposed images as illustrated in Figure 1.24. In many instances, the degraded observation $g(x, y)$ can be modeled as a two-dimensional convolution of the true image $f(x, y)$ and the point-spread function (also called

Fig. 1.24 Blind separation of binary text (black and white) images from a single overlapped image [757].

the blurring function) $h(x, y)$ of a linear shift-invariant system plus some additive noise $n(x, y)$. That is, $g(x, y) = f(x, y) * h(x, y) + n(x, y)$. In many situations, the point-spread function $h(x, y)$ is known explicitly. The goal of the general blind deconvolution problem is to recover convolved signals, when only a noisy version of their convolution is available along with some or no partial information about either signal. In practice, all blind deconvolution algorithms require some partial information to be known and some conditions to be satisfied. Our main interest concerns image enhancement, where the degradation involves a convolution process. Blind deconvolution is a technique that permits recovery of the target object from a set of "blurred" images in the presence of a poorly determined or unknown point spread function (PSF). Regular linear and non-linear deconvolution techniques require a known PSF. In many situations, the point-spread function is known explicitly prior to the image restoration process. In these cases, the recovery of the image is known as the classical linear image restoration problem. This problem has been thoroughly studied and a long list of restoration methods for this situation includes numerous well-known techniques, a few examples of which are inverse filtering, Wiener filtering, subspace filtering and least-squares filtering. However, there are numerous situations in which the point-spread function is not explicitly known, and the true image must be identified directly from the observed image $g(x, y)$ by using partial or no information about the true image and the point-spread function. In these cases, we have a more difficult problem of blind deconvolution of images. For the "blind" case a set of multiple images (data cube) of the same target object is preferable, each having dissimilar PSF's. The blind deconvolution algorithm would be then able to

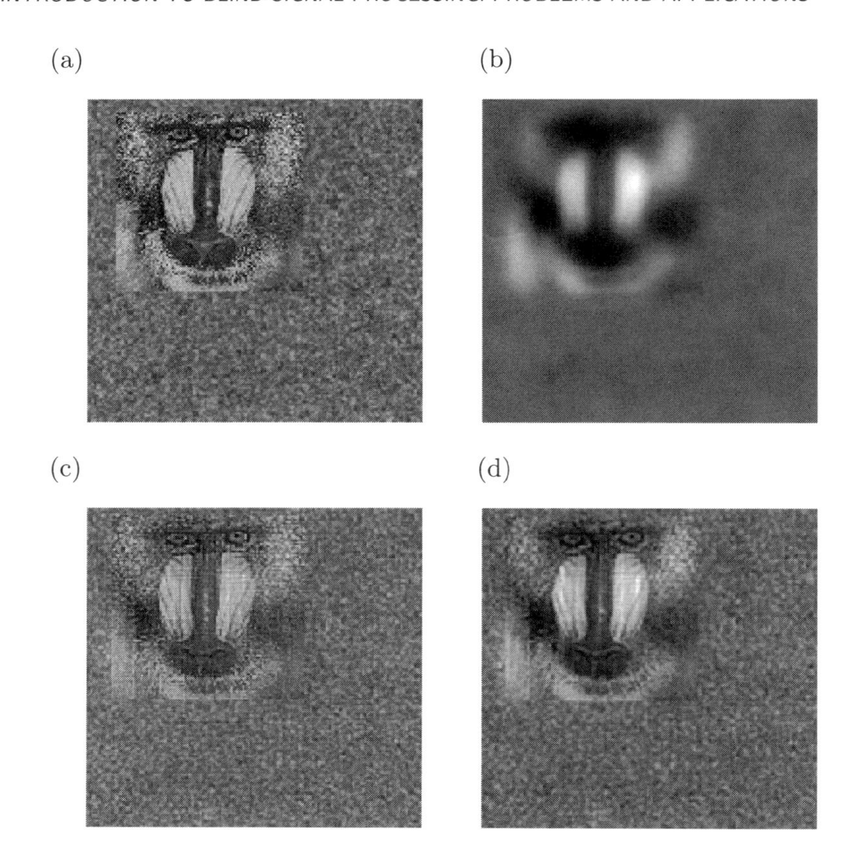

Fig. 1.25 Illustration of image restoration problem: (a) Original image (unknown), (b) distorted (blurred) available image, (c) restored image using blind deconvolution approach, (d) final restored image obtained after smoothing (post-processing) [323, 324].

restore not only the target object but also the PSFs. A good estimate of the PSF is helpful for quicker convergence but is not necessary.

The algorithmic way of processing and analyzing digital images has developed powerful means to interpret specific kinds of images, but failed to provide general image understanding methods that work on all kinds of images. This is mostly due to the fact that every image can be interpreted in many ways, as long as we do not know anything about what we expect to be in it. Thus, we need to build models about the expected contents of images in order to be able to "understand" them. There are many successful applications of image processing; but they are almost always fragile in the sense that it is difficult to adapt them to slightly different forms of imagery or to slightly different circumstances. The aim of Image Understanding is to address this fundamental problem by providing a set of image processing competences within an architecture that can observe the performance of each process, reflect on them, and choose to use/reject certain processes.

Obviously, there is a wide gap between the nature of images and descriptions. It is the bridging of this gap that has kept researchers very busy over the last two decades in the fields of Artificial Intelligence, Scene Analysis, Image Analysis, Image Processing, and Computer Vision. Nowadays we summarize these fields as "Image Understanding" research.

In order to make the link between image data and domain descriptions, an intermediate level of description is introduced. It generally contains geometric information. Processing usually starts with some image processing, where noise and distortion are reduced and certain important aspects of the imagery are emphasized. Then, events are extracted from the images that characterize the information needed for description. Typically, these events are such as blobs, edges, lines, corners and regions. They are stored at the intermediate level of abstraction. These are referred to in the literature as "features". Such descriptions are free of domain information - they are not specifically objects or entities of the domain of understanding, but they contain spatial and other information. It is the spatial/geometric (and other) information that can be analyzed in terms of the domain in order to interpret the images.

Image understanding is one of the most important and difficult tasks on the way towards what is known as artificial intelligence (AI). There is no working system yet which comes close to the capabilities of the human visual system. Some reasons are:

- Biological systems cannot be easily imitated.
- Specialized problem solving methods can hardly be generalized.
- The computational power needed for real-time digital image analysis exceeds the capacity of even the best workstation.

BSP algorithms, especially ICA/PCA, are promising approaches to Image Understanding. One of the ideas of a transform based image/signal description is to expand a signal by using a set of transform basis functions. A well-suited signal description allows us to extract characteristic signal properties which can be used for a variety of signal processing tasks, such as signal estimation, signal compression, or signal analysis. The suitability of an image transform in this context is connected to the efficiency of the transform in representing a given image, i.e. how many coefficients does a transform need to represent the image. The measure for efficiency is the sparseness of the transform coefficients, represented by the decay of the ordered coefficients from a given transform. The local singularities are characterized by location, orientation, and spatial extension. Finding a suitable signal transform for the description of linear singularities is the key for an analysis of the underlying information contained within natural images.

The question arises: How to efficiently describe images which contain a linear or nonlinear mixture of very different signal components. The application of classical signal transforms (such as the Fourier or wavelet transform) to such images is limited since there is no single dominant signal component that can be efficiently estimated with one transform. The idea of ICA or related decomposition approaches is to decompose the image to basic independent components and to start with a large set of independent components. For the image description, only those components that contribute to a sparse description are used. We want to have a small (sparse) number of large coefficients that condense the image information.

The reason for desiring a sparse representation is that under certain assumptions it will also reduce statistical dependencies among units: This provides a more efficient representation of the image structure.

One of the specific goals of research is to understand the coding strategies used by the human visual system for accomplishing tasks such as object recognition and scene analysis. In a task such as face recognition, much of the important information may be contained in the high-order relationships among the image pixels. ICA and related decomposition/separation techniques are able to recover signal components out of signal mixtures. Moreover, ICA/BSS image decomposition allows us to efficiently represent signal components in images. It also allows us to determine the "interesting" signal components in images (see Fig.1.19). Therefore, ICA image decomposition is a promising tool for image analysis, reconstruction, and classification, as well as for feature detection and image indexing. Statistically independent basis images (e.g., for the faces) can be viewed as a set of independent (facial) features. Unlike PCA basis vectors, the ICA basis images are spatially localized. The representation consists of the coefficients for the linear combinations of basis images that comprised each image. Theories of sensory coding based on the idea of maximizing information transmission while eliminating statistical redundancy from the raw sensory signals have been successful in explaining several properties of neural responses in the visual system such as receptive fields in the visual cortex.

The long-term goal of the Image Understanding research is to develop computational theories and techniques for use in artificial vision systems for which the performance matches or exceeds that of humans, by analyzing sequence of images in space, time and frequency domains.

2

Solving a System of Linear Algebraic Equations and Related Problems

A problem well stated is a problem half solved.

—(C.F. Kettering)

A problem adequately stated is a problem well on it's way to being solved.

—(R. Buckminster Fuller)

In modern signal and image processing fields such as biomedical engineering, computer tomography (image reconstruction from projections), automatic control, robotics, speech and communication, linear parametric estimation, models such as auto-regressive moving-average (ARMA) and linear prediction (LP) have been extensively utilized. In fact, such models can be mathematically described by an overdetermined system of linear algebraic equations. Such systems of equations are often contaminated by noise or errors, thus the problem of finding an optimal and robust with respect noise solution arises. On the other hand, wide classes of extrapolation, reconstruction, estimation, approximation, interpolation and inverse problems can be converted to minimum norm problems of solving underdetermined systems of linear equations. Generally speaking, in signal processing applications, the overdetermined system of linear equations describes filtering, enhancement, deconvolution and identification problems, while the underdetermined case describes inverse and extrapolation problems. This chapter provides a tutorial in the problem of solving large overdetermined and underdetermined systems of linear equations, especially when there is an uncertainty in parameter values and/or the systems are contaminated by noise. A special emphasis is placed in on-line fast adaptive and iterative algorithms for arbitrary noise statistics. This chapter also gives several illustrative examples that demonstrate the characteristics of the developed algorithms.

2.1 FORMULATION OF THE PROBLEM FOR SYSTEMS OF LINEAR EQUATIONS

Let us assume that we want to solve a large set of linear algebraic equations written in scalar form as

$$\sum_{j=1}^{n} h_{ij}\, s_j = x_i, \qquad (i = 1, 2, ..., m) \tag{2.1}$$

or in matrix form

$$\mathbf{H}\,\mathbf{s} = \mathbf{x}. \tag{2.2}$$

Here, $\mathbf{s}$ is the n-dimensional unknown vector, $\mathbf{x}$ is the m-dimensional sensor or measurement vector and $\mathbf{H} = [h_{ij}]$ is the $m \times n$ real, typically full column rank matrix with known elements.[1] Note that the number of equations is generally not restricted to $m = n$; it can be less than, equal to or greater than the number of variables, i.e., the components of $\mathbf{s}$. If $m < n$, then the system of equations is called underdetermined, and if $m > n$, then the system of equations is called overdetermined. Of course, such a system of equations may have a unique solution $\mathbf{s}_*$, an infinite number of solutions or no exact solution may exist. In practice, for linear estimation problems, a system of linear (overdetermined) equations is formulated in a more general form as

$$\mathbf{H}\,\mathbf{s} = \mathbf{x} - \mathbf{e} = \mathbf{x}_{true}, \tag{2.3}$$

where $\mathbf{H} = [h_{ij}] \in \mathbb{R}^{m\times n}$ is the matrix model, $\mathbf{x} \in \mathbb{R}^m$ is the vector of observations or measurements, $\mathbf{e} \in \mathbb{R}^m$ is the vector of unknown noise or measurement errors[2], $\mathbf{x}_{true} \in \mathbb{R}^m$ is the vector of true but unknown values and $\mathbf{s} \in \mathbb{R}^n$ is the vector of the system parameters or sources to be estimated or computed. From a practical point of view, it is usually desirable to find a (minimal norm) solution $\mathbf{s}_*$, if a solution exists, or to find an approximate solution which comes as close as possible to the original one, subject to a suitable optimality criterion if no exact solution exists. The problem can be formulated as an optimization problem:

Find a vector $\mathbf{s} \in \mathbb{R}^n$ that minimizes the scalar objective (cost) function

$$J_p(\mathbf{s}) = \|\mathbf{x} - \mathbf{H}\,\mathbf{s}\|_p = \|\mathbf{e}(\mathbf{s})\|_p, \quad p \geq 1, \tag{2.4}$$

where the residual error vector $\mathbf{e}$ for a given vector $\mathbf{s}$

$$\mathbf{e}(\mathbf{s}) = [e_1(\mathbf{s}), e_2(\mathbf{s}), ..., e_m(\mathbf{s})]^T \tag{2.5}$$

has the components

$$e_i(\mathbf{s}) = x_i - \mathbf{h}_i^T \mathbf{s} = x_i - \sum_{j=1}^{n} h_{ij}\, s_j, \quad (i = 1, 2, ..., m) \tag{2.6}$$

[1] In contrast to other chapters in this book, we assume here that matrix $\mathbf{H}$ is known *a priori* or can be estimated. In the next chapters, we will explain how such a matrix can be identified or estimated.

[2] In some applications *a priori* knowledge of the statistical nature of the error (noise) is available. Typically, it is assumed that noise is zero-mean and has a Gaussian distribution.

and $\|\mathbf{e}\|_p$ is the p-norm of the vector $\mathbf{e}$. The p-norm solution $\mathbf{s}_*$ satisfies the equation

$$\mathbf{H}\,\mathbf{s}_* + \mathbf{e}(\mathbf{s}_*) = \mathbf{x}. \tag{2.7}$$

Thus, $\mathbf{s}_*$ minimizes the p-norm of the residual vector $\mathbf{e}(\mathbf{s}) = \mathbf{x} - \mathbf{H}\,\mathbf{s}$ with respect to the vector $\mathbf{s}$, i.e., the following relation holds:

$$\|\mathbf{x} - \mathbf{H}\,\mathbf{s}_*\|_p \leq \|\mathbf{x} - \mathbf{H}\,\mathbf{s}\|_p \quad \forall \mathbf{s} \in \mathbb{R}^n. \tag{2.8}$$

It is important to note that the cost function 2.4) is a convex and continuous function of the vectors $\mathbf{x}$, $\mathbf{s}$ and matrix $\mathbf{H}$. The convexity property is essential since it ensures that any local minimum of J_p is also a global minimum.

For the optimization problem (2.4), there are three special cases which are important in practice:

(a) For $p = 1$, the problem is referred to as the 1-norm or least absolute deviation (LAD) problem.

(b) For $p = 2$, the problem is called the 2-norm or linear least-squares (LS) problem.

(c) For $p = \infty$, the problem is referred to as the Chebyshev (infinity - norm) or minimax problem.

A proper choice of the norm depends on the specific application and the distribution of the errors within the data $\mathbf{x}$. If the error distributions have sharply defined transitions (such as the uniform distribution), then the Chebyshev (infinity) norm may be the most suitable choice. For the Laplacian error distribution one can use the 1-norm. For example, in cases where the data being analyzed contain a few outliers (wild points or gross errors in the data) the minimum 1-norm solution is preferable since it tends to ignore bad data points. For the normal distribution of the errors, the best choice would be the 2-norm. It should be noted that, in the special case of zero noise, the matrix $\mathbf{H}$ is square and nonsingular, $\|\mathbf{e}(\mathbf{s}_*)\|_p = 0$ and all the three special cases ($p = 1, 2, \infty$) mentioned above are equivalent, i.e., they provide the same unique solution as $\mathbf{s}_* = \mathbf{H}^{-1}\,\mathbf{x}$. However, in general, for the noisy case for all the three problems above, the solutions are different. Moreover, both the 1-norm and infinity-norm solutions are not necessarily unique [276].

2.2 LEAST-SQUARES PROBLEMS

2.2.1 Basic Features of the Least-Squares Solution

The linear least-squares (LS) problem is a special case of the nonlinear least-squares problem (NLS) which is probably the most fundamental operation in signal processing. It is basic to Fourier analysis, deconvolution, correlation, optimum parameter estimation in Gaussian noise, linear prediction and many other signal processing methods (cf. next chapters).
The linear least-squares problem associated with the problem (2.4) can be formulated as

follows:
Find the vector $\mathbf{s} \in \mathbb{R}^n$ that minimizes the cost (energy) function

$$J(\mathbf{s}) = \frac{1}{2} \|\mathbf{x} - \mathbf{H}\,\mathbf{s}\|_2^2 = \frac{1}{2} \mathbf{e}^T \mathbf{e} = \frac{1}{2} \sum_{i=1}^{m} e_i^2, \tag{2.9}$$

where

$$e_i(\mathbf{s}) = x_i - \mathbf{h}_i^T\,\mathbf{s} = x_i - \sum_{j=1}^{n} h_{ij}\,s_j. \tag{2.10}$$

The cost function achieves the global minimum when its gradient equals zero:

$$\nabla J(\mathbf{s}) = \mathbf{H}^T\,(\mathbf{x} - \mathbf{H}\,\mathbf{s}) = \mathbf{0}. \tag{2.11}$$

Hence, $\mathbf{s}_* = (\mathbf{H}^T\,\mathbf{H})^{-1}\,\mathbf{H}^T\,\mathbf{x}$ and $\mathbf{H}^T \mathbf{e}(\mathbf{s}_*) = \mathbf{0}$, where $\mathbf{e}(\mathbf{s}_*) = \mathbf{x} - \mathbf{H}\,\mathbf{s}_*$.

Remark 2.1 *The cost function $J(\mathbf{s})$ has an interesting statistical interpretation. When the noise vector $\mathbf{e}$ is drawn from the Gaussian distribution with zero-mean and unity variance, the cost function $J(\mathbf{s})$ is proportional to the negative of the logarithm of the likelihood. Hence, minimizing $J(\mathbf{s})$ is equivalent to maximizing the log likelihood.*

The solutions of the LS problem can be grouped into three categories:

(i) $\mathbf{H} \in \mathbb{R}^{n \times n}$, rank$[\mathbf{H}] = n = m$ (exactly determined case): a unique solution

$$\mathbf{s}_* = \mathbf{H}^{-1}\,\mathbf{x} \text{ exists with } J(\mathbf{s}_*) = 0, \tag{2.12}$$

(ii) $\mathbf{H} \in \mathbb{R}^{m \times n}$, rank$[\mathbf{H}] = n < m$ (overdetermined case): an exact solution of the problem (2.2) generally does not exist, but the least-squares error solution can be expressed uniquely as

$$\mathbf{s}_* = (\mathbf{H}^T\,\mathbf{H})^{-1}\,\mathbf{H}^T\,\mathbf{x} = \mathbf{H}^+\,\mathbf{x}, \tag{2.13}$$

with

$$J(\mathbf{s}_*) = \frac{1}{2} \mathbf{x}^T\,(\mathbf{I} - \mathbf{H}\,\mathbf{H}^+)\,\mathbf{x} \geq 0, \tag{2.14}$$

where $\mathbf{H}^+$ is the Moore-Penrose pseudo-inverse,

(iii) $\mathbf{H} \in \mathbb{R}^{m \times n}$, rank$[\mathbf{H}] = m < n$ (underdetermined case): the solution of the problem (2.2) is not unique, but the LS problem can give the minimum 2-norm $\|\mathbf{s}\|_2^2$ unique solution [683]:

$$\mathbf{s}_* = \mathbf{H}^T\,(\mathbf{H}\,\mathbf{H}^T)^{-1}\,\mathbf{x} = \mathbf{H}^+\,\mathbf{x}, \tag{2.15}$$

with

$$J(\mathbf{s}_*) = 0 \tag{2.16}$$

and thereby the resulting value of the norm leads to

$$\|\mathbf{s}_*\|_2^2 = \mathbf{x}^T\,(\mathbf{H}\,\mathbf{H}^T)^{-1}\,\mathbf{x}. \tag{2.17}$$

Remark 2.2 *We can straightforwardly derive formula (2.15) by formulating the Lagrange function as*

$$L(\mathbf{s}, \boldsymbol{\lambda}) = \frac{1}{2}\mathbf{s}^T\mathbf{s} + \boldsymbol{\lambda}^T(\mathbf{x} - \mathbf{H}\mathbf{s}), \tag{2.18}$$

where $\boldsymbol{\lambda} = [\lambda_1, \lambda_2, \ldots, \lambda_n]^T$ is the vector of Lagrangian multipliers. For the optimal solution, the gradient of the Lagrange function becomes zero

$$\nabla_{\mathbf{s}} L(\mathbf{s}, \boldsymbol{\lambda}) = \mathbf{s}_* - \mathbf{H}^T\boldsymbol{\lambda}_* = \mathbf{0}. \tag{2.19}$$

Hence, we have $\boldsymbol{\lambda}_* = (\mathbf{H}\mathbf{H}^T)^{-1}\mathbf{H}\,\mathbf{s}_*$ and $\mathbf{s}_* = \mathbf{H}^T\boldsymbol{\lambda}_* = \mathbf{H}^T(\mathbf{H}\mathbf{H}^T)^{-1}\mathbf{x} = \mathbf{H}^+\mathbf{x}$.

It should be noted that generalized (the Moore-Penrose) pseudo-inverse matrix for the underdetermined case is $\mathbf{H}^+ = \mathbf{H}^T(\mathbf{H}\mathbf{H}^T)^{-1}$, $(n \geq m)$, while for the overdetermined case is $\mathbf{H}^+ = (\mathbf{H}^T\mathbf{H})^{-1}\mathbf{H}^T$, $(m \geq n)$ under assumption that the matrix $\mathbf{H}$ is full row rank or full column rank, respectively.

In the general case, when rank $(\mathbf{H}) < \min(m, n)$, i.e., matrix $\mathbf{H}$ may be rank deficient, we seek the minimum norm solution $\mathbf{s}_*$, which minimizes both $\|\mathbf{s}\|_2$ and $\|\mathbf{x} - \mathbf{H}\,\mathbf{s}\|_2$.

Assuming that the matrix $\mathbf{H}$ is not ill-conditioned, we can use (2.13) in a straightforward manner to find the parameters of the least-squares solution. For ill-conditioned problems, direct inversion of the matrix $\mathbf{H}^T\mathbf{H}$ may cause the noise to be amplified to a degree which is unacceptable in practice. In such cases, more robust solutions to the estimation problem need to be found.

2.2.2 Weighted Least-Squares and Best Linear Unbiased Estimation

In some applications (for $m > n$), it is reasonable to satisfy some of the more important equations at the expense of the others, i.e., it is required that some of the equations in (2.1) are nearly satisfied while "larger" errors are acceptable in the remaining equations. In such a case, a more general weighted least-squares problem can be formulated:

Minimize the cost function

$$J(\mathbf{s}) = \frac{1}{2}(\mathbf{x} - \mathbf{H}\,\mathbf{s})^T\,\boldsymbol{\Sigma}_e\,(\mathbf{x} - \mathbf{H}\,\mathbf{s}) = \frac{1}{2}\|\boldsymbol{\Sigma}_e^{1/2}\mathbf{e}\|_2^2, \tag{2.20}$$

where $\boldsymbol{\Sigma}_e$ is an $m \times m$ positive definite (typically diagonal) matrix which reflects a weighted error.

Remark 2.3 *The cost function (2.20) corresponds to a negative likelihood when the zero-mean error vector $\mathbf{e}$ is drawn from the Gaussian distribution with covariance matrix $\boldsymbol{\Sigma}_e^{-1}$.*

In the case that the matrix $\boldsymbol{\Sigma}_e$ is diagonal, i.e., $\boldsymbol{\Sigma}_e = \text{diag}(\sigma_{1e}^2, \sigma_{2e}^2, \ldots, \sigma_{me}^2)$, the 2-norm $\frac{1}{2}\|\boldsymbol{\Sigma}_e^{1/2}\mathbf{e}\|_2^2 = \frac{1}{2}\sum_{i=1}^m \sigma_{ie}^2 e_i^2$ will be minimized instead of $\frac{1}{2}\|\mathbf{e}\|_2^2$. Using such a transformation, it is easy to show that the optimal weighted least-squares solution has the form[3]

$$\mathbf{s}_* = \mathbf{s}_{\text{WLS}} = (\mathbf{H}^T\,\boldsymbol{\Sigma}_e\,\mathbf{H})^{-1}\,\mathbf{H}^T\,\boldsymbol{\Sigma}_e\,\mathbf{x} \tag{2.21}$$

[3]Matrix $\mathbf{H}^T\boldsymbol{\Sigma}_e\mathbf{H}$ in (2.21) must be nonsingular for its inverse to exist.

and its minimum LS error is

$$J(\mathbf{s}_*) = \mathbf{x}^T \left[\boldsymbol{\Sigma}_e - \boldsymbol{\Sigma}_e \mathbf{H} (\mathbf{H}^T \boldsymbol{\Sigma}_e \mathbf{H})^{-1} \mathbf{H}^T \boldsymbol{\Sigma}_e\right] \mathbf{x}. \tag{2.22}$$

It turns out that an optimal and natural choice for the weighting matrix $\boldsymbol{\Sigma}_e$ is the inverse of the covariance matrix of the noise under the assumptions that the noise is zero-mean with the positive definite covariance matrix $\mathbf{R}_{\mathbf{ee}}$ and $\mathbf{H}$ is deterministic.[4] Assuming that $\boldsymbol{\Sigma}_e = \mathbf{R}_{\mathbf{ee}}^{-1}$ is known or can be estimated, we obtain the so called BLUE (the Best Linear Unbiased Estimator)

$$\mathbf{s}_* = \mathbf{s}_{\mathrm{BLUE}} = (\mathbf{H}^T \mathbf{R}_{\mathbf{ee}}^{-1} \mathbf{H})^{-1} \mathbf{H}^T \mathbf{R}_{\mathbf{ee}}^{-1} \mathbf{x}, \tag{2.23}$$

which also minimizes the mean-square error.

The matrix $\mathbf{R}_{\mathbf{ee}}^{-1}$ emphasizes (amplifies) the contributions of the precise measurements and suppresses the contributions of the imprecise or noisy measurements. The above results can be summarized in the form of the following well known Gauss-Markov Theorem [683]:

Theorem 2.1 Gauss-Markov Estimator *If a system is described by a set of linear equations*

$$\mathbf{x} = \mathbf{H}\,\mathbf{s} + \mathbf{e}, \tag{2.24}$$

where $\mathbf{H}$ *is a known* $m \times n$ *matrix,* $\mathbf{s}$ *is an* $n \times 1$ *vector of parameters to be estimated, and* $\mathbf{e}$ *is an* $m \times 1$ *arbitrary distributed noise vector with zero-mean and known covariance matrix* $\mathbf{R}_{\mathbf{ee}} = E\{\mathbf{e}\mathbf{e}^T\}$, *then the BLUE of* $\mathbf{s}$ *is*

$$\mathbf{s}_{BLUE} = (\mathbf{H}^T \mathbf{R}_{\mathbf{ee}}^{-1} \mathbf{H})^{-1} \mathbf{H}^T \mathbf{R}_{\mathbf{ee}}^{-1} \mathbf{x}, \tag{2.25}$$

where $\mathbf{s}_{BLUE} = [\hat{s}_1, \hat{s}_2, \ldots, \hat{s}_n]^T$ *and the minimum variance of* $\hat{s}_j$ *is written as*

$$var(\hat{s}_j) = E\{\hat{s}_j^2\} = \left[\mathbf{H}^T \mathbf{R}_{\mathbf{ee}}^{-1} \mathbf{H}\right]_{jj}. \tag{2.26}$$

If the noise has the same variance, i.e., $\mathbf{R}_{\mathbf{ee}} = \sigma_e^2 \mathbf{I}_m$, the BLUE reduces to the standard Least-Squares formula

$$\mathbf{s}_{\mathrm{LS}} = (\mathbf{H}^T \mathbf{H})^{-1} \mathbf{H}^T \mathbf{x}. \tag{2.27}$$

2.2.3 Basic Network Structure-Least-Squares Criteria

To formulate the above problems (2.1)-(2.3) in terms of nonlinear adaptive filter and artificial neural networks (ANNs), the key step is to construct an appropriate computational cost (energy) function $J(\mathbf{s})$ such that the lowest energy state corresponds to the optimal solution $\mathbf{s}_*$ [276]. The derivations of the energy function enable us to transform the minimization problem into a set of differential or difference equations, on the basis of which

[4]In some applications, the matrix $\mathbf{H}$ varies in time.

we design ANN architectures with appropriate connection weights (synaptic strengths) and input excitations [276].

There are many ways to connect neuron-like computing units (cells) into large-scaled neural networks. These different patterns of connections between the cells are called architectures or circuit structures. The purpose of this section is to review known circuit structures and to propose some new configurations with improved performance and/or with a reduced set of such computing units. Using a general gradient approach for minimization of the energy function, the problem formulated by Eq. (2.9) can be mapped to a set of differential equations (an initial value problem) written in the matrix form as

$$\frac{d\mathbf{s}}{dt} = -\boldsymbol{\mu}\,\nabla J(\mathbf{s}) = \boldsymbol{\mu}\,\mathbf{H}^T(\mathbf{x} - \mathbf{H}\,\mathbf{s}) = \boldsymbol{\mu}\,\mathbf{H}^T\,\mathbf{e}, \tag{2.28}$$

where $\boldsymbol{\mu} = [\mu_{ij}]$ is an $n \times n$ positive definite matrix which is often chosen to be a diagonal one. The specific choice of the coefficients μ_{ij} must ensure both the stability of the differential equations and an appropriate convergence speed to the stationary solution (equilibrium) state. It is straightforward to prove that the system of differential Eqs. (2.28) is stable (i.e., it always has an asymptotically stable solution) since

$$\frac{dJ}{dt} = \sum_{j=1}^{n} \frac{\partial J(\mathbf{s})}{\partial s_j} \frac{ds_j}{dt} = -\left(\nabla J(\mathbf{s})\right)^T \boldsymbol{\mu}\,\nabla J(\mathbf{s}) \le 0 \tag{2.29}$$

under the condition that the matrix $\boldsymbol{\mu}$ is positive definite, and in the absence of round-off errors in the full rank matrix $\mathbf{H}$.

2.2.4 Iterative Parallel Algorithms for Large and Sparse Systems

The system of differential equations (2.28) can be easily and directly converted to the parallel iterative algorithm as [276]

$$\mathbf{s}(k+1) = \mathbf{s}(k) + \boldsymbol{\eta}(k)\,\mathbf{H}^T[\mathbf{x} - \mathbf{H}\,\mathbf{s}(k)], \tag{2.30}$$

where $\boldsymbol{\eta}$ is a symmetric positive definite matrix with upper bounded eigenvalues to ensure the stability of the algorithm. In the special case, for $\boldsymbol{\eta}(k) = \eta_k\,\mathbf{I}$ the algorithm is sometimes called the Landweber algorithm, which is known to converge to a LS solution of $\mathbf{H}\,\mathbf{s} = \mathbf{x}$, whenever the learning rate η_k is chosen so that $\mathbf{I} - \eta_k \mathbf{H}^T\,\mathbf{H}$ is nonnegative definite. In other words, the learning rate η_k should satisfy the constraint $0 < \eta_k \le 1/\lambda_{max}$, where λ_{max} is the maximum eigenvalue of $\mathbf{H}^T\mathbf{H}$. Generally, the algorithm converges faster if the η_k is near the upper limit. In order to apply the algorithm efficiently, we must first estimate the largest eigenvalue of $\mathbf{H}^T\mathbf{H}$, to determine the upper limit. By suitably rescaling the system of linear equations, we can accelerate the convergence and easily estimate the upper limit. For example, let $\mathbf{H}$ be an $m \times n$ matrix normalized so that the Euclidean norm of each row is unity. Furthermore, let r_j be the number of non-zero entries in the j-th column of $\mathbf{H}$ and r_{max} be maximum of the $\{r_j\}$. Then, the maximum eigenvalue of the matrix $\mathbf{H}^T\mathbf{H}$ does not exceed r_{max} [114]. This property allows us to take $\eta_k = 1/r_{max}$, which, for very sparse matrices $\mathbf{H}$, can considerably accelerate the convergence rate.

A further increase of convergence speed can be obtained by applying the block iterative or ordered subset version of the algorithm [114, 163]. We obtain the block iterative algorithms by partitioning the set $\{i = 1, 2, \ldots, m\}$ into (not necessary disjoint) subsets S_t, $t = 1, 2, \ldots, N_t$, for $N_t \geq 1$. At each iteration we select the current block and then we use only those data x_i with i in the current block. The use of blocks tends to increase the upper bounds of the learning rate and thereby increases the convergence speed.

The block iterative algorithm, called the BI-ART [163] (block iterative- algebraic reconstruction technique) can be written as

$$\mathbf{s}(k+1) = \mathbf{s}(k) + \eta_k \sum_{i \in S_t} \frac{x_i - \mathbf{h}_i^T \mathbf{s}(k)}{\mathbf{h}_i^T \mathbf{h}_i} \mathbf{h}_i, \tag{2.31}$$

where in each iterative step the sum is taken only over the subset S_t.

In extreme case for $N_t = 1$, we obtain the Kaczmarz algorithm [163, 276] (also called the row-action-projection method) which iterates through the set of equations in a periodic fashion, and can be written as

$$\boxed{\mathbf{s}(k+1) = \mathbf{s}(k) + \eta_k \frac{x_i - \mathbf{h}_i^T \mathbf{s}(k)}{\mathbf{h}_i^T \mathbf{h}_i} \mathbf{h}_i, \qquad i = k \text{ modulo } (m+1)} \tag{2.32}$$

where, $0 < \eta_k < 2$ and at each iteration, we use only one row of $\mathbf{H}$ and a corresponding component of $\mathbf{x}$ successively. In other words, the index i is taken modulo $(m+1)$, i.e., the equations are processed in a cyclical order.

The Kaczmarz algorithm[5], developed in 1937, has relatively low computational complexity and converges quickly to an exact solution if the system of equations is consistent without having to explicitly invert the matrix $\mathbf{H}^T\mathbf{H}$ or $\mathbf{H}\mathbf{H}^T$. This is important from a practical point of view, especially when $\mathbf{H}$ has a large number of rows. For inconsistent systems, the algorithm may fail to converge for the fixed learning rate and it can generate limit cycles, i.e., the solution fluctuates in the vicinity of the least-squares solution as has been shown by Amari [14] and Tanabe [114]. To remove this drawback, we can gradually reduce the learning rate η_k to zero [14].

A more elegant way, which does not easily generalize to other algorithms, is to simply apply the Kaczmarz algorithm twice. The procedure is sometimes called DART (double ART) [114]. Let us assume that for any $\mathbf{x}$, the inconsistent system of equations satisfies the equation $\mathbf{H}\,\mathbf{s}_* + \mathbf{e}(\mathbf{s}_*) = \mathbf{x}$ after convergence. The minimal error vector $\mathbf{e}(\mathbf{s}_*)$ is the orthogonal projection of $\mathbf{x}$ onto the null space of the matrix transformation $\mathbf{H}^T$. Therefore, if $\mathbf{H}^T\,\mathbf{e}(\mathbf{s}_*) = \mathbf{0}$ then $\mathbf{s}_*$ is a minimizer of $\|\mathbf{x} - \mathbf{H}\,\mathbf{s}\|_2$. The DART procedure is summarized as follows: first, we apply the Kaczmarz algorithm to $\mathbf{H}\,\mathbf{e} = \mathbf{0}$, with initial conditions $\mathbf{e}(0) = \mathbf{x}$. After convergence, the optimal error $\mathbf{e}(\mathbf{s}_*)$, will be the member of the null space closest to $\mathbf{x}$. In the second step, we apply the Kaczmarz algorithm to the consistent system of linear equations $\mathbf{H}\,\mathbf{s} = \mathbf{x} - \mathbf{e}(\mathbf{s}_*) = \mathbf{H}\,\mathbf{s}_*$.

[5]The Kaczmarz algorithm has been rediscovered over the years in many applications, for example, as the Widrow-Hoff NLMS (normalized least-mean-square) algorithm for adaptive array processing or the ART (algebraic reconstruction technique) in the field of medical image reconstruction in computerized tomography.

For sparse matrices, the convergence speed can be accelerated significantly without the use of blocks, through overrelaxation.

In general, the iterative algorithm takes the form

$$\mathbf{s}(k+1) = \mathbf{s}(k) + \overline{\boldsymbol{\eta}}(k)\,\mathbf{D}\,\mathbf{H}^T(\mathbf{x} - \mathbf{H}\,\mathbf{s}(k)), \tag{2.33}$$

where $\overline{\boldsymbol{\eta}}(k)$ is a diagonal matrix containing the learning rates and $\mathbf{D}$ is a diagonal and positive definite matrix ensuring the normalization of actual errors $x_i - \mathbf{h}_i^T\mathbf{s}(k)$. The diagonal matrix $\mathbf{D}$ can take different forms depending on the data structure and applications. Typically, for large, sparse and unstructured problems it can take the form [163]

$$\mathbf{D} = \operatorname{diag}\left\{\frac{1}{\|\mathbf{h}_1\|_{\mathbf{r}}^2}, \frac{1}{\|\mathbf{h}_2\|_{\mathbf{r}}^2}, \ldots, \frac{1}{\|\mathbf{h}_m\|_{\mathbf{r}}^2}\right\}, \tag{2.34}$$

where $\mathbf{h}_i^T$ is the i-th row of $\mathbf{H}$ and $\|\mathbf{h}_i\|_{\mathbf{r}}^2$ denotes a weighted norm of the vector $\mathbf{h}_i$. For a large sparse matrix, the norm is usually defined as $\|\mathbf{h}_i\|_{\mathbf{r}}^2 = \sum_{j=1}^n r_j h_{ij}^2$, where r_j is the number of non-zero elements h_{ij} of column j of $\mathbf{H}$. In such a case, the above iterative formula, which is called the CAV (Component Averaging) algorithm [163], can be written in vector form as

$$\boxed{\mathbf{s}(k+1) = \mathbf{s}(k) + \overline{\eta}(k)\sum_{i=1}^m \frac{x_i - \mathbf{h}_i^T\,\mathbf{s}(k)}{\sum_{j=1}^n r_j\,h_{ij}^2}\,\mathbf{h}_i} \tag{2.35}$$

with $0 < \overline{\eta}(k) < 2$.

2.2.5 Iterative Algorithms with Non-negativity Constraints

In many applications, some constraints may be imposed to obtain valid solutions for the system of linear equations $\mathbf{H}\,\mathbf{s} = \mathbf{x}$. For example, in the reconstruction of medical images in computer tomography, it is convenient to transform an arbitrary system of linear equations to equivalent systems in which both the matrix $\mathbf{H}$ and vector $\mathbf{x}$ have only nonnegative entries.

Remark 2.4 *There is no loss of generality in considering here only systems of linear equations in which all the entries of matrix $\mathbf{H} \in \mathbb{R}^{n\times m}$ are nonnegative. Such transformation can be done as follows [114]: Suppose that $\mathbf{H}\,\mathbf{s} = \mathbf{x}$ is an arbitrary (real) system of linear equation with the full-rank matrix $\mathbf{H}$. After rescaling of some equations, if necessary, we may assume that all x_i are non-negative and that for each j the column sum $\sum_{i=1}^m h_{ij}$ is non-zero. Now, we can redefine $\mathbf{H}$ and $\mathbf{s}$ as follows: $\tilde{h}_{kj} = h_{kj}/\sum_{i=1}^m h_{ij}$ and $\tilde{s}_j = s_j(\sum_i h_{ij})$. After such transformation, the new matrix $\tilde{\mathbf{H}}$ has column sums equal to one and the vector $\mathbf{x} = \tilde{\mathbf{H}}\,\tilde{\mathbf{s}}$ is unchanged. Since the sums of all columns are unity, we have $\sum_{j=1}^n \tilde{s}_j = s_+ = \sum_{i=1}^m x_i = x_+$. We can always find such positive coefficient β such that the matrix defined as $\mathbf{B} = \tilde{\mathbf{H}} + \beta\overline{\mathbf{1}}$ has only nonnegative entries, where $\overline{\mathbf{1}}$ is an $m \times n$ matrix whose entries are all unity. Hence, we have $\mathbf{B}\,\tilde{\mathbf{s}} = \tilde{\mathbf{H}}\,\tilde{\mathbf{s}} + (\beta\,s_+)\,\mathbf{1}$, where $\mathbf{1}$ is a vector whose entries are all one. Thus, the new system of equations to solve is $\mathbf{B}\,\tilde{\mathbf{s}} = \mathbf{x} + (\beta\,x_+)\,\mathbf{1} = \bar{\mathbf{x}}$.*

We often made a further assumption that the column-sums of the matrix $\mathbf{B}$ *are all unity. To achieve this, we make one additional renormalization: replace* b_{kj} *with* $\bar{h}_{kj} = b_{kj}/(\sum_{i=1}^{m} b_{ij}$ *and* $\tilde{s}_j$ *with* $\bar{s}_j = \tilde{s}_j \sum_{i=1}^{m} b_{ij}$*. Note that the vector* $\mathbf{B}\tilde{\mathbf{s}}$ *is identical to* $\bar{\mathbf{H}}\,\bar{\mathbf{s}}$ *and the new matrix* $\bar{\mathbf{H}}$ *is nonnegative and has column sums equal to one.*

Let us consider the problem solving a system of linear equations $\bar{\mathbf{H}}\,\mathbf{s} = \bar{\mathbf{x}}$ with nonnegative elements $\bar{h}_{ij}$ and $\bar{x}_i$ and the constraints $\mathbf{s} \geq \mathbf{0}$ (i.e., $s_j \geq 0,\ \forall j$) [114, 874]. Without loss of generality, we assume that the system is normalized so that

$$\bar{\mathbf{H}}^T \mathbf{1} = \mathbf{1}, \tag{2.36}$$

where $\mathbf{1}$ is a vector of all ones. This indicates that all the columns of $\bar{\mathbf{H}}$ are normalized to have their 1-norm equal unity.

To solve the problem, we can apply the standard LS criterion: minimize $J(\mathbf{s}) = \|\bar{\mathbf{x}} - \bar{\mathbf{H}}\,\mathbf{s}\|^2$ subject to the constraints $\mathbf{s} \geq \mathbf{0}$ or alternatively by applying the Shannon entropy type penalty term as $J(\mathbf{s}) = \|\bar{\mathbf{x}} - \bar{\mathbf{H}}\,\mathbf{s}\|^2 + \alpha \sum_{j=1}^{n} s_j \log s_j$ with $s_j \geq 0$.

Non-negativity can alternatively be enforced by choosing the Kullback-Leibler distance or maximum likelihood functional [114, 874]. The use of the maximum likelihood functional is justified by the assumption of Poisson noise, which is a typical case in medical image reconstruction [874].

To find a solution, we minimize the Kullback-Leibler distance defined as

$$\begin{aligned} KL(\bar{\mathbf{x}}\,||\,\bar{\mathbf{H}}\,\mathbf{s}) &= \sum_{i=1}^{m} KL(\bar{x}_i,\ \bar{\mathbf{h}}_i^T\,\mathbf{s}) \\ &= \sum_{i=1}^{m} \bar{x}_i \log \frac{\bar{x}_i}{\bar{\mathbf{h}}_i^T \mathbf{s}} + \bar{\mathbf{h}}_i^T \mathbf{s} - \bar{x}_i, \end{aligned} \tag{2.37}$$

subject to $\mathbf{s} \geq \mathbf{0}$, where $KL(a, b) = a \log(a/b) + b - a$, $KL(0, b) = b$ and $Kl(a, 0) = +\infty$ for positive scalars a and b, log denotes the natural logarithm and, by definition, $0 \log 0 = 0$. It is straightforward to check that the above cost function can be simplified to a likelihood function [874]

$$J(\mathbf{s}) = \sum_{i=1}^{m} \left[\bar{\mathbf{h}}_i^T \mathbf{s} - \bar{x}_i \log(\bar{\mathbf{h}}_i^T \mathbf{s})\right] \tag{2.38}$$

subject to $\mathbf{s} \geq \mathbf{0}$.

The above constrained optimization problem can be easily transformed into an unconstrained minimization problem by introducing a simple parametrization $\mathbf{s} = \exp(\mathbf{u})$, that is, $s_j = \exp(u_j),\ \ \forall j$. With this parametrization and taking into account that $\bar{\mathbf{H}}^T\,\mathbf{1} = \mathbf{1}$, we are able to evaluate the gradient of the cost function as follows [874]

$$\begin{aligned} \nabla_{\mathbf{u}} J(\mathbf{s}) = \mathbf{D}_{\mathbf{s}}\, \nabla_{\mathbf{s}} J(\mathbf{s}) &= \mathbf{D}_{\mathbf{s}}\, \bar{\mathbf{H}}^T\, \mathbf{D}_{\bar{\mathbf{H}}\mathbf{s}}^{-1}\, (\bar{\mathbf{H}}\,\mathbf{s} - \bar{\mathbf{x}}) \\ &= \mathbf{D}_{\mathbf{s}}\, \bar{\mathbf{H}}^T\, (\mathbf{1} - \mathbf{D}_{\bar{\mathbf{H}}\mathbf{s}}^{-1}\, \bar{\mathbf{x}}) = \mathbf{s} - \mathbf{D}_{\mathbf{s}}\, \bar{\mathbf{H}}^T\, \mathbf{D}_{\bar{\mathbf{H}}\mathbf{s}}^{-1}\, \bar{\mathbf{x}}, \end{aligned} \tag{2.39}$$

where $\mathbf{D}_{\mathbf{s}} = \mathrm{diag}\{\mathbf{s}\} = \mathrm{diag}\{s_1, s_2, \ldots, s_n\}$ and $\mathbf{D}_{\bar{\mathbf{H}}\mathbf{s}} = \mathrm{diag}\{\bar{\mathbf{H}}\,\mathbf{s}\}$ =diag$\{\bar{\mathbf{h}}_1^T \mathbf{s}, \bar{\mathbf{h}}_2^T \mathbf{s}, \ldots, \bar{\mathbf{h}}_n^T \mathbf{s}\}$. Hence , by setting the gradient to zero, we obtain the fixed point algorithm, which is sometimes called EMML (expectation maximization maximum likelihood) algorithm [114, 874]

$$\boxed{\mathbf{s}(k+1) = \mathbf{D}_{\mathbf{s}}(k)\, \bar{\mathbf{H}}^T\, \mathbf{D}_{\bar{\mathbf{H}}\mathbf{s}}^{-1}(k)\, \bar{\mathbf{x}}} \tag{2.40}$$

where $\mathbf{D}_{\mathbf{s}}(k) = \text{diag}\{\mathbf{s}(k)\}$ and $\mathbf{D}_{\bar{\mathbf{H}}\mathbf{s}}(k) = \text{diag}\{\bar{\mathbf{H}}\,\mathbf{s}(k)\}$. The EMML algorithm can be written in scalar form as

$$s_j(k+1) = s_j(k) \sum_{i=1}^{m} \bar{h}_{ij}\, \frac{\bar{x}_i}{\bar{\mathbf{h}}_i^T \mathbf{s}(k)}, \qquad (j = 1, 2, \ldots, n). \tag{2.41}$$

Had we not normalized $\mathbf{H}$ so as to have the columns of $\bar{\mathbf{H}}$ sum to one, the EMML algorithm would have had the iterative step

$$s_j(k+1) = \frac{s_j(k)}{\sum_i^m h_{ij}} \sum_{i=1}^{m} h_{ij}\, \frac{x_i}{\mathbf{h}_i^T \mathbf{s}(k)}, \qquad (j = 1, 2, \ldots, n). \tag{2.42}$$

An algorithm closely related to the EMML algorithm is the SMART (Simultaneous Multiplicative Algebraic Reconstruction Technique) developed and analyzed by Byrne [114]:

$$\begin{aligned} s_j(k+1) &= s_j(k) \exp\left(\sum_{i=1}^{m} \bar{h}_{ij} \log\left(\frac{\bar{x}_i}{\bar{\mathbf{h}}_i^T \bar{\mathbf{s}}(k)}\right) \right) \\ &= s_j(k) \prod_{i=1}^{m} \left(\frac{\bar{x}_i}{\bar{\mathbf{h}}_i^T \mathbf{s}(k)} \right)^{\bar{h}_{ij}}, \qquad (j = 1, 2, \ldots, n) \end{aligned} \tag{2.43}$$

which can be considered as the minimization of the Kullback-Leibler distance

$$KL(\bar{\mathbf{H}}\,\mathbf{s}||\mathbf{x}) = \sum_{i=1}^{m} KL(\bar{\mathbf{h}}_i^T\,\mathbf{s},\ \bar{x}_i), \tag{2.44}$$

over the nonnegative orthant.

In the consistent case (that is, when there is a vector $\mathbf{s} \geq \mathbf{0}$ which satisfies $\bar{\mathbf{x}} = \bar{\mathbf{H}}\,\mathbf{s}$), both the SMART and EMML algorithms converge to the nonnegative solution that minimizes $KL(\mathbf{s}||\mathbf{s}(0))$. When there are no such nonnegative vectors, SMART converges to the unique nonnegative minimizer of the cost function $KL(\bar{\mathbf{H}}\,\mathbf{s}||\bar{\mathbf{x}})$ for which $KL(\mathbf{s}||\mathbf{s}(0))$ is minimized, while EMML converges to the unique minimizer of $KL(\bar{\mathbf{x}}||\bar{\mathbf{H}}\,\mathbf{s})$. It is interesting to note that in the case when entries of the initial vector $\mathbf{s}(0)$ are all equal, SMART converges to the solution for which the Shannon entropy $J_S = -\sum_{i=1}^{n} s_j \log s_j$, is maximized [114].

2.2.6 Robust Criteria and Iteratively Reweighted Least-Squares Algorithm

Although the ordinary (standard) least-squares (2-norm) criterion discussed in the previous section is optimal for a Gaussian error distribution, it nevertheless, provides very poor estimates of the vector $\mathbf{s}_*$ in the presence of large errors (called outliers) or spiky noise.[6]

[6] For simplicity, we ensure that all the errors are confined to the observation (sensor) vector $\mathbf{x}$.

In order to mitigate the influence of outliers (i.e., to provide a more reliable and robust estimate of the unknown vector $\mathbf{s}$), we can employ the iteratively reweighted least-squares criterion (also called the robust least-squares criterion). According to this criterion, we wish to solve the following minimization problem:

Find the vector $\mathbf{s} \in \mathbb{R}^n$ that minimizes the energy function

$$J_\rho(\mathbf{s}) = \sum_{i=1}^{m} \rho[e_i(\mathbf{s})], \tag{2.45}$$

where $\rho[e]$ is a given (usually convex) function called the weighting or loss function and its derivative $\Psi(e) = d\rho(e)/de$ is called the influence function.[7] Typical robust loss functions and the corresponding influence functions are summarized in Table 2.1.

Remark 2.5 *From a statistical point of view this corresponds to the negative of the* log *likelihood, when the noise vector* $\mathbf{e}$ *has a distribution expressed by the pdf*

$$p(\mathbf{e}) = c \prod_{i=1}^{m} \exp(-\rho(e_i)). \tag{2.46}$$

Note that by taking $\rho(e_i) = e_i^2/2$, we obtain the ordinary linear least-squares problem considered in the previous section.[8] However, in order to reduce the influence of the outliers, other weighting functions should be chosen. One of the most popular weighting (loss) functions is the logistic function [276]

$$\rho_L[e] = \beta^2 \log(\cosh(e/\beta)), \tag{2.47}$$

where β is a problem dependent parameter, which is called the cut-off parameter. The iteratively reweighted least-squares problem given by Eq. (2.45), often used in robust statistics, is usually solved numerically by repeatedly solving a weighted least-squares problem. We will attack this problem by mapping the minimization problem (2.45) into a system of differential equations. For simplicity, in our further considerations, let us assume that the weighting function is the logistic function given by Eq. (2.47). Applying the gradient approach for the minimization of the energy function (2.45), we obtain

$$\frac{ds_j}{dt} = \mu_j \left(\sum_{i=1}^{m} h_{ij}\, \Psi_i \left[x_i - \sum_{p=1}^{n} h_{ip}\, s_p \right] \right) \quad (j = 1, 2, \dots, n), \tag{2.48}$$
$$\text{with} \quad s_j(0) = s_j^{(0)},$$

[7] It is important to note that the energy function so defined is convex if the loss functions are convex. Therefore the problem of convergence to a local minimum does not arise.

[8] The ordinary least-squares error criterion equally weights all the modelling errors and may produce a biased parameters estimation, if the observed data are contaminated by impulsive noise or large isolated errors.

Table 2.1 Basic robust loss functions $\rho(e)$ and the corresponding influence functions $\Psi(e) = d\rho(e)/de$.

Name	Loss Function $\rho(e)$	Influence Functions $\Psi(e)$
Logistic	$\rho_L = \frac{1}{\beta} \log(\cosh(\beta e))$	$\Psi_L = \tanh(\beta e)$
Huber	$\rho_H = \begin{cases} e^2/2, & for \ \ \lvert e\rvert \le \beta; \\ \beta\lvert e\rvert - \frac{\beta^2}{2}, & otherwise \end{cases}$	$\Psi_H = \begin{cases} e, & for \ \ \lvert e\rvert \le \beta; \\ \beta sign(e), & otherwise \end{cases}$
L_p	$\rho_{Lp} = \frac{1}{p}\lvert e\rvert^p$	$\Psi_{Lp} = \lvert e\rvert^{p-1} sign(e)$
Cauchy	$\rho_C = \frac{\sigma^2}{2} \log\left[1 + (\frac{e}{\sigma})^2\right]$	$\Psi_C = \frac{e}{1 + (\frac{e}{\sigma})^2}$
Geman, McCulre	$\rho_G = \frac{1}{2}\frac{e^2}{\sigma^2 + e^2}$	$\Psi_G = \frac{\sigma^2 e}{(\sigma^2 + e^2)^2}$
Welsh	$\rho_W = \frac{\sigma^2}{2}\left[1 - \exp(-(\frac{e}{\sigma})^2)\right]$	$\Psi_W = e\exp(-(\frac{e}{\sigma})^2)$
Fair	$\rho_F = \sigma^2\left[\frac{\lvert e\rvert}{\sigma} - log(1 + \frac{\lvert e\rvert}{\sigma})\right]$	$\Psi_F = \frac{e}{1 + \frac{\lvert e\rvert}{\sigma}}$
$L_1 - L_2$	$\rho_{L12} = 2(\sqrt{1 + e^2/2} - 1)$	$\Psi_{L12} = \frac{e}{\sqrt{1 + e^2/2}}$
Talvar	$\rho_{Ta} = \begin{cases} e^2/2, & for \ \ \lvert e\rvert \le \beta; \\ \beta^2/2, & otherwise \end{cases}$	$\Psi_{Ta} = \begin{cases} e, & for \ \ \lvert e\rvert < \beta; \\ 0, & otherwise \end{cases}$
Hampel	$\rho_{Ha} = \begin{cases} \frac{\beta^2}{\pi}(1 - \cos(\frac{\pi e}{\beta})), & for \ \ \lvert e\rvert \le \beta; \\ \frac{2\beta^2}{\pi}, & otherwise \end{cases}$	$\Psi_{Ha} = \begin{cases} \beta \sin(\frac{\pi e}{\beta}), \\ 0 \end{cases}$

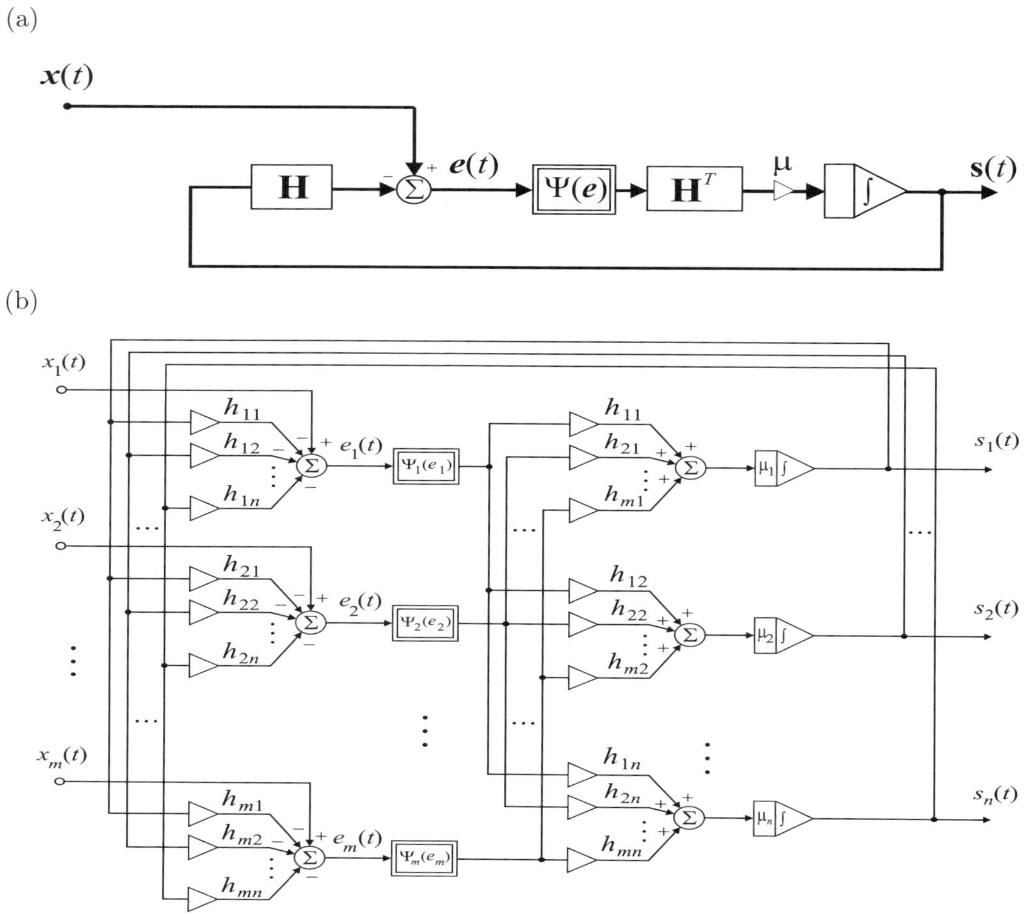

Fig. 2.1 Architecture of the Amari-Hopfield continuous-time (analog) model of recurrent neural network (a) block diagram, (b) detailed architecture.

where $\Psi_i[e_i]$ is the sigmoidal activation function described as

$$\Psi_i[e_i] = \frac{\partial \rho_L[e_i(\mathbf{s})]}{\partial e_i} = \beta \tanh(e_i/\beta). \tag{2.49}$$

The above system of differential equations can be rewritten in the compact matrix form

$$\frac{d\mathbf{s}}{dt} = \boldsymbol{\mu}\,\mathbf{H}^T\,\Psi(\mathbf{e}), \tag{2.50}$$

where $\mathbf{e} = \mathbf{x} - \mathbf{H}\,\mathbf{s}$ and $\boldsymbol{\mu} = \mathrm{diag}\{\mu_1, \mu_2, \ldots, \mu_n\}$. The above system of differential equations can be implemented by a flexible Amari-Hopfield neural network shown in Fig.2.1

[276]. Note that the above system of equations has a form similar to that given by Eqs. (2.28) and that they can be easily implemented by a similar network architecture under the assumption that the sigmoidal nonlinearities are incorporated in the first (sensor) layer of neurons. The use of the sigmoidal nonlinearities in the first layer of neurons is essential, since they compress large residuals $e_i(\mathbf{s})$, that is, their absolute values are prevented from being greater than the prescribed cut-off parameter β. Therefore, in comparison with the ordinary least-squares implementation, a more robust solution, which is less sensitive to large errors (outliers) is obtained.

The above solution is not equivariant with respect to scale. Thus, the residual errors e_i should be standardized by means of some estimate of the standard deviation σ (see Table 2.1). One possibility is to use the median absolute deviation

$$\hat{\sigma} = C\,\mathrm{med}\{|e_i - \mathrm{med}\{|e_i|\}|\}, \tag{2.51}$$

where $C = 1.4826$ if Gaussian noise is assumed.

The system of differential equations (2.50) achieves equilibrium point if the gradient of the cost function (2.45) will be zero, i.e.,

$$\mathbf{H}^T \mathbf{D_e}(\mathbf{x} - \mathbf{H}\,\mathbf{s}) = \mathbf{0}, \tag{2.52}$$

where a diagonal matrix $\mathbf{D_e}$ has the diagonal entries $d_{ii} = \Psi(e_i)/e_i$. Solving this equation, we get

$$\mathbf{s} = (\mathbf{H}^T \mathbf{D_e} \mathbf{H})^{-1} \mathbf{H}^T \mathbf{D_e} \mathbf{x}. \tag{2.53}$$

The above equation may be recognized as the weighted least squares problem. However, in this case the diagonal matrix $\mathbf{D_e}$ is the function of the residual error vector $\mathbf{e}$ and vector od source signals $\mathbf{s}$. Therefore, it has not any explicit solution and can be solved by the Iteratively Rewieghted Least Squares (IRLS) algorithm:

Outline of the IRLS Algorithms

---

1. $\mathbf{s}(0) = (\mathbf{H}^T \mathbf{H})^{-1} \mathbf{H}^T \mathbf{x},$
2. $\mathbf{e}(k) = \mathbf{x} - \mathbf{H}\,\mathbf{s}(k),$
3. $\hat{\sigma} = C\,\mathrm{med}\,\{|e_i(k) - \mathrm{med}\,\{|e_i(k)|\}|\},$
4. $d_{ii}(k) = \Psi(\mathbf{e}_i(k)/\hat{\sigma})/(e_i(k)/\hat{\sigma}),$
5. $\mathbf{s}(k) = \mathbf{H}^T \mathbf{D_e}(k))\mathbf{H})^{-1} \mathbf{H}^T \mathbf{D_e}(k))\mathbf{x},$
6. Iterate until convergence. (2.54)

2.2.7 Tikhonov Regularization and SVD

The noise in the measurements x_i, in combination with the ill-conditioning of matrix $\mathbf{H}$, means that the exact solution of the standard LS problem (2.9) usually deviates strongly from the noise-free solution and therefore is often worthless. To alleviate the problem, we can apply regularization. The Russian mathematician Tikhonov was probably the first who studied the concepts of regularization [276]. The idea behind this technique is to define a criterion to select an approximate solution from a set of admissible solutions. The basic feature of the regularization is a compromise between fidelity to data and fidelity to some *a priori* information about the solution. In other words, the regularization method imposes a weak smoothness constraint on a set of possible solutions.

According to the regularization theory, a regularized energy function (i.e., the function to be minimized) is a weighted sum of two (or even more) terms:

$$J(\mathbf{s}, \alpha) = J_d(\mathbf{s}) + \alpha\, J_s(\mathbf{s}), \tag{2.55}$$

where J_d is the data energy and J_s is the smoothness constraint (also called stabilizer energy) [276].

Remark 2.6 *This formulation has the following Bayesian statistical interpretation. Let us assume that the true signal* $\mathbf{s}$ *has the prior distribution*

$$\omega(\mathbf{s}) = \exp\{-\alpha J_s(\mathbf{s})\}. \tag{2.56}$$

Then the joint distribution of $\mathbf{s}$ *and* $\mathbf{x}$ *can be written as*

$$p(\mathbf{x}, \mathbf{s}) = c \exp\{-J_d(\mathbf{s}) - \alpha J_s(\mathbf{s})\}. \tag{2.57}$$

Hence, the criterion of maximizing a posteriori distribution $p(\mathbf{s}\,|\,\mathbf{x})$ *of the source signals* $\mathbf{s}$ *given* $\mathbf{x}$ *is equivalent to minimizing* $J(\mathbf{s}, \alpha)$.

For our linear least-squares problem (cf. Eq. (2.9)), a regularized solution can simply be defined as the solution of the following problem:

$$\mathbf{s}(\alpha) = \arg \min_{\mathbf{s} \in \mathbb{R}^n} J(\mathbf{s}, \alpha), \tag{2.58}$$

where

$$J(\mathbf{s}, \alpha) = \frac{1}{2} \left(\|\mathbf{x} - \mathbf{H}\,\mathbf{s}\|_2^2 + \alpha\, \|\mathbf{s}\|_2^2 \right), \quad \text{with} \quad \alpha > 0. \tag{2.59}$$

Thus, in this case, the smoothness constraint energy is the squared 2-norm of the vector $\mathbf{s}$. Here, the regularization parameter α controls the "smoothness" of the regularized solution.[9]

Applying a standard gradient descent approach to the cost function (2.59), we obtain a system of differential equations with leaky integrators [276]

$$\frac{d\mathbf{s}}{dt} = \boldsymbol{\mu} \left[\mathbf{H}^T \left(\mathbf{x} - \mathbf{H}\,\mathbf{s} \right) - \alpha\, \mathbf{s} \right]. \tag{2.60}$$

[9]Choosing the regularization parameter α for an ill-posed problem is an art based on good heuristic and *a priori* knowledge of the noise in the observations.

It should be noted that the minimization of the energy function $J(\mathbf{s}, \alpha)$ (2.59), with respect to $\mathbf{s}$, is equivalent to the solution of the normal equation

$$(\mathbf{H}^T \mathbf{H} + \alpha \mathbf{I})\mathbf{s} = \mathbf{H}^T \mathbf{x}. \tag{2.61}$$

It can be shown that the condition number of the realized matrix $\tilde{\mathbf{H}} = \mathbf{H}^T \mathbf{H} + \alpha \mathbf{I}$ is given by

$$\text{cond}(\tilde{\mathbf{H}}) = \frac{\sigma_{max}^2 + \alpha}{\sigma_{min}^2 + \alpha}, \tag{2.62}$$

where σ_{max} and σ_{min} are respectively the maximum and the minimum singular values of the matrix $\mathbf{H}^T \mathbf{H}$. Thus, the condition number of the regularized matrix $\tilde{\mathbf{H}} = \mathbf{H}^T \mathbf{H} + \alpha \mathbf{I}$ can be much lower than that of the matrix $\mathbf{H}^T \mathbf{H}$ (i.e., $\tilde{\mathbf{H}}$ for $\alpha = 0$). For example, for the setting $\sigma_{max} = 1$, $\sigma_{min} = 0.1$ and $\alpha = 0.1$, the condition number is improved by a factor of 10 (from 100 down to 10).

The solution of Eq. (2.61) can be interpreted by the use of the singular value decomposition (SVD) theory. Assume that the $m \times n$, matrix $\mathbf{H}$, with rank n $(m \geq n)$ has the following SVD:

$$\mathbf{H} = \mathbf{U}\,\mathbf{\Sigma}\,\mathbf{V}^T = \sum_{i=1}^{n} \sigma_i \mathbf{u}_i \mathbf{v}_i^T, \tag{2.63}$$

where both $\mathbf{U} = [\mathbf{u}_1, \mathbf{u}_2, \ldots, \mathbf{u}_m] \in \mathbb{R}^{m \times m}$ and $\mathbf{V} = [\mathbf{v}_1, \mathbf{v}_2, \ldots, \mathbf{v}_n] \in \mathbb{R}^{n \times n}$ are orthogonal matrices and $\mathbf{\Sigma}$ is a pseudo-diagonal m by n matrix whose top n rows contain $\text{diag}\{\sigma_1, \sigma_2, \ldots, \sigma_n\}$ (with ordered diagonal entries $\sigma_1 \geq \sigma_2 \geq \cdots \geq \sigma_n$) and whose bottom $(m - n)$ rows are all zero. It can be shown that if $\{\mathbf{u}_i\}$ and $\{\mathbf{v}_i\}$ are the columns of $\mathbf{U}$ and $\mathbf{V}$, respectively, then the minimum norm least-squares Tikhonov regularized solution to Eq. (2.61) can be approximated by

$$\mathbf{s}_*(\alpha) = \sum_{i=1}^{n} \frac{\sigma_i^2}{\sigma_i^2 + \alpha} \frac{\mathbf{u}_i^T \mathbf{x}}{\sigma_i} \mathbf{v}_i = \sum_{i=1}^{n} \frac{\beta_i}{\sigma_i + \alpha/\sigma_i} \mathbf{v}_i, \tag{2.64}$$

where $\beta_i = \mathbf{u}_i^T \mathbf{x}$ and $\alpha > 0$ is the regularization parameter. It is interesting to note that, if the singular values σ_i of the matrix $\mathbf{H}$ are much larger than the regularization parameter α, then the regularization has little effect on the final (optimal) solution. However, if one of the singular values σ_i is much smaller than α, the corresponding term in Eq. (2.64) can be expressed as

$$\frac{\beta_i\, \mathbf{v}_i}{\sigma_i + \alpha/\sigma_i} \cong \frac{\sigma_i}{\alpha} \beta_i\, \mathbf{v}_i \quad (\sigma_i \ll \alpha). \tag{2.65}$$

Note that this term approaches zero as σ_i tends to zero. This demonstrates the required continuity in the solution of a real physical system. Note also that such continuity of the solution cannot be achieved if $\alpha = 0$. We then see that the role of the regularization parameter α is to damp or filter the terms in the sum corresponding to the singular values σ_i smaller than α. Hence, in any practical application, α will always satisfy $\sigma_n \leq \alpha < \sigma_i$, where σ_i corresponds to the significant gap in the singular values spectrum [276, 920].

In contrast to the Tikhonov regularized solution (2.64), the true solution for the noise free problem is

$$\mathbf{s}_{true} = \sum_{i=1}^{n} \frac{\beta_i - \epsilon_i}{\sigma_i} \mathbf{v}_i, \tag{2.66}$$

where $\epsilon_i = \mathbf{u}_i^T \mathbf{e}$ represents the unknown noise component. The goal of the optimal regularization is to produce a solution as close as possible to the true solution. In other words, the close optimal value of the regularized parameter can be obtained by minimizing the distance $\tilde{J}(\alpha) = \|\mathbf{s}_*(\alpha) - \mathbf{s}_{true}\|^2$, which after some mathematical operation leads to the following algebraic equation [920]:

$$f(\alpha) = \sum_{i=1}^{n} \frac{\alpha(\mathbf{u}_i^T \mathbf{x})^2}{(\sigma_i^2 + \alpha)^3} - \frac{(\mathbf{u}_n^T \mathbf{x})^2}{(\sigma_n^2 + \alpha)^2} - \hat{\sigma}_e^2 \sum_{i=1}^{n-1} \frac{\mathbf{1}}{(\sigma_i^2 + \alpha)^2}, \tag{2.67}$$

where $\hat{\sigma}_e^2$ is the estimated variance of noise. As has been shown by O'Leary, finding the zero of such function gives an approximation of the Tikhonov regularization parameter α close to its optimal value [920].

An alternative method for the regularization of (2.9) is the truncated SVD approach, in which we discard the smallest singular value simply by truncating the sum in Eq. (2.64) at some $r < n$.

In practice, it appears that instead of keeping $\|\mathbf{s}\|_2^2$ small as in Eq. (2.60), it is often more effective to keep $\|\mathbf{L}\mathbf{s}\|_2^2$ small, where $\mathbf{L}$ is a suitably chosen matrix (typically, $\mathbf{L} = \mathbf{I}$).

We now state a generalized regularized least-squares solution, which corresponds to the minimum of the cost function

$$\min_{\mathbf{s} \in \mathbb{R}^n} \left\{ \frac{1}{2} \left(\|\mathbf{\Sigma}_e(\mathbf{x} - \mathbf{H}\,\mathbf{s})\|_2^2 + \alpha \,\|\mathbf{L}(\mathbf{s} - \bar{\mathbf{s}}_*)\|_2^2 \right) \right\}, \tag{2.68}$$

where $\bar{\mathbf{s}}_*$ is the expected mean value of the estimated vector $\hat{\mathbf{s}}$ (typically it is assumed that $\bar{\mathbf{s}}_* = \mathbf{0}$), both $\mathbf{\Sigma}_e$ and $\mathbf{L}^T\mathbf{L}$ are positive definite weighting matrices. The matrix $\mathbf{L}$ is the regularization matrix, which is an identity matrix or a discrete approximation to some derivative operator. Typical examples of such $\mathbf{L}$ are: the 1st derivative approximation $\mathbf{L}_1 \in \mathbb{R}^{(n-1)\times n}$ and the 2nd derivative approximation $\mathbf{L}_2 \in \mathbb{R}^{(n-2)\times n}$ given by

$$\mathbf{L}_1 = \begin{bmatrix} 1 & -1 & & & 0 \\ & 1 & -1 & & \\ \vdots & & \ddots & \ddots & \\ 0 & & & 1 & -1 \end{bmatrix}, \qquad \mathbf{L}_2 = \begin{bmatrix} -1 & 2 & -1 & & & 0 \\ & -1 & 2 & -1 & & \\ & & \ddots & \ddots & \ddots & \\ 0 & & & -1 & 2 & -1 \end{bmatrix}. \tag{2.69}$$

It is straightforward to show that the regularized solution (called the generalized Tikhonov regularized solution) can be written in the form

$$\hat{\mathbf{s}}_\alpha = \left(\mathbf{H}^T \mathbf{\Sigma}_e \mathbf{H} + \alpha \mathbf{L}^T \mathbf{L}\right)^{-1} \left(\mathbf{H}^T \mathbf{\Sigma}_e \mathbf{x} + \alpha \mathbf{W} \bar{\mathbf{s}}_*\right). \tag{2.70}$$

Such regularization has a close relationship with the Bayesian approach, where we use prior information in addition to the available data for the solution of the problem. In fact, it can be shown that if the errors $\mathbf{e}$ are jointly Gaussian with zero-mean, the elements of vector $\mathbf{s}$ are jointly Gaussian random variables with mean $\bar{\mathbf{s}}_*$ and the covariance matrix $\mathbf{R_{ss}} = E\{\mathbf{s}(k)\mathbf{s}^T(k)\}$ (where $\mathbf{s}(k)$ is the k-th observation of the vector $\mathbf{s}$), then the solution

$$\hat{\mathbf{s}}_\alpha = (\mathbf{H}^T\mathbf{R}_{\mathbf{ee}}^{-1}\mathbf{H} + \mathbf{R}_{\mathbf{ss}}^{-1})^{-1}(\mathbf{H}^T\mathbf{R}_{\mathbf{ee}}^{-1}\mathbf{x} + \mathbf{R}_{\mathbf{ss}}^{-1}\bar{\mathbf{s}}_*) \tag{2.71}$$

minimizes the Bayesian mean square estimation criterion $E\{\|\mathbf{s} - \hat{\mathbf{s}}\|^2\}$. It should be noted that the above formula simplifies to the ordinary BLUE (Gauss-Markov minimum variance) estimate:

$$\hat{\mathbf{s}}_{\text{BLUE}} = (\mathbf{H}^T\mathbf{R}_{\mathbf{ee}}^{-1}\mathbf{H})^{-1}\mathbf{H}^T\mathbf{R}_{\mathbf{ee}}^{-1}\mathbf{x} \tag{2.72}$$

by setting $\mathbf{R}_{\mathbf{ss}}^{-1} = \mathbf{0}$. This setting corresponds to an "infinite" variance of the parameters, that is, there is no assumption on the properties of the parameters.

In fact, for this case ($\mathbf{\Sigma}_e = \mathbf{R}_{\mathbf{ee}}^{-1}$), the minimization problem (2.68) is equivalent to the solution of the normal equation

$$(\mathbf{H}^T\mathbf{R}_{\mathbf{ee}}^{-1}\mathbf{H} + \alpha\,\mathbf{L}^T\mathbf{L})\,\mathbf{s} = \mathbf{H}^T\mathbf{R}_{\mathbf{ee}}^{-1}\,\mathbf{x}, \tag{2.73}$$

which corresponds to the maximum *a posteriori* (MAP) estimates [589].

In the subspace regularization approach, we combine two approaches: subspace method (SVD) and the Tikhonov regularization by setting $\mathbf{R}_{\mathbf{ss}}^{-1} = \alpha\mathbf{H}^T(\mathbf{I} - \mathbf{H}_\mathcal{S}\mathbf{H}_\mathcal{S}^T)\mathbf{H}$, where $\mathbf{H}_\mathcal{S}$ contains the n first principal eigenvectors of the data covariance matrix $\mathbf{R_{xx}} = E\{\mathbf{x}\mathbf{x}^T\} = \mathbf{U\Sigma U}^T$ associated with the n largest singular values [676].

Similarly, in the cost function (2.68) we select $\mathbf{L} = (\mathbf{I} - \mathbf{H}_\mathcal{S}\mathbf{H}_\mathcal{S}^T)^{1/2}\mathbf{H}$ in order to keep the second term $\|\mathbf{Ls}\|$ small for all expected $\bar{\mathbf{s}}_*$. Since

$$\mathbf{L}^T\mathbf{L} = \mathbf{H}^T(\mathbf{I} - \mathbf{H}_\mathcal{S}\mathbf{H}_\mathcal{S}^T)\mathbf{H}, \tag{2.74}$$

the desired subspace regularized solution can be written in the form

$$\hat{\mathbf{s}}_\alpha = \left[\mathbf{H}^T\mathbf{R}_{\mathbf{ee}}^{-1}\mathbf{H} + \alpha\mathbf{H}^T(\mathbf{I} - \mathbf{H}_\mathcal{S}\mathbf{H}_\mathcal{S}^T)\mathbf{H}\right]^{-1}\mathbf{H}^T\mathbf{R}_{\mathbf{ee}}^{-1}\mathbf{x}. \tag{2.75}$$

The above approach is closely related to Bayesian estimation. In this approach, the second-order statistics of the set of the observations is used to form an *a priori* information model for the regularization [676].

2.3 LEAST ABSOLUTE DEVIATION (1-NORM) SOLUTION OF SYSTEMS OF LINEAR EQUATIONS

The use of the minimum 1-norm or least absolute deviation (LAD) can often provide a useful alternative to the minimum 2-norm (least-squares) or the infinity- (Chebyshev) norm solution of a system of linear and nonlinear equations, especially in signal processing applications [276]. These applications span the areas of deconvolution, state space estimation, inversion and parameter estimation. The (LAD) solutions of systems of linear algebraic equations have certain properties not shared by the ordinary least-squares solutions, such as:

(1) The minimum 1-norm solution of an overdetermined system of linear equations always exists though the 1-norm solution is not necessarily unique in contrast to the minimum 2-norm solution, where the solution is always unique when the matrix $\mathbf{H}$ has a full rank.

(2) The minimum 1-norm solutions are robust to outliers (under condition that they are relatively few in numbers), that it, the solution is resistant (insensitive) to some large changes in the data. It is an extremely useful property when the data are known to be contaminated by occasional "wild points" (outliers) or spiky noise.

(3) For fitting a number of data points by a constant, the 1-norm estimate can be interpreted as the *median* while the interpretation of the 2-norm estimate is the *mean*.

(4) The minimum 1-norm solutions are in general sparse, in the sense that they have a small number of non-zero components in the underdetermined case (see Section 2.5).

(5) Minimum 1-norm problems are equivalent to linear programming problems and vice versa. Linear programming problems may also be formulated as minimum 1-norm problems, while linear least-squares problems can be considered as a special case of the quadratic programming problem [276].

2.3.1 Neural Network Architectures Using a Smooth Approximation and Regularization

The most straightforward approach for solving the problem of the least absolute value problem is to approximate the absolute value functions $|e_i(\mathbf{s})|$ $(i = 1, 2, ..., m)$ by smooth differentiable functions, for example

$$J_\rho(\mathbf{s}) = \sum_{i=1}^{m} \rho[e_i(\mathbf{s})], \tag{2.76}$$

where

$$\rho[e_i(\mathbf{s})] = \frac{1}{\gamma} \log(\cosh(\gamma e_i(\mathbf{s}))), \tag{2.77}$$

with $\gamma \gg 1$. To ensure a good approximation, the coefficient γ must be sufficiently large (although its actual value is not critical). Applying the gradient approach, we obtain the associated system of differential equations

$$\frac{ds_j}{dt} = \mu_j \sum_{i=1}^{m} h_{ij} \, \Psi_i[e_i(\mathbf{s})], \tag{2.78}$$

where

$$\Psi_i[e_i(\mathbf{s})] = \frac{\partial \rho[e_i(\mathbf{s})]}{\partial e_i(\mathbf{s})} = \tanh(\gamma e_i(\mathbf{s})), \tag{2.79}$$

$$e_i(\mathbf{s}) = \mathbf{x}_i - \sum_{j=1}^{n} h_{ij}\, s_j, \quad \mu_j = \frac{1}{\tau_j} > 0.$$

It should be noted that, for large values of the gain parameter γ (typically $\gamma > 50$), the sigmoidal activation function $\Psi_i[e_i(\mathbf{s})]$ approximates the sign (hard-limiter) function quite well and such a network is able to find a solution which approaches the minimum 1-norm solution as $\gamma \to \infty$. On the other hand, for small values of γ (typically $0.1 < \gamma < 1$) the activation function is almost linear over a wide range and the network is able to solve approximately the least-squares (a minimum 2-norm) problem.

In fact, by controlling the gain parameter γ (i.e., by changing its value over a wide range, typically $0.1 < \gamma < 1000$) the network is able to solve a system of linear equations in the minimum p-norm sense with $1 < p < 2$.[10] However, in order to achieve the exact 1-norm solution, it is necessary that the gain γ approaches infinity. Unfortunately, it is difficult to control large values of the parameter γ. This is inconvenient from a practical implementation point of view, since an infinite gain is in fact often responsible for various parasitic effects, such as parasitic oscillations that decrease the final accuracy. To avoid this problem (especially the ill-conditioned problems), we have developed a modified cost function with regularization [276]

$$J(\mathbf{s}, \alpha) = \sum_{i=1}^{m} \left(\rho[e_i(\mathbf{s})] + \alpha_j\, \vartheta_i(s_j)\right), \tag{2.80}$$

where $\vartheta(s)$ is a convex function, typically $\vartheta(s) = s^2$. The minimization of the above energy function leads to the set of differential equations

$$\begin{aligned} \frac{ds_j}{dt} &= \mu_j \left[\sum_{i=1}^{m} h_{ij}\, \Psi_i(e_i) - \alpha_i\, \varphi(s_j)\right], \\ e_i(t) &= x_i(t) - \sum_{j=1}^{n} h_{ij}\, s_j(t), \end{aligned} \tag{2.81}$$

where $\varphi_i(s_j) = d\vartheta_j/ds_j$. On the basis of the above system of differential equations, we can easily realize an appropriate neural network called the Amari-Hopfield network illustrated in Fig.2.2. Such a network will force the residuals $e_i(\mathbf{s})$ with the smallest absolute values to tends to zero, while other residuals with large values (corresponding to outliers) will be inhibited (suppressed). The algorithm given by Eqs. (2.81) makes it possible to obtain an approximate minimum 1-norm solution, even when the gain parameter γ has a relatively low value (typically $\gamma > 80$) [276].

[10] Strictly speaking, the activation function for the p-norm problem is given as $\Psi_i[e_i(\mathbf{s})] = |e_i(\mathbf{s})|^{p-1}\,\mathrm{sign}[e_i(\mathbf{s})]$ $(1 \le p \le \infty)$.

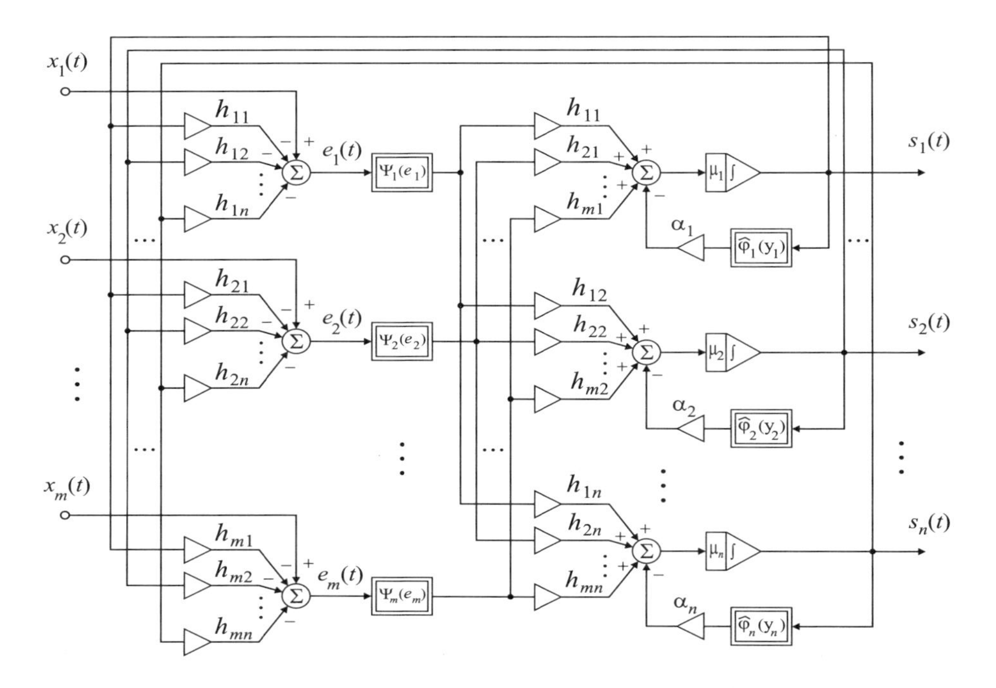

Fig. 2.2 Detailed architecture of the Amari-Hopfield continuous-time (analog) model of recurrent neural network with regularization.

2.3.2 Neural Network Model for LAD Problem Exploiting Inhibition Principles

Inhibition plays an important role of self-regulatory control in biologically plausible neural networks and also in many artificial neural networks (ANN), mainly in various decision-making and selection tasks [276]. The extremal form of inhibition is the Winner-Take-All (WTA) function. In general, the function of an inhibition sub-network is to suppress some signals (e.g., the strongest signals) while allowing other signals to be transmitted for further processing. Our goal in this section is to employ this mechanism explicitly for solving the LAD problem [276].

For an overdetermined linear system of equations (cf. Eq. (2.3)), in general, it is impossible to satisfy all the m equations exactly, but there may exist a point $\mathbf{s}_*$ in the n-space satisfying at least the n equations. This can be formulated by the following fundamental theorem [116, 276]:

Theorem 2.2 *For any matrix* $\mathbf{H} \in \mathbb{R}^{m \times n}$ *and vector* $\mathbf{x} \in \mathbb{R}^m$*, there exits a vector* $\mathbf{s}_* \in \mathbb{R}^n$*, with* $m > n$ *which minimizes 1-norm* $\|\mathbf{x} - \mathbf{H}\,\mathbf{s}\|_1$ *(i.e., the sum of residual error magnitudes) such that the optimal residual error vector* $\mathbf{e}(\mathbf{s}_*) = \mathbf{x} - \mathbf{H}\mathbf{s}_*$ *has at least n zero components. Furthermore, if the row vectors of the extended matrix* $\bar{\mathbf{H}} = [\mathbf{x}\ \mathbf{H}] \in \mathbb{R}^{m \times (n+1)}$ *satisfy the*

Haar condition[11] *then there exists a vector* $\mathbf{s}_*$ *which minimizes* $\|\mathbf{x} - \mathbf{H}\,\mathbf{s}\|_1$ *such that the residual vector has exactly* n *zero components.*

From this theorem, it follows that the minimum 1-norm solution $\mathbf{s}_{\text{LAD}} = \mathbf{s}_*$ of the overdetermined $m \times n$ system interpolates at least n points of the m (with $m > n$) observed or measured data points, assuming $\mathbf{H}$ is of full rank. We can say that the minimum 1-norm solution is the median solution. On the other hand, the ordinary minimum 2-norm LS solution is the mean solution, since it tries to satisfy most of the equations in the set, but this solution will not usually solve any of these equations exactly. The 1-norm criterion has the capability of ignoring a few outliers (bad data points) while taking into account the majority of the data points which usually reflect the true nature of the data [116].

From theorem 2.2, we can design a brute force method according to which we determine solutions for all possible subset of determined n equations taken from the set of all m equations $\mathbf{H}\,\mathbf{s} = \mathbf{x}$ and evaluates the 1-norm of the associated residual error vector $\mathbf{e} = \mathbf{x} - \mathbf{H}\,\mathbf{s}$ corresponding to each of these solutions. Unfortunately, this brute force method for large scale problem is inefficient since the number of subsets on n equations in a system of $m > n$ equations expressed as $m!/(n!(m-n)!)$ is huge [116].

Now we will propose simplified heuristic procedure which is much more efficient. The main idea is based on Theorem 2.2 and on the assumption that in most cases the 2-norm solution is very rough approximation of 1-norm solution. By using the ordinary least-squares technique, we can solve the LAD problem iteratively in several steps by eliminating equations with largest residuals. In the first stage, we compute all the residuals $e_i(\mathbf{s}_{\text{LS}})$ $(i = 1, 2, \ldots, m)$ and gradually select, from the m set of equations only the n equations corresponding to the n residuals which are the smallest in absolute value, the rest of the equations are ignored (or inhibited). In the next stage, we use the reduced number of determined equation to estimate the vector $\mathbf{s}_{\text{LAD}}$. The heuristic algorithm is summarized as follows:

Algorithm Outline: LAD Solution Using Multi-stage LS Procedure

---

Step 1. Compute the LS solution as

$$\mathbf{s}_{\text{LS}} = (\mathbf{H}\,\mathbf{H}^T)^{-1}\,\mathbf{H}^T\,\mathbf{x} \tag{2.82}$$

and on the basis of residual vector $\mathbf{e}(\mathbf{s}) = \mathbf{x} - \mathbf{H}\,\mathbf{s}_{\text{LS}}$ select the reduced set of equations corresponding to the modulus of the smallest residuals. After eliminating some equation, we obtain reduced set of equations $\mathbf{H}_{rj}\,\mathbf{s} = \mathbf{x}_{rj}$, where $\mathbf{H}_{rj} \in \mathbb{R}^{m_j \times n}$ with $n \leq m_j < m$.

Step 2. Estimate the new vector the vector $\mathbf{s}_{rj}$:

$$\mathbf{s}_{jr} = [\mathbf{H}_{rj}^T\,\mathbf{H}_{rj}]^{-1}\mathbf{H}_{rj}^T\,\mathbf{x}_{rj} = \mathbf{H}_{rj}^+\,\mathbf{x}_{rj} \tag{2.83}$$

[11]Let there be given set of vectors $\{\mathbf{a}_1, \mathbf{a}_2, \ldots, \mathbf{a}_m\}$ contained in vector space $\mathbb{R}^{n\times 1}$ (or $\mathbb{R}^{1\times n}$) with $m \geq n$. This set of vectors satisfy the Haar condition if every subset n of these vectors forms a linearly independent set.

is the reduced matrix obtained by removing from the matrix $\mathbf{H}$ certain rows corresponding to the largest magnitudes of residual error.

Step 3. Repeat Step 1 and 2 until at least $(m-n)$ (or the specified number of equations) from the original system of linear equations are removed.

After eliminating the $(m-n)$ equations one by one, we obtain the reduced set of determined equations in the form $\mathbf{H}_r\,\mathbf{s} = \mathbf{x}_r$, where $\mathbf{H}_r$ is a nonsingular $n \times n$ reduced matrix and $\mathbf{x}_r \in \mathbb{R}^n$ is a reduced sensor vector.

Example 2.1 Let us consider the following ill-conditioned LAD problem:

Minimize $J(\mathbf{s}, \mathbf{H}) = \|\mathbf{x} - \mathbf{H}\,\mathbf{s}\|_1$,

where

$$\mathbf{H}^T = \begin{bmatrix} 1.1 & 4.0 & 7.0 & -3.1 & 5.5 \\ 2.0 & 5.1 & 8.0 & -4.0 & 6.9 \\ 3.1 & 6.0 & 8.9 & -5.0 & 8.1 \end{bmatrix}$$

and $\mathbf{x} = [-2 \;\; 1.1 \;\; 4.2 \;\; -0.5 \;\; 2.1]^T$.

In the first step, we can obtain the LS (2-norm) solution and the corresponding vector of the residuals as

$$\begin{aligned} \mathbf{s}_{\text{LS}} &= (\mathbf{H}^T\,\mathbf{H})^{-1}\,\mathbf{H}^T\,\mathbf{x} = [1.6673 \;\; 1.4370 \;\; -2.1198]^T, \\ \mathbf{e}_{\text{LS}} &= \mathbf{x} - \mathbf{H}\,\mathbf{s}_{\text{LS}} = [0.1368 \;\; -0.1795 \;\; -0.1015 \;\; -0.1821 \;\; 0.1844]^T. \end{aligned}$$

Since the residuals corresponding to the fourth and fifth row have the largest amplitude, we can remove them and compute the optimal LAD solution as

$$\mathbf{s}_{\text{LAD}} = \mathbf{s}_* = \mathbf{H}_r^{-1}\,\mathbf{x}_r = [2 \;\; 1 \;\; -2]^T, \tag{2.84}$$

where $\mathbf{x}_r = [-2 \;\; 1.1 \;\; 4.2]^T$ and

$$\mathbf{H}_r = \begin{bmatrix} 1.1 & 2 & 3.1 \\ 4 & 5.1 & 6 \\ 7 & 8 & 8.9 \end{bmatrix}.$$

In many cases it is impossible explicitly to find the LAD solution in one step. The best performance is obtained if we reduce the number of equations gradually on-by-one.

The above simple algorithm can easily be implemented on-line by the Amari-Hopfield neural network shown in Fig.2.2, with the adaptive activation functions $\Psi_i(e_i)$ (taking one of two forms: e_i or 0). In the first phase of the computation, all the activation functions are $\Psi_i(e_i) = e_i$, thus both the LS estimation $\mathbf{s}_{\text{LS}}$ and the corresponding residuals $\mathbf{e}(\mathbf{s}) = \mathbf{x} - \mathbf{H}\,\mathbf{s}_{\text{LS}}$ are simultaneously estimated automatically by the neural network. In the second phase of the computation, the inhibition control circuit (not explicitly shown on Fig.2.2) selects the $(m-n)$ largest modulus of the residuals $e_i(\mathbf{s}_{\text{LS}})$ from the set of all the m residuals $e_i(\mathbf{s}_{\text{LS}})$ and inhibits the corresponding $(m-n)$ hidden neurons by switching their activation functions to $\Psi_p(e_i) = 0$, allowing the smallest n residuals to be further processed in the network. In this way, in the second phase of the computation, only n equations are selected for which the residuals are minimized to zero, while the rest of the equations are simply discarded. The inhibition control subnetwork can be realized on the basis of the Winner-Take-All principle. Firstly, the circuit selects the largest signal $|e_i(\mathbf{s}_{\text{LS}})|$ which is immediately inhibited and the corresponding switch is opened. Then, the procedure is sequentially repeated for $(n-m)$ times for the rest of the signals $|e_i(\mathbf{s})|$.

It is to be emphasized the above presented algorithm does not have guaranteed convergence to an optimal minimum 1-norm solution although typically converge to a good approximate solution at a very small computation expense. Therefore, we need some criterion or a measure which gives us guarantee that obtained solution is optimal. The following theorem provides the such measure [116].

Theorem 2.3 *Let the full column rank matrix $\mathbf{H} \in \mathbb{R}^{m \times n}$ with $m > n$ satisfies the Haar condition and let the reduced submatrix $\mathbf{H}_r \in \mathbb{R}^{n \times n}$ designate the n rows of matrix $\mathbf{H}$ associated with n zero elements of the residual error vector $\mathbf{e}(\mathbf{s}_*) = \mathbf{x} - \mathbf{H}\mathbf{s}_*$. Furthermore, let submatrix $\mathbf{H}_2 \in \mathbb{R}^{(m-n)\times n}$ designate the $(m-n)$ rows of the matrix $\mathbf{H}$ for which the residual error components are not equal zero, that is $\mathbf{e}_2 = \mathbf{x}_2 - \mathbf{H}_2\,\mathbf{s}_* \neq \mathbf{0}$ represents set of inconsistent equations, while $\mathbf{e}_1(\mathbf{s}_*) = \mathbf{x}_r - \mathbf{H}_r\,\mathbf{s}_*$. If the reduced matrix $\mathbf{H}_r$ is nonsingular then the vector $\mathbf{s}_*$ is a unique minimum 1-norm solution $(J(\mathbf{s}) = \|\mathbf{x} - \mathbf{H}\,\mathbf{s}\|_1)$, if and only if all the components of the vector*

$$\mathbf{g} = [\mathbf{H}_r^T]^{-1}\,\mathbf{H}_2^T\,\text{sign}(\mathbf{x}_2 - \mathbf{H}_2\mathbf{s}_*) \tag{2.85}$$

have magnitudes strictly less than one (i.e., $\|\mathbf{g}\|_\infty < 1$). However, if the elements of the vector $\mathbf{g}$ have magnitudes of less than or equal one the solution is optimal but not unique. In the case when $\|\mathbf{g}\|_\infty > 1$ the solution is not optimal.

In a degenerate (rank deficient) case, i.e., if more than n elements of the residual error vector $\mathbf{e}(\mathbf{s}_*) = \mathbf{x} - \mathbf{H}\,\mathbf{s}_*$ is equal to zero, we can use the following theorem formulated by Cadzow [116]

Theorem 2.4 *Let the full column rank matrix $\mathbf{H} \in \mathbb{R}^{m \times n}$ with $m > n$ satisfies the Haar condition and let a vector $\mathbf{s}_* \in \mathbb{R}^n$ will be a degenerate solution of the system of overdetermined equations $\mathbf{x} \approx \mathbf{H}\,\mathbf{s}$ so that its associated residual error vector $\mathbf{e}(\mathbf{s}) = \mathbf{x} - \mathbf{H}\,\mathbf{s}_*$ has $n_0 > n$ zero elements. Furthermore, let the reduced submatrix $\mathbf{H}_r \in \mathbb{R}^{n_0 \times n}$ designates the n_0 rows of the matrix $\mathbf{H}$ associated with n_0 zero elements of the residual error vector and $\mathbf{H}_2 \in \mathbb{R}^{(m-n_0)\times n}$ corresponds to the entities of the remaining subset of $(m - n_0)$ equations which have nonzero residuals. If the all n by n submatrices $\mathbf{H}_{rj}$ of the reduced matrix $\mathbf{H}_r$ are nonsingular then the vector $\mathbf{s}_*$ is a minimum 1-norm solution of the system of overdetermined equations $\mathbf{x} \approx \mathbf{H}\,\mathbf{s}$ if and only if all the components of the vectors*

$$\mathbf{g}_j = [\mathbf{H}_{rj}^T]^{-1}\,\mathbf{H}_2^T\,\text{sign}(\mathbf{x}_2 - \mathbf{H}_2\mathbf{s}_*) \tag{2.86}$$

for $1 \leq j \leq n_0!/(n!(n_0 - n)!)$ *have magnitudes less than or equal one (i.e.,* $||\mathbf{g}||_\infty \leq 1$*).*

While both the LS and LAD formulation of the estimation problem and their solution are commonly employed in practice, it is worth mentioning of not only their usefulness, but also their limitations [276]. These techniques are able to provide unbiased estimates of the coefficient vector in an ergodic environment, only when the entries of matrix $\mathbf{H}$ are known precisely. In other words, an implicit assumption employed in the LS and LAD estimation procedures and many of their variations is that the noise (errors) contained in the data matrix $\mathbf{H}$ are negligible. In other words, we have attributed so far that all the uncertainty about the system parameters to the noise contained in the sensor signals $\mathbf{x}$.

2.4 TOTAL LEAST-SQUARES AND DATA LEAST-SQUARES PROBLEMS

2.4.1 Problems Formulation

In the previous sections of this chapter, we have considered the case where only vector $\mathbf{x}$ is contaminated by the error and the matrix $\mathbf{H}$ is known precisely. The total least-squares approach is suitable for solving estimation problems that can be formulated as a system of over-determined linear equations of the form $\mathbf{Hs} \approx \mathbf{x}$, in which *both* the entries of data matrix $\mathbf{H}$ and the sensor vector $\mathbf{x}$ are contaminated by noise, i.e., we have the system of linear equations

$$(\mathbf{H}_{true} + \mathbf{N})\,\mathbf{s} \approx \mathbf{x}_{true} + \mathbf{n} = \mathbf{x} \tag{2.87}$$

where the true data (the data matrix $\mathbf{H}_{true}$ and the measurement vector $\mathbf{x}_{true}$) are unknown. In contrast to TLS, in LS problems it is assumed, that the data matrix $\mathbf{H}_{true}$ is known precisely, i.e., the noise matrix $\mathbf{N} \in \mathbb{R}^{m \times n}$ is zero or negligibly small and only the measurement vector $\mathbf{x}$ is contaminated by an unknown noise $\mathbf{n}$, while, in the data least squares (DLS) problem, it is assumed that noise $\mathbf{n}$ is zero or negligibly small and only the noise contained in the matrix $\mathbf{N}$ exists. In the TLS problem, it is assumed that the noise has zero-mean and a Gaussian distribution.

In many signal processing applications the TLS problem is reformulated as an approximate linear regression problem of the form

$$x_i \approx \mathbf{h}_i^T \mathbf{s}; \qquad (i = 1, 2, \dots, m) \tag{2.88}$$

where $\mathbf{h}_i^T = [h_{i1}, h_{i2}, \dots, h_{in}]$ is the i-th row of $\mathbf{H}$.

The TLS solution that eliminates the effects of certain types of noise in the signals can be shown to be related to a lower-rank approximation of the augmented matrix $\bar{\mathbf{H}} = [\mathbf{x} \;\; \mathbf{H}]$. Based on this result, we show how the unknown vector $\mathbf{s}$ can be estimated from noisy data. This section is concerned with the estimation algorithms that are designed to alleviate the effects of noise present in *both* the input and the sensor signals. We will show that the total least-squares estimation procedure can produce unbiased estimates in the presence of certain types of noise disturbances in the signals. This procedure will then be extended to the case of arbitrary noise distribution. There is a large class of problems requiring on-line estimation of the signals and the parameters of the underlying systems.

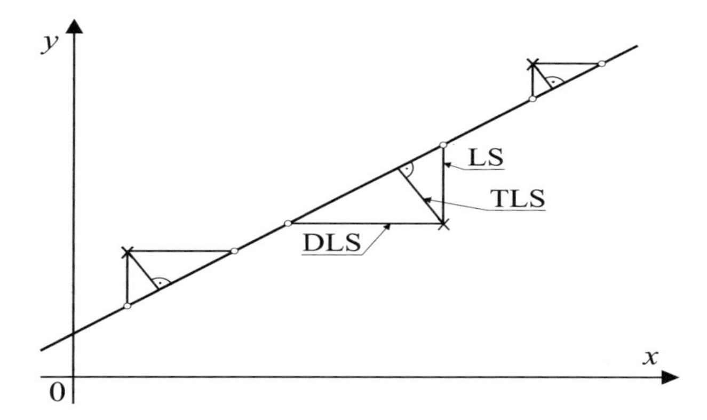

Fig. 2.3 This figure illustrates the optimization criteria employed in the total least-squares (TLS), least-squares (LS) and data least-squares (DLS) estimation procedures for the problem of finding a straight line approximation to a set of points. The TLS optimization assumes that the measurements of the x *and* y variables are in error, and seeks an estimate such that the sum of the squared values of the perpendicular distances of each of the points from the straight line approximation is minimized. The LS criterion assumes that only the measurements of the y variable is in error, and therefore the error associated with each point is parallel to the y axis. Therefore the LS minimizes the sum of the squared values of such errors. The DLS criterion assumes that only the measurements of the x variable is in error.

2.4.1.1 A Historical Overview of the TLS Problem The total least-squares (TLS) method was independently derived in several areas of science, and is known to statisticians as the *orthogonal regression* or the *error-in-variables* problem. The error-in-variables problem has a long history in statistical literature. Pearson in 1901 [575] solved the two-variable model fitting problem that may be formulated as follows: given a set of points (x_k, y_k) for $k = 1, 2 \ldots, m$, we wish to find the optimal straight line

$$y = mx + c \tag{2.89}$$

that minimizes the sum of squared perpendicular distances between the points in the set and the straight line. Figure 2.3 describes this problem graphically. Pearson solved this problem and expressed the modelling errors associated with the points in terms of the mean, standard deviation and correlation coefficients of the data. In the classical least-squares problem, we wish to find values of slope and intercept (m, c) which minimize the sum of the squared distances between y_k and its predicted values. In other words, the LS solution results from minimizing the sum of squared values of the vertical distances between the line and the measurements y_k. It assumes that the variables x_k are error free and all the noise is contained in y_k. The Data Least-Squares (DLS) algorithms are another class of estimation techniques based on the assumption that points y_k are error-free and all the noise is contained only in the measurements x_k. Therefore, the DLS algorithm attempts to minimize the sum of squared distances between the line and the measurements x_k along the horizontal axis. For example, the DLS solution is useful in equalization problems that involve certain types of deconvolution models. Unlike both the LS and DLS approaches, the TLS algorithms assume that both x_k and y_k are contaminated by noise. Consequently, we can consider the standard LS and DLS algorithms as special cases of the extended TLS technique.

Example 2.2 Let us determine the optimal line $y = mx + c$ by the ordinary LS, TLS, DLS criteria and also the solution to the minimum 1-norm and infinity-norm problem, for the data points:

$$(x_k, y_k) = (1, 2),\ (2, 1.5),\ (3, 3),\ (4, 2.5),\ (5, 3.5)$$

This problem is equivalent to solving a system of linear equations with respect to $\mathbf{s}$: $\mathbf{H}\,\mathbf{s} \approx \mathbf{x}$ or equivalently to minimize $\|\mathbf{x} - \mathbf{H}\,\mathbf{s}\|$, where $\mathbf{x} = \left[\begin{array}{ccccc} 2 & 1.5 & 3 & 2.5 & 3.5 \end{array}\right]^T$, $\mathbf{s} = [m \;\; c]^T$ and data (input) matrix

$$\mathbf{H}^T = \left[\begin{array}{ccccc} 1 & 2 & 3 & 4 & 5 \\ 1 & 1 & 1 & 1 & 1 \end{array}\right].$$

Fig.2.4 illustrates solutions of this problem using different criteria.

The problem of fitting a straight line to a noisy data set has been generalized to that of fitting a hyper-plane to noisy, higher-dimensional data [575, 576].

In the field of numerical analysis, the total least-squares problem was first introduced by Golub and Van Loan in 1980 [494], studied extensively and refined by Van Huffel, Vandevalle, Lemmerling *et. al.* [575, 576], Hansen and O'Leary, and many other researchers [575, 576].

There are many applications that require on-line adaptive computation of the parameters $\mathbf{s}$ for a system model. Adaptive algorithms that employ the TLS formulation and their extensions have been developed and analyzed by Amari and Kawanabe, Mathews, Cichocki, Unbehauen, Xu, Oja and Douglas along with many others [36, 575, 576, 276, 823, 1303, 287, 288].

2.4.2 Total Least-Squares Estimation

The TLS solution explicitly recognizes that both the input matrix $\mathbf{H}$ and the sensor vector $\mathbf{x}$ may be contaminated by noise. Let $\mathbf{H}_{true} = \mathbf{H} - \mathbf{N}$ represent the noise-free input matrix, and let $\mathbf{x}_{true} = \mathbf{x} - \mathbf{n}$ represent the noise-free desired response vector. Here, we do not consider any possible relationships that might exist among the elements of $\mathbf{H}$. Such constraints may be incorporated into the TLS formulation, but this will result in a considerable increase in the complexity of the solution. The TLS procedure attempts to estimate both the noise matrix $\mathbf{N}$ and the noise vector $\mathbf{n}$ to satisfy the exact solution of the system of linear equations:

$$(\mathbf{H} - \mathbf{N})\,\mathbf{s}_{\mathrm{TLS}} = (\mathbf{x} - \mathbf{n})\,. \tag{2.90}$$

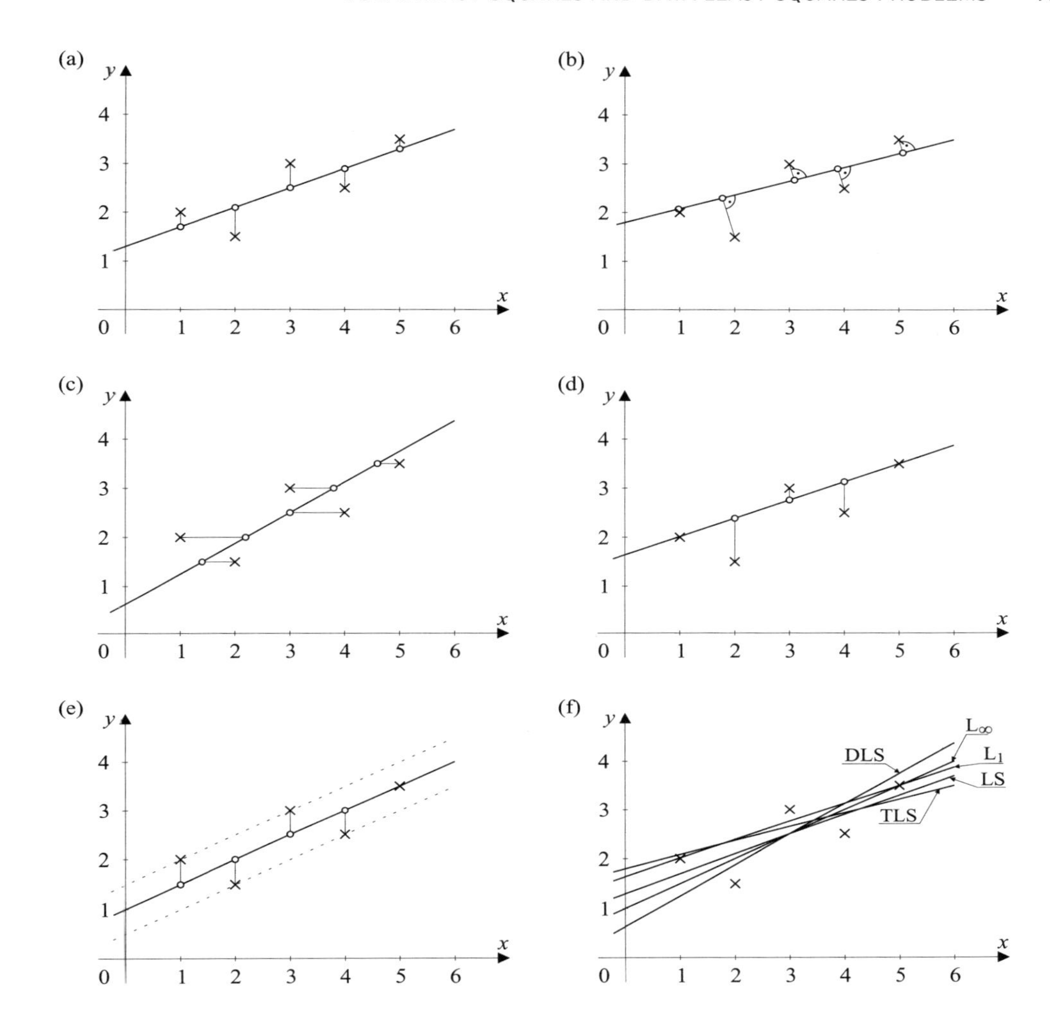

Fig. 2.4 Straight line fit for the five points marked by 'x' using the: (a) LS (L_2 -norm), (b) TLS, (c) DLS, (d) L_1-norm, (e) L_∞ -norm, and (f) combined results.

Generally, there may be many choices of $\mathbf{N}$ and $\mathbf{n}$ that satisfy (2.90). Among all such choices, we select $\mathbf{N}$ and $\mathbf{n}$ such that[12]

$$\| \begin{bmatrix} \mathbf{n} & \mathbf{N} \end{bmatrix} \|_{\mathrm{F}}^2 = \sum_{i=1}^{m} n_i^2 + \sum_{i=1}^{m} \sum_{j=1}^{n} n_{ij}^2 \tag{2.91}$$

[12]We have assumed, that the signals are real-valued. The extension to complex-valued data is straightforward.

is minimized, where n_{ij} is the (i,j)-th element of $\mathbf{N}$ and n_i is the i-th element of $\mathbf{n}$. To solve the above problem, we rewrite (2.90) as

$$(\overline{\mathbf{H}} - [\begin{array}{cc} \mathbf{n} & \mathbf{N} \end{array}]) \left[\begin{array}{c} -1 \\ \mathbf{s}_{\mathrm{TLS}} \end{array} \right] = \mathbf{0}, \tag{2.92}$$

where $\overline{\mathbf{H}} = [\mathbf{x},\ \mathbf{H}]$. In the above equation, $\mathbf{0}$ is an m-dimensional vector filled with all zeros. In general, the augmented input matrix $\overline{\mathbf{H}}$ is full rank due to the presence of noise. If we assume that $m > n+1$, the rank of $\overline{\mathbf{H}}$ is $(n+1)$.

The problem of finding $\mathbf{n}$ and $\mathbf{N}$ can be recast as that of finding the smallest perturbation of the augmented input matrix $\overline{\mathbf{H}}$ that results in a rank-n matrix.

Let us expand the matrix $\overline{\mathbf{H}}$ using the singular value decomposition as

$$\overline{\mathbf{H}} = [\begin{array}{cc} \mathbf{x} & \mathbf{H} \end{array}] = \sum_{i=1}^{n+1} \sigma_i \mathbf{u}_i \mathbf{v}_i^T, \tag{2.93}$$

where σ_i's are the singular values of $\overline{\mathbf{H}}$, arranged in the descending order of magnitude, $\mathbf{u}_i$'s and $\mathbf{v}_i$'s are respectively the left singular vectors containing m elements each, and the right singular vectors containing $(n+1)$ elements each. We assume that we have selected the singular vectors such that they have unit length, and that the sets $\{\mathbf{u}_i;\quad i = 1, 2, \ldots, n+1\}$ and $\{\mathbf{v}_i\ ;\ i = 1, 2, \ldots, n+1\}$ contain orthogonal elements so that $\mathbf{u}_i^T \mathbf{u}_j = 0$ and $\mathbf{v}_i^T \mathbf{v}_j = 0$ for $i \neq j$. It is well-known that the rank-n approximation of $\overline{\mathbf{H}}$ introducing the least amount of perturbation to its entries is given by [575]

$$\widehat{\mathbf{H}} = \sum_{i=1}^{n} \sigma_i \mathbf{u}_i \mathbf{v}_i^T. \tag{2.94}$$

Moreover, the error matrix $[\begin{array}{cc} \mathbf{n} & \mathbf{N} \end{array}]$ is given by

$$[\begin{array}{cc} \mathbf{n} & \mathbf{N} \end{array}] = \sigma_{n+1} \mathbf{u}_{n+1} \mathbf{v}_{n+1}^T. \tag{2.95}$$

Taking (2.92) and (2.94) into account, we can write

$$\widehat{\mathbf{H}} \left[\begin{array}{c} -1 \\ \mathbf{s}_{\mathrm{TLS}} \end{array} \right] = \left[\sum_{i=1}^{n} \sigma_i \mathbf{u}_i \mathbf{v}_i^T \right] \left[\begin{array}{c} -1 \\ \mathbf{s}_{\mathrm{TLS}} \end{array} \right] = \mathbf{0}. \tag{2.96}$$

Since $\mathbf{v}_{n+1}$ is orthogonal to the rest of the vectors: $\mathbf{v}_1$, $\mathbf{v}_2$, ..., $\mathbf{v}_n$, the TLS solution for the coefficient vector given by

$$\left[\begin{array}{c} -1 \\ \mathbf{s}_{\mathrm{TLS}} \end{array} \right] = -\frac{\mathbf{v}_{n+1}}{v_{n+1,1}}, \tag{2.97}$$

(where $v_{n+1,1}$ is the first non-zero entry of the right singular vector $\mathbf{v}_{n+1}$) satisfies (2.96). Thus, the total least-squares solution is described by the right singular vector corresponding to the smallest singular value of the augmented input matrix $\overline{\mathbf{H}}$. An efficient approach to computing this singular vector is to find the vector $\mathbf{v}$ that minimizes the cost function[13]

$$J(\mathbf{v}) = \frac{\mathbf{v}^T \overline{\mathbf{H}}^T \overline{\mathbf{H}} \mathbf{v}}{\| \mathbf{v} \|_2^2}, \tag{2.98}$$

and then to normalize the resulting vector so that

$$\begin{bmatrix} -1 \\ \mathbf{s}_{\text{TLS}} \end{bmatrix} = -\frac{\mathbf{v}_{opt}}{v_{opt,1}}, \tag{2.99}$$

where $\mathbf{v}_{opt}$ denotes the solution to the optimization problem in (2.98), and $v_{opt,1}$ is the first entry of $\mathbf{v}_{opt}$. Simple calculations will show that the vector $\mathbf{v}$ minimizing $J(\mathbf{v})$ is identical to the $(n+1)$-th right singular vector of $\overline{\mathbf{H}}$, and choosing $\mathbf{v} = \mathbf{v}_{n+1}$ will provide the minimum value of $J(\mathbf{v})$ given by

$$J(\mathbf{v}_{opt}) = \sigma_{n+1}^2. \tag{2.100}$$

It is also straightforward to show that the optimum choice for $\mathbf{v}$ corresponds to the eigenvector corresponding to the smallest eigenvalue of the matrix $\overline{\mathbf{H}}^T \overline{\mathbf{H}}$. Thus, a numerical method based on SVD or minor component analysis (MCA) for finding the TLS estimate of the coefficients can be applied. The above derivation assumes that the smallest singular value of the augmented data matrix $\overline{\mathbf{H}}$ is unique. If this is not the case, the TLS problem has an infinite number of solutions. To uniquely define $\mathbf{s}_{\text{TLS}}$ in such situations, usually a solution is chosen for which $\| \mathbf{s}_{\text{TLS}} \|^2$ is the smallest among all the possibilities.

When the noise in the entries of the augmented data matrix $\overline{\mathbf{H}}$ belongs to independent and identically distributed (i.i.d.) Gaussian processes with zero-mean, it can be shown that the TLS solution obtained by minimizing the cost function in (2.98) is the maximum likelihood estimate of the coefficient vector. When the noise sequences satisfy the i.i.d. condition, the standard TLS estimate is unbiased. In other words, we obtain an unbiased estimate if both the noise variance of the vector $\mathbf{x}$ (sensor signals) and the data matrix $\mathbf{H}$ are the same, i.e., $(\sigma_n^2 = \sigma_N^2)$.

Consequently, even though the standard total least-squares approach described above results in unbiased estimates of the parameters of the system model, it does not necessarily provide a good estimate of the signals of interest. In order to obtain better estimates of the parameters, one may use an augmented data matrix $\overline{\mathbf{H}}$ with the number of columns $n' >> n+1$ (where $(n+1)$ is the minimum number of columns needed by the TLS approach) and then approximate this matrix with a rank-n matrix. The noise matrix estimated using such an approximation has a rank larger than one. Therefore it provides a better estimate than the one provided by the standard TLS algorithm described above.

[13] Scaling the coefficient vector by a scalar multiplier does not change the cost function. Consequently, we can also formulate this problem equivalently as that of minimizing

$$J(\mathbf{v}) = \mathbf{v}^T \overline{\mathbf{H}}^T \overline{\mathbf{H}} \mathbf{v}$$

subject to $\| \mathbf{v} \|_2^2 = 1$.

2.4.3 Adaptive Generalized Total Least-Squares

The standard (ordinary) TLS is a method which gives an improved unbiased estimator only when both the noise (errors) in the data matrix $\mathbf{H}$ and the sensor vector $\mathbf{x}$ are i.i.d. and exhibit the same variance. However, in practice, the data matrix $\mathbf{H}$ and the observation vector $\mathbf{x}$ represent different physical quantities and are therefore usually subject to different noise or error levels. The generalized TLS (GTLS) problem deals with the case where the data errors $\Delta h_{ij} = n_{ij}$ are i.i.d. with zero-mean and variance σ_N^2 (i.e., $\mathbf{R}_{NN} = \sigma_N^2 \mathbf{I}_n$) and where the observation (sensor) vector components $\Delta x_i = n_i$ are also i.i.d. with zero-mean and variance $\sigma_n^2 \neq \sigma_N^2$.

There are many situations in which the parameters of the underlying system in the estimation problem are time-varying, while the input signals are corrupted by uncorrelated noise. Even in situations where the characteristics of the operating environment do not vary over time, either adaptive or iterative solutions are often sought, because the singular value decomposition-based solutions tend to be computationally expensive and such methods usually do not exploit the special structures or sparsity of the system model to reduce the computational complexity. In this section, we discuss how the traditional adaptive filtering algorithms such as the least-mean-square (LMS) algorithm can be modified to account for the presence of the additive i.i.d. noise in both the sensor signals and the mixing (data) matrix.

Solving the generalized TLS problem consists in finding the vector $\mathbf{s}$ which minimizes [576, 288]

$$\gamma \, \|\Delta \mathbf{H}\|_F^2 + (1-\gamma) \, \|\Delta \mathbf{x}\|_F^2, \qquad where \qquad \frac{1-\gamma}{\gamma} = \beta = \frac{\sigma_n^2}{\sigma_N^2}. \tag{2.101}$$

and $\Delta \mathbf{H}$ and $\Delta \mathbf{x}$ refer to perturbations of the matrix $\mathbf{H}$ and sensor vector $\mathbf{x}$, respectively. By changing the parameter γ in the range $[0, 1]$, we obtain the special cases: the standard LS, TLS and DLS problems. The parameter $\gamma = 0$ ($\beta = \infty$) yields the standard LS formulation since in this case $\sigma_N^2 = 0$, whereas $\gamma = 0.5$ gives the standard TLS formulation since $\sigma_N^2 = \sigma_n^2$, and finally $\gamma = 1$ ($\beta = 0$) results in the DLS formulation with $\sigma_n^2 = 0$.

Let us first consider the standard mean square error cost function formulated as

$$\widetilde{J}(\mathbf{s}) = E\{\mathbf{e}^T(k)\mathbf{e}(k)\}, \tag{2.102}$$

where the error vector $\mathbf{e}$ is defined as

$$\mathbf{e}(k) = \mathbf{x} - \mathbf{H}\,\mathbf{s}(k) = (\mathbf{x}_{true} + \mathbf{n}) - (\mathbf{H}_{true} + \mathbf{N})\,\mathbf{s}(k), \tag{2.103}$$

where $\mathbf{H}_{true}$ and $\mathbf{x}_{true}$ are unknown true parameters. The cost function (2.102) can be evaluated as follows

$$\begin{aligned} \widetilde{J}(\mathbf{s}) &= E\{(\mathbf{x}_{true} - \mathbf{H}_{true}\,\mathbf{s})^T(\mathbf{x}_{true} - \mathbf{H}_{true}\,\mathbf{s})\} + \sigma_n^2 + \mathbf{s}^T\,\mathbf{R}_{NN}\,\mathbf{s} \\ &= E\{(\mathbf{x}_{true} - \mathbf{H}_{true}\,\mathbf{s})^T(\mathbf{x}_{true} - \mathbf{H}_{true}\,\mathbf{s})\} + \sigma_N^2\,(\beta + \mathbf{s}^T\,\mathbf{s}) \end{aligned} \tag{2.104}$$

on the assumption that noise components are uncorrelated i.i.d. and $\mathbf{R}_{NN} = \sigma_N^2 \mathbf{I}_n$.

It is obvious that minimizing the cost function $\widetilde{J}(\mathbf{s})$ with respect to the vector $\mathbf{s}$ will yield a biased solution, since noise components are functions of $\mathbf{s}$. To avoid this problem, we can use the modified mean square error cost function formulated as the generalized TLS problem which can be further reformulated as the following optimization problem [276, 288]:

Minimize the cost function

$$\begin{aligned} J(\mathbf{s}) &= \frac{1}{2} \frac{E\{\mathbf{e}^T(k)\mathbf{e}(k)\}}{\beta + \mathbf{s}^T\, \mathbf{s}} \\ &= \frac{1}{2} \frac{E\{(\mathbf{x}_{true} - \mathbf{H}_{true}\, \mathbf{s}(k))^T (\mathbf{x}_{true} - \mathbf{H}_{true}\, \mathbf{s}(k))\}}{\beta + \mathbf{s}^T\, \mathbf{s}} + \sigma_N^2. \end{aligned} \tag{2.105}$$

The above cost function removes the effect of noise, assuming that the power ratio of the noise components $\beta = \sigma_n^2/\sigma_N^2$ is known, since the last term in (2.105) is independent of $\mathbf{s}$.

To derive the iterative adaptive algorithm, we represent the cost function as

$$J(\mathbf{s}) = \sum_{i=1}^{m} J_i(\mathbf{s}), \tag{2.106}$$

where

$$J_i(\mathbf{s}) = E\{\varepsilon^2(k)\} = \frac{1}{2} \frac{E\{e_i^2(k)\}}{\beta + \mathbf{s}^T\, \mathbf{s}}$$

and $e_i = x_i - \mathbf{h}_i^T\, \mathbf{s}$ ($\mathbf{h}_i^T$ denotes the i-th row of $\mathbf{H}$).

Then the instantaneous gradient components can be evaluated as

$$\frac{d\varepsilon^2}{d\mathbf{s}} = \frac{e_i(k)\, \mathbf{h}_i}{\beta + \mathbf{s}^T\, \mathbf{s}} - \frac{e_i^2(k)\, \mathbf{s}}{[\beta + \mathbf{s}^T\, \mathbf{s}]^2}. \tag{2.107}$$

Hence, the iterative discrete-time algorithm exploiting the gradient descent approach can be written as

$$\mathbf{s}(k+1) = \mathbf{s}(k) + \eta(k)\, \widetilde{e}_i(k)\, [\mathbf{h}_i + \widetilde{e}_i(k)\, \mathbf{s}(k)], \quad i = k \text{ modulo } (m+1), \tag{2.108}$$

where

$$\widetilde{e}_i(k) = \frac{e_i(k)}{\beta + \mathbf{s}^T(k)\, \mathbf{s}(k)} = \frac{x_i - \mathbf{h}_i^T \mathbf{s}(k)}{\beta + \mathbf{s}^T(k)\, \mathbf{s}(k)}. \tag{2.109}$$

Since the term $(\beta + \mathbf{s}^T(k)\, \mathbf{s}(k))^{-1}$ is always positive, it can therefore, be absorbed by the positive learning rate, thus the algorithm can be represented in a simplified form as

$$\boxed{\mathbf{s}(k+1) = \mathbf{s}(k) + \eta(k)\, e_i(k)\, [\mathbf{h}_i + \widetilde{e}_i(k)\, \mathbf{s}(k)], \quad i = k \text{ modulo } (m+1)} \tag{2.110}$$

Remark 2.7 *It should be noted that the index i is taken modulo $(m+1)$, i.e., the rows $\mathbf{h}_i$ of matrix $\mathbf{H}$ and elements x_i of vector $\mathbf{x}$ are selected and processed in a cyclical order. In* *other words, after the first m iterations, for the $(m+1)$th iteration, we revert back to the first row of $\mathbf{H}$ and the first component of $\mathbf{x}$. We continue with the $(m+2)$-nd iteration using the second row of $\mathbf{H}$ and second component of $\mathbf{x}$, and so on, repeating the cycle every m iterations. Moreover, it is interesting to note that the above algorithm simplifies to the standard LMS algorithm when $\beta = \infty$, while it becomes the standard DLS algorithm for $\beta = 0$.*

Using the concept of component averaging (say, for a block of all indices i in one iteration cycle) and by applying self-normalization as in the Kaczmarz or NLMS algorithms, we can easily derive a novel GTLS iterative formula for the sparse matrix $\mathbf{H}$ as

$$\boxed{\mathbf{s}(k+1) = \mathbf{s}(k) + \overline{\eta}(k) \sum_{i=1}^{m} \frac{x_i - \mathbf{h}_i^T \mathbf{s}(k)}{\sum_{j=1}^{n} r_j h_{ij}^2} \left[\mathbf{h}_i + \frac{x_i - \mathbf{h}_i^T \mathbf{s}(k)}{\beta + \mathbf{s}^T(k)\, \mathbf{s}(k)} \, \mathbf{s}(k) \right],} \tag{2.111}$$

where $0 < \overline{\eta}(k) < 2$ is the normalized learning rate (relaxation parameter) and r_j is the number of non-zero elements h_{ij} of the column j.

2.4.4 Extended TLS for Correlated Noise Statistics

We noted earlier that the TLS solution is unbiased when the noise in $\overline{\mathbf{H}}$ belongs to an i.i.d. process. When the noise is i.i.d. Gaussian-distributed this is also the maximum likelihood solution. Unfortunately, when the noise is correlated in turn, the estimates are no longer guaranteed to be unbiased. In such a case a modification is needed to make the TLS approach useful, especially when the input signal is corrupted by non-i.i.d. noise sequences. For the purpose of the derivation, we will assume that the noise samples are Gaussian distributed with zero-mean. If the noise is non-Gaussian, the procedure described here will still result in unbiased estimates of the system model parameters. However, these estimates will no longer satisfy the maximum likelihood property. Let the statistical expectation of the product of the augmented input matrix $\overline{\mathbf{H}}$ with its own transpose be given by [823]

$$E\left\{\overline{\mathbf{H}}^T \overline{\mathbf{H}}\right\} = \mathbf{R}_{\mathbf{HH}} + \bar{\mathbf{R}}_{\mathbf{NN}}, \tag{2.112}$$

where

$$\mathbf{R}_{\mathbf{HH}} = E\left\{[\mathbf{x}_{true} \;\; \mathbf{H}_{true}]^T \, [\mathbf{x}_{true} \;\; \mathbf{H}_{true}]\right\} \tag{2.113}$$

and

$$\bar{\mathbf{R}}_{\mathbf{NN}} = E\left\{[\mathbf{n} \;\; \mathbf{N}]^T \, [\mathbf{n} \;\; \mathbf{N}]\right\} \tag{2.114}$$

respectively represent the autocorrelation matrices of the unbiased by noise signal component and the zero-mean noise component. We have assumed that the two components are uncorrelated with each other. Let $\tilde{\mathbf{H}}$ denote the transformation of the augmented input matrix given by

$$\tilde{\mathbf{H}} = \overline{\mathbf{H}} \, \bar{\mathbf{R}}_{\mathbf{NN}}^{-1/2}. \tag{2.115}$$

The (scaled) autocorrelation matrix of $\tilde{\mathbf{H}}$ is then

$$E\left\{\tilde{\mathbf{H}}^T\tilde{\mathbf{H}}\right\} = \bar{\mathbf{R}}_{\mathbf{NN}}^{-1/2}\,\mathbf{R}_{\mathbf{HH}}\,\bar{\mathbf{R}}_{\mathbf{NN}}^{-1/2} + \mathbf{I}. \tag{2.116}$$

Thus, the transformed input matrix is corrupted by a noise process that is i.i.d. Gaussian with zero-mean. Therefore, the maximum likelihood estimate $\tilde{\mathbf{s}}_{\mathrm{ETLS}}$ of the coefficient vector for the transformed input matrix is given by the solution of the optimization problem

$$\min_{\tilde{\mathbf{s}}} J(\tilde{\mathbf{s}}) = \frac{\tilde{\mathbf{s}}^T\tilde{\mathbf{H}}^T\tilde{\mathbf{H}}\,\tilde{\mathbf{s}}}{\tilde{\mathbf{s}}^T\tilde{\mathbf{s}}}. \tag{2.117}$$

Obviously, this solution is unbiased. Let $\hat{\mathbf{s}}_{\mathrm{ETLS}}$ denote the coefficient vector for $\overline{\mathbf{H}}$, obtained by appropriately transforming the optimal solution $\tilde{\mathbf{s}}_{\mathrm{ETLS}}$ of (2.117). Since

$$\tilde{\mathbf{H}}\,\tilde{\mathbf{s}}_{\mathrm{ETLS}} = \overline{\mathbf{H}}\,\bar{\mathbf{R}}_{\mathbf{NN}}^{-1/2}\,\tilde{\mathbf{s}}_{\mathrm{ETLS}}, \tag{2.118}$$

we conclude that $\hat{\mathbf{s}}_{\mathrm{ETLS}}$, which is the optimal solution for the correlated noise problem, is related to $\tilde{\mathbf{s}}_{\mathrm{ETLS}}$ through the transformation

$$\hat{\mathbf{s}}_{\mathrm{ETLS}} = \bar{\mathbf{R}}_{\mathbf{NN}}^{-1/2}\,\tilde{\mathbf{s}}_{\mathrm{ETLS}}. \tag{2.119}$$

The entries for the augmented regression vector $[x_i \quad \mathbf{h}_i^T]^T$ are obtained by an appropriate scaling of $\hat{\mathbf{s}}_{\mathrm{ETLS}}$ and given as

$$\begin{bmatrix} -1 \\ \mathbf{s}_{\mathrm{ETLS}} \end{bmatrix} = -\frac{\hat{\mathbf{s}}_{\mathrm{ETLS}}}{\hat{s}_{\mathrm{ETLS},1}}, \tag{2.120}$$

where $\hat{s}_{\mathrm{ETLS},1}$ is the first non-zero element of $\hat{\mathbf{s}}_{\mathrm{ETLS}}$. Since scaling the solution vector does not change the cost function, after substituting (2.119) and (2.120) in (2.117), we can state the optimization problem for the extended TLS approach as [823]

$$\min_{\mathbf{s}} J(\mathbf{s}) = \frac{\begin{bmatrix} -1 \\ \mathbf{s} \end{bmatrix}^T \overline{\mathbf{H}}^T\overline{\mathbf{H}} \begin{bmatrix} -1 \\ \mathbf{s} \end{bmatrix}}{\begin{bmatrix} -1 \\ \mathbf{s} \end{bmatrix}^T \bar{\mathbf{R}}_{\mathbf{NN}} \begin{bmatrix} -1 \\ \mathbf{s} \end{bmatrix}}. \tag{2.121}$$

The solution to the above optimization problem is given by the generalized eigenvector[14] corresponding to the smallest generalized eigenvalue of the matrix pencil $(\overline{\mathbf{H}}^T\overline{\mathbf{H}},\ \bar{\mathbf{R}}_{\mathbf{NN}})$.

[14]Given two square matrices $\mathbf{G}$ and $\mathbf{H}$, the generalized eigenvector of the matrix pencil $(\mathbf{G}, \mathbf{H})$ is a vector $\mathbf{v}$ that satisfies the equality $\mathbf{G}\mathbf{v} = \lambda\,\mathbf{H}\mathbf{v}$, where the constant λ is known as a generalized eigenvalue. Assuming that the inverse of the matrix $\mathbf{H}$ exists, the generalized eigenvectors of the matrix pencil $(\mathbf{G}, \mathbf{H})$ are the eigenvectors of $\mathbf{H}^{-1}\mathbf{G}$.

2.4.4.1 Choice of $\bar{\mathbf{R}}_{\mathbf{NN}}$ in Some Practical Situations The coefficient estimate obtained by solving the optimization problem (2.121), can be shown to be unbiased for any noise correlation matrix $\bar{\mathbf{R}}_{\mathbf{NN}}$. However, we need to have *a priori* knowledge of the noise correlation matrix in order to implement the procedure. Fortunately, we need to use only a scaled version of the noise correlation matrix, since the minimization of the cost function

$$J(\mathbf{s}) = \frac{\begin{bmatrix} -1 \\ \mathbf{s} \end{bmatrix}^T \overline{\mathbf{H}}^T \overline{\mathbf{H}} \begin{bmatrix} -1 \\ \mathbf{s} \end{bmatrix}}{\begin{bmatrix} -1 \\ \mathbf{s} \end{bmatrix}^T c\bar{\mathbf{R}}_{\mathbf{NN}} \begin{bmatrix} 1 \\ -\mathbf{s} \end{bmatrix}}, \tag{2.122}$$

where c is a scalar positive constant, gives the same solution as the optimization problem in (2.121). There are many situations in which we can provide an estimate of such scaled noise correlation matrices. Some of these situations are:

- *Uncorrelated Noise in the Input Signals:* In many estimation problems, the matrix $\mathbf{H}$ or equivalently each of its column vector $\mathbf{h}_i$ are corrupted by the additive i.i.d. noise with variance σ_N^2, and the vector of sensor signals $\mathbf{x}$ is contaminated by independent noise with variance σ_n^2. In such cases, we can use

  $$c\bar{\mathbf{R}}_{\mathbf{NN}} = \text{diag}\{\beta, 1, 1, \ldots, 1\}, \tag{2.123}$$

  where $\beta = \sigma_n^2/\sigma_N^2$ is the ratio of the variance of the noise sequences associated with x_i and $\mathbf{h}_i$.

- *Data Least-Squares (DLS) problem:* In a variety of the estimation problems that belong to this class, the sensor signal x_i contains no noise and the noise in $\mathbf{h}_i$ can be reliably modelled as i.i.d. An appropriate choice of $c\bar{\mathbf{R}}_{\mathbf{NN}}$ in this case can be

  $$c\bar{\mathbf{R}}_{\mathbf{NN}} = \text{diag}\{0, 1, 1, \ldots, 1\}. \tag{2.124}$$

- *Least-Squares (LS) Problems:* In this situation, we assume that the regression vector $\mathbf{h}_i$ is noise-free and that only x_i are corrupted by noise. Then, choosing

  $$c\bar{\mathbf{R}}_{\mathbf{NN}} = \text{diag}\{1, 0, 0, \ldots, 0\} \tag{2.125}$$

  results in the standard least-squares problem formulation.

2.4.5 An Illustrative Example - Fitting a Straight Line to a Set of Points

Example 2.3 Consider the problem of estimating the slope and the intercept of a straight line that fits the points $(1, 2.0)$, $(2, 1.5)$, $(3, 3.0)$, $(4, 2.5)$, $(5, 3.5)$. The augmented data matrix for this problem is given by

$$\overline{\mathbf{H}} = \begin{bmatrix} 2.0 & 1 & 1 \\ 1.5 & 2 & 1 \\ 3.0 & 3 & 1 \\ 2.5 & 4 & 1 \\ 3.5 & 5 & 1 \end{bmatrix}.$$

The entries in the third column of this matrix are all fixed to ones because the y-intercept of the line does not depend on the independent variable. The estimated autocorrelation matrix of the data matrix $\overline{\mathbf{H}}$ is given by

$$\hat{\mathbf{R}}_{\mathbf{HH}} = \overline{\mathbf{H}}^T \overline{\mathbf{H}} = \begin{bmatrix} 33.75 & 41.50 & 12.50 \\ 41.50 & 55.00 & 15.00 \\ 12.50 & 15.00 & 5.00 \end{bmatrix}.$$

The least-squares estimate of the slope and the intercept may be found as

$$\begin{bmatrix} m_{\text{LS}} \\ c_{\text{LS}} \end{bmatrix} = \begin{bmatrix} 55.00 & 15.00 \\ 15.00 & 5.00 \end{bmatrix}^{-1} \begin{bmatrix} 41.50 \\ 12.50 \end{bmatrix} = \begin{bmatrix} 0.40 \\ 1.30 \end{bmatrix}.$$

The total least-squares solution assumes that there is noise in all the entries of the matrix $\overline{\mathbf{H}}$, and the solution is obtained from the eigenvector corresponding to the smallest eigenvalue of $\hat{\mathbf{R}}_{\mathbf{HH}}$. This estimate can be calculated as

$$\begin{bmatrix} m_{\text{TLS}} \\ c_{\text{TLS}} \end{bmatrix} = \begin{bmatrix} 0.2660 \\ 1.7958 \end{bmatrix}.$$

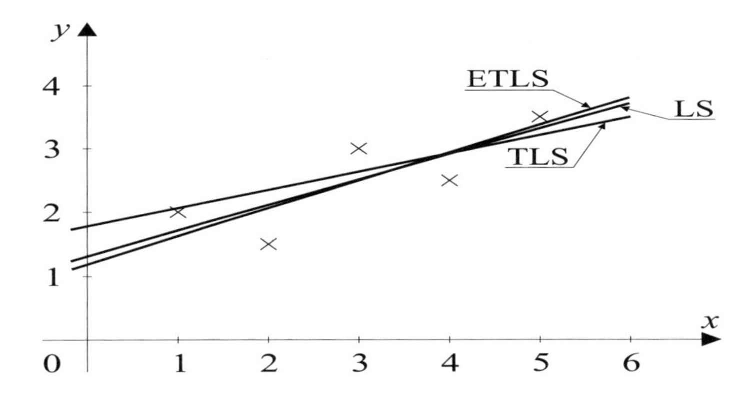

Fig. 2.5 Straight lines fit for the five points marked by 'x' using the LS, TLS and ETLS methods.

A drawback of the above solution is that we have assumed that the last column of $\overline{\mathbf{H}}$ is in error, even though this column can be exactly specified in our problem. Consequently, here we can utilize the extended TLS approach to estimate the parameters. Let us assume that the noise in the first two columns of the data matrix $\overline{\mathbf{H}}$ are uncorrelated with each other. Then, a scaled autocorrelation matrix for the noise is given by diag$\{1,\ 1,\ 0\}$. The extended TLS solution for the slope and intercept of the straight line is specified by the generalized eigenvector corresponding to the smallest generalized eigenvalue of the matrix pencil $\left(\hat{\mathbf{R}}_{HH}, \text{diag}\{1,\ 1,\ 0\}\right)$, and is given by

$$\begin{bmatrix} m_{\text{ETLS}} \\ c_{\text{ETLS}} \end{bmatrix} = \begin{bmatrix} 0.4332 \\ 1.2003 \end{bmatrix}.$$

Figure 2.5 shows the plots of the straight lines estimated using the three approaches: LS, TLS and ETLS.

2.5 SPARSE SIGNAL REPRESENTATION AND MINIMUM 1-NORM SOLUTION

In the previous sections, we have considered the problems of solving overdetermined systems of linear equations. The problem of underdetermined systems of linear equations can be usually formulated as the following constrained optimization problem:

Minimize

$$J_p(\mathbf{s}) = \|\mathbf{s}\|_p \tag{2.126}$$

subject to the constraint

$$\mathbf{H}\,\mathbf{s} = \mathbf{x},$$

where $\mathbf{H} \in \mathbb{R}^{m \times n}$ (with $m < n$).

The above formulated optimization problem arises in many applications such as electromagnetic and biomagnetic inverse problems, time-frequency representation, neural and speech coding, spectral estimation, direction of arrival estimation and failure diagnosis [276, 504, 1002].

For $p = 1$ the problem can be formulated as the standard linear programming problem and several efficient and powerful algorithms can be applied. For the standard 2-norm, the problem is usually called the minimum energy solution whereas for the infinity-norm, it is called the minimum amplitude solution. For $p \leq 1$, the problem provides a sparse representation of the vector $\mathbf{s}$ and is sometimes called the minimum fuel solution. The term sparse representation or solution usually refers to a solution with $(m - n)$ or more zero entries in the vector $\mathbf{s}$. This property can be summarized in the form of the following theorem.

Theorem 2.5 *For any matrix $\mathbf{H} \in \mathbb{R}^{m \times n}$ and vector $\mathbf{x} \in \mathbb{R}^m$, with $m < n$, there exits a vector $\mathbf{s}_* \in \mathbb{R}^n$, which minimizes 1-norm $\|\mathbf{s}\|_1$ subject to the constraints $\mathbf{H}\,\mathbf{s} = \mathbf{x}$ such that the optimal vector $\mathbf{s}_*$ has at most m non zero components. Furthermore, if the column vectors of the extended matrix $\bar{\mathbf{H}} = [\mathbf{x}\ \mathbf{H}] \in \mathbb{R}^{m \times (n+1)}$ satisfy the Haar condition then there exists a vector $\mathbf{s}_*$ which minimizes the 1-norm $\|\mathbf{s}\|_1$ subject to the constraints $\mathbf{H}\,\mathbf{s} = \mathbf{x}$ that has exactly m nonzero components.*

The minimum p-norm solution or minimum fuel problem of the optimization problem 2.126 for $p \leq 1$ is closely related to overcomplete signal representation and the best basis selection (matching pursuit) problems [698, 1002]. In the overcomplete signal representation problem, we search for an efficient overcomplete dictionary to represent the signal. To solve the problem, a given signal is decomposed into a number of optimal basis components which can be found from an overcomplete basis dictionary via some optimization algorithms, such as matching pursuit and basis pursuit. The problem of basis selection, i.e., choosing a proper subset of vectors from the given dictionary naturally arises in the overcomplete representation of signals. In other words, in the problem of the best basis selection, it is necessary to identify or select a few columns $\mathbf{h}_i$ of matrix $\mathbf{H}$ that best represent the sensor vector $\mathbf{x}$. This corresponds to finding a solution to (2.126) for $p \leq 1$ with a few non-zero entries [504, 698, 1001, 1002].

From theorem 2.5, we can design a brute force method according to which we determine solutions for all possible subset of determined m equations by removing each time the $(n-m)$ columns from the matrix $\mathbf{H}$ and evaluates the 1-norm solution of the vectors $\mathbf{s}_{j*} = \mathbf{H}_r^{-1}\mathbf{x}$ for the each of these set of equations. Unfortunately, this brute force method for large scale problem is inefficient. In fact, finding an optimal (smallest) basis set of vectors is NP hard and requires a combinatorial search [1002]. For example, if we were interested in selecting m vectors $\mathbf{h}_i$ that best represent sensor data $\mathbf{x}$, this would require searching over $n!/(n-m)!m!$ possible ways in which the basis set can be chosen as the best solution. This search cost is prohibitive for large values of n, making combinatoric approaches non-feasible [1001, 1002].

The main objective of this section is to present several efficient and robust algorithms which enable us to find the suboptimal solutions for the minimum fuel problem and its generalizations, especially when the data are corrupted by noise.

2.5.1 Approximate Solution of Minimum p-norm Problem Using Iterative LS Approach

Intuitively, in order to find the minimum p-norm solution with $0 \leq p \leq 1$, i.e., possibly the most sparse representation of the vector $\mathbf{s}$, we must optimally select some columns of the matrix $\mathbf{H}$. Alternatively, using a neural network representation, we should impose some 'competition' between the columns of matrix $\mathbf{H}$ to represent optimally and sparsely the data vector $\mathbf{x}$. Due to this competition, certain columns will get emphasized, while others will be de-emphasized. In the end, at most m columns will survive to represent $\mathbf{x}$, while the rest or at least $(n-m)$ will be ignored or neglected, thereby providing a sparse solution.

The minimum energy (2-norm) solution is usually a rough approximation of the 1-norm solution. However, in contrast to the 1-norm solution, the minimum 2-norm solution will not provide a sparse representation. It rather has a tendency to spread the energy among a large number of entries of $\mathbf{s}$, instead by putting all the energy (concentrating it) into just a few entries. The minimum energy problem can be easily solved explicitly using

$$\mathbf{s}_{\mathrm{LS}} = \mathbf{H}^{+}\,\mathbf{x},$$

where $\mathbf{H}^{+} = \mathbf{H}^T\,(\mathbf{H}\,\mathbf{H}^T)^{-1}$ denotes the Moore-Penrose generalized pseudo-inverse. The solution has a number of computational advantages, but does not provide a desirable sparse solution.

Exploiting these properties and theorem 2.5, we propose the following approximative multiple (at least two) stage algorithm based on the iterative minimum energy solution:

Algorithm Outline: Approximate Procedure for Sparse Solution

Step 1. Estimate the minimum 2-norm solution of the problem (2.126) as

$$\mathbf{s}_{\mathrm{LS}} = \mathbf{H}^T\,(\mathbf{H}\,\mathbf{H}^T)^{-1}\,\mathbf{x} = \mathbf{H}^{+}\,\mathbf{x}, \tag{2.127}$$

where $\mathbf{H}^{+} \in \mathbb{R}^{n\times m}$ is the Moore-Penrose pseudo-inverse matrix of $\mathbf{H}$. On the basis of vector $\mathbf{s}_{\mathrm{LS}}$, we remove certain columns of the matrix $\mathbf{H}$ corresponding to the smallest modulus of the components of the vector $\mathbf{s}_{\mathrm{LS}}$. Then, we set the components of the vector $\mathbf{s}_{\mathrm{LS}}$ to zero as a partial solution of the minimum 1-norm problem.

Step 2. Estimate the remaining components of the vector $\mathbf{s}_1$:

$$\mathbf{s}_{1r} = \mathbf{H}_r^T(\mathbf{H}_r\,\mathbf{H}_r^T)^{-1}\mathbf{x} = \mathbf{H}_r^{+}\,\mathbf{x}, \tag{2.128}$$

where $\mathbf{H}_r \in \mathbb{R}^{m\times r}$ (with $r \geq m$) is the reduced matrix obtained by removing from the matrix $\mathbf{H}$ certain columns corresponding to the smallest magnitude and $\mathbf{s}_{1r} \in \mathbb{R}^r$.

Step 3. Repeat Step 1 and 2 until at least $(n-m)$ or specified by the user the number of columns from the original matrix $\mathbf{H}$ are removed.

The algorithm will be illustrated by a simple example.

Example 2.4 Let us consider the following minimum fuel problem:

Minimize $||\mathbf{s}||_1$ subject to the constraint $\mathbf{H}\,\mathbf{s} = \mathbf{x}$,

where

$$\mathbf{H} = \begin{bmatrix} 2 & 3 & -1 & 10 & 21 & 44 & -9 & 1 & -1 \\ 1 & 2 & 2 & 8 & 15 & 35 & 8 & -3 & 1 \\ 3 & 1 & 1 & 6 & 16 & 53 & -7 & 2 & 2 \end{bmatrix}$$

and $\mathbf{x} = [118\ \ 77\ \ 129]^T$.

It is impossible to find the 1-minimum solution in one step. In the first step, we obtain the minimum energy (2-norm) solution as

$$\begin{aligned} \mathbf{s}_{LS} &= \mathbf{H}^T\,(\mathbf{H}\,\mathbf{H}^T)^{-1}\,\mathbf{x} \qquad (2.129) \\ &= [0.131\ 0.086\ \ -0.104\ 0.302\ 0.795\ 2.022\ \ -0.9373\ 0.222\ 0.037]^T. \end{aligned}$$

Since the components s_1, s_2, s_3, s_4, s_8 and s_9 have the smallest magnitudes, we set them to zero and remove the corresponding columns (i.e., $[1, 2, 3, 4, 8, 9]$) of the matrix $\mathbf{H}$ which yields its reduced version:

$$\mathbf{H}_r = \begin{bmatrix} 21 & 44 & -9 \\ 15 & 35 & 8 \\ 16 & 53 & -7 \end{bmatrix}.$$

In the second step, we compute the remaining (in general non-zero) components of the vector $\mathbf{s}_1$ as

$$\mathbf{s}_{1r} = \mathbf{H}_r^{-1}\,\mathbf{x} = [1\ \ 2\ \ -1]^T.$$

Thus, the minimum 1-norm solution finally takes the sparse form as:

$$\mathbf{s}_{1*} = [0\ \ 0\ \ 0\ \ 0\ \ 1\ \ 2\ \ -1\ \ 0\ \ 0]^T.$$

Remark 2.8 *It is important to note in order to ensure a good performance of the algorithm in general a multistage procedure should be applied, so in each stage only one column is removed. As the criterion for removing the specific column we chosen magnitude of components of a vector* $\mathbf{s}_{LS}$ *in each iteration stage. Alternative criterion is the* 1*-norm of the vector* $\mathbf{s}$ *such that removing selected column(s) causes the largest possible decreasing of the* $\|\mathbf{s}\|_1$.

In many signal processing applications, the sensor (observed) vector is available at a number of time instants, as in multiple measurements or recordings, thus, the system of linear equations $\mathbf{H}\,\mathbf{s}(k) = \mathbf{x}(k), \quad (k = 1, 2 \ldots, N)$ can be written in a compact aggregated matrix form as

$$\mathbf{H}\,\mathbf{S} = \mathbf{X}, \tag{2.130}$$

where $\mathbf{S} = [\mathbf{s}(1), \mathbf{s}(2), \ldots, \mathbf{s}(N)]$ and $\mathbf{X} = [\mathbf{x}(1), \mathbf{x}(2), \ldots, \mathbf{x}(N)]$.

Our objective is to find a sparse representation of the matrix $\mathbf{S}$. However, we require that individual columns of $\mathbf{S}$ not only have a sparse structure but also share a common structure and have a common sparsity profile; that is, possibly a small number of rows $\mathbf{s}_j = [s_j(1), s_j(2), \ldots, s_j(N)]$ $(j = 1, 2, \ldots, n)$ of the matrix $\mathbf{S}$ have non-zero entries. In such a case, we can extend or modify the proposed algorithm as follows:

Algorithm Outline: Extended Algorithm for Sparse Solution with Multiple Observed Vectors

---

Step 1. Estimate the minimum 2-norm solution of the problem (2.130) as

$$\mathbf{S}_{\rm LS} = \mathbf{H}^T\,(\mathbf{H}\,\mathbf{H}^T)^{-1}\,\mathbf{X} = \mathbf{H}^+\,\mathbf{X}, \tag{2.131}$$

where $\mathbf{H}^+ \in \mathbb{R}^{n\times m}$ is the Moore-Penrose pseudo-inverse of $\mathbf{H}$ and $\mathbf{S}_{\rm LS} \in \mathbb{R}^{n\times N}$ is the matrix of estimated sources $s_j(k)$.
Then, we remove certain columns of the matrix $\mathbf{H}$ corresponding to the smallest value of norm[15] $\|\mathbf{s}_j\|$ of the row vectors $\mathbf{s}_j = [s_j(1), s_j(2), \ldots, s_j(N)]$ of the matrix $\mathbf{S}_{\rm LS}$. Next, certain components of these row vectors are set to zero if they are below some threshold value as a partial solution to the minimum fuel problem. In this stage, we can remove at least $(n - m)$ columns of $\mathbf{H}$.

Step 2. Estimate the remaining components of the matrix $\mathbf{S}$:

$$\mathbf{S}_{1r} = \mathbf{H}_r^+\,\mathbf{X} = \mathbf{H}_r^T(\mathbf{H}_r\mathbf{H}_r^T)^{-1}\,\mathbf{X}, \tag{2.132}$$

where $\mathbf{S}_{1r} \in \mathbb{R}^{r\times N}$ is a required partial solution and $\mathbf{H}_r$ is the reduced version of the matrix $\mathbf{H}$ (with removed certain columns of $\mathbf{H}$ corresponding to the smallest norms of row vectors of the matrix $\mathbf{S}_{\rm LS}$).

[15]The choice of norm $\|\mathbf{s}_j\|$ depends on the noise distribution, e.g., for Gaussian noise, the optimal is 2-norm, and for Laplacian (impulsive noise) the 1-norm, whereas for uniformly distributed noise infinity-norm is the best choice.

Step 3 Repeat Step 2 and 3 until at least $(n-m)$ or the required number of columns from the original matrix $\mathbf{H}$ are removed.

2.5.2 Uniqueness and Optimal Solution for Sparse Representation

It is to be emphasized the above presented algorithms do not have guaranteed convergence to an optimal minimum 1-norm solution although typically converge to a good approximate solution at a very small computation expense. Therefore, we need some criterion or a measure which gives us guarantee that obtained minimum 1-norm (or more generally minimum p-norm) solution is an optimal. The following theorem provides the such criterion.

Theorem 2.6 *Let us consider the system of linear underdetermined equations* $\mathbf{Hs} = \mathbf{x}$ *with the degenerate (rank deficient) matrix* $\mathbf{H} \in \mathbb{R}^{m\times n}$ *with* $m < n$ *Furthermore, let* $\mathbf{H}_r \in \mathbb{R}^{m\times m_0}$ *designate the* m_0 *columns of the matrix* $\mathbf{H}$ *associated with* $m_0 \le m$ *zero elements of the desired vector* $\mathbf{s}_*$. *Furthermore, let a submatrix* $\mathbf{H}_2 \in \mathbb{R}^{m\times(n-m_0)}$ *designate the* $(n-m_0)$ *columns of the matrix* $\mathbf{H}$ *which are associated with the zero components of the vector* $\mathbf{s}_*$. *If the reduced matrix* $\mathbf{H}_r$ *is the full column rank then the vector* $\mathbf{s}_*$ *is a unique minimum 1-norm solution* $\|\mathbf{s}\|_1$ *subject to the constraint* $\mathbf{x} = \mathbf{Hs}$ *if and only if all the components of the vector*

$$\mathbf{g} = \mathbf{H}_2^T\ [\mathbf{H}_r^+]^T\, \mathrm{sign}(\mathbf{H}_r^+\mathbf{x}) \tag{2.133}$$

with $\mathbf{H}_r^+ = [\mathbf{H}_r^T\mathbf{H}_r]^{-1}\mathbf{H}_r^T)$ *have magnitudes strictly less than one (i.e.,* $\|\mathbf{g}\|_\infty < 1$*). However, if the elements of the vector* $\mathbf{g}$ *have magnitudes of less than or equal one the solution is optimal but not unique.*

2.5.3 FOCUSS Algorithms

An alternative algorithm for the minimum fuel problem, called FOCUSS (FOCal Underdetermined System Solver) has been proposed by Gorodnitsky and Rao [504] and extended and generalized by Kreutz-Delgado and Rao [698, 1001, 1002].
Let us consider the following constrained optimization problem [698, 1001, 1002]:

minimize $J_\rho(\mathbf{s}) = \sum_{j=1}^n \rho|s_j|$

subject to $\mathbf{H}\,\mathbf{s} = \mathbf{x}$,

where the cost function $J_\rho(\mathbf{s})$ (often called the diversity measure) is some measure of sparsity of signals and it can take various forms [1002]:

1. The generalized p-norm

$$J_p(\mathbf{s}) = \mathrm{sign}(p)\sum_{j=1}^n |s_j|^p, \tag{2.134}$$

where $p \le 1$ and is selected by the user.

2. The Gaussian entropy diversity measure

$$J_G(\mathbf{s}) = H_G(\mathbf{s}) = \sum_{j=1}^{n} \log |s_j|^2. \tag{2.135}$$

3. The Shannon entropy diversity measure

$$J_S(\mathbf{s}) = H_S(\mathbf{s}) = -\sum_{j=1}^{n} \tilde{s}_j \log |\tilde{s}_j|, \tag{2.136}$$

where the components $\tilde{s}_j$ can take different forms, e.g. $\tilde{s}_j = |s_j|$, $\tilde{s}_j = |s_j|/\|\mathbf{s}\|_2$, $\tilde{s}_j = |s_j|/\|\mathbf{s}\|_1$ or $\tilde{s}_j = s_j$ for $s_j \geq 0$.

4. Renyi entropy diversity measure

$$J_R(\mathbf{s}) = H_R(\mathbf{s}) = \frac{1}{1-p} \log \sum_{j=1}^{n} (\tilde{s}_j)^p, \tag{2.137}$$

where $\tilde{s}_j = s_j/\|\mathbf{s}\|_1$ and $p \neq 1$.

It should be noted that, for $p = 1$, we obtain the formulation of the standard linear programming problem in which at least $(n - m)$ components are zero. Choosing the above diversity measures, we can obtain a sparser solution than for the minimum 1-norm solution (corresponding to $p = 1$) (i.e., more than $(n-m)$ entries in the vector $\mathbf{s}$ are zero). Moreover, the solution can be more robust with respect to the additive noise. The general diversity measures based on the negative norm or Gaussian, Shannon and Renyi entropies ensure that a relatively large number of entries s_j tend to be very small, albeit usually of non-zero amplitude. In such cases, we use a small threshold below which the entries are set to zero.

To minimize the generalized p norm diversity measure $J_p(\mathbf{s})$ in (2.134), subject to the equality constraint $\mathbf{H}\,\mathbf{s} = \mathbf{x}$, we define the Lagrangian $L(\mathbf{s}, \boldsymbol{\lambda})$ as

$$L(\mathbf{s}, \boldsymbol{\lambda}) = J_p(\mathbf{s}) + \boldsymbol{\lambda}\,(\mathbf{x} - \mathbf{H}\,\mathbf{s}), \tag{2.138}$$

where $\boldsymbol{\lambda} \in \mathbb{R}^n$ is a vector of Lagrange multipliers [698, 1001, 1002].

The stationary points of the Lagrangian function above can be evaluated as follows

$$\nabla_{\mathbf{s}} L(\mathbf{s}_* \boldsymbol{\lambda}_*) = \nabla_{\mathbf{s}} J_p(\mathbf{s}) - \mathbf{H}^T \boldsymbol{\lambda}_* = \mathbf{0}, \tag{2.139}$$

$$\nabla_{\boldsymbol{\lambda}} L(\mathbf{s}_* \boldsymbol{\lambda}_*) = \mathbf{x} - \mathbf{H}\,\mathbf{s}_* = \mathbf{0}, \tag{2.140}$$

where the gradient of the p norm can be expressed as

$$\nabla_{\mathbf{s}} J_p(\mathbf{s}) = |p|\; \mathbf{D}_{|\mathbf{s}|}^{-1}(\mathbf{s})\, \mathbf{s} \tag{2.141}$$

and $\mathbf{D}_{|\mathbf{s}|}(\mathbf{s}) \in \mathbb{R}^{n \times n}$ is a diagonal matrix with the entries $d_j = |s_j|^{2-p}$. Solving the above equations by simple mathematical operations, we obtain

$$\boldsymbol{\lambda}_* = |p| \left(\mathbf{H}\,\mathbf{D}_{|\mathbf{s}|}(\mathbf{s}_*)\,\mathbf{H}^T\right)^{-1} \mathbf{x}, \tag{2.142}$$

$$\begin{aligned} \mathbf{s}_* &= |p|^{-1}\, \mathbf{D}_{|\mathbf{s}|}(\mathbf{s}_*)\, \mathbf{H}^T \boldsymbol{\lambda}_* \\ &= \mathbf{D}_{|\mathbf{s}|}(\mathbf{s}_*)\, \mathbf{H}^T \left(\mathbf{H}\,\mathbf{D}_{|\mathbf{s}|}(\mathbf{s}_*)\mathbf{H}^T\right)^{-1} \mathbf{x}. \end{aligned} \tag{2.143}$$

The equation (2.143) is not in a convenient form for computation since the desired vector $\mathbf{s}_*$ is implicitly in the right side of the equation. However, it suggests that an iterative algorithm for estimation of the optimal vector $\mathbf{s}_*$ is given as

$$\mathbf{s}(k+1) = \mathbf{D}_{|\mathbf{s}|}(k)\,\mathbf{H}^T\left(\mathbf{H}\,\mathbf{D}_{|\mathbf{s}|}(k)\,\mathbf{H}^T\right)^{-1}\mathbf{x}, \tag{2.144}$$

where $\mathbf{D}_{|\mathbf{s}|}(k) = \operatorname{diag}\{|s_1(k)|^{2-p}, |s_1(k)|^{2-p}, \ldots, |s_n(k)|^{2-p}\}$. The above algorithm, called the generalized FOCUSS algorithm can be expressed in a more compact form [504]:

$$\mathbf{s}(k+1) = \tilde{\mathbf{D}}_{|\mathbf{s}|}(k)\left(\mathbf{H}\,\tilde{\mathbf{D}}_{|\mathbf{s}|}(k)\right)^{+}\mathbf{x}, \tag{2.145}$$

where the superscript $(\cdot)^+$ denotes the Moore-Penrose pseudo-inverse and $\tilde{\mathbf{D}}_{|\mathbf{s}|}(k) = \mathbf{D}^{1/2}_{|\mathbf{s}|}(k) = \operatorname{diag}\{|s_1|^{1-\frac{p}{2}}(k), |s_2|^{1-\frac{p}{2}}(k), \ldots, |s_n|^{1-\frac{p}{2}}(k)\}$. It should be noted that the matrix $\mathbf{D}_{|\mathbf{s}|}$ exists for all $\mathbf{s}$ and even for a negative p. For $p = 2$, the matrix $\mathbf{D}_{|\mathbf{s}|} = \mathbf{I}$ and the Focuss algorithm simplifies to the standard LS or the minimum 2-norm solution $\mathbf{s}_* = \mathbf{H}^T(\mathbf{H}\,\mathbf{H}^T)^{-1}\mathbf{x}$. For another special case $p = 0$, the diagonal matrix $\tilde{\mathbf{D}}_{|\mathbf{s}|} = \operatorname{diag}\{|s_1|, |s_2|, \ldots, |s_n|\}$. In order to derive rigorously the algorithm for $p = 0$, instead of (2.134) we should use the Gaussian entropy (2.135), for which the gradient can be expressed as

$$\nabla_{\mathbf{s}} J_G(\mathbf{s}) = 2\,\mathbf{D}_G^{-1}\,\mathbf{s}, \tag{2.146}$$

where $\mathbf{D}_G(\mathbf{s}) = \operatorname{diag}\{|s_1|^2, |s_2|^2, \ldots, |s_n|^2\}$.

For noisy data, we can use a more robust regularized Focuss algorithm in the form:

$$\mathbf{s}(k+1) = \mathbf{D}_{|\mathbf{s}|}(k)\,\mathbf{H}^T\left(\mathbf{H}\,\mathbf{D}_{|\mathbf{s}|}(k)\,\mathbf{H}^T + \alpha(k)\,\mathbf{I}\right)^{-1}\mathbf{x}, \tag{2.147}$$

where $\alpha(k) \geq 0$ is the Tikhonov regularization parameter depending on the noise level [698, 1002].

Finally, it is worthy of mention that in order to solve the minimum p-norm problem as in (2.130) for the case of multiple sensor vectors, we can formulate the following generalized constrained optimization problem [698, 1002]:

Minimize

$$J_p(\mathbf{S}) = \operatorname{sign}(p)\sum_{j=1}^{n}(\|\mathbf{s}_j\|_2)^p, \qquad p \leq 1, \tag{2.148}$$

subject to the constraints $\mathbf{H}\,\mathbf{S} = \mathbf{X}$,

where $\mathbf{s}_j = [s_j(1), s_j(2), \ldots, s_j(N)]^T$ and $\|\mathbf{s}_j\|_2 = (\sum_{l=1}^{N}|s_j(l)|^2)^{1/2}$.

Similar to the previous case, we can derive the Focuss algorithm for the multiple sensor vectors as

$$\mathbf{S}(k+1) = \mathbf{D}_{\|\mathbf{s}\|}(k)\,\mathbf{H}^T\left(\mathbf{H}\,\mathbf{D}_{\|\mathbf{s}\|}(k)\,\mathbf{H}^T\right)^{-1}\mathbf{X}, \tag{2.149}$$

where $\mathbf{D}_{\|\mathbf{s}\|}(k) = \operatorname{diag}\{d_1(k), d_2(k), \ldots, d_n(k)\}$ with $d_j(k) = \|\mathbf{s}_j\|^{2-p}(k)$. The algorithm can be considered as a natural generalization of the Focuss algorithm (2.144). and initialized by using the minimum Frobenius norm solution [1001, 1002]. Alternatively for noisy data, we can use the Tikhonov regularization technique, the truncated SVD or a modified L-curve approach for noisy data [698, 1001, 1002].

3

Principal/Minor Component Analysis and Related Problems

I want to get the structural problems out of the way first, so I can get to what matters more.
—(John McPhee)

3.1 INTRODUCTION

Neural networks with unsupervised learning algorithms organize themselves in such a way that they can detect or extract useful features, regularities, correlations of data or signals or separate or decorrelate some signals with little or no prior knowledge of the desired results.[1] Normalized (constrained) Hebbian and anti-Hebbian learning rules are simple variants of basic unsupervised learning algorithms; in particular, learning algorithms for principal component analysis (PCA), singular value decomposition (SVD) and minor component analysis (MCA) belong to this class of unsupervised rules [349, 907, 1202].

PCA is perhaps one of the oldest and the best-known techniques in multivariate analysis and data mining. It was introduced by Pearson, who used it in a biological context and further developed by Hotelling in works done on psychometry. PCA was also developed independently by Karhunen in the context of probability theory and was subsequently generalized by Loève [349]. Recently, many efficient and powerful adaptive algorithms have been developed for PCA, SVD and MCA and their extensions [15, 907, 914, 57, 275]. The main objective of this chapter is to derive and present an overview of the most important algorithms.

[1]It is generally believed that the shape of the receptive fields in the visual cortex is determined by some form of unsupervised learning.

3.2 BASIC PROPERTIES OF PCA

3.2.1 Eigenvalue Decomposition

The purpose of principal component analysis (PCA) is to derive a relatively small number of decorrelated linear combinations (principal components) of a set of random zero-mean variables while retaining as much of the information from the original variables as possible.
Among the objectives of Principal Components Analysis are the following.

1. dimensionality reduction;
2. determination of linear combinations of variables;
3. feature selection: the choosing of the most useful variables;
4. visualization of multidimensional data;
5. identification of underlying variables;
6. identification of groups of objects or of outliers.

PCA has been widely studied and used in pattern recognition and signal processing. In fact it is important in many engineering and scientific disciplines, e.g., in data compression, feature extraction, noise filtering, signal restoration and classification [349]. PCA is used widely in data mining as a data reduction technique. In image processing and computer vision PCA representations have been used for solving problems such as face and object recognition, tracking, detection, background modelling, parameterizing shape, appearance and motion [1202, 706].

Often the principal components (PCs) (i.e., directions on which the input data have the largest variances) are regarded as important, while those components with the smallest variances called minor components (MCs) are regarded as unimportant or associated with noise. However, in some applications, the MCs are of the same importance as the PCs, for example, in curve and surface fitting or total least squares (TLS) problems [1303, 276].

Generally speaking, PCA is related and motivated by the following two problems:

1. Given random vectors $\mathbf{x}(k) \in \mathbb{R}^m$, with finite second order moments and zero mean, find the reduced n-dimensional ($n < m$) linear subspace that minimizes the expected distance of $\mathbf{x}$ from the subspace. This problem arises in the area of data compression where the task is to represent all the data with a reduced number of parameters while assuring minimum distortion due to projection.

2. Given random vectors $\mathbf{x}(k) \in \mathbb{R}^m$, find the n-dimensional linear subspace that captures most of the variance of the data $\mathbf{x}$. This problem is related to feature extraction, where the objective is to reduce the dimension of the data while retaining most of its information content.

It turns out that both problems have the same optimal solution (in the sense of least-squares error) which is based on the second order statistics, in particular, on the eigen structure of the data covariance matrix[2]. PCA can be converted to the eigenvalue problem of the covariance matrix of $\mathbf{x}$ and it is essentially equivalent to the Karhunen-Loève transform used in image and signal processing. In other words, PCA is a technique for computation of eigenvectors and eigenvalues for the estimated covariance matrix[3]

$$\widehat{\mathbf{R}}_{\mathbf{x}\,\mathbf{x}} = E\{\mathbf{x}(k)\,\mathbf{x}^T(k)\} = \mathbf{V}\,\mathbf{\Lambda}\,\mathbf{V}^T \in \mathbb{R}^{m\times m}, \tag{3.1}$$

where $\mathbf{\Lambda} = \operatorname{diag}\{\lambda_1, \lambda_2, ..., \lambda_m\}$ is a diagonal matrix containing the m eigenvalues and $\mathbf{V} = [\mathbf{v}_1, \mathbf{v}_2, \ldots, \mathbf{v}_m] \in \mathbb{R}^{m\times m}$ is the corresponding orthogonal or unitary matrix consisting of the unit length eigenvectors referred to as principal eigenvectors.

The Karhunen-Loéve-transform determines a linear transformation of an input vector $\mathbf{x}$ as

$$\mathbf{y}_{\mathrm{P}} = \mathbf{V}_{\mathcal{S}}^T\,\mathbf{x}, \tag{3.2}$$

where
$\mathbf{x} = [x_1(k), x_2(k), \ldots, x_m(k)]^T$ is the zero-mean input vector, $\mathbf{y}_{\mathrm{P}} = [y_1(k), y_2(k), \ldots, y_n(k)]^T$ is the output vector called the vector of principal components (PCs), and $\mathbf{V}_{\mathcal{S}} = [\mathbf{v}_1, \mathbf{v}_2, \ldots, \mathbf{v}_n]^T \in \mathbb{R}^{m\times n}$ is the set of signal subspace eigenvectors, with the orthonormal vectors $\mathbf{v}_i = [v_{i1}, v_{i2}, \ldots, v_{im}]^T$, (i.e., $(\mathbf{v}_i^T\mathbf{v}_j = \delta_{ij})$ for $j \leq i$, (δ_{ij} is the Kronecker delta)). The vectors $\mathbf{v}_i$ $(i = 1, 2, \ldots, n)$ are eigenvectors of the covariance matrix, while the variances of the PCs y_i are the corresponding principal eigenvalues. On the other hand, the $(m - n)$ minor components are given by

$$\mathbf{y}_M = \mathbf{V}_{\mathcal{N}}^T\,\mathbf{x}, \tag{3.3}$$

where $\mathbf{V}_{\mathcal{N}} = [\mathbf{v}_m, \mathbf{v}_{m-1}, \ldots, \mathbf{v}_{m-n+1}]$ consists of the $(m - n)$ eigenvectors associated with the smallest eigenvalues.

Therefore, the basic problem we try to solve is the standard eigenvalue problem which can be formulated by the equations

$$\mathbf{R}_{\mathbf{x}\,\mathbf{x}}\mathbf{v}_i = \lambda_i\mathbf{v}_i, \qquad (i = 1, 2, \ldots, n) \tag{3.4}$$

where $\mathbf{v}_i$ are the eigenvectors, λ_i are the corresponding eigenvalues and $\mathbf{R}_{\mathbf{x}\,\mathbf{x}} = E\{\mathbf{x}\,\mathbf{x}^T\}$ is the covariance matrix of zero-mean signal $\mathbf{x}$ and E is the expectation operator. Note that Eq.(3.4) can be written in matrix form $\mathbf{V}^T\,\mathbf{R}_{\mathbf{x}\,\mathbf{x}}\,\mathbf{V} = \mathbf{\Lambda}$, where $\mathbf{\Lambda}$ is the diagonal matrix of eigenvalues of the covariance matrix $\mathbf{R}_{\mathbf{x}\,\mathbf{x}}$.

In the standard numerical approach for extracting the principal components, first the covariance matrix $\mathbf{R}_{\mathbf{x}\,\mathbf{x}} = E\{\mathbf{x}\,\mathbf{x}^T\}$ is computed and then its eigenvectors and (corresponding) associated eigenvalues are determined by one of the known numerical algorithms. However, if the input data vectors have a large dimension (e.g., 1000 elements), then the covariance matrix $\mathbf{R}_{\mathbf{x}\,\mathbf{x}}$ becomes very large (10^6 entries) and it may be difficult to compute the required eigenvectors.

[2]If signals are zero mean, the covariance and correlation matrices are identical.
[3]The covariance matrix is the correlation matrix of the vector with the mean removed. Since, we consider zero mean signals, both matrices are equivalent.

A neural network approach with adaptive learning algorithms enables us to find the eigenvectors and the associated eigenvalues directly from the input vectors $\mathbf{x}(k)$ without a need to compute or estimate the very large covariance matrix $\mathbf{R_{xx}}$. Such an approach will be especially useful for nonstationary input data, i.e., in cases of tracking slow changes of correlations in the input data (signals) or in updating eigenvectors with new samples. Computing the sample covariance matrix itself is very costly. Furthermore, the direct diagonalization of a matrix or eigenvalue decomposition can be extremely costly since this operation is of complexity $\mathcal{O}(m^3)$. Most of the adaptive algorithms presented in this chapter do not require the sample covariance matrix to be computed and have low complexity.

3.2.2 Estimation of Sample Covariance Matrices

In practice, the ideal covariance matrix $\mathbf{R_{xx}}$ is not available. We can only have an estimate $\widehat{\mathbf{R}}_{\mathbf{xx}}$ of $\mathbf{R_{xx}}$ called the sample covariance matrix based on a finite number of samples:

$$\widehat{\mathbf{R}}_{\mathbf{xx}} = \frac{1}{N} \sum_{k=1}^{N} \mathbf{x}(k)\, \mathbf{x}^T(k) \,. \tag{3.5}$$

We assume that the covariance matrix does not change (or changes very slowly) over the length of the block. Alternatively, we can use the Moving Average (MA) approach for an on-line estimation of the sample covariance matrix as follows:

$$\boxed{\widehat{\mathbf{R}}_{\mathbf{xx}}^{(k)} = (1-\eta_0)\, \widehat{\mathbf{R}}_{\mathbf{xx}}^{(k-1)} + \eta_0\, \mathbf{x}(k)\, \mathbf{x}^T(k)} \tag{3.6}$$

where $\eta_0 > 0$ is a learning rate (and $(1-\eta_0)$ is a forgetting factor) to be chosen according to the stationarity of the signal (typically $0.01 \le \eta_0 \le 0.1$).

Alternatively, in real time applications, the sample covariance matrix can be recursively updated as

$$\begin{aligned} \widehat{\mathbf{R}}_N &= \frac{1}{N} \sum_{l=k-N+1}^{k} \mathbf{x}(l)\, \mathbf{x}^T(l) = \frac{1}{N}\left[\sum_{l=k-N+1}^{k-1} \mathbf{x}(l)\, \mathbf{x}^T(l) + \mathbf{x}(k)\, \mathbf{x}^T(k)\right] \\ &= \frac{N-1}{N} \widehat{\mathbf{R}}_{N-1} + \frac{1}{N} \mathbf{x}(k)\, \mathbf{x}^T(k), \end{aligned} \tag{3.7}$$

where $\widehat{\mathbf{R}}_N$ denotes the estimated covariance matrix at k-th data instant so that

$$\widehat{\mathbf{R}}_{N-1} = \frac{1}{N-1} \sum_{l=k-N+1}^{k-1} \mathbf{x}(l)\, \mathbf{x}^T(l) \,.$$

The recursive update can be formulated in a more general form as

$$\widehat{\mathbf{R}}_N = \alpha\, \widehat{\mathbf{R}}_{N-1} + \triangle \widehat{\mathbf{R}}, \tag{3.8}$$

where α is a parameter in the range $(0, 1]$ and $\triangle \hat{\mathbf{R}}$ is a symmetric matrix of rank much less than that of $\widehat{\mathbf{R}}_{N-1}$. While working with stationary signals, we usually use rank-1 update with $\alpha = (N-1)/N$ and $\triangle \widehat{\mathbf{R}} = (1/N)\, \mathbf{x}(k)\, \mathbf{x}^T(k)$, where $\mathbf{x}(k)$ is the data vector at k-th instant. On the other hand, in the nonstationary case, rank-1 updating is carried out by choosing $0 < \alpha \ll 1$ and $\triangle\widehat{\mathbf{R}} = \mathbf{x}(k)\,\mathbf{x}^T(k)$. Alternatively, in the nonstationary case, we can use the rank-2 updating is achieved with $\alpha = 1$ and $\triangle\widehat{\mathbf{R}} = \mathbf{x}(k)\,\mathbf{x}^T(k) - \mathbf{x}(k-N+1)\,\mathbf{x}^T(k-N+1)$, where N is the sliding window length over which the covariance matrix is computed. The term $\widehat{\mathbf{R}}_{N-1}$ may be thought of as a prediction of $\mathbf{R}$ based on $N-1$ observations and $\mathbf{x}(k)\,\mathbf{x}^T(k)$ may be thought of as an instantaneous estimate of $\mathbf{R}$.

3.2.3 Signal and Noise Subspaces - Automatic Choice of Dimensionality for PCA

A very important problem arising in many application areas is determination of the dimension of the signal and noise subspaces. In other words, a central issue in PCA is choosing the number of principal components to be retained [845]. To solve this problem, we usually exploit a fundamental property of PCA: It projects the input data $\mathbf{x}(k)$ from their original m-dimensional space onto an n-dimensional output subspace $\mathbf{y}(k)$ (typically, with $n \ll m$), thus performing a dimensionality reduction which retains most of the intrinsic information in the input data vectors. In other words, the principal components $y_i(k) = \mathbf{v}_i^T\mathbf{x}(k)$ are estimated in such a way that, for $n \ll m$, although the dimensionality of data is strongly reduced, the most relevant information is retained in the sense that the original input data $\mathbf{x}$ can be reconstructed from the output data (signals) $\mathbf{y}$ by using the transformation $\hat{\mathbf{x}} = \mathbf{V}_S\,\mathbf{y}$, that minimizes a suitable cost function. A commonly used criterion is the minimization of mean squared error $\|\mathbf{x} - \mathbf{V}_S^T\,\mathbf{V}_S\,\mathbf{x}\|_2^2$.

PCA enables us to divide observed (measured), sensor signals: $\mathbf{x}(k) = \mathbf{x}_s(k) + \boldsymbol{\nu}(k)$ into two subspaces: the *signal subspace* corresponding to principal components associated with the largest eigenvalues called principal eigenvalues: $\lambda_1, \lambda_2, ..., \lambda_n$, $(m > n)$ and associated eigenvectors $\mathbf{V}_s = [\mathbf{v}_1, \mathbf{v}_2, \ldots, \mathbf{v}_n]$ called the principal eigenvectors and the *noise subspace* corresponding to the minor components associated with the eigenvalues $\lambda_{n+1}, ..., \lambda_m$. The subspace spanned by the n first eigenvectors $\mathbf{v}_i$ can be considered as an approximation of the noiseless signal subspace. One important advantage of this approach is that it enables not only a reduction in the noise level, but also allows us to estimate the number of sources on the basis of distribution of eigenvalues. However, a problem arising from this approach, is how to correctly set or estimate the threshold which divides eigenvalues into the two subspaces, especially when the noise is large (i.e., the SNR is low).

Let us assume that we model the vector $\mathbf{x}(k) \in \mathbb{R}^m$ as

$$\mathbf{x}(k) = \mathbf{H}\,\mathbf{s}(k) + \boldsymbol{\nu}(k), \tag{3.9}$$

where $\mathbf{H} \in \mathbb{R}^{m\times n}$ is a full column rank mixing matrix with $m > n$, $\mathbf{s}(k) \in \mathbb{R}^n$ is a vector of zero-mean Gaussian sources with the nonsingular covariance matrix $\mathbf{R}_{\mathbf{s}\mathbf{s}} = E\{\mathbf{s}(k)\mathbf{s}^T(k)\}$ and $\boldsymbol{\nu}(k) \in \mathbb{R}^m$ is a vector of Gaussian zero-mean i.i.d. noise modelled by the covariance matrix $\mathbf{R}_{\boldsymbol{\nu}\boldsymbol{\nu}} = \sigma_\nu^2\mathbf{I}_m$, furthermore, random vectors $\{\mathbf{s}(k)\}$ and $\{\boldsymbol{\nu}(k)\}$ are uncorrelated [769].

Remark 3.1 *The model given by Eq. (3.9) is often referred to as probabilistic PCA, and has been introduced in the machine learning context [1018, 1148]. Moreover, such a model* *can be also considered as a special form of Factor Analysis (FA) with isotropic noise [1148]. The only difference is that, in FA the noise covariance matrix is a general diagonal matrix.*

For the model (3.9) and under the above assumptions the covariance matrix of $\mathbf{x}(k)$ can be written as

$$\begin{aligned}
\mathbf{R_{xx}} &= E\{\mathbf{x}(k)\,\mathbf{x}^T(k)\} = \mathbf{H}\,\mathbf{R_{s\,s}}\,\mathbf{H}^T + \sigma_\nu^2 \mathbf{I}_m \\
&= [\mathbf{V}_\mathcal{S}, \mathbf{V}_\mathcal{N}] \begin{bmatrix} \mathbf{\Lambda}_\mathcal{S} & \mathbf{0} \\ \mathbf{0} & \mathbf{\Lambda}_\mathcal{N} \end{bmatrix} [\mathbf{V}_\mathcal{S}, \mathbf{V}_\mathcal{N}]^T \\
&= \mathbf{V}_\mathcal{S} \mathbf{\Lambda}_\mathcal{S} \mathbf{V}_\mathcal{S}^T + \mathbf{V}_\mathcal{N} \mathbf{\Lambda}_\mathcal{N} \mathbf{V}_\mathcal{N}^T, \qquad (3.10)
\end{aligned}$$

where $\mathbf{H}\,\mathbf{R_{s\,s}}\,\mathbf{H}^T = \mathbf{V}_\mathcal{S} \mathbf{\Lambda}_\mathcal{S} \mathbf{V}_\mathcal{S}^T$ is a rank-n matrix, $\mathbf{V}_\mathcal{S} \in \mathbb{R}^{m \times n}$ contains the eigenvectors associated with n principal (signal+noise subspace) eigenvalues of $\mathbf{\Lambda}_\mathcal{S} = \text{diag}\{\lambda_1 \geq \lambda_2 \cdots \geq \lambda_n\}$ in a descending order. Similarly, the matrix $\mathbf{V}_\mathcal{N} \in \mathbb{R}^{m \times (m-n)}$ contains the $(m - n)$ (noise) eigenvectors that correspond to noise eigenvalues $\mathbf{\Lambda}_\mathcal{N} = \text{diag}\{\lambda_{n+1}, \ldots, \lambda_m\} = \sigma_\nu^2 \mathbf{I}_{m-n}$. This means that, theoretically, the $(m - n)$ smallest eigenvalues of $\mathbf{R_{xx}}$ are equal to σ_ν^2, so we can determine the dimension of the signal subspace from the multiplicity of the smallest eigenvalues under the assumption that the variance of the noise is relatively low and we have a perfect estimate of the covariance matrix. However, in practice, we estimate the sample covariance matrix from a limited number of samples and the smallest eigenvalues are usually different, so the determination of the dimension of the signal subspace is usually not an easy task.

Instead of setting the threshold between the signal and noise eigenvalues by using some heuristic procedure or a rule of thumb, we can use one of the two well-known information theoretic criteria, namely, Akaike's information criterion (AIC), the minimum description length (MDL) and Beyesian Information Criterion (BIC) [665, 1260, 845]. To do this, we compute the probability of the data for each possible dimension. Akaike's information theoretic criterion (AIC) selects the model that minimizes the cost function [769]

$$AIC = -2 \log(p(\mathbf{x}(1), \mathbf{x}(2), \ldots, x(N)|\hat{\mathbf{\Theta}})) + 2n, \qquad (3.11)$$

where $p(\mathbf{x}(1), \mathbf{x}(2), \ldots, x(N)|\hat{\mathbf{\Theta}})$ is a parameterized probability density function, $\hat{\mathbf{\Theta}}$ is the maximum likelihood estimator of a parameter vector $\mathbf{\Theta}$, and n is the number of free adjusted parameters.

The minimum description length (MDL) criterion selects the model that instead minimizes

$$MDL = -\log(p(\mathbf{x}(1), \mathbf{x}(2), \ldots, x(N)|\hat{\mathbf{\Theta}})) + \frac{1}{2} n \log N. \qquad (3.12)$$

Assuming that the observed vectors $\{\mathbf{x}(k)\}_{k=1}^N$ are zero-mean, i.i.d. Gaussian random vectors, it can be shown [1260] that the dimension of the signal subspace can be estimated by taking the value of $n \in \{1, 2, \ldots, m\}$ for which

$$AIC(n) = -2N(m-n) \log \varrho(n) + 2n(2m-n), \qquad (3.13)$$

$$MDL(n) = -N(m-n) \log \varrho(n) + 0.5n(2m-n) \log N \qquad (3.14)$$

is minimized. Here, N is the number of the data vectors $\mathbf{x}(k)$ used in estimating the data covariance matrix $\mathbf{R}_{\mathbf{xx}}$, and

$$\varrho(n) = \frac{(\lambda_{n+1}\lambda_{n+2}\cdots\lambda_m)^{\frac{1}{m-n}}}{\frac{1}{m-n}(\lambda_{n+1}+\lambda_{n+2}+\cdots+\lambda_m)} \tag{3.15}$$

is the ratio of the geometric mean of the $(m-n)$ smallest PCA eigenvalues to their arithmetic mean. The estimate $\hat{n}$ of the number of terms (sources) is chosen so it minimizes either the AIC or MDL criterion.

Unfortunately, both criteria provide only rough estimates (of the number of sources) that are rather very sensitive to variations in the SNR and the number of available data samples [769]. Another problem with the AIC and MDL criteria given above is that they have been derived by assuming that the data vectors $\mathbf{x}(k)$ have a Gaussian distribution [1260]. This is done for mathematical tractability, by making it possible to derive closed form expressions. The Gaussianity assumption does not usually hold exactly in the BSS and other blind signal processing applications. Instead of setting the threshold between the signal and noise eigenvalues, one might even suppose that the AIC and MDL criteria cannot be used in the BSS or ICA problems, because there we assume that the source signals $s_i(k)$ are non-Gaussian. However, it should be noted that the components of the data vectors $\mathbf{x}(k) = \mathbf{H}\mathbf{s}(k) + \boldsymbol{\nu}(k)$ are mixtures of the sources, and therefore often have distributions that are not so far from a Gaussian one [665].

Recently Minka proposed very efficient criterion for estimation of the true dimensionality of the data on basis of Bayesian model selection [845]. The estimate involve an integral over the Stiefel manifold which is approximated by Laplace method. We refer to this criterion as MInka Bayesian model Selection ($MIBS$) which can be summarized as follows: Find a index n for $1 \le n \le m$ such that the cost function is maximized [845]

$$MIBS(n) = p(\mathbf{X}|n) \approx p_n\Big(\prod_{j=1}^{n}\lambda_j\Big)^{-N/2}\tilde{\sigma}_n^{-N(m-n)}|\mathbf{A}_n|^{-1/2}(2\pi)^{(d_n+n)/2}N^{-n/2}, \tag{3.16}$$

where

$$\begin{aligned} p_n &= 2^{-n}\prod_{i=1}^{n}\Gamma\Big(\frac{m-i+1}{2}\Big)\,\pi^{-(m-i+1)/2}, \\ |\mathbf{A}_n| &= \prod_{i=1}^{n}\prod_{j=i+1}^{m}(\hat{\lambda}_j^{-1}-\hat{\lambda}_i^{-1})(\lambda_i-\lambda_j)N, \\ \tilde{\sigma}_n^2 &= \Big(\sum_{j=n+1}^{m}\lambda_j\Big)/(m-n), \qquad d_n = mn - n(n+1)/2 \end{aligned}$$

and $\hat{\lambda}_j$ are identical with λ_j expect for $j > n$ where $\hat{\lambda}_j = \tilde{\sigma}_n$. In order to estimate the latent dimensionality of the data $\mathbf{X}$, we choose the value n that maximizes the approximation to the model evidence $p(\mathbf{X}|n)$ [845].

An approximation of the MIBS leads to the Bayesian Information Criterion (BIC) which neglects all terms which do not grow with N:

$$BIC(n) = \Big(\prod_{j=1}^{n} \lambda_j\Big)^{-N/2} \tilde{\sigma}_n^{-N(m-n)/2} N^{-(d_n+n)/2}. \tag{3.17}$$

Minka's experiments and our own, have shown that his criterion is quite robust in respect of distribution of sources and a remarkably consistent even with relative few data points. In practical experiments, the $MIBS$ criterion have quite often performed very well in estimating the true number of sources n in BSS and ICA problems and even for non-Gaussian sources accurate model selection is still feasible.

The following two conditions may lead to correct estimation of sources. Firstly, the number of mixtures must be larger than the number of sources. If the number of sources is equal to the number of sensors, all criteria inevitably underestimate n by one. The second condition is that there must be at least a small amount of noise. This also guarantees that the eigenvalues $\lambda_{n+1}, \lambda_{n+2}, \ldots, \lambda_m$, corresponding to noise, are nonzero. It is obvious that zero eigenvalues cause numerical difficulties in formulas (3.13), (3.14) and (3.16).

3.2.4 Basic Properties of PCA

It is straightforward to obtain the following properties for principal components (PCs) $y_i(k) = \mathbf{v}_i^T \mathbf{x}(k)$:

1. The factor $y_1(k) = \mathbf{v}_1^T \mathbf{x}(k)$ is the first principal component of $\mathbf{x}(k)$ if the variance of $y_1(k)$ is maximally large under the constraint that the norm of vector $\mathbf{v}_1$ is constant [907]. Then the weight vector $\mathbf{v}_1$ maximizes the following criterion

$$J_1(\mathbf{v}_1) = E\{y_1^2\} = E\{\mathbf{v}_1^T \mathbf{R_{xx}} \mathbf{v}_1\}, \tag{3.18}$$

subject to the constraint $\|\mathbf{v}_1\|_2 = 1$. This criterion can be extended to n principal components (n can be any integer between 1 and m) as

$$J_n(\mathbf{v}_1, \mathbf{v}_2, \ldots, \mathbf{v}_n) = E\{\sum_{i=1}^{n} y_i^2\} = E\{\sum_{i=1}^{n} (\mathbf{v}_i^T \mathbf{x})^2\} = \sum_{i=1}^{n} \mathbf{v}_i^T \mathbf{R_{xx}} \mathbf{v}_i, \tag{3.19}$$

subject to the constraints $\mathbf{v}_i^T \mathbf{v}_j = \delta_{ij}$.

2. The PCs have zero mean values

$$E\{y_i\} = 0, \qquad \forall i. \tag{3.20}$$

3. Different PCs are mutually uncorrelated

$$E\{y_i y_j\} = \delta_{ij} \lambda_j, \qquad (i, j = 1, 2, \ldots, n). \tag{3.21}$$

4. The variance of the i-th PC is equal to the i-th eigenvalue of the covariance matrix $\mathbf{R_{xx}}$

$$var\{y_i\} = \sigma_{y_i}^2 = E\{y_i^2\} = E\{(\mathbf{v}_i^T \mathbf{x})^2\} = \mathbf{v}_i^T \mathbf{R_{xx}} \mathbf{v}_i = \lambda_i. \tag{3.22}$$

5. The PCs are hierarchically organized with respect to decreasing values of their variances

$$\sigma_{y_1}^2 \geqslant \sigma_{y_2}^2 \geqslant \cdots \geqslant \sigma_{y_n}^2, \tag{3.23}$$

i.e., $\lambda_1 \geqslant \lambda_2 \geqslant \cdots \geqslant \lambda_n$.

6. Best approximation property: for the mean-square error of the approximation

$$\widehat{\mathbf{x}} = \sum_{i=1}^{n} y_i \mathbf{v}_i = \sum_{i=1}^{n} \mathbf{v}_i \mathbf{v}_i^T \mathbf{x}, \qquad n < m, \tag{3.24}$$

we have

$$E\{\|\mathbf{x} - \widehat{\mathbf{x}}\|_2^2\} = E\{\| \sum_{i=n+1}^{m} y_i \mathbf{v}_i \|_2^2\} = \sum_{i=n+1}^{m} E\{|y_i|^2\} = \sum_{i=n+1}^{m} \lambda_i. \tag{3.25}$$

Taking into account that $\lambda_1 \geqslant \lambda_2 \geqslant \cdots \geqslant \lambda_n$, it is obvious that an approximation with these eigenvectors $\mathbf{v}_1, \mathbf{v}_2, \ldots, \mathbf{v}_n$, corresponding to the largest eigenvalues, leads to the minimal mean square error.

3.3 EXTRACTION OF PRINCIPAL COMPONENTS USING OPTIMAL COMPRESSION–RECONSTRUCTION PRINCIPLE

One of the simplest and intuitively understandable approaches to the derivation of adaptive algorithms for PCA is based on self-association (also called self-supervising or the replicator principle) [15, 275, 276]. According to this approach, we first compress the data vector $\mathbf{x}(k)$ to one variable $y_1(k) = \mathbf{v}_1^T \mathbf{x}(k)$ and next we attempt to reconstruct the original data from $y_1(k)$ by using the transformation $\hat{\mathbf{x}}(k) = \mathbf{v}_1 y_1(k)$. Let us assume, that we wish to extract principal components (PCs) sequentially by employing the self-supervising principle (replicator) and a cascade (hierarchical) neural network architecture [275, 271, 276].
Let us consider a simple processing unit (see Fig.3.1)

$$y_1(k) = \mathbf{v}_1^T \mathbf{x}(k) = \sum_{p=1}^{m} v_{1p} x_p(k), \tag{3.26}$$

which extracts the first principal component, with $\lambda_1 = E\{y_1^2(k)\}$. Strictly speaking, the factor $y_1(k)$ is called the first principal component of $\mathbf{x}(k)$, if the variance of $y_1(k)$ is maximally large under the constraint that the principal vector $\mathbf{v}_1$ has unit length.

The vector $\mathbf{v}_1 = [v_{11}, v_{12}, \ldots, v_{1m}]^T$ should be determined in such a way so that the reconstruction vector $\hat{\mathbf{x}}(k) = \mathbf{v}_1 y_1(k)$ will reproduce (reconstruct) the input training vectors $\mathbf{x}(k)$ as correctly as possible, according to a suitable optimization criterion. In general, the loss (cost) function is expressed as

$$J_1(\mathbf{v}_1) = E\{\|\mathbf{x}(k) - \mathbf{v}_1 \mathbf{v}_1^T \mathbf{x}(k)\|_2^2\} \cong \sum_{k=1}^{N} \gamma^{N-k} \|\mathbf{x}(k) - \mathbf{v}_1 \mathbf{v}_1^T \mathbf{x}(k)\|_2^2, \tag{3.27}$$

where γ is the forgetting factor.

Let us consider the simplified (continuous time) version of the cost function which can be written as

$$\tilde{J}_1(\mathbf{v}_1) \stackrel{\triangle}{=} \frac{1}{2}||\mathbf{e}_1(t)||_2^2 = \sum_{p=1}^{m} e_{1p}^2(t), \tag{3.28}$$

with

$$\mathbf{e}_1(t) \stackrel{\triangle}{=} \mathbf{x}(t) - \mathbf{v}_1 y_1(t), \quad y_1(t) \stackrel{\triangle}{=} \mathbf{v}_1^T \mathbf{x}(t).$$

The formulation of the computational cost function $J_1(\mathbf{v}_1)$ is a key step in our approach, because this enables us to transform the minimization problem into a set of differential or difference equations, which then determines the adaptive learning algorithm [275, 276].

The minimization of the cost function (3.28), according to the standard gradient descent approach [275] for vector $\mathbf{v}_1$, leads to the following set of differential equations:

$$\frac{dv_{1p}}{dt} = -\mu_1 \frac{\partial J_1(\mathbf{v}_1)}{\partial v_{1p}} = \mu_1 \left[y_1(t) e_{1p}(t) + x_p(t) \sum_{h=1}^{m} v_{1h}(t) e_{1h}(t) \right], \; (p = 1, 2, \ldots, m) \tag{3.29}$$

which can be written in matrix form as

$$\frac{d\mathbf{v}_1}{dt} = \mu_1 [y_1 \mathbf{e}_1 + \mathbf{x}\mathbf{v}_1^T \mathbf{e}_1], \tag{3.30}$$

for any $\mathbf{v}_1(0) \neq \mathbf{0}$, $\mu_1 > 0$.
The above learning rule can be further simplified as

$$\frac{d\mathbf{v}_1}{dt} = \mu_1 y_1 \mathbf{e}_1 = \mu_1 y_1 [\mathbf{x} - \mathbf{v}_1 y_1], \tag{3.31}$$

since the second term on the right hand side in Equation (3.30), which can be written as $\mathbf{x}\mathbf{v}_1^T \mathbf{e}_1 = \mathbf{x}\mathbf{v}_1^T(\mathbf{x} - \mathbf{v}_1 y_1) = \mathbf{x}(1 - \mathbf{v}_1^T \mathbf{v}_1) y_1$, tends quickly to zero as $\mathbf{v}_1^T \mathbf{v}_1$ tends to 1 with $t \to \infty$ and can therefore be neglected. This feature has also been confirmed by extensive computer simulations.

It is interesting to note that the discrete-time realization of the learning rule:

$$\mathbf{v}_1(k+1) = \mathbf{v}_1(k) + \eta_1(k)\, y_1(k) [\mathbf{x}(k) - \mathbf{v}_1(k)\, y_1(k)], \qquad (k = 0, 1, 2, \ldots) \tag{3.32}$$

is in a form known as the Oja algorithm [907].

Remark 3.2 *It is not well known, however, that the paper by Amari [15] was the first paper in which a normalized Hebbian learning was used to perform PCA [15]:*

$$\tilde{\mathbf{v}}_1(k+1) = \mathbf{v}_1(k) + \eta_1(k)\, y_1(k)\, \mathbf{x}(k), \tag{3.33}$$

$$\mathbf{v}_1(k+1) = \frac{\tilde{\mathbf{v}}_1(k+1)}{\| \tilde{\mathbf{v}}_1(k+1) \|_2} \tag{3.34}$$

and if the normalization step is combined with the Hebbian rule, we obtain the learning algorithm (3.32).

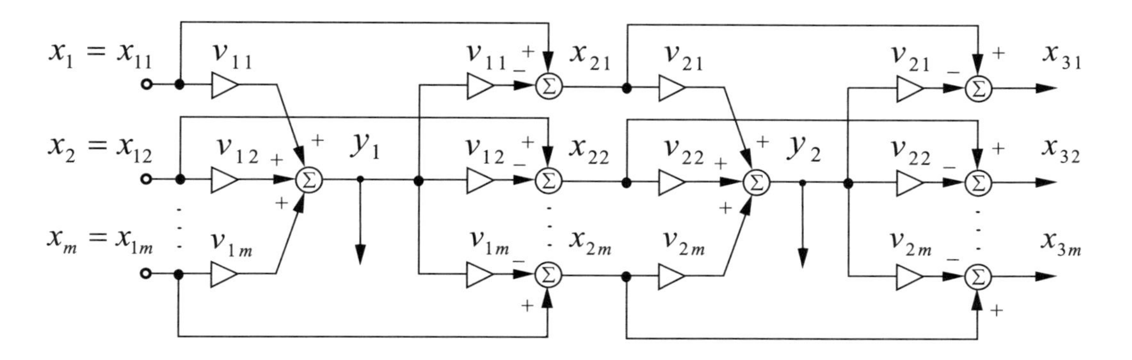

Fig. 3.1 Sequential extraction of principal components.

The above learning rule can be extended for an arbitrary number of PCs using the self-supervising principle and the cascade hierarchical neural network shown in Fig.3.1.

In what follows, we will discuss a deflation approach for sequential extraction of principal components, corresponding to real-valued zero-mean signals without estimating directly the large covariance matrix. We will extract principal components sequentially as long as the eigenvalues λ_i are larger than some suitably chosen threshold. We assume that minor components for $i > n$ correspond to additive noise.

The learning algorithm for the extraction of the second PC corresponding to the second largest eigenvalue $\lambda_2 = E\{y_2^2(k)\}$ is similar to that used for the extraction of the first principal component. However, we do not apply the extraction process directly to the input data $\mathbf{x}_1(k) = \mathbf{x}(k)$ but to the residual error

$$\mathbf{e}_1(k) \triangleq \mathbf{x}_2(k) = \mathbf{x}_1(k) - \hat{\mathbf{x}}_1(k) = \mathbf{x}_1(k) - \mathbf{v}_1 y_1(k)$$

and $y_2(k) \triangleq \mathbf{v}_2^T \mathbf{e}_1(k)$ (not $y_2(k) = \mathbf{v}_2^T \mathbf{x}(k)$ as usually is assumed). It follows that the learning rule for the $i - th$ PC can be written in the general form as follows

$$\mathbf{v}_i(k+1) = \mathbf{v}_i(k) + \eta_i(k)\, y_i(k)\, \mathbf{x}_{i+1}(k), \tag{3.35}$$

where

$$\mathbf{e}_i = \mathbf{x}_{i+1} \triangleq \mathbf{x}_i - \mathbf{v}_i y_i, \quad y_i \triangleq \mathbf{v}_i^T \mathbf{x}_i, \quad \mathbf{x}_1(k) \triangleq \mathbf{x}(k).$$

The extracted output signals $y_i(k)$ after applying the above learning procedure will be decorrelated with decreasing values of the variances $\lambda_i = E\{y_i^2(k)\}, \quad (i = 1, 2, ..., n)$.

In the accelerated version of the above algorithm, key roles are played by the learning rates $\eta_i(k) \geq 0$ and the forgetting factor γ. If the learning rate is too large, the algorithm is unstable. Otherwise, if it is a fixed or an exponentially decreasing parameter, the convergence speed of the algorithm may be very slow [275, 271, 276].

In order to increase convergence speed, we can minimize the cost function (3.27) by employing the recursive least-squares (RLS) or Kalman filtering approach for optimal updating

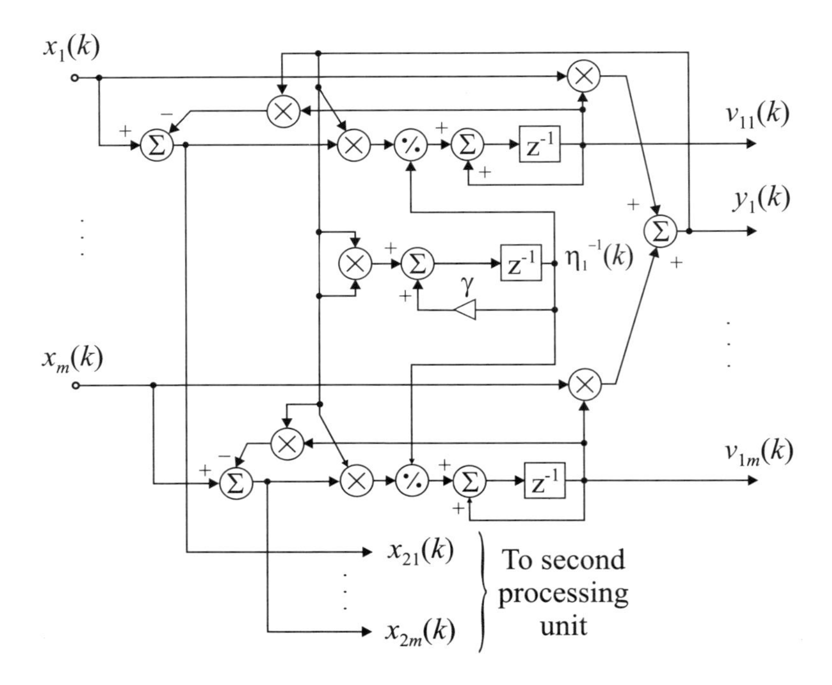

Fig. 3.2 On-line on chip implementation of fast RLS learning algorithm for the principal component estimation.

of the learning rate η_i [275, 349, 1312] (see Fig. 3.2):

$$\mathbf{x}_1(k) = \mathbf{x}(k), \;\; \eta_i^{-1}(0) = 2\max\{\|\mathbf{x}_i(k)\|_2^2\} = 2\,\mathbf{x}_{i,max}, \tag{3.36}$$

$$\mathbf{v}_i(0) = \mathbf{x}_{i,max}/\|\mathbf{x}_{i,max}\|_2, \quad (i = 1, 2, \ldots, n), \tag{3.37}$$

$$y_i(k) = \mathbf{v}_i^T(k)\mathbf{x}_i(k), \tag{3.38}$$

$$\mathbf{v}_i(k+1) = \mathbf{v}_i(k) + \frac{y_i(k)}{\eta_i^{-1}(k)}[\mathbf{x}_i(k) - y_i(k)\mathbf{v}_i(k)], \tag{3.39}$$

$$\eta_i^{-1}(k+1) = \gamma\eta_i^{-1}(k) + |y_i(k)|^2, \tag{3.40}$$

$$\mathbf{x}_{i+1}(k) = \mathbf{x}_i(k) - y_i(k)\mathbf{v}_{i*}, \tag{3.41}$$

where γ is the forgetting factor (typically, $0.9 \le \gamma \le 0.99$) and $\mathbf{v}_{i*}$ means vector $\mathbf{v}_i(k)$ after achieving convergence.

The above fast algorithm can be generalized for Principal Subspace Analysis (PSA) and nonlinear PCA.

3.4 BASIC COST FUNCTIONS AND ADAPTIVE ALGORITHMS FOR PCA

3.4.1 The Rayleigh Quotient – Basic Properties

Most of the adaptive algorithms for PCA and MCA (minor component analysis) can be defined directly or indirectly by using the Rayleigh quotient (RQ) of the specific covariance matrix as the cost function.

The Rayleigh quotient $r(\mathbf{v})$ is defined for $\mathbf{v} \neq \mathbf{0}$, as

$$r(\mathbf{v}) = r(\mathbf{v}, \mathbf{R_{xx}}) = \frac{\mathbf{v}^T \mathbf{R_{xx}} \mathbf{v}}{\mathbf{v}^T \mathbf{v}}, \tag{3.42}$$

where $\mathbf{R_{xx}} = E\{\mathbf{x}\mathbf{x}^T\}$. The Rayleigh quotient has the following important properties:

1. Stationarity and critical points:

$$\lambda_1 = \max r(\mathbf{v}, \mathbf{R_{xx}}) \tag{3.43}$$

$$\lambda_m = \min r(\mathbf{v}, \mathbf{R_{xx}}), \tag{3.44}$$

where λ_1 and λ_m denote the largest and smallest eigenvalues of the covariance matrix $\mathbf{R_{xx}}$.

More generally, the critical points and critical values of $r(\mathbf{v}, \mathbf{R_{xx}})$ are the eigenvectors and eigenvalues of $\mathbf{R_{xx}}$. Let the eigenvalues of the covariance matrix be ordered as

$$\lambda_1 \geqslant \lambda_2 \geqslant \cdots \geqslant \lambda_m. \tag{3.45}$$

2. Homogeneity:

$$r(\alpha\, \mathbf{v}, \beta\, \mathbf{R_{xx}}) = \beta\, r(\mathbf{v}, \mathbf{R_{xx}}) \qquad \forall \alpha \neq 0, \;\; \beta \neq 0. \tag{3.46}$$

3. Translation invariance:

$$r(\mathbf{v}, \mathbf{R_{xx}} - \alpha \mathbf{I}) = r(\mathbf{v}, \mathbf{R_{xx}}) - \alpha. \tag{3.47}$$

4. Minimal residual:

$$\| (\mathbf{R_{xx}} - r(\mathbf{v}, \mathbf{R_{xx}})\mathbf{I})\, \mathbf{v} \| \leqslant \|(\mathbf{R_{xx}} - \alpha \mathbf{I})\mathbf{v}\|_2 \tag{3.48}$$

$$\forall \mathbf{v} \neq 0 \quad \text{and any scalar coefficient } \alpha. \tag{3.49}$$

5. Orthogonality:

$$\mathbf{v} \perp (\mathbf{R_{xx}} - r(\mathbf{v}, \mathbf{R_{xx}})\mathbf{I})\, \mathbf{v}. \tag{3.50}$$

6. The Hessian matrix of the Rayleigh quotient is:

$$\mathbf{H}_r(\mathbf{v}, \mathbf{R_{xx}}) = \left[\frac{\partial^2 r(\mathbf{v}, \mathbf{R_{xx}})}{\partial v_i \partial v_j} \right] = \frac{2}{\|\mathbf{v}\|_2^2} \left(\mathbf{R_{xx}} - r(\mathbf{v})\, \mathbf{I}\right) \left(\mathbf{I} - \frac{4}{\|\mathbf{v}\|_2^2} \mathbf{v}\mathbf{v}^T \right). \tag{3.51}$$

For the eigenvalues $\lambda_1, \lambda_2, \ldots, \lambda_m$ and the corresponding eigenvectors $\mathbf{v}_1, \mathbf{v}_2, \ldots, \mathbf{v}_m$, the Hessian can be expressed as [287]

$$\mathbf{H}_r(\mathbf{v}_i) = 2\,(\mathbf{R_{xx}} - \lambda_i \mathbf{I}) \tag{3.52}$$

$$\mathbf{H}_r(\mathbf{v}_i)\mathbf{v}_j = \begin{cases} 0, & i = j \\ 2\,(\lambda_j - \lambda_i)\mathbf{v}_j, & i \neq j \end{cases} \tag{3.53}$$

$$\det \mathbf{H}_r(\mathbf{v}_i) = \det[\mathbf{R_{xx}} - \lambda_i \mathbf{I}] = 0 \tag{3.54}$$

$$\text{i.e., } \mathbf{H}_r \text{ is singular for any eigenvector } \mathbf{v}_i. \tag{3.55}$$

The Hessian matrix $\mathbf{H}_r$ has the same eigenvectors as $\mathbf{R_{xx}}$ but with different eigenvalues.

Remark 3.3 *It should be noted that it is not practical to use the Newton or quasi Newton method for minimization of the Rayleigh quotient, since the Hessian matrix $\mathbf{H}_r$ is singular at the extremum points and so the inverse matrix does not exist.*

3.4.2 Basic Cost Functions for Computing Principal and Minor Components

The maximum and minimum eigenvalues of the covariance matrix $\mathbf{R_{xx}} = E\{\mathbf{x}\mathbf{x}^T\}$ can be found as the extrema of the Rayleigh quotient, so the following basic cost can be used

$$J_2(\mathbf{v}) = r(\mathbf{v}) = \frac{\mathbf{v}^T \mathbf{R_{xx}} \mathbf{v}}{\mathbf{v}^T \mathbf{v}}, \qquad \mathbf{v} \in \mathbb{R}^m \quad \|\mathbf{v}\|_2 \neq 0. \tag{3.56}$$

To find these extrema, we can compute the gradient as

$$\nabla_{\mathbf{v}} r(\mathbf{v}) = \frac{\partial r(\mathbf{v})}{\partial \mathbf{v}} = 2\,\frac{\mathbf{R_{xx}}\mathbf{v}(\mathbf{v}^T\mathbf{v}) - \mathbf{v}(\mathbf{v}^T\mathbf{R_{xx}}\mathbf{v})}{(\mathbf{v}^T\mathbf{v})^2}, \tag{3.57}$$

from which it follows that the stationary points corresponding to $\nabla_{\mathbf{v}} r(\mathbf{v}) = \mathbf{0}$ satisfy

$$\mathbf{R_{xx}}\mathbf{v} = \frac{\mathbf{v}^T \mathbf{R_{xx}} \mathbf{v}}{\mathbf{v}^T \mathbf{v}}\, \mathbf{v}. \tag{3.58}$$

This equation can be satisfied if $\mathbf{v}$ is a unit length eigenvector of $\mathbf{R_{xx}}$ with corresponding eigenvalue $\lambda = \mathbf{v}^T \mathbf{R_{xx}} \mathbf{v}$. Obviously, the minimum will correspond to the minimal eigenvalue of $\mathbf{R_{xx}}$ and the maximum of $r(\mathbf{v})$ corresponds to the maximum eigenvalue. In the general case, the zeros of $\nabla_{\mathbf{v}} r(\mathbf{v})$ correspond to the eigenvectors of $\mathbf{R_{xx}}$, which can be assumed to have unit length.

The above unconstrained optimization problem can also be formulated as a constrained one:

Maximize $\widetilde{\mathbf{J}}_3(\mathbf{v}) = \mathbf{v}^T \mathbf{R_{xx}} \mathbf{v}$

subject to the constraints $\mathbf{v}^T \mathbf{v} = 1$.

The Lagrangian for this constrained problem is

$$J_3(\mathbf{v}, \lambda) = \mathbf{v}^T \mathbf{R_{xx}} \mathbf{v} + \lambda(1 - \mathbf{v}^T \mathbf{v}), \tag{3.59}$$

where $\lambda \in \mathbb{R}$ is a scalar Lagrange multiplier. A necessary and sufficient condition for a stationary point corresponding to an eigenvalue of $\mathbf{R_{xx}}$ is

$$\nabla_{\mathbf{v}} J_3(\mathbf{v}, \lambda) = \mathbf{0} \;\text{ and }\; \nabla_{\lambda} J_3(\mathbf{v}, \lambda) = 0 \tag{3.60}$$

and corresponds to

$$\mathbf{R_{xx}} \mathbf{v} = \mathbf{v}\lambda, \qquad \mathbf{v}^T \mathbf{v} = 1. \tag{3.61}$$

Alternatively, instead of the Lagrangian, we can employ the penalty method and formulate the following cost function

$$J_4(\mathbf{v}) = \mathbf{v}^T \mathbf{R_{xx}} \mathbf{v} - \alpha(1 - \mathbf{v}^T \mathbf{v})^2, \tag{3.62}$$

where α is a positive penalty coefficient.

Another class of objective functions, based on an information-theoretic criterion has been recently proposed as[4] [1, 3]

$$J_5(\mathbf{v}) = \log \frac{\mathbf{v}^T \mathbf{R_{xx}} \mathbf{v}}{\mathbf{v}^T \mathbf{v}} \tag{3.63}$$

and

$$J_6(\mathbf{v}) = \log(\mathbf{v}^T \mathbf{R_{xx}} \mathbf{v}) - \mathbf{v}^T \mathbf{v}. \tag{3.64}$$

Another important and relative simple cost function for PCA, that is not based on the Raleigh quotient is:

$$J_7((\mathbf{v}, c) = E\{\|\mathbf{x} - \mathbf{v}\, c\|_2\}, \tag{3.65}$$

where c is a scalar [349].

Various cost functions used for derivation of PCA algorithms are summarized in Table 3.1.

It is interesting to note that minimization of the above cost functions lead to adaptive algorithms written in the general form:

$$\mathbf{v}_1(k+1) = \mathbf{v}_1(k) + \eta_1(k) F\left[\mathbf{v}_1(k), \mathbf{R}_{\mathbf{xx}}^{(k)}\right], \tag{3.66}$$

where the function F can take various forms (see Table 3.2). The covariance matrix can be estimated on-line as

$$\mathbf{R}_{\mathbf{xx}}^{(k)} = (1 - \eta_0)\, \mathbf{R}_{\mathbf{xx}}^{(k-1)} + \eta_0\, \mathbf{x}(k) \mathbf{x}^T(k), \tag{3.67}$$

where η_0 is the learning rate.

[4]It is noteworthily, that any nonlinear monotonic transformation of the Raleigh quotient (e.g., $\log(r(\mathbf{v}))$) will have the same minimum as the standard cost function for PCA.

Table 3.1 Basic cost functions whose maximization leads to adaptive PCA algorithms.

1.	$J_1(\mathbf{v}) = -E\{\|\mathbf{x} - \mathbf{v}\mathbf{v}^T\mathbf{x}\|_2^2\} \cong -\sum_{k=1}^{N} \gamma^{N-k} \|\mathbf{x}(k) - \mathbf{v}\mathbf{v}^T\mathbf{x}(k)\|_2^2$
2.	$J_2(\mathbf{v}) = (\mathbf{v}^T\mathbf{R}_{\mathbf{x}\mathbf{x}}\mathbf{v})/(\mathbf{v}^T\mathbf{v})$
3.	$J_3(\mathbf{v}) = \mathbf{v}^T\mathbf{R}_{\mathbf{x}\mathbf{x}}\mathbf{v} + \lambda(\mathbf{v}^T\mathbf{v} - 1)$
4.	$J_4(\mathbf{v}) = \mathbf{v}^T\mathbf{R}_{\mathbf{x}\mathbf{x}}\mathbf{v} - \alpha(\mathbf{v}^T\mathbf{v} - 1)^2$
5.	$J_5(\mathbf{v}) = \log(\mathbf{v}^T\mathbf{R}_{\mathbf{x}\mathbf{x}}\mathbf{v})/(\mathbf{v}^T\mathbf{v})$
6.	$J_6(\mathbf{v}) = \log(\mathbf{v}^T\mathbf{R}_{\mathbf{x}\mathbf{x}}\mathbf{v}) - \mathbf{v}^T\mathbf{v}$
7.	$J_7(\mathbf{v}) = E\{\|\mathbf{x} - \mathbf{v}\, c\|_2\}$

3.4.3 Fast PCA Algorithm Based on the Power Method

Alternative fast algorithms for PCA can be derived by using the power method and properties of the Rayleigh quotient [1003, 1018]. Assuming that the principal eigenvector $\mathbf{v}_1$ has unit length, i.e., $\mathbf{v}_1^T\mathbf{v}_1 = 1$, we can estimate it using the following iterations

$$\mathbf{v}_1(l+1) = \frac{\mathbf{R}_{\mathbf{x}\mathbf{x}}\mathbf{v}_1(l)}{\mathbf{v}_1^T(l)\mathbf{R}_{\mathbf{x}\mathbf{x}}\mathbf{v}_1(l)}. \tag{3.68}$$

Taking into account that $y_1^{(l)}(k) = \mathbf{v}_1^T(l)\mathbf{x}(k)$ and $\widehat{\mathbf{R}}_{\mathbf{x}\mathbf{x}} = \langle\mathbf{x}(k)\mathbf{x}^T(k)\rangle$, we can then use the following simplified formula

$$\mathbf{v}_1(l+1) = \frac{\sum_{k=1}^{N} y_1^{(l)}(k)\mathbf{x}(k)}{\sum_{k=1}^{N} [y_1^{(l)}(k)]^2} \tag{3.69}$$

or more generally, for a number of higher PCs, we use the deflation approach as

$$\mathbf{v}_i(l+1) = \frac{\sum_{k=1}^{N} y_i^{(l)}(k)\mathbf{x}_i(k)}{\sum_{k=1}^{N} [y_i^{(l)}(k)]^2}, \qquad (i = 1, 2, \ldots, n) \tag{3.70}$$

where $y_i^{(l)}(k) = \mathbf{v}_i^T(l)\mathbf{x}_i(k)$. After convergence of the vector $\mathbf{v}_i(l)$ to $\mathbf{v}_{i*}$, we perform the deflation as: $\mathbf{x}_{i+1} = \mathbf{x}_i - \mathbf{v}_{i*}y_i, \quad \mathbf{x}_1 = \mathbf{x}$.

Table 3.2 Basic adaptive learning algorithms for principal component analysis (PCA).

No.	Learning Algorithm	Notes, References
1.	$\tilde{\mathbf{v}}_1(k+1) = \mathbf{v}_1(k) + \eta_1 \mathbf{v}_1^T(k)\mathbf{x}(k)\mathbf{x}(k)$ $\mathbf{v}_1(k+1) = \tilde{\mathbf{v}}_1(k+1)/\lVert\tilde{\mathbf{v}}_1(k+1)\rVert_2^2$	Amari (1978) [15]
2.	$\Delta\mathbf{v}_1 = \eta_1\left[\mathbf{R_{xx}}\mathbf{v}_1 - \mathbf{v}_1\mathbf{v}_1^T\mathbf{R_{xx}}\mathbf{v}_1\right]$ $\cong \eta_1[y_1\mathbf{x} - \lvert y_1\rvert^2\mathbf{v}_1]$	Oja (1982) [915]
3.	$\Delta\mathbf{v}_1 = \eta_1\left[\mathbf{R_{xx}}\mathbf{v}_1(\mathbf{v}_1^T\mathbf{v}_1) - \mathbf{v}_1\mathbf{v}_1^T\mathbf{R_{xx}}\mathbf{v}_1\right]$ $\cong \eta_1[y_1\mathbf{x}\mathbf{v}_1^T\mathbf{v}_1 - y_1^2\mathbf{v}_1]$	Chen, Amari (1998) [183]
4.	$\Delta\mathbf{v}_1 = \eta_1\left[y_1\mathbf{x} - \lVert\mathbf{v}_1\rVert_2^2\mathbf{v}_1\right]$	Yuille *et al.* (1994) [1335]
5.	$\Delta\mathbf{v}_1 = \eta_1 g(\mathbf{v}_1^T\mathbf{v}_1)\left[\mathbf{R_{xx}}\mathbf{v}_1 - \mathbf{v}_1(\mathbf{v}_1^T\mathbf{R_{xx}}\mathbf{v}_1/\mathbf{v}_1^T\mathbf{v}_1)\right]$ $g(\mathbf{v}_1^T\mathbf{v}_1) = 1$ or $\mathbf{v}_1^T\mathbf{v}_1$ or $(\mathbf{v}_1^T\mathbf{v}_1)^{-1}$	Luo *et al.* (1996), Chatterje (1999) [171]
6.	$\Delta\mathbf{v}_1 = \eta_1\left[\mathbf{R_{xx}}\mathbf{v}_1 - \mathbf{v}_1\mathbf{v}_1^T\mathbf{R_{xx}}\mathbf{v}_1 - \mathbf{v}_1(1 - \mathbf{v}_1^T\mathbf{v}_1)\right]$ $\cong \eta_1[y_1\mathbf{x} - \lvert y_1\rvert^2\mathbf{v}_1 - \mathbf{v}_1(1 - \mathbf{v}_1^T\mathbf{v}_1)]$	Abed-Meraim, Douglas, Hua, Chatterje (1999) [172]
7.	$\Delta\mathbf{v}_1 = \eta_1\left[2\mathbf{R_{xx}}\mathbf{v}_1 - \mathbf{v}_1\mathbf{v}_1^T\mathbf{R_{xx}}\mathbf{v}_1 - \mathbf{R_{xx}}\mathbf{v}_1\mathbf{v}_1^T\mathbf{v}_1)\right]$ $\cong \eta_1[2y_1\mathbf{x} - \lvert y_1\rvert^2\mathbf{v}_1 - y_1\mathbf{v}_1^T\mathbf{v}_1\mathbf{x})$	Abed-Meraim, Douglas, [3, 405] Hua (1999) [572]
8.	$\Delta\mathbf{v}_1 = \eta_1 y_1 \Psi_1(\mathbf{x} - y_1\mathbf{v}_1)$	Robust Algorithm, Cichocki - Unbehauen (1993) [275]
9.	$\mathbf{v}_1 = \dfrac{\mathbf{R_{xx}}\mathbf{v}_1}{\mathbf{v}_1^T\mathbf{R_{xx}}\mathbf{v}_1}$ $\cong \dfrac{\sum_{k=1}^N y_1(k)\mathbf{x}(k)}{\sum_{k=1}^N y_1^2(k)}$	Rao, Principe (2000) [1003] Roweis (1998) [1018, 1148]
10.	$y_i(k) = \mathbf{v}_i^T(k)\mathbf{x}_i(k)$ $\eta_i^{-1}(k) = \gamma\eta_i^{-1}(k-1) + \lvert y_i(k)\rvert^2$ $\mathbf{v}_i(k+1) = \dfrac{y_i(k)}{\eta_i^{-1}(k)}\left[[\mathbf{x}_i(k) - y_i(k)\mathbf{v}_i(k)\right]$ $\mathbf{x}_{i+1}(k+1) = \mathbf{x}_i(k) - y_i(k)\mathbf{v}_{i*}$ $\mathbf{x}_1(k) = \mathbf{x}(k),\ \ \eta_i^{-1}(0) = \sigma_{yi}^2 = E\{\lvert y_i\rvert^2\}$	Fast RLS Algorithm, Cichocki, Kasprzak, Skarbek [275, 263] Yang (1995) [1312]

The above fast PCA algorithm can be rigorously derived in a slightly modified form by minimizing the cost function:

$$\begin{aligned} J_1(\mathbf{v}, y^{(l)}) &= \sum_{k=1}^{N} \|\mathbf{x}(k) - y^{(l)}\mathbf{v}\|_2^2 \qquad (3.71) \\ &= \sum_{k=1}^{N} \|\mathbf{x}(k)\|_2^2 + \|\mathbf{v}\|_2^2 \sum_{k=1}^{N} [y^{(l)}(k)]^2 - 2\mathbf{v}^T \sum_{k=1}^{N} y^{(l)}(k)\mathbf{x}(k), \end{aligned}$$

subject to the constraint $\|\mathbf{v}\|_2 = 1$. The above cost function achieves equilibrium when the gradient of J_2 is zero, i.e., at

$$\mathbf{v}_* = \frac{\sum_{k=1}^N y_*(k)\mathbf{x}(k)}{\sum_{k=1}^N y_*^2(k)}. \tag{3.72}$$

This suggests the following iteration formula [1018, 1148]:

$$\mathbf{v}(l+1) = \frac{\sum_{k=1}^N y^{(l)}(k)\mathbf{x}(k)}{\| \sum_{k=1}^N y^{(l)}(k)\mathbf{x}(k)\|_2}. \tag{3.73}$$

An outline of the fast PCA algorithm:

1. Initialization: set $\mathbf{v}_1(0) \neq \mathbf{0}$ for $l = 0$

2. Set $y_1^{(l)}(k) = \mathbf{v}_1^T(l)\mathbf{x}(k), \quad (k = 1, 2, \ldots, N)$

3. Compute

$$\mathbf{v}_1(l+1) = \frac{\sum_{k=1}^N y_1^{(l)}(k)\mathbf{x}(k)}{\| \sum_{k=1}^N y_1^{(l)}(k)\mathbf{x}(k)\|_2}$$

4. Stop if $\left(1 - \frac{J_1(\mathbf{v}_1^{(l+1)})}{J_1(\mathbf{v}_1^{(l)})}\right)$ is less than a certain small threshold ε. Otherwise, let $l := l+1$ and go to step 2.

It should be noted that the convergence rate of the power algorithm depends on a ratio λ_2/λ_{max}, where λ_2 is the second largest eigenvalue of $\mathbf{R_{xx}}$. This ratio is generally smaller than one, allowing adequate convergence of the algorithm. However, if the eigenvalue $\lambda_1 = \lambda_{max}$ has one or more other eigenvalues of $\mathbf{R_{xx}}$ close by, in other words, when λ_1 belongs to a cluster of eigenvalues then the ratio can be very close to one, causing very slow convergence, and in consequence, the estimated eigenvector $\mathbf{v}$ may be inaccurate. For multiple eigenvalues, the power method fails to converge.

3.4.4 Inverse Power Iteration Method

The drawback of the power method can be partially overcome by applying the power method to the matrix $\mathbf{T_{xx}} = (\mathbf{R_{xx}} - \sigma \mathbf{I})^{-1}$ instead of $\mathbf{R_{xx}}$, where σ is a positive coefficient, called shift, specified by the user. This method converges to the eigenvector corresponding to the eigenvalue λ_j closest to σ rather λ_{max}. This method is called the inverse power iteration and can be formulated as the following algorithm:

$$\mathbf{v}_T(l+1) = \frac{\mathbf{T_{xx}} \mathbf{v}_T(l)}{\|\mathbf{T_{xx}} \mathbf{v}_T(l)\|_2} = \frac{\mathbf{z}^{(l)}}{\|\mathbf{z}^{(l)}\|_2}, \tag{3.74}$$

where $\mathbf{z}^{(l)} = [\mathbf{R_{xx}} - \sigma \mathbf{I}]^{-1} \mathbf{v}(l)$. The stop criterion can be established as: $\|\mathbf{z}^{(l)} - \lambda_T(l) \mathbf{v}(l)\|_2 \leq \varepsilon$, where $\lambda_T = \mathbf{v}_T(l) \mathbf{z}^{(l)}$ and ε is a small threshold. After convergence $\mathbf{v}_j = \mathbf{v}_T / \lambda_T$ and $\lambda_j = \sigma + 1/\lambda_T$.

The inverse power method converges if $\mathbf{v}_T(0)$ is not perpendicular to $\mathbf{v}_j$. The convergence rate is $|(\lambda_j - \sigma)/(\lambda_i - \sigma)|$, where λ_i is an eigenvalue of $\mathbf{R_{xx}}$ such that $|\lambda_i - \sigma|^{-1}$ is the second largest eigenvalue of $\mathbf{T_{xx}} = (\mathbf{R_{xx}} - \sigma \, \mathbf{I})^{-1}$ in magnitude. The algorithm is particularly effective when we have a good approximation to an eigenvalue for which we want to compute the eigenvector. By choosing σ very close to a desired eigenvalue, the algorithm can converge very quickly [55]. One advantage of the inverse power method is its ability to converge to any desired eigenvalue closest to σ. For other efficient and fast algorithms on eigenvalue problems the reader should refer to the excellent book [55].

3.5 ROBUST PCA

The learning algorithms discussed in the previous sections (e.g., (3.29), (3.30)) are optimal only for a Gaussian distribution of the input data and they are rather sensitive to impulsive noise or outliers.

Remark 3.4 *It is well known that standard PCA is optimal in the sense of Mean Square Error (MSE). However, the estimation based on MSE is rather sensitive to non Gaussian noise or outliers, so it is not a robust estimator. It is interesting to note that choosing the* 1*-norm (or more generally robust criteria) instead of the* 2*-norm cost function (3.71), we may obtain a more robust estimation of components when signals are corrupted by noise or outliers.*

Many approaches can be taken to increase the robustness of PCA with respect to noise and outliers. Firstly, outlying measurements can be eliminated from the data; secondly outliers can be suppressed or modified by replacing them with more appropriate values; and finally, more robust criteria can be applied. For example, a highly robust and efficient method employs the Minimum Covariance Determinant (MCD) estimator developed and refined by Rousseeuw [1017]. The MCD looks for the subset of K observations (samples) having the smallest determinant of the covariance matrix computed from that subset. Typically, $K \approx 3N/4$ or $K \approx N/2$, where N is the total number of samples. The main disadvantage of the MCD method is that is not suitable for large scale problem (in such applications as image processing and computer vision) in which the estimation of high dimension covariance matrix is not feasible [341]. Moreover, this method discard the entire data vectors. In image processing applications a whole image would be eliminated, just because a few outlying pixels. In this section, we describe in more detail method based on robust M-estimator.

In order to derive a robust algorithm, we can formulate a cost function as:

$$J_{1\rho}(\mathbf{v}_1) \triangleq \boldsymbol{\rho}(\mathbf{e}_1) = \sum_{p=1}^{m} \rho_p(e_{1p}), \tag{3.75}$$

and $\rho_i(e_{1i})$ are real, typically convex functions known in statistics as "robust loss functions." In order to reduce the influence of outliers many different robust loss functions $\rho(e)$ have been proposed. Here, we give only four examples (see Table 2.1) [275]:

1. The absolute value function (i.e., 1-norm criterion)

$$\rho_A(e) = |e| \tag{3.76}$$

2. Huber's function

$$\rho_H(e) = \begin{cases} e^2/2 & \text{for} \quad |e| \le \beta, \\ \beta|e| - \beta^2/2 & \text{for} \quad |e| > \beta, \end{cases} \tag{3.77}$$

3. Talvar's function

$$\rho_T(e) = \begin{cases} e^2/2 & \text{for} \quad |e| \le \beta, \\ \beta^2/2 & \text{for} \quad |e| > \beta, \end{cases} \tag{3.78}$$

4. The logistic function

$$\rho_L(e) = \beta^2 \log(\cosh(e/\beta)), \tag{3.79}$$

where $\beta > 0$ is a problem dependent parameter, called the cut-off parameter (typically $1 \le \beta \le 3$). Typical robust loss (cost) functions and their influence (activation) functions defined by

$$\Psi_p(e_p) \triangleq \frac{\partial \rho_p}{\partial e_p} \tag{3.80}$$

are collected in Table 2.1.

Generally speaking, a suitable choice of the loss function depends on the distribution of the input vector $\mathbf{x}(t)$. Applying a standard gradient descent approach to the energy function (3.75) after some mathematical manipulations, we obtain a learning algorithm (generalization of Equation (3.29)).

$$\frac{dv_{1p}}{dt} = \mu_1 \left[y_1 \Psi_p(e_{1p}) + x_p \sum_{h=1}^{m} v_{1h} \Psi_h(e_{1h}) \right], \tag{3.81}$$

where $\mu_1(t) > 0$ and

$$\Psi_p(e_{1p}) \triangleq \frac{\partial \rho_p(e_1)}{\partial e_{1p}}.$$

The above learning algorithm can be written in matrix form as

$$\frac{d\mathbf{v}_1}{dt} = \mu_1 \left[y_1 \boldsymbol{\Psi}(\mathbf{e}_1) + \mathbf{x}\mathbf{v}_1^T \boldsymbol{\Psi}(\mathbf{e}_1) \right], \tag{3.82}$$

where

$$\boldsymbol{\Psi}(\mathbf{e}_1) \triangleq [\Psi_1(e_{11}), \Psi_2(e_{12}), \dots, \Psi_n(e_{1m})]^T.$$

The simplified (approximated) version of the above learning rule takes the form

$$\frac{d\mathbf{v}_1}{dt} = \mu_1 y_1 \boldsymbol{\Psi}(\mathbf{e}_1) = \mu_1 y_1 \boldsymbol{\Psi}[\mathbf{x} - \mathbf{v}_1 y_1], \tag{3.83}$$

with $\mu_1 > 0$.

By using the self-supervising principle and a cascade hierarchical neural network the above learning rules (3.81), (3.82) and (3.83) can be extended to obtain a number of (higher) PCs. In other words, the learning algorithm for the extraction of the second PC (y_2) corresponding to the second largest eigenvalue $\lambda_2 = E\{y_2 y_2^*\}$, is performed in a similar way as the first component, but we apply the extraction process not directly to the input data $\mathbf{x}$ but to the available errors

$$\tilde{\mathbf{x}_1} \triangleq \mathbf{e}_1 = \mathbf{x} - \hat{\mathbf{x}} = (\mathbf{x} - \mathbf{v}_1 y_1), \tag{3.84}$$

and

$$y_2 \triangleq \mathbf{v}_2^T \mathbf{e}_1. \tag{3.85}$$

In general, the sequence of cost functions can be formulated as

$$J_\rho(\mathbf{v}_i) \triangleq \boldsymbol{\rho}(\mathbf{e}_i) = \sum_{p=1}^{m} \rho_p(e_{ip}), \qquad (i = 1, 2, \dots, n) \tag{3.86}$$

where $\mathbf{e}_i = \mathbf{e}_{i-1} - \mathbf{v}_i y_i$, $y_i = \mathbf{v}_i^T \mathbf{e}_{i-1}$, with $\mathbf{e}_0(t) \triangleq \mathbf{x}(t)$. The minimization of these cost functions by the gradient descent technique leads to an adaptive learning algorithm

$$\frac{d\mathbf{v}_i}{dt} = \mu_i \left[y_i \boldsymbol{\Psi}(e_i) + \mathbf{e}_{i-1} \mathbf{v}_i^T \boldsymbol{\Psi}(\mathbf{e}_i) \right], \tag{3.87}$$

for any $\mathbf{v}_i(0) \neq \mathbf{0}$, $(i = 1, 2, \dots, n)$, where

$$\begin{aligned}
\mu_i(t) &> 0, \\
\boldsymbol{\Psi}(\mathbf{e}_i) &= [\Psi_1(e_{i1}), \Psi_2(e_{i2}), \dots, \Psi_m(e_{im})]^T, \\
\Psi_p(e_{ip}) &= \frac{\partial \rho_p(e_{ip})}{\partial e_{ip}}, \qquad \text{(e.g.,} \qquad \Psi_p(e_{ip}) = \tanh(e_{ip}/\beta)), \\
\mathbf{e}_0(t) &= \mathbf{x}(t).
\end{aligned}$$

Usually, the second term in the right hand side of Equation (3.87) is relatively small and can be neglected[5]. Thus, yielding a simplified version of the learning algorithm for the extraction of the first m PCs:

$$\frac{d\mathbf{v}_i(t)}{dt} = \mu_i(t)\, y_i(t)\boldsymbol{\Psi}[\mathbf{e}_i(t)], \tag{3.88}$$

or in a discrete-time form as

$$\mathbf{v}_i(k+1) = \mathbf{v}_i(k) + \eta_i(k) y_i(k)\boldsymbol{\Psi}[\mathbf{e}_i(k)] \tag{3.89}$$

with $\mathbf{v}_i(0) \neq \mathbf{0}$, $\eta_i(k) > 0$.

3.6 ADAPTIVE LEARNING ALGORITHMS FOR SEQUENTIAL MINOR COMPONENTS EXTRACTION

In contrast to principal components which are directions in which the data have the largest variances the minor components are directions in which the data have the smallest variances. In other words, the minor component analysis (MCA) is the eigenvalue decomposition (EVD) problem of finding the smallest eigenvalues and corresponding eigenvectors.

One would expect that algorithms similar to PCA could be applied, however a simple change of sign of the learning rate causes most of the algorithms to be numerically unstable. Therefore, special stabilizing terms are usually introduced to provide stability for the algorithms. For example, the Amari/Oja learning rule for PCA can be modified for MCA as follows

$$\mathbf{v}(k+1) = \mathbf{v}(k) - \eta(k)[y(k)\mathbf{x}(k) - y^2(k)\mathbf{v}(k) + (\mathbf{v}^T(k)\mathbf{v}(k) - 1)\mathbf{v}(k)], \tag{3.90}$$

where $\mathbf{v}$ is the eigenvector corresponding to the smallest eigenvalue $\lambda_{min} = E\{y^2\} = E\{(\mathbf{v}^T\mathbf{x})^2\} < 1$. It should be noted that the auxiliary penalty term $(\mathbf{v}^T\mathbf{v} - 1)\mathbf{v}$ is added ensuring stability of the algorithm by forcing vector $\mathbf{v}$ to tend towards unit length ($\|\mathbf{v}\|_2 = 1$).

A wide class of MCA algorithms can be derived from the unconstrained minimization problem

$$\min_{\mathbf{v}\in\mathbb{R}^m} r(\mathbf{v})/2 \quad \text{with} \quad r(\mathbf{v}) = \frac{\mathbf{v}^T\mathbf{R}_{\mathbf{xx}}\mathbf{v}}{\mathbf{v}^T\mathbf{v}}. \tag{3.91}$$

Applying the gradient descent approach directly, we obtain a system of nonlinear ordinary differential equations (ODE)

$$\frac{d\mathbf{v}}{dt} = -\mu\,\nabla_v r(\mathbf{v}) = -\mu\,\frac{\partial r(\mathbf{v})}{\partial \mathbf{v}} = -\mu\left[\frac{\mathbf{R}_{\mathbf{xx}}\mathbf{v}\mathbf{v}^T\mathbf{v} - \mathbf{v}(\mathbf{v}^T\mathbf{R}_{\mathbf{xx}}\mathbf{v})}{(\mathbf{v}^T\mathbf{v})^2}\right], \tag{3.92}$$

[5] In fact the second term can be omitted if the actual error $\mathbf{e}_i$ is small compared to the excitation input vector $\mathbf{e}_{i-1}$.

where $\mu(t) > 0$ is a learning rate.

An important property of the above flow is its isonormal property, i.e., the norm of the vector $\mathbf{v}(t)$ is constant over time. This is straightforward to prove due to

$$\frac{d\|\mathbf{v}\|_2^2}{dt} = 2\mathbf{v}^T \frac{d\mathbf{v}}{dt} = 0, \tag{3.93}$$

with initial conditions $\|\mathbf{v}(0)\|_2 \neq \mathbf{0}$. Hence, $\mathbf{v}(t)$ is constant over time with

$$\|\mathbf{v}(t)\|_2 = \|\mathbf{v}(0)\|_2, \qquad \forall t \geqslant 0. \tag{3.94}$$

Without loss of generality, we can assume that $\|\mathbf{v}(0)\|_2 = 1$. This means that the state vector of the above nonlinear system (3.93) evolves on a sphere with unit radius. In fact, the term $\mathbf{v}^T\mathbf{v}$ can be neglected or formally absorbed by a positive learning rate $\widetilde{\mu}(t)$, that is,

$$\frac{d\mathbf{v}}{dt} = -\widetilde{\mu} \left(\mathbf{R_{xx}v} - \mathbf{v}\frac{\mathbf{v}^T\mathbf{R_{xx}v}}{\mathbf{v}^T\mathbf{v}} \right). \tag{3.95}$$

The flow is isonormal and it will converge to the eigenvector corresponding to the minimal eigenvalue of the covariance matrix $\mathbf{R_{xx}}$. This flow can also be interpreted as a special case of Brockett's double bracket flow [105, 106].

On the basis of equation (3.95), several extensions or modifications have been proposed in the literature, which can be written in the general form as follows:

$$\frac{d\mathbf{v}}{dt} = -\mu(t)\, g(\mathbf{v}^T\mathbf{v}) \left(\mathbf{R_{xx}v} - \frac{\mathbf{v}^T\mathbf{R_{xx}v}}{\mathbf{v}^T\mathbf{v}}\mathbf{v} \right) \tag{3.96}$$

or equivalently

$$\frac{d\mathbf{v}}{dt} = -\mu(t)\, \frac{g(\mathbf{v}^T\mathbf{v})}{\mathbf{v}^T\mathbf{v}} \left[\mathbf{R_{xx}vv}^T\mathbf{v} - (\mathbf{v}^T\mathbf{R_{xx}v})\mathbf{v} \right], \tag{3.97}$$

where $g(\mathbf{v}^T\mathbf{v})$ can take various forms, e.g., $(\mathbf{v}^T\mathbf{v}), \mathbf{1}, (\mathbf{v}^T\mathbf{v})^{-1}$.

The discrete-time algorithm can be written in its simplest forms as:

$$\mathbf{v}(k+1) = \mathbf{v}(k) - \eta(k) \left[\mathbf{R}_{\mathbf{xx}}^{(k)}\mathbf{v}(k) - \frac{\mathbf{v}^T(k)\mathbf{R}_{\mathbf{xx}}^{(k)}\mathbf{v}(k)}{\mathbf{v}^T(k)\mathbf{v}(k)}\mathbf{v}(k) \right] \tag{3.98}$$

and its on-line version

$$\mathbf{v}(k+1) = \mathbf{v}(k) - \eta(k) g(\|\mathbf{v}(k)\|_2^2) \left[y(k)\mathbf{x}(k) - \frac{y^2(k)}{\mathbf{v}^T(k)\mathbf{v}(k)}\mathbf{v}(k) \right], \tag{3.99}$$

where $y(k) = \mathbf{v}^T\mathbf{x}(k)$.

Unfortunately, due to numerical approximation the above discrete-time algorithms are unstable (i.e., they can diverge after a large number of iterations unless the normalization to unit length is performed every few iterations). To prevent this instability, the learning

Table 3.3 Basic adaptive learning algorithms for minor component analysis (MCA).

No.	Learning Algorithm	Notes, References
1.	$\bar{\mathbf{v}}_n(k+1) = \mathbf{v}_n(k) - \eta_n(k) y_n(k) \left[\mathbf{x}(k) - y_n(k)\mathbf{v}_n(k)\right]$ $\mathbf{v}_n(k+1) = \bar{\mathbf{v}}_n(k+1)/\lVert\bar{\mathbf{v}}_n(k+1)\rVert_2$ or $\Delta\mathbf{v}_n(k) = -\eta_n(k)\left[y_n(k)\mathbf{x}(k) - y_n^2(k)\mathbf{x}(k) + \right.$ $\left. + \mathbf{v}_n(k)(\mathbf{v}_n^T(k)\mathbf{v}_n(k) - 1)\right]$	Oja (1992) [907]
2.	$\Delta\mathbf{v}_n(k) = -\eta_n(k) g(\mathbf{v}_n^T\mathbf{v}_n)\left[y_n(k)\mathbf{x}(k) - \frac{y_1^2(k)}{\lVert\hat{\mathbf{v}}_n(k)\rVert_2^2}\mathbf{v}_n(k)\right]$ where $g(\mathbf{v}^T\mathbf{v}) = 1$ or $\mathbf{v}_n^T\mathbf{v}_n$ or $(\mathbf{v}_n^T\mathbf{v}_n)^{-1}$	Oja (1991) [907], Luo *et al.* (1997), Cirrincione (1998) [287]
3.	$\Delta\mathbf{v}_n(k) = -\eta_n(k)\left[y_n(k)\mathbf{x}(k) - \frac{y_1^2(k)}{\lVert\mathbf{v}_n(k)\rVert_2^4}\mathbf{v}_n(k)\right]$	Xu (1994) [1303, 1304]
4.	$\Delta\mathbf{v}_n(k) = -\eta_n(k) y_n(k)\left[\lVert\mathbf{v}_n\rVert_2^2\mathbf{x}(k) - y_n(k)\mathbf{v}_n(k)\right]$	Chen, Amari, Lin [183] (1998)
5.	$\Delta\mathbf{v}_n(k) = -\eta_n(k) y_n(k)\left[\lVert\mathbf{v}_n\rVert_2^4\mathbf{x}(k) - y_n(k)\mathbf{v}_n(k)\right]$	Douglas, Kung, [405] Amari (1998) [183]
6.	$\Delta\mathbf{v}_n(k) = -\eta_n(k)\left[y_n(k)\mathbf{x}(k) + \log(\lVert\mathbf{v}_n(k)\rVert_p)\mathbf{v}_n(k)\right]$	
7.	$\Delta\mathbf{v}_n(k) = -\eta_n(k)\left[y_n(k)\mathbf{x}(k) + (d - \lVert\mathbf{v}_1(k)\rVert_p)\mathbf{v}_n(k)\right]$ where $d > \lambda_{max}$	Zhang-Leung (1997)
8.	$\Delta\mathbf{v}_n(k) = -\eta_n(k)\left[\mathbf{v}_n(k) - y_n(k)\mathbf{x}(k)\lVert\mathbf{v}_n\rVert_p\right]$	
9.	$y_i(k) = \mathbf{v}_i^T(k)\mathbf{x}_i(k)$ $\eta_i^{-1}(k) = \gamma\eta_i^{-1}(k-1) + \lvert y_i(k-1)\rvert^2, \quad \eta_i^{-1}(0) = E\{\lvert y_i\rvert^2\}$ $\bar{\mathbf{v}}_i(k+1) = \mathbf{v}_i(k) - \frac{y_i(k)}{\eta_i^{-1}(k)}\left[\mathbf{x}_i(k) - y_i(k)\mathbf{v}_i(k)\right]$ $\bar{\mathbf{v}}_i := \bar{\mathbf{v}}_i - \sum_{j=n}^{i+1}(\bar{\mathbf{v}}_i^T\bar{\mathbf{v}}_j)\bar{\mathbf{v}}_j, \quad \mathbf{v}_i = \bar{\mathbf{v}}_i/\lVert\bar{\mathbf{v}}_i)\rVert_2$ $\mathbf{x}_{i-1}(k) = \mathbf{x}_i(k) + \gamma_n y_i(k)\mathbf{v}_{i*}, \quad \mathbf{x}_n = \mathbf{x}(k)$ $for \;\; i = m, m-1, \ldots, m-n+1$	Sakai and Shimizu (1997) [1030]

rate $\eta(k)$ must decay exponentially to zero or we need to orthonormalize the vector $\mathbf{v}(k)$ after every few iterations by formally taking:

$$\mathbf{v}(k) := \frac{\mathbf{v}(k)}{\|\mathbf{v}(k)\|_2}. \tag{3.100}$$

After extraction of the first minor component, in order to extract the next minor component, instead of eliminating the vector $\mathbf{v} = \mathbf{v}_n$ from the sample covariance matrix, we attempt to make it the greatest principal component of the new covariance matrix defined as

$$\mathbf{R}_{\mathbf{xx}}^{(2)} = \mathbf{R}_{\mathbf{xx}}^{(1)} + \gamma_n \mathbf{v}_n \mathbf{v}_n^T, \tag{3.101}$$

where $\mathbf{R}_{\mathbf{xx}}^{(1)} = E\{\mathbf{x}\mathbf{x}^T\} = \sum_{i=1}^{n} \lambda_i \mathbf{v}_i \mathbf{v}_i^T$ and γ_n is a fixed constant larger that λ_1.

All the above algorithms for MCA are rather slow and their convergence speed strongly depends on the learning rate $\eta(k)$.

Recently, Sakai and Shimizu extended and modified a fast PCA RLS algorithm (3.36) - (3.41) for MCA [1030] (see also ([275]):

$$y_i(k) = \mathbf{v}_i^T(k) y_i(k), \tag{3.102}$$

$$\overline{\mathbf{v}}_i(k+1) = \mathbf{v}_i(k) - \frac{y_i(k)}{\eta_i^{-1}(k)} [\mathbf{x}_i(k) - y_i(k)\mathbf{v}_i(k)], \tag{3.103}$$

$$\eta_i^{-1}(k+1) = \gamma \eta_i^{-1}(k) + |y_i(k)|^2, \tag{3.104}$$

$$\overline{\mathbf{v}}_i := \overline{\mathbf{v}}_i - \sum_{j=m}^{i+1} (\overline{\mathbf{v}}_i^T \overline{\mathbf{v}}_j) \overline{\mathbf{v}}_j, \quad \mathbf{v}_i(k) = \overline{\mathbf{v}}_i(k) \left(\overline{\mathbf{v}}_i^T(k) \overline{\mathbf{v}}_i(k)\right)^{-1/2} \tag{3.105}$$

$$\mathbf{x}_{i-1}(k) = \mathbf{x}_i + \gamma_m y_i(k) \mathbf{v}_i(k) \qquad i = m, m-1, \ldots, \tag{3.106}$$

$$\mathbf{x}_n(k) = \mathbf{x}(k), \tag{3.107}$$

where $\gamma_m > \lambda_1$.

The main difference between this algorithm and the RLS PCA algorithm (3.36) - (3.41) lies in changing the sign of the learning rate, the orthonormalization of vector $\mathbf{v}_i(k)$ in each iteration step and different deflation procedure, which shifts the already extracted minor components to the principal components.

3.7 UNIFIED PARALLEL ALGORITHMS FOR ESTIMATING PRINCIPAL COMPONENTS, MINOR COMPONENTS AND THEIR SUBSPACES

In the previous sections, we have presented simple fast local algorithms which enables us to extract principal and minor components sequentially one by one. In this section, we will present a more general and unified approach to estimate principal and minor components in parallel. Moreover, the algorithms discussed can also be used for principal subspace analysis (PSA) and minor subspace analysis (MSA). When we are interested only in the subspace spanned by the n largest or smallest eigenvectors, we do not need to identify the respective eigenvectors $\mathbf{v}_i$. Any set of $\tilde{\mathbf{v}}_i$'s, which spans the same subspace as the $\mathbf{v}_i$s, is sufficient.

Indeed, when there is multiplicity in the eigenvalues, say $\lambda_1 = \lambda_2$, we cannot obtain $\mathbf{v}_1$ and $\mathbf{v}_2$ uniquely, but we can obtain the subspace spanned by $\mathbf{v}_1$ and $\mathbf{v}_2$. Such a problem is called the principal subspace or minor subspace extraction.

All these problems (PCA, PSA, MCA and MSA) are very similar, including sequential extraction ($n = 1$) as a special case. Therefore, it is desirable to use a general unified principle applicable to all of these problems and obtain algorithms in a unified way. The principle should only elucidate the structure of the PCA and MCA problems, and therby explain most of the algorithms proposed so far, but also guide us in developing unified algorithms.

3.7.1 Cost Function for Parallel Processing

We first explain the idea intuitively. Let $\{\lambda_1, \dots, \lambda_m\}$ and $\{d_1, \cdots, d_m\}$ be two sets of m positive numbers, where $d_1 > \cdots > d_m > 0$. Consider the problem to maximize (or minimize) the sum

$$S = \sum_{i=1}^{m} \lambda_{i'} d_i \tag{3.108}$$

by rearranging the order of $\{\lambda_1, \dots, \lambda_m\}$ as $\{\lambda_{1'}, \dots, \lambda_{m'}\}$, where $\{1', \dots, m'\}$ is a permutation of $\{1, 2, \dots, m\}$. It is straightforward to see that S is maximized when $\{\lambda_{i'}\}$ is arranged in a decreasing order (that is, $\lambda_{1'} > \cdots > \lambda_{m'}$) and is minimized when $\lambda_{1'} < \cdots < \lambda_{m'}$.

Brockett generalized this idea to a matrix calculation [105, 106]. Let $\mathbf{V} = [\mathbf{v}_1, \dots, \mathbf{v}_m]$ be an orthogonal matrix whose columns satisfy

$$\mathbf{v}_i^T \mathbf{v}_j = \delta_{ij} \quad \left(\mathbf{V}^T \mathbf{V} = \mathbf{I}_m\right). \tag{3.109}$$

Let us put

$$J(\mathbf{V}) = \operatorname{tr}\left(\mathbf{D}\mathbf{V}^T \mathbf{R}_{\mathbf{xx}} \mathbf{V}\right) = \operatorname{tr}\left(\mathbf{V}\mathbf{D}\mathbf{V}^T \mathbf{R}_{\mathbf{xx}}\right) \tag{3.110}$$

where $\mathbf{D} = \operatorname{diag}(d_1, \dots, d_m)$. When $\mathbf{V}$ consists of m eigenvectors of $\mathbf{R}_{\mathbf{xx}}$, $\mathbf{V} = [\mathbf{v}_{1'}, \dots, \mathbf{v}_{m'}]$,

$$\mathbf{V}^T \mathbf{R}_{\mathbf{xx}} \mathbf{V} = \operatorname{diag}(\lambda_{1'}, \dots, \lambda_{m'}) \tag{3.111}$$

and Eq. (3.110) reduces to $J(\mathbf{V}) = \sum d_i \lambda_{i'}$. When $\mathbf{V}$ is a general orthogonal matrix, $\mathbf{V}^T \mathbf{R}_{\mathbf{xx}} \mathbf{V}$ is not a diagonal matrix. However, the following proposition holds.

Proposition 3.1 *The cost function $J(\mathbf{V})$ is maximized when*

$$\mathbf{V} = [\mathbf{v}_1, \dots, \mathbf{v}_m] \tag{3.112}$$

and minimized when

$$\mathbf{V} = [\mathbf{v}_m, \dots, \mathbf{v}_1] \tag{3.113}$$

provided the eigenvalues satisfy $\lambda_1 > \cdots > \lambda_m$. $J(\mathbf{V})$ has no local minima nor local maxima except for the global ones.

Now consider the case where $d_1 > \cdots > d_n > d_{n+1} = \cdots = d_m = 0$. In such a case,

$$\begin{aligned}\mathbf{V}\mathbf{D}\mathbf{V}^T &= [\mathbf{v}_1, \ldots, \mathbf{v}_n, \mathbf{v}_{n+1}, \ldots, \mathbf{v}_m] \begin{bmatrix} d_1 & & & \\ & \ddots & & \\ & & d_n & \\ & & & 0 \end{bmatrix} \\ & \quad [\mathbf{v}_1, \ldots, \mathbf{v}_n, \mathbf{v}_{n+1}, \ldots, \mathbf{v}_m]^T \\ &= [\mathbf{v}_1, \ldots, \mathbf{v}_n] \begin{bmatrix} d_1 & & \\ & \ddots & \\ & & d_n \end{bmatrix} [\mathbf{v}_1, \ldots, \mathbf{v}_n]^T, \end{aligned} \tag{3.114}$$

so that the last $(m-n)$ columns of $\mathbf{V}$ automatically vanish. Let us use n arbitrary mutually orthogonal unit vectors $\mathbf{w}_1, \ldots, \mathbf{w}_m$ and put

$$\mathbf{W} = [\mathbf{w}_1, \ldots, \mathbf{w}_n]. \tag{3.115}$$

We have

$$J(\mathbf{W}) = \operatorname{tr}\left(\mathbf{D}\mathbf{W}^T\mathbf{R}_{\mathbf{xx}}\mathbf{W}\right) = \operatorname{tr}\left(\mathbf{W}\mathbf{D}\mathbf{W}^T\mathbf{R}_{\mathbf{xx}}\right), \tag{3.116}$$

which plays the same role as $J(\mathbf{V})$. It is immediately visible that $J(\mathbf{W})$ is maximized (minimized) when $\mathbf{W}$ consists of the eigenvectors of the n largest (smallest) eigenvalues, and there are no local maxima nor minima except for the true solution. So this cost function is applicable for both PCA and MCA as well as PSA and MSA.

When $n = 1$, $J(\mathbf{w})$ reduces to

$$J(\mathbf{w}) = \mathbf{w}^T\mathbf{R}_{\mathbf{xx}}\mathbf{w} \tag{3.117}$$

under the condition that $d_1 = 1$, $\mathbf{w}^T\mathbf{w} = 1$. Hence, this is equivalent to the Rayleigh quotient or its constrained version.

When $d_1 = d_2 = \cdots = d_n = d$, $J(\mathbf{W})$ is maximized (minimized) if $\mathbf{W}$ is composed of the eigenvectors of the n largest (smallest) eigenvalues. There are no other local maxima or minima. Only the subspace, not the exact eigenvectors, can be extracted by maximizing (minimizing) this cost function.

3.7.2 Gradient of $J(\mathbf{W})$

The matrix $\mathbf{W} \in \mathbb{R}^{n\times m}$ satisfies

$$\mathbf{W}^T\mathbf{W} = \mathbf{I}_m. \tag{3.118}$$

A set of such matrices is called the Stiefel manifold $\boldsymbol{O}_{m,n}$. We calculate the gradient of $J(\mathbf{W})$ in $\boldsymbol{O}_{m,n}$. Let $d\mathbf{W}$ be a small change of $\mathbf{W}$, and the corresponding change in $J(\mathbf{W})$ is $dJ = J(\mathbf{W} + d\mathbf{W}) - J(\mathbf{W})$. We have

$$dJ = \operatorname{tr}\left(d\mathbf{W}\ \mathbf{D}\ \mathbf{W}^T\ \mathbf{R}_{\mathbf{xx}}\right) + \operatorname{tr}\left(\mathbf{W}\ \mathbf{D}\ d\mathbf{W}^T\ \mathbf{R}_{\mathbf{xx}}\right). \tag{3.119}$$

From (3.118), we have

$$d\mathbf{W}^T\mathbf{W} + \mathbf{W}^Td\mathbf{W} = 0 \tag{3.120}$$

Hence, from

$$dJ = \frac{\partial J}{\partial \mathbf{W}} \cdot d\mathbf{W}, \tag{3.121}$$

where $\mathbf{A} \cdot \mathbf{B}$ is $\sum_{i,j} a_{ij} b_{ij} = \text{tr}(\mathbf{A}\mathbf{B}^T)$ and the gradient is a matrix given by

$$\begin{aligned} \frac{\partial J}{\partial \mathbf{W}} &= -\mathbf{W}\mathbf{D}\mathbf{W}^T \mathbf{R}_{\mathbf{xx}} \mathbf{W} + \mathbf{R}_{\mathbf{xx}} \mathbf{W}\mathbf{D}\mathbf{W}^T \mathbf{W} \\ &= -\mathbf{W}\mathbf{D}\mathbf{W}^T \mathbf{R}_{\mathbf{xx}} \mathbf{W} + \mathbf{R}_{\mathbf{xx}} \mathbf{W}\mathbf{D}. \end{aligned} \tag{3.122}$$

The gradient method for obtaining the principal components is written as

$$\Delta \mathbf{W}_P = \eta \left(\mathbf{R}_{\mathbf{xx}} \mathbf{W}_P \mathbf{D} - \mathbf{W}_P \mathbf{D} \mathbf{W}_P^T \mathbf{R}_{\mathbf{xx}} \mathbf{W}_P \right) \tag{3.123}$$

and, for the minor components,

$$\Delta \mathbf{W}_M = -\eta \left(\mathbf{R}_{\mathbf{xx}} \mathbf{W}_M \mathbf{D} - \mathbf{W}_M \mathbf{D} \mathbf{W}_M^T \mathbf{R}_{\mathbf{xx}} \mathbf{W}_M \right), \tag{3.124}$$

where $\mathbf{W}_P = \mathbf{V}_\mathcal{S} = [\mathbf{w}_1, \dots, \mathbf{w}_n]$ and $\mathbf{W}_M = \mathbf{V}_\mathcal{N} = [\mathbf{w}_m, \dots, \mathbf{w}_{n+1}]$.

Let us consider the special case with $n = 1$. By putting $d_1 = 1$, Eq. (3.123) reduces to

$$\Delta \mathbf{w} = \eta \left(\mathbf{R}_{\mathbf{xx}} \mathbf{w} - \mathbf{w}\mathbf{w}^T \mathbf{R}_{\mathbf{xx}} \mathbf{w} \right), \tag{3.125}$$

which is a well-known algorithm. Its on-line version is

$$\Delta \mathbf{w} = \eta \left(y\mathbf{x} - y^2 \mathbf{w} \right), \tag{3.126}$$

where we replaced the covariance matrix $\mathbf{R}_{\mathbf{xx}}$ by its instantaneous version $\mathbf{x}^T\mathbf{x}$, and $y = \mathbf{w}^T\mathbf{x}$. This is the classic algorithm found by Amari (1978) and Oja (1982) in which the constraint term imposing $\mathbf{w}^T\mathbf{w} = 1$ is treated separately [15, 907, 911, 1].

In the algorithm given by Brockett $n = m$, while in the one developed by Xu $n < m$ [106, 1303, 1304]. If we put $d_1 = \cdots = d_n = 1$, we obtain the subspace algorithm.

How does the algorithm (3.124) for extracting minor components work? It is obtained from the same cost function, but it uses minimization instead of maximization. Hence, the MCA algorithm changes the sign of the gradient. However, computer simulations show that it does not work. We have shown that $J(\mathbf{W})$ has only one maximum and one minimum. This remained a puzzle for many years and has attracted a lot of attention. This phenomena might be related to the stability of the algorithms. We need a more detailed stability analysis in order to elucidate the structure. Table 3.4 summarizes several parallel algorithms for PCA and PSA while Table 3.5 summarizes algorithms for MSA/MCA.

3.7.3 Stability Analysis

It is easier to replace a finite time difference equation by its continuous time version for analyzing stability. The continuous time versions of (3.123) and (3.124) are

$$\frac{d\mathbf{W}(t)}{dt} = \mu \left(\mathbf{R}_{\mathbf{xx}} \mathbf{W}\mathbf{D} - \mathbf{W}\mathbf{D}\mathbf{W}^T \mathbf{R}_{\mathbf{xx}} \mathbf{W} \right), \tag{3.127}$$

$$\frac{d\mathbf{W}(t)}{dt} = -\mu \left(\mathbf{R}_{\mathbf{xx}} \mathbf{W}\mathbf{D} - \mathbf{W}\mathbf{D}\mathbf{W}^T \mathbf{R}_{\mathbf{xx}} \mathbf{W} \right), \tag{3.128}$$

Table 3.4 Parallel adaptive algorithms for PSA/PCA.

No.	Learning algorithm	Notes, References
1.	$\Delta \mathbf{W} = \eta \left[\mathbf{x}\mathbf{y}^H - \mathbf{W}\mathbf{y}\mathbf{y}^H \right]$	Oja-Karhunen (1985) [911]
2.	$\Delta \mathbf{W} = \eta \left[\mathbf{R}_{\mathbf{xx}}\mathbf{W}\mathbf{D} - \mathbf{W}\mathbf{D}\mathbf{W}^H\mathbf{R}_{\mathbf{xx}}\mathbf{W}\mathbf{D} \right]$ $\cong \eta \left[\mathbf{x}\mathbf{y}^H\mathbf{D} - \mathbf{W}\mathbf{D}\mathbf{y}\mathbf{y}^H\mathbf{D} \right]$	Brockett (1991) [106]
3.	$\Delta \mathbf{W} = \eta \left[\mathbf{R}_{\mathbf{xx}}\mathbf{W}\mathbf{W}^H - \mathbf{W}\mathbf{W}^H\mathbf{R}_{\mathbf{xx}} \right] \mathbf{W}$ $\cong \eta \left[\mathbf{x}\mathbf{y}^H\mathbf{W}^H\mathbf{W} - \mathbf{W}\mathbf{y}\mathbf{y}^H \right]$	Chen, Amari (1998) [183]
4.	$\Delta \mathbf{W} = \eta \left[\mathbf{R}_{\mathbf{xx}}\mathbf{W}\mathbf{D}\mathbf{W}^H - \mathbf{W}\mathbf{D}\mathbf{W}^H\mathbf{R}_{\mathbf{xx}} \right] \mathbf{W}$	Chen, Amari (2001) [182]

For the diagonal matrix $\mathbf{D}$ with positive-valued strictly decreasing entries the Brockett and Chen-Amari algorithms perform parallel adaptive PCA [106, 182, 183].

where we omitted suffixes P and M.

We assume here that $\mathbf{W}$ is a general $n \times m$ matrix, not necessarily belonging to $\boldsymbol{O}_{m,n}$, and we put

$$\boldsymbol{U}(t) = \mathbf{W}^T(t)\mathbf{W}(t). \tag{3.129}$$

We can prove that $\boldsymbol{U}(t)$ is invariant under the gradient dynamics (3.127) and (3.128), that is

$$\frac{d}{dt}\boldsymbol{U}(t) = \mathbf{0} \tag{3.130}$$

when $\mathbf{W}(t)$ changes under the dynamics of (3.127) or (3.128). This can be proved by the direct calculations,

$$\begin{aligned} \frac{d}{dt}\boldsymbol{U}(t) &= \frac{d\mathbf{W}^T}{dt}\mathbf{W} + \mathbf{W}^T\frac{d\mathbf{W}}{dt} \\ &= \mu\,(\mathbf{I} - \mathbf{W}^T\mathbf{W})(\mathbf{D}\mathbf{W}^T\mathbf{R}_{\mathbf{xx}}\mathbf{W} + \mathbf{W}^T\mathbf{R}_{\mathbf{xx}}\mathbf{W}\mathbf{D}) = \mathbf{0} \end{aligned} \tag{3.131}$$

with the initial condition: $\boldsymbol{U}(0) = \mathbf{W}^T(0)\mathbf{W}(0) = \mathbf{I}$. Therefore, when $\mathbf{W}(0)$ at time $t = 0$ belongs to the Stiefel manifold $\boldsymbol{O}_{m,n}$, it holds, $\boldsymbol{U}(0) = \mathbf{I}$, and hence, $\boldsymbol{U}(t) = \mathbf{I}$ for any t, implying that $\mathbf{W}(t)$ always belongs to $\boldsymbol{O}_{m,n}$. Since any global extremum gives the true solution of n principal or minor components, the dynamical equation (3.123) or (3.124) should converge to the true solution, provided $\boldsymbol{W}(t)$ always belongs to $\boldsymbol{O}_{m,n}$.

If computer simulations show that (3.124) does not work for minor components extraction, the dynamics (3.124) defined in $\boldsymbol{O}_{m,n}$ are stable in $\boldsymbol{O}_{m,n}$ but unstable when extended to the entire space of $\mathbb{R}^{m \times n}$. In other words, if $\mathbf{W}(t)$ deviates to $\mathbf{W}(t) + d\mathbf{W}$ outside $\boldsymbol{O}_{m,n}$ due to noise or numerical round-off, the deviation grows and $\mathbf{W}(t)$ escapes from $\boldsymbol{O}_{m,n}$.

To show the above scenario, let $\delta\mathbf{W}$ be a small deviation of $\mathbf{W} \in \boldsymbol{O}_{m,n}$ in any direction. It is known that the deviation in general can be decomposed into three terms

$$\delta\mathbf{W} = \mathbf{W}\delta\boldsymbol{S} + \mathbf{W}\delta\boldsymbol{G} + \boldsymbol{N}\delta\boldsymbol{B}, \tag{3.132}$$

where $\delta\boldsymbol{S}$ is an $n \times n$ skew symmetric matrix, $\delta\boldsymbol{G}$ is an $n \times n$ symmetric matrix, $\delta\boldsymbol{B}$ is an $(m-n) \times n$ matrix, and $\boldsymbol{N}$ is an $m \times (m-n)$ matrix whose $(m-n)$ columns are orthogonal to the columns of $\boldsymbol{W}$. The current $\mathbf{W}$ consists of n orthonormal vectors $\{\mathbf{w}_1, \ldots, \mathbf{w}_n\}$ which define the n-dimensional subspace. The first term $\mathbf{W}\delta\boldsymbol{S}$ represents a change of $\mathbf{w}_i$ to $\mathbf{w}_i + \delta\mathbf{w}_i$, keeping the subspace spanned by the $\mathbf{w}_i$s invariant although its orthonormal basis vectors change. The third term $\boldsymbol{N}\delta\boldsymbol{B}$ alters $\mathbf{w}_i$ from the subspace, so that it alters the directions of the subspace spanned by $\mathbf{W}$ but still staying in $\boldsymbol{O}_{m,n}$. The second term $\mathbf{W}\delta\boldsymbol{G}$ destroys the orthonormality of $\mathbf{W}$, so that it represents changes of $\mathbf{W}$ in the direction orthogonal to $\boldsymbol{O}_{m,n}$. Hence, when the dynamics are restricted only to inside $\boldsymbol{O}_{m,n}$, this term is always 0.

When $\mathbf{W}(t)$ deviates to $\mathbf{W}(t) + \delta\mathbf{W}(t)$, how does such change $\delta\mathbf{W}(t)$ develop through the dynamics. The dynamics of $\delta\mathbf{W}(t)$ are given by the variational equation

$$\begin{aligned} \frac{d}{dt}\,\delta\,\mathbf{W}(t) &= \pm\,\mu\,\delta\,(\mathbf{R_{xx}}\mathbf{W}\mathbf{D} - \mathbf{W}\mathbf{D}\mathbf{W}^T\mathbf{R_{xx}}\mathbf{W}) \\ &= \pm\,\mu,\,(\mathbf{R_{xx}}\delta\,\mathbf{W}\mathbf{D} - \delta\,\mathbf{W}\mathbf{D}\mathbf{W}^T\mathbf{R_{xx}}\mathbf{W} - \mathbf{W}\mathbf{D}\delta\,\mathbf{W}^T\mathbf{R_{xx}}\mathbf{W}). \end{aligned} \tag{3.133}$$

We analyze the variational equation in the neighborhood of the true solution, where the variation $\delta\mathbf{W}$ is not only inside $\boldsymbol{O}_{m,n}$ but also in the orthogonal directions.

The results are summarized as follows (see Chen and Amari, 2001; Chen *et al.* 1998) [182, 183].

1. The variational equations are stable at the true solution with respect to changes $\delta\boldsymbol{S}$ and $\delta\boldsymbol{B}$ inside $\boldsymbol{O}_{m,n}$, when all d_is are different and all λ_i s are different.

2. When some d_is or some λ_is are equal, the variational equations are stable with respect to $\delta\boldsymbol{B}$ but only neutrally stable with respect to $\delta\boldsymbol{S}$.

3. The variational equation is stable at the true solution concerning changes in $\delta\boldsymbol{G}$ for principal component extraction (3.127), but is unstable for minor component extraction (3.128).

Result 1 shows that the gradient dynamics (3.127) and (3.128) are successful in obtaining the principal and minor components, respectively, in parallel, provided $\mathbf{W}(t)$ is exactly controlled to belong to $\boldsymbol{O}_{n,m}$.

Result 2 shows that the algorithms are successful for extracting principal and minor subspaces, under the same condition when some d_is or λ_is are equal.

Result 3 shows that algorithm (3.127) is successful in extracting principal components, but algorithm (3.128) fails in extracting minor components due to the instability of the

Table 3.5 Adaptive parallel MSA/MCA algorithms for complex valued data.

No.	Learning rule	Notes, References
1.	$\Delta \mathbf{W} = -\eta(\mathbf{R_{xx}}\mathbf{W}\mathbf{W}^H - \mathbf{W}\mathbf{W}^H\mathbf{R_{xx}})\mathbf{W} =$ $\cong -\eta[\mathbf{x}\mathbf{y}^H\mathbf{W}^H\mathbf{W} - \mathbf{W}\mathbf{y}\mathbf{y}^H]$	Chen, Amari, Lin (1998) [183]
2.	$\Delta \mathbf{W} = -\eta(\mathbf{R_{xx}}\mathbf{W}\mathbf{D}\mathbf{W}^H - \mathbf{W}\mathbf{D}\mathbf{W}^H\mathbf{R_{xx}})\mathbf{W}+$ $+\mathbf{W}(\mathbf{D} - \mathbf{W}^H\mathbf{D}\mathbf{W})$	Chen, Amari (2001) [182]
3.	$\Delta \mathbf{W} = -\eta\,[\mathbf{R_{xx}}\mathbf{W}\mathbf{W}^H\mathbf{W}\mathbf{W}^H\mathbf{W} - \mathbf{W}\mathbf{W}^H\mathbf{R_{xx}}\mathbf{W}] =$ $\cong -\eta\,[\mathbf{x}\,\mathbf{y}^H\,\mathbf{W}^H\,\mathbf{W}\,\mathbf{W}^H\,\mathbf{W} - \mathbf{W}\,\mathbf{y}\,\mathbf{y}^H]$	Douglas, Kung, Amari (1998) [405]
4.	$\mathbf{y}(k) = \mathbf{W}^H(k)\mathbf{x}(k)$ $\hat{\mathbf{x}}(k) = \mathbf{W}(k)\mathbf{y}(k)$ $\mathbf{e}(k) = \mathbf{x}(k) - \hat{\mathbf{x}}(k)$ $\alpha(k) = (1 + \eta^2\,\|\mathbf{e}(k)\|_2^2\|\mathbf{y}(k)\|_2^2)^{-1/2}$ $\beta(k) = (\alpha(k) - 1)/\|\mathbf{y}(k)\|_2^2$ $\bar{\mathbf{e}}(k) = -\beta(k)\,\hat{\mathbf{x}}(k)/\eta + \alpha(k)\,\mathbf{e}(k)$ $(\Delta\mathbf{W}(k) = -\eta\,\bar{\mathbf{e}}(k)\mathbf{y}^H(k))$ or $\mathbf{u}(k) = \bar{\mathbf{e}}(k)\,/\|\bar{\mathbf{e}}(k)\|_2$ $\mathbf{z}(k) = \mathbf{W}^H(k)\,\mathbf{u}(k)$ $\Delta\mathbf{W}(k) = -2\mathbf{u}(k)\,\mathbf{z}^H(k)$	Orthogonal Algorithm [1] Abed-Meraim *et al.* (2000)

For $\mathbf{D} = \mathbf{I}$ the Chen-Amari algorithm can perform MSA [182, 183], however, for matrix $\mathbf{D}$ with positive strictly decreasing entries, the algorithm performs stable MCA.

directions orthogonal to $\boldsymbol{O}_{n,m}$. Such deviations can be caused by noise or numerical round-off, so $\boldsymbol{W}(t)$ should be adjusted in each step to satisfy (3.118).

3.7.4 Unified Stable Algorithms

In order to overcome the instability of minor components extraction, Chen *et al.* (1998) added a term which forces $\mathbf{W}(t)$ to return to $\boldsymbol{O}_{m,n}$ [183]. The algorithms of principal and minor components extraction differ only in the signs:

$$\frac{d\mathbf{W}}{dt} = \mu\left(\mathbf{R_{xx}}\mathbf{W}\mathbf{D}\mathbf{W}^T\mathbf{W} - \mathbf{W}\mathbf{D}\mathbf{W}^T\mathbf{R_{xx}}\mathbf{W}\right) \tag{3.134}$$

for PCA,

$$\frac{d\mathbf{W}}{dt} = -\mu\left(\mathbf{R_{xx}}\mathbf{W}\mathbf{D}\mathbf{W}^T\mathbf{W} - \mathbf{W}\mathbf{D}\mathbf{W}^T\mathbf{R_{xx}}\mathbf{W}\right) \tag{3.135}$$

for MCA. Here,

$$\mathbf{W}^T(t)\mathbf{W}(t) = \mathbf{I}_n \tag{3.136}$$

is not necessarily guaranteed in the course of dynamics, although we choose the initial value to satisfy $\mathbf{W}^T(0)\mathbf{W}(0) = \mathbf{I}_n$. When (3.118) holds, (3.134) is the same as Xu's algorithm [1303, 1304]. However, it cannot be applied to MCA just by changing the sign, while (3.135) works well for MCA.

However, the dynamics of (3.134) and (3.135) are neutrally stable with respect to changes in $\mathbf{W}\delta\boldsymbol{G}$. Douglas *et al.* performed detailed numerical simulations and showed that the discretized version of the algorithm does not work when $\mathbf{R_{xx}}$ is replaced by $\mathbf{x}(t)\mathbf{x}^T(t)$ [404, 405, 406]. They proposed to strengthen the term to return to $\boldsymbol{O}_{m,n}$.

Taking this into account, Chen and Amari proposed the following algorithms [182]

$$\frac{d\mathbf{W}}{dt} = \mu\left[(\mathbf{R_{xx}}\mathbf{W}\mathbf{D}\mathbf{W}^T\mathbf{W} - \mathbf{W}\mathbf{D}\mathbf{W}^T\mathbf{R_{xx}}\mathbf{W}) + \mathbf{W}(\mathbf{D} - \mathbf{W}\mathbf{D}\mathbf{W}^T)\right] \tag{3.137}$$

for PCA, and

$$\frac{d\mathbf{W}}{dt} = -\mu\left[(\mathbf{R_{xx}}\mathbf{W}\mathbf{D}\mathbf{W}^T\mathbf{W} - \mathbf{W}\mathbf{D}\mathbf{W}^T\mathbf{R_{xx}}\mathbf{W}) + \mathbf{W}(\mathbf{D} - \mathbf{W}\mathbf{D}\mathbf{W}^T)\right] \tag{3.138}$$

for MCA.

It is interesting to show that almost all algorithms proposed so far are induced by modifying the penalty term. See the discussion by Chen and Amari for details [182].

Remark 3.5 *Let λ_0 be an upper bound of all λ_i that is a constant larger than λ_1. When we know a bound λ_0, we can define*

$$\bar{\mathbf{R}}_{\mathbf{xx}} = \lambda_0\,\mathbf{I} - \mathbf{R_{xx}}. \tag{3.139}$$

Then, the eigenvalues of $\bar{\mathbf{R}}_{\mathbf{xx}}$ are $\lambda_0 - \lambda_1, \lambda_0 - \lambda_2, \ldots, \lambda_0 - \lambda_m$. Hence, by performing PCA on $\bar{\mathbf{R}}_{\mathbf{xx}}$, we can obtain minor components and their eigenvectors. This was pointed out by Chen, Amari and Murata [184].

It is well known that the principal components of $\mathbf{R}_{\mathbf{xx}}^{-1}$ correspond to the minor components of $\mathbf{R_{xx}}$. Hence any PCA algorithm can be used for MCA if we can calculate $\mathbf{R}_{\mathbf{xx}}^{-1}$, but will cost an additional matrix inversion.

3.8 SINGULAR VALUE DECOMPOSITION IN RELATION TO PCA AND FUNDAMENTAL MATRIX SUBSPACES

The singular value decomposition (SVD) is a tool of both practical and theoretical importance in signal processing and identification problems. The SVD of a real valued matrix

$\mathbf{X} = [\mathbf{x}(1), \mathbf{x}(2), \ldots, \mathbf{x}(N)] \in \mathbb{R}^{m \times N}$ $(m \geq n)$ is given by

$$\mathbf{X} = \mathbf{U}\,\mathbf{\Sigma}\,\mathbf{V}^T, \tag{3.140}$$

where $\mathbf{U} \in \mathbb{R}^{m \times m}$ and $\mathbf{V} \in \mathbb{R}^{N \times N}$ are orthogonal matrices and $\mathbf{\Sigma} \in \mathbb{R}^{m \times N}$ is a pseudo-diagonal matrix whose top n rows contain $\mathbf{\Sigma}_{\mathcal{S}} = \text{diag}\{\sigma_1, \ldots, \sigma_n\}$ (with non-negative diagonal entries ordered from the largest to the smallest) and whose bottom $(m-n)$ rows are zero. Note that only n singular values σ_i are non zero and for the full rank matrix $\mathbf{X}$, $n = m$.

For noiseless data, we can use the following decomposition

$$\mathbf{X} = [\mathbf{U}_{\mathcal{S}}, \mathbf{U}_{\mathcal{N}}] \begin{bmatrix} \mathbf{\Sigma}_{\mathcal{S}} & \mathbf{0} \\ \mathbf{0} & \mathbf{0} \end{bmatrix} [\mathbf{V}_{\mathcal{S}}, \mathbf{V}_{\mathcal{N}}]^T, \tag{3.141}$$

where $\mathbf{U}_{\mathcal{S}} = [\mathbf{u}_1, \ldots, \mathbf{u}_n] \in \mathbb{R}^{m \times n}$, $\mathbf{\Sigma}_{\mathcal{S}} = \text{diag}\{\sigma_1, \ldots, \sigma_n\}$ and $\mathbf{U}_{\mathcal{N}} = [\mathbf{u}_{n+1}, \ldots, \mathbf{u}_m]$. The set of matrices $\{\mathbf{U}_{\mathcal{S}}, \mathbf{\Sigma}_{\mathcal{S}}, \mathbf{V}_{\mathcal{S}}\}$ represents this signal subspace and the set of matrices $\{\mathbf{U}_{\mathcal{N}}, \mathbf{\Sigma}_{\mathcal{N}}, \mathbf{V}_{\mathcal{N}}\}$ represents the null subspace or, in practice for noisy data, the noise subspace. The n columns of $\mathbf{U}$ corresponding to these non-zero singular values that span the column space of $\mathbf{X}$ are called the left singular vectors. Similarly, the n columns of $\mathbf{V}$ are called the right singular vectors and they span the row space of $\mathbf{X}$. Using these terms, the SVD of $\mathbf{X}$ can be written in a more compact size:

$$\mathbf{X} = \mathbf{U}_{\mathcal{S}}\,\mathbf{\Sigma}_{\mathcal{S}}\,\mathbf{V}_{\mathcal{S}}^T = \sum_{i=1}^{n} \sigma_i \mathbf{u}_i \mathbf{v}_i^T. \tag{3.142}$$

and we also have

$$\begin{aligned} \mathbf{X}\mathbf{v}_i &= \sigma_i \mathbf{u}_i, \\ \mathbf{X}^T\mathbf{u}_i &= \sigma_i \mathbf{v}_i. \end{aligned} \tag{3.143}$$

Perturbation theory for the SVD is partially based on the link between the SVD and the PCA and eigenvalue decomposition. It is evident that from the SVD of matrix $\mathbf{X} = \mathbf{U}\,\mathbf{\Sigma}\,\mathbf{V}^T$ with rank $n \leq m \leq N$, we have

$$\mathbf{X}\mathbf{X}^T = \mathbf{U}\mathbf{\Sigma}_1^2\mathbf{U}^T, \tag{3.144}$$

$$\mathbf{X}^T\mathbf{X} = \mathbf{V}\,\mathbf{\Sigma}_2^2\,\mathbf{V}^T, \tag{3.145}$$

where $\mathbf{\Sigma}_1 = \text{diag}\{\sigma_1, \ldots, \sigma_m\}$ and $\mathbf{\Sigma}_2 = \text{diag}\{\sigma_1, \ldots, \sigma_N\}$. This means that the singular values of $\mathbf{X}$ are the positive square roots of the eigenvalues of $\mathbf{X}\mathbf{X}^T$ and the eigenvectors $\mathbf{U}$ of $\mathbf{X}\mathbf{X}^T$ are the left singular vectors of $\mathbf{X}$. Note that if $m < N$, the matrix $\mathbf{X}^T\mathbf{X}$ will contain at least $N - m$ additional eigenvalues that are not included as singular values of $\mathbf{X}$.

As we discussed earlier, an estimate $\hat{\mathbf{R}}_{\mathbf{xx}}$ of the covariance matrix corresponding to a set of observed vectors $\mathbf{x}(k) \in \mathbb{R}^m$ may be computed as $\hat{\mathbf{R}}_{\mathbf{xx}} = (1/N)\sum_{k=1}^{N} \mathbf{x}(k)\mathbf{x}^T(k)$. An alternate and equivalent way of computing $\widehat{\mathbf{R}}_{\mathbf{xx}}$ is to form a data matrix $\mathbf{X} = [\mathbf{x}(1), \mathbf{x}(2), \ldots, \mathbf{x}(N)]$ $\in \mathbb{R}^{m \times N}$ and represent the estimated covariance matrix by

$$\hat{\mathbf{R}}_{\mathbf{xx}} = \frac{1}{N}\mathbf{X}\mathbf{X}^T. \tag{3.146}$$

Hence, the eigenvectors of the sample covariance matrix $\hat{\mathbf{R}}_{\mathbf{xx}}$ are the left singular vectors $\mathbf{U}$ of $\mathbf{X}$ and the singular values σ_i of $\mathbf{X}$ are the positive square roots of the eigenvalues of $\hat{\mathbf{R}}_{\mathbf{xx}}$.

From this discussion it follows that all the algorithms discussed in this chapter for PCA and MCA can be applied (after some simple tricks) to the SVD of arbitrary matrix $\mathbf{X} = [\mathbf{x}(1), \mathbf{x}(2), \ldots, \mathbf{x}(N)]$ without any need to directly compute or estimate the covariance matrix. The opposite is also true; the PCA or EVD of the covariance matrix $\hat{\mathbf{R}}_{\mathbf{xx}}$ can be performed via the SVD numerical algorithms. However, for large matrices $\mathbf{X}$, the SVD algorithms become usually more costly, than the relatively efficient and fast PCA adaptive algorithms. Several reliable and efficient numerical algorithms for the SVD do however, exist [305].

Now suppose that the data matrix $\mathbf{X}$ is perturbed by some noise matrix $\boldsymbol{\mathcal{N}}$, such that $\tilde{\mathbf{X}} = \mathbf{X} + \boldsymbol{\mathcal{N}}$. The entries of $\boldsymbol{\mathcal{N}}$ are generated by an uncorrelated, zero mean, white noise process with variance $\sigma_{\mathcal{N}}^2$ so that the covariance matrix of noise is given by $E\{\boldsymbol{\mathcal{N}}\boldsymbol{\mathcal{N}}^T/N\} = \sigma_{\mathcal{N}}^2 \mathbf{I}_m$. Under these conditions, we have [1217]

$$E\{\tilde{\mathbf{X}}\tilde{\mathbf{X}}^T/N\} = E\{\mathbf{X}\mathbf{X}^T/N\} + \sigma_{\mathcal{N}}^2 \mathbf{I}_m, \tag{3.147}$$

so that for large p, the SVD of the noisy matrix $\tilde{\mathbf{X}}$ is approximated by

$$\tilde{\mathbf{X}} \approx \mathbf{U}(\boldsymbol{\Sigma}_1^2 + N\sigma_{\mathcal{N}}^2 \mathbf{I}_m)^{1/2} \tilde{\mathbf{V}}^T \tag{3.148}$$

for some orthogonal matrix $\tilde{\mathbf{V}}$. This expression shows that, for large N and small noise variance $\sigma_{\mathcal{N}}^2$, the subspace spanned by the left singular vectors and singular values of the perturbed covariance matrix $E\{\mathbf{X}\mathbf{X}^T/N\}$ is relatively insensitive to the added perturbations in the entries of the matrix $\mathbf{X}$. Therefore, the SVD is a robust and numerically reliable approach. Moreover, the singular values of $\tilde{\mathbf{X}}$ increase by an amount approximately equal to $\sigma_{\mathcal{N}}\sqrt{N}$ while the left singular vectors remain the same as for a noiseless matrix $\mathbf{X}$. Furthermore, the matrix $\tilde{\mathbf{X}}$ is now a full rank one and its $(m - n)$ smallest singular values are no longer zero, but now equal to $\sigma_{\mathcal{N}}\sqrt{N}$. In theory, we can recover the noiseless matrix $\mathbf{X}\mathbf{X}^T$ by subtracting the term $N\sigma_{\mathcal{N}}^2 \mathbf{I}_m$ from $\boldsymbol{\Sigma}_1$. However, it is impossible to recover matrix $\mathbf{V}$ or $\mathbf{X}$ because the length of the columns of $\mathbf{V}$ is equal to N and hence these vectors do not participate in the averaging effect of increasing N [1217].

3.9 MULTISTAGE PCA FOR BLIND SOURCE SEPARATION OF COLORED SOURCES

It is easy to show that, under some mild conditions, we can perform blind separation of source signals with a temporal structure using two-stage or multistage PCA.

Let us consider the instantaneous mixing model:

$$\mathbf{x}(k) = \mathbf{H}\,\mathbf{s}(k) + \boldsymbol{\nu}(k), \tag{3.149}$$

where $\mathbf{x}(k) \in \mathbb{R}^m$ is an available vector of sensor signals, $\mathbf{H} \in \mathbb{R}^{m \times n}$ is an unknown full rank mixing matrix, with $m \geq n$, $\mathbf{s}(k) \in \mathbb{R}^n$ is the vector of colored source signals and $\boldsymbol{\nu}(k) \in \mathbb{R}^m$ is a vector of independent white noise signals.

In the first stage, we perform the standard PCA for the vector $\mathbf{x}(k)$, using formal eigenvalue decomposition of the covariance matrix [349]:

$$\mathbf{R_{xx}} = \mathbf{V}\,\boldsymbol{\Lambda}\,\mathbf{V}^T. \tag{3.150}$$

On the basis of dominant (largest) eigenvalues, we perform the spatial whitening procedure

$$\overline{\mathbf{x}}(k) = \mathbf{Q}\,\mathbf{x}(k) = \boldsymbol{\Lambda}_{\mathcal{S}}^{-1/2}\,\mathbf{V}_{\mathcal{S}}^T\,\mathbf{x}(k), \tag{3.151}$$

where $\boldsymbol{\Lambda}_{\mathcal{S}} = \text{diag}\{\lambda_1, \lambda_2, \ldots, \lambda_n\}$ with $\lambda_1 \geq \lambda_2 \geq \cdots \geq \lambda_n$ and $\mathbf{V}_{\mathcal{S}} = [\mathbf{v}_1, \mathbf{v}_2, \ldots, \mathbf{v}_n] \in \mathbb{R}^{n\times m}$.

In the second stage, we can perform PCA for a new vector of signals defined by [349, 217, 218]

$$\widetilde{\mathbf{x}}(k) = \overline{\mathbf{x}}(k) + \overline{\mathbf{x}}(k-p), \tag{3.152}$$

where p is an arbitrary time delay (typically, $p = 1$). It is interesting to note that the covariance matrix of the vector $\widetilde{\mathbf{x}}(k)$ can be expressed as

$$\mathbf{R}_{\widetilde{\mathbf{x}}\,\widetilde{\mathbf{x}}} = \mathbf{R}_{\widetilde{\mathbf{x}}}(0) = E\{\widetilde{\mathbf{x}}(k)\,\widetilde{\mathbf{x}}^T(k)\} = 2\,\mathbf{R}_{\overline{\mathbf{x}}}(0) + \mathbf{R}_{\overline{\mathbf{x}}}(p) + \mathbf{R}_{\overline{\mathbf{x}}}^T(p), \tag{3.153}$$

where

$$\mathbf{R}_{\overline{\mathbf{x}}\,\overline{\mathbf{x}}} = \mathbf{R}_{\overline{\mathbf{x}}}(0) = E\{\overline{\mathbf{x}}(k)\,\overline{\mathbf{x}}^T(k)\} = \mathbf{A}\,\mathbf{R_{ss}}\,\mathbf{A}^T = \mathbf{I} \tag{3.154}$$

under the assumption that $\mathbf{A} = \mathbf{QH}$ is orthogonal and $\mathbf{R_{ss}} = \mathbf{I}$ and

$$\mathbf{R}_{\overline{\mathbf{x}}}(p) = E\{\overline{\mathbf{x}}(k)\,\overline{\mathbf{x}}^T(k-p)\} = \mathbf{A}\,\mathbf{R_s}(p)\,\mathbf{A}^T. \tag{3.155}$$

Hence, we obtain the matrix decomposition

$$\mathbf{R}_{\widetilde{\mathbf{x}}\,\widetilde{\mathbf{x}}} = \mathbf{A}\,\mathbf{D}(p)\,\mathbf{A}^T = \mathbf{V}_{\widetilde{\mathbf{x}}}\,\boldsymbol{\Lambda}_{\widetilde{\mathbf{x}}}\,\mathbf{V}_{\widetilde{\mathbf{x}}}^T, \tag{3.156}$$

where the $\mathbf{D}(p)$ is a diagonal matrix expressed as

$$\mathbf{D}(p) = 2\,\mathbf{I} + \mathbf{R_s}(p) + \mathbf{R_s}^T(p),$$

with diagonal elements $d_{ii}(p) = 2(1+E\{s_i(k)\,s_i(k-p)\}$. If the diagonal elements are distinct (i.e., $E\{s_i(k)\,s_i(k-p)\} \neq E\{s_j(k)\,s_j(k-p)\}, \ \ \forall i \neq j$), then the eigenvalue decomposition is unique up to the permutation and sign of the eigenvectors. Thus, the mixing matrix can be estimated as $\mathbf{H} = \mathbf{Q}^+\,\mathbf{V}_{\widetilde{\mathbf{x}}}$ and the source signals can be estimated as

$$\hat{\mathbf{s}}(k) = \mathbf{V}_{\widetilde{\mathbf{x}}}^T\,\overline{\mathbf{x}}(k) = \mathbf{V}_{\widetilde{\mathbf{x}}}^T\,\mathbf{Q}\,\mathbf{x}(k). \tag{3.157}$$

If some of the eigenvalues of the diagonal matrix $\boldsymbol{\Lambda}_{\widetilde{\mathbf{x}}}$ are very close to each other the performance of separation can be poor. In such cases, we can try to repeat the last step for different time delays until all eigenvalues are distinct.

The above described procedure belongs to a wide class of second order statistics (SOS) techniques [1158, 84, 349, 217, 218]. More advanced and improved algorithms for BSS of colored sources with different auto-correlation functions based on SOS and spatio-temporal blind decorrelation will be described in Chapter 4.

It should be noted that the above PCA algorithm can perform only BSS of colored sources with different temporal structure or equivalently with different power spectra. In order to perform ICA, we can alternatively apply nonlinear PCA which minimizes the following cost function [589]

$$J(\mathbf{W}) = E\{\|\overline{\mathbf{x}} - \mathbf{W}\mathbf{g}(\mathbf{W}^T\,\overline{\mathbf{x}})\|_2^2\} \tag{3.158}$$

or equivalently

$$\begin{aligned} J(\mathbf{w}_1, \mathbf{w}_2, \ldots, \mathbf{w}_n) &= E\{\|\overline{\mathbf{x}} - \sum_{i=1}^{n} \mathbf{w}_i\, g_i(y_i)\|_2^2\} \\ &= \sum_{i=1}^{n} E\{[y_i - g_i(y_i)]^2\}, \end{aligned} \tag{3.159}$$

where $\overline{\mathbf{x}} = \mathbf{Q}\,\mathbf{x} = \mathbf{W}\,\mathbf{y}$, $\mathbf{y} = \mathbf{W}^T\overline{\mathbf{x}}$, $\mathbf{w}_i$ is the i-th vector of the orthogonal matrix $\mathbf{W}$, $y_i(k) = \mathbf{w}_i^T\,\overline{\mathbf{x}}(k)$ and $g_i(y_i)$ are suitably chosen nonlinear functions, e.g., $g_i(y_i) = y_i + \mathrm{sign}(y_i)y_i^2$ or $g_i(y_i) = \tanh(\beta y_i)$.

There are at least several algorithms that can perform efficient minimization of the above cost function in order to estimate the separating matrix $\mathbf{W}$. The modified recursive least squares (RLS) method leads to the following algorithm [589]

$$\mathbf{q}(k) = \mathbf{g}[\mathbf{y}(k)] = \mathbf{g}[\mathbf{W}^T(k)\,\overline{\mathbf{x}}(k)], \tag{3.160}$$

$$\mathbf{e}(k) = \overline{\mathbf{x}}(k) - \mathbf{W}(k)\,\mathbf{q}(k), \tag{3.161}$$

$$\mathbf{m}(k) = \frac{\mathbf{P}(k)\,\mathbf{q}(k)}{\gamma + \mathbf{q}^T(k)\,\mathbf{P}(k)\,\mathbf{q}(k)}, \tag{3.162}$$

$$\mathbf{P}(k+1) = \frac{1}{\gamma}\mathrm{Tri}\,[\mathbf{P}(k) - \mathbf{m}(k)\,\mathbf{q}^T(k)\,\mathbf{P}^T(k)], \tag{3.163}$$

$$\mathbf{W}(k+1) = \mathbf{W}(k) + \mathbf{e}(k)\,\mathbf{m}^T(k), \tag{3.164}$$

with nonzero initial conditions, typically $\mathbf{W}(0) = \mathbf{P}(0) = \mathbf{I}_n$; where $\overline{\mathbf{x}} = \mathbf{Q}\,\mathbf{x}$, γ is the forgetting factor and Tri means that only the upper triangular part of the matrix is computed and then its transpose is copied to the lower triangular part, ensuring that the resulting matrix is symmetric.

Alternatively, it is possible to estimate the vectors $\mathbf{w}_i(k)$ in a sequential manner using an algorithm similar to (3.36)-(3.41) as

$$\overline{\mathbf{x}}_1(k) = \overline{\mathbf{x}}(k), \;\; \eta_i^{-1}(0) = 2\max\{\|\overline{x}_i(k)\|_2^2\} = 2\,\|\overline{x}_{i,max}\|_2^2, \tag{3.165}$$

$$\mathbf{w}_i(0) = \overline{\mathbf{x}}_{i,max}/\|\overline{\mathbf{x}}_{i,max}\|_2, \tag{3.166}$$

$$q_i(k) = g_i[y_i(k)] = g_i[\mathbf{w}_i^T(k)\,\overline{\mathbf{x}}_i(k)], \tag{3.167}$$

$$\mathbf{w}_i(k+1) = \mathbf{w}_i(k) + \frac{q_i(k)}{\eta_i^{-1}(k)}[\overline{\mathbf{x}}_i(k) - q_i(k)\,\mathbf{w}_i(k)], \tag{3.168}$$

$$\eta_i^{-1}(k+1) = \gamma\,\eta_i^{-1}(k) + |q_i(k)|^2, \tag{3.169}$$

$$\overline{\mathbf{x}}_{i+1}(k) = \overline{\mathbf{x}}_i(k) - y_i(k)\,\mathbf{v}_{i*}, \tag{3.170}$$

Appendix A. Basic Neural Networks Algorithms for Real and Complex-Valued PCA

Many researchers have modified Amari and Oja's algorithms to extract the true PCs for real-valued data or signals [907], [435]. The main purpose of this Appendix is to review and summarize some of those algorithms closely related to the subspace rule and then to generalize them for complex-valued signals. We will restrict our considerations to four popular learning algorithms[6] namely:

1. **Sanger's Generalized Hebbian Algorithm (GHA)**

$$\frac{d\mathbf{V}}{dt} = \mu \left[\mathbf{x}\,\mathbf{y}^T - \mathbf{V}\; UT(\mathbf{y}\,\mathbf{y}^T)\right], \tag{A.1}$$

where $UT(\cdot)$ means the Upper Triangular operation, i.e., it sets the lower diagonal elements of its matrix argument to zero. Sanger's GHA learning algorithm can be written in a scalar form as [1037]

$$\frac{dv_{ip}}{dt} = \mu_i\, y_i \left[x_p - \sum_{k=1}^{i} v_{kp} y_k\right], \tag{A.2}$$

where $\mu_i > 0$, $(i = 1, 2, \ldots, n;\;\; p = 1, 2, \ldots, m)$.

2. **The stochastic Gradient Ascent (SGA)** proposed by Oja [907] can be formulated as

$$\frac{dv_{ip}}{dt} = \mu_i\, y_i \left[x_p - v_{ip} y_i - \alpha \sum_{k=1}^{i-1} v_{kp} y_k\right], \tag{A.3}$$

where $\alpha > 1$ typically $\alpha = 2$.

3. **The Weighted Subspace Algorithm (WSA)** proposed by Oja, Ogawa and Wangviwattana [916, 907]

$$\frac{d\mathbf{V}}{dt} = \boldsymbol{\mu} \left[\mathbf{x}\,\mathbf{y}^T - \mathbf{V}\,\mathbf{y}\,\mathbf{y}^T \boldsymbol{\Theta}\right], \tag{A.4}$$

where $\boldsymbol{\Theta} = \mathrm{diag}[\theta_1, \theta_2, \ldots, \theta_n]$ with $\theta_i < \theta_{i+1}$, i.e., Θ is a diagonal matrix with positive and strictly decreasing entries.

In a scalar form, the WSA learning rule can take the form [916]

$$\frac{dv_{ip}}{dt} = \mu_i\, y_i \left[x_p - \theta_i \sum_{k=1}^{n} v_{kp} y_p\right], \tag{A.5}$$

where $(i = 1, 2, \ldots, n;\;\; p = 1, 2, \ldots, m)$, $\mu_i > 0$, $0 < \theta_1 < \theta_2 < \cdots < \theta_n$.

[6] All these algorithms have been developed only for real-valued data.

4. **Brockett's Algorithm [105, 106]** for extracting principal components can be written as

$$\frac{d\mathbf{V}}{dt} = \boldsymbol{\mu}\left[\mathbf{x}\,\mathbf{y}^T\,\mathbf{D} - \mathbf{V}\,\mathbf{D}\,\mathbf{y}\,\mathbf{y}^T\,\mathbf{D}\right], \tag{A.6}$$

where $\mathbf{D}$ is a diagonal matrix with positive and strictly decreasing entries, i.e.,

$$\mathbf{D} = \mathrm{diag}[d_1, d_2, \ldots, d_n], \qquad \text{with} \quad d_1 > d_2 > \cdots > d_n > 0.$$

Brockett's Algorithm can also be written in a scalar form as

$$\frac{dv_{ip}}{dt} = \hat{\mu}_i\, y_i \left[x_p - \sum_{k=1}^{n} \alpha_{ki} v_{kp} y_k\right], \tag{A.7}$$

where

$$\alpha_{ki} = \begin{cases} \frac{d_k}{d_i} < 1 & \text{for} \quad k < i, \\ 1 & \text{for} \quad k = i, \\ \frac{d_k}{d_i} > 1 & \text{for} \quad k > i, \end{cases}$$

and $\hat{\mu}_i = \mu_i\, d_i > 0$.

It is interesting to note that the above four learning algorithms can be written in the "generalized form" given by Equation (A.7). For example, in the case of the GHA algorithm, coefficients α_{ki} in (A.7) can be defined by

$$\alpha_{ki} = \begin{cases} 0 & \text{if} \quad k \leq i, \\ 1 & \text{if} \quad k > i. \end{cases} \tag{A.8}$$

Analogously, for the WSA algorithm, we have $\alpha_{ki} = \theta_i$, for any k.

The above algorithms can be extended or generalized for the extraction of the PCs of complex-valued signals. For example, we can derive the Brockett's algorithm for complex-valued signals as follows [106].

Let us consider a three layer self-supervising linear neural network with matrix transformations: $\tilde{\mathbf{y}} = \mathbf{D}\,\mathbf{V}^H\,\mathbf{x}$, $\hat{\mathbf{x}} = \mathbf{V}\,\tilde{\mathbf{y}}$ where $\mathbf{V} \in C^{m \times n}$ is a feed-forward matrix of complex-valued synaptic weights v_{ip} and $\mathbf{D}$ is a real-valued diagonal matrix with strictly decreasing entries providing a scaling (or inducing asymmetry) for outputs $\tilde{\mathbf{y}}$. For such a network, we can formulate a standard cost function as

$$J_c = \frac{1}{2}\left\|\mathbf{e}^R\right\|_2^2 + \frac{1}{2}\left\|\mathbf{e}^I\right\|_2^2, \tag{A.9}$$

where $\mathbf{e} \triangleq \mathbf{e}^R + j\mathbf{e}^I = \mathbf{x} - \hat{\mathbf{x}} = (\mathbf{I} - \mathbf{V}\,\mathbf{D}\,\mathbf{V}^H)\,\mathbf{x}$. Using the back-propagation gradient descent method, the minimization of the above cost function leads to a general learning rule

$$\begin{aligned} \frac{d\mathbf{V}}{dt} &= \boldsymbol{\mu}\,[\mathbf{e}\,\mathbf{y}^H\,\mathbf{D} + \mathbf{x}\,\mathbf{e}^H\,\mathbf{V}\,\mathbf{D}] \\ &= \boldsymbol{\mu}\,[\mathbf{x}\,\mathbf{y}^H\,\mathbf{D} - \mathbf{V}\,\mathbf{D}\,\mathbf{y}\,\mathbf{y}^H\,\mathbf{D}]. \end{aligned} \tag{A.10}$$

Table A.1 Fast implementations of PSA algorithms for complex-valued signals and matrices.

No.	Learning rule	Notes, References
1.	$\Delta \mathbf{W}(k) = \bar{\mathbf{e}}(k)\mathbf{z}^H(k)$	PAST Algorithm
	$\mathbf{z}(k) = \frac{1}{\alpha}\mathbf{Q}(k-1)\mathbf{y}(k)$	Yang (1995) [1312]
	$\mathbf{y}(k) = \mathbf{W}^H(k-1)\mathbf{x}(k)$	
	$\gamma(k) = [1+\mathbf{y}^H(k)\mathbf{z}(k)]^{-1}$	
	$\bar{\mathbf{e}}(k) = \gamma(k)[\bar{\mathbf{x}}(k) - \mathbf{W}(k-1)\mathbf{y}(k)]$	
	$\mathbf{Q}(k) = \frac{1}{\alpha}\mathbf{Q}(k-1) - \gamma(k)\mathbf{z}(k)\mathbf{z}^H(k)$	
2.	$\Delta \mathbf{W}(k) = \widetilde{\mathbf{e}}(k)\mathbf{z}^H(k)$	OPAST Algorithm
	$\beta(k) = \frac{1}{\\|\mathbf{z}(k)\\|_2^2} \quad \frac{1}{\sqrt{1+\\|\bar{\mathbf{e}}(k)\\|_2^2\\|\mathbf{z}(k)\\|_2^2}} - 1 \Big)$	Abed-Meraim, Chkeif,
	$\widetilde{\mathbf{e}}(k) = \beta(k)\,\mathbf{W}(k-1)\mathbf{z}(k) + \;\; 1+\beta(k)\,\\|\bar{\mathbf{z}}(k)\\|_2^2 \;\; \bar{\mathbf{e}}(k)$	and Hua (2000) [1]

Assuming again that the second term on the right hand side in Eq. (A.10) is small and can be neglected, we obtain a generalized form of Brockett's algorithm for complex-valued signals as

$$\frac{d\mathbf{V}}{dt} = \boldsymbol{\mu}\,[\mathbf{x}\,\mathbf{y}^H\mathbf{D} - \mathbf{V}\,\mathbf{D}\,\mathbf{y}\,\mathbf{y}^H\,\mathbf{D}], \tag{A.11}$$

with $\mathbf{V}(0) \neq \mathbf{0}$. In a scalar form, the above algorithm becomes

$$\frac{dv_{ip}}{dt} = \hat{\mu}_i\, y_i \left[x_p^* - \sum_{k=1}^{n} y_k^* v_{kp}\alpha_{ki}\right], \qquad (i=1,2,\ldots,n;\;\; p=1,2,\ldots,m,\;\; n\le m) \tag{A.12}$$

where

$$\alpha_{ki} = d_k/d_i, \qquad \hat{\mu}_i = \mu_i d_i > 0$$

(i.e., the parameter d_i is absorbed by the learning rate $\hat{\mu}_i(t) > 0$), and $d_1 \geq d_2 \geq \cdots \geq d_n$. In the special case of $\alpha_{ki} = 1$ $(\forall k,\ i)$, (A.12) simplifies to the subspace learning algorithm.

Appendix B. Hierarchical Neural Network for Complex-valued PCA

Assume that the process $\mathbf{x}(k) \in \mathbf{C}^m$ comprises a zero-mean sequence whose covariance matrix is defined as $\mathbf{R}_{\mathbf{x}\mathbf{x}} = E\{\mathbf{x}\,\mathbf{x}^H\}$ and we wish to estimate (extract) its complex-valued eigenvectors $\mathbf{v}_i$ and corresponding PCs on-line [275, 271, 276]. Employing a self-supervising principle and hierarchical neural network architecture, we shall extract the PCs sequentially. Let us temporarily assume that we need to extract only the first[7] PC (y_1) by a linear single neuron

$$y_1 = \mathbf{v}_1^T \mathbf{x} = \sum_{p=1}^{m} v_{1p} \, x_p(t). \tag{B.1}$$

The vector $\mathbf{v}_1$ should be determined in such a way so that the reconstructed vector

$$\hat{\mathbf{x}} = \mathbf{v}_1^* y_1, \tag{B.2}$$

will reproduce the input vector $\mathbf{x}(t)$ as well as possible, according to a suitable optimization criterion.

For this purpose let us define a complex-valued instantaneous error vector as

$$\begin{aligned} \mathbf{e}_1(t) &= [e_{11}(t), e_{12}(t), \ldots, e_{1m}(t)]^T \\ &\triangleq \mathbf{x}(t) - \hat{\mathbf{x}}(t) = \mathbf{x}(t) - \mathbf{v}_1^* y_1(t) \\ &= \left(\mathbf{I} - \mathbf{v}_1 \mathbf{v}_1^H\right) \mathbf{x}(t) = \mathbf{e}_1^R(t) + j\mathbf{e}_1^I(t), \end{aligned} \tag{B.3}$$

where $\mathbf{I}$ is an identity matrix, $\mathbf{e}_1^R$ is the real part and $\mathbf{e}_1^I$ is the imaginary part of the error vector $\mathbf{e}_1(t)$ and $j = \sqrt{-1}$.

In order to find the optimal value of the vector $\mathbf{v}_1$, we can define a standard 2-norm cost function

$$\begin{aligned} E_1(\mathbf{v}_1) &= \frac{1}{2} \left[\|\mathbf{e}_1^R\|_2^2 + \|\mathbf{e}_1^I\|_2^2 \right] \\ &= \frac{1}{2} \left[\sum_{p=1}^{m} (e_{1p}^R)^2 + \sum_{p=1}^{m} (e_{1p}^I)^2 \right] \end{aligned} \tag{B.4}$$

where e_{1p}^R is the pth element of $\mathbf{e}_1^R$ etc.

The minimization of the cost function (B.4), according to the standard gradient descent approach for the real and imaginary parts of the vector $\mathbf{v}_1 = \mathbf{v}_1^R + j\mathbf{v}_1^I$, leads to the following

[7] The first PC y_1 corresponds to the largest eigenvalue $\lambda_1 = E\{y_1 y_1^*\} = E\{|y_1|^2\}$ of the covariance matrix $\mathbf{R_{xx}} = E\{\mathbf{x}\mathbf{x}^H\}$, $y_1 = \mathbf{v}_1^T \mathbf{x}$, where $(\cdot)^*$ denotes complex conjugate and $(\cdot)^H$ means the complex conjugate transpose or Hermitian operation.

set of differential equations:

$$\begin{aligned}\frac{dv_{1p}^R}{dt} &= -\mu_1 \frac{\partial E_1(\mathbf{v}_1)}{\partial v_{1p}^R} \\ &= \mu_1 \left\{ (e_{1p}^R y_1^R + e_{1p}^I y_1^I) + x_p^R \sum_{h=1}^{m} (e_{1h}^R v_{1h}^R - e_{1h}^I v_{1h}^I) + x_p^I \sum_{h=1}^{m} (e_{1h}^R v_{1h}^I + e_{1h}^I v_{1h}^R) \right\},\end{aligned} \tag{B.5}$$

$$\begin{aligned}\frac{dv_{1p}^I}{dt} &= -\mu_1 \frac{\partial E_1(\mathbf{v}_1)}{\partial v_{1p}^I} \\ &= \mu_1 \left\{ (e_{1p}^R y_1^I - e_{1p}^I y_1^R) + x_p^R \sum_{h=1}^{m} (e_{1h}^R v_{1h}^I + e_{1h}^I v_{1h}^R) - x_p^I \sum_{h=1}^{m} (e_{1h}^R v_{1h}^R - e_{1h}^I v_{1h}^I) \right\},\end{aligned} \tag{B.6}$$

where $\mu_1 > 0$ is the learning rate and $y_1 \triangleq y_1^R + jy_1^I$, $e_{1p} \triangleq e_{1p}^R + je_{1p}^I$, $(p = 1, 2, \ldots, m)$. Combining the above equations (B.5) and (B.6) and taking into account that $v_{1p} \triangleq v_{1p}^R + jv_{1p}^I$, we obtain the learning algorithm

$$\frac{dv_{1p}(t)}{dt} = \mu_1(t) \left[y_1(t) e_{1p}^*(t) + x_p^*(t) \sum_{h=1}^{m} v_{1h}(t) e_{1h}(t) \right], \qquad (p = 1, 2, \ldots, m) \tag{B.7}$$

which can be written in matrix form as

$$\frac{d\mathbf{v}_1}{dt} = \mu_1 \, [y_1 \mathbf{e}_1^* + \mathbf{x}^* \mathbf{v}_1^T \mathbf{e}_1], \tag{B.8}$$

for any $\mathbf{v}_1(0) \neq \mathbf{0}$, $\mu_1(t) > 0$.

The above learning rule can be further simplified to

$$\begin{aligned}\frac{d\mathbf{v}_1}{dt} &= \mu_1 \, y_1 \mathbf{e}_1^* \\ &= \mu_1 \, y_1 [\mathbf{x} - \mathbf{v}_1^* y_1]^* \\ &= \mu_1 \, y_1 [\mathbf{x}^* - \mathbf{v}_1 y_1^*] \\ &= \mu_1 \, \mathbf{v}_1^T \mathbf{x} [\mathbf{I} - \mathbf{v}_1 \mathbf{v}_1^H] \mathbf{x}^*,\end{aligned} \tag{B.9}$$

since the second term in Equation (B.8), which can be written as

$$\mathbf{x}^* \mathbf{v}_1^T \mathbf{e}_1 = \mathbf{x}^* \mathbf{v}_1^T (\mathbf{x} - \mathbf{v}_1^* y_1) = \mathbf{x}^* (1 - \mathbf{v}_1^H \mathbf{v}_1) y_1, \tag{B.10}$$

tends quickly to zero as $\mathbf{v}_1^H \mathbf{v}_1$ tends to 1 with $t \to \infty$; it can therefore be neglected.

It is interesting to note that the discrete-time realization of the learning algorithm:

$$\mathbf{v}_1(k+1) = \mathbf{v}_1(k) + \eta_1(k) \, y_1(k) [\mathbf{x}^*(k) - \mathbf{v}_1(k) y_1^*(k)], \qquad (k = 0, 1, 2, \ldots) \tag{B.11}$$

is known as the Oja learning rule for complex-valued data.

4

Blind Decorrelation and Second Order Statistics for Robust Blind Identification

There are very few human beings who receive the truth, complete and staggering, by instant illumination. Most of them acquire it fragment by fragment, on a small scale, by successive developments, cellularly, like a laborious mosaic.

—(Anais Nin; 1903-1977)

Temporal, spatial and spatio-temporal decorrelations play important roles in signal processing. These techniques are based only on second-order statistics (SOS). They are the basis for modern subspace methods of spectrum analysis and array processing and are often used in a preprocessing stage in order to improve convergence properties of adaptive systems, to eliminate redundancy or to reduce noise. Spatial decorrelation or prewhitening is often considered a necessary (but not sufficient) condition for stronger stochastic independence criteria. After prewhitening, the BSS or ICA tasks usually become somewhat easier and well-posed (less ill-conditioned), because the subsequent separating (unmixing) system is described by an orthogonal matrix for real-valued signals and a unitary matrix for complex-valued signals and weights. Furthermore, spatio-temporal and time-delayed decorrelation can be used to identify the mixing matrix and to perform blind source separation of colored sources. In this chapter, we will discuss and analyze a number of efficient and robust adaptive and batch algorithms for spatial whitening, orthogonalization, spatio-temporal and time-delayed blind decorrelation. Moreover, we discuss several promising robust algorithms for blind identification and blind source separation of nonstationary and/or colored sources.

4.1 SPATIAL DECORRELATION - WHITENING TRANSFORMS

4.1.1 Batch Approach

Some adaptive algorithms for blind separation require prewhitening (also called sphering or normalized spatial decorrelation) of mixed (sensor) signals. A random, zero-mean vector $\mathbf{y}$ is said to be *white* if its covariance matrix is an identity matrix, i.e., $\mathbf{R_{yy}} = E\{\mathbf{y}\,\mathbf{y}^T\} = \mathbf{I}_n$ or $E\{y_i y_j\} = \delta_{ij}$, where δ_{ij} is the Kronecker delta. In *whitening*, the sensor vectors $\mathbf{x}(k)$ are pre-processed using the following transformation (see Fig. 4.1):

$$\mathbf{y}(k) = \mathbf{W}\,\mathbf{x}(k). \tag{4.1}$$

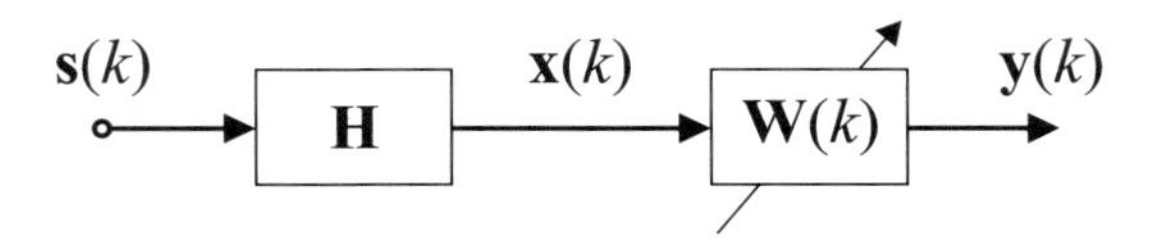

Fig. 4.1 Basic model for blind spatial decorrelation of sensor signals.

Here $\mathbf{y}(k)$ denotes the whitened vector, and $\mathbf{W}$ is an $n \times m$ whitening matrix. If $m > n$, where n is known in advance, $\mathbf{W}$ simultaneously reduces the dimension of the data vectors from m to n. In whitening, the matrix $\mathbf{W}$ is chosen so that the covariance matrix $\mathrm{E}\{\mathbf{y}(k)\,\mathbf{y}(k)^T\}$ becomes the unit matrix $\mathbf{I}_n$. Thus, components of the whitened vectors $\mathbf{y}(k)$ are mutually uncorrelated and have unit variance, i.e.,

$$\mathbf{R_{yy}} = E\left\{\mathbf{y}\,\mathbf{y}^T\right\} = E\left\{\mathbf{W}\,\mathbf{x}\,\mathbf{x}^T\,\mathbf{W}^T\right\} = \mathbf{W}\mathbf{R_{xx}}\mathbf{W}^T = \mathbf{I}_n. \tag{4.2}$$

Fig. 4.2 illustrates three basic transformations of sensor signals: prewhitening, PCA and ICA.

Generally, the sensor signals are mutually correlated, i.e., the covariance matrix $\mathbf{R_{xx}} = E\left\{\mathbf{x}\mathbf{x}^T\right\}$ is a full (not diagonal) matrix. It should be noted that the matrix $\mathbf{W} \in \mathbb{R}^{n\times m}$ is not unique, since multiplying it by an arbitrary orthogonal matrix from the left still preserves property (4.2).

Since the covariance matrix of sensor signals $\mathbf{x}(k)$ is usually symmetric positive definite, it can be decomposed as follows

$$\mathbf{R_{xx}} = \mathbf{V_x}\boldsymbol{\Lambda}_\mathbf{x}\mathbf{V}_\mathbf{x}^T = \mathbf{V_x}\boldsymbol{\Lambda}_\mathbf{x}^{1/2}\boldsymbol{\Lambda}_\mathbf{x}^{1/2}\mathbf{V}_\mathbf{x}^T, \tag{4.3}$$

where $\mathbf{V_x}$ is an orthogonal matrix and $\boldsymbol{\Lambda}_\mathbf{x} = \mathrm{diag}\{\lambda_1, \lambda_2, \dots, \lambda_n\}$ is a diagonal matrix with positive eigenvalues $\lambda_1 \geq \lambda_2 \geq \cdots \geq \lambda_n > 0$. Hence, under the condition that the covariance matrix is positive definite[1], the required decorrelation matrix $\mathbf{W}$ (also called a

[1] If the covariance matrix is positive semi-definite, we can take only positive eigenvalues and associated eigenvectors.

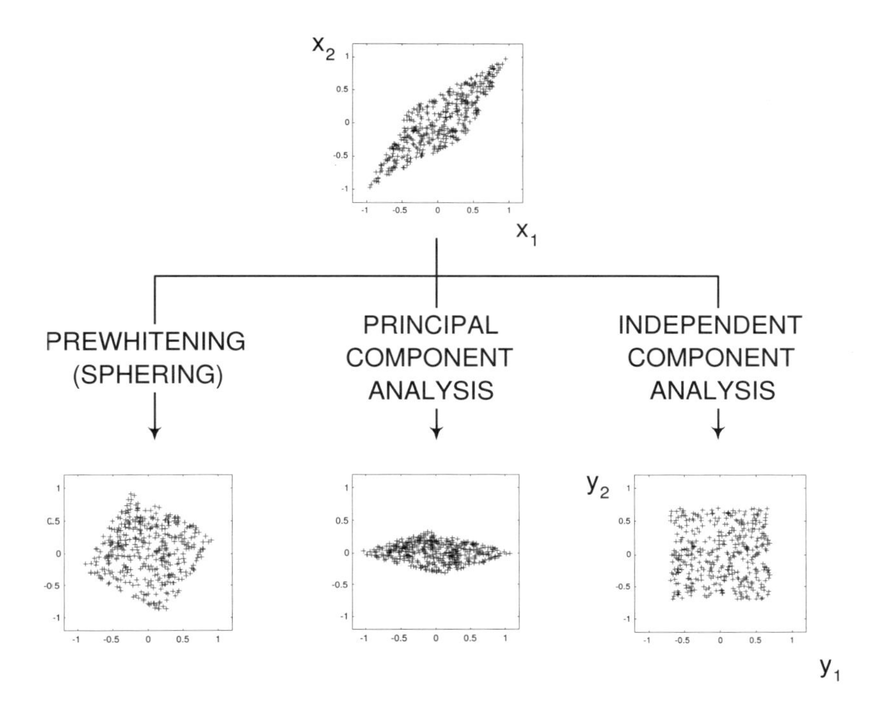

Fig. 4.2 Scatter plots illustrating of basic transformations for two sensor signals with uniform distributions.

whitening matrix or Mahalanobis transform) can be computed as follows

$$\boxed{\mathbf{W} = \mathbf{\Lambda}_{\mathbf{x}}^{-1/2}\, \mathbf{V}_{\mathbf{x}}^T = \text{diag}\left\{ \frac{1}{\sqrt{\lambda_1}}, \frac{1}{\sqrt{\lambda_2}}, \ldots, \frac{1}{\sqrt{\lambda_n}} \right\} \mathbf{V}_{\mathbf{x}}^T} \tag{4.4}$$

or

$$\boxed{\mathbf{W} = \mathbf{U}\mathbf{\Lambda}_{\mathbf{x}}^{-1/2}\mathbf{V}_{\mathbf{x}}^T,} \tag{4.5}$$

where $\mathbf{U}$ is an arbitrary orthogonal matrix. This can be easily verified by substituting (4.4) or (4.5) into (4.2):

$$\mathbf{R_{yy}} = E\left\{\mathbf{y}\,\mathbf{y}^T\right\} = \mathbf{\Lambda}_{\mathbf{x}}^{-1/2}\mathbf{V}_{\mathbf{x}}^T\mathbf{V}_{\mathbf{x}}\mathbf{\Lambda}\mathbf{V}_{\mathbf{x}}^T\mathbf{V}_{\mathbf{x}}\mathbf{\Lambda}_{\mathbf{x}}^{-1/2} = \mathbf{I}_n, \tag{4.6}$$

or

$$\mathbf{R_{yy}} = \mathbf{U}\mathbf{\Lambda}_{\mathbf{x}}^{-1/2}\mathbf{V}_{\mathbf{x}}^T\mathbf{V}_{\mathbf{x}}\mathbf{\Lambda}_{\mathbf{x}}\mathbf{V}_{\mathbf{x}}^T\mathbf{V}_{\mathbf{x}}\mathbf{\Lambda}_{\mathbf{x}}^{-1/2}\mathbf{U}^T = \mathbf{I}_n. \tag{4.7}$$

Alternatively, we can apply the Cholesky decomposition

$$\mathbf{R_{xx}} = \mathbf{L}\mathbf{L}^T, \tag{4.8}$$

where $\mathbf{L}$ is a lower triangular matrix. The whitening (decorrelation) matrix in this case is

$$\boxed{\mathbf{W} = \mathbf{U}\mathbf{L}^{-1},} \tag{4.9}$$

where $\mathbf{U}$ is an arbitrary orthogonal matrix, since

$$\mathbf{R}_{\mathbf{yy}} = E\left\{\mathbf{y}\,\mathbf{y}^T\right\} = \mathbf{W}\mathbf{R}_{\mathbf{xx}}\mathbf{W}^T = \mathbf{U}\mathbf{L}^{-1}\mathbf{L}\mathbf{L}^T\left(\mathbf{L}^{-1}\right)^T\mathbf{U}^T = \mathbf{I}_n. \tag{4.10}$$

In the special case when $\mathbf{x}(k) = \boldsymbol{\nu}(k)$ is colored Gaussian noise with $\mathbf{R}_{\boldsymbol{\nu}\boldsymbol{\nu}} = E\left\{\boldsymbol{\nu}\boldsymbol{\nu}^T\right\} \neq \sigma_\nu^2\,\mathbf{I}_n$, the whitening transform converts it into a white noise (i.i.d.) process.

4.1.2 Optimization Criteria for Adaptive Blind Spatial Decorrelation

In the previous section, we described some simple numerical or batch methods to estimate decorrelation matrix $\mathbf{W}$. Now, we consider optimization criteria that enable us to derive adaptive algorithms. Let us consider a mixing system

$$\mathbf{x}(k) = \mathbf{H}\,\mathbf{s}(k) \tag{4.11}$$

and a decorrelation system as depicted on Fig. 4.1

$$\mathbf{y}(k) = \mathbf{W}\,\mathbf{x}(k), \tag{4.12}$$

where matrices $\mathbf{H}$ and $\mathbf{W}$ are n by n nonsingular matrices. Our objective is to find a simple adaptive algorithm for estimation of decorrelation matrix $\mathbf{W}$, such that the covariance matrix of the output signals will be a diagonal matrix, i.e.,

$$\mathbf{R}_{\mathbf{yy}} = E\left\{\mathbf{y}\,\mathbf{y}^T\right\} = \boldsymbol{\Lambda}, \tag{4.13}$$

where $\boldsymbol{\Lambda} = \operatorname{diag}\{\lambda_1, \ldots, \lambda_n\}$ is a diagonal matrix, typically, $\boldsymbol{\Lambda} = \mathbf{I}_n$. It should be noted that the output signals will be mutually uncorrelated if all the cross-correlations are zero:

$$r_{ij} = E\{y_i y_j\} = 0, \quad \text{for all} \quad i \neq j, \tag{4.14}$$

with non-zero autocorrelations

$$r_{ii} = E\left\{y_i^2\right\} = \lambda_i > 0. \tag{4.15}$$

The natural minimization criterion can be formulated as the p-norm

$$J_p(\mathbf{W}) = \frac{1}{p}\sum_{i=1}^{n}\sum_{\substack{j=1\\ j\neq i}}^{n} |r_{ij}|^p, \tag{4.16}$$

subject to the constraints $r_{ii} \neq 0,\ \forall i$, typically, $r_{ii} = 1,\ \forall i$.

The following cases are especially interesting:
1-norm (Absolute value criterion),
2-norm (Frobenius norm),
∞-norm (Chebyshev norm).

The problem is how to derive adaptive algorithms based on these criteria. The present chapter studies only the Frobenius norm.

4.1.3 Derivation of Equivariant Adaptive Algorithms for Blind Spatial Decorrelation

We can derive an adaptive learning algorithm using the following criterion:
Minimize the global cost function:

$$J_2(\mathbf{W}) = \frac{1}{4}\sum_{i=1}^{n}\sum_{j=1}^{n}\left(E\{y_i y_j\} - \lambda_i \delta_{ij}\right)^2 = \frac{1}{4}\left\|E\{\mathbf{y}\,\mathbf{y}^T\} - \mathbf{\Lambda}\right\|_F^2, \tag{4.17}$$

where $\|\mathbf{A}\|_F$ denotes the Frobenius norm of matrix $\mathbf{A}$. In order to derive an adaptive learning algorithm, we use the following transformation:

$$\begin{aligned}\mathbf{R_{yy}} &= E\{\mathbf{y}\,\mathbf{y}^T\} = E\{\mathbf{W}\,\mathbf{x}\mathbf{x}^T\,\mathbf{W}^T\} = E\left\{\mathbf{W}\,\mathbf{H}\,\mathbf{s}\,\mathbf{s}^T\,(\mathbf{W}\,\mathbf{H})^T\right\}\\ &= \mathbf{G}\,\mathbf{R_{ss}}\,\mathbf{G}^T = \mathbf{G}\,\mathbf{G}^T,\end{aligned} \tag{4.18}$$

where $\mathbf{G} = \mathbf{W}\,\mathbf{H}$ is the global transformation matrix from $\mathbf{s}$ to $\mathbf{y}$, and we assumed without loss of generality that $\mathbf{R_{ss}} = E\{\mathbf{s}\,\mathbf{s}^T\} = \mathbf{I}_n$. The optimization criterion can be written in the form:

$$J_2(\mathbf{W}) = \frac{1}{4}\left\|\mathbf{G}\,\mathbf{G}^T - \mathbf{\Lambda}\right\|_F^2 = \frac{1}{4}\mathrm{tr}[(\mathbf{G}\,\mathbf{G}^T - \mathbf{\Lambda})\,(\mathbf{G}\,\mathbf{G}^T - \mathbf{\Lambda})]. \tag{4.19}$$

Applying the standard gradient descent approach and the chain rule, we have

$$\frac{dg_{ij}}{dt} = -\mu\,\frac{\partial J_2(\mathbf{W})}{\partial g_{ij}} = -\mu\sum_{k=1}^{n}\sum_{p=1}^{n}\frac{\partial J_2}{\partial r_{kp}}\,\frac{\partial r_{kp}}{\partial g_{ij}}, \tag{4.20}$$

where we use the continuous-time version of the learning rule. Taking into account that $\mathbf{R_{yy}} = \mathbf{G}\,\mathbf{G}^T$, we obtain

$$\begin{aligned}\frac{dg_{ij}}{dt} &= \frac{\mu}{2}\left[\sum_{k=1}^{n}\lambda_k\,\frac{\partial r_{kk}}{\partial g_{ij}} - \sum_{k=1}^{n}\sum_{p=1}^{n} r_{kp}\,\frac{\partial r_{kp}}{\partial g_{ij}}\right]\\ &= \frac{\mu}{2}\left[2\lambda_i\,g_{ij} - \sum_{k=1}^{n} r_{ik}\,g_{kj} - \sum_{p=1}^{n} r_{pi}\,g_{pj}\right].\end{aligned} \tag{4.21}$$

The above formula can be simplified by taking into account that the output covariance matrix $\mathbf{R_{yy}}$ is symmetric, i.e. $r_{ij} = r_{ji}$, as

$$\frac{dg_{ij}}{dt} = \mu\left[\lambda_i\,g_{ij} - \sum_{k=1}^{n} r_{ik}\,g_{kj}\right] \qquad (i,j = 1,2,\ldots,n) \tag{4.22}$$

or in more compact matrix form

$$\frac{d\mathbf{G}}{dt} = \mu\,(\mathbf{\Lambda} - \mathbf{R_{yy}})\,\mathbf{G} = \mu\,\left(\mathbf{\Lambda} - \mathbf{G}\,\mathbf{G}^T\right)\mathbf{G}. \tag{4.23}$$

Taking into account that $\mathbf{G} = \mathbf{W}\,\mathbf{H}$ and assuming that $\mathbf{H}$ is only very slowly varying in time, i.e., $d\mathbf{H}/dt \approx \mathbf{0}$, we have

$$\frac{d\mathbf{W}}{dt}\mathbf{H} = \mu\,(\boldsymbol{\Lambda} - \mathbf{R_{yy}})\,\mathbf{W}\,\mathbf{H}. \tag{4.24}$$

Hence

$$\boxed{\frac{d\mathbf{W}}{dt} = \mu\,(\boldsymbol{\Lambda} - \mathbf{R_{yy}})\,\mathbf{W}.} \tag{4.25}$$

Using the simple Euler formula, the corresponding discrete-time algorithm can be written as

$$\boxed{\mathbf{W}(l+1) = \mathbf{W}(l) + \eta_l\,\left[\boldsymbol{\Lambda} - \mathbf{R}^{(l)}_{\mathbf{yy}}\right]\mathbf{W}(l).} \tag{4.26}$$

The covariance matrix $\mathbf{R_{yy}}$ can be estimated as follows:

$$\widehat{\mathbf{R}}^{(l)}_{\mathbf{yy}} = \langle \mathbf{y}\,\mathbf{y}^T \rangle = \frac{1}{N}\sum_{k=0}^{N-1} \mathbf{y}^{(l)}(k)\,[\mathbf{y}^{(l)}(k)]^T, \tag{4.27}$$

where $\mathbf{y}^{(l)}(k) = \mathbf{W}(l)\,\mathbf{x}(k)$.

Alternatively, we can apply the moving average (MA) approach to estimate matrix $\mathbf{W}$ on-line as

$$\begin{aligned}\mathbf{W}(k+1) &= \mathbf{W}(k) + \eta(k)\,\left[\boldsymbol{\Lambda} - \widehat{\mathbf{R}}^{(k)}_{\mathbf{yy}}\right]\mathbf{W}(k)\\ &= \mathbf{W}(k) + \eta(k)\,\left[\boldsymbol{\Lambda} - \mathbf{W}(k)\widehat{\mathbf{R}}^{(k)}_{\mathbf{xx}}\mathbf{W}^T(k)\right]\mathbf{W}(k),\end{aligned} \tag{4.28}$$

where $0 < \eta(k) \leq 0.5$ and

$$\widehat{\mathbf{R}}^{(k)}_{\mathbf{yy}} = (1-\eta_0)\,\widehat{\mathbf{R}}^{(k-1)}_{\mathbf{yy}} + \eta_0\,\mathbf{y}(k)\,\mathbf{y}^T(k), \tag{4.29}$$

where $\eta_0 \in (0, 1]$ is a fixed step size (learning rate). For on-line learning, the covariance matrix can be very roughly estimated simply by neglecting the expectation operator as

$$\widehat{\mathbf{R}}_{\mathbf{yy}} \cong \mathbf{y}(k)\,\mathbf{y}^T(k), \tag{4.30}$$

thus, the discrete-time, on-line algorithm (4.26) simplifies as

$$\boxed{\Delta\mathbf{W}(k) = \mathbf{W}(k+1) - \mathbf{W}(k) = \eta(k)\,\left[\boldsymbol{\Lambda} - \mathbf{y}(k)\,\mathbf{y}^T(k)\right]\mathbf{W}(k).} \tag{4.31}$$

A functional diagram illustrating implementation of the discrete-time on-line learning algorithm (4.31) is shown in Fig. 4.3.

It is interesting to note that a similar algorithm can be derived by using an information-theoretic criterion (see Chapter 6 for more detail)

$$\boxed{J(\mathbf{W}) = -\frac{1}{2}\left[\log\left(\det\left(\mathbf{W}\mathbf{W}^T\right)\right) - \sum_{i=1}^{n} E\left\{|y_i|^2\right\}\right],} \tag{4.32}$$

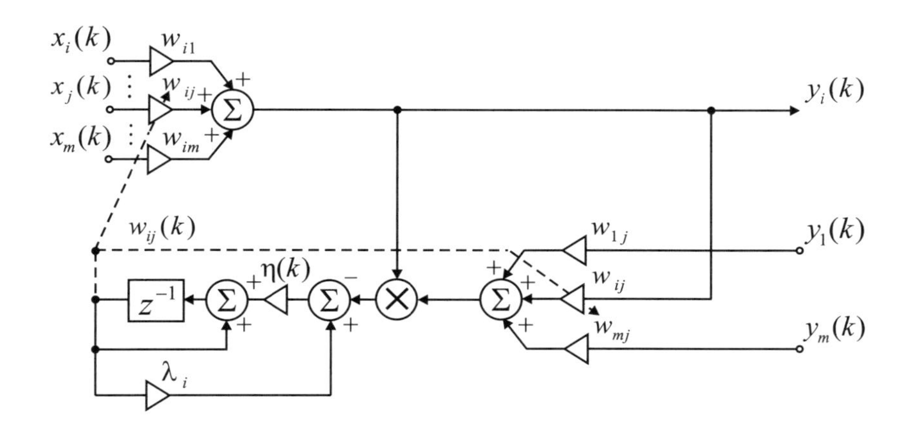

Fig. 4.3 Block diagram illustrating implementation of the learning algorithm (4.31).

where $\mathbf{W} \in \mathbb{R}^{n \times m}$ with $m \geq n$. The first term of the cost function makes sure that the decorrelation matrix $\mathbf{W}$ is full rank and thus, prevents all the outputs from decaying to zero. The second term assures that the output signals will be mutually uncorrelated by minimizing the global energy of these signals. The gradient components of the cost function can be computed as follows

$$\frac{\partial \log \det \mathbf{W}\mathbf{W}^T}{\partial \mathbf{W}} = 2\left(\mathbf{W}\mathbf{W}^T\right)^{-1}\mathbf{W} \tag{4.33}$$

and

$$\frac{\partial E\left\{|y_i|^2\right\}}{\partial w_{ij}} = \frac{\partial E\left\{|y_i|^2\right\}}{\partial y_i}\frac{\partial y_i}{\partial w_{ij}} = 2\left\langle y_i x_j \right\rangle \qquad (i,j = 1,2,\ldots,n) \tag{4.34}$$

or equivalently

$$\frac{\partial \sum_{i=1}^{n} E\left\{|y_i|^2\right\}}{\partial \mathbf{W}} = 2\left\langle \mathbf{y}\,\mathbf{x}^T \right\rangle. \tag{4.35}$$

Hence, applying the standard gradient descent approach, we obtain the learning algorithm expressed in matrix form as

$$\frac{d\mathbf{W}}{dt} = -\mu\left[\frac{\partial J}{\partial \mathbf{W}}\right] = \mu\,\left[(\mathbf{W}\,\mathbf{W}^T)^{-1}\mathbf{W} - \left\langle \mathbf{y}\,\mathbf{x}^T \right\rangle\right]. \tag{4.36}$$

In order to avoid matrix inversion, we can apply the Atick-Redlich formula [45]:

$$\boxed{\frac{d\mathbf{W}}{dt} = -\mu\mathbf{W}\left[\frac{\partial J}{\partial \mathbf{W}}\right]^T\mathbf{W} = \mu\left(\mathbf{I} - \left\langle \mathbf{y}\,\mathbf{y}^T \right\rangle\right)\mathbf{W}.} \tag{4.37}$$

Alternatively, we can use Amari's natural gradient (NG) formula to obtain the same final algorithm [20] (see Appendix A and Chapter 6 for theoretical explanation)

$$\boxed{\frac{d\mathbf{W}}{dt} = -\mu \frac{\partial J}{\partial \mathbf{W}} \mathbf{W}^T \mathbf{W} = \mu \left(\mathbf{I} - \langle \mathbf{y}\,\mathbf{y}^T \rangle\right) \mathbf{W}.} \tag{4.38}$$

The corresponding discrete-time on-line algorithm can be written as:

$$\mathbf{W}(k+1) = \mathbf{W}(k) + \eta(k) \left[\mathbf{I} - \mathbf{y}(k)\,\mathbf{y}^T(k)\right] \mathbf{W}(k). \tag{4.39}$$

Remark 4.1 *It is interesting to note that the above algorithms (4.38) and (4.39) converge when global matrix* $\mathbf{G} = \mathbf{W}\,\mathbf{H}$ *becomes a matrix satisfying relations*

$$\mathbf{G}\,\mathbf{G}^T = \mathbf{G}^T\,\mathbf{G} = \mathbf{I}_n \tag{4.40}$$

or $\mathbf{G}^{-1} = \mathbf{G}^T$, *so the matrix* $\mathbf{G}$ *is orthogonal. Indeed, multiplying equation (4.39) by the mixing matrix* $\mathbf{H}$ *from the right hand side, we get:*

$$\mathbf{W}(k+1)\,\mathbf{H} \stackrel{df}{=} \mathbf{G}(k+1) = \mathbf{G}(k) + \eta(k) \left[\mathbf{I} - \mathbf{G}(k) \left\langle \mathbf{s}(k)\,\mathbf{s}^T(k)\right\rangle \mathbf{G}^T(k)\right] \mathbf{G}(k). \tag{4.41}$$

Assuming, without loss of generality, that the autocorrelation matrix $\mathbf{R_{ss}} = \left\langle \mathbf{s}(k)\,\mathbf{s}^T(k)\right\rangle$ *is the identity matrix, it is evident that a learning algorithm, employing the rule (4.41), reaches an equilibrium when the matrix* $\mathbf{G}(k)$ *becomes orthogonal, i.e.,* $\mathbf{G}^{-1} = \mathbf{G}^T$.

Moreover, the above algorithm possesses the so-called "equivariant property" such that its average performance does not depend on the eigenvalues of the covariance matrix $\mathbf{R_{xx}}$.

The algorithm (4.38) can be expressed in terms of the entries of a nonsingular mixing matrix $\mathbf{H} = [h_{ij}] \in \mathbb{R}^{n \times n}$. Assuming that $\mathbf{W}\,\mathbf{H} = \mathbf{I}_n$, we have

$$\frac{d\mathbf{H}}{dt} = -\mathbf{H}\,\frac{d\mathbf{W}}{dt}\,\mathbf{H}. \tag{4.42}$$

Hence, we obtain a local biologically plausible (normalized Hebbian) algorithm

$$\frac{d\mathbf{H}}{dt} = -\mu \mathbf{H} \left[\mathbf{I} - \langle \mathbf{y}\,\mathbf{y}^T \rangle\right] = \mu \left[\langle \mathbf{x}\,\mathbf{y}^T \rangle - \mathbf{H}\right] \tag{4.43}$$

or in scalar form

$$\boxed{\frac{dh_{ij}}{dt} = \mu \left[\langle x_i\, y_j \rangle - h_{ij}\right] \qquad (i, j = 1, 2, \ldots, n).} \tag{4.44}$$

4.1.4 Simple Local Learning Rule

The learning rules discussed in the previous section can be considerably simplified, if we can assume that the decorrelation matrix $\mathbf{W}$ is symmetric positive definite.[2] To this end,

[2] It is always possible to decorrelate vector $\mathbf{x}$ by using a symmetric positive definite matrix $\mathbf{W}$ by taking $\mathbf{U} = \mathbf{V}$ in (4.5).

we can use a stable simple gradient formula

$$\frac{d\mathbf{W}}{dt} = -\mu \frac{\partial J}{\partial \mathbf{W}} \mathbf{W}^{T/2} \mathbf{W}^{1/2} = -\mu \frac{\partial J}{\partial \mathbf{W}} \mathbf{W}^T = \mu \left[\mathbf{\Lambda} - \langle \mathbf{y}\, \mathbf{y}^T \rangle \right] \tag{4.45}$$

or equivalently

$$\frac{d\mathbf{W}}{dt} = -\mu \mathbf{W} \frac{\partial J}{\partial \mathbf{W}} = \mu \left[\mathbf{\Lambda} - \langle \mathbf{y}\, \mathbf{y}^T \rangle \right]. \tag{4.46}$$

It should be noted that $\mathbf{W}(t)$ will be symmetric positive definite if $\mathbf{W}(0)$ is positive definite (typically $\mathbf{W}(0) = \mathbf{I}$), since $\widehat{\mathbf{R}}_{\mathbf{yy}} = \langle \mathbf{y}\, \mathbf{y}^T \rangle$ is also symmetric in each iteration step.
The above formula can be written in scalar form as

$$\frac{dw_{ij}}{dt} = \mu \left(\delta_{ij}\, \lambda_i - \langle y_i y_j \rangle \right) \qquad (i, j = 1, 2, \ldots, n). \tag{4.47}$$

The discrete-time, on-line, local learning algorithm can be written as

$$\boxed{\mathbf{W}(k+1) = \mathbf{W}(k) + \eta(k) \left(\mathbf{\Lambda} - \mathbf{y}(k)\, \mathbf{y}^T(k) \right)} \tag{4.48}$$

or in scalar form (see Fig. 4.4) as

$$\boxed{w_{ij}(k+1) = w_{ij}(k) + \eta(k) \left[\delta_{ij}\, \lambda_i - y_i(k) y_j(k) \right] \qquad (i, j = 1, 2, \ldots, n).} \tag{4.49}$$

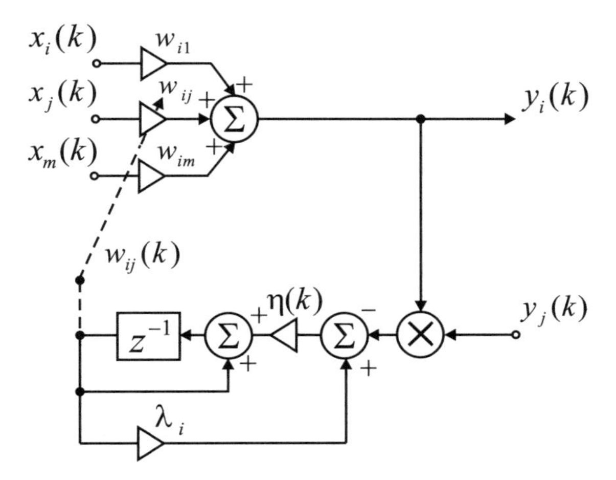

Fig. 4.4 Implementation of the local learning rule (4.48) for the blind decorrelation.

In addition to the merit that the algorithm (4.48) is much simpler to implement than (4.39), the local signal requirements of the algorithm in (4.48) make it ideal for hardware and VLSI implementations. However, performances of (4.39) and (4.48) are not the same, and convergence speed of the local algorithm is usually much slower. In order to improve convergence properties, a multi-layer neural network can be employed as it has been shown in [260, 262, 390] and described in detail in the next chapters. Furthermore, the learning rate should be suitably chosen. A theoretical performance comparison of these two algorithms is discussed in [390].

The update in (4.48) has an interesting property that it converges also for a suitable sequence of *negative* step sizes $\eta(k)$ provided $\mathbf{W}(0)$ is a negative semi-definite matrix. To see this result, multiply both sides of (4.48) by (-1). By defining $\tilde{\mathbf{W}}(k) = -\mathbf{W}(k)$, $\tilde{\mathbf{y}}(k) = -\mathbf{y}(k) = \tilde{\mathbf{W}}(k)\,\mathbf{x}(k)$, and $\tilde{\eta}(k) = -\eta(k)$,

$$\tilde{\mathbf{W}}(k+1) \quad = \quad \tilde{\mathbf{W}}(k) + \tilde{\eta}(k)\,(\mathbf{\Lambda} - \tilde{\mathbf{y}}(k)\tilde{\mathbf{y}}^T(k)). \tag{4.50}$$

This algorithm is algebraically-equivalent to that in (4.48), and thus the coefficient matrix $\tilde{\mathbf{W}}(k)$ tends towards the solution obtained by $-\mathbf{W}(k)$ in the original algorithm. The convergence conditions on $\tilde{\eta}(k)$ are the same as those for $-\eta(k)$ in the original algorithm.

Summarizing, the local learning rule can be formulated in a more general form as

$$\boxed{\mathbf{W}(k+1) = \mathbf{W}(k) \pm \eta(k)\left(\mathbf{\Lambda} - \langle \mathbf{y}(k)\,\mathbf{y}^T(k)\rangle\right),} \tag{4.51}$$

where $\eta(k) > 0$ is the learning step, $\mathbf{W}(0)$ is a symmetric positive definite matrix and $\mathbf{\Lambda}$ is a diagonal positive definite matrix (typically, an identity matrix).

4.1.5 Gram-Schmidt Orthogonalization

The above adaptive algorithms for spatial decorrelation (whitening) are highly redundant in the sense that each processing unit is connected to all inputs so that the decorrelating matrix $\mathbf{W}$ is generally a full one. Blind decorrelation can also be performed by imposing some constraints on matrix $\mathbf{W}$, e.g., forcing it to be a lower triangular matrix with unit entries on the main diagonal. Let us consider the following simple cost function

$$J(\mathbf{W}) = \frac{1}{2}\sum_{i=1}^{n} E\{y_i^2\}, \tag{4.52}$$

with the constraint that the decorrelation matrix is a lower triangular matrix.

Applying the standard gradient descent method leads to a simple adaptive algorithm called Gram-Schmidt orthogonalization

$$y_i(k) \quad = \quad x_i(k) + \sum_{j \le i-1} w_{ij}(k) x_j(k) = \mathbf{w}_i^T(k)\mathbf{x}(k), \tag{4.53}$$

$$w_{ij}(k+1) \quad = \quad w_{ij}(k) - \eta(k)\,\langle y_i(k) x_j(k)\rangle\,, \quad j \le i-1, \quad i \ge 2, \tag{4.54}$$

with $y_1(k) = x_1(k)$ and $w_{11} = 1$. It should be noted that this set of weight vectors is not unique.

Alternatively, we can use the batch Gram-Schmidt orthogonalization as follows:

$$\mathbf{w}_1 \quad = \quad \mathbf{e}_1, \tag{4.55}$$

$$\mathbf{w}_i \quad = \quad \mathbf{e}_i - \sum_{j=1}^{i-1} \alpha_{ij}\,\mathbf{w}_j, \qquad (i = 2, \ldots, n) \tag{4.56}$$

where $\mathbf{e}_i = [0, \ldots 1, \ldots, 0]^T$ is the unit vector with 1 in the i-th place and

$$\alpha_{ij} = \frac{\mathbf{e}_i^T \, \mathbf{R_{xx}} \, \mathbf{w}_j}{\mathbf{w}_j^T \, \mathbf{R_{xx}} \, \mathbf{w}_j}, \qquad (j = 1, 2, \ldots, i-1; \;\; i = 1, 2, \ldots, n). \tag{4.57}$$

One advantage of the Gram-Schmidt orthogonalization algorithm over the eigenvalue decomposition approach is its lower computational complexity.

4.1.6 Blind Separation of Decorrelated Sources Versus Spatial Decorrelation

All the prewhitening rules can be used in the context of neural separating algorithms. The algorithm which iteratively applies either the rule (4.39) or (4.51) achieves the equilibrium point when the output signal covariance matrix becomes

$$\mathbf{R_{yy}} = E\{\mathbf{y}\,\mathbf{y}^T\} = E\{\mathbf{W}\mathbf{x}\mathbf{x}^T\mathbf{W}^T\} = \mathbf{W}E\{\mathbf{x}\mathbf{x}^T\}\mathbf{W}^T = \mathbf{W}\mathbf{R_{xx}}\mathbf{W}^T = \mathbf{I}_n. \tag{4.58}$$

Hence, assuming that $\mathbf{W}$ is a symmetric matrix, we get the equilibrium point at

$$\mathbf{W}_* = \mathbf{R}_{\mathbf{xx}}^{-\frac{1}{2}} = \mathbf{V_x}\mathbf{\Lambda}_{\mathbf{x}}^{-1/2}\mathbf{V}_{\mathbf{x}}^T, \tag{4.59}$$

where $\mathbf{V_x}$ is an orthogonal matrix and $\mathbf{\Lambda_x}$ is the diagonal matrix obtained by eigenvalue decomposition of the covariance matrix: $\mathbf{R_{xx}} = \mathbf{V_x}\mathbf{\Lambda_x}\mathbf{V}_{\mathbf{x}}^T$. This means that the output signals $y_i(k)$ will be mutually orthogonal with unit variances. In general, spatial decorrelation is not sufficient to perform instantaneous blind source separation from linear mixtures.

Remark 4.2 *It is interesting to note that for a special case when a mixing matrix* $\mathbf{H}$ *is nonsingular and symmetric, blind spatial decorrelation (whitening) algorithms with a symmetric whitening matrix* $\mathbf{W}$ *perform directly blind signal separation, since* $\mathbf{R_{xx}} = \mathbf{H}^2$ *under weak assumption that* $\mathbf{R_{ss}} = \mathbf{I}$ *(i.e., the sources are spatially uncorrelated and have unit variance), hence*

$$\boxed{\mathbf{W}_* = \widehat{\mathbf{H}}^{-1} = \widehat{\mathbf{R}}_{\mathbf{xx}}^{-1/2} = \mathbf{V_x}\mathbf{\Lambda}_{\mathbf{x}}^{-1/2}\mathbf{V}_{\mathbf{x}}^T.} \tag{4.60}$$

4.1.7 Bias Removal for Noisy Data

It should be noted that when the sensor signals $\mathbf{x}(k)$ are noisy such that $\mathbf{x}(k) = \hat{\mathbf{x}}(k) + \boldsymbol{\nu}(k)$ and $\hat{\mathbf{x}}(k) = \mathbf{H}\,\mathbf{s}(k)$, and $\hat{\mathbf{y}}(k) = \mathbf{W}(k)\,\hat{\mathbf{x}}(k)$ are noiseless estimates of the sensor and output vectors, respectively, it is easy to show that the additive noise $\boldsymbol{\nu}(k)$ within $\mathbf{x}(k)$ introduces a bias in the estimated decorrelation matrix $\mathbf{W}$. The covariance matrix of the output can be evaluated as

$$\mathbf{R_{yy}} = E\{\mathbf{y}(k)\,\mathbf{y}^T(k)\} = \mathbf{W}\,\boldsymbol{R}_{\hat{\mathbf{x}}\hat{\mathbf{x}}}\,\mathbf{W}^T + \mathbf{W}\,\mathbf{R}_{\boldsymbol{\nu}\boldsymbol{\nu}}\,\mathbf{W}^T, \tag{4.61}$$

where $\mathbf{R}_{\hat{\mathbf{x}}\hat{\mathbf{x}}} = E\{\hat{\mathbf{x}}(k)\,\hat{\mathbf{x}}^T(k)\}$ and $\mathbf{R}_{\boldsymbol{\nu}\boldsymbol{\nu}} = E\{\boldsymbol{\nu}(k)\,\boldsymbol{\nu}^T(k)\}$.

Assuming that the sample covariance matrix of the noise can be estimated (e.g., $\widehat{\mathbf{R}}_{\boldsymbol{\nu}\boldsymbol{\nu}} = \hat{\sigma}_\nu^2 \mathbf{I}$), modified adaptive learning algorithms (cf. (4.51) and (4.39)) employing bias removal can take the following forms:

$$\boxed{\Delta \mathbf{W}(k) = \eta(k)[\mathbf{I} - \widehat{\mathbf{R}}_{\mathbf{yy}}^{(k)} + \mathbf{W}(k)\,\widehat{\mathbf{R}}_{\boldsymbol{\nu}\boldsymbol{\nu}}\,\mathbf{W}^T(k)]} \tag{4.62}$$

and

$$\boxed{\Delta \mathbf{W}(k) = \eta(k)[\mathbf{I} - \widehat{\mathbf{R}}_{\mathbf{yy}}^{(k)} + \mathbf{W}(k)\,\widehat{\mathbf{R}}_{\boldsymbol{\nu}\boldsymbol{\nu}}\,\mathbf{W}^T(k)]\,\mathbf{W}(k).} \tag{4.63}$$

where $\widehat{\mathbf{R}}_{\mathbf{yy}}^{(k)} = (1-\eta_0)\,\widehat{\mathbf{R}}_{\mathbf{yy}}^{(k-1)} + \eta_0\,\mathbf{y}(k)\,\mathbf{y}^T(k)$.

The stochastic gradient version of the on-line global algorithm (4.39) for $\mathbf{R}_{\boldsymbol{\nu}\boldsymbol{\nu}} = \sigma_\nu^2\,\mathbf{I}$ is

$$\boxed{\Delta \mathbf{W}(k) = \eta(k)[\mathbf{I} - \mathbf{y}(k)\,\mathbf{y}^T(k) + \sigma_\nu^2\,\mathbf{W}(k)\,\mathbf{W}^T(k)]\mathbf{W}(k).} \tag{4.64}$$

4.1.8 Robust Prewhitening - Batch Algorithm

For data corrupted by noise, instead of the adaptive unbiased algorithms discussed in the previous section, we can attempt to apply a batch procedure called robust prewhitening based on the subspace approach. Using the subspace technique, we can relatively easily estimate the variance of noise and number of sources in the simplest case, when the covariance matrix of noise can be modelled as $\mathbf{R}_{\boldsymbol{\nu}\boldsymbol{\nu}} = \sigma_\nu^2\,\mathbf{I}_m$ and the variance of noise is relatively small (that is, the SNR is relatively high above some threshold).

The Algorithm Outline: Robust Prewhitening for $m > n$

---

1. Compute the sample covariance matrix: $\widehat{\mathbf{R}}_{\mathbf{xx}} = \mathbf{H}\,\widehat{\mathbf{R}}_{\mathbf{ss}}\,\mathbf{H}^T + \hat{\sigma}_\nu^2\,\mathbf{I}_m = \mathbf{H}\,\mathbf{H}^T + \hat{\sigma}_\nu^2\,\mathbf{I}_m$, which holds asymptotically under the assumption of independent sources with the unit variances and uncorrelated white noise.

2. Compute the eigenvalue decomposition:

$$\begin{aligned}
\widehat{\mathbf{R}}_{\mathbf{xx}} &= \mathbf{V}_x \boldsymbol{\Lambda}_x \mathbf{V}_x^T = [\mathbf{V}_\mathcal{S}, \mathbf{V}_\mathcal{N}] \begin{bmatrix} \boldsymbol{\Lambda}_\mathcal{S} & \mathbf{0} \\ \mathbf{0} & \boldsymbol{\Lambda}_\mathcal{N} \end{bmatrix} [\mathbf{V}_\mathcal{S}, \mathbf{V}_\mathcal{N}]^T \\
&= \mathbf{V}_\mathcal{S} \boldsymbol{\Lambda}_\mathcal{S} \mathbf{V}_\mathcal{S}^T + \mathbf{V}_\mathcal{N} \boldsymbol{\Lambda}_\mathcal{N} \mathbf{V}_\mathcal{N}^T,
\end{aligned} \tag{4.65}$$

   where $\mathbf{V}_\mathcal{S} \in \mathbb{R}^{m \times n}$ contains the eigenvectors associated with n principal eigenvalues of $\boldsymbol{\Lambda}_\mathcal{S} = \text{diag}\{\lambda_1 \geq \lambda_2 \cdots \geq \lambda_n\}$ in a descending order. Similarly, the matrix $\mathbf{V}_\mathcal{N} \in \mathbb{R}^{m \times (m-n)}$ contains $(m-n)$ noise eigenvectors that correspond to noise eigenvalues $\boldsymbol{\Lambda}_\mathcal{N} = \text{diag}\{\lambda_{n+1} \geq \cdots \geq \lambda_m\}$, with $\lambda_n > \lambda_{n+1}$. Usually, it is required that $\lambda_n >> \lambda_{n+1}$.

3. Estimate $\hat{\sigma}_\nu^2$ by computing the mean value of $(m-n)$ minor eigenvalues and the rank of the matrix $\mathbf{H}$. This can be done by using the distribution of eigenvalues by

detecting the gap between them or applying the AIC or MDL criteria (see Chapter 3 for details).

4. Define the whitening matrix[3]

$$\begin{aligned} \mathbf{W} &= \widehat{\mathbf{\Lambda}}_S^{-1/2} \mathbf{V}_S^T = (\mathbf{\Lambda}_S - \hat{\sigma}_\nu^2 \mathbf{I}_n)^{-1/2} \mathbf{V}_S^T \\ &= \operatorname{diag}\left\{ \frac{1}{\sqrt{(\lambda_1 - \hat{\sigma}_\nu^2)}}, \ldots, \frac{1}{\sqrt{(\lambda_n - \hat{\sigma}_\nu^2)}} \right\} \mathbf{V}_S^T \end{aligned} \tag{4.66}$$

and the prewhitened sensor vector: $\mathbf{y} = \mathbf{W}\,\mathbf{x}$.

Remark 4.3 *It should be noted that for noisy data ($\mathbf{x}(k) = \mathbf{H}\,\mathbf{s}(k) + \boldsymbol{\nu}(k)$) the above described whitening transform ($\mathbf{y}(k) = \mathbf{W}\,\mathbf{x}(k) = \mathbf{W}\,\mathbf{H}\,\mathbf{s}(k) + \mathbf{W}\,\boldsymbol{\nu}(k)$) can amplify the noise rather than suppress it, especially when $m = n$ and/or the mixing matrix $\mathbf{H}$ is ill-conditioned. For the ill-conditioned $\mathbf{H}$, some eigenvalues $\lambda_n, \lambda_{n-1}, \ldots$ are very small. Enhancement of noise will be different in different channels depending on the distribution of the eigenvalues. In such cases, to alleviate the problem, we can apply the regularization approach discussed in Chapter 2. Instead of (4.66) we can use the following formula for $m \geq n$:*

$$\boxed{\mathbf{W} = \operatorname{diag}\left\{ \sqrt{\frac{\lambda_1}{\lambda_1^2 + \hat{\sigma}_\nu^2}}, \ldots, \sqrt{\frac{\lambda_n}{\lambda_n^2 + \hat{\sigma}_\nu^2}} \right\} \mathbf{V}_S^T,} \tag{4.67}$$

where $\hat{\sigma}_\nu^2$ is the estimated noise variance.

For Gaussian noise, instead of the standard covariance matrix $\mathbf{R}_{\mathbf{xx}}$, we can employ fourth-order matrix cumulants which are insensitive to arbitrary Gaussian noise [886].

4.2 SECOND ORDER STATISTICS BLIND IDENTIFICATION BASED ON EVD AND GEVD

4.2.1 Mixing Model

In this section, we will discuss basic methods that jointly exploit the second order statistics (correlation matrices for different time delays) and temporal structure of sources. We show how the problem of blind identification of mixing matrix can be converted to the standard eigenvalue decomposition (EVD), generalized eigenvalue decomposition (GEVD) and simultaneous diagonalization (SD) problems.

[3]Such an operation is sometimes called "quasi-whitening", because it performs whitening not on the noisy sensor signals but rather on the estimated noise-free data.

We consider the case where sources may have arbitrary distributions but non-vanishing temporal correlations. More precisely, let us consider the simple mixing model where the m-dimensional observation (sensor) vector $\mathbf{x}(k) \in \mathbb{R}^m$ is assumed to be generated by

$$\mathbf{x}(k) = \mathbf{H}\,\mathbf{s}(k) + \boldsymbol{\nu}(k), \tag{4.68}$$

where $\mathbf{H} \in \mathbb{R}^{m \times n}$ is an unknown full column rank mixing matrix, $\mathbf{s}(k)$ is an n-dimensional source vector (which is also unknown and $m \geq n$), and $\boldsymbol{\nu}(k)$ is an additive noise vector that is assumed to be statistically independent of $\mathbf{s}(k)$.

The task of blind identification or equivalently blind source separation (BSS) entails estimation of the mixing matrix $\mathbf{H}$ or its pseudo inverse separating (unmixing) matrix $\mathbf{W} = \mathbf{H}^{+}$ in order to estimate the original source signals $\mathbf{s}(k)$, given only a finite number of observation data, $\{\mathbf{x}(k)\}$, $k = 1, \ldots, N$. Recall that two indeterminacies cannot be resolved in BSS without some *a priori* knowledge: Scaling and permutation ambiguities. Thus, if the estimate of the mixing matrix $\widehat{\mathbf{H}}$ satisfies $\mathbf{G} = \mathbf{W}\,\mathbf{H} = \widehat{\mathbf{H}}^{+}\,\mathbf{H} = \mathbf{P}\,\mathbf{D}$, where $\mathbf{G}$ is a global transformation which combines the mixing and separating system, $\mathbf{P}$ is some permutation matrix and $\mathbf{D}$ is some nonsingular scaling diagonal matrix, then $(\widehat{\mathbf{H}}, \widehat{\mathbf{s}})$ and $(\mathbf{H},\, \mathbf{s})$ are said to be related by a waveform-preserving relation [1159]. A key factor in BSS is the assumption about the statistical properties of sources such as statistical independence. That is the reason why BSS is often confused with independent component analysis (ICA). In this chapter, we exploit some weaker conditions for separation of sources assuming that they have temporal structures with different autocorrelation functions or equivalently different power spectra and/or they are nonstationary with time varying variances. Methods that exploit either the temporal structure of sources (mainly the second-order correlations) or the nonstationarity of sources, lead to the second-order statistics (SOS) based BSS methods. In contrast to the higher-order statistics (HOS) based BSS methods, all SOS based methods do not have to infer the probability distributions of sources or nonlinear activation functions [228, 230].

In this section and the next, we describe several batch methods that exploit spatio-temporal decorrelation to estimate (or identify) the mixing matrix in the presence of spatially correlated but temporally white noise (which might not necessarily be Gaussian). Moreover, we show that for a suitable set of time-delayed correlations of the observation data, we can find a robust (with respect to additive noise) estimate of the separating matrix $\mathbf{H}$. Throughout this section and the next the following assumptions are made unless stated otherwise:

(AS1) The mixing matrix $\mathbf{H}$ is of full column rank.

(AS2) Sources are spatially uncorrelated with different autocorrelation functions but are temporally correlated (colored) stochastic signals with zero-mean.

(AS3) Sources are stationary signals and/or second-order nonstationary signals in the sense that their variances are time varying.

(AS4) Additive noises $\{\nu_i(k)\}$ are independent of source signals. They can be spatially correlated but temporally white, i.e.,

$$E\{\boldsymbol{\nu}(k)\boldsymbol{\nu}^T(k-p)\} = \delta_{p0}\mathbf{R}_{\boldsymbol{\nu}}(p), \tag{4.69}$$

where δ_{p0} is the Kronecker symbol and $\mathbf{R}_{\boldsymbol{\nu}}$ is an arbitrary $m \times m$ matrix.

4.2.2 Basic Principles: Simultaneous Diagonalization and Eigenvalue Decomposition

Taking into account the above assumptions, it is straightforward to check if the correlation matrices of the vector $\mathbf{x}(k)$ of sensor signals satisfy

$$\begin{aligned} \mathbf{R}_{\mathbf{x}}(0) &= E\{\mathbf{x}(k)\mathbf{x}^T(k)\} = \mathbf{H}\,\mathbf{R}_{\mathbf{s}}(0)\,\mathbf{H}^T + \mathbf{R}_{\boldsymbol{\nu}}(0), &(4.70)\\ \mathbf{R}_{\mathbf{x}}(p) &= E\{\mathbf{x}(k)\mathbf{x}^T(k-p)\} = \mathbf{H}\,\mathbf{R}_{\mathbf{s}}(p)\,\mathbf{H}^T, &(4.71) \end{aligned}$$

for some non-zero time lag p. It follows from the assumption (AS2) that both $\mathbf{R}_{\mathbf{s}}(0) = E\{\mathbf{s}(k)\mathbf{s}^T(k)\}$ and $\mathbf{R}_{\mathbf{s}}(p) = E\{\mathbf{s}(k)\mathbf{s}^T(k-p)\}$ are non-zero distinct diagonal matrices.

In the case of overdetermined mixtures (more sensors than sources) when the covariance matrix of the noise has the special form $\mathbf{R}_{\boldsymbol{\nu}\boldsymbol{\nu}} = \mathbf{R}_{\boldsymbol{\nu}}(0) = E\{\boldsymbol{\nu}(k)\boldsymbol{\nu}^T(k)\} = \sigma_\nu^2 \mathbf{I}_m$, the noise variance σ_ν^2 can be estimated for relatively high SNR (signal to noise ratio) from the least singular value of $\mathbf{R}_{\mathbf{x}}(0)$ (or the average of minor $(m-n)$ singular values of $\mathbf{R}_{\mathbf{x}}(0)$) and the unbiased covariance matrix $\mathbf{R}_{\mathbf{x}}(0)$ can be estimated as

$$\bar{\mathbf{R}}_{\mathbf{x}}(0) = \mathbf{R}_{\mathbf{x}}(0) - \sigma_\nu^2\,\mathbf{I}_m = \mathbf{H}\,\mathbf{R}_{\mathbf{s}}(0)\,\mathbf{H}^T. \tag{4.72}$$

In order to estimate the mixing matrix $\mathbf{H}$ up to its re-scaled and permuted version, we can perform simultaneous diagonalization of two covariance matrices: $\widehat{\mathbf{R}}_{\mathbf{x}}(0)$ and $\widehat{\mathbf{R}}_{\mathbf{x}}(p)$, according to (4.71) and (4.72).

For the sake of simplicity, simultaneous diagonalization will be explained first, for the case where the number of sensors is equal to the number of sources ($m = n$). This is[4] performed in two steps: orthogonalization followed by unitary transformation as shown below

(1) First, the covariance matrix $\widehat{\bar{\mathbf{R}}}_{\mathbf{x}}(0) = (1/N)\sum_{k=1}^{N}(\mathbf{x}(k)\,\mathbf{x}^T(k)) - \hat{\sigma}_\nu^2\,\mathbf{I}_n$ is estimated and its EVD is performed as $\widehat{\bar{\mathbf{R}}}_{\mathbf{x}}(0) = \mathbf{V}_{\mathbf{x}}\,\boldsymbol{\Lambda}_{\mathbf{x}}\,\mathbf{V}_{\mathbf{x}}^T$. Then the standard whitening is realized by a linear transformation:

$$\overline{\mathbf{x}}(k) = \mathbf{Q}\,\mathbf{x}(k) = \boldsymbol{\Lambda}_{\mathbf{x}}^{-\frac{1}{2}}\,\mathbf{V}_{\mathbf{x}}^T\,\mathbf{x}(k), \tag{4.73}$$

where $\mathbf{Q} = \boldsymbol{\Lambda}_{\mathbf{x}}^{-\frac{1}{2}}\,\mathbf{V}_{\mathbf{x}}^T$. Hence, we have

$$\begin{aligned} \widehat{\mathbf{R}}_{\overline{\mathbf{x}}}(0) &= \frac{1}{N}\sum_{k=1}^{N}\overline{\mathbf{x}}(k)\,\overline{\mathbf{x}}^T(k) = \mathbf{Q}\,\widehat{\mathbf{R}}_{\mathbf{x}}(0)\,\mathbf{Q}^T = \mathbf{I}_n, &(4.74)\\ \widehat{\mathbf{R}}_{\overline{\mathbf{x}}}(p) &= \frac{1}{N}\sum_{k=1}^{N}\overline{\mathbf{x}}(k)\,\overline{\mathbf{x}}^T(k-p) = \mathbf{Q}\,\widehat{\mathbf{R}}_{\mathbf{x}}(p)\,\mathbf{Q}^T. &(4.75) \end{aligned}$$

[4]In contrast to the joint diagonalization problem, where we attempt to diagonalize (approximately) an arbitrary number of matrices, in the simultaneous diagonalization, the task is to diagonalize simultaneously (exactly) only two matrices.

(2) Second, an orthogonal transformation is applied to diagonalize the matrix $\widehat{\mathbf{R}}_{\overline{\mathbf{x}}}(p)$. The eigenvalue decomposition of $\widehat{\mathbf{R}}_{\overline{\mathbf{x}}}(p)$ has the form

$$\widehat{\mathbf{R}}_{\overline{\mathbf{x}}}(p) = \mathbf{V}_{\overline{\mathbf{x}}} \mathbf{\Lambda}_{\overline{\mathbf{x}}} \mathbf{V}_{\overline{\mathbf{x}}}^T. \tag{4.76}$$

Simultaneously, based on (4.71) and (4.75), we obtain

$$\widehat{\mathbf{R}}_{\overline{\mathbf{x}}}(p) = \mathbf{Q}\, \widehat{\mathbf{R}}_{\mathbf{x}}(p)\, \mathbf{Q}^T = \mathbf{Q}\, \mathbf{H}\, \widehat{\mathbf{R}}_{\mathbf{s}}(p)\, \mathbf{H}^T\, \mathbf{Q}^T. \tag{4.77}$$

Hence, if the diagonal matrix $\mathbf{\Lambda}_{\overline{\mathbf{x}}}$ has distinct eigenvalues then the mixing matrix can be estimated uniquely (up to the sign and permutation matrices) (see Theorem 4.1 as given below)

$$\widehat{\mathbf{H}} = \mathbf{Q}^{-1} \mathbf{V}_{\overline{\mathbf{x}}} = \mathbf{V}_{\mathbf{x}}\, \mathbf{\Lambda}_{\mathbf{x}}^{1/2}\, \mathbf{V}_{\overline{\mathbf{x}}}. \tag{4.78}$$

Simultaneous diagonalization of two symmetric matrices can be carried out without going through the two-step procedure, by converting the problem to the generalized eigenvalue decomposition (GEVD) [1158, 217]. In fact, the problem can be easily converted to the standard eigenvalue problem which can be formulated for the nonsingular mixing matrix $\mathbf{H}$ as (see Eqs. (4.71)-(4.72)):

$$\widehat{\mathbf{R}}_{\mathbf{x}}^{-1}(0)\widehat{\mathbf{R}}_{\mathbf{x}}(p) = (\mathbf{H}^T)^{-1}\mathbf{R}_{\mathbf{s}}^{-1}(0)\mathbf{R}_{\mathbf{s}}(p)\mathbf{H}^T = \mathbf{V}\mathbf{\Lambda}\mathbf{V}^{-1}, \tag{4.79}$$

or equivalently, the generalized eigenvalue problem:

$$\widehat{\mathbf{R}}_{\mathbf{x}}(p)\mathbf{V} = \widehat{\mathbf{R}}_{\mathbf{x}}(0)\mathbf{V}\mathbf{\Lambda} \tag{4.80}$$

under the condition that $\mathbf{\Lambda} = \mathbf{R}_{\mathbf{s}}^{-1}(0)\mathbf{R}_{\mathbf{s}}(p)$ has distinct eigenvalues. Then, the mixing matrix $\mathbf{H}$ can be estimated based on the eigenvectors of the GEVD (4.80) as

$$\widehat{\mathbf{H}} = (\mathbf{V}^T)^{-1} = \mathbf{V}^{-T}, \tag{4.81}$$

up to arbitrary scaling and permutation of columns.

The following Theorem [1158, 217] explains and summarizes these basic results.

Theorem 4.1 *Let $\mathbf{\Lambda}_1$, $\mathbf{\Lambda}_2$, $\mathbf{D}_1$, $\mathbf{D}_2 \in \mathbb{R}^{n\times n}$ be diagonal matrices with non-zero diagonal entries and additionally matrices $\mathbf{\Lambda}_1$ and $\mathbf{D}_1$ are positive definite. Suppose that $\mathbf{G} \in \mathbb{R}^{n\times n}$ satisfies the following decompositions:*

$$\mathbf{D}_1 = \mathbf{G}\, \mathbf{\Lambda}_1\, \mathbf{G}^T, \tag{4.82}$$

$$\mathbf{D}_2 = \mathbf{G}\, \mathbf{\Lambda}_2\, \mathbf{G}^T. \tag{4.83}$$

Then, the matrix $\mathbf{G}$ is a generalized permutation matrix[5] if $\mathbf{D}_1^{-1}\mathbf{D}_2$ and $\mathbf{\Lambda}_1^{-1}\mathbf{\Lambda}_2$ have distinct diagonal entries.

[5]The generalized permutation matrix is defined as $\mathbf{G} = \mathbf{P}\mathbf{D}$, where $\mathbf{P}$ is a standard permutation matrix and $\mathbf{D}$ is any nonsingular diagonal matrix.

Proof. From (4.82), there exists an orthogonal matrix $\mathbf{U}$ such that

$$\left(\mathbf{G}\,\mathbf{\Lambda}_1^{\frac{1}{2}}\right) = \left(\mathbf{D}_1^{\frac{1}{2}}\right)\mathbf{U}. \tag{4.84}$$

Hence,

$$\mathbf{G} = \mathbf{D}_1^{\frac{1}{2}}\mathbf{U}\mathbf{\Lambda}_1^{-\frac{1}{2}}. \tag{4.85}$$

Substitute (4.85) into (4.83) to obtain

$$\mathbf{D}_1^{-1}\mathbf{D}_2 = \mathbf{U}\mathbf{\Lambda}_1^{-1}\mathbf{\Lambda}_2\mathbf{U}^T. \tag{4.86}$$

Since the right-hand side of (4.86) is the eigen-decomposition of the matrix $\mathbf{D}_1^{-1}\mathbf{D}_2$, the diagonal elements of $\mathbf{D}_1^{-1}\mathbf{D}_2$ and $\mathbf{\Lambda}_1^{-1}\mathbf{\Lambda}_2$ are the same. From the assumption that the diagonal elements of $\mathbf{D}_1^{-1}\mathbf{D}_2$ are distinct, the orthogonal matrix $\mathbf{U}$ must have the form $\mathbf{U} = \mathbf{P}\boldsymbol{S}_{\boldsymbol{g}}$, where $\mathbf{P}$ is a permutation matrix and $\boldsymbol{S}_{\boldsymbol{g}}$ is a diagonal matrix whose diagonal elements are either $+1$ or -1. Hence, we have

$$\begin{aligned}\mathbf{G} &= \mathbf{D}_1^{\frac{1}{2}}\mathbf{P}\boldsymbol{S}_{\boldsymbol{g}}\mathbf{\Lambda}_1^{-\frac{1}{2}} \\ &= \mathbf{P}\mathbf{P}^T\mathbf{D}_1^{\frac{1}{2}}\mathbf{P}\boldsymbol{S}_{\boldsymbol{g}}\mathbf{\Lambda}_1^{-\frac{1}{2}} \\ &= \mathbf{P}\mathbf{D}_0,\end{aligned} \tag{4.87}$$

where $\mathbf{D}_0$ is a diagonal matrix expressed as

$$\mathbf{D}_0 = \mathbf{P}^T\mathbf{D}_1^{\frac{1}{2}}\mathbf{P}\boldsymbol{S}_{\boldsymbol{g}}\mathbf{\Lambda}_1^{-\frac{1}{2}}. \tag{4.88}$$

Remark 4.4 *For successful source separation, we may choose any time delay p for which $\widehat{\mathbf{R}}_{\mathbf{x}}^{-1}(0)\widehat{\mathbf{R}}_{\mathbf{x}}(p)$ has non-zero distinct eigenvalues. We have found, by extensive experiments, that for typical real world signals a good choice is usually $p = 1$. It is also possible to choose a linear combination $\sum_i \alpha_i\widehat{\mathbf{R}}_{\mathbf{x}}(i)$ instead of $\widehat{\mathbf{R}}_{\mathbf{x}}(p)$.*

It is also important to note, that instead of using the generalized eigenvalue decomposition, we can use the standard eigenvalue decomposition (EVD) or equivalently the singular value decomposition (SVD) in a two stage procedure described in detail below [1158].

The Algorithm Outline: Two-stage EVD/SVD for the case where number of sensors is more than the number sources

---

1. Estimate the correlation matrix of sensor signals as

$$\widehat{\mathbf{R}}_{\mathbf{x}}(0) = \frac{1}{N}\sum_{k=1}^{N}\mathbf{x}(k)\,\mathbf{x}^T(k). \tag{4.89}$$

2. Compute the EVD (or equivalently SVD) of $\widehat{\mathbf{R}}_{\mathbf{x}}(0)$ as (see section 4.1.8)

$$\begin{aligned}\widehat{\mathbf{R}}_{\mathbf{x}}(0) &= \mathbf{U}_{\mathbf{x}}\,\mathbf{\Sigma}_{\mathbf{x}}\,\mathbf{V}_{\mathbf{x}}^T = \mathbf{V}_{\mathbf{x}}\mathbf{\Lambda}_{\mathbf{x}}\mathbf{V}_{\mathbf{x}}^T \\ &= \mathbf{V}_{\mathcal{S}}\mathbf{\Lambda}_{\mathcal{S}}\mathbf{V}_{\mathcal{S}}^T + \mathbf{V}_{\mathcal{N}}\mathbf{\Lambda}_{\mathcal{N}}\mathbf{V}_{\mathcal{N}}^T,\end{aligned} \tag{4.90}$$

where $\mathbf{V}_{\mathcal{S}} = [\mathbf{v}_1, \mathbf{v}_2, \ldots, \mathbf{v}_n] \in \mathbb{R}^{m \times n}$ contains the eigenvectors associated with n principal eigenvalues of $\mathbf{\Lambda}_{\mathcal{S}} = \text{diag}\{\lambda_1 \geq \lambda_2 \cdots \geq \lambda_n\}$ in descending order. Similarly, matrix $\mathbf{V}_{\mathcal{N}} \in \mathbb{R}^{m \times (m-n)}$ contains $(m-n)$ noise eigenvectors that correspond to noise eigenvalues $\mathbf{\Lambda}_{\mathcal{N}} = \text{diag}\{\lambda_{n+1} \geq \cdots \geq \lambda_m\}$, with $\lambda_n > \lambda_{n+1}$. It should be noted that eigenvalues have usually the typical relationship $\lambda_1 \geq \lambda_2 \geq \cdots \lambda_n > \lambda_{n+1} \approx \cdots \approx \lambda_m$, $m > n$. This means that the last $(m-n)$ non-significant (minor) eigenvalues correspond to noise subspace and the first significant (principal) eigenvalues correspond to signal plus noise subspace. Estimate the number of sources n from the number of most significant singular values.

3. Estimate also the variance σ_ν^2 of the white noise as the mean value of the $(m-n)$ least significant eigen (or singular) values.

4. Perform a robust (with respect to the white noise) prewhitening transformation as

$$\overline{\mathbf{x}}(k) = \widehat{\mathbf{\Lambda}}_{\mathcal{S}}^{-1/2} \mathbf{V}_{\mathcal{S}}^T \mathbf{x}(k) = \mathbf{Q}\, \mathbf{x}(k), \tag{4.91}$$

where $\widehat{\mathbf{\Lambda}}_{\mathcal{S}} = \text{diag}\{(\lambda_1 - \hat{\sigma}_\nu^2), (\lambda_2 - \hat{\sigma}_\nu^2), \ldots, (\lambda_n - \hat{\sigma}_\nu^2)\})$.

5. Estimate the covariance matrix of the vector $\overline{\mathbf{x}}(k)$ for specific time delay $p \neq 0$ (typically, $p = 1$ gives best results) and perform the SVD of the covariance matrix:

$$\widehat{\mathbf{R}}_{\overline{\mathbf{x}}}(p) = \frac{1}{N} \sum_{k=1}^{N} \overline{\mathbf{x}}(k)\, \overline{\mathbf{x}}^T(k-p) = \mathbf{U}_{\overline{\mathbf{x}}}\, \mathbf{\Sigma}_{\overline{\mathbf{x}}}\, \mathbf{V}_{\overline{\mathbf{x}}}^T. \tag{4.92}$$

6. Check whether for a specific time delay p, all singular values of the diagonal matrix $\mathbf{\Sigma}_{\overline{\mathbf{x}}}$ are distinct. If not, repeat step 4 for a different time delay p.

   If singular values are distinct and sufficiently far away from each other, then, we can estimate successfully the mixing matrix as

$$\widehat{\mathbf{H}} = \mathbf{Q}^+\, \mathbf{U}_{\overline{\mathbf{x}}} = \mathbf{V}_{\mathcal{S}}\, \widehat{\mathbf{\Lambda}}_{\mathcal{S}}^{1/2}\, \mathbf{U}_{\overline{\mathbf{x}}} \tag{4.93}$$

   and if necessary, noisy source signals[6] as

$$\mathbf{y}(k) = \widehat{\mathbf{s}}(k) = \mathbf{U}_{\overline{\mathbf{x}}}^T\, \overline{\mathbf{x}}(k) = \mathbf{U}_{\overline{\mathbf{x}}}^T\, \widehat{\mathbf{\Lambda}}_{\mathcal{S}}^{-1/2}\, \mathbf{V}_{\mathcal{S}}^T\, \mathbf{x}(k). \tag{4.94}$$

It should be noted that if both covariance matrices $\widehat{\mathbf{R}}_{\mathbf{x}}(0)$ and $\widehat{\mathbf{R}}_{\overline{\mathbf{x}}}(p)$ are symmetric positive definite, then $\mathbf{U}_{\mathbf{x}} = \mathbf{V}_{\mathbf{x}}$ and $\mathbf{U}_{\overline{\mathbf{x}}} = \mathbf{V}_{\overline{\mathbf{x}}}$, respectively and the SVD and PCA/EVD techniques are equivalent.

[6]The estimated sources will be recovered without cross-talking due to unbiased estimation of the unmixing matrix. However, they will be corrupted by additive noise since the noise is projected from the sensor signals by the linear transformation (4.94). In order to remove noise, we need to apply methods described in Chapters 1 and 8.

The above procedure is a modified and optimized version of the algorithm called the AMUSE (Algorithm for Multiple Unknown Signals Extraction) [1158, 850]. Usually, for single sample time delay $p = 1$, the above algorithm successfully separates colored sources with different power spectra shapes. This means that in such a case the eigenvalues of the time-delayed covariance matrix are distinct. The main disadvantage of this algorithm is that its accuracy quickly deteriorates in the presence of additive noise.

The AMUSE algorithm for BSS of colored sources can be naturally extended to the ICA of independent non Gaussian source signals if instead of the standard time-delayed covariance matrices $\mathbf{R}_{\overline{\mathbf{x}}}(p)$, we use the contracted quadricovariance matrices defined as

$$\begin{aligned}\mathbf{C}_{\overline{\mathbf{x}}}(\mathbf{E}) &= \mathbf{C}_{\overline{\mathbf{x}}}\{\overline{\mathbf{x}}^T(k)\,\mathbf{E}\,\overline{\mathbf{x}}(k)\,\overline{\mathbf{x}}(k)\,\overline{\mathbf{x}}^T(k)\} \\ &= E\{\overline{\mathbf{x}}^T(k)\,\mathbf{E}\,\overline{\mathbf{x}}(k)\,\overline{\mathbf{x}}(k)\,\overline{\mathbf{x}}^T(k)\} - \mathbf{R}_{\overline{\mathbf{x}}}(0)\,\mathbf{E}\,\mathbf{R}_{\overline{\mathbf{x}}}(0) \\ &\quad - \mathrm{tr}(\mathbf{E}\,\mathbf{R}_{\overline{\mathbf{x}}}(0))\,\mathbf{R}_{\overline{\mathbf{x}}}(0) - \mathbf{R}_{\overline{\mathbf{x}}}(0)\,\mathbf{E}^T\,\mathbf{R}_{\overline{\mathbf{x}}}(0), \end{aligned} \tag{4.95}$$

where $\mathbf{R}_{\overline{\mathbf{x}}}(0) = E\{\overline{\mathbf{x}}(k)\,\overline{\mathbf{x}}^T(k)\}$ [7] and $\mathbf{E} \in \mathbb{R}^{n \times n}$ is some freely chosen matrix called eigenmatrix (typically, $\mathbf{E} = \mathbf{I}_n$ or $\mathbf{E} = \mathbf{e}_q\,\mathbf{e}_q^T$, where $\mathbf{e}_q$ are the column vectors of some unitary matrix) [129, 130].

It can be shown that such a matrix has the following eigenvalue decomposition (EVD):

$$\mathbf{C}_{\overline{\mathbf{x}}}(\mathbf{E}) = \mathbf{U}\,\mathbf{\Lambda}_{\mathbf{E}}\,\mathbf{U}^T, \tag{4.96}$$

with $\mathbf{\Lambda}_{\mathbf{E}} = \mathrm{diag}\{\lambda_1 \mathbf{u}_1^T \mathbf{E}\,\mathbf{u}_1, \ldots, \lambda_n \mathbf{u}_n^T \mathbf{E}\,\mathbf{u}_n\}$, $\lambda_i = \kappa_4(s_i) = E\{s_i^4\} - 3E^2\{s_i^2\}$ is the kurtosis of the zero-mean i-th source and $\mathbf{u}_i$ is the i-th column of the orthogonal eigenvector matrix $\mathbf{U}$. Hence, if the EVD of $\mathbf{C}_{\overline{\mathbf{x}}}(\mathbf{E}) = \mathbf{U}\,\mathbf{\Lambda}_{\mathbf{E}}\,\mathbf{U}^T = \widehat{\mathbf{A}}\,\mathbf{C}_{\mathbf{s}}(\mathbf{E})\,\widehat{\mathbf{A}}^T$ is unique in the sense that all eigenvalues of $\mathbf{\Lambda}_{\mathbf{E}}$ are distinct, we can estimate the mixing matrix $\widehat{\mathbf{A}} = \mathbf{QH} = \mathbf{U}$. In the special case for $\mathbf{E} = \mathbf{I}_n$ these conditions are satisfied if the source signals have different values of kurtosis. The above procedure is called FOBI (Fourth-Order Blind Identification) [129, 130, 879, 589].

Remark 4.5 *The main advantage of using fourth order quadricovariance matrices is their theoretical insensitivity to an arbitrary Gaussian noise. Furthermore, HOS-based techniques enable us to identify the mixing system when sources are i.i.d. and mutually independent. However, it should be emphasized that standard time-delayed covariance matrices can be estimated accurately with far fewer data samples than their higher order counterparts. In such cases, when the number of available samples is relatively low, working with SOS-based instead of HOS-based techniques is advantageous, especially in a time-varying environment.*

The above algorithms based on the time-delayed covariance matrices and symmetric EVD/SVD and GEVD are probably the simplest batch algorithms for blind identification and blind separation of sources with temporal structure. However, their robustness with respect to noise and performance can be poor, especially when the additive noise is large or we are not able to estimate precisely the covariance matrix of the noise. In order to alleviate this problem, we can use two covariance matrices: $\widehat{\mathbf{R}}_{\mathbf{x}}(p_1)$ and $\widehat{\mathbf{R}}_{\mathbf{x}}(p_2)$ for non-zero time

[7] For the prewhitened data, we have $\mathbf{R}_{\overline{\mathbf{x}}}(0) = \mathbf{I}_n$.

delays ($p_1 \neq p_2 \neq 0$). Since the noise vector was assumed to be temporally white, the covariance matrices $\widehat{\mathbf{R}}_{\mathbf{x}}(p_1)$ and $\widehat{\mathbf{R}}_{\mathbf{x}}(p_2)$ are not affected by the noise vector, i.e.,

$$\begin{aligned}\widehat{\mathbf{R}}_{\mathbf{x}}(p_1) &= \mathbf{H}\widehat{\mathbf{R}}_{\mathbf{s}}(p_1)\mathbf{H}^T, \\ \widehat{\mathbf{R}}_{\mathbf{x}}(p_2) &= \mathbf{H}\widehat{\mathbf{R}}_{\mathbf{s}}(p_2)\mathbf{H}^T\end{aligned}$$

for any time delay different from zero. Thus, it is possible to obtain a robust estimate of the unmixing matrix, regardless of the probability distributions and the spatial structure of the noise vector [217, 218]. However, to perform GEVD or EVD with robust prewhitening one of the matrices $\mathbf{R}_{\mathbf{x}}(p_1)$ or $\mathbf{R}_{\mathbf{x}}(p_2)$ must be positive definite. This is not guaranteed for all time delays. Therefore, a new problem arises; how to select an optimal time delay, such that at least one of the covariance matrices is symmetric positive definite. Furthermore, the described algorithms exploit only two different correlation matrices of the observation vector, so their performance is degraded if some eigenvalues of $\mathbf{R}_{\mathbf{s}}(p)$ are close to each other. In order to avoid these drawbacks, we should use rather a larger set of time-delayed correlation matrices for various time lags as explained in the future sections.

4.3 IMPROVED SOS BLIND IDENTIFICATION ALGORITHMS BASED ON SYMMETRIC EVD/SVD

There is a current trend in ICA/BSS to investigate the "average eigen-structure" of a large set of data matrices formed as functions of available data (typically, covariance or cumulant matrices for different time delays). In other words, the objective is to extract reliable information (e.g., estimation of sources and/or the mixing matrix) from the eigen-structure of a possibly large set of data matrices [153]. However, since in practice we only have a finite number of samples of signals corrupted by noise, the data matrices do not exactly share the same eigen-structure. Furthermore, it should be noted that determining the eigen-structure on the basis of one or even two data matrices leads usually to poor or unsatisfactory results because such matrices, based usually on an arbitrary choice, may have some degenerate eigenvalues which leads to loss of information contained in other data matrices. Therefore, from a statistical point of view, in order to provide robustness and accuracy, it is necessary to consider the average eigen-structure by taking into account simultaneously a possibly large set of data matrices [151, 152, 153, 1216]. In this section and the next, we will describe several approaches that exploit average eigen-structure in order to estimate reliable sources and mixing matrix.

4.3.1 Robust Orthogonalization of Mixing Matrices for Colored Sources

Let us consider the standard mixing model:

$$\mathbf{x}(k) = \mathbf{H}\,\mathbf{s}(k) + \boldsymbol{\nu}(k), \tag{4.97}$$

where $\mathbf{x}(k) \in \mathbb{R}^m$ is the available vector of sensor signals, $\mathbf{H} \in \mathbb{R}^{m \times n}$ is the full column rank mixing matrix and $\mathbf{s}(k) \in \mathbb{R}^n$ is the vector of temporally correlated sources.

We formulate the robust orthogonalization problem as follows: Find a linear transformation $\overline{\mathbf{x}}(k) = \mathbf{Q}\,\mathbf{x}(k) \in \mathbb{R}^n$ such that the global mixing matrix, defined as $\mathbf{A} = \mathbf{Q}\,\mathbf{H} \in \mathbb{R}^{n\times n}$, will be orthogonal and unbiased by the additive white noise $\boldsymbol{\nu}(k)$.

Such robust orthogonalization is an important pre-processing step in a variety of BSS methods. It ensures that the global mixing matrix is orthogonal. The conventional whitening exploits the zero time-lag covariance matrix $\mathbf{R}_{\mathbf{xx}} = \mathbf{R}_{\mathbf{x}}(0) = E\{\mathbf{x}(k)\,\mathbf{x}^T(k)\}$, so that the effect of the additive white noise can not be removed if the covariance matrix of the noise can not be precisely estimated, especially in the case when the number of sensors is equal to the number of sources.

The idea of the robust orthogonalization is to search for such a linear combination of several (typically, from 5 to 50) symmetric time-delayed covariance matrices, i.e.,

$$\overline{\mathbf{R}}_{\mathbf{x}}(\boldsymbol{\alpha}) = \sum_{i=1}^{K} \alpha_i\, \tilde{\mathbf{R}}_{\mathbf{x}}(p_i), \tag{4.98}$$

where matrix $\overline{\mathbf{R}}_{\mathbf{x}}$ is positive definite and insensitive to the additive white noise [1155]. A proper choice of coefficients $\{\alpha_i\}$ induces that the matrix $\overline{\mathbf{R}}_{\mathbf{x}}$ will be symmetric positive definite. It should be noted that matrices $\tilde{\mathbf{R}}_{\mathbf{x}}(p_i) = [\mathbf{R}_{\mathbf{x}}(p_i) + \mathbf{R}_{\mathbf{x}}^T(p_i)]/2$ are symmetric but not necessarily positive definite, especially for a large time delay p_i.

The practical implementation of the algorithm for data corrupted by white noise is given below [1155, 82, 230].

The Algorithm Outline: Robust Orthogonalization

---

1. Estimate a set of time-delayed covariance matrices of sensor signals for preselected time delays $(p_1, p_2, \ldots, p_K)$ and construct an $m\times mK$ matrix $\boldsymbol{\mathcal{R}} = [\tilde{\mathbf{R}}_{\mathbf{x}}(p_1) \cdots \tilde{\mathbf{R}}_{\mathbf{x}}(p_K)]$, where $\tilde{\mathbf{R}}_{\mathbf{x}}(p_i) = (\langle \mathbf{x}(k)\mathbf{x}^T(k-p)\rangle + \langle \mathbf{x}(k-p)\mathbf{x}^T(k)\rangle)/2$.

   Then, compute the singular value decomposition (SVD) of $\boldsymbol{\mathcal{R}}$, i.e.,

   $$\boldsymbol{\mathcal{R}} = \mathbf{U}\,\boldsymbol{\Sigma}\,\mathbf{V}^T, \tag{4.99}$$

   where $\mathbf{U} = [\mathbf{U}_s, \mathbf{U}_\nu] \in \mathbb{R}^{m\times m}$ (with $\mathbf{U}_s = [\mathbf{u}_1, \ldots \mathbf{u}_n] \in \mathbb{R}^{m\times n}$) and $\mathbf{V} \in \mathbb{R}^{mK\times mK}$ are orthogonal matrices, and $\boldsymbol{\Sigma}$ is an $m \times mK$ matrix whose left n columns contain $\operatorname{diag}\{\sigma_1, \sigma_2, \ldots, \sigma_n\}$ (with non-increasing singular values) and whose right $(mK - n)$ columns are zero. The number of unknown sources n can be detected by inspecting the singular values as explained in the previous section under the assumption that the noise covariance matrix is modelled as $\mathbf{R}_{\boldsymbol{\nu}} = \sigma_\nu^2 \mathbf{I}_m$ and the variance of noise is relative low, i.e., $\sigma_\nu^2 \ll \sigma_n^2$.

2. For $i = 1, 2, \ldots, K$, compute

   $$\mathbf{R}_i = \mathbf{U}_{\mathbf{s}}^T \tilde{\mathbf{R}}_{\mathbf{x}}(p_i) \mathbf{U}_{\mathbf{s}}. \tag{4.100}$$

3. Choose any non-zero initial vector of parameters $\boldsymbol{\alpha} = [\alpha_1, \alpha_2, \ldots, \alpha_K]^T$.

4. Compute

$$\overline{\mathbf{R}} = \sum_{i=1}^{K} \alpha_i \, \mathbf{R}_i. \tag{4.101}$$

5. Compute the EVD decomposition of $\overline{\mathbf{R}}$ and check if $\overline{\mathbf{R}}$ is positive definite or not. If $\overline{\mathbf{R}}$ is positive definite, go to Step 7. Otherwise, go to Step 6.

6. Choose an eigenvector $\mathbf{u}$ corresponding to the smallest eigenvalue[8] of $\overline{\mathbf{R}}$ and update $\boldsymbol{\alpha}$ via replacing $\boldsymbol{\alpha}$ by $\boldsymbol{\alpha} + \boldsymbol{\delta}$, where

$$\boldsymbol{\delta} = \frac{\left[\mathbf{u}^T \mathbf{R}_1 \mathbf{u} \cdots \mathbf{u}^T \mathbf{R}_K \mathbf{u}\right]^T}{\|[\mathbf{u}^T \mathbf{R}_1 \mathbf{u} \cdots \mathbf{u}^T \mathbf{R}_K \mathbf{u}]\|}. \tag{4.102}$$

   Go to step 4.

7. Compute symmetric positive definite matrix

$$\overline{\mathbf{R}}_{\mathbf{x}}(\boldsymbol{\alpha}_*) = \sum_{i=1}^{K} \alpha_i \, \tilde{\mathbf{R}}_{\mathbf{x}}(p_i), \tag{4.103}$$

   and perform SVD or symmetric EVD of $\overline{\mathbf{R}}_{\mathbf{x}}$,

$$\overline{\mathbf{R}}_{\mathbf{x}}(\boldsymbol{\alpha}_*) = [\mathbf{U}_{\mathcal{S}}, \mathbf{U}_{\mathcal{N}}] \begin{bmatrix} \boldsymbol{\Sigma}_{\mathcal{S}} & \mathbf{0} \\ \mathbf{0} & \boldsymbol{\Sigma}_{\mathcal{N}} \end{bmatrix} [\mathbf{V}_{\mathcal{S}}, \mathbf{V}_{\mathcal{N}}]^T, \tag{4.104}$$

   where $(\boldsymbol{\alpha}_*)$ is the set of parameters α_i after the algorithm achieves convergence, i.e., positive definiteness of the matrix $\overline{\mathbf{R}}$, $\mathbf{U}_{\mathcal{S}}$ contains the eigenvectors associated with n principal singular values of $\boldsymbol{\Sigma}_{\mathcal{S}} = \text{diag}\{\sigma_1, \sigma_2, \ldots, \sigma_n\}$.

8. The robust orthogonalization transformation is performed by

$$\overline{\mathbf{x}}(k) = \mathbf{Q}\,\mathbf{x}(k), \tag{4.105}$$

   where $\mathbf{Q} = \boldsymbol{\Sigma}_{\mathcal{S}}^{-\frac{1}{2}} \mathbf{U}_{\mathcal{S}}^T$.

Some remarks and comments are now in order:

- The robust orthogonalization algorithm converges globally for any non-zero initial condition of $\boldsymbol{\alpha}$ under the assumption that all sources have different autocorrelation functions which are linearly independent or equivalently have distinct power spectra. Moreover, it converges in a finite number of steps [1155].

[8] If the smallest eigenvalue has some multiplicity, take any vector $\mathbf{u}$ corresponding to the smallest eigenvalue.

- In the ideal noiseless case, the last $(m-n)$ singular values of $\overline{\mathbf{R}}_{\mathbf{x}}(\alpha)$ are equal to zero, thus $\mathbf{\Sigma}_{\mathcal{N}} = \mathbf{0}$.

- In the case of $m = n$ (equal numbers of sources and sensors), steps 1 and 2 are not necessary. Simply, we let $\mathbf{R}_i = \tilde{\mathbf{R}}_{\mathbf{x}}(p_i) = (\mathbf{R}_{\mathbf{x}}(p_i) + \mathbf{R}_{\mathbf{x}}^T(p_i))/2$.

- For $m > n$ the linear transformation $\overline{\mathbf{x}}(k) = \mathbf{Q}\,\mathbf{x}(k)$ besides orthogonalization enables us to estimate the number of sources, i.e., the orthogonalization matrix reduces the array of sensor signals to an n-dimensional vector, thus the number of sources can be estimated, under conditions that the SNR is relatively high.

- By defining a new mixing matrix as $\mathbf{A} = \mathbf{Q}\,\mathbf{H}\mathbf{D}^{1/2}$, where $\mathbf{D} = \sum_{i=1}^{L} \alpha_i\, \tilde{\mathbf{R}}_{\mathbf{s}}(p_i)$ is a diagonal (scaling) matrix with positive entries, it is straightforward to show that $\mathbf{A}\mathbf{A}^T = \mathbf{I}_n$, thus, the matrix $\mathbf{A}$ is orthogonal. This orthogonality condition is necessary for performing separation of signals using the symmetric EVD or Joint Diagonalization approaches. It should be noted that in contrast to the standard prewhitening procedure for our robust orthogonalization, generally $E\{\overline{\mathbf{x}}\,\overline{\mathbf{x}}^T\} = \mathbf{D}_{\overline{\mathbf{x}}} \neq \mathbf{I}_n$. We have $\overline{\mathbf{x}} = \mathbf{A}\,\tilde{\mathbf{s}} + \mathbf{Q}\,\mathbf{n}$, where $\tilde{\mathbf{s}} = \mathbf{D}^{-1/2}\,\mathbf{s}$, but due to the scaling indeterminacy of the sources, we may write in the sequel that $\overline{\mathbf{x}} = \mathbf{A}\,\mathbf{s} + \tilde{\boldsymbol{\nu}}$ $\quad(\tilde{\boldsymbol{\nu}} = \mathbf{Q}\,\boldsymbol{\nu})$. The diagonal elements of $\mathbf{D}$ are positive, due to the positive definiteness of $\overline{\mathbf{R}}_{\mathbf{x}}(\boldsymbol{\alpha})$ [450].

Several extensions and improvements to the above presented robust orthogonalization algorithm are possible, especially, if the noise is not completely white (i.e., the noise has white and colored components) and/or the number of available samples is relatively small.

First of all, instead of a simple shift (time delay) operator, we can use generalized delay operators or more generally suitably designed filters. In other words, instead of the standard time-delayed sampled covariance matrices $\widehat{\mathbf{R}}_{\mathbf{x}}(p_i) = \langle \mathbf{x}(k)\mathbf{x}^T(k-p_i)\rangle$, we can use the generalized sampled covariance matrices of the form

$$\widehat{\mathbf{R}}_{\mathbf{x}\,\widetilde{\mathbf{x}}}(B_i) = \frac{1}{N}\sum_{k=1}^{N} \mathbf{x}(k)\widetilde{\mathbf{x}}_{B_i}^T(k), \qquad (i = 1, 2, \ldots, K) \tag{4.106}$$

where vector $\widetilde{\mathbf{x}}_{B_i}(k) = B_i(z)[\mathbf{x}(k)] = \sum_p b_{ip}\mathbf{x}(k-p)$ is a filtered version of the vector $\mathbf{x}(k)$ and $B_i(z)$ denotes transfer function of a suitably designed filter or generalized time-delay operator[9]. It should be noted that in general any set of FIR (finite impulse response) or stable IIR (infinite impulse response) filter may be used in the preprocessing stage. However, we propose to use banks of bandpass filters possibly with overlapping band-passes covering a bandwidth of all source signals but with different central frequencies as is illustrated in Fig. 4.5. For example, we can use simple second-order IIR bandpass filters with transfer characteristics

$$B_i(z) = z^{-q_i}(1-r_i)\frac{(\omega_{ci}z^{-1}/(r_i + r_i^2)) - 1}{1 - \omega_{ci}z^{-1} + r_i^2 z^{-2}}, \tag{4.107}$$

[9]In the simplest case, $B_i(z) = z^{-i}$. Generalized delay operator of the first-order has the following form $B_1(z) = \frac{\alpha + \beta z^{-1}}{1 + \gamma z^{-1}}$, where α, β and γ are suitably chosen coefficients.

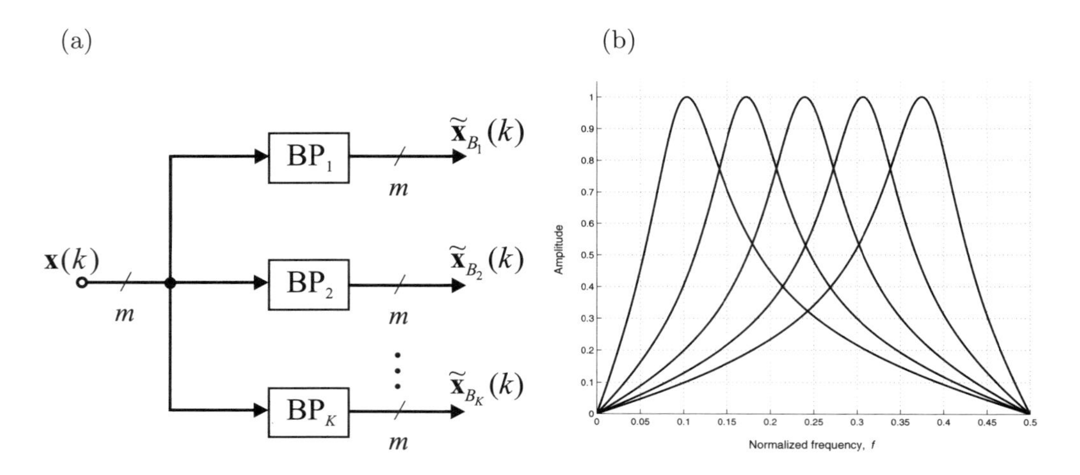

Fig. 4.5 Illustration of processing of signals by using a bank of bandpass filters: (a) Filtering the vector $\mathbf{x}(k)$ of sensor signals by a bank of sub-band filters, (b) typical frequency characteristics of bandpass filters.

where $\omega_{ci}(k) = 2r_i \cos(2\pi f_{ci} k)$ with the center frequency f_{ci} and the parameter r_i related to the frequency bandwidth by relationship $B_{wi} = (1 - r_i)/2$. The suitably designed bank of bandpass filters enables us to remove efficiently wide-band noise, which is out of the bandwidth of the source signals.[10] If the bandwidth of source signals is known approximately, we can use only $K = n$ bandpass filters with bandwidths consistent with the bandwidth of the sources.

In order to ensure the symmetry of the generalized covariance matrices, we will use the matrices defined as: $\widetilde{\mathbf{R}}_{\mathbf{x}\,\widetilde{\mathbf{x}}}(B_i) = (E\{\mathbf{x}(k)\,\widetilde{\mathbf{x}}_{B_i}^T(k)\} + E\{\widetilde{\mathbf{x}}_{B_i}(k)\,\mathbf{x}^T(k)\})/2$, where the vectors $\widetilde{\mathbf{x}}_{B_i}(k) = [\widetilde{x}_1(k), \ldots, \widetilde{x}_m(k)]^T$ represent sub-band filtered versions of the vector $\mathbf{x}(k)$.

For nonstationary source signals and stationary noise and/or interference, we can adopt an alternative approach, based on the concept of differential correlation matrices, defined as [219]

$$\delta \mathbf{R}_{\mathbf{x}}(T_i, T_j, p_l) = \mathbf{R}_{\mathbf{x}}(T_i, p_l) - \mathbf{R}_{\mathbf{x}}(T_j, p_l), \tag{4.108}$$

where T_i and T_j are two non-overlapping time windows of the same size and $\mathbf{R}_{\mathbf{x}}(T_i, p_l)$ denotes the time-delayed correlation matrix for the time window T_i. It should be noted that such defined differential time-delayed correlation matrices are insensitive to stationary signals. In order to perform robust orthogonalization for nonstationary sources, we divide the sensor data $\mathbf{x}(k)$ into K non-overlapping blocks (time windows T_i) and estimate the set of differential matrices $\delta\tilde{\mathbf{R}}_{\mathbf{x}}(T_i, T_j, p_l)$ for $i = 1, \ldots, K$, $j > i$ and $l = 1, \ldots, M$ (typically, $M = 5$ and $K = 10$ and the number of samples in each block is 100).

[10] It is important that the bandpass of the filters possibly match bandwidths corresponding to the highest energy of individual sources.

In the next step, we formulate the composite differential matrix defined as

$$\overline{\delta \mathbf{R}_{\mathbf{x}}}(\boldsymbol{\alpha}) = \sum_{ijl} \alpha_{ijl}\, \delta \tilde{\mathbf{R}}_{\mathbf{x}}(T_i, T_j, p_l) \qquad (i = 1, 2, \ldots, K;\; j > i;\; l = 1, 2, \ldots, M) \quad (4.109)$$

and using the approach described above, we can estimate the set of coefficients α_{ijl} for which the matrix $\overline{\delta \mathbf{R}_{\mathbf{x}}}(\boldsymbol{\alpha})$ is positive definite. In the last step, we perform the symmetric EVD of the positive definite matrix $\overline{\delta \mathbf{R}_{\mathbf{x}}}(\boldsymbol{\alpha}_*)$ and compute the orthogonalization matrix $\mathbf{Q}$ (cf. Eqs. (4.101)-(4.105)).

For sensor signals corrupted by any Gaussian noise, instead of the time-delayed covariance matrices, we can use the fourth order quadricovariance matrices defined as:

$$\begin{aligned}
\mathbf{C}_{\mathbf{x}}(p, \mathbf{E}_q) &= \mathbf{C}_{\mathbf{x}}\, \{(\mathbf{x}^T(k-p)\, \mathbf{E}_q\, \mathbf{x}(k-p))\, \mathbf{x}(k)\, \mathbf{x}^T(k)\} \\
&= E\{(\mathbf{x}^T(k-p)\, \mathbf{E}_q\, \mathbf{x}(k-p))\, \mathbf{x}(k)\, \mathbf{x}^T(k)\} - \mathbf{R}_{\mathbf{x}}(p)\, \mathbf{E}_q\, \mathbf{R}_{\mathbf{x}}^T(p) \\
&\quad - \operatorname{tr}(\mathbf{E}_q\, \mathbf{R}_{\mathbf{x}}(0))\, \mathbf{R}_{\mathbf{x}}(0) - \mathbf{R}_{\mathbf{x}}(p)\, \mathbf{E}_q^T\, \mathbf{R}_{\mathbf{x}}^T(p),
\end{aligned} \quad (4.110)$$

where $\mathbf{R}_{\mathbf{x}}(0) = E\{\mathbf{x}(k)\, \mathbf{x}^T(k)\} = E\{\mathbf{x}(k-p)\, \mathbf{x}^T(k-p)\}$, $\mathbf{R}_{\mathbf{x}}(p) = E\{\mathbf{x}(k)\, \mathbf{x}^T(k-p)\}$ and $\mathbf{E}_q \in \mathbb{R}^{n \times n}$ is any matrix, typically, $\mathbf{E}_q = \mathbf{I}$ or $\mathbf{E}_q = \mathbf{u}_q \mathbf{u}_q^T$, where $\mathbf{u}_p$ is the p-th vector of some orthogonal matrix $\mathbf{U}$. [11] Our objective is to find such a set of matrices $\mathbf{E}_q$ and time-delays p such that the quadricovariance matrix (4.110) (or a linear combination of several such matrices) becomes positive definite.

4.3.2 An Improved Algorithm Based on GEVD

Using robust matrix orthogonalization, we can develop several improved and extended algorithms based on the EVD/SVD or GEVD. In this section, we will discuss an improved algorithm based on GEVD or the matrix pencil proposed by Choi *et al.* [217, 229].

The set of all matrices of the form $\mathbf{R}_1 - \lambda \mathbf{R}_2$ (with some parameter λ) is said to be a *matrix pencil.* Frequently, we encounter the case where $\mathbf{R}_1$ is symmetric and $\mathbf{R}_2$ is symmetric and positive definite. Pencils of this variety are referred to as *symmetric definite pencils* [494].

Theorem 4.2 *If* $\mathbf{R}_1 - \lambda \mathbf{R}_2$ *is a symmetric definite pencil (i.e. both matrices are symmetric and* $\mathbf{R}_2$ *is positive definite), then there exists a nonsingular matrix* $\mathbf{V} = [\mathbf{v}_1, \ldots, \mathbf{v}_n]$ *which performs simultaneous diagonalization of* $\mathbf{R}_1$ *and* $\mathbf{R}_2$*:*

$$\mathbf{V}^T \mathbf{R}_1 \mathbf{V} = \mathbf{D}_1, \quad (4.111)$$

$$\mathbf{V}^T \mathbf{R}_2 \mathbf{V} = \mathbf{D}_2, \quad (4.112)$$

if the diagonal matrix $\mathbf{D}_1\, \mathbf{D}_2^{-1}$ *has distinct entries. Moreover, the problem can be converted to GEVD:* $\mathbf{R}_1\, \mathbf{V} = \mathbf{R}_2\, \mathbf{V}\, \boldsymbol{\Lambda}$*, where* $\boldsymbol{\Lambda} = \operatorname{diag}\{\lambda_1, \lambda_2, \ldots, \lambda_n\} = \mathbf{D}_1\, \mathbf{D}_2^{-1}$ *(or equivalently* $\mathbf{R}_1 \mathbf{v}_i = \lambda_i \mathbf{R}_2 \mathbf{v}_i$ *for* $i = 1, \ldots, n$*), if all eigenvalues* $\lambda_i = \frac{d_i(\mathbf{R}_1)}{d_i(\mathbf{R}_2)}$ *are distinct.*

[11] The matrix $\mathbf{U}$ can be estimated by EVD of the simplified contracted quadricovariance matrix for $p = 0$ and $\mathbf{E}_q = \mathbf{I}$ as $\mathbf{C}_{\mathbf{x}}(0, \mathbf{I}) = \mathbf{C}_{\mathbf{x}}\, \{(\mathbf{x}^T(k)\, \mathbf{x}(k))\, \mathbf{x}(k)\, \mathbf{x}^T(k)\} = E\{(\mathbf{x}^T(k)\, \mathbf{x}(k))\, \mathbf{x}(k)\, \mathbf{x}^T(k)\} - 2\, \mathbf{R}_{\mathbf{x}}(0)\, \mathbf{R}_{\mathbf{x}}(0) - \operatorname{tr}(\mathbf{R}_{\mathbf{x}}(0))\, \mathbf{R}_{\mathbf{x}}(0) = \mathbf{U}\, \boldsymbol{\Lambda}_{\mathbf{I}}\, \mathbf{U}^T$.

It is apparent from Theorem 4.2 that $\mathbf{R}_1$ should be symmetric and $\mathbf{R}_2$ should be symmetric and positive definite so that the generalized eigenvector $\mathbf{V}$ can be a valid solution (in the sense that $\widehat{\mathbf{H}} = (\mathbf{V}^T)^{-1}$) on the condition that all the generalized eigenvalues λ_i are distinct. Unfortunately, for some time delays, the covariance matrices $\mathbf{R_x}(p_1) = \mathbf{R}_1$ and $\mathbf{R_x}(p_2) = \mathbf{R}_2$ cannot be positive definite. Moreover, due to some noise and numerical error they cannot be symmetric. Thus, we might have a numerical problem in the calculation of generalized eigenvectors, which can be complex-valued in such cases [229, 230].

Remark 4.6 *In fact, the positiveness of matrix* $\mathbf{R}_2 = \overline{\mathbf{R}}_\mathbf{x}$ *is not absolutely necessary. If* $\mathbf{R}_1$ *and* $\mathbf{R}_2$ *are symmetric and* $\mathbf{R}_2$ *is not positive definite, then we can try to construct a positive definite matrix* $\mathbf{R}_3 = \beta_1 \mathbf{R}_1 + \beta_2 \mathbf{R}_2$ *for some choice of real coefficients* β_1, β_2*, and next to solve equivalent generalized symmetric eigen problem* $\mathbf{R}_1 \mathbf{V} = \mathbf{R}_3 \mathbf{V} \boldsymbol{\Lambda}$ *in the sense that the eigenvectors of the pencils* $\mathbf{R}_1 - \lambda \mathbf{R}_2$ *and* $\mathbf{R}_1 - \lambda \mathbf{R}_3$ *are identical. The eigenvalues* λ_i *of* $\mathbf{R}_1 - \lambda \mathbf{R}_2$ *and the eigenvalues* $\tilde{\lambda}_i$ *of* $\mathbf{R}_1 - \lambda \mathbf{R}_3$ *are related by* $\lambda_i = \beta_2 \tilde{\lambda}_i / (1 - \beta_1 \tilde{\lambda}_i)$*.*

Let us consider two time-delayed covariance matrices $\mathbf{R_x}(p_1)$ and $\mathbf{R_x}(p_2)$ for non-zero time lags p_1 and p_2. For the requirement of symmetry, we replace $\mathbf{R_x}(p_1)$ and $\mathbf{R_x}(p_2)$ by $\tilde{\mathbf{R}}_\mathbf{x}(p_1)$ and $\tilde{\mathbf{R}}_\mathbf{x}(p_2)$ that are defined by

$$\tilde{\mathbf{R}}_\mathbf{x}(p_1) = \frac{1}{2} \left\{ \mathbf{R_x}(p_1) + \mathbf{R}_\mathbf{x}^T(p_1) \right\}, \tag{4.113}$$

$$\tilde{\mathbf{R}}_\mathbf{x}(p_2) = \frac{1}{2} \left\{ \mathbf{R_x}(p_2) + \mathbf{R}_\mathbf{x}^T(p_2) \right\}. \tag{4.114}$$

Then, the pencil $\tilde{\mathbf{R}}_\mathbf{x}(p_1) - \lambda \tilde{\mathbf{R}}_\mathbf{x}(p_2)$ is a symmetric pencil. In general, the matrix $\tilde{\mathbf{R}}_\mathbf{x}(p_2)$ is not positive definite. Therefore, instead of $\tilde{\mathbf{R}}_\mathbf{x}(p_2)$ for a single time delay, we consider a linear combination of several time-delayed covariance matrices:

$$\overline{\mathbf{R}}_\mathbf{x}(\boldsymbol{\alpha}) = \sum_{i=1}^{K} \alpha_i \, \tilde{\mathbf{R}}_\mathbf{x}(p_i). \tag{4.115}$$

The set of coefficients $\{\alpha_i\}$ is chosen in such a way that the matrix $\overline{\mathbf{R}}_\mathbf{x}(\boldsymbol{\alpha})$ is positive definite, as described in the previous section. Hence, the pencil $\mathbf{R_x}(p_1) - \lambda \overline{\mathbf{R}}_\mathbf{x}(\boldsymbol{\alpha})$ is a symmetric definite pencil and its generalized eigenvectors are calculated without any numerical problem.

This method referred to as *Improved GEVD (Matrix Pencil) Method* is summarized below [217, 229].

Algorithm Outline: The Improved GEVD (Matrix Pencil) Algorithm

---

1. Compute $\mathbf{R}_1 = \tilde{\mathbf{R}}_\mathbf{x}(p_1)$ for some time lag $p_1 \neq 0$ (typically, $p = 1$) and calculate a symmetric positive definite matrix $\mathbf{R}_2 = \overline{\mathbf{R}}_\mathbf{x}(\boldsymbol{\alpha}) = \sum_{i=1}^{K} \alpha_i \, \tilde{\mathbf{R}}_{\mathbf{x}\tilde{\mathbf{x}}}(p_i)$ by using the robust orthogonalization method (employing a time-delay operator, bank of bandpass filters or differential correlation matrices).

2. Find the generalized eigenvector matrix $\mathbf{V}$ for the generalized eigen value decomposition (GEVD)

$$\tilde{\mathbf{R}}_{\mathbf{x}}(p_1)\mathbf{V} = \overline{\mathbf{R}}_{\mathbf{x}}(\boldsymbol{\alpha})\,\mathbf{V}\,\boldsymbol{\Lambda}. \tag{4.116}$$

3. The mixing matrix is given by $\widehat{\mathbf{H}} = (\mathbf{V}^T)^{-1}$ on the condition that all eigenvalues are real and distinct.

4.3.3 An Improved Two-stage Symmetric EVD/SVD Algorithm

Instead of GEVD approach, we can use the standard symmetric EVD or SVD in a two-stage (or more) procedure. Let us assume, that the sensor signals are corrupted by the additive white noise and the number of sources is generally unknown (with the number of sensors larger or equal to the number of sources).

Using a set of covariance matrices and the robust orthogonalization described above, we can implement the following algorithm.

The Algorithm Outline: Robust EVD/SVD Algorithm

---

1. Perform the robust orthogonalization transformation $\overline{\mathbf{x}}(k) = \mathbf{Q}\,\mathbf{x}(k)$ using one of the methods described in the previous section, such that the global mixing matrix $\mathbf{A} = \mathbf{Q}\,\mathbf{H}$ is orthogonal.

2. Compute the linear combination of a set of the time-delayed covariance matrices of the vector $\overline{\mathbf{x}}(k)$ for a set of time delays $p_i \neq 0$ (or alternatively using a bank of bandpass filters)

$$\overline{\mathbf{R}}_{\overline{\mathbf{x}}}(\boldsymbol{\beta}) = \sum_{i=1}^{M} \beta_i\,\tilde{\mathbf{R}}_{\overline{\mathbf{x}}}(p_i), \tag{4.117}$$

where a set of coefficients β_i can be randomly chosen.

3. Perform SVD (or equivalently EVD) as

$$\overline{\mathbf{R}}_{\overline{\mathbf{x}}}(\boldsymbol{\beta}) = \mathbf{U}_{\overline{\mathbf{x}}}\boldsymbol{\Sigma}_{\overline{\mathbf{x}}}\mathbf{U}_{\overline{\mathbf{x}}}^T \tag{4.118}$$

and check whether for the specific set of parameters β_i and p_i all singular values of the diagonal matrix $\boldsymbol{\Sigma}_{\overline{\mathbf{x}}}$ are distinct. If not, repeat steps 2 and 3 for a different set of parameters. If the singular values are distinct and sufficiently far away from each other, then, we can estimate (unbiased by white noise) the mixing matrix as

$$\widehat{\mathbf{H}} = \mathbf{Q}^{+}\,\mathbf{U}_{\overline{\mathbf{x}}} \tag{4.119}$$

and/or if necessary estimate (noisy) colored source signals as

$$\mathbf{y}(k) = \widehat{\mathbf{s}}(k) = \mathbf{U}_{\overline{\mathbf{x}}}^T\,\overline{\mathbf{x}}(k) = \mathbf{U}_{\overline{\mathbf{x}}}^T\,\mathbf{Q}\,\mathbf{x}(k). \tag{4.120}$$

4.3.4 Blind Separation and Identification Using Bank of Bandpass Filters and Robust Orthogonalization

Instead of using a linear combination of a set of covariance matrices for various time delays, we can use a single generalized covariance matrix $\widetilde{\mathbf{R}}_{\overline{\mathbf{x}}\,\widetilde{\overline{\mathbf{x}}}} = (E\{\overline{\mathbf{x}}(k)\,\widetilde{\overline{\mathbf{x}}}(k)^T\} + E\{\widetilde{\overline{\mathbf{x}}}(k)\,\overline{\mathbf{x}}^T(k)\})/2$, where the vector $\widetilde{\overline{\mathbf{x}}}(k) = [\widetilde{\overline{x}}_1(k), \ldots, \widetilde{\overline{x}}_n(k)]^T$ represents a filtered version of the vector $\overline{\mathbf{x}}$. More precisely, each signal $\widetilde{\overline{x}}_j(k) = B(z)\overline{x}_j(k) = \sum_p b_p \overline{x}_j(k-p)$ is a sub-bandpass filtered version of signal $\overline{x}_j$. It should be noted that all filters have identical transfer functions $B(z)$ for each channel and each source signal should have a frequency range located, at least partially, in the bandwidth of the filters. Detailed implementation of the algorithm is given below.

The Algorithm Outline: Robust SVD with Bank of Band-Pass Filters

---

1. Perform the robust orthogonalization transformation ($\overline{\mathbf{x}}(k) = \mathbf{Q}\,\mathbf{x}(k)$), for example, by computing the SVD of a symmetric positive definite matrix

$$\overline{\mathbf{R}}_{\mathbf{x}\,\widetilde{\mathbf{x}}}(\boldsymbol{\alpha}) = \sum_{i=1}^{K} \alpha_i\, \widetilde{\mathbf{R}}_{\mathbf{x}\,\widetilde{\mathbf{x}}}(B_i) = \mathbf{U}_S\, \boldsymbol{\Sigma}_S\, \mathbf{V}_S^T, \tag{4.121}$$

such that the global mixing matrix ($\mathbf{A} = \mathbf{Q}\,\mathbf{H} \in \mathbb{R}^{n \times n}$) is orthogonal.

2. Generate the vector $\widetilde{\overline{\mathbf{x}}}(k) = [\widetilde{\overline{x}}_1(k), \ldots, \widetilde{\overline{x}}_n(k)]^T$, defined as $\widetilde{\overline{\mathbf{x}}}(k) = B(z)\overline{\mathbf{x}}(k) = \sum_{p=1}^{L} b_p \overline{\mathbf{x}}(k-p)$, by passing the signals $\overline{x}_j(k)$ through the bandpass filter $B(z)$. The estimate the symmetric generalized covariance matrix defined as

$$\widetilde{\mathbf{R}}_{\overline{\mathbf{x}}\,\widetilde{\overline{\mathbf{x}}}} = \frac{1}{2}(E\{\overline{\mathbf{x}}(k)\,\widetilde{\overline{\mathbf{x}}}(k)^T\} + E\{\widetilde{\overline{\mathbf{x}}}(k)\,\overline{\mathbf{x}}^T(k)\}). \tag{4.122}$$

3. Perform SVD (or equivalently EVD) of the symmetric covariance matrix $\widetilde{\mathbf{R}}_{\overline{\mathbf{x}}\,\widetilde{\overline{\mathbf{x}}}}$

$$\widetilde{\mathbf{R}}_{\overline{\mathbf{x}}\,\widetilde{\overline{\mathbf{x}}}} = \mathbf{U}_{\overline{\mathbf{x}}} \boldsymbol{\Sigma}_{\overline{\mathbf{x}}} \mathbf{U}_{\overline{\mathbf{x}}}^T \tag{4.123}$$

and check whether for a specific set of filter parameters ($B(z) = \sum_{p=1}^{L} b_p z^{-1}$) all singular values of the diagonal matrix $\boldsymbol{\Sigma}_{\overline{\mathbf{x}}}$ are distinct. If not, repeat steps 2 and 3 for a different set of filter parameters. If the singular values are distinct and sufficiently far away from each other then, we can estimate (unbiased by the noise) the mixing matrix as

$$\widehat{\mathbf{H}} = \mathbf{Q}^+\, \mathbf{U}_{\overline{\mathbf{x}}} = \mathbf{U}_S\, (\boldsymbol{\Sigma}_S)^{1/2}\, \mathbf{U}_{\overline{\mathbf{x}}} \tag{4.124}$$

and/or the noisy source signals as

$$\mathbf{y}(k) = \widehat{\mathbf{s}}(k) = \mathbf{U}_{\overline{\mathbf{x}}}^T\, \overline{\mathbf{x}}(k) = \mathbf{U}_{\overline{\mathbf{x}}}^T\, (\boldsymbol{\Sigma}_S)^{-1/2}\, \mathbf{U}_S^T \mathbf{x}(k). \tag{4.125}$$

4.4 JOINT DIAGONALIZATION - ROBUST SOBI ALGORITHMS

In the previous section, we implemented the average eigen-structure by taking linear combination of several covariance matrices and applying the standard EVD or SVD. An alternative approach to EVD/SVD is to apply the approximate joint diagonalization procedure [97, 151, 153, 564, 433, 1380]. The objective of this procedure is to find the orthogonal matrix $\mathbf{U}$ which diagonalizes a set of matrices [112, 866, 1380]:

$$\mathbf{M}_i = \mathbf{U}\mathbf{D}_i\mathbf{U}^T + \boldsymbol{\varepsilon}_i, \qquad (i = 1, 2, \ldots, L) \tag{4.126}$$

where $\mathbf{M}_i \in \mathbb{R}^{n\times n}$ are data matrices (for example, time-delayed covariance and/or cumulant matrices), the $\mathbf{D}_i$ are diagonal and real, and $\boldsymbol{\varepsilon}_i$ represent additive errors or noise matrix (as small as possible). If $L > 2$ for arbitrary matrices $\mathbf{M}_i$, the problem becomes overdetermined and generally we can not find an exact diagonalizing matrix $\mathbf{U}$ with $\boldsymbol{\varepsilon}_i = \mathbf{0}$, $\forall i$.

A natural and common criterion for the Joint Approximative Diagonalization (JAD) is the least-squares (LS) approach which can be formulated as the minimization of a general cost function [1216, 1261]:

$$J(\mathbf{U}, \mathbf{D}_i) = \sum_{i=1}^{L} \|\mathbf{M}_i - \mathbf{U}\mathbf{D}_i\mathbf{U}^T\|_F^2. \tag{4.127}$$

It should be noted that the minimization proceeds not only over the orthogonal matrix $\mathbf{U}$, but also over the set of diagonal matrices $\mathbf{D}_i$, since they are also unknown. Thus, the problem can be solved by, the so called Alternating Least Squares (ALS) technique. In the ALS technique, we alternatively minimize over one component set, keeping the other component set fixed. In particular, assume that at the k-th iteration, we have an estimate $\mathbf{U}_k$. The next step is to minimize $J(\mathbf{U}_k, \mathbf{D}_i)$ with respect to $\mathbf{D}_i$ [1216].

It can be shown that the problem of estimating the orthogonal matrix $\mathbf{U}$ can be converted to the problem of minimization of the following cost function [1261]

$$\begin{aligned}\tilde{J}(\mathbf{U}) &= -\sum_{i=1}^{L}\sum_{j=1}^{n} |\mathbf{u}_j^T\mathbf{M}_i\mathbf{u}_j|^2 \\ &= -\sum_{i=1}^{L} \|\operatorname{diag}\{\mathbf{U}^T\mathbf{M}_i\mathbf{U}\}\|_F^2,\end{aligned} \tag{4.128}$$

where $\|\operatorname{diag}(\cdot)\|$ denotes the norm of the vector built from the diagonal entries of the matrix.

The above criterion can be formulated in a slightly different form as

$$\bar{J}(\mathbf{U}) = \sum_{i=1}^{L} \mathbf{off}\{\mathbf{U}^T\mathbf{M}_i\mathbf{U}\}, \tag{4.129}$$

where

$$\mathbf{off}\{\mathbf{M}\} = \sum_{i\neq j} |m_{ij}|^2.$$

In order to improve convergence of the optimization procedure, we can attempt to find a good (close to optimum) initial estimate of $\mathbf{U}$ by applying the eigenvalue decomposition for two selected data matrices as

$$\mathbf{M}_p \mathbf{M}_q^{-1} = \mathbf{U}_0 \mathbf{\Lambda}_p \mathbf{\Lambda}_q^{-1} \mathbf{U}_0^{-1}. \tag{4.130}$$

It is known from our previous discussion that such initialization is possible if the inverse of matrix $\mathbf{M}_q$ exists and the eigenvalues $\mathbf{\Lambda}_{pq} = \mathbf{\Lambda}_p \mathbf{\Lambda}_q^{-1}$ are real and distinct. However, due to noise and numerical errors, the eigenvalues $\mathbf{\Lambda}_{pq}$ may become complex-valued. We have shown earlier that if either $\mathbf{M}_p$ or $\mathbf{M}_q$ is positive definite, then eigenvalues of $\mathbf{\Lambda}_{pq}$ are real. Thus, to avoid the problem, we can search among all the data matrices $\{\mathbf{M}_i\}$ for a matrix that is symmetric positive definite and then use this matrix in our initialization step [1216].

The above criteria assume that the matrix $\mathbf{U}$ is orthogonal and sensor data are preprocessed using robust whitening or orthogonalization procedure.

Recently, Pham proposed a different criterion for a set of symmetric positive definite matrices $\{\mathbf{M}_i\}$, which does not require any prewhitening and the diagonalizing matrix $\mathbf{W}$ is simultaneously a separating matrix [964]. Using the Hadamard inequality $\det \mathbf{M} \leq \det(\operatorname{diag} \mathbf{M})$ for a symmetric positive definite matrix $\mathbf{M}$, with equality if and only if $\mathbf{M}$ is the diagonal, he proposed the cost function

$$J(\mathbf{W}) = \sum_{i=1}^{L} \gamma_i \left[\log \det \operatorname{diag}\{\mathbf{W}\mathbf{M}_i \mathbf{W}^T\} - \log \det(\mathbf{W}\mathbf{M}_i \mathbf{W}^T) \right], \tag{4.131}$$

where γ_i are positive weighting coefficients and $\operatorname{diag}\{\cdot\}$ denotes the diagonal matrix with the same diagonal as its argument. One advantage of such a cost function is that its minimization directly leads to estimation of the separating matrix without orthogonalization or prewhitening. In practice, we usually want to avoid this prewhitening since it deteriorates the performance of the whole process, since the bias or error in preprocessing stage cannot be corrected in the following separation (rotation) stage. However, the drawback of this approach is that it requires the set of data matrices to be symmetric and positive definite. Therefore, we need to find linear combinations of data matrices that are positive definite what increases the computational complexity.

An important advantage of the Joint Approximate Diagonalization (JAD) is that several numerically efficient algorithms exist for its computation, including Jacobi techniques (one sided and two sided), Alternating Least Squares (ALS), PARAFAC (Parallel Factor Analysis) and subspace fitting techniques employing the efficient Gauss-Newton optimization [151, 1216, 1261].

The matrices $\mathbf{M}_i$ can take different forms. In one of the simplest cases, for colored sources with distinct power spectra (or equivalently, different autocorrelation functions), we can use time-delayed covariance matrices, i.e.,

$$\mathbf{M}_i = \mathbf{R}_{\overline{\mathbf{x}}}(p_i) = E\{\overline{\mathbf{x}}(k)\, \overline{\mathbf{x}}^T(k - p_i)\}. \tag{4.132}$$

In such a case, we obtain the second order blind identification (SOBI) algorithm developed first by Belouchrani *et al.* [84, 80]. It should be noted that for prewhitened sensor signals or orthogonalized mixing matrix $\mathbf{A} = \mathbf{Q}\,\mathbf{H}$, we have

$$\mathbf{R}_{\overline{\mathbf{x}}}(p_i) = \mathbf{Q}\, \mathbf{R}_{\mathbf{x}}(p_i)\, \mathbf{Q}^T = \mathbf{A}\, \mathbf{R}_{\mathbf{s}}(p_i)\, \mathbf{A}^T = \mathbf{U}\, \mathbf{D}_i\, \mathbf{U}^T, \qquad (i = 1, 2, \ldots L). \tag{4.133}$$

Thus, the orthogonal mixing matrix can be estimated as $\widehat{\mathbf{A}} = \mathbf{Q}\,\widehat{\mathbf{H}} = \mathbf{U}$ up to irrelevant scaling and permutation of columns on the condition that at least one diagonal matrix $\mathbf{D}_i(p_i)$ has distinct diagonal entries. The source signals can be estimated as $\hat{\mathbf{s}}(k) = \mathbf{U}^T\,\mathbf{Q}\,\mathbf{x}(k)$ and the mixing matrix is estimated as $\widehat{\mathbf{H}} = \mathbf{Q}^{+}\,\mathbf{U}^T$. It should be noted that it is rather difficult to determine *a priori* a single time lag p for which the diagonal matrix $\mathbf{D}(p)$ has distinct diagonal elements. The JAD reduces the probability of un-identifiability of a mixing matrix caused by an unfortunate choice of time lag p. The Robust Second Order Blind Identification (RSOBI) algorithm can be summarized as follows [82].

The Algorithm Outline: Robust SOBI Algorithm

---

1. Perform robust orthogonalization $\overline{\mathbf{x}}(k) = \mathbf{Q}\,\mathbf{x}(k)$, according to the algorithm described in section 4.3.1.

2. Estimate the set of covariance matrices:

$$\widehat{\mathbf{R}}_{\overline{\mathbf{x}}}(p_i) = (1/N)\sum_{k=1}^{N} \overline{\mathbf{x}}(k)\,\overline{\mathbf{x}}^T(k-p_i) = \mathbf{Q}\,\widehat{\mathbf{R}}_{\mathbf{x}}(p_i)\,\mathbf{Q}^T \tag{4.134}$$

   for a preselected set of time lags $(p_1, p_2, \ldots, p_L)$ or bandpass filters B_i.

3. Perform JAD: $\mathbf{R}_{\overline{\mathbf{x}}}(p_i) = \mathbf{U}\mathbf{D}_i\mathbf{U}^T, \quad \forall i$, i.e., estimate the orthogonal matrix $\mathbf{U}$ using one of the available numerical algorithms [153, 298, 300, 433, 564, 1216, 1261].

4. Estimate the source signals as

$$\widehat{\mathbf{s}}(k) = \mathbf{U}^T\,\mathbf{Q}\,\mathbf{x}(k) \tag{4.135}$$

   and the mixing matrix as

$$\widehat{\mathbf{H}} = \mathbf{Q}^{+}\,\mathbf{U}. \tag{4.136}$$

Remark 4.7 *It should be noted that the sampled covariance matrices* $\widehat{\mathbf{R}}_{\overline{\mathbf{x}}\,\widetilde{\overline{\mathbf{x}}}} = \left\langle \overline{\mathbf{x}}(k)\widetilde{\overline{\mathbf{x}}}_{B_i}^T(k)\right\rangle$, *with* $\widetilde{\overline{\mathbf{x}}}_{B_i}^T(k) = B_i(z)[\overline{\mathbf{x}}(k)]$, *employing the bandpass filters* $B_i(z)$, *can be very ill-conditioned, especially, if the filters have very narrow band-passes. Therefore, joint diagonalization may not work properly for such covariance matrices or the separation performance can be poor. To avoid this problem, we can jointly diagonalize the following composite sampled data matrices*

$$\overline{\mathbf{R}}_{\widetilde{\overline{\mathbf{x}}}}(q) = \sum_{i=1}^{M} \tilde{\mathbf{R}}_{\overline{\mathbf{x}}\,\widetilde{\overline{\mathbf{x}}}}(B_{iq}) = \frac{1}{2}\sum_{i=1}^{M}\left(\left\langle \overline{\mathbf{x}}(k)\widetilde{\overline{\mathbf{x}}}_{B_{iq}}^T(k)\right\rangle + \left\langle \widetilde{\overline{\mathbf{x}}}_{B_{iq}}(k)\overline{\mathbf{x}}^T(k)\right\rangle\right), \tag{4.137}$$

where $\widetilde{\overline{\mathbf{x}}}_{B_{iq}}(k) = [B_{iq}(z)]\overline{\mathbf{x}}(k) = [z^{-q}B_i(z)]\overline{\mathbf{x}}(k), \quad (q = 1, 2, \ldots, L)$.

4.4.1 The Modified SOBI Algorithm for Nonstationary Sources: SONS Algorithm

In this section, we describe a very flexible and efficient algorithm developed by Choi and Cichocki referred to as SONS (Second-Order Nonstationary Source separation) [217, 218, 229, 219]. The method jointly exploits the nonstationarity and the temporal structure of the sources under the assumption that additive noise is white or the undesirable interference and noise are stationary signals.

The main idea of the SONS algorithm is to exploit the nonstationarity of signals by partitioning the prewhitened sensor data into non-overlapping blocks (time windows T_i), for which we estimate time-delayed covariance matrices. We consider the case where source signals have time varying variances and non-vanishing temporal correlations. Moreover, we assume that undesirable additive noise is white or stationary. It follows from the assumptions (AS1)-(AS4) given in section 4.2, that we have

$$\mathbf{R}_{\mathbf{x}}(T_i, p_l) = \mathbf{A}\mathbf{R}_{\mathbf{s}}(T_i, p_l)\mathbf{A}^T, \qquad \forall i, l \tag{4.138}$$

and

$$\delta\mathbf{R}_{\mathbf{x}}(T_i, T_j, p_l) = \mathbf{A}\delta\mathbf{R}_{\mathbf{s}}(T_i, T_j, p_l)\mathbf{A}^T, \qquad \forall i, l, j > i, \tag{4.139}$$

where $\mathbf{A} \in \mathbb{R}^{n\times n}$ is an orthogonal mixing matrix, p_l's are time lags. The index T_i denotes the i-th time window. The discrete-time differential correlation matrix is defined as [219]

$$\delta\mathbf{R}_{\mathbf{x}}(T_i, T_j, p_l) = \delta\mathbf{R}_{\mathbf{x}}(T_i, p_l) - \delta\mathbf{R}_{\mathbf{x}}(T_j, p_l), \qquad (i \neq j). \tag{4.140}$$

The Algorithm Outline: SONS Algorithm with Robust Orthogonalization [218]

---

1. The robust orthogonalization method (described in section 4.3.1) is applied to obtain the whitened vector $\overline{\mathbf{x}}(k) = \mathbf{Q}\,\mathbf{x}(k)$. In the robust orthogonalization step, we use all available data points.

2. Divide the spatial whitened sensor data $\{\overline{\mathbf{x}}(k)\}$ into L non-overlapping blocks (time windows T_i) and estimate the set of covariance matrices $\tilde{\mathbf{R}}_{\overline{\mathbf{x}}}(T_i, p_l)$ for $i = 1, \dots, L$ and $l = 1, \dots, M$. In other words, at each time-windowed data frame, we compute M different time-delayed covariance matrices of $\overline{\mathbf{x}}(k)$ (typically, good performance is obtained for $M = 1$, $L = 20$, with $10 - 200$ samples in each block.

3. Find an orthogonal matrix $\mathbf{U}$ for all $\{\mathbf{R}_{\overline{\mathbf{x}}}(T_i, p_l)\}$ using the joint approximate diagonalization method in [153], which satisfies

$$\mathbf{U}^T\,\mathbf{R}_{\overline{\mathbf{x}}}(T_i, p_l)\,\mathbf{U} = \mathbf{D}_{i,l}, \tag{4.141}$$

   where $\{\mathbf{D}_{i,l}\}$ is a set of diagonal matrices.

4 The mixing matrix is computed as $\widehat{\mathbf{H}} = \mathbf{Q}^{+}\,\mathbf{U}$.

Remark 4.8 *Instead of the sampled covariance matrices* $\mathbf{R}_{\overline{\mathbf{x}}}(T_i, p_l)$, *with* $i = 1, 2, \ldots, L$ *and* $p_l = 1, 2, \ldots, M$, *we can attempt to jointly diagonalize the sampled differential correlation matrices* $\delta\mathbf{R}_{\overline{\mathbf{x}}}(T_i, T_j, p_l)$, *with* $j > i$ *and* $p_l = 0, 1, \ldots, M-1$. *However, such matrices can be very ill-conditioned. To improve the conditioning, we can diagonalize the following composite sampled matrices rather than the previous ones:*

$$\overline{\mathbf{R}}_{\overline{\mathbf{x}}}(p_l) = \sum_{i=1}^{L} \sum_{j>i} \delta\mathbf{R}_{\overline{\mathbf{x}}}(T_i, T_j, p_l), \qquad (p_l = 0, 1, \ldots, M-1). \tag{4.142}$$

4.4.2 Computer Simulation Experiments

Computer simulations show that the SONS algorithm is very robust with respect to temporally white noise, regardless of the probability distributions of noises. In fact, SONS is a generalization of SOBI [78, 82] to the case of nonstationary sources. The SONS algorithm is applicable to the case of nonstationary sources including nonstationary i.i.d. and/or temporally correlated (colored) sources while the SOBI algorithm is able to separate or extract only colored sources. In fact, through numerical experiments, we have confirmed the robustness with respect to noise and useful behavior of the SONS algorithm in a variety of cases: (1) the case where several nonstationary Gaussian sources exist and they lack temporal correlation; (2) the case where additive noises are spatially correlated but temporally white Gaussian processes; (3) the case where measurement noises have white uniform distributions.

In order to measure the performance of algorithms, we use the performance index (PI) defined by

$$\mathrm{PI} = \frac{1}{n(n-1)} \sum_{i=1}^{n} \left\{ \left(\sum_{k=1}^{n} \frac{|g_{ik}|}{\max_j |g_{ij}|} - 1 \right) + \left(\sum_{k=1}^{n} \frac{|g_{ki}|}{\max_j |g_{ji}|} - 1 \right) \right\}, \tag{4.143}$$

where g_{ij} is the (i,j)-element of the global system matrix $\mathbf{G} = \mathbf{W}\,\mathbf{H}$. The term $\max_j g_{ij}$ represents the maximum value among the elements in the ith row vector of $\mathbf{G}$. Similarly, the term represents $\max_j g_{ji}$ does the maximum value among the elements in the ith column vector of $\mathbf{G}$. When the perfect separation is achieved, the performance index is zero. In practice, a performance index around 10^{-2} indicates quite a good performance. Figure 4.6 shows typical performances of the several algorithms discussed so far in this chapter. At high SNR, all tested algorithms perform very well. At low SNR, one can observe that the RSOBI with robust orthogonalization outperforms the standard SOBI with standard whitening. SONS gives better performance than the RSOBI algorithm in most SNR ranges. In $0 - 6$dB range SONS is worse than the RSOBI. The advantage of SONS over RSOBI with robust orthogonalization lies in the fact that the first method works even for the case of nonstationary sources with identical spectral shapes, whereas the latter does not.

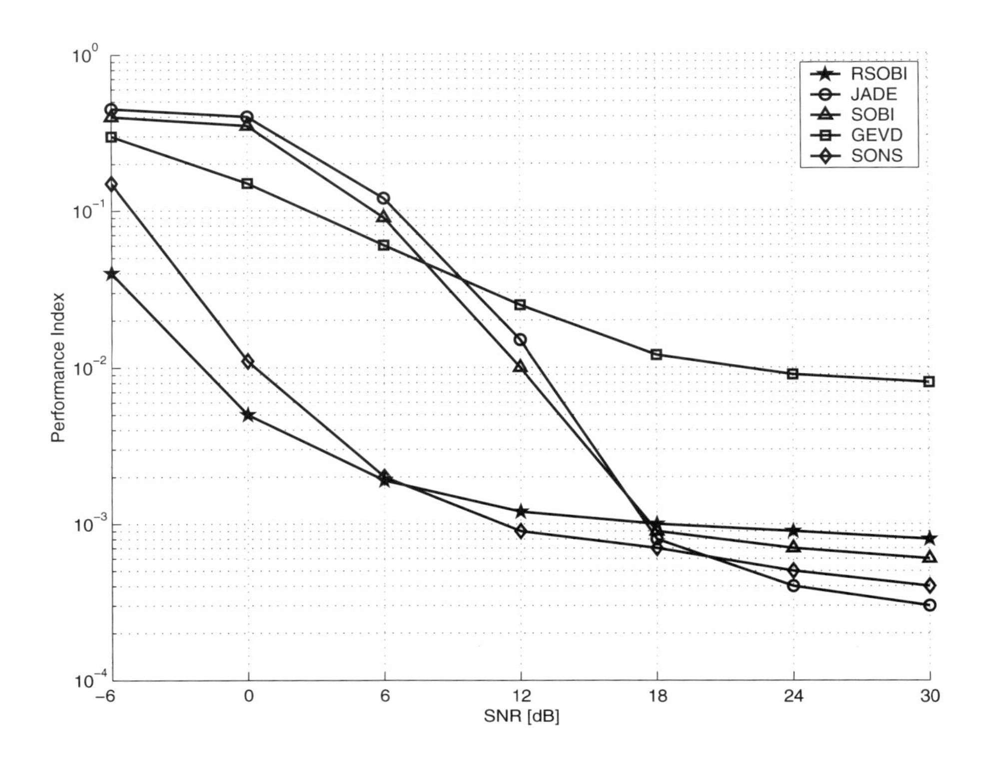

Fig. 4.6 Comparison of performance of various algorithms as a function of signal to noise ratio (SNR) [217, 229].

4.4.3 Possible Extensions of Joint Approximate Diagonalization Technique

In order to improve performance and/or to extend the JAD approach to various kinds of source signals several extensions and generalizations have been proposed [1216, 1328, 1330, 1331].

For example, instead of standard covariance matrices $\widehat{\mathbf{R}}_{\overline{\mathbf{x}}}(p_i) = (1/N)\sum_{k=1}^{N} \overline{\mathbf{x}}(k)\overline{\mathbf{x}}^T(k-p_i)$, we can use the generalized covariance matrices of the form

$$\mathbf{M}_i = \widehat{\mathbf{R}}_{\overline{\mathbf{x}}\,\widetilde{\overline{\mathbf{x}}}}(B_i) = \frac{1}{N}\sum_{k=1}^{N} \overline{\mathbf{x}}(k)\widetilde{\overline{\mathbf{x}}}_{B_i}^T(k), \tag{4.144}$$

where vector $\widetilde{\overline{\mathbf{x}}}_{B_i}(k) = B_i(z)[\overline{\mathbf{x}}(k)] = \sum_p b_{ip}\overline{\mathbf{x}}(k-p)$ is a filtered version of the vector $\overline{\mathbf{x}}(k)$ and $B_i(z)$ denotes the transfer function of a filter or a generalized time-delay operator.

Choosing the entries of the covariance matrix $\mathbf{R}_{\overline{\mathbf{x}}\,\widetilde{\overline{\mathbf{x}}}}(p_l) = [r_{ij}(p_l)]_{n\times n}$ as

$$r_{ij}(p_l) = \frac{1}{N}\sum_{k=1}^{N} \overline{x}_i(k)\,\overline{x}_j^*(k-p_l), \tag{4.145}$$

leads to the extended SOBI algorithm for complex-valued signals.

Other choices for the entries r_{ij} of $\mathbf{R}_{\overline{\mathbf{x}}}(p_l, \alpha)$ can be

$$r_{ij}(p_l, \alpha) = \frac{1}{N} \sum_{k=1}^{N} \overline{x}_i(k)\, \overline{x}_j^*(k - p_l) \exp(-2\pi j \alpha k), \tag{4.146}$$

which leads to an algorithm in which sources with different cyclostationarity[12] properties can be separated [1330, 1331].

The choice

$$r_{ij}(k, f) = \sum_{m,l=-L}^{L} \phi(m,l)\, \overline{x}_i(k+m+l)\, \overline{x}_j^*(k+m-l) \exp(-4\pi j f k), \tag{4.147}$$

for various k- time and f- frequency indices and specific smoothing kernel $\phi(m,l)$, leads to the time-frequency JAD algorithm proposed by Belouchrani and Amin [79, 80].

Instead of covariance or generalized covariance matrices, we can employ higher order statistics (HOS), i.e., cumulant matrices. For example, by performing the joint approximate diagonalization for fourth-order $n \times n$ cumulant matrices $\mathbf{M}_{ij} = \mathbf{C}_{\overline{\mathbf{x}}}(i,j), \;\; \forall i,j = 1,2,\ldots,n;$ whose (m,l)-th element is given by [151, 153, 1261, 1380]

$$c_{ml}(i,j) = cum\{\overline{x}_i(k), \overline{x}_j^*(k), \overline{x}_m(k), \overline{x}_l^*(k)\}, \tag{4.148}$$

we obtain the JADE (Joint Approximate Diagonalization of Eigen-matrices) algorithm for ICA (see Appendix C and Chapter 8 for more detail).

4.4.4 Comparison of the Joint Approximate Diagonalization and Symmetric Eigenvalue Decomposition Approaches

Although the JAD approach usually gives a better performance, especially for noisy data with low SNR, the symmetric EVD approach has several important advantages that are worth mentioning.

- Both approaches: The EVD and JAD are batch processing algorithms in the sense that the entire data set or block of data is collected and processed at once. However, the EVD approach needs to perform only one average diagonalization, instead of performing many joint diagonalizations of data matrices simultaneously, thus the symmetric EVD algorithm is generally less numerical complex than the JAD algorithms.

- The EVD controls explicitly whether a separation of the sources is performed successfully by monitoring eigenvalues, which must be distinct.

- Recently, several efficient algorithms have been developed for EVD with high convergence speed (even cubic convergence) such as the power method, PCA RLS algorithm and conjugate gradient on the Stiefel manifold [1018].

[12]The signal $x(k)$ is said to be cyclostationary if its correlation function is cyclic, e.g. with period q, i.e., the following relation holds $E\{x(k)\, x^*(k+p)\} = E\{x(k+q)\, x^*(k+p+q)\}$, for all k, p.

4.5 CANCELLATION OF CORRELATION

4.5.1 Standard Estimation of Mixing Matrix and Noise Covariance Matrix

The concept of spatial decorrelation also called correlation cancelling plays an important role in signal processing [922]. Consider two zero-mean vector signals $\mathbf{x}(k) \in \mathbb{R}^m$ and $\mathbf{s}(k) \in \mathbb{R}^n$ that are related by the linear transformation

$$\mathbf{x}(k) = \mathbf{H}\,\mathbf{s}(k) + \mathbf{e}(k), \tag{4.149}$$

where $\mathbf{H} \in \mathbb{R}^{m\times n}$ is an unknown full column rank mixing matrix and $\mathbf{e}(k) \in \mathbb{R}^m$ is a vector of zero-mean error, interference or noise depending on application. Generally, vectors $\mathbf{s}(k)$, $\mathbf{x}(k)$ are correlated, i.e., $\mathbf{R_{xs}} = E\left\{\mathbf{x}\,\mathbf{s}^T\right\} \neq \mathbf{0}$ but the error or noise $\mathbf{e}$ is uncorrelated with $\mathbf{s}$. Hence, our objective is to find the matrix $\mathbf{H}$ such that the new pair of vectors $\mathbf{e} = \mathbf{x} - \mathbf{H}\,\mathbf{s}$ and $\mathbf{s}$ are no longer correlated with each other, i.e.,:

$$\mathbf{R_{es}} = E\left\{\mathbf{e}\,\mathbf{s}^T\right\} = E\left\{(\mathbf{x} - \mathbf{H}\,\mathbf{s})\,\mathbf{s}^T\right\} = \mathbf{0}. \tag{4.150}$$

The cross-correlation matrix can be written as

$$\mathbf{R_{es}} = E\left\{\mathbf{x}\,\mathbf{s}^T - \mathbf{H}\,\mathbf{s}\,\mathbf{s}^T\right\} = \mathbf{R_{xs}} - \mathbf{H}\,\mathbf{R_{ss}}. \tag{4.151}$$

Hence the optimal mixing matrix for cancelling correlation can be expressed as

$$\mathbf{H}_{opt} = \mathbf{R_{xs}}\,\mathbf{R_{ss}}^{-1} = E\left\{\mathbf{x}\,\mathbf{s}^T\right\}\left(E\left\{\mathbf{s}\,\mathbf{s}^T\right\}\right)^{-1}. \tag{4.152}$$

It should be noted that the same result is obtained by minimizing the mean square error cost function:

$$\begin{aligned} J(\mathbf{e}) &= \frac{1}{2}E\{\mathbf{e}^T\mathbf{e}\} = E\{(\mathbf{x} - \mathbf{H}\,\mathbf{s})^T\,(\mathbf{x} - \mathbf{H}\,\mathbf{s})\} \\ &= \frac{1}{2}\left(E\{\mathbf{x}^T\,\mathbf{x}\} - E\{\mathbf{s}^T\,\mathbf{H}^T\,\mathbf{x}\} - E\{\mathbf{x}^T\mathbf{H}\,\mathbf{s}\} + E\{\mathbf{s}^T\,\mathbf{H}^T\,\mathbf{H}\,\mathbf{s}\}\right). \end{aligned} \tag{4.153}$$

By computing the gradient of the cost function $J(\mathbf{e})$ with respect to $\mathbf{H}$, we obtain

$$\frac{\partial J(\mathbf{e})}{\partial \mathbf{H}} = -E\{\mathbf{x}\,\mathbf{s}^T\} + \mathbf{H}\,E\{\mathbf{s}\,\mathbf{s}^T\}. \tag{4.154}$$

Hence, applying the standard gradient descent approach, we obtain an on-line adaptive algorithm for the estimation of the mixing matrix

$$\Delta\widehat{\mathbf{H}}(k) = -\eta\,\frac{\partial J(\mathbf{e})}{\partial \widehat{\mathbf{H}}} = \eta\left(\mathbf{R_{xs}} - \widehat{\mathbf{H}}(k)\,\mathbf{R_{ss}}\right). \tag{4.155}$$

Assuming that the optimum matrix $\mathbf{H}_{opt}$ is achieved when the gradient is zero, we have

$$\boxed{\mathbf{H}_{opt} = \mathbf{R_{xs}}\,\mathbf{R_{ss}}^{-1}}, \tag{4.156}$$

where the minimum value for the error (or noise) covariance matrix can be estimated as

$$\boxed{\mathbf{R_{ee}} = E\{\mathbf{e}\mathbf{e}^T\} = \mathbf{R_{xx}} - \mathbf{R_{xs}}\,\mathbf{R_{ss}^{-1}}\,\mathbf{R_{sx}}} \tag{4.157}$$

assuming that noise is spatially colored but independent of the source signals. In the simplest case, when noise is spatially white, the covariance matrix of the noise takes the form: $\mathbf{R_{ee}} = \sigma_e^2 \mathbf{I}$, where the variance of noise can be estimated for relative high SNR as

$$\hat{\sigma}_e^2 = \text{minimum eigenvalue of } \mathbf{R_{xx}}$$

or alternatively

$$\hat{\sigma}_e^2 = \text{mean of } \; (m-n) \; \text{ minor eigenvalues of } \mathbf{R_{xx}} \in \mathbb{R}^{m \times m}, \qquad m > n.$$

In the case, when the noise covariance matrix is known or it can be estimated, we can obtain an alternative formula (for $m \geq n$) by simple manipulations of Eqs. (4.156)-(4.157)

$$\boxed{\widehat{\mathbf{H}} = (\mathbf{R_{xx}} - \mathbf{R_{ee}})\;\mathbf{R_{sx}^{+}},} \tag{4.158}$$

where $\mathbf{R_{sx}^{+}}$ is the Moore-Penrose pseudo-inverse matrix of $\mathbf{R_{sx}}$. Hence, neglecting the noise covariance matrix $\mathbf{R_{ee}}$, we obtain the Wiener filter equation:

$$hat\mathbf{s} = \mathbf{R_{sx}}\mathbf{R_{xx}^{+}}\,\mathbf{x}, \tag{4.159}$$

which minimizes the mean square error $E\{\|\mathbf{s} - \hat{\mathbf{s}}\|_2^2\}$. The fundamental problem in this method is to obtain an estimator for the cross-covariance matrix $\mathbf{R_{sx}}$ when the vector $\mathbf{s}$ is not available.

4.5.2 Blind Identification of Mixing Matrix Using the Concept of Cancellation of Correlation

In a blind separation scenario, both the mixing matrix $\mathbf{H}$ and the source signals $\mathbf{s}$ are unknown. In this case, we need to formulate a modified estimation function in the form (see Chapter 10 and [22] for more detail)

$$\boxed{\mathbf{F}(\mathbf{H}, \mathbf{s}) = \widetilde{\mathbf{R}}_{\tilde{\mathbf{e}}\tilde{\mathbf{s}}} = E\{\widetilde{\mathbf{e}}\,\widetilde{\mathbf{s}}^T\},} \tag{4.160}$$

where $\widetilde{\mathbf{e}}(k) = \widetilde{\mathbf{x}}(k) - \widehat{\mathbf{H}}\,\widetilde{\mathbf{s}}$ and $\widetilde{\mathbf{x}} = [\widetilde{x}_1, \widetilde{x}_2, \ldots, \widetilde{x}_m]^T$ and $\widetilde{\mathbf{s}} = [\widetilde{s}_1, \widetilde{s}_2, \ldots, \widetilde{s}_n]^T$ are filtered versions of sensor signals $\mathbf{x}(k)$ and estimated sources $\mathbf{s}(k)$ respectively. More precisely, all the sensor signals and also the estimated source signals are filtered through filters with identical transfer functions $B(z) = \sum_{p=1}^{L} b_p\, z^{-1}$, i.e.,

$$\widetilde{x}_j(k) = \sum_{p=1}^{L} b_p x_j(k-p), \qquad (j = 1, 2, \ldots, m) \tag{4.161}$$

and

$$\widetilde{s}_i(k) = \sum_{p=1}^{L} b_p \, \hat{s}_i(k-p), \qquad (i = 1, 2, \ldots, n). \tag{4.162}$$

The choice of the filter $B(z)$ depends on the statistics of sources and additive noise (for more detail see the next chapter). Generally, the approach presented in this section is useful when source signals are colored, i.e., they have a temporal structure.

It is straightforward to check that the above function satisfies the basic properties of an estimation function [22].

From (4.160) we can obtain the formula for estimation of the mixing matrix

$$\boxed{\widehat{\mathbf{H}} = \widehat{\mathbf{R}}_{\widetilde{\mathbf{x}}\,\widetilde{\mathbf{s}}} \, \widehat{\mathbf{R}}_{\widetilde{\mathbf{s}}\,\widetilde{\mathbf{s}}}^{-1}.} \tag{4.163}$$

Let us consider first the case, when the number of sensors is larger than the number of sources ($m > n$) and the sources are colored, that is, they have a temporal structure and the covariance matrix $\mathbf{R_{ee}}$ of the additive zero-mean and uncorrelated noise is known or can be estimated. In such a case, we propose the following algorithm which is robust with respect to noise.

The Algorithm Outline: BLUE Algorithm for Blind Identification and Source Separation

---

1. Make arbitrary non-zero initialization of the mixing matrix $\widehat{\mathbf{H}}$ and estimate the sources using the BLUE formula (see in Chapter 2 Eq. (2.23))

$$\hat{\mathbf{s}}(k) = (\widehat{\mathbf{H}}^T \, \widehat{\mathbf{R}}_{\mathbf{ee}}^{-1} \, \widehat{\mathbf{H}})^{-1} \, \widehat{\mathbf{H}}^T \, \widehat{\mathbf{R}}_{\mathbf{ee}}^{-1} \, \mathbf{x}(k), \tag{4.164}$$

   where $\widehat{\mathbf{R}}_{\mathbf{ee}}$ is the estimated covariance matrix of noise uncorrelated with sources.

2. Compute the mixing matrix $\widehat{\mathbf{H}}$ on the basis of estimated sources in Step 1 as

$$\widehat{\mathbf{H}} = \widehat{\mathbf{R}}_{\widetilde{\mathbf{x}}\,\widetilde{\mathbf{s}}} \, \widehat{\mathbf{R}}_{\widetilde{\mathbf{s}}\,\widetilde{\mathbf{s}}}^{-1}, \tag{4.165}$$

   where $\widehat{\mathbf{R}}_{\widetilde{\mathbf{x}}\,\widetilde{\mathbf{s}}} = \frac{1}{N} \sum_{k=1}^{N} \widetilde{\mathbf{x}}(k) \widetilde{\mathbf{s}}^T(k)$ and $\widehat{\mathbf{R}}_{\widetilde{\mathbf{s}}\,\widetilde{\mathbf{s}}} = \frac{1}{N} \sum_{k=1}^{N} \widetilde{\mathbf{s}}(k) \widetilde{\mathbf{s}}^T(k)$.

3. Repeat Steps 1 and 2 alternately, until convergence is achieved.

Remark 4.9 *The above two-phase procedure is similar to the expectation maximization (EM) scheme: (i) Freeze the entries of the mixing matrix* $\mathbf{H}$ *and learn new statistics (i.e., the actual vector of the estimated source signal; (ii) freeze the covariance and cross-correlation matrices and learn the entries of the mixing matrix, then go back to (i) and repeat. Hence, in* *phase (i) our algorithm estimates source signals, whereas in phase (ii) it learns the statistics of the sources and estimates the mixing matrix.*

Let us consider now a more challenging task when the number of sensors is less than the number of sources ($m < n$). Assuming that all sources have a sparse representation, we can apply the robust FOCUSS algorithm discussed in Chapter 2. The idea is similar to the above procedure: First, after initialization, we estimate a sparse representation of sources, then an estimate of the mixing matrix, $\widehat{\mathbf{H}} = \widehat{\mathbf{R}}_{\tilde{\mathbf{x}}\,\tilde{\mathbf{s}}}\widehat{\mathbf{R}}_{\tilde{\mathbf{s}}\,\tilde{\mathbf{s}}}^{-1}$, which consequently produces a new estimate of the sparse source vector $\widehat{\mathbf{s}}(k)$ obtained, for example, by using the FOCUSS algorithm. Iterations are conducted until the algorithm converges.

Alternatively, we can use the following adaptive algorithm.

The Algorithm Outline: FOCUSS Adaptive Algorithm for Blind Identification and Estimation of Sparse Sources

---

1. After initialization, we estimate sparse source signals using the Focuss algorithm as

$$\hat{\mathbf{s}}_{l+1}(k) = \mathbf{D}_{|\mathbf{s}|}(l)\,\widehat{\mathbf{H}}_k^T \left(\widehat{\mathbf{H}}_k\,\mathbf{D}_{|\mathbf{s}|}(l)\,\widehat{\mathbf{H}}_k^T + \alpha_l\,\mathbf{I}\right)^{-1} \mathbf{x}(k), \qquad \forall k,\ l = 1, 2, \ldots, \tag{4.166}$$

   where $\mathbf{D}_{|\mathbf{s}|}(l) = \mathrm{diag}\{|\hat{s}_{1l}|, |\hat{s}_{2l}|, \ldots, |\hat{s}_{nl}|\}$, $\hat{s}_{jl}$ denotes estimation of $\hat{s}_j(k)$ in j-th internal iteration, and α_l is a suitably chosen regularization parameter [1001].

2. Estimate iteratively the mixing matrix $\mathbf{H}$ as

$$\widehat{\mathbf{H}}_{k+1} = \widehat{\mathbf{H}}_k - \eta_k \left[\widehat{\mathbf{H}}_k\widehat{\mathbf{R}}_{\tilde{\mathbf{s}}\,\tilde{\mathbf{s}}} - \widehat{\mathbf{R}}_{\tilde{\mathbf{x}}\,\tilde{\mathbf{s}}} - \beta\gamma_k\,\widehat{\mathbf{H}}_k\right], \tag{4.167}$$

   where $\beta > 0$ is scaling factor and $\gamma_k = \mathrm{tr}[\widehat{\mathbf{H}}_k^T(\widehat{\mathbf{H}}_k\widehat{\mathbf{R}}_{\tilde{\mathbf{s}}\,\tilde{\mathbf{s}}} - \widehat{\mathbf{R}}_{\tilde{\mathbf{x}}\,\tilde{\mathbf{s}}})]$ is a forgetting factor, which ensures that the Frobenius norm of the matrix $\widehat{\mathbf{H}}$ is kept bounded during the iteration process in the sense that $|||\mathbf{H}_k|||_F^2 \approx \beta^{-1}$ (see Appendix B for proof). Such constraints on entries of the mixing matrix prevent the trivial solution $\widehat{\mathbf{H}} = \mathbf{0}$ and ensure the stability of the algorithm.

   Repeat Steps 1 and 2 alternately, until convergence is achieved.

Example 4.1 Figure 4.7 illustrates the performance of the blind identification and estimation of sparse images in the case when the number of observations is less than the number of sources. Three sparse images shown in Fig. 4.7 (a) are mixed by the full row rank ill-conditioned mixing matrix $\mathbf{H} \in \mathbb{R}^{2\times 3}$. In this way, we obtained two superimposed images shown in Fig. 4.7 (b). Using the algorithm (4.166)-(4.167), we reconstructed approximately the original images as shown in Fig. 4.7 (c) from the superimposed (overlapped) images.

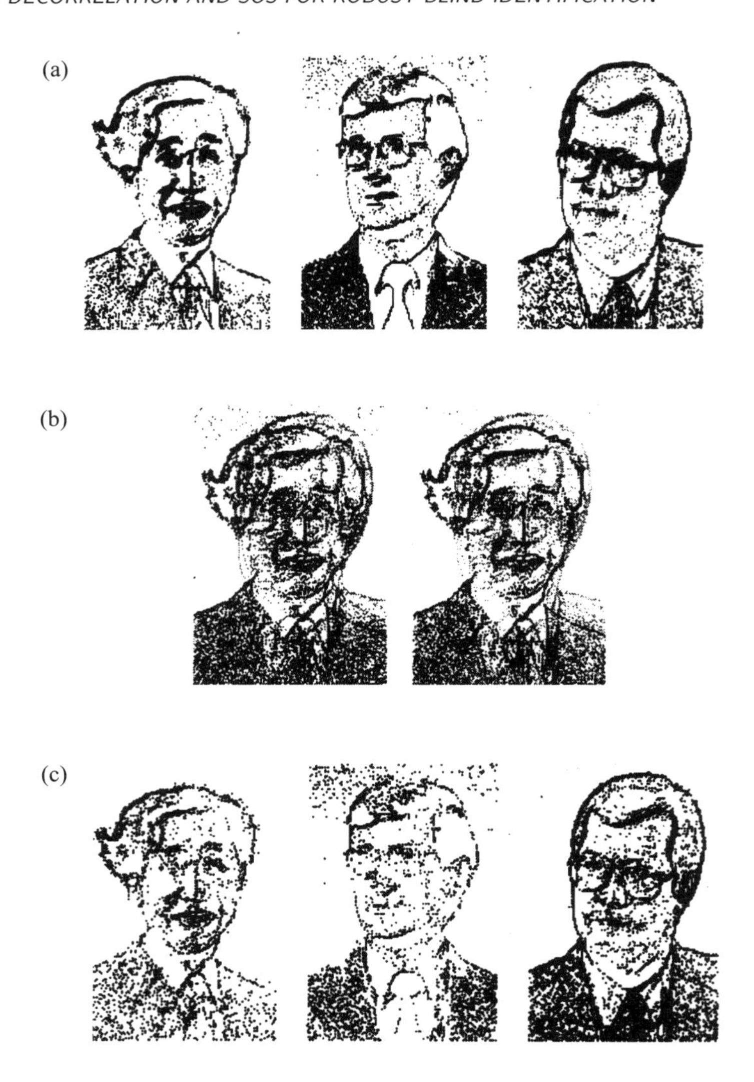

Fig. 4.7 Blind identification and estimation of sparse images: (a) Original sources, (b) mixed available images, (c) reconstructed images using the proposed algorithm (4.166)-(4.167).

Appendix A. Stability of Amari's Natural Gradient and Atick-Redlich Formula

Theorem A.3 *Consider a dynamical system, described by the following differential equation*

$$\frac{d\mathbf{W}}{dt} = -\boldsymbol{\mu}\frac{\partial J(\mathbf{W})}{\partial \mathbf{W}}\mathbf{W}^T\mathbf{W}, \tag{A.1}$$

where $J(\mathbf{W})$ is a lower bounded differentiable function and $\boldsymbol{\mu} \in \mathbb{R}^{n \times n}$ is a symmetric positive definite matrix of learning rates. Then, $J(\mathbf{W})$ is a Lyapunov function for the dynamical system (A.1).

Proof. [452] Denote by w_{ij}, μ_{ij} and b_{ij}, $(i, j = 1, ..., n)$ the elements of the matrices $\mathbf{W}, \boldsymbol{\mu}$ and $\mathbf{B} = \frac{\partial J(\mathbf{W})}{\partial \mathbf{W}} \mathbf{W}^T$ respectively. We calculate:

$$\begin{aligned}
\frac{dJ}{dt} &= \sum_{i,j=1}^{n} \frac{\partial J}{\partial w_{ij}} \frac{dw_{ij}}{dt} \\
&= -\sum_{i,j=1}^{n} \frac{\partial J}{\partial w_{ij}} \sum_{l=1}^{n} \mu_{il} \sum_{r=1}^{n} \sum_{k=1}^{n} \frac{\partial J}{\partial w_{lk}} w_{rk} w_{rj} \\
&= -\sum_{i,r=1}^{n} b_{ir} \sum_{l=1}^{n} \mu_{il} b_{lr} \\
&= -\sum_{r=1}^{n} \mathbf{b}_r^T \boldsymbol{\mu} \mathbf{b}_r \leq 0,
\end{aligned} \tag{A.2}$$

where $\mathbf{b}_r$ denotes the r-th column vector of $\mathbf{B}$. It is easy to see that zero is achieved if and only if $\mathbf{b}_r = \mathbf{0}$ for every $r = 1, ..., n$, i.e., when $d\mathbf{W}/dt = \mathbf{0}$. According to Lyapunov's theorem, this means the continuous trajectory of the natural gradient algorithm converges to a stationary point. In the special case, we can use a scalar learning rate by putting $\boldsymbol{\mu} = \mu_0 \mathbf{I}$, $\mu_0 > 0$.

In contrast to NG, for the gradient formula proposed by Atick and Redlich (see Eq. 4.37) [45], we can formulate more restrictive conditions for stability [452]:

Theorem A.4 *Consider a dynamical system, described by the following differential equation*

$$\frac{d\mathbf{W}}{dt} = -\mu \mathbf{W} \left[\frac{\partial J(\mathbf{W})}{\partial \mathbf{W}} \right]^T \mathbf{W}, \tag{A.3}$$

where $J(\mathbf{W})$ is a lower bounded, differentiable cost function and $\mu > 0$ is a learning rate. Suppose that the matrix $\mathbf{B} = \frac{\partial J}{\partial \mathbf{W}} \mathbf{W}^T$ is a symmetric matrix, then $J(\mathbf{W})$ is a Lyapunov function for the dynamical system (A.3).

Proof. Denote by w_{ij} and b_{ij}, $(i, j = 1, ..., n)$ elements of the matrices $\mathbf{W}$ and $\mathbf{B}$ respectively. We calculate:

$$\begin{aligned}
\frac{dJ}{dt} &= \sum_{i,j=1}^{n} \frac{\partial J}{\partial w_{ij}} \frac{dw_{ij}}{dt} \\
&= -\mu \sum_{i,j=1}^{n} \frac{\partial J}{\partial w_{ij}} \sum_{r=1}^{n} \sum_{k=1}^{n} w_{ki} \frac{\partial J}{\partial w_{rk}} w_{rj} \\
&= -\mu \sum_{i,r=1}^{n} b_{ri} b_{ir} \\
&= -\mu \sum_{i,r=1}^{n} b_{ir}^2 \leq 0
\end{aligned} \tag{A.4}$$

as zero is achieved if and only if $d\mathbf{W}/dt = \mathbf{0}$.

Using a nonholonomic basis $d\mathbf{X} = d\mathbf{W}\mathbf{W}^{-1}$, (A.1) and (A.3) become respectively

$$\frac{d\mathbf{X}}{dt} = -\mu \frac{\partial J}{\partial \mathbf{W}} \mathbf{W}^T \tag{A.5}$$

and

$$\frac{d\mathbf{X}}{dt} = -\mu \mathbf{W} \left[\frac{\partial J}{\partial \mathbf{W}} \right]^T. \tag{A.6}$$

Putting $\mathbf{B} = \frac{\partial J}{\partial \mathbf{W}} \mathbf{W}^T$, we have $\mathbf{B} = \frac{\partial J}{\partial \mathbf{X}}$, we obtain

$$\frac{dJ(\mathbf{W}(t))}{dt} = -\mu \operatorname{tr}(\mathbf{B}^T \mathbf{B}) = -\mu \sum_{i,j=1}^{n} b_{ij}^2 \leq 0, \tag{A.7}$$

as equality is achieved if and only if $\partial J / \partial \mathbf{W} = \mathbf{0}$ (assuming that $\mathbf{W}$ is nonsingular).

Analogously, for (A.6), we obtain

$$\frac{dJ(\mathbf{W}(t))}{dt} = -\mu \operatorname{tr}(\mathbf{B}\mathbf{B}) = -\mu \sum_{i,j=1}^{n} b_{ij} b_{ji}. \tag{A.8}$$

Equation (A.7) shows that J is a Lyapunov function. The trace in (A.8) is not always positive. Let us decompose $\mathbf{B}$ as

$$\mathbf{B} = \mathbf{B}_S + \mathbf{B}_A,$$

where $\mathbf{B}_S$ is a symmetric matrix and $\mathbf{B}_A$ is an antisymmetric matrix, respectively. Then, we have

$$\operatorname{tr}(\mathbf{B}\mathbf{B}) = \sum_{i,j=1}^{n} b_{ij} b_{ji} = \|\mathbf{B}_S\|^2 - \|\mathbf{B}_A\|^2. \tag{A.9}$$

This gives a sufficient condition for convergence of the Atick-Redlich algorithm, that is, $\|\mathbf{B}_S\|_F > \|\mathbf{B}_A\|_F$ for any $\mathbf{W}$ (the matrices $\mathbf{B}$, $\mathbf{B}_S, \mathbf{B}_A$ depend on $\mathbf{W}$).

Appendix B. Gradient Descent Learning Algorithms with Bounded Frobenius Norm of the Separating Matrix

In order to ensure convergence of some learning algorithms and to provide their practical implementations it sometimes is necessary to restrict the entries of separating matrices to a bounded subset. Such bound can be imposed by a gradient descent learning system by keeping the norm of the separating matrix or estimated mixing matrix bounded or fixed (invariant) during the learning process. In this appendix, we present a theorem which proposes how to solve this problem.

Theorem B.5 *The learning rule*

$$\frac{d\,\mathbf{W}(t)}{d\,t} = \mu(t)\ [\mathbf{F}(\mathbf{y}(t)) - \beta\gamma(t)\mathbf{I}_n]\ \mathbf{W}(t), \tag{B.1}$$

where $\beta > 0$ *is a scaling factor and* $\gamma(t) = trace\left(\mathbf{W}^T(t)\mathbf{F}(\mathbf{y}(t))\mathbf{W}(t)\right) > 0$*, stabilizes the Frobenius norm of* $\mathbf{W}(t)$ *such* $||\mathbf{W}(t)||_F^2 = \mathrm{tr}(\mathbf{W}^T(t)\mathbf{W}(t)) \approx \beta^{-1}$.

Proof. It is straightforward to show that

$$\frac{d\,\mathrm{tr}(\mathbf{W}^T\mathbf{W})}{dt} = -2\,\mu(t)\beta\gamma(t)\left[\mathrm{tr}(\mathbf{W}^T\mathbf{W}) - \beta^{-1})\right]. \tag{B.2}$$

Denote $z(t) = ||\mathbf{W}||_F^2 = \mathrm{tr}\left(\mathbf{W}^T(t)\mathbf{W}(t)\right)$, and consider a differential equation of the form

$$\frac{dz}{dt} = -2\,\mu(t)\,\gamma(t)\beta\left(z(t) - \beta^{-1})\right).$$

The above differential equation has for an initial condition $z(0) = \beta^{-1} + \delta$ (where δ is a small coefficients representing perturbation) the following solution

$$z(t) = \beta^{-1} + \delta\exp(-2\int_0^t \mu(t)\,\gamma(t)\beta\,dt). \tag{B.3}$$

Since $\mu(t)$, $\gamma(t)$ and β are assumed to be nonnegative, the exponential term in (B.3) decays to zero keeping the norm $||\mathbf{W}(t)||_F^2 = \mathrm{tr}(\mathbf{W}^T(t)\mathbf{W}(t))$ close to value β^{-1} which prevents the norm from exploding. □

Remark Theorem B.5 does not discuss the stability problem of the corresponding learning algorithms, which depend on a cost function and a learning rate but only states that by introducing the forgetting factor we are able to bound entries of the matrix $\mathbf{W}$ in such way that its Frobenius norm is bounded or even fixed.

In the next corollaries from Theorem B.5, we present stabilizing modifications of some known algorithms.

Corollary B.1 *The modified natural gradient descent learning algorithm with a forgetting factor described as*

$$\frac{d\,\mathbf{W}(t)}{d\,t} = -\mu(t)\,[\frac{\partial J(\mathbf{W})}{\partial\mathbf{W}}\,\mathbf{W}^T(t)\,\mathbf{W}(t) + \beta\gamma(t)\,\mathbf{W}(t)], \tag{B.4}$$

where $J(\mathbf{W})$ *is the cost function,* $\nabla_{\mathbf{W}} J = \frac{\partial J(\mathbf{W})}{\partial \mathbf{W}}$ *denotes its gradient with respect to the nonsingular matrix* $\mathbf{W} \in \mathbb{R}^{n \times n}$, $\mu > 0$ *is the learning rate and*

$$\gamma(t) = -\mathrm{tr}\left(\mathbf{W}^T(t)\,\frac{\partial J(\mathbf{W})}{\partial \mathbf{W}}\,\mathbf{W}^T(t)\,\mathbf{W}(t)\right) > 0 \tag{B.5}$$

is a forgetting factor, which ensures that the Frobenius norm of the matrix $\mathbf{W}(t)$ *is bounded.*

Proof. Take $F(\mathbf{y}) = -\frac{\partial J(\mathbf{W})}{\partial \mathbf{W}}\,\mathbf{W}^T$ in Theorem B.5. □

Corollary B.2 *The stochastic gradient descent learning algorithm*

$$\frac{d\,\mathbf{W}(t)}{dt} = -\mu(t)\left[\frac{\partial J(\mathbf{W})}{\partial \mathbf{W}} + \beta\gamma(t)\mathbf{W}(t)\right], \tag{B.6}$$

where $J(\mathbf{W})$ *is the cost function,* $\nabla_{\mathbf{W}} J(\mathbf{W}) = \frac{\partial J(\mathbf{W})}{\partial \mathbf{W}}$ *denotes its gradient with respect to the nonsingular matrix* $\mathbf{W} \in \mathbb{R}^{n \times n}$, $\mu(t) > 0$ *is the learning rate and*

$$\gamma(t) = -\mathrm{tr}\left(\mathbf{W}^T(t)\,\frac{\partial J(\mathbf{W})}{\partial \mathbf{W}}\right) > 0 \tag{B.7}$$

is a forgetting factor, which ensures that the Frobenius norm of the matrix $\mathbf{W}(t)$ *is stable.*

Corollary B.3 *The modified Atick-Redlich descent learning algorithm with forgetting factor*

$$\frac{d\,\mathbf{W}(t)}{dt} = -\mu(t)\Big[\mathbf{W}(t)\left[\frac{\partial J(\mathbf{W})}{\partial \mathbf{W}}\right]^T \mathbf{W}(t) + \beta\gamma(t)\mathbf{W}(t)\Big], \tag{B.8}$$

where $J(\mathbf{W})$ *is a differentiable cost function,* $\nabla_{\mathbf{W}} J = \frac{\partial J(\mathbf{W})}{\partial \mathbf{W}}$ *denotes its gradient with respect to the nonsingular matrix* $\mathbf{W} \in \mathbb{R}^{n \times n}$, $\mu > 0$ *is the learning rate and*

$$\gamma(t) = -\mathrm{tr}\left(\mathbf{W}^T(t)\,\mathbf{W}(t)\left[\frac{\partial J(\mathbf{W})}{\partial \mathbf{W}}\right]^T \mathbf{W}(t)\right) > 0 \tag{B.9}$$

is a forgetting factor, which ensures that the Frobenius norm of the matrix $\mathbf{W}(t)$ *is bounded.*

Appendix C. The JADE Algorithm

The JADE (Joint Approximate Diagonalization of Eigenmatrices) algorithm can be considered a natural extension or generalization of the SOBI and FOBI algorithms [151, 153, 1261, 1380].

In JADE, in contrast to the FOBI algorithm, we jointly diagonalize a set of n^2 (or less) contracted quadricovariance matrices defined as:

$$\begin{aligned}\mathbf{M}_{pq}\left(\mathbf{E}_{pq}\right) &= \mathbf{C}_{\overline{\mathbf{x}}}\left\{(\overline{\mathbf{x}}^T(k)\,\mathbf{E}_{pq}\,\overline{\mathbf{x}}(k))\,\overline{\mathbf{x}}(k)\,\overline{\mathbf{x}}^T(k)\right\} = E\{(\overline{\mathbf{x}}^T(k)\,\mathbf{E}_{pq}\,\overline{\mathbf{x}}(k))\,\overline{\mathbf{x}}(k)\,\overline{\mathbf{x}}^T(k)\} \\ &\quad -\mathbf{R}_{\overline{\mathbf{x}}}(0)\,\mathbf{E}_{pq}\,\mathbf{R}_{\overline{\mathbf{x}}}(0) - \mathrm{tr}(\mathbf{E}_{pq}\,\mathbf{R}_{\overline{\mathbf{x}}}(0))\,\mathbf{R}_{\overline{\mathbf{x}}}(0) - \mathbf{R}_{\overline{\mathbf{x}}}(0)\,\mathbf{E}_{pq}^T\,\mathbf{R}_{\overline{\mathbf{x}}}(0)\end{aligned} \tag{C.1}$$

for all $1 \leq p, q \leq n$, where $\mathbf{R}_{\overline{\mathbf{x}}}(0) = E\{\overline{\mathbf{x}}(k)\, \overline{\mathbf{x}}^T(k)\}$, $\mathbf{E}_{pq} \in \mathbb{R}^{n \times n}$ is a set of matrices called eigen-matrices. It has been shown that eigen-matrices $\mathbf{E}_{pq}$ should satisfy the following conditions:

$$\mathbf{C}_{\overline{\mathbf{x}}}(\mathbf{E}_{pq}) = \lambda_{pq}\, \mathbf{E}_{pq}, \qquad \mathrm{tr}(\mathbf{E}_{pq}\, \mathbf{E}_{kl}^T) = \delta(p,q,k,l), \qquad (1 \leq p, q \leq n) \tag{C.2}$$

where $\delta(p,q,k,l) = 1$ for $p = q = k = l$, and 0 otherwise. Each $\mathbf{E}_{pq}$ is the eigen-matrix and the real scalar λ_{pq} is the corresponding eigenvalue. Only n non-zero λ_{pq} eigenvalues exist [151].

There are several techniques to select the eigen-matrices $\mathbf{E}_{pq}$ that satisfy the above relations. In the ideal case, we can choose $\mathbf{E}_{pq} = \mathbf{e}_p\, \mathbf{e}_q^T$, where $\mathbf{e}_p$ denotes the n-dimensional vector with 1 at the pth position and 0 elsewhere. However, this method creates a large number of n^2 matrices such that the problem cannot be computationally solved for $n > 40$. We can reduce the number of matrices to $n(n+1)/2$ by selecting the following matrices [886]:

$$\mathbf{E}_{pq} = \begin{cases} \mathbf{e}_p \mathbf{e}_q^T, & \text{for} \quad p = k, \\ (\mathbf{e}_p \mathbf{e}_q^T + \mathbf{e}_q \mathbf{e}_p^T)/\sqrt{2}, & \text{for} \quad p < q, \\ \mathbf{0}, & \text{for} \quad p > q. \end{cases} \tag{C.3}$$

If the number of matrices is still too large, we can reduce it by selecting only L matrices $\mathbf{C}_{\overline{\mathbf{x}}}(\mathbf{E}_{pq})$ with the largest squared sums of their diagonal elements. The number L is selected by a user depending on required performance and computation speed (typically, $L \geq n$).

An alternative approach is to generate only n (where n is the number of sources) quadricovariance matrices $\mathbf{C}_{\overline{\mathbf{x}}}(\mathbf{E}_p)$, by estimating the eigen-matrices $\mathbf{E}_p = \hat{\mathbf{u}}_p \hat{\mathbf{u}}_p^T$ (for $p = 1, 2, \ldots, n$), where $\mathbf{u}_p$ is the p-th column of the orthogonal matrix $\widehat{\mathbf{U}}$ which estimates diagonalizing matrix. The exact diagonalizing matrix is, of course, not known in advance, but we can roughly estimate it by EVD of the special quadricovariance matrix for $\mathbf{E} = \mathbf{I}_n$ as

$$\begin{aligned} \mathbf{C}_{\overline{\mathbf{x}}}(\mathbf{I}) &= E\{\overline{\mathbf{x}}^T(k)\, \overline{\mathbf{x}}(k)\, \overline{\mathbf{x}}(k)\, \overline{\mathbf{x}}^T(k)\} - 2\, \mathbf{R}_{\overline{\mathbf{x}}}(0)\, \mathbf{R}_{\overline{\mathbf{x}}} - \mathrm{tr}(\mathbf{R}_{\overline{\mathbf{x}}}(0))\, \mathbf{R}_{\overline{\mathbf{x}}}(0) \\ &= \widehat{\mathbf{U}}\, \mathbf{\Lambda}_{\mathbf{I}}\, \widehat{\mathbf{U}}^T, \end{aligned} \tag{C.4}$$

where $\mathbf{\Lambda}_{\mathbf{I}} = \mathrm{diag}\{\kappa_4(s_1), \kappa_4(s_2), \ldots, \kappa_4(s_n)\}$.

It is easy to check, that $\mathbf{E}_p = \mathbf{u}_p \mathbf{u}_p^T$ satisfies the conditions in (C.2). We have, in fact,

$$\begin{aligned} \mathbf{C}_{\overline{\mathbf{x}}}(\mathbf{E}_p) &= \mathbf{C}_{\overline{\mathbf{x}}}(\mathbf{u}_p \mathbf{u}_p^T) = \mathbf{U}\ \mathrm{diag}\{\lambda_1 \mathbf{u}_1^T \mathbf{u}_p \mathbf{u}_p^T \mathbf{u}_1, \ldots, \lambda_n \mathbf{u}_n^T \mathbf{u}_p \mathbf{u}_p^T \mathbf{u}_n\}\, \mathbf{U}^T \\ &= \mathbf{U}\, \mathrm{diag}\{0, \ldots, 0, \lambda_p, 0, \ldots, 0\}\, \mathbf{U}^T = \lambda_p \mathbf{u}_p \mathbf{u}_p^T = \lambda_p\, \mathbf{E}_p. \end{aligned} \tag{C.5}$$

It is also straightforward to verify, that $\mathrm{tr}(\mathbf{E}_p\, \mathbf{E}_q) = \delta_{p,q}$ and $\mathbf{C}_{\overline{\mathbf{x}}}(\mathbf{u}_p \mathbf{u}_q^T) = \mathbf{0}$ for $p \neq q$. This means, that we can reduce considerably the number of diagonalized quadricovariance matrices to n, which makes the JADE algorithm computationally attractive [151, 886].

The Algorithm Outline: Robust and Efficient JADE Algorithm

---

1. Apply the robust prewhitening or orthogonalization method (described in section 4.3.1 and section 4.1.8) to obtain the whitened (pre-processed) vector $\overline{\mathbf{x}}(k) = \mathbf{Q}\,\mathbf{x}(k)$. Preferably, for data corrupted by Gaussian noise, use a set of quadricovariance matrices.

2. Perform EVD of the sampled contracted quadricovariance matrix

$$\begin{aligned} \mathbf{C}_{\overline{\mathbf{x}}}(\mathbf{I}) &= \frac{1}{N}\sum_{k=1}^{N}[\overline{\mathbf{x}}^T(k)\,\overline{\mathbf{x}}(k)\;\overline{\mathbf{x}}(k)\,\overline{\mathbf{x}}^T(k)]\} - 2\,\widehat{\mathbf{R}}_{\overline{\mathbf{x}}}(0)\widehat{\mathbf{R}}_{\overline{\mathbf{x}}}(0) - \operatorname{tr}(\widehat{\mathbf{R}}_{\overline{\mathbf{x}}}(0))\,\widehat{\mathbf{R}}_{\overline{\mathbf{x}}}(0) \\ &= \widehat{\mathbf{U}}\,\mathbf{\Lambda}_{\mathbf{I}}\,\widehat{\mathbf{U}}^T, \end{aligned} \tag{C.6}$$

where $\widehat{\mathbf{R}}_{\overline{\mathbf{x}}}(0) = \frac{1}{N}\sum_{k=1}^{N}[\overline{\mathbf{x}}(k)\,\overline{\mathbf{x}}^T(k)]\}$ and $\widehat{\mathbf{U}} = [\hat{\mathbf{u}}_1, \hat{\mathbf{u}}_2, \ldots, \mathbf{u}_n]$.

3. Estimate the n sampled contracted quadricovariance matrices:

$$\begin{aligned} \mathbf{C}_{\overline{\mathbf{x}}}(\mathbf{E}_p) &= \frac{1}{N}\sum_{k=1}^{N}[\overline{\mathbf{x}}^T(k)\,\mathbf{E}_p\overline{\mathbf{x}}(k)\;\overline{\mathbf{x}}(k)\,\overline{\mathbf{x}}^T(k)] - \widehat{\mathbf{R}}_{\overline{\mathbf{x}}}(0)\,\mathbf{E}_p\,\widehat{\mathbf{R}}_{\overline{\mathbf{x}}}(0) \\ &\quad - \operatorname{tr}(\mathbf{E}_p\,\widehat{\mathbf{R}}_{\overline{\mathbf{x}}}(0))\,\widehat{\mathbf{R}}_{\overline{\mathbf{x}}}(0) - \widehat{\mathbf{R}}_{\overline{\mathbf{x}}}\,\mathbf{E}_p^T\,\widehat{\mathbf{R}}_{\overline{\mathbf{x}}} \end{aligned} \tag{C.7}$$

for $\mathbf{E}_p = \hat{\mathbf{u}}_p\hat{\mathbf{u}}_p^T, \quad p = 1, 2, \ldots, n.$

4. Find an orthogonal joint diagonalizing orthogonal matrix $\mathbf{U}$ for all n matrices $\{\mathbf{C}_{\overline{\mathbf{x}}}(\mathbf{E}_p)\}$ using one of the available joint approximate diagonalization numerical methods.

5. Estimate the mixing matrix as $\widehat{\mathbf{H}} = \mathbf{Q}^+\widehat{\mathbf{A}} = \mathbf{Q}^+\,\mathbf{U}$.

Appendix D. The MATLAB Implementation of the Robust SOBI Algorithm

```
 function [W]= sobiro(X,n,p),
%  Robust Second Order Blind Identification SOBIRO.
%  This function was created on the basis of publications
%  of Belouchrani et al., F. Cardoso et al.,
%  A. Belouchrani - A. Cichocki, S. Choi - A. Cichocki,
%  S. Cruces, A. Cichocki - S. Amari and P. Georgiev - A. Cichocki.
%  (C)
%  [W]=sobiro(X,n,p) estimates a separating matrix W of dimension
%  n by m  or  m by m  and a  matrix  of  estimated  sources S of
%  dimension n by N or m by N.
%  > X: Data matrix of dimension m by N representing observed sensor
%  signals,
%  > m: sensor number,
%  > n: source number by default m=n,
```

```
%  > N: sample number,
%  > p: number of the covariance matrices to be diagonalized,
%  Remark: for noisy data it is recommended to take at least p=100.
%
  [m,N]=size(X);
 if nargin==1,
   n=m;
   p=100; % Number of the time delayed covariance matrices.
 end; if nargin==2,
   p=100 ;
 end;
  Xz=X-(mean(X')'*ones(1,N)); % Removing  mean value
  Rxx=(Xz(:,1:N-1)*Xz(:,2:N)')/(N-1); % Estimation of sample covariance
% matrix for the time delay p=1 in order to reduce influence of a white
% noise.
%
 [Ux,Dx,Vx]=svd(Rxx);
  Dx=diag(Dx);
%n=10;
 if n<m, % under assumption of additive white noise and
       %when the number of sources are known or can a priori estimated
   Dx=Dx-real((mean(Dx(n+1:m))));
   Q= diag(real(sqrt(1./Dx(1:n))))*Ux(:,1:n)';
 else   % under assumption of no additive noise and when the
       % number of sources is unknown
   n=max(find(Dx>1e-14)); %Detection the number of sources
   Q= diag(real(sqrt(1./Dx(1:n))))*Ux(:,1:n)';
 end;
  Xb=Q*Xz; % prewhitened data
% Estimation of the time delayed covariance matrices:
  k=1; pn=p*n; for u=1:n:pn, k=k+1;
  Rxp=Xb(:,k:N)*Xb(:,1:N-k+1)'/(N-k+1);
  M(:,u:u+n-1)=norm(Rxp,'fro')*Rxp;
 end;
% Approximate joint diagonalization:
 eps=1/sqrt(N)/100; encore=1; U=eye(n); while encore, encore=0;
 for p=1:n-1,
   for q=p+1:n,
% Givens rotations:
     g=[ M(p,p:n:pn)-M(q,q:n:pn)  ;
        M(p,q:n:pn)+M(q,p:n:pn)  ;
        i*(M(q,p:n:pn)-M(p,q:n:pn))];
     [Ucp,D] = eig(real(g*g')); [la,K]=sort(diag(D));
     angles=Ucp(:,K(3)); angles=sign(angles(1))*angles;
     c=sqrt(0.5+angles(1)/2);
```

```
      sr=0.5*(angles(2)-j*angles(3))/c; sc=conj(sr);
      asr = abs(sr)>eps ;
      encore=encore | asr ;
    if asr , % Update of the M and U matrices:
      colp=M(:,p:n:pn); colq=M(:,q:n:pn);
      M(:,p:n:pn)=c*colp+sr*colq; M(:,q:n:pn)=c*colq-sc*colp;
      rowp=M(p,:); rowq=M(q,:);
      M(p,:)=c*rowp+sc*rowq; M(q,:)=c*rowq-sr*rowp;
      temp=U(:,p);
      U(:,p)=c*U(:,p)+sr*U(:,q); U(:,q)=c*U(:,q)-sc*temp;
     end  %% if
    end  %% q loop
   end  %% p loop
  end  %% while
% Estimation of the separating matrix W and the sources S:
 W=U'*Q; % Estimated separating matrix
 S= W*X; % Estimated sources.
```

5

Statistical Signal Processing Approach to Blind Signal Extraction

A large brain, like large government, may not be able to do simple things in a simple way.
—(Donald O. Hebb)

The problem of blind signal extraction (BSE) has received wide attention in various fields such as biomedical signal analysis and processing (EEG, MEG, fMRI), geophysical data processing, data mining, wireless communications, speech and image recognition and enhancement.

In this chapter, we will discuss a large family of unconstrained optimization criteria, from which we derive learning algorithms that can extract a single source signal from a linear mixture of source signals. One can repeat this process to extract the original source signals one by one. To prevent the newly extracted source signal from being extracted again in the next processing unit, we employ another unconstrained optimization criterion that uses information about this signal. From this criterion, we then derive a learning rule that deflates from the mixture of the newly extracted signal. By virtue of blind extraction and deflation processing, the described cascade neural network can cope with a practical case where the number of mixed signals is equal to or larger than the unknown number of sources. We prove that the proposed criteria both for blind extraction and deflation processing have no spurious equilibria. In addition, the proposed criteria, in most cases, do not require whitening of mixed signals. Using computer simulation experiments, we also confirm the validity and performance of the developed learning algorithms. In this chapter, we adopt a neural network approach. There are three main objectives to this chapter:

1. To present simple neural networks (processing units) and propose unconstrained extraction and deflation criteria that do not require either *a priori* knowledge of source

signals or the whitening of mixed signals. These criteria lead to simple, efficient, purely local and biologically plausible learning rules (e.g., Hebbian/anti-Hebbian type learning algorithms).

2. To prove that the proposed criteria have no spurious equilibria. In other words, most learning rules discussed in this chapter reach desired solutions, regardless of initial conditions (see appendixes for proof).

3. To demonstrate with computer simulations the validity and high performance of the presented neural networks and associated learning algorithms.

We will use two different models and approaches. The first approach is based on higher order statistics (HOS) which assumes that the sources are mutually statistically independent and non-Gaussian (at most only one can be Gaussian). For independence criteria, we will use some measures of non-Gaussianity. The second approach, based on the second order statistics (SOS) assumes that source signals have some temporal structure, i.e., the sources are colored with different autocorrelation functions or equivalently have different spectra shapes. Special emphasis will be given to blind source extraction (BSE) in the case when sensor signals are corrupted by additive noise.

5.1 INTRODUCTION AND PROBLEM FORMULATION

Mixing and filtering processes of the unknown input sources $s_j(t)$ $(j = 1, 2, ..., n)$ may have different mathematical or physical models, depending on the specific applications. In this chapter, we will focus mainly on the simplest cases when m mixed signals $x_i(t)$ are linear combinations of n (typically $m \geq n$) unknown, zero mean source signals $s_j(t)$ that are either statistically independent and/or have different temporal structures. They are written as

$$x_i(t) = \sum_{j=1}^{n} h_{ij}\, s_j(t) \qquad (i = 1, 2, \ldots, m) \tag{5.1}$$

or in matrix notation

$$\mathbf{x}(t) = \mathbf{H}\,\mathbf{s}(t), \tag{5.2}$$

where $\mathbf{x}(t) = [x_1(t), x_2(t), \ldots, x_m(t)]^T$ is a sensor vector, $\mathbf{s}(t) = [s_1(t), s_2(t), \ldots, s_n(t)]^T$ is a vector of source signals assumed to be zero mean and statistically independent, and $\mathbf{H}$ is an unknown full column rank $m \times n$ mixing matrix. It is assumed that only the sensor vector $\mathbf{x}(t)$ is available to use. It is desired to develop algorithms for estimation of primary sources and/or identification of the mixing matrix $\mathbf{H}$ with some intrinsic ambiguities such as arbitrary permutations and scaling factors (see Chapter 1 for more detail).

There are two principal approaches to solve this problem. The first approach is to separate all sources simultaneously. In the second approach, we extract sources one by one sequentially rather than separating them all simultaneously. In many applications, a large number of sensors (electrodes, microphones or transducers) is available but only a very few source signals are the subject of interest. For example, in EEG or MEG, we observe typically more than 64 sensor signals but only a few source signals are considered interesting, and the rest are considered to be interfering noise. Another example is the cocktail party problem; it is usually desired to extract the voice of a specific person rather than separate all the available source signals from an array of microphones. For such applications, it is essential to develop reliable, robust and effective learning algorithms which enable us to extract only a small number of source signals that are potentially interesting and contain useful information.

Before we begin to explain the derivation of learning algorithms for blind source extraction (BSE), let us recall some of the advantages of this approach. The blind signal extraction approach has the following advantages over simultaneous blind separation [248, 273]:

1. Signals can be extracted in a specified *order* according to the stochastic features of the source signals, (e.g., the order determined by absolute values of generalized normalized kurtosis, some measures of sparseness, non-Gaussianity, smoothness or linear predictability.) The blind extraction can be considered as a generalization of PCA (principal components analysis), where decorrelated output signals (principal components) are extracted according to decreasing order of their variance. Analogously, independent components can be ordered according for example to a decreasing absolute value of normalized kurtosis which is a measure of non-Gaussianity or according to any higher order normalized moment or cumulant.

2. The approach is very flexible, because many different criteria based on HOS and SOS can be applied for extraction of a wide spectrum of sources, such as i.i.d. sources, colored Gaussian, sparse sources, nonstationary sources, smooth sources with relative high measure of predictability, etc. In fact, in each stage of extraction, we can use a different criterion and the corresponding algorithm to extract sources with specific features.

3. Only "interesting" signals need to be extracted. For example, if the source signals are mixed with a large number of noise sources or interferences, we may extract only signals with some desired stochastic properties.

   In EEG/MEG signal processing it is often desired to extract the so called evoked potentials with non-symmetric distributions from symmetrically distributed noises and interferences.

4. The learning algorithms developed for BSE are local and biologically plausible. In fact, the learning algorithms derived in this chapter can be considered as extensions or modifications of the Hebbian/anti-Hebbian learning rule. Typically, they are simpler than algorithms for simultaneous blind source separation.

In summary, blind signal extraction is a useful approach when it is desired to extract several source signals with specific stochastic properties from a large number of mixtures. Extraction of a single source is closely related to the problem of blind deconvolution discussed in chapters 9 through 12 [545, 610, 606, 1081].

On the other hand, the sequential blind extraction approach may give poorer performance in comparison to the simultaneous blind separation approach discussed in the following

(a)

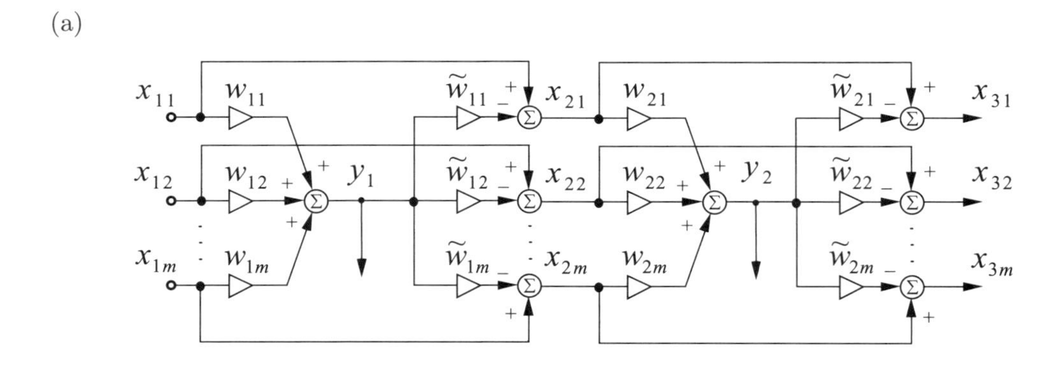

(b)

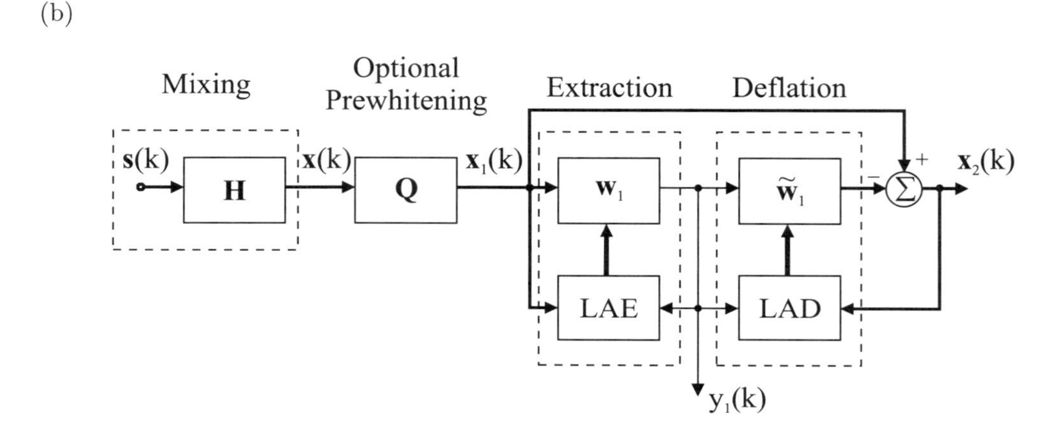

Fig. 5.1 Block diagrams illustrating: (a) Sequential blind extraction of sources and independent components, (b) implementation of extraction and deflation principles. LAE and LAD mean learning algorithm for extraction and deflation, respectively.

chapters for some ill-conditioned problems due to accumulation of error during deflation procedures. Furthermore, some of the blind extraction approaches need some preprocessing of sensor data, such as prewhitening or matrix orthogonalization.

5.2 LEARNING ALGORITHMS USING KURTOSIS AS A COST FUNCTION

Sequential blind source extraction (BSE) can be performed by using a cascade neural network similar to the one used for the extraction of principal components (see Fig. 5.1). However, in contrast to PCA, optimization criteria for BSE are different. For the principal component extraction, we apply an optimization criterion that ensures the best possible reconstruction of vector $\mathbf{x}(k)$ after its compression using a single processing unit.

In order to extract independent source signals, we use different criteria, e.g., maximization of the absolute value of normalized kurtosis which is a measure of the deviation of the extracted source signal from Gaussianity.

5.2.1 A Cascade Neural Network for Blind Extraction of Non-Gaussian Sources with Learning Rule Based on Normalized Kurtosis

A single processing unit (artificial neuron) is used in the first step to extract one independent source signal with the specified stochastic properties. In the next step, a deflation technique is used in order to eliminate the already extracted signals from the mixtures.

Let us assume that observed (sensor) signals are prewhitened (sphered), for example, by using the standard PCA technique. Then, the transformed sensor signals satisfy the condition[1]

$$E\{\mathbf{x}_1 \mathbf{x}_1^T\} = \mathbf{I}_n, \tag{5.3}$$

where $\mathbf{x}_1 = \overline{\mathbf{x}} = \mathbf{Q}\mathbf{x}$ and a new global $n \times n$ mixing matrix $\mathbf{A} = \mathbf{Q}\,\mathbf{H}$ is orthogonal, that is, $\mathbf{A}\mathbf{A}^T = \mathbf{A}^T\mathbf{A} = \mathbf{I}_n$. Hence the ideal $n \times n$ separating matrix is $\mathbf{W}_* = \mathbf{A}^{-1} = \mathbf{A}^T$ for $m = n$.

Let us consider a single processing unit, as shown in Fig. 5.1 described by

$$y_1 = \mathbf{w}_1^T \mathbf{x}_1 = \sum_{i=1}^{m} w_{1i} x_{1i}. \tag{5.4}$$

The unit successfully extracts a zero-mean source signal, e.g. the jth signal, if $\mathbf{w}_1(\infty) = \mathbf{w}_{1*}$ satisfying the relation $\mathbf{w}_{1*}^T \mathbf{A} = \mathbf{e}_j^T$, where $\mathbf{e}_j$ denotes the jth column of an $n \times n$ nonsingular diagonal matrix.

As a cost function for minimization, we may employ [273, 1144]

$$\boxed{\mathcal{J}_1(\mathbf{w}_1) = -\frac{1}{4}\left|\kappa_4(y_1)\right| = -\frac{\beta}{4}\kappa_4(y_1)}, \tag{5.5}$$

where $\kappa_4(y_1)$ is the normalized kurtosis defined for zero-mean signals by

$$\boxed{\kappa_4(y_1) = \frac{E\{|y_1|^4\}}{E^2\{|y_1|^2\}} - 3} \tag{5.6}$$

and the parameter β determines the sign of the kurtosis of the extracted signal, i.e.,

$$\beta = \begin{cases} -1, & \text{for extraction of a source signal with negative kurtosis,} \\ +1, & \text{for extraction of a source signal with positive kurtosis.} \end{cases} \tag{5.7}$$

[1]Time indices are omitted for brevity unless this not cause confusion.

We do not employ any further constraints (such as normalization of output signal to unit variance), since we used the normalized kurtosis. We will show in Appendix A that such a cost function has no spurious equilibria if the sources are mutually independent and they a non-Gaussian distribution (see Appendix A). In other words, we will prove that the cost function (5.5) does not have local minima corresponding to spurious (undesired) equilibria of the dynamics. Each local minimum corresponds to one extracted source signal. Therefore, successful extraction of a source signal from a mixture is always guaranteed, regardless of initial conditions. In addition, use of the normalized criteria makes it possible to eliminate any constraints that were required for the normalized criteria in [338, 589].

Remark 5.1 *Intuitively, the use of kurtosis as a cost function can be justified as follows: According to Central Limit Theorem (a classical result in probability theory), the distribution of a sum of independent random variables tends towards a Gaussian distribution (under certain mild conditions). This means that a sum of several variables typically has a distribution that is closer to Gaussian than any of the original random variables. Therefore, roughly speaking, our objective is to find such vector* $\mathbf{w}_1$ *that maximizes the non-Gaussianity of the output variable* $y_1 = \mathbf{w}_1^T \mathbf{x}$. *On the other hand, it should be noted that the absolute value of the normalized kurtosis may be considered as one of the simplest measures of non-Gaussianity of the extracted signal* y_1. *Furthermore, kurtosis measures the flatness or peakedness of a distribution of signals. A distribution with negative kurtosis is called sub-Gaussian, platykurtic or short-tailed (e.g., uniform). A distribution with positive kurtosis is referred to as super-Gaussian, leptokurtic or long-tailed (e.g., Laplacian), and a zero-kurtosis distribution is named mesokurtic (e.g., Gaussian).*

Applying the standard gradient descent approach to minimize the cost function (5.5), we have

$$\frac{d\mathbf{w}_1}{dt} = -\mu_1 \frac{\partial \mathcal{J}_1(\mathbf{w}_1)}{\partial \mathbf{w}_1} = \mu_1 \beta \frac{m_4(y_1)}{m_2^3(y_1)} \left[\frac{m_2(y_1)}{m_4(y_1)} E\{y_1^3 \mathbf{x}_1\} - E\{y_1 \mathbf{x}_1\} \right], \tag{5.8}$$

where $\mu_1 > 0$. It should be noted that the term $E\{|y|^4\}/E^3\{|y|^2\} = m_4(y_1)/m_2^3(y_1)$ is always positive, so it can be absorbed by the learning rate μ_1 as $\tilde{\mu}_1 = \frac{m_4(y_1)}{m_2^3(y_1)} \mu_1 > 0$. The moments $m_q(y_1) = E\{y_1(t)^q\}$, for $q = 2$ and 4, can be estimated on-line using the following moving averaging (MA) formula

$$\frac{d\hat{m}_q(y_1(t))}{dt} = \mu_0 \left[y_1^q(t) - \hat{m}_q(y_1(t)) \right], \quad (q = 2, 4). \tag{5.9}$$

Now applying the stochastic approximation technique, we obtain an on-line learning formula [245, 247]:

$$\boxed{\frac{d\mathbf{w}_1}{dt} = \mu_1(t)\, \varphi(y_1(t))\, \mathbf{x}_1(t),} \tag{5.10}$$

where $\mu_1(t) > 0$ is a learning rate and

$$\varphi(y_1) = \beta \frac{\hat{m}_4(y_1)}{\hat{m}_2^3(y_1)} \left[\frac{\hat{m}_2(y_1)}{\hat{m}_4(y_1)} y_1^3 - y_1 \right] \tag{5.11}$$

is a nonlinear adaptive activation function. Since the positive term $\hat{m}_4(y_1)/\hat{m}_2(y_1)$ can be absorbed by the learning rate, we can also use the following approximation of nonlinear activation function:

$$\varphi_1(y_1) = \beta \left[\frac{\hat{m}_2(y_1)}{\hat{m}_4(y_1)} y_1^3 - y_1 \right] \tag{5.12}$$

or

$$\varphi_2(y_1) = \beta \left[\frac{1}{\hat{m}_4(y_1)} y_1^3 - \frac{1}{m_2(y_1)} y_1 \right]. \tag{5.13}$$

In general, the nonlinear activation function is not fixed but changes in shape during the learning process according to the statistics of the extracted signals [245, 273].

Remark 5.2 *It should be noted that in our approach, we use the normalized kurtosis. Moreover, for spiky signals with positive kurtosis (super-Gaussian signals), the nonlinear activation function closely approximates a sigmoidal function, which is not only robust, but also biologically plausible.*

In the special case, applying a simple approximation for the derivative using the Euler approximation, we obtain the discrete-time learning rule [248, 273]:

$$\boxed{\mathbf{w}_1(k+1) = \mathbf{w}_1(k) + \eta_1(k)\, \varphi_1(y_1(k))\mathbf{x}_1(k),} \tag{5.14}$$

where $\mathbf{x}_1$ is a vector of sensor signals and the nonlinear activation function φ_1 is evaluated by (5.12).

The higher-order moments m_2 and m_4 and the sign of the kurtosis $\kappa_4(y_1)$ can be estimated on-line by using the following averaging formula

$$\hat{m}_q(k) = (1 - \eta_0)\, \hat{m}_q(k-1) + \eta_0\, |y_1(k)|^q, \tag{5.15}$$

with $\eta_0 \in (0, 1]$ and $\hat{m}_q(k) \cong E\{|y_1(k)|^q\}$, $q = 2, 4$. The above cost function is similar to that already proposed by Shalvi and Weinstein [1068, 1070, 1199] and in an extended form by Inouye [606, 607] and Comon [301] for blind deconvolution problems [545]. However, instead of the standard kurtosis $(E\{y_1^4\} - 3E^2\{y_1^2\})$, we employ a normalized version which poses no constraint on the variance of the output signals and therefore considerably simplifies the optimization procedure. Furthermore, the normalized kurtosis is more robust with respect to outliers and spiky noise.

In some applications such as communications, source signals are typically sub-Gaussian, so they have negative kurtosis. In such cases, the nonlinear function can be simplified as

$$\varphi_1[y_1] = (y_1 - \alpha y_1^3) = (y_1 - f(y_1)), \tag{5.16}$$

where $\alpha = E\{|y|^2\}/E\{|y|^4\}$, $f(y) = y^3$ and $\beta = \text{sign}(\kappa_4\{y_1\}) = -1$.

The presented algorithms can be considered as the Blind Least Mean Square (BLMS) algorithm with a "blind" error signal equal to $e_1(k) = y_1(k) - f[y_1(k)]$. It should be noted that the BLMS learning algorithm resembles the standard LMS algorithm where error is represented by $e_1(k)$ (see Fig. 5.2). In fact, many techniques and approaches known from standard adaptive signal processing can be adopted to improve the performance of the BLMS algorithm.

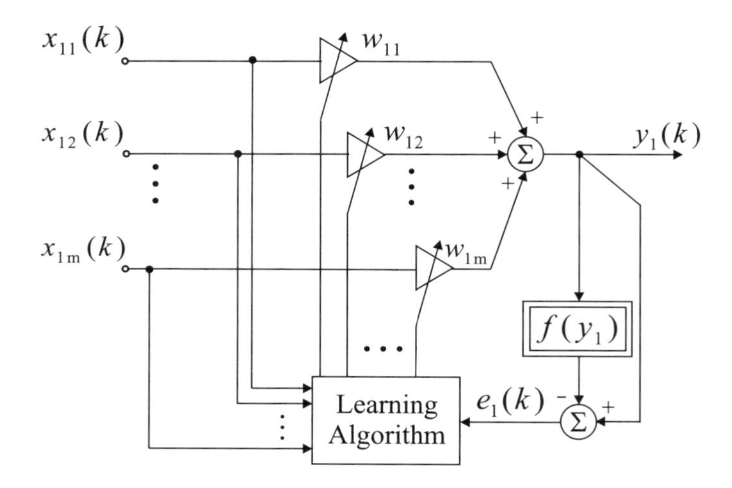

Fig. 5.2 Block diagram illustrating the blind LMS algorithm.

5.2.2 Algorithms Based on Optimization of Generalized Kurtosis

The above adaptive learning algorithm can easily be generalized by minimizing the following cost function

$$\boxed{\mathcal{J}_{pq}[\mathbf{w}_1] = -\frac{1}{p}\,|\kappa_{p,q}\{y_1\}| = -\frac{\beta}{p}\kappa_{p,q}\{y_1\},} \tag{5.17}$$

where $\kappa_{p,q}\{y_1\}$ is the generalized normalized kurtosis (or Gray's variable norm) [509, 715], defined as

$$\boxed{\kappa_{p,q}\{y_1\} = \frac{E\{|y_1|^p\}}{E^q\{|y_1|^{p/q}\}} - c_{pq},} \tag{5.18}$$

where c_{pq} is a positive constant, such that for the Gaussian distribution $\kappa_{p,q} = 0$ and p, q are chosen suitably positive (typically, $q = 2$ and $p = 1, 3, 4, 6$). In the special case for $p = 4$, $q = 2$ and $c_{pq} = 3$, the generalized kurtosis reduces to the standard normalized kurtosis.

Applying the standard gradient descent approach to minimize the cost function (5.17), we have

$$\begin{aligned}
\frac{d\mathbf{w}_1}{dt} &= -\mu_{pq}\,\frac{\partial \mathcal{J}_{pq}(\mathbf{w}_1)}{\partial \mathbf{w}_1} \\
&= \tilde{\mu}_{pq}\,\beta\left[\frac{E\{|y_1|^{p/q}\}}{E\{|y_1|^p\}}E\{\mathrm{sign}(y_1)|y_1|^{p-1}\mathbf{x}_1\} - E\{\mathrm{sign}(y_1)|y_1|^{p/q-1}\mathbf{x}_1\}\right],
\end{aligned} \tag{5.19}$$

where $\mu_{pq} > 0$ and $\tilde{\mu}_{pq} = \frac{E\{|y_1|^p\}}{E^{q+1}\{|y_1|^{p/q}\}}\mu_{pq} > 0$ are the learning rates.
Applying the stochastic approximation technique, we obtain an on-line learning formula

$$\boxed{\Delta\mathbf{w}_1(k) = \mathbf{w}_1(k+1) - \mathbf{w}_1(k) = \tilde{\eta}_{pq}(k)\,\varphi_1(y_1(k))\,\mathbf{x}_1(k),} \tag{5.20}$$

where $\varphi_1[y_1(k)]$ has the general form

$$\boxed{\varphi_1[y_1] = \mathrm{sign}(y_1)\left(|y_1|^{p/q-1} - \frac{E\{|y_1|^{p/q}\}}{E\{|y_1|^p\}}|y_1|^{p-1}\right)} \tag{5.21}$$

assuming for simplicity that all source signals are sub-Gaussian with negative kurtosis. In the special case, for sub-Gaussian signals, where $p = 1$ and $q = 1/2$, we obtain the modified Sato algorithm with nonlinearity [1041, 1042, 1172, 941]

$$\varphi_1[y_1] \quad = \quad y_1 - \frac{E\{|y_1|^2\}}{E\{|y_1|\}}\mathrm{sign}(y_1). \tag{5.22}$$

More generally, for sub-Gaussian signals, the choice of $q = 1/2$ produces the class of the Godard or constant modulus algorithms (CMA) (5.14) with an adaptive nonlinear activation function [942, 715, 270, 269, 33, 397]

$$\varphi_1[y_1] \quad = \quad (|y_1|^p - \gamma_p)\, y_1 |y_1|^{p-2}, \tag{5.23}$$

where

$$\gamma_p = \frac{E\{|\hat{s}_1(k)|^{2p}\}}{E\{|\hat{s}_1(k)|^p\}} = \mathrm{const} \tag{5.24}$$

assuming that the statistics of the estimated source signal $\hat{s}_1$ are known. However, in general, when the statistics of source signals are not known (or cannot be estimated) the parameter $R_p(\hat{s}_1(k))$ is not fixed but can be adapted during the learning process depending on the higher-order moments of the absolute values of the estimated output signal $y(k)$. In this case, the higher-order moments of the form $m_r(k) = E\{|y(k)|^r\}$ appearing in the nonlinearities, can be estimated on-line by using the moving average (MA) procedure.

It is interesting that the above algorithms can easily be extended to complex-valued signals by noting that in such a case, $y_1(k) = \mathbf{x}_1^H(k)\mathbf{w}_1(k)$ (where superscript H denotes the complex conjugate transpose or Hermitian operation) and $\mathrm{sign}[y_1(k)]$ is replaced by $y_1(k)/|y_1(k)|$. For example, the constant modulus algorithm (CMA) for $p = 4$, $q = 2$, for complex-valued coefficients and signals, takes the form:

$$\boxed{\mathbf{w}_1(k+1) = \mathbf{w}_1(k) \pm \eta(k)\left[y_1(k) - \frac{E\{|y_1|^2\}}{E\{|y_1|^4\}} y_1(k)|y_1(k)|^2\right]\mathbf{x}_1^*(k),} \tag{5.25}$$

where $\mathbf{x}_1^*$ is a vector of the complex conjugate sensor signals, where the plus sign is used for sub-Gaussian signals and the minus sign is used for super-Gaussian signals. The above algorithms can be considered as the BLMS (Blind Least Mean Square) algorithm with a "blind" error signal equal to $y_1(k) - \frac{\hat{m}_2(|y_1(k)|)}{\hat{m}_4(|y_1(k)|)} y_1(k)|y_1(k)|^2$. Many powerful and efficient techniques developed originally for standard LMS can also be applied to the above algorithms (see Fig. 5.2).

It should be noted that, in general, the activation function can not be fixed but is instead adaptive during the learning process. This observation was first made by several researchers [29, 221, 223]. Adapting the nonlinearities are important in a multichannel case, since signals may consist of a mixture of several sources with different distributions. The optimal choice for values of p and q depends on the statistics of the input signals, implying a trade-off between the tracking ability and the estimation accuracy of the algorithm [715]. Such methods also have a natural extension to multichannel blind deconvolution problems [225, 220, 941].

Remark 5.3 *It should be noted that the above method does not need prewhitening. However, for ill-conditioned problems (when a mixing matrix is ill-conditioned and/or source signals have different amplitude or variance), we can apply preprocessing (prewhitening) in the form of* $\mathbf{x}_1 = \mathbf{Q}\mathbf{x}$, *where the decorrelation matrix* $\mathbf{Q} \in \mathbb{R}^{n \times m}$ *ensures that the auto-correlation matrix* $\mathbf{R}_{\mathbf{x}_1\mathbf{x}_1} = E\{\mathbf{x}_1\mathbf{x}_1^T\} = \mathbf{I_n}$. *Prewhitening can simultaneously reduce the dimensional redundancy of the signals from* m *to* n, *if we select* $\mathbf{Q} \in \mathbb{R}^{n \times m}$. *It should be noted that after the decorrelation process, the new unknown mixing matrix defined as* $\mathbf{A} = \mathbf{Q}\,\mathbf{H}$, *is an orthogonal matrix satisfying the relation* $\mathbf{A}^T\mathbf{A} = \mathbf{I}_n$, *i.e.,* $\mathbf{a}_i^T\mathbf{a}_j = \delta_{ij}$, *where* $\mathbf{a}_i$ *is the i-th column vector of the global mixing matrix* $\mathbf{A}$ *(see Fig. 5.1 (b)).*

5.2.3 KuicNet Learning Algorithm

The KuicNet learning algorithm developed by Kung and Douglas uses the same cost function (normalized kurtosis) but under the constraint that the vector $\mathbf{w}_1$ has unit length, i.e., $||\mathbf{w}_1||_2^2 = 1$ [407]. Taking into account that

$$E\{y_1^2\} = E\{\mathbf{w}_1^T\mathbf{x}_1\mathbf{x}_1^T\mathbf{w}_1\} = \mathbf{w}_1^T\mathbf{R}_{\mathbf{x}_1\mathbf{x}_1}\mathbf{w}_1 = \mathbf{w}_1^T\mathbf{w}_1 = 1, \tag{5.26}$$

the cost function can be reformulated as

$$\boxed{\mathcal{J}_1(\mathbf{w}_1) = -\frac{\beta}{4}\frac{E\{y_1^4\}}{||\mathbf{w}_1||_2^4}.} \tag{5.27}$$

Applying the standard stochastic gradient approach, we obtain a simple learning rule

$$\mathbf{w}_1(k+1) = \mathbf{w}_1(k) + \eta\,\beta\,\left(y_1^3(k)\,\mathbf{x}_1(k) - y_1^4(k)\,\mathbf{w}_1(k)\right), \tag{5.28}$$

where $\eta > 0$ is the learning rate and $\beta = sign(\kappa_4(y_1))$.

It should be noted that the above KuicNet learning rule has a self-normalizing property, such that the vector $\mathbf{w}_1$ ($||\mathbf{w}_1(k)||_2 \approx 1$) approximately maintains its unit length. However, when the extracted source signal $y_1(k)$ has a negative value of kurtosis, the above algorithm is unstable (due to accumulation of error during iterative process) and the vector $\mathbf{w}_1$ must periodically be renormalized to unit length as follows

$$\mathbf{w}_1^+(k+1) = \mathbf{w}_1(k) + \eta\,\beta\,\left[y_1^3\,\mathbf{x}_1 - y_1^4\,\mathbf{w}_1\right], \tag{5.29}$$

where $\mathbf{w}_1(k+1) = \mathbf{w}_1^+(k+1)/||\mathbf{w}_1^+(k+1)||_2$.

Alternatively, we can use the ordered rotation KuicNet learning rule proposed recently by S. Douglas and S. Y. Kung [407]

$$\mathbf{w}_1(k+1) = \mathbf{w}_1(k) + \eta\,y_1^3(k)\left[\mathbf{x}_1(k) - y_1(k)\,\mathbf{w}_1(k)\right] \quad \text{for } \beta > 0, \tag{5.30}$$

$$\mathbf{w}_1(k+1) = \mathbf{w}_1(k) - \eta\,y_1^3\,\left[||\mathbf{w}_1||_2^4\mathbf{x}_1(k) - y_1(k)\,\mathbf{w}_1(k)\right] \quad \text{for } \beta < 0. \tag{5.31}$$

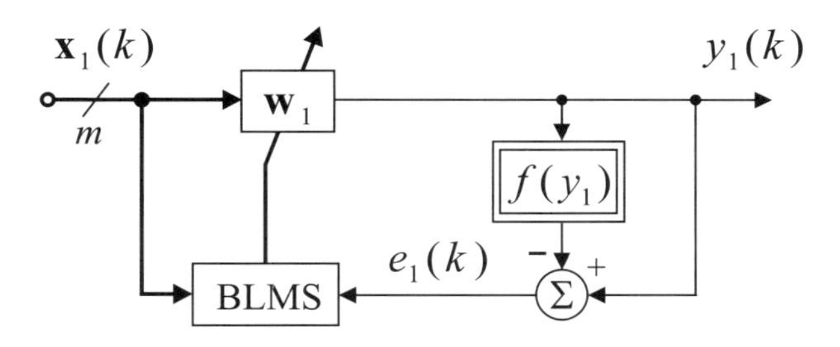

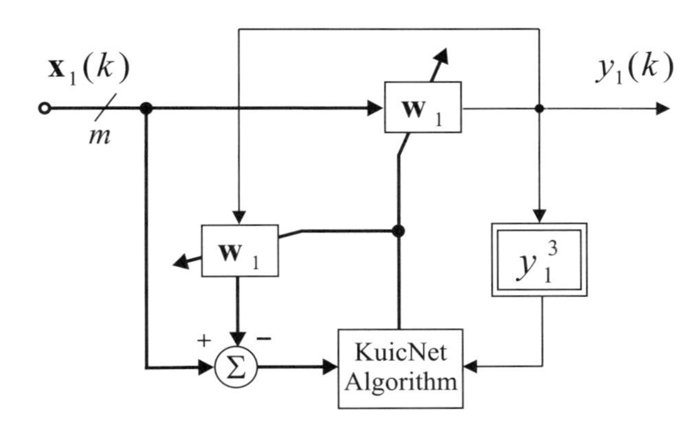

Fig. 5.3 Implementation of the BLMS and KuicNet algorithms.

Implementations of BLMS and KuicNet algorithms are shown in Fig. 5.3 (a) and (b).

5.2.4 Fixed-point Algorithms

Hyvärinen and Oja proposed a family of batch learning rules, called fixed-point or fast ICA algorithms for a hierarchical neural network that extracts source signals from their mixtures in a sequential fashion. In the hierarchical neural network, the output of the jth extraction processing unit is described as $y_j = \mathbf{w}_j^T \mathbf{x}_1$, where $\mathbf{w}_j = [w_{j1}, w_{j2}, \dots, w_{jn}]^T$. Contrary to the cascade neural network, the input vector for each processing unit of the hierarchical neural network is the same $\mathbf{x}_1 = \mathbf{Q}\,\mathbf{x}$ vector from the prewhitened sensor signals.

Let us consider as the cost function, the standard kurtosis for a zero mean signal y_1

$$\mathcal{J}(\mathbf{w}_1, y_1) = \kappa_4(y(\mathbf{w}_1)) = \frac{1}{4}\left[E\{y_1^4\} - 3E^2\{y_1^2\}\right], \tag{5.32}$$

where $y_1 = \mathbf{w}_1^T \mathbf{x}_1$ is the output of a single processing unit.

In order to find the optimal value of vector $\mathbf{w}_1$, we apply the following iteration rule [589, 453]:

$$\boxed{\mathbf{w}_1(l+1) = \frac{\nabla_{w_1} \kappa_4(\mathbf{w}_1(l))}{||\nabla_{w_1} \kappa_4(\mathbf{w}_1(l))||_2},} \tag{5.33}$$

where $\nabla_{w_1} \kappa_4(\mathbf{w}_1) = \partial \kappa_4(\mathbf{w}_1)/\partial \mathbf{w}_1$. Equivalently, we can apply the following formula:

$$\begin{aligned} \mathbf{w}_1^+(l+1) &= \nabla_{w_1} \kappa_4(\mathbf{w}_1(l)), \\ \mathbf{w}_1(l+1) &= \frac{\mathbf{w}_1^+(l+1)}{||\mathbf{w}_1^+(l+1)||_2}, \end{aligned} \tag{5.34}$$

which enforces that the vector $\mathbf{w}_1$ has unit length in each iteration step. The gradient of the cost function can be evaluated as

$$\nabla_{w_1} \kappa_4(\mathbf{w}_1) = \frac{\partial \kappa_4(\mathbf{w}_1)}{\partial \mathbf{w}_1} = E\{y_1^3 \mathbf{x}_1\} - 3E\{y_1^2\}E\{y_1 \mathbf{x}_1\}. \tag{5.35}$$

Assuming that sensor signals are prewhitened and the covariance matrix $\mathbf{R_x} = E\{\mathbf{x}_1 \mathbf{x}_1^T\} = \mathbf{I}$ and the vector $\mathbf{w}_1$ is normalized to unit length, (i.e., $E\{y_1^2\} = 1$), we obtain

$$\begin{aligned} \nabla_{w_1} \kappa_4(\mathbf{w}_1) &= E\{y_1^3 \mathbf{x}_1\} - 3E\{y_1^2\}E\{\mathbf{x}_1 \mathbf{x}_1^T \mathbf{w}\} \\ &= E\{y_1^3 \mathbf{x}_1\} - 3\mathbf{w}_1. \end{aligned} \tag{5.36}$$

Thus, the fixed point algorithm in its standard form can be written as

$$\begin{aligned} \mathbf{w}_1^+(l+1) &= \langle y_1^3 \mathbf{x}_1 \rangle - 3\mathbf{w}_1(l), \quad y_1 = \mathbf{w}_1^T(l)\mathbf{x}_1, \\ \mathbf{w}_1(l+1) &= \frac{\mathbf{w}_1^+(l+1)}{||\mathbf{w}_1^+(l+1)||_2}. \end{aligned} \tag{5.37}$$

In a similar way, we can derive the modified fixed point algorithm for generalized normalized kurtosis

$$\kappa_{p,q}\{y_1\} = \frac{1}{p}\left(\frac{E\{|y_1|^p\}}{E^q\{|y_1(k)|^{p/q}\}} - c_{pq}\right), \tag{5.38}$$

where c_{pq} is a positive constant, such that for the Gaussian distribution $\kappa_{p,q} = 0$.

It is straightforward to check that the gradient of the generalized normalized kurtosis with respect to vector $\mathbf{w}_1$, can be expressed as

$$\frac{\partial \kappa_{p,q}}{\partial \mathbf{w}_1} = \frac{E\{\text{sign}(y_1)|y_1|^{p-1}\mathbf{x}_1\}E^q\{|y_1|^{p/q}\} - E\{|y_1|^p\}E^{q-1}\{|y_1|^{p/q}\}E\{\text{sign}(y_1)|y_1|^{p/q-1}\mathbf{x}_1\}}{E^{2q}\{|y_1|^{p/q}\}}.$$

Thus, the new algorithm can take the form

$$\boxed{\mathbf{w}_1(l+1) = \frac{\nabla_{w_1} \kappa_{p,q}(\mathbf{w}_1(l))}{||\nabla_{w_1} \kappa_{p,q}(\mathbf{w}_1(l))||_2},} \tag{5.39}$$

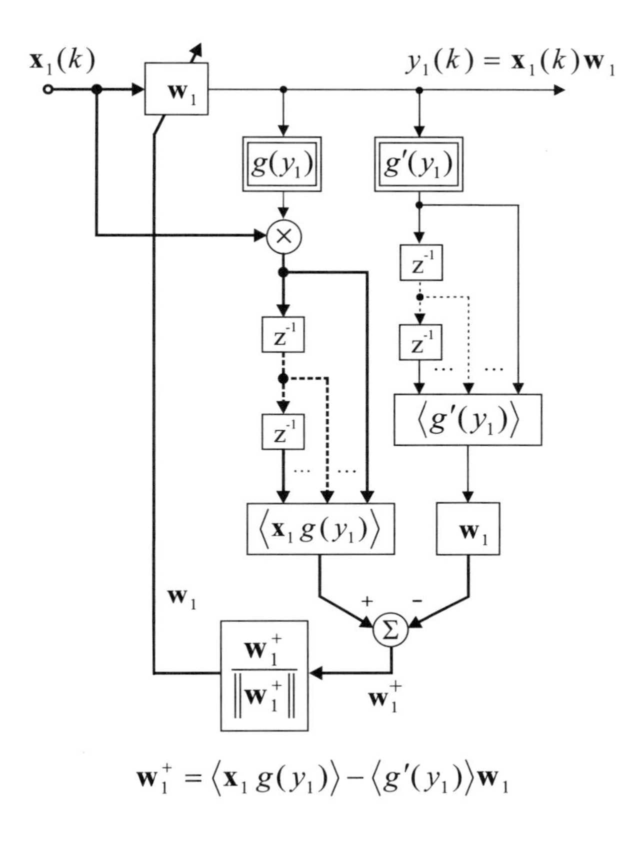

Fig. 5.4 Block diagram illustrating implementation of the generalized fixed-point learning algorithm developed by Hyvärinen-Oja [589]. $\langle\rangle$ indicates an averaging operator. In the special case of optimization of standard kurtosis $g(y_1) = y_1^3$ and $g'(y_1) = 3y_1^2$.

where

$$\nabla_{w_1} \kappa_{p,q}(\mathbf{w}_1(l)) = \frac{E\{|y_1|^p\}}{E^{q+1}\{|y_1|^{p/q}\}} E\{\text{sign}(y_1)|y_1|^{p-1}\mathbf{x}_1\} - E\{\text{sign}(y_1)|y_1|^{p/q-1}\mathbf{x}_1\}, \quad (5.40)$$

with $y_1 = \mathbf{w}_1^T(l)\mathbf{x}_1$. In the special case, for $p = 4$, $q = 2$, $c_{pq} = 3$, and prewhitened sensor data the above algorithm can be simplified to a form similar to the learning rule given by Eq. (5.37) as

$$\mathbf{w}_1^+(l+1) = \frac{\langle y_1^3 \mathbf{x}_1 \rangle}{\langle y_1^4 \rangle} - \mathbf{w}_1(l), \quad y_1 = \mathbf{w}_1^T(l)\mathbf{x}_1, \quad (5.41)$$

$$\mathbf{w}_1(l+1) = \frac{\mathbf{w}_1^+(l+1)}{||\mathbf{w}_1^+(l+1)||_2}. \quad (5.42)$$

The above algorithm is more robust to outliers and spiky noise than algorithm (5.37).

Let us consider an alternative derivation of the fixed point algorithm (5.37) in a slightly more general form. For this purpose, we formulate the following constrained optimization problem:

maximize $\mathcal{J}(\mathbf{w}_1) = E\{G(y_1)\}$,

subject to the constraint $E\{y_1^2\} = ||\mathbf{w}_1||_2^2 = 1$,

where $G(y)$ is a suitably chosen convex function (typically $G(y) = \log\cosh(\alpha y)/\alpha$). For signals with spiky noise, we can use more robust functions (see Chapter 2 and Table 2.1).

Assume now that observed (sensor) data are prewhitened (i.e., $E\{\mathbf{x}_1\mathbf{x}_1^T\} = \mathbf{I}$). According to Kuhn-Tucker conditions, the maxima of $\mathcal{J}(\mathbf{w}_1) = E\{G(y_1)\}$ (under the constraint $E\{|\mathbf{w}_1^T\mathbf{x}_1|^2\} = ||\mathbf{w}_1||_2^2 = 1$) are obtained at points $\mathbf{w}_1$ satisfying

$$\nabla E\{G(y_1)\} - \lambda \nabla E\{|\mathbf{w}_1^T\mathbf{x}_1|^2\} = 0, \tag{5.43}$$

where λ is a Lagrange multiplier. After a simple mathematical manipulation, we obtain

$$F(\mathbf{w}_1) = E\{\mathbf{x}_1 g(y_1)\} - \lambda \mathbf{w}_1 = \mathbf{0}, \tag{5.44}$$

where $g(y) = dG(y)/dy$. The Newton method can be used to efficiently solve Eq. (5.44). For this purpose the Jacobian matrix of $\nabla E\{G(y_1)\} = E\{\mathbf{x}_1 g(y_1)\}$ is evaluated as follows:

$$\nabla^2 E\{G(y_1)\} = E\{\mathbf{x}_1\,\mathbf{x}_1^T\, g'(y_1)\} \approx E\{\mathbf{x}_1\,\mathbf{x}_1^T\}\, E\{g'(y_1)\}. \tag{5.45}$$

Taking into account that the approximate Jacobian matrix

$$\mathbf{J} = (E\{g'(y_1)\} - \lambda)\,\mathbf{I} \tag{5.46}$$

is a diagonal matrix (and thus easy to invert), we obtain the following approximate Newton iteration:

$$\begin{aligned} \mathbf{w}_1^+ &= \mathbf{w}_1 - \mathbf{J}^{-1}F(\mathbf{w}_1) \\ &= \mathbf{w}_1 - \frac{E\{\mathbf{x}_1 g(y_1)\} - \lambda \mathbf{w}_1}{E\{g'(y_1)\} - \lambda}, \\ \mathbf{w}_1 &= \frac{\mathbf{w}_1^+}{||\mathbf{w}_1^+||_2}. \end{aligned} \tag{5.47}$$

Finally, by multiplying both sides of the above equation by the factor $(-E\{g'(y_1)\} + \lambda)$, the algorithm is simplified to the so called Fast-ICA algorithm as

$$\begin{aligned} \mathbf{w}_1^+ &= E\{\mathbf{x}_1 g(y_1)\} - E\{g'(y_1)\}\mathbf{w}_1, \\ \mathbf{w}_1 &= \frac{\mathbf{w}_1^+}{||\mathbf{w}_1^+||_2}, \end{aligned} \tag{5.48}$$

where $g'(y_1) = dg(y_1)/dy_1$. Fig. 5.4 illustrates implementation of the fast ICA batch learning algorithm for the extraction of the first source. In order to extract subsequent source signals, we can apply the deflation technique as described in the next section.

In theory, the Hyvärinen-Oja method (as shown in [589]) can successfully extract a wide class of non-Gaussian source signals. However, in practice, use of fixed nonlinearities and high sensitivity to accumulated errors of Gram-Schmidt-like orthogonalization (or decorrelation) for deflation procedure may lead, for a large number of sources, to rapid decrease the quality of the extracted signals. In the next section, we present a simple and robust cascade (sequential) extraction deflation procedure which avoids accumulation of error during the deflation procedure.

5.2.5 Sequential Extraction and Deflation Procedure

We now describe a simple and efficient deflation procedure. Figures 5.1 (a) and (b) illustrate the extraction and deflation process. The cascade neural network employs two different types of processing units, alternately connected with each other in a cascade fashion, one type for blind extraction and the other for deflation. The j-th extraction processing unit extracts a source signal from inputs that are linear mixtures of the remaining source signals yet to be extracted. The jth deflation processing unit then deflates (removes) the newly extracted source signal from the mixtures and feeds the resulting outputs to the next $(j+1)$-th extraction processing unit.

After the successful extraction of the first source signal $y_1(k) \approx s_i(k)$ $\left(i \in \overline{1,n}\right)$, we can apply the deflation procedure which removes previously extracted signals from the mixtures. This procedure may be recursively applied to extract all source signals sequentially. This means, that we require an on-line linear transformation given by (see Fig.5.1) [273]

$$\mathbf{x}_{j+1}(k) = \mathbf{x}_j(k) - \widetilde{\mathbf{w}}_j y_j(k), \qquad (j = 1, 2, \dots,) \tag{5.49}$$

where $\widetilde{\mathbf{w}}_j$ can be optimally estimated by minimization of the cost (energy) function

$$\mathcal{J}_j(\widetilde{\mathbf{w}}_j) = E\{\rho(\mathbf{x}_{j+1})\} = \frac{1}{2} E\{\sum_{p=1}^{m} x_{j+1,p}^2\}, \tag{5.50}$$

where $E\{\rho(\mathbf{x}_{j+1})\}$ is an objective function and $y_j(k) = \mathbf{w}_j^T \mathbf{x}_j(k)$. Intuitively speaking, such a cost (objective) function can be considered an energy function whose minimum is achieved when the extracted source signal y_j is eliminated from the mixture of sources. Note that $\tilde{\mathbf{w}}_j$ is different from $\mathbf{w}_j$. Minimization of the cost function (5.50) leads to the simple local type LMS learning rule [273, 272, 1144]

$$\boxed{\widetilde{\mathbf{w}}_j(k+1) = \widetilde{\mathbf{w}}_j(k) + \tilde{\eta}_j(k)\, y_j(k)\, \mathbf{x}_{j+1}(k), \qquad (j = 1, 2, \dots, n)} \tag{5.51}$$

where (as we will show later) $\widetilde{\mathbf{w}}_j$ is an estimation of the j-th column $\widehat{\mathbf{h}}_j$ of the identified mixing matrix $\widehat{\mathbf{H}}$, $y_j = \mathbf{w}_j^T\ \mathbf{x}_j$. The vector $\mathbf{w}_j$ can be estimated by using the following learning rule (similar to (5.21))

$$\mathbf{w}_j(k+1) = \mathbf{w}_j(k) + \eta_j(k)\, \varphi_j[y_j(k)]\, \mathbf{x}_j(k), \tag{5.52}$$

$$\varphi_j(y_j) = \frac{y_j}{|y_j|^2} \left(|y_j|^{p/q} - \frac{E\{|y_j|^{p/q}\}}{E\{|y_j|^p\}} |y_j|^p \right). \tag{5.53}$$

The procedure can be continued until all estimated source signals are recovered, that is, until the amplitude of each signal $\mathbf{x}_{j+1}$ reaches a preassigned threshold. This procedure means it is not required to know the number of source signals in advance; however, it is assumed that this number is constant and the mixing system is stationary (e.g., the sources do not "change" during the convergence of the algorithm). It can be proved that this deflation optimization procedure has no spurious (undesired) minima and, after the convergence, the algorithm (5.51) estimates one column of the mixing matrix $\mathbf{H}$ with scaling and permutation indeterminacy.

Alternatively, instead of an adaptive on-line algorithm, we can use a very simple and efficient batch one-step formula. Minimization of the mean squares cost function

$$\begin{aligned} \tilde{\mathcal{J}}_j(\widetilde{\mathbf{w}}_j) = E\{\mathbf{x}_{j+1}^T \mathbf{x}_{j+1}\} = \\ E\{\mathbf{x}_j^T \mathbf{x}_j\} - 2\widetilde{\mathbf{w}}_j^T E\{\mathbf{x}_j y_j\} + \widetilde{\mathbf{w}}_j^T \widetilde{\mathbf{w}}_j E\{y_j^2\}, \end{aligned} \tag{5.54}$$

with respect to $\widetilde{\mathbf{w}}_j$ leads to an alternative batch simple updating equation,

$$\boxed{\widetilde{\mathbf{w}}_j = \widehat{\mathbf{h}}_j = \frac{E\{\mathbf{x}_j y_j\}}{E\{y_j^2\}} = \frac{E\{\mathbf{x}_j \mathbf{x}_j^T\}\mathbf{w}_j}{E\{y_j^2\}},} \tag{5.55}$$

where $\widehat{\mathbf{h}}_j$ is, in fact, an estimated column of the mixing matrix $\mathbf{H}$, ignoring arbitrary scaling and permutation of columns ambiguities.

It is important to note that by prewhitening or performing PCA, for each processing unit the covariance matrices are turned into identity matrices, i.e.,

$$\mathbf{R}_{\mathbf{x}_j \mathbf{x}_j} = E\{\mathbf{x}_j \mathbf{x}_j^T\} = \mathbf{I} \qquad \forall j \tag{5.56}$$

and keeping $||\mathbf{w}_j||_2 = 1$ which implies $E\{y_j^2\} = 1$, we do not need to estimate any of the vectors $\widetilde{\mathbf{w}}_j$ in the deflation procedure since

$$\widetilde{\mathbf{w}}_j = \widehat{\mathbf{a}}_j = \mathbf{w}_j, \tag{5.57}$$

where $\widehat{\mathbf{a}}_j$ is the j-th vector of the estimated orthogonal mixing matrix $\widehat{\mathbf{A}} = \mathbf{Q}\widehat{\mathbf{H}}$ with $\mathbf{Q}$ prewhitening matrix.

In Appendices A and B, we prove that such algorithms converge to a desired solution, i.e., they can successfully extract sources that have nonzero kurtosis. More precisely, we can prove now that the family of extraction criteria discussed in the previous sections have no spurious equilibria and hence, successful extraction of a source signal can always be achieved, regardless of initial conditions. We then extend the theoretical results and perform a more general analysis for the jth extraction and deflation processing units. In this general analysis, it will be shown that the jth extraction processing unit can always successfully extract a source signal from a mixture of the remaining source signals and that the outputs of the jth deflation processing unit produce a new mixture which does not contain this newly extracted signal. Finally, based on these analytical results, in Appendix B, we describe implementation techniques for determining appropriate values of β, and we formulate criteria for terminating the extraction and deflation procedures.

5.3 ADAPTIVE ON-LINE ALGORITHMS FOR BLIND SIGNAL EXTRACTION OF TEMPORALLY CORRELATED SOURCES

In previous sections, we discussed algorithms based on the assumptions that "interesting" sources are independent and non-Gaussian. In this section and the next, we show how to relax these conditions assuming that sources are colored, i.e., they have different temporal structures and arbitrary distributions including Gaussian. In this section, we will derive a family of on-line adaptive learning algorithms for sequential blind extraction of arbitrarily distributed but generally not i.i.d. (independent and identically distributed) sources from their linear mixtures. The algorithms discussed in this section are computationally simple and efficient, and they only exploit second order statistics. Thus, in contrast to the algorithms described in the previous sections of this chapter they do not assume non-zero kurtosis for the sources. Thus, signals with low or even zero kurtosis (colored Gaussian) can be successfully extracted. Specifically, some biomedical source signals are characterized by extremely low values of normalized kurtosis and due to nonstationarities, their distribution may change in time. In fact, the algorithms discussed in the previous sections use some nonlinear activation functions whose optimal forms depend on the statistics of source signals. However, such statistics are usually unknown. These algorithms moreover, may have poor performance and relatively slow convergence speeds for small absolute values of normalized kurtosis.

Our main objective in this section is to derive an alternative class of algorithms for estimation of colored source signals sequentially, one-by-one, assuming that source signals are arbitrarily distributed and have different auto-correlation functions, i.e., $E\{s_i(k-p)\, s_i(k)\} \neq E\{s_j(k-p)\, s_j(k)\}$ for some time delays.

Let us assume that temporally correlated source signals are modelled by autoregressive processes (AR) as

$$\begin{aligned} s_j(k) &= \widetilde{s}_j(k) + \sum_{p=1}^{L} \widetilde{a}_{jp} s_j(k-p) \\ &= \widetilde{s}_j(k) + A_j(z) s_j(k), \end{aligned} \tag{5.58}$$

where $A_j(z) = \sum_{p=1}^{L} \widetilde{a}_{jp}\, z^{-p}$, $z^{-p}s(k) = s(k-p)$ and $\widetilde{s}_j(k)$ are i.i.d. unknown innovative processes. In practice, the AR model can be extended to more general models such as the Auto Regressive Moving Average (ARMA) model or the Hidden Markov Model (HMM) [22, 46, 589].

For ill-conditioned problems (when a mixing matrix is ill-conditioned and/or source signals have different amplitudes), we can apply optional preprocessing (prewhitening) to the

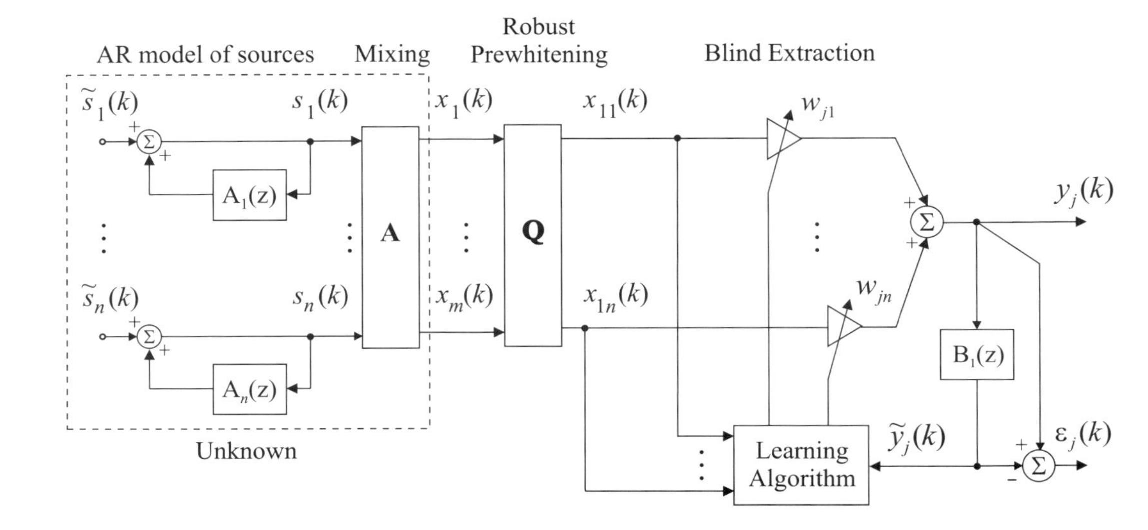

Fig. 5.5 Block diagram illustrating implementation of learning algorithm for temporally correlated sources.

sensor signals $\mathbf{x}$ in the form

$$\mathbf{x}_1 = \mathbf{Q}\mathbf{x},$$

where $\mathbf{Q} \in \mathbb{R}^{n \times m}$ is a decorrelation matrix ensuring that the correlation matrix $\mathbf{R}_{\mathbf{x}_1\mathbf{x}_1} = E\{\mathbf{x}_1\mathbf{x}_1^T\} = \mathbf{I}_n$ is an identity matrix (see Chapters 4 and 8 for more detail on robust algorithms with respect to noise).

To model temporal structures of source signals, we consider a linear neural network cascaded with an adaptive filter with the transfer function $B_1(z)$ (which estimates one $A_j(z)$) as illustrated in Fig. 5.5, where the input-output relations of the network and the filter are described, respectively, as follows:

$$y_j(k) = \mathbf{w}_j^T \mathbf{x}_1(k) = \sum_{i=1}^{n} w_{ji} x_{1i}(k) \tag{5.59}$$

and

$$\varepsilon_j(k) = [1 - B_j(z)]\, y_j(k), \qquad (j = 1, 2, \ldots, n) \tag{5.60}$$

where $B_j(z)$ is the transfer function of a suitably chosen filter.

Depending on specific applications and requirements, the filters $B_j(z)$ can take different forms to extract the source signals with specific stochastic properties, for example:

- The filter $B_j(z)$ can be a simple Finite Impulse Response (FIR) filter which performs a linear prediction of the output signal $y_j(k)$. (for more detail, see the next section).

- As a special case, we can use the simplest first order FIR filter (with a single unit delay) with one step prediction, i.e., only the first order predictor can be applied in the simplest case.

- We can also employ Infinite Impulse Response (IIR) or FIR band pass filters or banks of band pass filters in order to extract source signals with specific properties, i.e., with specific frequency bandwidth.

- Furthermore, the concept can be generalized by employing a nonlinear predictor instead of a simple linear predictor. For example, a multilayer perceptron (MLP) or radial basis function (RBF) network can be used as a nonlinear predictor. In the simplest case, a nonlinear predictor can be described as

$$\varepsilon_j(k) = \left[y_j(k) - \varphi\left(\sum_{p=1}^{L} b_p y_j(k-p) \right) \right], \tag{5.61}$$

  where $\varphi(y)$ is a suitably chosen adaptive or fixed nonlinear function [22, 276]. By employing a suitably designed nonlinear predictor, we can extend the class of extracted signals, e.g., it may be possible to extract close to white independent and/or colored signals with similar shapes of power spectra.

5.3.1 On-Line Algorithms for Blind Extraction Using a Linear Predictor

Let us assume for simplicity, that we want to extract only one source signal, e.g. $s_j(k)$, from the available sensor vector $\mathbf{x}(k)$. For this purpose, we employ a single processing unit described above as (see Fig. 5.6):

$$y_1(k) = \mathbf{w}_1^T \mathbf{x}(k) = \sum_{i=1}^{m} w_{1i} x_i(k), \tag{5.62}$$

$$\begin{aligned} \varepsilon_1(k) &= y_1(k) - \sum_{p=1}^{L} b_{1p} y_1(k-p) \\ &= \mathbf{w}_1^T \mathbf{x}(k) - \mathbf{b}_1^T \bar{\mathbf{y}}_1(k), \end{aligned} \tag{5.63}$$

where $\mathbf{w}_1 = [w_{11}, w_{12}, \ldots, w_{1m}]^T$, $\bar{\mathbf{y}}_1(k) = [y_1(k-1), y_1(k-2), \ldots, y_1(k-L)]^T$, $\mathbf{b}_1 = [b_{11}, b_{12}, \ldots, b_{1L}]^T$ and $B_1(z) = \sum_{p=1}^{L} b_{1p} z^{-p}$ is the transfer function of the corresponding FIR filter. It should be noted that the FIR filter can have a sparse representation. In particular, only one single processing unit, e.g. with delay p and $b_{1p} \neq 0$ can be used instead of L parameters. The processing unit has two outputs: $y_1(k)$ which estimates the extracted source signals, and $\varepsilon_1(k)$, which represents a prediction error or innovation, after passing the output signal $y_1(k)$ through FIR filter.

Our objective is to estimate optimal values of vectors $\mathbf{w}_1$ and $\mathbf{b}_1$, in such a way that the processing unit successfully extracts one of the sources. This is achieved if the global vector defined as $\mathbf{g}_1 = \mathbf{A}^T \mathbf{w}_1 = \left(\mathbf{w}_1^T \mathbf{A}\right)^T = c_j \mathbf{e}_j$ contains only one nonzero element, e.g. in the

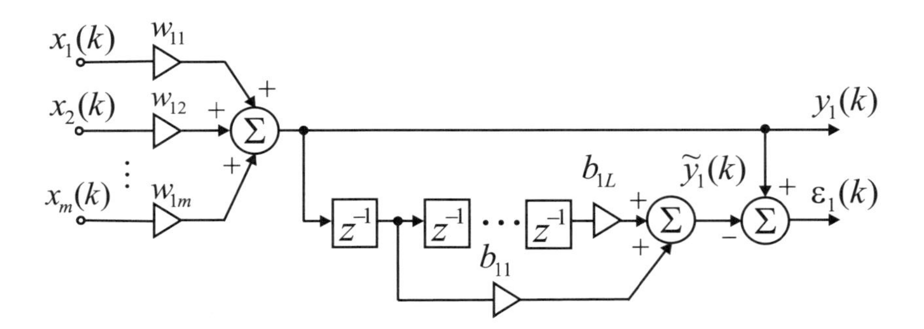

Fig. 5.6 The neural network structure of single extraction unit using a linear predictor.

j-th row, such that $y_1(k) = c_j s_j$, where c_j is an arbitrary nonzero scaling factor. For this purpose, we reformulate the problem as a minimization of the cost function

$$\mathcal{J}(\mathbf{w}_1, \mathbf{b}_1) = E\{\varepsilon_1^2\}. \tag{5.64}$$

The main motivation for applying such a cost function is the assumption that primary source signals (signals of interest) have temporal structures and can be modelled, e.g., by an autoregressive model [29, 60, 61, 268, 952].

According to the AR model of source signals, the filter output can be represented as $\varepsilon_1(k) = y_1(k) - \tilde{y}_1(k)$, where $\tilde{y}_1(k) = \sum_{p=1}^{L} b_{1p} y_1(k-p)$ is defined as an error or estimator of the innovation source $\tilde{s}_j(k)$. The mean square error $E\{\varepsilon_1^2(k)\}$ achieves a minimum $c_1^2 E\{\tilde{s}_j^2(k)\}$, where c_1 is a positive scaling constant, if and only if $y_1 = \pm c_1 s_j$ for any $j \in \{1, 2, \ldots, m\}$ or $y_1 = 0$ holds (see Appendix C). To prevent the latter trivial case, we need a constraint to bound $E\{y_1^2(k)\}$ e.g. to 1. We can formulate this constrained optimization criterion as

$$\text{minimize } \mathcal{J}_1(\mathbf{w}_1, \mathbf{b}_1) = \frac{1}{2} E\{\varepsilon_1^2\} + \frac{\beta_1}{4}(1 - E\{y_1^2\})^2, \tag{5.65}$$

where $\beta_1 > 0$ is the constant penalty factor. The standard stochastic gradient descent method leads to an on-line learning algorithm for the vector $\mathbf{w}_1$ and coefficients of the FIR adaptive filter b_{1p}, respectively,

$$\begin{aligned}
\Delta \mathbf{w}_1(k) &= \mathbf{w}_1(k+1) - \mathbf{w}_1(k) = -\eta_1 \frac{\partial \mathcal{J}_1(\mathbf{w}_1, \mathbf{b}_1)}{\partial \mathbf{w}_1} \\
&= -\eta_1(k) \left[\langle \varepsilon_1(k) \widetilde{\mathbf{x}}_1(k) \rangle - \gamma(k) \mathbf{w}(k) \right],
\end{aligned} \tag{5.66}$$

where $\gamma(k) = -\beta_1[1 - \hat{\sigma}_{y_1}^2(k)]$ is a forgetting factor and

$$\begin{aligned}
\Delta b_{1p}(k) &= b_{ip}(k+1) - b_{ip}(k) = -\widetilde{\eta}_1 \frac{\partial \mathcal{J}_1(\mathbf{w}_1, \mathbf{b}_1)}{\partial b_{1p}} \\
&= \widetilde{\eta}_1(k) \left\langle \varepsilon_1(k) y_1(k-p) \right\rangle,
\end{aligned} \tag{5.67}$$

where

$$\widetilde{\mathbf{x}}_1(k) = \mathbf{x}_1(k) - \sum_{p=1}^{L} b_{1p}\mathbf{x}_1(k-p) \tag{5.68}$$

and η_1 as well as $\widetilde{\eta}_1$ are learning rates.
The variance of output signal $\sigma_{y_1}^2 = E\{y_1(k)^2\}$ can be estimated on-line by using the standard moving averaging formula

$$\hat{\sigma}_{y_1}^2(k) = (1-\eta_0)\,\hat{\sigma}_{y_1}^2(k-1) + \eta_0\, y_1^2(k). \tag{5.69}$$

In this model, we exploit the temporal structure of signals rather than their statistical independence [272, 1105]. Intuitively speaking, the source signals s_j have less complexity than the mixed sensor signals x_j. In other words, the degree of temporal predictability of any source signal is higher than (or equal to) that of any mixture. For example, waveforms of a mixture of two sine waves with different frequencies are more complex or less predictable than either of the original sine waves. This means that applying the standard linear predictor model and minimizing the mean squared error $E\{\varepsilon^2\}$, which is measure of predictability, we can separate or extract signals with different temporal structures. More precisely, by minimizing the error, we maximize a measure of temporal predictability for each recovered signal [272, 268, 61].

It is interesting to note that there is some analogy between the measure of temporal predictability and the measure of non-Gaussianity. The central limit theorem ensures that the probability density function (pdf) of any mixture is closer to the Gaussian distribution than (or equal to) any of its component source signals. As some measure of non Gaussianity or statistical independence, we have used in the previous section the absolute value of the kurtosis and the generalized kurtosis. However, it should be noted, that these two measures: temporal linear predictability and non-Gaussianity based on kurtosis may lead to different results. Temporal predictability forces the extracted signal to be smooth and possibly less complex while the non-Gaussianity measure forces the extracted signals to be as independent as possible with sparse representation for sources that have positive kurtosis.

In Appendix C, we formulate and prove sufficient conditions in order to successfully extract source signals using the cost function $E\{\varepsilon_1^2\}$ subject to some constraints.

5.3.2 Neural Network for Multi-unit Blind Extraction

For extraction of multiple source signals, we present a neural network architecture (see Fig. 5.7) that connects, in a cascade fashion, extraction processing units and other processing units of different types for deflation as described in [273, 1144]. In this cascade architecture, a jth deflation processing unit deflates (eliminates) the newly extracted source signal y_j, (yielded by the jth extraction processing unit) from the mixtures $\mathbf{x}_j = [x_{j1}, x_{j2}, \dots, x_{jm}]^T$ and feeds the resulting new mixtures as outputs $\mathbf{x}_{j+1}$, to the next $(j+1)$th extraction processing unit which then extracts another source signal. It can be analytically shown by the following linear transformation that the resulting outputs $\mathbf{x}_{j+1}$ of the jth deflation

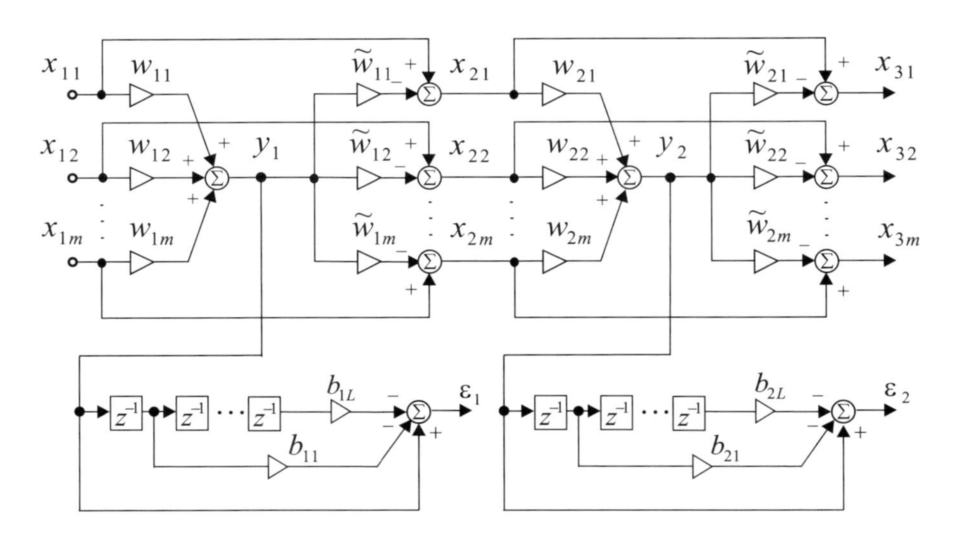

Fig. 5.7 The cascade neural network structure for multi-unit extraction.

processing unit do not include the already extracted signals $\{y_1, \dots, y_j\}$ [1144]

$$\mathbf{x}_{j+1}(k) \stackrel{\text{def}}{=} \mathbf{x}_j(k) - \tilde{\mathbf{w}}_j(k) y_j(k), \tag{5.70}$$

which minimizes the loss function

$$\widehat{\mathcal{J}}_j(\tilde{\mathbf{w}}_j) = \frac{1}{2} E\{\mathbf{x}_{j+1}^2\}, \tag{5.71}$$

where $\tilde{\mathbf{w}}_j = \left[\tilde{w}_{j1}, \tilde{w}_{j2}, \dots, \tilde{w}_{jm}\right]^T$ and $\mathbf{x}_j = \left[x_{j1}, x_{j2}, \dots, x_{jm}\right]^T$. In (5.70), $y_j = \mathbf{w}_j^T \mathbf{x}_j$ is the output of the jth extraction processing unit whose weights $\mathbf{w}_j = \left[w_{j1}, w_{j2}, \dots, w_{jm}\right]^T$ are updated according to the learning rule in (5.74).

For $\tilde{\mathbf{w}}_j$, we obtain the following updating rule by applying the stochastic gradient descent method to (5.71)

$$\Delta \widetilde{\mathbf{w}}_j(k) = -\widetilde{\eta}_j \, \frac{\partial \widehat{\mathcal{J}}_j(\widetilde{\mathbf{w}}_j)}{\partial \widetilde{\mathbf{w}}_j} = \widehat{\eta}_j(k) \, y_j(k) \, \mathbf{x}_{j+1}(k), \tag{5.72}$$

where $\widehat{\eta}_j(k) > 0$ is a learning rate.

Applying the standard stochastic gradient descent method to a generalized criterion of (5.65) for the jth extraction processing unit, i.e.,

$$\mathcal{J}_j\left(\mathbf{w}_j, \mathbf{b}_j\right) = \frac{1}{2} E\{\varepsilon_j^2\} + \frac{\beta_j}{4} (1 - E\{y_j^2\})^2, \tag{5.73}$$

we obtain the following adaptive on-line learning algorithms for the vectors $\mathbf{w}_j$ ($j = 1, 2, \dots, n$) and coefficients of the FIR adaptive filters b_{jp}, respectively,

$$\boxed{\Delta \mathbf{w}_j(k) = -\eta_j(k) \, \varepsilon_j(k) \, \widetilde{\mathbf{x}}_j(k) + \eta_j(k) \, \beta_j \, \{1 - m_2(y_j(k))\} \, y_j(k) \, \mathbf{x}_j(k)} \tag{5.74}$$

and

$$\boxed{\Delta b_{jp}(k) = -\widetilde{\eta}_j(k)\,\varepsilon_j(k)\,y_j(k-p),} \tag{5.75}$$

where

$$\varepsilon_j(k) = y_j(k) - \sum_{p=1}^{L} b_{jp} y_j(k-p) = [1 - B_j(z)]\, y_j(k),$$

$$\widetilde{\mathbf{x}}_j(k) = \mathbf{x}_j(k) - \sum_{p=1}^{L} b_{jp} \mathbf{x}_j(k-p) = [1 - B_j(z)]\, \mathbf{x}_j(k),$$

$$\Delta m_2(y_j(k)) = \bar{\eta}_j(k)\left[y_j^2(k) - m_2(y_j(k))\right],$$

$\beta_j > 0$ is a penalty factor, and $\eta_j(k)$, $\widetilde{\eta}_j(k)$, $\bar{\eta}_j(k)$ are learning rates [272].

5.4 BATCH ALGORITHMS FOR BLIND EXTRACTION OF TEMPORALLY CORRELATED SOURCES

The objective of this section is to derive alternative batch algorithms for extraction of colored sources with different autocorrelation functions. Let us consider the processing unit shown in Fig. 5.6 with the following constrained minimization problem:

Minimize the cost function

$$\mathcal{J}(\mathbf{w}_1, \mathbf{b}_1) \quad = \quad E\left\{\varepsilon_1^2\right\} \tag{5.76}$$

subject to the constraint $||\mathbf{w}_1||_2 = 1$,

where $\varepsilon_1(k) = y_1(k) - \widetilde{y}_1(k)$, $y_1 = \mathbf{w}_1^T \mathbf{x}_1$, $\mathbf{x}_1 = \mathbf{x}$, $\widetilde{y}_1(k) = \mathbf{b}_1^T \bar{\mathbf{y}}_1(k) = \sum_{p=1}^{L} b_{1p} y_1(k-p)$ and $\bar{\mathbf{y}}_1 = [y_1(k-1), y_2(k), \dots, y_1(k-L)]^T$.

The cost function can be evaluated as follows:

$$E\left\{\varepsilon_1^2\right\} = \mathbf{w}_1^T \widehat{\mathbf{R}}_{\mathbf{x}_1\mathbf{x}_1} \mathbf{w}_1 - 2\mathbf{w}_1^T \widehat{\mathbf{R}}_{\mathbf{x}_1\bar{\mathbf{y}}_1} \mathbf{b}_1 + \mathbf{b}_1^T \widehat{\mathbf{R}}_{\bar{\mathbf{y}}_1\bar{\mathbf{y}}_1} \mathbf{b}_1, \tag{5.77}$$

where $\widehat{\mathbf{R}}_{\mathbf{x}_1\mathbf{x}_1} \approx E\{\mathbf{x}_1\mathbf{x}_1^T\}$, $\widehat{\mathbf{R}}_{\mathbf{x}_1\bar{\mathbf{y}}_1} \approx E\{\mathbf{x}_1\bar{\mathbf{y}}_1^T\}$ and $\widehat{\mathbf{R}}_{\bar{\mathbf{y}}_1\bar{\mathbf{y}}_1} \approx E\{\bar{\mathbf{y}}_1\bar{\mathbf{y}}_1^T\}$, are estimators of true values of correlation and cross-correlation matrices: $\mathbf{R}_{\mathbf{x}_1\mathbf{x}_1}, \mathbf{R}_{\mathbf{x}_1\bar{\mathbf{y}}_1}, \mathbf{R}_{\bar{\mathbf{y}}_1\bar{\mathbf{y}}_1}$, respectively. In order to estimate vectors $\mathbf{w}_1$ and $\mathbf{b}_1$, we evaluate gradients of the cost function and equalize them to zero as follows:

$$\frac{\partial \mathcal{J}_1(\mathbf{w}_1, \mathbf{b}_1)}{\partial \mathbf{w}_1} \quad = \quad 2\widehat{\mathbf{R}}_{\mathbf{x}_1\mathbf{x}_1}\mathbf{w}_1 - 2\widehat{\mathbf{R}}_{\mathbf{x}_1\bar{\mathbf{y}}_1}\mathbf{b}_1 = \mathbf{0}, \tag{5.78}$$

$$\frac{\partial \mathcal{J}_1(\mathbf{w}_1, \mathbf{b}_1)}{\partial \mathbf{b}_1} \quad = \quad 2\widehat{\mathbf{R}}_{\bar{\mathbf{y}}_1\bar{\mathbf{y}}_1}\mathbf{b}_1 - 2\widehat{\mathbf{R}}_{\bar{\mathbf{y}}_1\mathbf{x}_1}\mathbf{w}_1 = \mathbf{0}. \tag{5.79}$$

Solving the above matrix equations, we obtain an iterative algorithm

$$\mathbf{w}_1^+ \quad = \quad \widehat{\mathbf{R}}_{\mathbf{x}_1\mathbf{x}_1}^{-1}\widehat{\mathbf{R}}_{\mathbf{x}_1\bar{\mathbf{y}}_1}\mathbf{b}_1, \quad \mathbf{w}_1 = \frac{\mathbf{w}_1^+}{||\mathbf{w}_1^+||_2}, \tag{5.80}$$

$$\mathbf{b}_1 \quad = \quad \widehat{\mathbf{R}}_{\bar{\mathbf{y}}_1\bar{\mathbf{y}}_1}^{-1}\widehat{\mathbf{R}}_{\bar{\mathbf{y}}_1\mathbf{x}_1}\mathbf{w}_1 = \widehat{\mathbf{R}}_{\bar{\mathbf{y}}_1\bar{\mathbf{y}}_1}^{-1}\widehat{\mathbf{R}}_{\bar{\mathbf{y}}_1 y_1}, \tag{5.81}$$

where the matrices $\widehat{\mathbf{R}}_{\bar{\mathbf{y}}_1 \bar{\mathbf{y}}_1}$ and $\widehat{\mathbf{R}}_{\bar{\mathbf{y}}_1 y_1}$ are estimated based on the parameters $\mathbf{w}_1$ obtained in the previous iteration step.
In order to avoid the trivial solution $\mathbf{w}_1 = \mathbf{0}$, we normalize the vector $\mathbf{w}_1$ to unit length in each iteration step as $\mathbf{w}_1(l+1) = \mathbf{w}_1^+(l+1) / \left\|\mathbf{w}_1^+(l+1)\right\|_2$ (which ensures that $E\{y_1^2\} = 1$).

Remark 5.4 *It should be emphasized here that in our derivation* $\widehat{\mathbf{R}}_{\bar{\mathbf{y}}_1 \bar{\mathbf{y}}_1}$ *and* $\widehat{\mathbf{R}}_{\bar{\mathbf{y}}_1 y_1}$ *are assumed to be independent of the vector* $\mathbf{w}_1(l+1)$, *i.e., they are estimated based on* $\mathbf{w}_1(l)$ *in the previous iteration step. This two-phase procedure is similar to the expectation maximization (EM) scheme: (i) Freeze the correlation and cross-correlation matrices and learn the parameters of the processing unit* $(\mathbf{w}_1, \mathbf{b}_1)$; *(ii) freeze* $\mathbf{w}_1$ *and* $\mathbf{b}_1$ *and learn new statistics (i.e., matrices* $\widehat{\mathbf{R}}_{\bar{\mathbf{y}}_1 y_1}$ *and* $\mathbf{R}_{\bar{\mathbf{y}}_1 \bar{\mathbf{y}}_1}$*) of the estimated source signal, then go back to (i) and repeat. Hence, in phase (i), our algorithm extracts a source signal, whereas in phase (ii) it learns the statistics of the source.*

The above algorithm can be considerably simplified. It should be noted that in order to avoid inversion of the autocorrelation matrix $\mathbf{R}_{\mathbf{x}_1\mathbf{x}_1}$ in each iteration step, we can perform the standard prewhitening or standard PCA as a preprocessing step and then normalize the sensor signals to unit variance. In such cases, $\widehat{\mathbf{R}}_{\mathbf{x}_1\mathbf{x}_1} = \mathbf{I}_n$ and the algorithm is simplified to [61]

$$\boxed{\mathbf{w}_1^+ = \widehat{\mathbf{R}}_{\mathbf{x}_1\bar{\mathbf{y}}_1}\mathbf{b}_1 = \widehat{\mathbf{R}}_{\mathbf{x}_1\tilde{y}_1}, \qquad \mathbf{w}_1 = \frac{\mathbf{w}_1^+}{||\mathbf{w}_1^+||_2},} \tag{5.82}$$

where $\widehat{\mathbf{R}}_{\mathbf{x}_1\tilde{y}_1} = \frac{1}{N}\sum_{k=1}^{N} \mathbf{x}_1(k)\,\tilde{y}_1(k)$.
It is interesting to note that the algorithm can be formulated in an equivalent form as

$$\boxed{\mathbf{w}_1(l+1) = \frac{\langle \mathbf{x}_1(k)\tilde{y}_1(k)\rangle}{\langle y_1^2(k)\rangle}.} \tag{5.83}$$

In order to reduce bias caused by white additive noise, we can modify the formula (5.83) as

$$\boxed{\mathbf{w}_1(l+1) = \frac{\langle \mathbf{x}_1(k)\tilde{y}_1(k)\rangle}{\langle y_1(k)\tilde{y}_1(k)\rangle}.} \tag{5.84}$$

The formulas (5.82) and (5.83) extend our basic simplified learning batch algorithm. The length of the FIR filter should be chosen sufficiently large. In our experiments we have shown that a value of 5-10 for L gives satisfactory results. However, as shown by our extensive computer simulations, in practice it is sufficient to use only a single delay unit with a suitably chosen delay q if some *a priori* information about source signals is available. The suitable choice of a single delay q depends on the autocorrelation function of the extracted source [60, 61, 268, 272].

Remark 5.5 *From (5.82)-(5.83) it follows that our algorithm is similar to the power method for finding the eigenvector* $\mathbf{w}_1$ *associated with the maximal eigenvalue of the matrix*

$\mathbf{R}_{\mathbf{x}_1}(\mathbf{b}_1) = E\{\sum_{p=1}^{L} b_{1p}\mathbf{x}_1(k)\mathbf{x}_1^T(k-p)\}$. *This observation suggests that it is not necessary to minimize the cost function with respect to parameters* $\{b_{1p}\}$ *but it is enough to choose an arbitrary set of them for which the largest eigenvalue is unique. More generally, if all eigenvalues of the generalized covariance matrix* $\mathbf{R}_{\mathbf{x}_1}(\mathbf{b}_1)$ *are distinct, then we can extract all sources simultaneously by estimating principal eigenvectors of* $\mathbf{R}_{\mathbf{x}_1}(\mathbf{b}_1)$ *(see Chapter 4 for more details).*

5.4.1 Blind Extraction Using a First Order Linear Predictor

The algorithms derived in the previous section can be further simplified, if we assume that the linear predictor has only one single time delay z^{-q}. In such a case, the cost function can be simplified as follows [61]

$$\mathcal{J}(\mathbf{w}_1, b_{1q}) = E\{\varepsilon_1^2\} = \mathbf{w}_1^T\, E[\mathbf{x}_1\, \mathbf{x}_1^T]\mathbf{w}_1 - 2b_{1q}\, E\{y_{1q}\, \mathbf{w}_1^T\, \mathbf{x}_1\} + b_{1q}^2\, E\{\mathbf{y}_{1q}^2\}, \tag{5.85}$$

where $y_{1q} = y_1(k-q) = \mathbf{w}_1^T\, \mathbf{x}(k-q)$. This cost function achieves a minimum when its gradient reaches zero with respect to $\mathbf{w}_1$ and b_{1q}. Thus, taking into account that $y_1 = \mathbf{w}_1^T\, \mathbf{x}_1$, we find

$$\frac{\partial \mathcal{J}(\mathbf{w}_1, b_{1q})}{\partial \mathbf{w}_1} = 2E\{\mathbf{x}_1\mathbf{x}_1^T\}\mathbf{w}_1 - 2b_{1q}\, E\{y_{1q}\mathbf{x}_1\} + 2b_{1q}^2\, E\{\mathbf{x}_1(k-q)\mathbf{x}_1^T(k-q)\}\mathbf{w}_1 = \mathbf{0}, \tag{5.86}$$

$$\frac{\partial \mathcal{J}(\mathbf{w}_1, b_{1q})}{\partial b_{1q}} = -2E\{y_{1q}y_1\} + 2b_{1q}\, E\{y_{1q}^2\} = 0. \tag{5.87}$$

Solving the above system of equations, we obtain

$$\mathbf{w}_1 = [E\{\mathbf{x}_1(k)\, \mathbf{x}_1^T(k)\} + \mathbf{b}_{1q}^2 E\{\mathbf{x}_1(k-q)\, \mathbf{x}_1^T(k-q)\}]^{-1} E\{y_1(k-q)\mathbf{x}_1(k)\} b_{1q}, \tag{5.88}$$

with $b_{1q} = E\{y_1(k-q)y_1(k)\}/E\{y_1^2(k-q)\}$.
This equation yields the following updating rule,

$$\mathbf{w}_1^+ = E\{\mathbf{x}_1\mathbf{x}_1^T\}^{-1} E\{y_1(k-q)\mathbf{x}_1(k)\} \frac{b_{1q}}{1 + b_{1q}^2}. \tag{5.89}$$

In order to avoid the trivial solution $\mathbf{w}_1 = \mathbf{0}$, we normalize the vector to unit length at each iteration step as $\mathbf{w}_1 = \mathbf{w}_1^+/||\mathbf{w}_1^+||_2$. With this, the term $b_{1q}/(1 + b_{1q}^2)$ can be disregarded. Moreover, without losing generality, we can assume that sensor data are prewhitened, thus $E[\mathbf{x}_1\mathbf{x}_1^T] = \mathbf{I}$. With this, (5.89) leads to a very simple learning rule,

$$\mathbf{w}_1^+(l+1) = \langle \mathbf{x}_1(k) y_1(k-q) \rangle = \frac{1}{N} \sum_{k=1}^{N} \mathbf{x}_1(k) y_1(k-q), \tag{5.90}$$

$$\mathbf{w}_1(l+1) = \frac{\mathbf{w}_1^+(l+1)}{||\mathbf{w}_1^+(l+1)||_2}, \tag{5.91}$$

where $y_1(k) = \mathbf{w}_1^T(l)\, \mathbf{x}_1(k)$, $\mathbf{x}_1(k) = \mathbf{Q}\, \mathbf{x}(k)$ and $y_{1q} = y_1(k-q) = \mathbf{w}_1^T(l)\, \mathbf{x}_1(k-q)$. The above algorithm can be formulated in the following simplified form (as shown by Barros and Cichocki in [61]):

$$\boxed{\mathbf{w}_1(l+1) = \frac{\langle \mathbf{x}_1(k)\, y_1(k-q) \rangle}{\langle y_1^2(k) \rangle}.} \tag{5.92}$$

Remark 5.6 *It is interesting to note that minimization of the cost function*

$$\mathcal{J}(\mathbf{w}) = \frac{1}{2} E\{\varepsilon^2\} = \frac{1}{2} E\{(y(k) - b_q\, y(k-q))^2\} \tag{5.93}$$

is equivalent to maximization of

$$\mathbf{w}^T \mathbf{R}_{\mathbf{x}_1}(q)\, \mathbf{w} \tag{5.94}$$

subject to the constraint $||\mathbf{w}||_2 = 1$, *where* $\mathbf{R}_{\mathbf{x}_1}(q) = E\{\mathbf{x}_1(k)\, \mathbf{x}_1^T(k-q)\}$.
In fact, the algorithm (5.91)-(5.92) is the well-known power method for finding the eigenvector corresponding to the maximal eigenvalue of the covariance matrix $\mathbf{R}_{\mathbf{x}_1}(q)$. *This means that the problem is equivalent to PCA or the eigenvalue problem of finding an eigenvector* $\mathbf{w}$ *corresponding to the largest eigenvalue of the covariance matrix* $\mathbf{R}_{\mathbf{x}_1}(q)$. *Thus, any efficient algorithm for the estimation of an extremum eigenvalue and associated eigenvector can be employed. The problem has a solution if this largest eigenvalue is distinct from the other eigenvalues. If the largest eigenvalue is multiple, we must choose another time delay or employ the linear predictor with many delays.*
It can easily be shown that if all the eigenvalues of the covariance matrix $\mathbf{R}_{\mathbf{x}_1}(q)$ *are distinct, then, for the prewhitened data* $\mathbf{x}_1$ *by applying an eigenvalue decomposition as*

$$\mathbf{R}_{\mathbf{x}_1}(q) = E\{\mathbf{x}_1(k)\, \mathbf{x}_1^T(k-q)\} = \mathbf{A}\, \mathbf{R}_{\mathbf{s}}(q)\, \mathbf{A}^T = \mathbf{V}\, \mathbf{\Lambda}\, \mathbf{V}^T, \tag{5.95}$$

we can estimate the global orthogonal mixing matrix $\widehat{\mathbf{A}} = \mathbf{Q}\widehat{\mathbf{H}} = \mathbf{V}$ *or equivalently the separating matrix* $\mathbf{W} = \mathbf{V}^T$ *(see Chapters 3 and 4 for more details).*

5.4.2 Blind Extraction of Sources Using Bank of Adaptive Bandpass Filters

For noisy data, instead of linear predictor, we can use a bandpass filter (or in a parallel way several processing units with a bank of bandpass filters) with fixed or adjustable center frequency and a bandpass bandwidth. The approach is illustrated in Fig. 5.8. By minimizing the cost function

$$\mathcal{J}(\mathbf{w}_j) = E\{\varepsilon_j^2\} \tag{5.96}$$

subject to the constraint $\|\mathbf{w}_j\|_2 = 1$, we obtain the on-line learning rule (see Section 5.3.1 for more details)

$$\Delta \mathbf{w}_j(k) = -\eta_1(k)\, [\langle \varepsilon_j(k) \widetilde{\mathbf{x}}_1(k) \rangle - \gamma_j(k) \mathbf{w}_1(k)] \tag{5.97}$$

where $\gamma_j(k) = -\beta_j\, [1 - \hat{\sigma}_{y_j}^2(k)]$ is a forgetting factor, $\widetilde{\mathbf{x}}_1(k) = \mathbf{x}_1(k) - B_j(z)\mathbf{x}_1(k)$ and $B_j(z)$ is the transfer function of the bandpass filters.

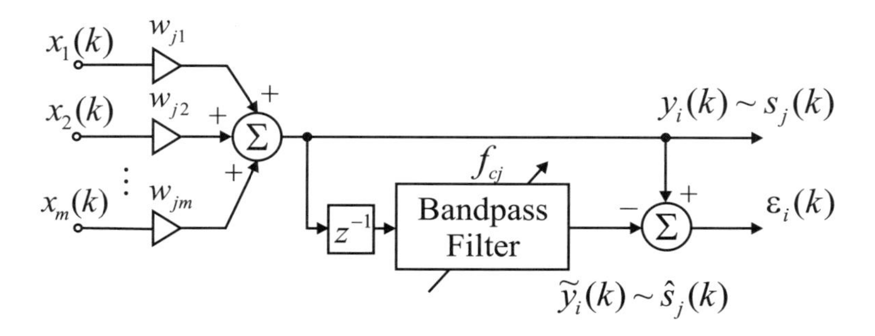

Fig. 5.8 The conceptual model of single processing unit for extraction of sources using an adaptive bandpass filter.

Analogous to the procedure presented in the previous section, for prewhitened sensor signals, we can derive a simple batch algorithm as

$$\boxed{\mathbf{w}_j(l+1) = \frac{\langle \mathbf{x}_1(k)\widetilde{y}_j(k)\rangle}{\langle y_j^2(k)\rangle}} \tag{5.98}$$

or alternatively

$$\mathbf{w}_j^+(l+1) = \langle \mathbf{x}_1(k)\widetilde{y}_j(k)\rangle = \frac{1}{N}\sum_{k=1}^{N} \mathbf{x}_1(k)\tilde{y}_j(k), \tag{5.99}$$

$$\mathbf{w}_j(l+1) = \frac{\mathbf{w}_j^+(l+1)}{||\mathbf{w}_j^+(l+1)||_2}, \tag{5.100}$$

where $y_j(k) = \mathbf{w}_j^T(l)\,\mathbf{x}_1(k)$, $\widetilde{y}_j(k) = B_j(z)y_j(k) = \tilde{\mathbf{x}}_1^T(k)\mathbf{w}_j(l)$. The above algorithms extract sources successfully if the covariance matrix $\mathbf{R}_{\mathbf{x}_1\tilde{\mathbf{x}}_1} = E\{\mathbf{x}_1\widetilde{\mathbf{x}}_1\}$ has a unique maximum eigenvalue.

The proposed algorithm (5.99)-(5.100) is insensitive to white noise and/or arbitrary distributed zero-mean noise which is beyond the bandwidth of the bandpass filter. Moreover, the processing unit is able to extract the filtered from noise version of a source signal if it is a narrow band signal.

As one of the simplest bandpass filters, we can use the second order IIR filter with transfer function (4.107) discussed in Chapter 4. An alternative realization of a bandpass filter with easy adjustable central frequency and bandwidth is the 4-th order Butterworth filter with the transfer function:

$$B(z) = \frac{b_0 + b_2 z^{-2} + b_4 z^{-4}}{1 + a_1\omega_c z^{-1}(a_2\omega_c^2 + a_2')z^{-2} + a_3\omega_c z^{-3} + a_4 z^{-4}} \tag{5.101}$$

where

$$\begin{aligned}
b_0 &= b_4 = 1/(d^2 + 2^{0.5}d + 1), \qquad b_2 = -2b_0, \\
b_1 &= b_3 = -4b_0, \qquad a_1 = -2d(2d + 2^{0.5})b_0, \\
a_2 &= 4d^2 b_0, \qquad a_2' = 2(d^2 - 1)b_0, \\
a_3 &= 2d(-2d + 2^{0.5})b_0, \quad a_4 = (d^2 - 2^{0.5}d + 1)b_0, \qquad d = \mathrm{cotan}(\pi B_w),
\end{aligned}$$

and

$$\omega_c = \frac{\cos(\pi(f_1 + f_2))}{\cos(\pi B_w)} < 1, \tag{5.102}$$

where f_1 and f_2 are normalized lower and higher cutoff frequencies and B_w is a normalized bandwidth. It should be noted that for the fixed (constant) bandwidth B_w, ω_c is the only center frequency f_c dependent parameter. It is worthwhile mentioning that the stability constraints on $B(z)$ are provided if

$$|\omega_c| < 1 \quad \text{and} \quad d > 0. \tag{5.103}$$

It should be noted that the second order bandpass filter (4.107) provides unity gain only at center frequency f_c alone and makes rather large distortions for source signals that are not pure sinusoids with some frequency variability. Therefore, by choosing a very narrow bandwidth, this filter can be applied for tracking and enhancement of a single sinusoid in white noise. In contrast, the fourth-order Butterworth filter has a flat characteristic around central frequency and can enhance an arbitrary narrow-band source signal with low distortion (see Fig. 5.9). By changing or adjusting the center frequency and bandwidth of the band pass filter, we can extract different narrow band sources using the some processing unit. We can also extract sources simultaneously by employing several processing units with bandpass filters with different bandwidths and center frequencies.

By maximizing the output power of the band pass filter, we can automatically adjust the center frequency to extract narrow-band sources located in a specific bandwidth. The filter output is given by

$$\begin{aligned}
\widetilde{y}(k) &= b_0 y(k) + b_2 y(k-2) + b_4 y(k-4) - a_1 \omega_c(k)\widetilde{y}(k-1) \\
&- (a_2 \omega_c^2(k) + a_2')\widetilde{y}(k-2) a_3 \omega_c(k)\widetilde{y}(k-3) - a_4 \widetilde{y}(k-4).
\end{aligned} \tag{5.104}$$

In order to find an optimal value of the parameter $\omega_c(k)$ for a specific bandwidth, we can maximize the cost function $E\{\widetilde{y}^2(k)\}$ using the gradient ascent procedure obtaining a simple learning rule for automatically estimating the optimal center frequency of a specific bandwidth

$$\omega_c(k+1) = \omega_c(k) + \eta(k)\widetilde{y}(k)\alpha(k), \tag{5.105}$$

where $\eta(k) = \eta/r(k)$, $r(k) = (1 - \eta_0)\, r(k-1) + \eta_0\, \alpha^2(k)$ and

$$\begin{aligned}
\alpha(k) &= \frac{\partial \widetilde{y}(k)}{\partial \omega_c(k)} = -a_1 \widetilde{y}(k-1) - 2a_2 \omega_c(k)\widetilde{y}(k-2) - a_3 \widetilde{y}(k-3) - a_1 \omega_c(k)\alpha(k-1) \\
&- (a_2' + a_2 \omega_c(k)^2)\alpha(k-2) - a_3 \omega_c(k)\alpha(k-3) - a_4 \alpha(k-4).
\end{aligned} \tag{5.106}$$

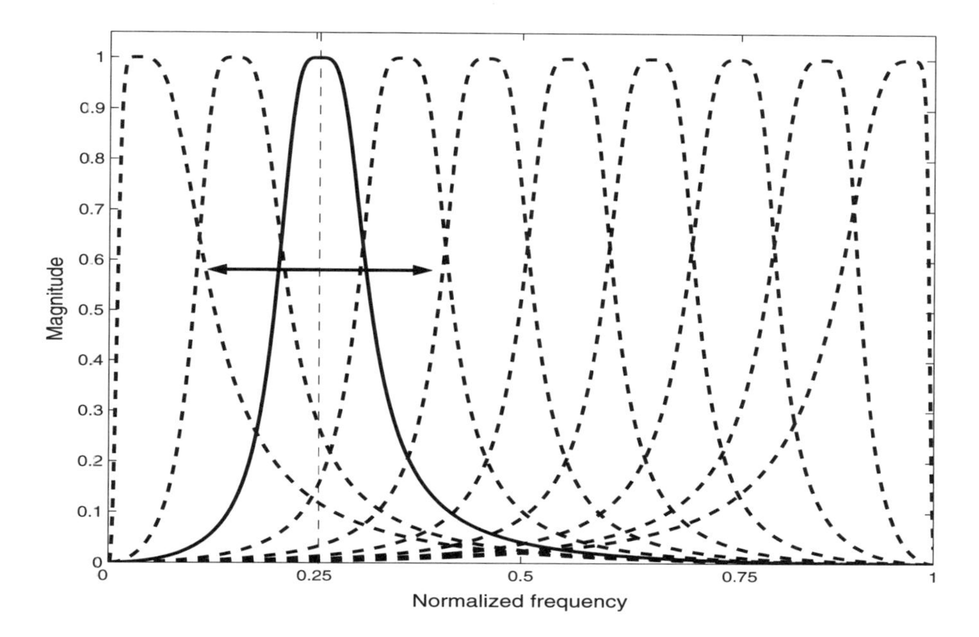

Fig. 5.9 Frequency characteristics of the 4-th order Butterworth bandpass filter with adjustable center frequency and fixed bandwidth.

Summarizing, the method presented in this section has several advantages:

- The method does not need a deflation procedure. One processing unit can extract all desired narrow-band sources sequentially one-by-one by adjusting the center frequency and bandwidth of the bandpass filter. Parallel extraction of an arbitrary group of sources is also possible by employing several bandpass filters with different characteristics.

- The algorithms are computationally very simple and efficient.

- The proposed algorithms are robust to additive noise, both white and narrow band colored noise. In contrast to other methods, the covariance matrix of noise does not need to be estimated or modelled.

5.4.3 Blind Extraction of Desired Sources Correlated with Reference Signals

In many applications, it is desired to extract independent components or separate sources with specific stochastic properties or features, but ignore other "uninteresting" sources and noises. Such extraction is possible if some *a priori* information about original sources is available.

For example, if the bandwidth of a desired narrow-band source signal is known, then we can apply a bandpass filter with the same specific bandwidth (passband) to extract the desired source (see Section 5.4.2 and learning rules (5.97)-(5.100)).

If a source signal is periodic or quasi periodic and its frequency f_q is known or can be estimated, then we can apply the learning rules (5.90) - (5.92) with time delay $q = 1/f_q$.

In many applications, such as biomedical applications (e.g., fMRI) some reference signals related to stimulus is explicitly available which can be exploited as some *a priori* knowledge. In such cases, it is usually desired to extract an independent source which is correlated as high as possible with the reference signal $r(k-\Delta)$, where Δ is a suitably chosen time delay. For this purpose, we can add to the cost functions discussed above an auxiliary penalty term $E\{r^2(k-\Delta)y^2(k)\}$. For example, we can use the following cost function (see Section 5.3 and 5.4)

$$J(\mathbf{w}) = \frac{1}{2}\, E\{\varepsilon^2\} + \frac{\beta}{4}\, (c - E\{y^2(k) r^2(k-\Delta)\})^2, \tag{5.107}$$

where $\varepsilon(k) = y(k) - \sum_{p=1}^{L} b_p y(k-p)$, $y(k) = \mathbf{w}^T \mathbf{x}_1(k)$, $\beta \geq 1$ is a penalty factor and c is suitably chosen positive constant (typically $c = 1$).

Minimization of the cost function according to the standard gradient descent method leads to a learning rule for the vector $\mathbf{w}$ as

$$\begin{aligned} \Delta\mathbf{w}(l) &= \mathbf{w}(l+1) - \mathbf{w}(l) = -\eta\, \frac{\partial J(\mathbf{w})}{\partial \mathbf{w}} \\ &= -\eta(l)\ \left[\langle \varepsilon(k) \widetilde{\mathbf{x}}_1(k)\rangle - \gamma(l) \left\langle y(k)\, r^2(k-\Delta) \mathbf{x}_1(k)\right\rangle\right], \end{aligned} \tag{5.108}$$

where

$$\gamma(l) = \beta\, [c - \left\langle y^2(k) r^2(k-\Delta)\right\rangle]$$

is a forgetting factor and $\widetilde{\mathbf{x}}_1(k) = \mathbf{x}(k) - \sum_{p=1}^{L} b_p \mathbf{x}(k-p)$.

5.5 A STATISTICAL APPROACH TO SEQUENTIAL EXTRACTION OF INDEPENDENT SOURCES

5.5.1 Log Likelihood and Cost Function

Let us consider the problem of extracting one source from a statistical point of view. Let $\overline{\mathbf{x}} = \mathbf{x}_1 = \mathbf{Q}\, \mathbf{x}$ be the vector of prewhitened sensor signals $\mathbf{x} = \mathbf{H}\, \mathbf{s}$, i.e,

$$\mathbf{x}_1 = \mathbf{A}\, \mathbf{s} = \sum_{j=1}^{n} s_j\, \mathbf{a}_j, \tag{5.109}$$

where $\mathbf{A} = \mathbf{Q}\,\mathbf{H}$ and $\mathbf{a}_1, \mathbf{a}_2, \ldots, \mathbf{a}_n$ are the column vectors of the orthogonal mixing matrix $\mathbf{A}$. Let the probability density function of $\mathbf{s}$ be

$$q(\mathbf{s}) = \prod_{j=1}^{n} q_j\left(s_j\right), \tag{5.110}$$

where $q_j\left(s_j\right)$ is the p.d.f. of s_j.

We assume without loss of generality, that source signals are independent and have unit variance, i.e., $E\{s_j^2\} = 1$. Then, $\mathbf{A}$ is an $n \times n$ orthogonal matrix, satisfying $\mathbf{a}_i^T \mathbf{a}_j = \delta_{ij}$.

We extract only one source signal by

$$y_1 = \mathbf{w}^T \mathbf{x}_1. \tag{5.111}$$

To this end, we formulate the problem in such a way that all source signals are extracted by $\mathbf{w}_1, \mathbf{w}_2, \ldots, \mathbf{w}_n$, although we are interested only in a single source,

$$y_j = \mathbf{w}_j^T \mathbf{x}_1, \quad (j = 1, 2, \ldots, n). \tag{5.112}$$

or

$$\mathbf{y} = \mathbf{W}\,\mathbf{x}_1, \tag{5.113}$$

where the true sources are extracted when

$$\mathbf{W} = \mathbf{A}^T \tag{5.114}$$

or $\mathbf{w}_j = \mathbf{a}_j$ (ignoring irrelevant scaling and permutation of columns of matrix).

Because $\mathbf{s} = \mathbf{W}\mathbf{x}_1$ and $\det|\mathbf{W}| = 1$, the probability density function of $\mathbf{x}_1$ is given by

$$\begin{aligned} p\left(\mathbf{x}_1; \mathbf{W}\right) &= \det|\mathbf{W}| q\left(\mathbf{W}\mathbf{x}_1\right) = q\left(\mathbf{W}\mathbf{x}_1\right) \\ &= \prod_{j=1}^{m} q_j\left(\mathbf{w}_j^T \mathbf{x}_1\right). \end{aligned} \tag{5.115}$$

Hence, the log likelihood

$$\rho\left(\mathbf{x}_1; \mathbf{W}\right) = \log p\left(\mathbf{x}_1; \mathbf{W}\right) \tag{5.116}$$

is decomposed as

$$\rho\left(\mathbf{x}_1; \mathbf{W}\right) = \sum_{j=1}^{n} \log q_j\left(\mathbf{w}_j^T \mathbf{x}_1\right) = \sum_{j=1}^{m} \rho_j\left(\mathbf{x}_1; \mathbf{w}_j\right), \tag{5.117}$$

where each term depends only on one $\mathbf{w}_j$. In order to extract one independent source signal, e.g., $y_1 = \mathbf{w}_1^T \mathbf{x}_1$, the maximum likelihood method searches for $\mathbf{w}_1$ that maximizes the log likelihood.

The problem is hence formulated as follows[2]:

$$\text{minimize } \mathcal{J}(\mathbf{w}) = -E\{\log(q_1(y)\} \tag{5.118}$$

$$\text{subject to } \ ||\mathbf{w}||_2^2 = 1. \tag{5.119}$$

[2]For simplicity, in further considerations in this section, we omit indexes, i.e, $y_1 = y$, $\mathbf{x}_1 = \mathbf{x}$ and $\mathbf{w}_1 = \mathbf{w}$.

This is a similar formulation to the extended fixed point algorithm [589].

The Lagrangian function can be formulated as

$$L(\mathbf{w}, \lambda) = -E\{\log(q_1(y)\} + \lambda(\|\mathbf{w}\|_2^2 - 1). \tag{5.120}$$

Using Lagrange's Theorem, we obtain the condition at the equilibrium point $\mathbf{w} = \mathbf{w}_*$

$$\nabla_{\mathbf{w}} L(\mathbf{w}, \lambda) = -\frac{\partial E\{\log(q_1(y)\}}{\partial \mathbf{w}} + 2\lambda \mathbf{w} = \mathbf{0} \tag{5.121}$$

or

$$E\{\varphi(y)\mathbf{x}\} + 2\lambda \mathbf{w} = \mathbf{0}, \tag{5.122}$$

where

$$\varphi(y) = -\frac{d\rho_1(\mathbf{x}_1, \mathbf{w})}{dy} = -\frac{d \log q_1(y)}{dy}. \tag{5.123}$$

By multiplying Eq. (5.122) by $\mathbf{w}^T$ and taking into account that $\|\mathbf{w}\|_2^2 = 1$, we obtain

$$\lambda = -\frac{1}{2} E\{\varphi(y) y\} \tag{5.124}$$

and

$$E\{\varphi(y)\mathbf{x}\} = E\{\varphi(y)y\}\mathbf{w}. \tag{5.125}$$

Hence, we can obtain a new simple batch fixed point algorithm

$$\boxed{\mathbf{w}(l+1) = \frac{E\{\varphi(\mathbf{w}^T(l)\mathbf{x})\, \mathbf{x}\}}{E\{\varphi(\mathbf{w}^T(l)\mathbf{x})\, y\}}}. \tag{5.126}$$

5.5.2 Learning Dynamics

Furthermore, a similar analysis is applicable to all gradient methods where various cost functions $\mathcal{J}(\mathbf{w})$ are introduced from different considerations (see Table 5.1).

Applying the standard gradient descent method to the Lagrangian function (5.120), we obtain the batch learning rule:

$$\begin{aligned} \Delta \mathbf{w}(l) &= \mathbf{w}(l+1) - \mathbf{w}(l) = -\eta \frac{\partial L(\mathbf{w}, \lambda)}{\partial \mathbf{w}} \\ &= -\eta \left[E\{\varphi(y)\, \mathbf{x}\} - E\{\varphi(y)\, y\}\, \mathbf{w}(l) \right], \end{aligned} \tag{5.127}$$

where expectation terms can be estimated as follows:

$$E\{\varphi(y)\mathbf{x}\} \approx (1/N) \sum_{k=1}^{N} \varphi[\mathbf{w}^T(l)\mathbf{x}(k)]\mathbf{x}(k)$$

and

$$E\{\varphi(y)y\} \approx (1/N) \sum_{k=1}^{N} \varphi[\mathbf{w}^T(l)\mathbf{x}(k)][\mathbf{w}^T(l)\mathbf{x}(k)]$$

and the learning rate is bounded as [322]

$$0 < \eta < \frac{2}{E\{\varphi'(y) - y\varphi(y)\}}. \tag{5.128}$$

Hence, applying a stochastic approximation, we obtain a simple on-line learning rule:

$$\boxed{\Delta\mathbf{w}(k) = -\eta(k) \left[\varphi(y(k))\, \mathbf{x}(k) - y(k)\varphi(y(k))\mathbf{w}(k)\right].} \tag{5.129}$$

However, when the step-size η is not so small, we need to normalize $\mathbf{w}+\Delta\mathbf{w}$ at each iteration step.

The present method is designed to extract source signals whose probability distribution is $q_1(s_1)$. The problem is that we do not know the exact source distribution $q_1(s_1)$. However, even when we misfit $q_1(s_1)$, or even when we use an arbitrary function $\varphi(y)$, the true $\mathbf{w}$ is the equilibrium point of the dynamics (5.129). This is because the right-hand side of (5.129) gives an estimating function of $\mathbf{w}$, which will be explained in Chapter 10.

5.5.3 Equilibrium of Dynamics

We show here directly that $\mathbf{w}_1 = \mathbf{w}$, or more generally any $\mathbf{w}_i = \mathbf{a}_i$, is an equilibrium point of the dynamics, regardless of φ. Then, the equilibrium of (5.129) satisfies

$$E\{\varphi(y)\mathbf{x} - y\varphi(y)\mathbf{w}\} = \mathbf{0}. \tag{5.130}$$

Because $\mathbf{x} = \sum s_j \mathbf{w}_j$ and

$$\begin{aligned} y = \mathbf{w}^T\mathbf{x} &= \mathbf{w}^T \left(\sum s_j \mathbf{w}_j\right) \\ &= \sum g_j s_j = \mathbf{g}^T\mathbf{s}, \end{aligned} \tag{5.131}$$

by putting

$$\beta_j(\mathbf{g}) = E\{s_j \varphi\left(\sum g_i s_i\right)\}, \tag{5.132}$$

we have

$$E\{\varphi(y)\mathbf{x}\} = \sum \beta_j \mathbf{w}_j, \tag{5.133}$$

$$E\{y\varphi(y)\mathbf{w}\} = \left(\sum g_j \beta_j\right)\left(\sum g_i \mathbf{w}_i\right). \tag{5.134}$$

Hence, any equilibrium $\mathbf{w}$ should satisfy

$$\boldsymbol{\beta}(\mathbf{g}) = \left(\sum g_j \beta_j\right) \mathbf{g}, \tag{5.135}$$

which implies

$$\boldsymbol{\beta} = c\mathbf{g} \tag{5.136}$$

for some constant c. It is straightforward to see that $\mathbf{g} = \mathbf{e}_j = [0, \dots, 0, 1, 0, \dots, 0]^T$ satisfies this condition. That is, $\mathbf{g} = (\mathbf{w}^T \mathbf{A})^T = \mathbf{e}_j$ is an equilibrium.

Alternatively, we can show this feature by reformulating the optimization problem (5.118) as

$$\text{minimize } \mathcal{J}(\mathbf{g}) = -E\{\log(q_1(y)\} \tag{5.137}$$

$$\text{subject to } ||\mathbf{g}||_2^2 = 1, \tag{5.138}$$

in terms of the global vector defined as $\mathbf{g} = \mathbf{A}^T \mathbf{w}$ and taking into account that $\mathbf{y} = \mathbf{w}^T \mathbf{x} = \mathbf{w}^T \mathbf{A}\, \mathbf{s} = \mathbf{g}^T \mathbf{s}$. A Lagrangian function can be formulated as

$$L(\mathbf{g}, \lambda) = -E\{\log(q_1(\mathbf{g}^T \mathbf{s})\} + \lambda(\|\mathbf{g}\|_2^2 - 1), \tag{5.139}$$

with $\lambda = (1/2)E\{\varphi(y)y\}$. Using Lagrange's Theorem, we obtain

$$E\{\varphi(\mathbf{g}^T \mathbf{s})\, \mathbf{s}\} = E\{\varphi(y)\, y\}\, \mathbf{g}. \tag{5.140}$$

It is straightforward to see that the above equation is satisfied for $\mathbf{g} = \mathbf{e}_j = [0, \dots, 0, 1, 0, \dots, 0]$ if $y = s_j$.

By choosing the nonlinear activation functions $\varphi(y)$ adequately, the algorithm can extract the source whose probability distribution is close to the one that gives $\varphi(y)$. However, there may exist local minima. There are two special functions φ which guarantee global convergence. One is derived from the 4th order cumulant (kurtosis) [339, 340]. The tensorial property of cumulants guarantees this, as we have proved in Appendix A. The other is the one derived from minimization of entropy or the negative log likelihood.

$$J(\mathbf{w}) = -E\{\log q_i \left(\mathbf{w}^T \mathbf{x}\right)\}. \tag{5.141}$$

This was proved by Cruces *et al.* by using the entropy inequality [323]. They further gave another interpretation of the log likelihood [322, 323]. Let $q(y) = q_1 \left(\mathbf{w}_1^T \mathbf{x}_1\right)$ and $q_G(y)$ be the standard Gaussian distribution. Then,

$$\begin{aligned} K\left[q(y) \;||\; q_G(y)\right] &= E_{q(y)} \left\{\log \frac{q(y)}{q_G(y)}\right\} \\ &= E\{\log q(y)\} + \frac{1}{2} \log(2\pi e) \end{aligned} \tag{5.142}$$

is the Kullback-Leibler divergence from $q(y)$ to the Gaussian $q_G(y)$. Hence, this is a measure of the non-Gaussianity of $q(y)$. The minimization of the negative log likelihood is exactly the same as maximization of non-Gaussianity. Gaussianity is increased by mixing signals because of the central limit theorem. Hence, as the entropy increases, higher-order cumulants decrease at the same time.

Remark 5.7 *When we do not approximately know the pdf of the estimated sources signal $q(y)$, we can use an adaptive method of estimating $q(y)$. We can use a parametric form $q(y; \mathbf{\Theta})$ of the density function, and modify a set of parameters $\mathbf{\Theta}$ at each step by (see Chapter 6 for more detail)*

$$\Delta\mathbf{\Theta} = \eta_{\Theta}\, \frac{\partial \log q\,(y; \mathbf{\Theta})}{\partial \mathbf{\Theta}}. \tag{5.143}$$

5.5.4 Stability of Learning Dynamics and Newton's Method

Now, lets consider the stability analysis when the activation function $\varphi(y)$ is properly chosen. This would lead to Newton's method automatically. The continuous-time averaged version of learning dynamics is much easier to analyze, so we use the continuous-time version of (5.129) given by

$$\frac{d\mathbf{w}}{dt} = -\mu\, \boldsymbol{f}(\mathbf{x}, \mathbf{w}), \tag{5.144}$$

where

$$\boldsymbol{f}(\mathbf{x}, \mathbf{w}) = E\{\varphi(y)\mathbf{x} - y\varphi(y)\mathbf{w}\}, \tag{5.145}$$

$$y = \mathbf{w}^T \mathbf{x}. \tag{5.146}$$

The variational equation, which shows how a small deviation $\delta\mathbf{w}$ develops in the time course of the dynamics, is

$$\frac{d}{dt}\delta\mathbf{w}(t) = -\mu \frac{\partial \boldsymbol{f}(\mathbf{x}, \mathbf{w})^T}{\partial \mathbf{w}} \delta\mathbf{w}. \tag{5.147}$$

The stability of the algorithm is determined by the eigenvalues of the Hessian matrix

$$\boldsymbol{\mathcal{K}}(\mathbf{w}) = \frac{\partial \boldsymbol{f}(\mathbf{x}, \mathbf{w})}{\partial \mathbf{w}} \tag{5.148}$$

at the equilibrium $\mathbf{w} = \mathbf{a}_1$.

For $\lambda = E\{y\varphi(y)\}$, we have

$$\boldsymbol{\mathcal{K}}(\mathbf{w}) = E\{\mathbf{w} \left(\frac{\partial \lambda}{\partial \mathbf{w}}\right)^T + \left(\lambda \mathbf{I} - \varphi'(y)\mathbf{x}\,\mathbf{x}^T\right)\}. \tag{5.149}$$

At the equilibrium $\mathbf{w} = \mathbf{a}_1$, we have

$$\begin{aligned} E\{\varphi'(y)\mathbf{x}\,\mathbf{x}^T\} &= E\{\varphi'(s_1) \sum_{j=1}^{n} s_j^2\, \mathbf{a}_j\, \mathbf{a}_j^T\} \\ &= E\{\varphi'(s_1)\, s_1^2\}\, \mathbf{a}_1\, \mathbf{a}_1^T + E\{\varphi'(s_1)\} \left(\mathbf{I} - \mathbf{a}_1 \mathbf{a}_1^T\right), \end{aligned} \tag{5.150}$$

because $E\{s_j^2\}$, $\sum_{j=2}^{m} \mathbf{a}_j\, \mathbf{a}_j^T = \mathbf{I} - \mathbf{a}_1\, \mathbf{a}_1^T$. Hence, the variational equation at $\mathbf{w} = \mathbf{a}_1$,

$$\frac{d}{dt}\delta\mathbf{w} = \boldsymbol{\mu}\boldsymbol{\mathcal{K}}\left(\mathbf{a}_1\right)\delta\mathbf{w}, \tag{5.151}$$

can be written as

$$\frac{d}{dt}\delta\mathbf{w} = \delta c_1 \mathbf{a}_1 + c_2 \delta\mathbf{w}. \tag{5.152}$$

The first term on the right-hand side is in the direction of the solution $\mathbf{w} = \mathbf{a}_1$, so that it enlarges or shrinks the magnitude of $\mathbf{w}$, but it is ineffective because $||\mathbf{w}||_2^2 = 1$. The change in the direction orthogonal to $\mathbf{a}_1$ is given by the second term

$$\frac{d}{dt}\delta\mathbf{w} = -\chi\, \delta\, \mathbf{w}, \tag{5.153}$$

where

$$\chi = E\{\varphi'(s_1)\} - E\{s_1\varphi(s_1)\}. \tag{5.154}$$

When φ is given by (5.123),

$$E\{s_1\varphi(s_1)\} = \int -s\frac{d}{ds}q_1(s)ds = \int q_1(s)ds = 1. \tag{5.155}$$

On the other hand,

$$E\{\varphi'(s_1)\} = E\{-\frac{d^2}{ds^2}\log q_1(s)\} = \mathcal{G} \tag{5.156}$$

is the Fisher information on the distribution of the source signal s_1. Since s_1 is a normalized, we easily have

$$\mathcal{G} \geq \infty, \tag{5.157}$$

where the equality holds when $q_1(s)$ is Gaussian. Hence, stability of the dynamics is proved when φ is derived from $q_1(s)$.

For an arbitrary chosen φ, the stability condition is given by

$$\boxed{E\{s_1\varphi(s_1)\} < E\{\varphi'(s_1)\}.} \tag{5.158}$$

The above analysis shows that

$$\mathcal{K}(\mathbf{a}_1) = c\mathbf{I} + c_1\mathbf{a}_1\mathbf{a}_1^T. \tag{5.159}$$

Hence, its inverse has a similar form

$$\mathcal{K}^{-1}(\mathbf{a}_1) = \tilde{c}\mathbf{I} + \tilde{c}_1\mathbf{a}_1\mathbf{a}_1^T. \tag{5.160}$$

Newton's algorithm corresponding to (5.129) is given by

$$\Delta\mathbf{w} = -\eta\,\mathcal{K}^{-1}(\mathbf{w})\,\frac{\partial \boldsymbol{f}(\mathbf{w})}{\partial \mathbf{w}}. \tag{5.161}$$

However, by multiplying $\mathcal{K}^{-1}(\mathbf{w})$ and (5.129), we have exactly the same form of the updating equation. Hence, the dynamics of a single source extraction (5.129) is by itself equivalent to Newton's method [3], proving its efficiency.

5.6 A STATISTICAL APPROACH TO TEMPORALLY CORRELATED SOURCES

The temporally correlated but spatially independent source signals $\mathbf{s}(1), \mathbf{s}(2), \ldots, \mathbf{s}(N)$ are statistically modelled as follows. For an estimated source signal, e.g., $y_1(k) = \hat{s}_1(k) = \hat{s}(k)$ that is the first component of $\hat{\mathbf{s}}(k)$, we assume an AR model

$$\hat{s}(k) = \varepsilon(k) + \sum_{p=1}^{L} b_p\hat{s}(k-p). \tag{5.162}$$

[3] Assuming that we are sufficiently close to the equilibrium $\mathbf{w} = \mathbf{h}_1$.

or

$$[1 - B(z)]\, \hat{s}(k) = \varepsilon(k). \tag{5.163}$$

Here, $B(z) = \sum_{p=1}^{L} b_p z^{-p}$, $\varepsilon(k)$, $k = 1, 2, \ldots$, is a sequence of an independent innovation signal. Let $q(\varepsilon)$ represent its probability density function. Then, for a large number of samples N, the joint probability density of $s(1), s(2), \ldots, s(N)$ is the same as the probability density of the corresponding $\varepsilon(1), \varepsilon(2), \ldots, \varepsilon(N)$. Hence, it is written as

$$\begin{aligned} p\left\{s(1), s(2), \ldots, s(N); \mathbf{b}\right\} &= \prod_{k=1}^{N} q\left(\varepsilon(k)\right) \\ &= \prod_{k=1}^{N} q\left([1 - B(z)]\, s(k)\right), \end{aligned} \tag{5.164}$$

where $\mathbf{b} = [b_1, b_2, \ldots b_L]^T$. The observed prewhitened signals are given by

$$\mathbf{x}_1(k) = \mathbf{A}\, \mathbf{s}(k) = \mathbf{Q}\, \mathbf{H}\, \mathbf{s}(k) = \sum_{j=1}^{n} s_j(k)\, \mathbf{a}_j. \tag{5.165}$$

We recover the source signal by

$$\mathbf{y}(k) = \hat{\mathbf{s}}(k) = \mathbf{W}\, \mathbf{x}_1(k), \tag{5.166}$$

where $\mathbf{W} = \mathbf{A}^T$ is the true solution. We use a sequential method for extracting signals

$$y_i(k) = \mathbf{w}_i^T \mathbf{x}_1(k) \tag{5.167}$$

one by one. The total log likelihood of the observed signal sequence $\mathbf{x}_1(1), \ldots, \mathbf{x}_1(N)$ is decomposed as

$$\begin{aligned} &\log p\left\{\mathbf{x}_1(1), \mathbf{x}_1(2), \ldots, \mathbf{x}_1(N); \mathbf{W}, B_1, \ldots, B_n\right\} \\ &= \sum_{j,k} \log q_j \left\{\mathbf{w}_j^T \left(1 - B_j\right) \mathbf{x}_1(k)\right\} \end{aligned} \tag{5.168}$$

as before. Hence, for the maximum likelihood method, the cost function is

$$\mathcal{J}(\mathbf{w}, \mathbf{b}) = -\log q \left\{\mathbf{w}^T \widetilde{\mathbf{x}}(k)\right\} \tag{5.169}$$

under the constraint $||\mathbf{w}||_2^2 = 1$, where

$$\widetilde{\mathbf{x}}(k) = [1 - B(z)]\, \mathbf{x}_1(k) = 1 - \sum_{p=1}^{L} b_p\, \mathbf{x}_1(k - p). \tag{5.170}$$

The on-line gradient learning algorithm is derived by

$$\begin{aligned} \Delta \mathbf{w}(k) &= -\eta\, \frac{\partial \mathcal{J}(\mathbf{w}, \mathbf{b})}{\partial \mathbf{w}} \\ &= -\eta\, \left[\varphi[\varepsilon(k)] \tilde{\mathbf{x}}(k) - \gamma \mathbf{w}(k)\right], \end{aligned} \tag{5.171}$$

where

$$\gamma = \beta(\langle y^2 \rangle - 1) \tag{5.172}$$

is a forgetting factor (see Section 5.3.1), and

$$\varphi(\varepsilon) = -\frac{d}{d\varepsilon} \log q(\varepsilon), \tag{5.173}$$

is a nonlinear activation function, and

$$\varepsilon(k) = [1 - B(z)]\mathbf{w}^T \mathbf{x}_1(k) = [1 - B(z)]\, y(k),$$

is the innovation signal. When the innovation ε is Gaussian, we have a linear activation function,

$$\varphi(\varepsilon) = \varepsilon. \tag{5.174}$$

The linear φ works well in the case of temporally correlated sources, but a nonlinear φ may be more robust.

Since the temporal structure of source signals is not exactly known, we use an adaptive method to select $B(z)$. The update rule of the filter is derived from the likelihood as

$$\Delta b_p = -\tilde{\eta} \frac{\partial \mathcal{J}(\mathbf{w}, \mathbf{b})}{\partial b_p} \tag{5.175}$$

$$= \tilde{\eta}\, \varphi\, [\varepsilon(k)]\; y(k-p). \tag{5.176}$$

The true solution $\mathbf{w} = \mathbf{a}_1$ and $B(z) = B_1(z)$ is an equilibrium of dynamics (5.171) and (5.176) whatever φ is chosen. We can prove this in the same way as we did in the previous section. The stability analysis can also be done similarly, giving the same stability condition where y is replaced by $\tilde{y}$. The sequential update rule (5.171) is automatically turned into Newton's method (assuming that we are sufficiently close to equilibrium $\mathbf{w} = \mathbf{a}_1$).

5.7 ON-LINE SEQUENTIAL EXTRACTION OF CONVOLVED AND MIXED SOURCES

The criteria and algorithms discussed in the previous sections for blind signal extraction from an instantaneous mixture can be relatively easily extended or generalized to the problem of extraction of convolved and mixed independent sources. In this section, we illustrate this by a simple extension of the standard Godard-type blind equalization algorithm that is able to extract multiple source signals from their unknown convolutive mixtures [213, 214, 215].

5.7.1 Formulation of the Problem

In multichannel blind deconvolution, an m dimensional vector of received signals $\mathbf{x}(k) = [x_1(k), x_2(k), \ldots, x_m(k)]^T$ is assumed to be generated from an n dimensional vector of spatially independent, temporally i.i.d. unknown source signals $\mathbf{s}(k) = [s_1(k), s_2(k), \ldots, s_n(k)]^T$

using multi-variate linear time invariant filters, i.e.,

$$\mathbf{x}(k) = \sum_{p=-\infty}^{\infty} \mathbf{H}_p \, \mathbf{s}(k-p) = [\mathbf{H}(z)] \, \mathbf{s}(k) \tag{5.177}$$

or equivalently in scalar form

$$x_i(k) = \sum_{p=-\infty}^{\infty} \sum_{j=1}^{n} h_{ijp} \, s_j(k-p), \qquad (i = 1, 2, \ldots, m) \tag{5.178}$$

where $\mathbf{H}(z) = \sum_{p=-\infty}^{\infty} \mathbf{H}_p z^{-p}$ is an unknown $(m \times n)$ polynomial matrix with $m \geq n$, and z^{-p} is the delay operator such that $z^{-p}\mathbf{s}(k) = \mathbf{s}(k-p)$. The task of multichannel deconvolution is to recover the source signals $\mathbf{s}(k)$ from the received signals $\mathbf{x}(k)$, up to a scaled, permuted, and delayed version of source signals, i.e., the estimates of sources, $\mathbf{y}(k) = \hat{\mathbf{s}}(k) = \mathbf{P} \, \boldsymbol{\Lambda} \, \mathbf{D}(z) \, \mathbf{s}(k)$, where $\mathbf{P} \in \mathbb{R}^{n \times n}$ is a permutation matrix, $\boldsymbol{\Lambda} \in \mathbb{R}^{n \times n}$ is a nonsingular scaling diagonal matrix, and $\mathbf{D}(z)$ is a diagonal matrix whose ith diagonal element is given by z^{-d_i}.

5.7.2 Extraction of Single i.i.d. Source Signal

Let us consider an FIR equalizer whose first processing unit with the output $y_1(k)$ is described by

$$y_1(k) = \sum_{i=1}^{m} \sum_{p} w_{1ip} \, x_i(k-p), \tag{5.179}$$

where $\{w_{1ip}\}$ are the FIR equalizer coefficients and $\{x_i(k)\}$ is the ith sensor output.

A single source can be extracted by minimization of Godard criterion which is described by [213, 371]

$$J_1 = \frac{1}{4} E\{[|y_1(k)|^2 - \gamma_2]^2\}, \tag{5.180}$$

where γ is some positive constant, typically $\gamma = 1$. For the constant modulus (CM) signals γ can be chosen as $\gamma_2 = E\{|\hat{y}_1|^4\}/E\{|\hat{y}_1|^2\}$. Using the stochastic gradient descent method, one can derive the updating rule for the FIR equalizer coefficients w_{1ip} in the form of

$$\begin{aligned} w_{1ip}(k+1) &= w_{1ip}(k) - \eta_1(k) \, \frac{\partial J_1}{\partial w_{1ip}} \\ &\approx w_{1ip}(k) + \eta_1(k) \, \varphi_1(y_1(k)) \, x_i^*(k-p), \end{aligned} \tag{5.181}$$

where the complex conjugate variables are denoted by superscript $*$, $\eta_1(k) > 0$ is the learning rate, and the nonlinear activation function $\varphi_1(y_1(k))$ is given by

$$\varphi_1(y_1) = \gamma_2 \, y_1 - y_1 |y_1|^2. \tag{5.182}$$

Table 5.1 Cost functions for sequential blind source extraction one by one, $y = \mathbf{w}^T\mathbf{x}$. (Some criteria require prewhitening of sensor data, i.e., $\mathbf{R_{xx}} = \mathbf{I}$ or $\mathbf{A}\mathbf{A}^T = \mathbf{I}$).

No.	Cost function $J(y, \mathbf{w})$	Remarks
1.	$-E\{\log p(y)\}$ s.t. $\|\|\mathbf{w}\|\|_2 = 1$	Entropy
2.	$-\frac{1}{4}\|E\{\|y\|^4\} - 3E^2\{\|y^2\|\}\|$ s.t. $\|\|\mathbf{w}\|\|_2 = 1$	Kurtosis
3.	$-\sum_\beta \alpha_\beta \|C_{1+\beta}(y)\|$ s.t. $\|\mathbf{w}\|_2 = 1, \alpha_\beta \geq 0$ where $C_{1+\beta}(y) = \text{Cum}\{\underbrace{y(k), y(k), \ldots, y(k)}_{1+\beta}\}$	Sum of cumulants
4.	$-\|(E\{\|y\|^p\}/E^{p/q}\{\|y\|^q\}) - c_{pq}\|$	Generalized normalized kurtosis
5.	$-\text{Cum}\{y(k), y(k), y(k-p), y(k-p)\} =$ $= -(E\{y^2(k)y^2(k-p)\} - E\{y^2(k)\}E\{y^2(k-p)\}$ $-2E^2\{y(k)y(k-p)\})$ s.t. $\|\|\mathbf{w}\|\|_2 = 1$	Self cumulant with time delay
6.	$E\{\varepsilon^2\}$ s.t. $\|\|\mathbf{w}\|\|_2 = 1$, where $\varepsilon(k) = y(k) - \sum_{p=1}^{L} b_p y(k-p)$, particularly $b_1 = 1$ and $b_p = 0$ for $p \neq 1$	Linear predictor
7.	$E\{\Psi(y(k))\}$, e.g., $\Psi[y(k)] = \frac{1}{2\beta}(\|y\|^\beta - \gamma_\beta)^2$ $\gamma_\beta = E\{\|\hat{y}\|^{2\beta}\}/E\{\|\hat{y}\|^\beta\}$	Constant modulus criteria

A similar activation function can be obtained by using the normalized kurtosis or the generalized normalized kurtosis as optimization criteria. Table 5.1 shows typical criteria for blind source extraction and blind equalization.

For a doubly-infinite FIR channel, the only existing minima of (5.180) correspond to the points where a single source is extracted, provided that the source signal is sub-Gaussian (negative kurtosis)[4] [1195, 1196] whenever the equalizer is doubly-infinite. For a finite order FIR channel, there also exists a finite order FIR equalizer under some mild conditions on the FIR channel [220, 225, 1195, 1196] (in this case, the number of sensors should be greater than the number of sources). Thus, using (5.180) one can extract a single source successfully. The algorithm in this case can written in the vector form as

$$\mathbf{w}_{1i}(k+1) = \mathbf{w}_{1i}(k) \pm \eta_1(k)\, \varphi(y_1(k))\, \bar{\mathbf{x}}_i^*(k), \tag{5.183}$$

where $\mathbf{w}_{1i} = [w_{1i0}, w_{1i1}, \dots, w_{1iM}]^T$, $\bar{\mathbf{x}}_i^*(k) = [x_i^*(k), x_i^*(k-1), \dots, x_i^*(k-M)]^T$ and the $\pm$ sign is chosen opposite to the sign of the source kurtosis.

It should be noted that by applying the extended Godard criterion (5.180) several times, the same sources might be extracted at different outputs even under different initial conditions. To avoid this problem, Inouye [606], Papadias and Paulraj [941, 942] introduced auxiliary constraints and extra processing which would spatio-temporally decorrelate the extracted signals. This leads to a relatively complicated iterative algorithm that requires knowledge of the number of source signals in advance.

5.7.3 Extraction of Multiple i.i.d. Sources

We discuss now an on-line approach to extract multiple source signals one by one using the cascaded connections of modules which consist of equalization units (a processing unit extracting a single source) and deflation units (a processing unit eliminating contribution of extracted signals to mixtures) [213].

Without loss of generality, we can assume that the first extracted signal $y_1(k)$ corresponds to the first source signal $s_1(k)$, i.e., $y_1(k) = c_1 s_1(k-d_1)$. The deflation unit coefficients $\{\widetilde{w}_{1ip}\}$ are updated to minimize the energy (cost) function given by

$$\rho = \frac{1}{2} E\{\sum_{i=1}^{m} |x_{2i}(k)|^2\}, \tag{5.184}$$

where

$$x_{2i}(k) = x_i(k) - \sum_p \widetilde{w}_{1ip}(k) y_1(k-p). \tag{5.185}$$

Applying the stochastic gradient descent method, the updating rule for $\{\widetilde{w}_{1ip}\}$ is given by

$$\widetilde{w}_{1ip}(k+1) = \widetilde{w}_{1ip}(k) - \eta_1\, x_{2i}(k)\, y_1^*(k-p). \tag{5.186}$$

In order to show that the learning algorithm (5.186) is able to eliminate the contribution of the first extracted signal $y_1(k)$ is given by

$$y_1(k) = c_1 s_1(k-d_1), \tag{5.187}$$

[4] For super-Gaussian signals (positive kurtosis), the updating rule is derived from maximization of (5.180) instead of minimization.

from the observation $\mathbf{x}(k)$, we investigate the stationary points of the averaged version of (5.186). If the learning algorithm (5.186) approaches a steady state, we have

$$E\{x_{2i}(k)y_1^*(k-p)\} = E\{x_i(k)y_1^*(k-p)\} - \widetilde{w}_{1i\,p}(k)E\{|y_1(k-p)|^2\} = 0. \tag{5.188}$$

Then, $\widetilde{w}_{1i\,p}(k)$ is given by

$$\widetilde{w}_{1i\,p}(k) = \frac{E\{x_i(k)y_1^*(k-p)\}}{E\{|y_1(k-p)|^2\}}, \quad (i = 1, 2, \ldots, m). \tag{5.189}$$

Using the fact that $y_1(k)$ is the first extracted source signal and assuming that all source signals are i.i.d., we have [213]

$$E\{x_i(k)y_1^*(k-p)\} = E\{\sum_q \sum_{j=1}^{m} h_{ijq}s_j(k-q)c_1 s_1^*(k-p-d_1)\} = c_1 h_{i1,p+d_1}\sigma_{s_1}^2, \tag{5.190}$$

where $\sigma_{s_1}^2 = E\{|s_1(k)|^2\}$. Using the result (5.190), the $\widetilde{w}_{1i\,p}(k)$ defined in (5.189) becomes

$$\widetilde{w}_{1i\,p}(k) = \frac{h_{i1,p+d_1}}{c_1}, \quad (i = 1, 2, \ldots, m). \tag{5.191}$$

Thus, after deflation processing is done, the input to the equalization unit in the next module $\{x_{1i}(k)\}$ is

$$\begin{aligned} x_{2i}(k) &= x_i(k) - \sum_p \frac{h_{i1,p+d_1}}{c_1} c_1 s_1(k-p-d_1) \\ &= x_i(k) - \sum_q h_{i1,q}s_1(k-q). \end{aligned} \tag{5.192}$$

Therefore, we can see that the deflation algorithm given in (5.186) can eliminate the contribution of the first source signal $s_1(k)$ to the received signals $\{x_i(k)\}$. The deflated mixture $x_{1i}(k)$ is fed into the next module in order to extract the 2nd source signal. Subsequently, it generates a mixture by eliminating the contribution of the 2nd extracted signal. By continuing this procedure until the output of the module converges to a small value pre-specified (which means all source signals are extracted), we can successfully extract all source signals [213, 214]. We should emphasize that any other blind equalization algorithm instead of the Godard algorithm, can be applied similarly. Note that a similar deflation approach to multichannel blind deconvolution has been introduced by Inouye [606], Chi [198, 199, 200] and Tugnait [1195, 1196]. In [1196], batch-type deflation processing, using the equation (5.189) was applied in order to cancel the contribution of already extracted signals.

5.7.4 Extraction of Colored Sources from Convolutive Mixture

The procedure described in the previous section allows us to extract i.i.d. sources. Moreover, to extract several unknown sources, we need to apply a rather computationally involved deflation procedure.

Using the concepts presented in section 5.4, we easily extend derived algorithms for extraction of multiple colored sources form their multichannel convolutive mixture by employing suitable designed linear predictors (LP's) or a bank of bandpass filters (BPF's).

For example, using the concept of a bank of bandpass filters (cf. Eqs. (5.80)-(5.84) and (5.98)- (5.100)) we can use the following novel iterative algorithm:

$$\mathbf{w}_{ji}(l+1) = \widehat{\mathbf{R}}_{\mathbf{x}_i\mathbf{x}_i}^{-1} \frac{\langle \widetilde{y}_j(k)\bar{\mathbf{x}}_i^*(k)\rangle}{\langle y_j(k) y_j^*(k)\rangle}, \tag{5.193}$$

where

$$y_j(k) = \sum_{i=1}^{m}\sum_{p=1}^{M} w_{jip}\, x_i(k-p), \tag{5.194}$$

$$\widetilde{y}_j(k) = [B_j(z)]\, y_j(k), \tag{5.195}$$

$\mathbf{w}_{ji} = [w_{ji0}, w_{ji1}, \ldots, w_{jiM}]^T$ and $\bar{\mathbf{x}}_i^*(k) = [x_i^*(k), x_i^*(k-1), \ldots, x_i^*(k-M)]^T$.

The advantage of the proposed approach is that we can avoid the deflation procedure by using bandpass filters with different frequency characteristics. Furthermore, the algorithm allows us to extract colored sources with different shape spectra.

5.8 COMPUTER SIMULATIONS: ILLUSTRATIVE EXAMPLES

We now illustrate the performance of selected algorithms presented in this chapter. In each example presented in this section, source signals are mixed with a mixing matrix $\mathbf{H}$ whose elements are randomly selected in the range [-1, 1]. All weights are initialized with random values in the range [-0.1, 0.1].

To show qualitatively the performance of the presented algorithms, we use a performance index which is defined at the ith extraction processing unit by

$$PI_i = \frac{1}{m}\left(\sum_{j=1}^{m} \frac{\widehat{g}_{ij}^2}{\widehat{g}_{ij^*}^2} - 1\right),$$

where

$$\widehat{\mathbf{g}}_i = \mathbf{w}_i^T \bar{\mathbf{H}}_i = [\widehat{g}_{i1}, \widehat{g}_{i2}, \ldots, \widehat{g}_{im}],$$

$$\widehat{g}_{ij^*} = \max\{\widehat{g}_{ij}\} \qquad \text{for } j = 1, 2, \ldots, m,$$

$$\bar{\mathbf{H}}_i = (\mathbf{I} - \widehat{\mathbf{w}}_{i-1}\mathbf{w}_{i-1}^T)\bar{\mathbf{H}}_{i-1},$$

and

$$\bar{\mathbf{H}}_1 = \begin{cases} \mathbf{QH}, & \text{when mixed signals are whitened} \\ \mathbf{H}, & \text{when mixed signals are NOT whitened.} \end{cases}$$

The smaller the value of PI_i, the better the quality of the extracted source signal as compared to the original source signal at the ith extraction processing unit.

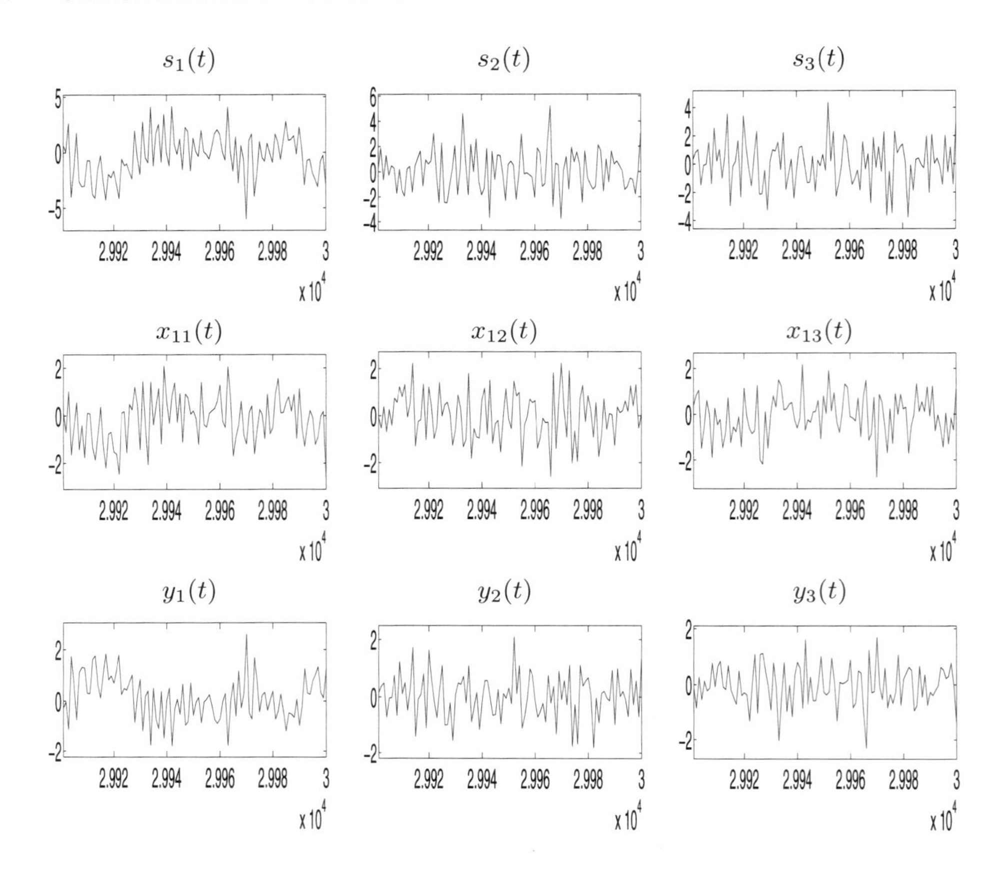

Fig. 5.10 Computer simulation results for mixture of three colored Gaussian signals, where s_j, x_{1j}, and y_j stand for the j-th source signals, whitened mixed signals, and extracted signals, respectively. Sources signals were extracted by employing the learning algorithm (5.74)-(5.75) with $L = 5$ [1142].

5.8.1 Extraction of Colored Gaussian Signals

Example 5.1 Three colored Gaussian signals are used here. Each colored Gaussian signal is generated by passing Gaussian sequences with variance 1 through an FIR filter of length 20, whose elements were randomly chosen between -1 and 1[5]. The normalized kurtosis of the resulting signals s_1, s_2, and s_3 are close to zero, i.e., 0.02, -0.02, and -0.06, respectively. These signals are mixed with the randomly chosen mixing matrix

$$\mathbf{H} = \begin{pmatrix} 0.82 & -0.90 & -0.62 \\ -0.54 & -0.84 & 0.69 \\ -0.52 & 0.28 & -0.65 \end{pmatrix}.$$

[5]Colored Gaussian signals used in subsequent examples are also generated in the same fashion.

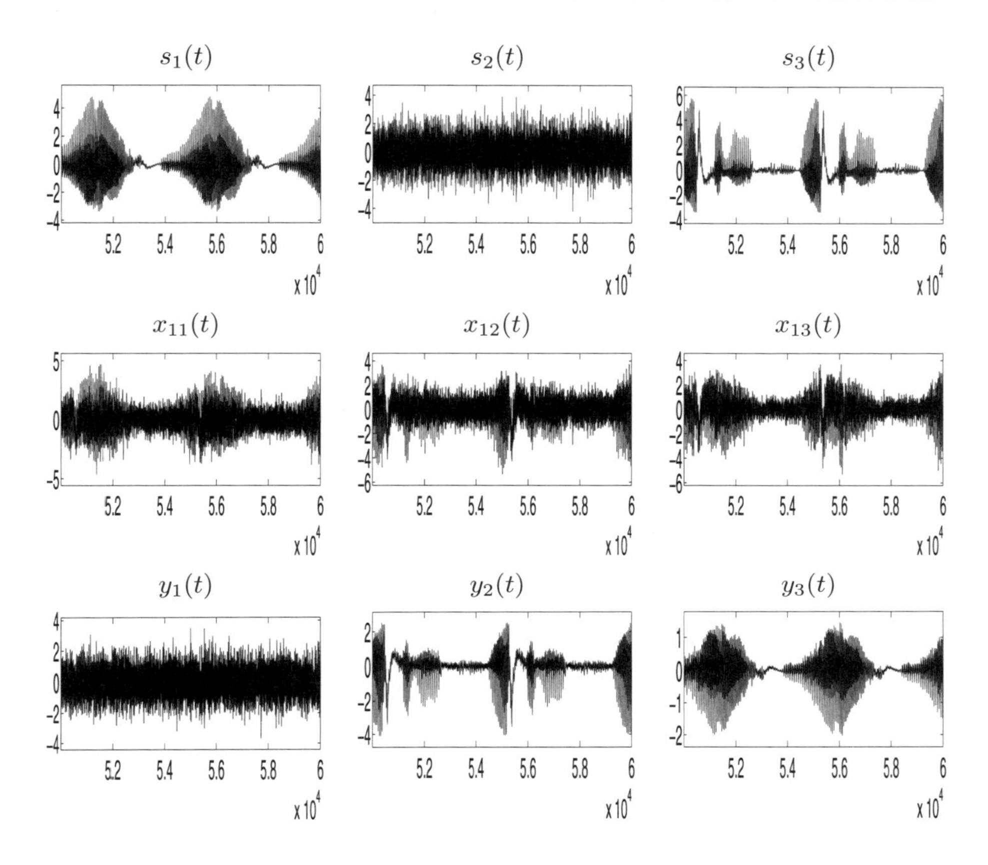

Fig. 5.11 Computer simulation results for mixture of natural speech signals and a colored Gaussian noise, where s_j and x_{1j}, stand for the j-th source signal and mixed signal, respectively. The signals y_j was extracted by using the neural network shown in Fig. 5.7 and the associated learning algorithm (5.92) with $q = 1, 5, 12$.

In order to extract source signals, we applied the algorithm (5.74) - (5.75) with $L = 5$ and the learning rates $\eta = 0.005$. Fig. 5.10 shows, from top to bottom, the original source signals $\mathbf{s} = [s_1, s_2, s_3]^T$, the whitened mixed signals $\mathbf{x}_1 = [x_{11}, x_{12}, x_{13}]^T$, and the extracted signals $\mathbf{y} = [y_1, y_2, y_3]^T$. The performance indices are $PI_1 = 0.00002$, $PI_2 = 0.00006$, and $PI_3 = 0.00011$. Visual comparison of the original source signals and the extracted signals (with $y_1 = -s_1$, $y_2 = s_3$, and $y_3 = -s_2$), together with the performance indexes, confirms the validity of the proposed algorithms.

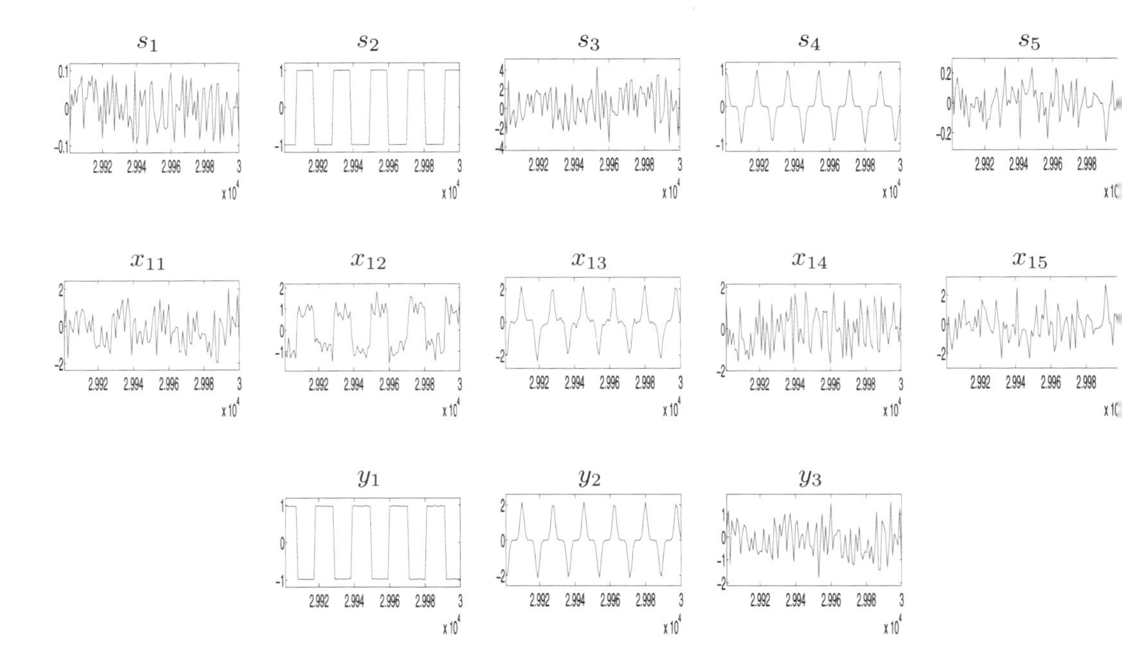

Fig. 5.12 Computer simulation results for mixture of three non-i.i.d. signals and two i.i.d. random sequences, where s_j, x_{1j}, and y_j stand for the j-th source signals, mixed signals, and extracted signals, respectively. The learning algorithm (5.82) with $L = 10$ was employed [1142].

5.8.2 Extraction of Natural Speech Signals from Colored Gaussian Signals

Example 5.2 Two natural speech signals[6], i.e., an English word */hello/* (s_1 with normalized kurtosis $= 3.44$) and a Japanese word */moshimoshi/* (s_3 with normalized kurtosis $= 6.13$), and a colored Gaussian signal (s_2 with normalized kurtosis $= -0.003$) are mixed by the same mixing matrix used in the above example. The original source signals and the whitened mixed signals are shown in the top and middle rows of Fig. 5.11, respectively. In order to extract source signals, we applied the batch algorithm (5.83) with $L = 10$. The extracted signals are shown in the bottom row of Fig. 5.11 which reveals that the recovered signals (with $y_1 = s_2$, $y_2 = -s_3$, and $y_3 = -s_1$) are very close to the original sources. The performance indices are $PI_1 = 0.0037$, $PI_2 = 0.0045$, and $PI_3 = 0.0042$.

5.8.3 Extraction of Colored and White Sources

Example 5.3 For this example, three non-i.i.d. signals, i.e., a sub-Gaussian signal s_2, a colored Gaussian signal s_3, and a super-Gaussian signal s_4, are mixed with two i.i.d.

[6]It is known that speech signals have temporal structures.

Fig. 5.13 Computer simulation results for mixture of three 512 × 512 image signals, where s_j and x_{1j} stand for the j-th original image and mixed image, respectively. y_1 represents the image extracted by the extraction processing unit shown in Fig. 5.6. The learning algorithm (5.92) with $q = 1$ was employed [61, 1142].

sequences, one being a uniform random noise s_1, and the other being a Gaussian random noise s_5. The normalized kurtosis of s_1, s_2, s_3, s_4, and s_5 are $-1.22, -2.00, -0.04, 0.41$, and 0.07, respectively. These signals are mixed with the randomly generated mixing matrix $\mathbf{H}$. Our aim here is to show an interesting property of the algorithm (5.82), which is able to extract only colored sources. Specifically, from mixtures of non-i.i.d. signals (or signals with temporal structures) and i.i.d. signals, only non-i.i.d. signals can be extracted using the second order statistics (SOS) algorithms.

Fig. 5.12 shows, from top to bottom, the original source signals $\mathbf{s} = [s_1, s_2, s_3, s_4, s_5]^T$, the whitened mixed signals $\mathbf{x}_1 = [x_{11}, x_{12}, x_{13}, x_{14}, x_{15}]^T$, and the estimated source signals from the first three extraction processing units, $\mathbf{y} = [y_1, y_2, y_3]^T$. The performance indices are $PI_1 = 0.0030$, $PI_2 = 0.0001$, and $PI_3 = 0.0077$. Both the performance indices as well as a visual comparison between the original source signals and the extracted signals (with $y_1 = -s_2$, $y_2 = -s_4$, and $y_3 = -s_3$) confirm the validity of the aforementioned conjecture.

5.8.4 Extraction of Natural Image Signal from Interferences

Example 5.4 In this section, we further test performance of the SOS algorithms using image signals for a case where the number of sensors is greater than the number of sources. Three 512×512 images are used, including an image for s_2. This image has temporal correlations when scanned in one dimension. The other two images are interferences artificially generated from Gaussian i.i.d. noises s_1 and binary i.i.d. sequences s_3. The normalized kurtosis of images s_1, s_2, and s_3 are 0.02, 0.31 and -2.00, respectively. These image signals are mixed with the randomly chosen non-square mixing matrix

$$\mathbf{H} = \begin{pmatrix} -0.97 & -0.16 & 0.68 \\ 0.49 & 0.69 & -0.96 \\ -0.11 & 0.05 & 0.36 \\ 0.87 & -0.59 & -0.24 \\ -0.07 & 0.34 & 0.66 \end{pmatrix}.$$

Fig. 5.13 shows, from top to bottom, the original images $\mathbf{s} = [s_1, s_2, s_3]^T$, the mixed images $\mathbf{x}_1 = [x_{11}, x_{12}, x_{13}, x_{14}, x_{15}]^T$, and the extracted signal at the first processing unit y_1 using the learning algorithm (5.92) with $q = 1$. The performance index at the first processing unit is 0.00004. As can be seen from Fig. 5.13, this unit has successfully extracted the natural image.

5.9 CONCLUDING REMARKS

In this chapter, we have presented a rather large number of algorithms for blind source extraction (BSE). The reader faced with the problem of BSE or BSS/ICA is justified in being puzzled as to which algorithm to use. Unfortunately, there is not a general valid answer. The right choice may very well depend on the nature and statistical properties of sources and the specific applications. If source signals are mutually independent i.i.d. signals, methods based on the HOS, especially, maximization of the absolute value of kurtosis may give the best results. However, this approach fails to extract sources with small values of kurtosis or colored Gaussian signals. For extraction of colored sources, especially smooth signals with high degree of predictability, the best performance can be obtained by using the SOS approach, especially based on linear predictor and eigenvalue decomposition using, for example, the power method or any other efficient method for estimation of eigenvectors. The SOS methods fail to extract white or i.i.d. sources. Unsymmetrical distributed sources can be extracted by employing skewness instead of kurtosis. For extraction of very sparse sources, we should use higher order cumulants or generalized kurtosis with $p = 6$. Also cumulants criteria are useful for independent signals corrupted by arbitrary Gaussian noise. For colored sensor signals buried in white arbitrary distributed noise, the SOS robust algorithm should be used. If we have a mixture of different sources (i.i.d. and colored), the best results may be achieved by a combination of several algorithms in the sense that in each stage we use a different algorithm depending on the kind of source signal desired to be extracted.

Appendix A. Global Convergence of Algorithms for Blind Source Extraction Based on Kurtosis

Assuming that source signals s_j are independent, the following property holds [340]

$$\kappa_4(y_1) = \kappa_4(\mathbf{g}_1^T \mathbf{s}) = \sum_{j=1}^{m} \kappa_4(s_j) g_{1j}^4, \tag{A.1}$$

where g_{1j} is the jth element of the row vector $\mathbf{g}_1^T = \mathbf{w}_1^T \mathbf{A}$. Next, let us assume throughout this section without loss of generality that every source signal has unit variance (or unit second order cumulant), i.e.,

$$\kappa_2(s_j) = \sigma^2(s_j) = E\{s_j^2\} = 1, \qquad \forall j. \tag{A.2}$$

This assumption is always feasible because source signals have zero-mean and differences in variance (power) can be absorbed in the mixing matrix $\mathbf{A}$.

Lemma A.1 *Consider the optimization problem [453]:*

Maximize

$$J(\mathbf{v}) = \sum_{j=1}^{n} d_j v_j^2 = \mathbf{v}^T \mathbf{D}\, \mathbf{v}, \qquad \|\mathbf{v}\|_2 = c > 0 \tag{A.3}$$

where $\mathbf{D}$ is a diagonal matrix: $\mathbf{D} = diag\{d_1, ..., d_n\}$. Assume that there exists only one index j_0 such that $d_{j_0} = \max_{1 \leq j \leq n} d_j$. Then, every local maximum is global and the point of global maximum is exactly the vector $\pm c\mathbf{e}_{j_0}$, where $\mathbf{e}_{j_0} = (0, ...0, 1, 0, ..., 0), (1$ is in the j_0-th place).

Proof. Applying a Lagrangian approach, for a point of local maximum $\mathbf{v}_* = (v_{1*}, ..., v_{n*})$, we write:

$$d_j v_{j*} - \lambda v_{j*} = 0, \quad (j = 1, 2, \ldots, n) \tag{A.4}$$

where λ is a Lagrange multiplier.
Multiplying (A.4) by v_{j*} and summing, we obtain:

$$J_{max} = \lambda c^2,$$

where J_{max} represents the value of J at the local maximum. Hence

$$\lambda = \frac{J_{max}}{c^2}. \tag{A.5}$$

From (A.4), we obtain

$$v_{j*}(d_j - \frac{J_{max}}{c^2}) = 0,$$

whence either $v_{j*} = 0$, or $J_{max} = d_j c^2$ for every $j = 1, ..., n$. Therefore, if $\mathbf{v}_*$ is a global maximum, then $\mathbf{v}_* = \pm c\mathbf{e}_{j_0}$.
We shall prove that every local maximum is global.
According to the second order optimality condition, a point $\mathbf{v}_*$ is a local maximum if

$$\mathbf{c}^T \nabla_v^2 L(\mathbf{v}_*)\mathbf{c} < 0 \quad \forall \mathbf{c} \in K(\mathbf{v}_*) = \{\mathbf{c} : \mathbf{c}^T \mathbf{v}_* = 0\}, \;\; \mathbf{c} \neq 0,$$

where

$$L(\mathbf{v}) = \sum_{j=1}^{n} d_j v_j^2 - \lambda(\|\mathbf{v}\|_2^2 - c^2)$$

is the Lagrangian function.
In our case, we obtain

$$\mathbf{c}^T \nabla_v^2 L(\mathbf{v}_*)\mathbf{c} = \sum_{j=1}^{n} (d_j - \lambda) c_j^2; \tag{A.6}$$

$$K(\pm c\mathbf{e}_j) = \{\mathbf{c} : c_j = 0\}.$$

We conclude that the quadratic form (A.6) is negative definite at $\mathbf{v}_*$ for $\mathbf{c} \in K(\mathbf{v}_*)$ if and only if $\lambda = d_{j_0}$ and $\mathbf{v}_* = \pm c\mathbf{e}_{j_0}$. $\square$

In a similar way, we can prove the following Lemma [453].

Lemma A.2 *Consider the optimization problem:*

Minimize (maximize) $\sum_{j=1}^{n} \kappa_j v_j^p$
subject to $||\mathbf{v}||_2 = c > 0$*, where* $p > 2$ *is even.*

Denote $I^+ = \{j \in \{1, 2, \ldots, n\} : \kappa_j > 0\}$*,* $I^- = \{j \in \{1, 2, \ldots, n\} : \kappa_j < 0\}$ *and* $\mathbf{e}_j = (0, ..., 0, 1, 0, ..., 0)$*, (1 is the* $j-th$ *place). Assume that* $I^+ \neq \emptyset$ *and* $I^- \neq \emptyset$*.*

Then the points of local minimum are exactly the vectors $\mathbf{v}_{j*}^{\pm} = \pm c\mathbf{e}_j, j \in I^-$ *and the points of local maximum are exactly the vectors* $\mathbf{v}_{j*}^{\pm} = \pm c\mathbf{e}_j, j \in I^+$*.*

Using Lemma (A.2), we are able to formulate and prove the following Theorem [453].

Theorem A.1 *Let* $\mathbf{w}_1$ *be a local minimum of the cost function given by (5.5). Then, the output signal* $y_1 = \mathbf{w}_1^T \mathbf{x}$ *recovers one of the source signals,* y_1 *converges to one of the desired solutions, i.e.,* $\pm c s_{1^*}$ *if and only if* $\beta\kappa_4(s_{1^*}) > 0$*, where* $y_1 = \mathbf{w}_1^T \mathbf{x}_1$*,* $\mathbf{x}_1 = \mathbf{A}\mathbf{s}$*,* $\mathbf{g}_1^T = \mathbf{w}_1^T \mathbf{A}$*, and* $\mathbf{A}$ *is full column rank and* $s_{1^*} \in \{s_1, s_2, \ldots, s_m\}$*.*

Proof. According to the property in $(A.1)$, minimization of (5.5) is equivalent to maximization of

$$\mathcal{J}_1(\mathbf{g}_1) = \frac{\beta}{4} \frac{\sum_{j=1}^{n} \kappa_4(s_j) g_{1j}^4}{(\sum_{j=1}^{n} m_2(s_j) g_{1j}^2)^2}. \tag{A.7}$$

Because of the assumption in $(A.2)$, the criterion $\tilde{\mathcal{J}}_1(\mathbf{g}_1)$ in $(A.7)$ reduces to

$$\tilde{\mathcal{J}}_1(\mathbf{g}_1) = \beta\frac{1}{4}\frac{\sum_{j=1}^{n}\kappa_4(s_j)g_{1j}^4}{\| \mathbf{g}_1 \|_2^4}. \tag{A.8}$$

Maximization of (A.8) is equivalent to the constraint maximization problem:

Maximize $\sum_{j=1}^{n}\beta\kappa_4(s_j)g_{1j}^4$ under the constraint $\|\mathbf{g}_1\|_2 = 1$. Applying the Lemma A.2, we finish the proof. □

Appendix B. Analysis of Extraction and Deflation Procedure

At the jth extraction and deflation processing units, we assume without loss of generality that elements s_j in $\mathbf{s}$ are permuted such that the first $j-1$ elements are the source signals that were previously extracted. We further assume that the column vectors $\mathbf{h}_i$ in the mixing matrix $\mathbf{H}$ are also permuted accordingly.

Theorem B.2 *Under notations of section 5.2.5 and assuming $E\{s_j^2\} = 1, \ \forall j$, we have*

$$\mathbf{x}_i(k) = \sum_{r=j+1}^{n} \mathbf{h}_r\, s_r(k), \tag{B.1}$$

where $\mathbf{x}_0(k) = \mathbf{x}(k)$, i.e., $\mathbf{x}_i = \mathbf{H}\mathbf{D}_i\,\mathbf{s}$, where $\mathbf{D}_i = \operatorname{diag}\{d_{i1}, d_{i2}, \ldots, d_{in}\}$, with $d_{ij} = 0$ for $j \leq i$ and $d_{ij} = 1$ for $j > i$.

Proof. We shall prove the Theorem by induction. For $j = 0$, the conclusion (B.1) is true. Assume that (B.1) is true for some j. Using Theorem A.1, we obtain $y_j = \pm s_{j+1}$ and for $j+1$, we obtain:

$$\begin{aligned}
\mathbf{x}_{j+1}(k) &= \mathbf{x}_j(k) - \widetilde{\mathbf{w}}_j y_j(k) = \sum_{r=j+1}^{n} \mathbf{h}_r s_r(k) - E\{\mathbf{x}_j y_j\} y_j(k) \\
&= \sum_{r=j+1}^{n} \mathbf{h}_r s_r(k) - E\{\sum_{r=j+1}^{n} \mathbf{h}_r s_r(k) s_{j+1}(k)\} s_{j+1}(k) = \sum_{r=j+2}^{n} \mathbf{h}_r s_r(k).
\end{aligned} \tag{B.2}$$

□

Next, we discuss two implementation issues: The first issue is how to choose or estimate a proper value of β on-line, and the second issue is when to terminate extraction and deflation procedures if it is necessary to extract all non-Gaussian signals.

Use of Analytical Results for Implementation

Implementation Issue 1

One of the main assumptions used both in the Theorem A.1 and the Theorem B.2 is that at the jth extraction and deflation processing units ($j = 1, 2, \dots$), there exists an index $r \leq j$ such that $\beta\kappa_4(s_r) > 0$. Let S_j denote the set of indices of the remaining, non-extracted source signals at the jth extraction processing unit. At the jth extraction processing unit, there are two possible cases where this assumption is violated, i.e.,

- Case I: β is set to -1 when $\kappa_4(s_r) > 0$, $\forall r \in S_j$, or
- Case II: β is set to +1 when $\kappa_4(s_r) < 0$, $\forall r \in S_j$.

Let us denote $\mathbf{g}_j = [g_{j1}, g_{j2}, \dots, g_{jn}]^T = (\mathbf{w}_j^T \mathbf{H}\mathbf{D}_j)^T$. Applying the property (A.1), we have

$$\kappa_4(y_j) = \sum_{r=j}^{n} \kappa_4(s_r) g_{jr}^4. \tag{B.3}$$

Based on this feature, we can avoid the above two cases by fixing β to 1 (or -1). Let $\widetilde{\kappa}_4\big(y_j(k)\big)$ denote the estimated value of the normalized 4th-order cumulant. At the jth extraction processing unit, if the condition $\beta\,\widetilde{\kappa}_4\big(y_j(k)\big) > 0$ holds for $k = \widetilde{T}_j$, where $\widetilde{T}_j$ is a specified time period, this means that $\kappa_4(s_r) > 0$ (or < 0), $\forall r \in S_j$. We then stop and restart the extraction process at the jth extraction processing unit with β being flipped to -1 (or 1). Alternatively, we can estimate the sign of $\kappa_4(y_j)$ on-line during the learning process, as done in [272, 273, 1143, 1144]. Another approach is to maximize the absolute value of the normalized kurtosis.

Implementation Issue 2

Here we discuss a terminating condition for extraction and deflation procedures. We need to consider this issue because we employ the assumption that the number of source signals is not known in advance and is less than or equal to the number of sensor signals, i.e., $n \leq m$. According to the Theorem B.2, at the ($j = n$)th deflation processing unit, $\mathbf{x}_{n+1} = \mathbf{H}\,\mathbf{D}_n\,\mathbf{s} = [0, 0, \dots, 0]^T$. As a result, in practice, due to error and/or additive noise, we can terminate extraction and deflation procedures when the amplitudes of all entries of the vector $\mathbf{x}_{j+1} = \mathbf{x}_j - \widetilde{\mathbf{w}}_j y_j$ are below a small given threshold or all the elements of the vector $\mathbf{x}_{j+1}$ are Gaussian signals, i.e., they have zero kurtosis.

Appendix C. Conditions for Extraction of Sources Using Linear Predictor Approach

The theoretical results for the learning model used in the linear predictor approach can be summarized in the following Theorem.

Theorem C.3 *Consider the minimization problem:*
Minimize

$$J(\mathbf{w}, \mathbf{b}) = E\{\varepsilon^2\}, \tag{C.1}$$

where $\varepsilon(k) = y(k) - \sum_{p=1}^{L} b_p y(k-p)$.
Assume that the following conditions are satisfied

1) $E\{s_j^2\} = 1$ *for* $j = 1, 2, \ldots, n$.

2) $\mathbf{R_s}(p,q) = E\{\mathbf{s}(k-p)\,\mathbf{s}(k-q)^T\} = E\{\mathbf{s}_p\,\mathbf{s}_q^T\}$ *are diagonal matrices for* $p, q = 0, 1, ..., L$ *such that* J *attains its global minimum with respect to* $\mathbf{w}$ *and the matrix*

$$\mathbf{R_s}(\mathbf{b}) = 2\sum_{p=1}^{L} b_p\,\mathbf{R_s}(p) - \sum_{p,q=1}^{L} b_p\,b_q\,\mathbf{R_s}(p,q) - \mathbf{I}_n$$

is diagonal with unique maximal diagonal elements for all coefficients b_p.
Then, $\mathbf{g}_* = \mathbf{H}^T\mathbf{w}_*$ *has the form* $\mathbf{g}_* = \pm c_*\mathbf{e}_j$, *where* $e_j = (0, ..., 1, ..., 0)$, *with* 1 *in the* j*-th place, for some index* $j \in \{1, ..., n\}$ *and some scalar* c_*.

The above Theorem implies that the cost function in (5.65) has no spurious minima. As a result, it is guaranteed that a desired solution can always be reached independent of initial conditions. In other words, a source signal having a unique temporal structure can always be extracted from the mixtures.

Proof. Let J given by Eq: (C.1) attain its minimum at the point $\mathbf{w}_*$ (such a point exists, since the cost function is quadratic).

Under the assumptions of the Theorem and taking into account that $y = \mathbf{w}^T\mathbf{x}$ and $\mathbf{x} = \mathbf{H}\,\mathbf{s}$, we obtain

$$J(\mathbf{w},\mathbf{b}) = E\{y^2\} - 2\sum_{p=1}^{L} b_p\,\mathbf{w}^T E\{\mathbf{x}\,\mathbf{x}_p^T\}\,\mathbf{w} + \sum_{p,q=1}^{L} b_p\,b_q\,\mathbf{w}^T E\{\mathbf{x}_p\,\mathbf{x}_q^T\}\,\mathbf{w}.$$

Since $E\{y^2\} = E\{\mathbf{w}^T\mathbf{x}\,\mathbf{x}^T\mathbf{w}\} = E\{\mathbf{w}^T\mathbf{H}\,\mathbf{s}\,\mathbf{s}^T\mathbf{H}^T\mathbf{w}\} = \mathbf{g}^T\mathbf{g} = \|\mathbf{g}\|_2^2$ the cost function can be expressed as

$$J(\mathbf{w},\mathbf{b}) = \mathbf{g}^T\mathbf{g} - 2\sum_{p=1}^{L} b_p\,\mathbf{g}^T\mathbf{R_s}(p)\,\mathbf{g} + \sum_{p,q=1}^{L} b_p\,b_q\,\mathbf{g}^T\mathbf{R_s}(p,q)\,\mathbf{g},$$

where the global vector $\mathbf{g} = \mathbf{H}^T\mathbf{w}$.

Thus, the problem converts to the following one:

Maximize $\mathbf{g}^T\mathbf{R}\,\mathbf{g}$, subject to $\mathbf{g}^T\mathbf{g} = c$.

Applying Lemma A.1, we finish the proof. □

6

Natural Gradient Approach to Independent Component Analysis

Two roads diverged in a wood, and I... took the one less travelled by, and that has made all the difference.

—Robert Frost "The Road Not Taken"

In this chapter, fundamental signal processing and information theoretic approaches are presented together with learning algorithms for the problem of adaptive blind source separation (BSS) and Independent Component Analysis (ICA). We discuss recent developments of adaptive learning algorithms based on the natural gradient approach in the general linear, orthogonal and Stiefel manifolds. Mutual information, Kullback-Leibler divergence, and several promising schemes are discussed and reviewed in this chapter, especially for signals with various unknown distributions and unknown number of sources. Emphasis is given to an information-theoretical and information-geometrical unifying approach, adaptive filtering models and associated on-line adaptive nonlinear learning algorithms. We discuss the optimal choice of nonlinear activation functions for various distributions, e.g., Gaussian, Laplacian, impulsive and uniformly-distributed signals based on a generalized-Gaussian-distributed model. Furthermore, a family of efficient and flexible algorithms that exploits nonstationarity of signals is also derived.

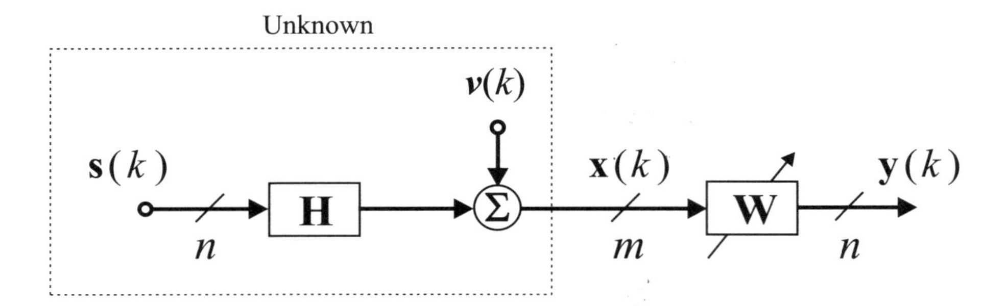

Fig. 6.1 Block diagram illustrating standard independent component analysis (ICA) and blind source separation (BSS) problem.

6.1 BASIC NATURAL GRADIENT ALGORITHMS

Let us consider the simple mixing system illustrated in Fig. 6.1 and described in vector-matrix form as

$$\mathbf{x}(k) = \mathbf{H}\,\mathbf{s}(k) + \boldsymbol{\nu}(k), \tag{6.1}$$

where $\mathbf{x}(k) = [x_1(k), \ldots, x_m(k)]^T$ is a noisy sensor vector, $\mathbf{s}(k) = [s_1(k), \ldots, s_n(k)]^T$ is a source signal vector, $\boldsymbol{\nu}(k) = [\nu_1(k), \ldots, \nu_m(k)]^T$ is a noise vector, and $\mathbf{H}$ is an unknown full rank $m \times n$ mixing matrix. It is assumed that only the sensor vector $\mathbf{x}(k)$ is available. The objective is to design a feed-forward or recurrent neural network and an associated adaptive learning algorithm that enable estimation of sources and/or identification of the mixing matrix $\mathbf{H}$ and/or the separating matrix $\mathbf{W}$ with good tracking abilities for time variable systems.

The present section is devoted to the analysis of learning algorithms for a typical but simple instantaneous blind source separation problem. Here, we assume that the number of source signals is known and is equal to the number of sensors (with $m = n$), so that both $\mathbf{H}$ and $\mathbf{W}$ are nonsingular $n \times n$ matrices. The source signals $s_i(k)$ are assumed to be mutually independent with zero-mean ($E\{s_i(k)\} = 0$). We also assume that the additive noise terms $\boldsymbol{\nu}(k)$ are negligible or reduced to be at negligible levels due to the preprocessing stated in the previous chapters. The following section gives a prototype mathematical analysis. We will then relax most of these constraints in later sections.

6.1.1 Kullback–Leibler Divergence - Relative Entropy as a Measure of Stochastic Independence

In order to obtain a good estimate $\mathbf{y} = \mathbf{W}\mathbf{x}$ of the source signals $\mathbf{s}$, we introduce an objective or loss function $\rho(\mathbf{y}, \mathbf{W})$ in terms of the estimated $\mathbf{y}$ and $\mathbf{W}$. The expected value of this function, called a risk function

$$\mathcal{R}(\mathbf{W}) = E\{\rho(\mathbf{y}, \mathbf{W})\}, \tag{6.2}$$

represents the measure of mutual independence of output signals $\mathbf{y}(k)$. In other words, the risk function $\mathcal{R}(\mathbf{W})$ should be minimized when the components of $\mathbf{y}$ become independent, that is, when $\mathbf{W}$ is a rescaled permutation of $\mathbf{H}^{-1}$. To achieve this minimization, we use the Kullback-Leibler divergence as a measure of independence. Let $p_y(\mathbf{y};\mathbf{W})$ be the probability density function of the random variable $\mathbf{y} = \mathbf{W}\mathbf{x} = \mathbf{W}\mathbf{H}\mathbf{s}$ and $q(\mathbf{y})$ denotes another probability density function of $\mathbf{y}$, in which all the y_i are statistically independent. In this case, $q(\mathbf{y})$ can be decomposed into a product form as

$$q(\mathbf{y}) = \prod_{i=1}^{n} q_i(y_i). \tag{6.3}$$

This independent distribution is called the reference function, which is arbitrary for the moment. We will use the Kullback-Leibler divergence between the distribution $p_y(\mathbf{y};\mathbf{W})$ of $\mathbf{y}$ obtained by the actual value of $\mathbf{W}$ and the reference distribution $q(\mathbf{y})$ as

$$\begin{aligned} \mathcal{R}(\mathbf{W}) &= E\{\rho(\mathbf{y},\mathbf{W})\} = K_{pq}(\mathbf{W}) = K\left[p_y(\mathbf{y};\mathbf{W})\|q(\mathbf{y})\right] \\ &= \int p_y(\mathbf{y};\mathbf{W}) \log \frac{p_y(\mathbf{y};\mathbf{W})}{q(\mathbf{y})} d\mathbf{y}. \end{aligned} \tag{6.4}$$

The Kullback-Leibler divergence is a natural measure of deviation for two probability distributions. Hence, $\mathcal{R}(\mathbf{W}) = K_{pq}(\mathbf{W})$ shows how far the distribution $p_y(\mathbf{y};\mathbf{W})$ is from the reference distribution.

When $q(\mathbf{y})$ is equal to the true distribution $p_s(\mathbf{s})$ of source signals, then $p_y(\mathbf{y},\mathbf{W}) = p_s(\mathbf{y})$ when $\mathbf{W} = \mathbf{H}^{-1}$, because $\mathbf{y} = \mathbf{W}\,\mathbf{H}\,\mathbf{s} = \mathbf{s}$ in this case. Hence, K_{pq} is an appropriate function. However, even when $q(\mathbf{y})$ is not equal to p_s, $\mathbf{W} = \mathbf{H}^{-1}$ is still a critical point of $\mathcal{R}(\mathbf{W}) = K_{pq}(\mathbf{y},\mathbf{W})$. This important property follows from information geometric relations between pdfs described by the Pythagorean Theorem [548].

It is interesting to note that most learning algorithms proposed from heuristic considerations can be explained in terms of the above cost function. It is remarkable that the entropy maximization [71, 1316], independent components analysis (ICA) [299, 28, 29], nonlinear PCA [914, 589], and the maximum likelihood approach [142, 865, 1316, 967] are formulated in the above framework, where the only difference lies in the choice of the reference function $q(\mathbf{y})$. If we choose the true distribution of the sources as q, we have the maximum likelihood approach. Note that the true distribution is unknown, in general; therefore, we need to estimate $q(\mathbf{y})$. If we choose the marginalized independent distribution of $p_y(\mathbf{y};\mathbf{W})$, this leads to ICA [28, 29, 299]. The entropy maximization uses nonlinear transformations $z_i = g_i(y_i)$ to maximize the joint entropy of $\mathbf{z}$. This is easily shown to be equivalent to choosing [71]

$$q_i(y_i) = \frac{d}{dy_i} g_i(y_i). \tag{6.5}$$

The Kullback-Leibler divergence always takes nonnegative values, achieving zero if and only if $p_y(\mathbf{y})$ and q($\mathbf{y}$) are the same distributions. This divergence is invariant with respect to the invertible (monotonic) nonlinear transformations of variables (y_i), including amplitude scaling and permutation in which the variables y_i are rescaled and rearranged. For the independent components analysis problem, we assume that $q(\mathbf{y})$ is the product of the distribution of independent variables y_i. It can be the product of the marginal pdfs of $\mathbf{y}$, in particular,

$$q(\mathbf{y}) = \tilde{p}(\mathbf{y}) = \prod_{i=1}^{n} p_i(y_i), \tag{6.6}$$

where $p_i(y_i)$ are the marginal probability density functions of y_i $(i = 1, 2, \ldots, n)$. The marginal pdf is defined by

$$p_i(y_i) = \int_{-\infty}^{\infty} p_y(\mathbf{y}) d\check{\mathbf{y}}^i, \tag{6.7}$$

where the integration is taken over $\check{\mathbf{y}}^i = [y_1, \ldots, y_{i-1}\ y_{i+1}, \ldots, y_n]^T$ left after removing variable y_i from $\mathbf{y}$.

The natural measure of independence can be formulated as

$$K_{pq} = E\{\rho(\mathbf{y}, \mathbf{W})\} = \int_{-\infty}^{\infty} p_y(\mathbf{y}) \log \frac{p_y(\mathbf{y})}{\prod_{i=1}^{n} q_i(y_i)} d\mathbf{y}, \tag{6.8}$$

where $\rho(\mathbf{y}, \mathbf{W}) = \log(p_y(\mathbf{y})/q(\mathbf{y}))$. The Kullback-Leibler divergence can be expressed in terms of mutual information as

$$K_{pq} = -H(\mathbf{y}) - \sum_{i=1}^{n} \int_{-\infty}^{\infty} p_y(\mathbf{y}) \log q_i(y_i) d\mathbf{y}, \tag{6.9}$$

where the differential entropy of the output signals $\mathbf{y} = \mathbf{W}\mathbf{x}$ is defined by

$$H(\mathbf{y}) = -\int_{-\infty}^{\infty} p_y(\mathbf{y}) \log p_y(\mathbf{y}) d\mathbf{y}. \tag{6.10}$$

When $q_i(y_i) = \tilde{p}_i(y_i)$, taking into account that $d\mathbf{y} = d\check{\mathbf{y}}^i dy_i$, the second term in Eq.(6.9) can be expressed by the marginal entropies as

$$\int_{-\infty}^{\infty} p_y(\mathbf{y}) \log \tilde{p}_i(y_i) d\mathbf{y} = \int_{-\infty}^{\infty} \log \tilde{p}_i(y_i) \int_{-\infty}^{\infty} p_y(\mathbf{y}) d\check{\mathbf{y}}^i dy_i$$

$$= \int_{-\infty}^{\infty} \tilde{p}_i(y_i) \log \tilde{p}_i(y_i) dy_i = E\{\log(\tilde{p}_i(y_i))\} = -H_i(y_i). \tag{6.11}$$

Hence, the Kullback-Leibler divergence can be expressed by the difference between $H(\mathbf{y})$ and the marginal entropies $H_i(y_i)$ as

$$K_{pq} = E\{\rho(\mathbf{y}, \mathbf{W})\} = -H(\mathbf{y}) + \sum_{i=1}^{n} H_i(y_i). \tag{6.12}$$

Assuming $\mathbf{y} = \mathbf{W}\mathbf{x}$, the differential entropy can be expressed by

$$H(\mathbf{y}) = H(\mathbf{x}) + \log|\det(\mathbf{W})|, \tag{6.13}$$

where $H(\mathbf{x}) = -\int_{-\infty}^{\infty} p_x(\mathbf{x}) \log p_x(\mathbf{x}) d\mathbf{x}$ is independent of the matrix $\mathbf{W}$. For general q, we obtain a simple cost (risk) function in the same way,

$$\mathcal{R} = K_{pq} = E\{\rho(\mathbf{y}, \mathbf{W})\} = -H(\mathbf{x}) - \log|\det(\mathbf{W})| - \sum_{i=1}^{n} E\{\log(q_i(y_i))\}. \tag{6.14}$$

It should be noted that the term $H(\mathbf{x})$ can be omitted since it is independent of the demixing matrix $\mathbf{W}$.

6.1.2 Derivation of Natural Gradient Basic Learning Rules

Since our target is to minimize the expectation of the loss function $\rho(\mathbf{y}, \mathbf{W})$ in (6.14), a simple idea is to use the ordinary stochastic gradient descent on-line learning algorithm given by

$$\Delta\mathbf{W}(k) = \mathbf{W}(k+1) - \mathbf{W}(k) = -\eta(k) \frac{\partial \rho\left[\mathbf{y}(k), \mathbf{W}\right]}{\partial \mathbf{W}}, \tag{6.15}$$

where $\eta(k)$ is a learning rate depending on k and $\partial\rho/\partial\mathbf{W}$ is an $n \times n$ gradient matrix whose entries are $\partial\rho/\partial w_{ij}$. We can calculate the gradient matrix $\partial\rho/\partial\mathbf{W}$ by component-wise differentiation. By simple differential matrix calculus, we obtain

$$\Delta\mathbf{W}(k) = \eta(k)\left[\mathbf{W}^{-T}(k) - \mathbf{f}[\mathbf{y}(k)]\,\mathbf{x}^T(k)\right], \tag{6.16}$$

where $\mathbf{W}^{-T}$ is the transpose of the inverse of $\mathbf{W}$ and $\mathbf{f}(\mathbf{y}) = [f_1(y_1), f_2(y_2), \ldots, f_n(y_n)]^T$ is the column vector whose i-th component is (see Table 6.1)

$$f_i(y_i) = -\frac{d \log q_i(y_i)}{dy_i} = -\frac{dq_i(y_i)/dy_i}{q_i(y_i)} = -\frac{q_i'(y_i)}{q_i(y_i)}, \tag{6.17}$$

where $q_i(y_i)$ are approximate (typically parametric) models of the pdf of source signals $\{s_i\}$. The gradient $-\partial\rho/\partial\mathbf{W}$ represents the steepest decreasing direction of function ρ when the parameter space is Euclidean. In the present case, the parameter space consists of all the nonsingular $n \times n$ matrices $\mathbf{W}$. This is a multiplicative group where the identity matrix $\mathbf{I}_n$ is the unit. Moreover, it is a manifold, so that it forms a Lie group. Amari *et al.* [28, 23, 24, 29] exploited this fact to introduce a natural Riemannian metric to the space of $\mathbf{W}$. They showed that the true steepest descent direction in the Riemannian space of parameters $\mathbf{W}$ is not $\partial\rho/\partial\mathbf{W}$ but

$$-\frac{\partial \rho(\mathbf{y}; \mathbf{W})}{\partial \mathbf{W}} \mathbf{W}^T \mathbf{W} = \left[\mathbf{I} - \mathbf{f}(\mathbf{y})\,\mathbf{y}^T\right] \mathbf{W}. \tag{6.18}$$

Hence, the learning algorithm takes the form [28, 29]

$$\boxed{\Delta \mathbf{W}(k) = -\eta(k) \frac{\partial \rho(\mathbf{y}, \mathbf{W})}{\partial \mathbf{W}} \mathbf{W}^T \mathbf{W} = \eta(k) \left[\mathbf{I} - \mathbf{f}[\mathbf{y}(k)]\, \mathbf{y}^T(k) \right] \mathbf{W}(k).} \tag{6.19}$$

This type of learning rule was first introduced in [279, 278, 277] and further developed and analyzed in [23, 24, 28, 29]. This is essentially the same as the relative gradient introduced independently by Cardoso and Laheld [148]. The equivariant property [148] holds in this learning rule, because the underlying Riemannian metric is based on the Lie group structure. Amari [20] showed that the natural gradient learning has a number of desirable properties including the Fisher efficiency.

Alternatively, in order to calculate the gradient of $\rho(\mathbf{y}, \mathbf{W})$ expressed by (6.14), we use the total differential $d\rho$ of ρ when $\mathbf{W}$ is changed from $\mathbf{W}$ to $\mathbf{W} + d\mathbf{W}$ [24, 29]. In the component form,

$$d\,\rho = \rho(\mathbf{y}, \mathbf{W} + d\mathbf{W}) - \rho(\mathbf{y}, \mathbf{W}) = \sum_{i,j} \frac{\partial \rho}{\partial w_{ij}} dw_{ij}, \tag{6.20}$$

where the coefficients $\partial \rho / \partial w_{ij}$ represent the gradient of ρ. Algebraic manipulation and differential calculus yields

$$d\,\rho = -\operatorname{tr}(d\mathbf{W}\mathbf{W}^{-1}) + \mathbf{f}^T(\mathbf{y})\, d\mathbf{y}, \tag{6.21}$$

where tr represents the trace of a matrix and $\mathbf{f}(\mathbf{y})$ is a column vector whose components are $f_i(y_i) = -q_i'(y_i)/q_i(y_i)$. From $\mathbf{y} = \mathbf{W}\mathbf{x}$, we have

$$d\mathbf{y} = d\mathbf{W}\mathbf{x} = d\mathbf{W}\mathbf{W}^{-1}\mathbf{y}. \tag{6.22}$$

Hence, we denote

$$d\mathbf{X} = d\mathbf{W}\mathbf{W}^{-1}, \tag{6.23}$$

whose components dx_{ij} are linear combinations of dw_{ij}. The differentials $\{dx_{ij}\}$ form a basis of the tangent space of nonsingular matrices $\mathbf{W}$, since they are linear combinations of the basis $\{dw_{ij}\}$. It should be noted that $d\mathbf{X} = d\mathbf{W}\mathbf{W}^{-1}$ is a non-integrable differential form, so that we do not have a matrix function $\mathbf{X}(\mathbf{W})$ which gives (6.23). Nevertheless, the nonholonomic basis $d\mathbf{X}$ has a definite geometrical meaning and is very useful. It is effective to analyze the differential in terms of $d\mathbf{X}$, since the natural Riemannian gradient [28, 29] is automatically implemented by it and the equivariant properties investigated in [148] automatically hold in this basis. It is straightforward to rewrite the results in terms of $d\mathbf{W}$ by using (6.23). The gradient $d\rho$ in (6.21) is expressed by the differential form

$$d\,\rho = -\operatorname{tr}(d\mathbf{X}) + \mathbf{f}^T(\mathbf{y})\, d\mathbf{X}\mathbf{y}. \tag{6.24}$$

This leads to the stochastic gradient learning algorithm:

$$\Delta \mathbf{X}(k) = -\eta(k) \frac{d\,\rho}{d\mathbf{X}} = \eta(k) \left[\mathbf{I} - \mathbf{f}(\mathbf{y}(k))\, \mathbf{y}^T(k) \right] \tag{6.25}$$

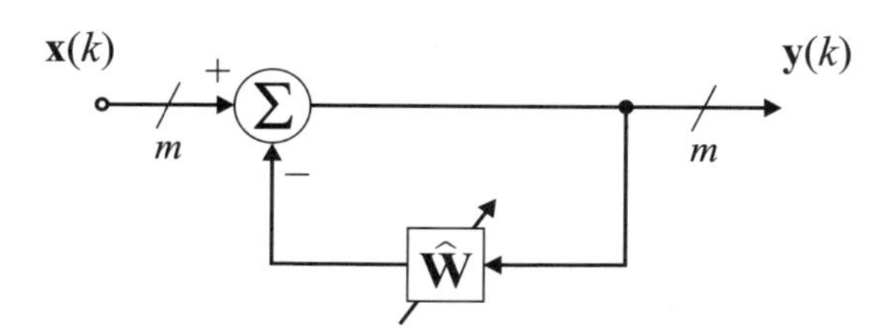

Fig. 6.2 Block diagram of a fully connected recurrent network.

in terms of $\Delta\mathbf{X}(k) = \Delta\mathbf{W}(k)\mathbf{W}^{-1}(k)$, or

$$\mathbf{W}(k+1) = \mathbf{W}(k) + \eta(k)\left[\mathbf{I} - \mathbf{f}(\mathbf{y}(k))\,\mathbf{y}^T(k)\right]\mathbf{W}(k) \tag{6.26}$$

in terms of $\mathbf{W}(k)$.

Applying the natural gradient (NG) rule (6.18), to the risk function (6.14), we obtain the batch algorithm

$$\boxed{\Delta\mathbf{W} = -\eta\frac{\partial K_{pq}}{\partial \mathbf{W}}\,\mathbf{W}^T\mathbf{W} = \eta\left[\mathbf{I} - \langle \mathbf{f}(\mathbf{y})\,\mathbf{y}^T\rangle\right]\mathbf{W},} \tag{6.27}$$

which can be converted to on-line algorithms by omitting the average (expectation) operation.

Remark 6.1 *We can derive similar algorithms for a fully recurrent neural network described by* $\mathbf{y}(k) = \mathbf{x}(k) - \widehat{\mathbf{W}}\mathbf{y}(k)$ *(see Fig. 6.2). Assuming that the demixing matrix* $\mathbf{W}$ *(giving* $\mathbf{y}(k) = \mathbf{W}\mathbf{x}(k)$*) is nonsingular, we have simple relations:* $\widehat{\mathbf{W}} = \mathbf{W}^{-1} - \mathbf{I}$ *and*

$$\frac{d\mathbf{W}}{dt}\mathbf{W}^{-1} = -\mathbf{W}\frac{d\mathbf{W}^{-1}}{dt}.$$

Hence, we obtain a learning rule first developed by Cichocki and Unbehauen [277]

$$\Delta\widehat{\mathbf{W}}(k) = -\eta(k)\left[\widehat{\mathbf{W}}(k) + \mathbf{I}\right]\left[\mathbf{I} - \langle \mathbf{f}[\mathbf{y}(k)]\,\mathbf{y}^T(k)\rangle\right] \tag{6.28}$$

or equivalently

$$\boxed{\Delta\widehat{\mathbf{H}}(k) = -\eta(k)\widehat{\mathbf{H}}(k)\left[\mathbf{I} - \langle \mathbf{f}[\mathbf{y}(k)]\,\mathbf{y}^T(k)\rangle\right],} \tag{6.29}$$

where $\widehat{\mathbf{H}}(k) = \mathbf{W}^{-1}(k) = \widehat{\mathbf{W}}(k) + \mathbf{I}$ *is an estimating matrix of the unknown mixing matrix* $\mathbf{H}$ *(up to permutations and scale factors).*

6.2 GENERALIZATIONS OF THE BASIC NATURAL GRADIENT ALGORITHM

6.2.1 Nonholonomic Learning Rules

Now, we may present some modifications and extensions of the basic natural gradient learning equation (6.19).

In order to resolve the indeterminacy of scales, the basic NG learning algorithms impose some constraints on the magnitudes of the recovered signals, e.g., $E\{f_i(y_i)y_i\} = 1$. However, when source signals are non-stationary and their average magnitudes change rapidly, the constraints force a rapid change in the magnitude of the separating matrix. This is the case with most applications (speech, music, biomedical signals, etc.). It is known that this might cause numerical instability in some cases. In order to resolve this difficulty, we introduce so-called nonholonomic constraints in the NG learning algorithm in this section [25]. This is motivated by the geometrical consideration that the directions of change in the separating matrix $\mathbf{W}$ should be orthogonal to the equivalence class of separating matrices due to the scaling indeterminacy. These constraints are proven to be nonholonomic, so that the proposed algorithm is able to adapt to rapid and/or intermittent changes in the magnitude of source signals. The proposed algorithm works well even when the number of sources is overestimated (assuming the sensor noise is negligibly small), because they amplify the null components not included in the sources.

Let us consider the NG learning algorithm (referred to as the nonholonomic NG algorithm)

$$\boxed{\Delta\mathbf{W}(k) = \eta(k)\left[\mathbf{\Lambda}(k) - \mathbf{f}\left[\mathbf{y}(k)\right]\mathbf{y}^T(k)\right]\mathbf{W}(k),} \tag{6.30}$$

where $\mathbf{\Lambda} = \mathrm{diag}\{\lambda_1, \lambda_2, \dots, \lambda_n\}$ is a diagonal positive definite matrix. Amari *et al.* [29] demonstrated that, when the magnitudes of source signals rapidly change over time or when some of them become zero for a while, the learning algorithm with entries $\lambda_{ii} = f(y_i)y_i$ works very well. These are nonholonomic constraints in the learning algorithm.

Another modification is to use a learning algorithm of the form [270, 277, 279]

$$\Delta\mathbf{W}(k) = \eta(k)\left[\mathbf{\Lambda}(k) - \mathbf{y}(k)\,\boldsymbol{g}^T[\mathbf{y}(k)]\right]\mathbf{W}(k). \tag{6.31}$$

This is the modified gradient [45, 270], which is the dual to the natural gradient descent (see Appendix B),

$$\begin{aligned}\Delta\mathbf{W}(k) &= -\eta(k)\,\mathbf{W}(k)\left[\frac{\partial K_{pq}(\mathbf{W})}{\partial\mathbf{W}}\right]^T\mathbf{W}(k)\\ &= \eta(k)\left[\mathbf{I} - \langle\mathbf{y}(k)\boldsymbol{g}^T[\mathbf{y}(k)]\rangle\right]\mathbf{W}(k).\end{aligned} \tag{6.32}$$

The nonlinearity $g_i(y_i)$ is the inverse of (dual to) the function $f_i(y_i) = -\{\log q(y_i)\}'$. The dynamics and the behavior of the nonholonomic learning algorithms are analyzed in [29] (see Appendix D).

More generally, Amari *et al.* have proposed the following learning rule [20, 24, 29]

$$\boxed{\Delta\mathbf{W}(k) = \eta(k)\left[\mathbf{\Lambda}(k) - \alpha\left\langle\mathbf{f}[\mathbf{y}(k)]\,\mathbf{y}^T(k)\right\rangle + \beta\left\langle\mathbf{y}(k)\,\boldsymbol{g}^T[\mathbf{y}(k)]\right\rangle\right]\mathbf{W}(k),} \tag{6.33}$$

where $\mathbf{\Lambda}$ is a diagonal matrix configured to eliminate the diagonal entries in the bracket of the right-hand side, and α and β are suitably chosen parameters to be adaptively determined. This guarantees the best performance, that is the most efficient online estimation, provided constants α and β are adequately determined. See [24] for details about α and β.

6.2.2 Natural Riemannian Gradient in Orthogonality Constraint

Let us assume that the observation vector $\mathbf{x}(k) = \mathbf{Q}\,\mathbf{H}\,\mathbf{s}(k) = \mathbf{A}\,\mathbf{s}(k)$ has already been whitened by preprocessing and that the source signals are normalized, i.e.,

$$\mathbf{R}_{\mathbf{x}\,\mathbf{x}} = E\{\mathbf{x}(k)\,\mathbf{x}^T(k)\} = \mathbf{I}_m, \tag{6.34}$$

$$\mathbf{R}_{\mathbf{s}\,\mathbf{s}} = E\{\mathbf{s}(k)\,\mathbf{s}^T(k)\} = \mathbf{I}_n. \tag{6.35}$$

When $m = n$, from (6.34) and (6.35), we have

$$\mathbf{A}\mathbf{A}^T = \mathbf{I}_n, \tag{6.36}$$

where the mixing matrix $\mathbf{A} = \mathbf{Q}\mathbf{H} \in \mathbb{R}^{n\times n}$ is orthogonal. In such a case, we may consider only the orthogonal matrices $\mathbf{W}$ (such that $\mathbf{W}^{-1} = \mathbf{W}^T$) for recovering the original signals by $\mathbf{y} = \mathbf{W}\,\mathbf{x}$. The nonholonomic basis is written as

$$d\mathbf{X} = \mathbf{dW}\,\mathbf{W}^{-1} = d\mathbf{W}\,\mathbf{W}^T, \tag{6.37}$$

and it is skew-symmetric,

$$\mathbf{dX} = -\mathbf{dX}^T, \tag{6.38}$$

which is easily shown by

$$\mathbf{0} = \mathbf{dI} = d(\mathbf{W}\mathbf{W}^T) = d\mathbf{W}\mathbf{W}^T + \mathbf{W}d\mathbf{W}^T = d\mathbf{X} + d\mathbf{X}^T. \tag{6.39}$$

This implies that the gradient

$$\frac{\partial \rho}{\partial \mathbf{X}} = \frac{\partial \rho}{\partial \mathbf{W}}\,\mathbf{W}^T \tag{6.40}$$

is also skew-symmetric. Hence, we have

$$\frac{\partial \rho}{\partial \mathbf{X}}\,\mathbf{W}^T = \mathbf{f}(\mathbf{y})\,\mathbf{y}^T - \mathbf{y}\,\mathbf{f}^T(\mathbf{y}). \tag{6.41}$$

The natural gradient learning algorithm for the separating matrix $\mathbf{W} \in \mathbb{R}^{n\times n}$ in the case where data are prewhitened, is given by

$$\begin{aligned}\mathbf{W}(k+1) &= \mathbf{W}(k) - \eta(k)\tilde{\nabla}\rho(\mathbf{y}, \mathbf{W}) \\ &= \mathbf{W}(k) - \eta(k)\,\frac{\partial \rho}{\partial \mathbf{X}}\mathbf{W}.\end{aligned} \tag{6.42}$$

It should be noted that the matrix $\mathbf{W}$ is (approximately)[1] orthogonal in each iteration step, so, $\mathbf{x} = \mathbf{W}^T\mathbf{y}$ and the above algorithm reduces to the following form

$$\mathbf{W}(k+1) = \mathbf{W}(k) - \eta(k)\left[\mathbf{f}(\mathbf{y}(k))\,\mathbf{y}^T(k) - \mathbf{y}(k)\,\mathbf{f}^T(\mathbf{y}(k))\right]\mathbf{W}(k). \tag{6.43}$$

[1]The matrix is precisely orthogonal for the continuous time version of the algorithm if $\mathbf{W}(0)$ is orthogonal. However, for the discrete-time algorithm, it is only approximately orthogonal under the condition that the learning rate $\eta(k)$ has a sufficiently small positive value, and the second term $\eta^2(k)\mathcal{O}$ can be neglected.

In practice, due to the skew-symmetry of the term $\mathbf{f}(\mathbf{y}(k))\mathbf{y}^T(k) - \mathbf{y}(k)\mathbf{f}^T(\mathbf{y}(k))$, decorrelation (or whitening) processing can be performed simultaneously with separation. By taking this into account, the algorithm becomes the EASI algorithm proposed by Cardoso and Laheld [148]

$$\Delta\mathbf{W}(k) = \eta(k)\left[\mathbf{I} - \mathbf{y}(k)\mathbf{y}^T(k) - \mathbf{f}(\mathbf{y}(k))\mathbf{y}^T(k) + \mathbf{y}(k)\mathbf{f}^T(\mathbf{y}(k))\right]\mathbf{W}(k). \tag{6.44}$$

The aforementioned algorithms belong to a class of on-line learning algorithms that are based on stochastic approximation. We can also consider batch versions of the algorithms by estimating a time average instead of an instantaneous realization. For example, the batch version of the algorithm (6.44) is given by

$$\boxed{\Delta\mathbf{W}(k) = \eta(k)\left[\mathbf{I} - \langle\mathbf{y}(k)\mathbf{y}^T(k)\rangle - \langle\mathbf{f}(\mathbf{y}(k))\mathbf{y}^T(k)\rangle + \langle\mathbf{y}(k)\mathbf{f}^T(\mathbf{y}(k))\rangle\right]\mathbf{W}(k),} \tag{6.45}$$

where $\langle\cdot\rangle$ denotes the time average operation.

Remark 6.2 *In the case of $m > n$, we may use $\mathbf{W} \in \mathbb{R}^{n\times m}$, where all the n rows of $\mathbf{W}$ are mutually orthogonal m-dimensional vectors, so that*

$$\mathbf{W}\mathbf{W}^T = \mathbf{I}_n \tag{6.46}$$

is satisfied. A set of such matrices is called the Stiefel manifold. The natural Riemannian gradient in the Stiefel manifold was calculated by Amari [29] as (see also Edelman, Arias and Smith [410])

$$\tilde{\nabla}\rho(\mathbf{y},\mathbf{W}) = \nabla\rho(\mathbf{y},\mathbf{W}) - \mathbf{W}\left[\nabla\rho(\mathbf{y},\mathbf{W})\right]^T\mathbf{W}. \tag{6.47}$$

Using this formula, the natural gradient of the loss function $\rho(\mathbf{y},\mathbf{W})$ is given by

$$\tilde{\nabla}\rho(\mathbf{y},\mathbf{W}) = \mathbf{f}(\mathbf{y})\,\mathbf{x}^T - \mathbf{y}\,\mathbf{f}^T(\mathbf{y})\mathbf{W}, \tag{6.48}$$

and we have

$$\Delta\mathbf{W} = -\eta\,\tilde{\nabla}\rho(\mathbf{y},\mathbf{W}). \tag{6.49}$$

6.2.2.1 Local Stability Analysis The stability conditions for the algorithms (6.27) and (6.44) were given by Amari *et al.* [24] and by Cardoso and Laheld [148], respectively. In this section, we focus on the algorithm (6.44) which employs the natural gradient in the Stiefel manifold.

Since the algorithm (6.44) was derived from the gradient $d\rho = \mathbf{f}^T(\mathbf{y})d\mathbf{W}\mathbf{x}$, we need to calculate its Hessian $d^2\rho$ to check the stability of stationary points. Amari *et al.* have shown that the calculation of the Hessian $d^2\rho$ is relatively easy if the modified differential coefficient matrix $d\mathbf{X} = d\mathbf{W}\mathbf{W}^{-1}$ is employed [24]. Noting that the modified differential coefficient matrix $d\mathbf{X}$ is skew-symmetric in the orthogonality constraint, we calculate the Hessian $d^2\rho$. By using the fact that $d\mathbf{X}$ is skew-symmetric, $d\rho$ can be written as

$$\begin{aligned} d\rho &= \mathbf{f}^T(\mathbf{y})d\mathbf{y} = \mathbf{f}^T(\mathbf{y})d\mathbf{X}\mathbf{y} \\ &= \sum_{i,j} f_i(y_i)dx_{ij}y_j = \sum_{i>j}\{f_i(y_i)y_j - f_j(y_j)y_i\}\,dx_{ij}. \end{aligned} \tag{6.50}$$

Then the Hessian $d^2\rho$ is calculated as

$$\begin{aligned} d^2\rho &= \sum_{i>j}\sum_{k}\{f_i'(y_i)dx_{ik}y_ky_j - f_j'(y_j)dx_{jk}y_ky_i \\ &\quad + f_i(y_i)dx_{jk}y_k - f_j(y_j)dx_{ik}y_k\ \}\, dx_{ij}. \end{aligned} \tag{6.51}$$

Taking into account the normalization constraint ($E\{y_i^2\} = 1$) and the skew-symmetry, $dx_{ij} = -dx_{ji}$, the Hessian at $\mathbf{W} = \mathbf{A}^{-1}$ (i.e. the desirable solution) is given by

$$E\{d^2\rho\} = \sum_{i>j} \left[E\left\{f_i'(y_i)\right\} + E\left\{f_j'(y_j)\right\} - E\left\{f_i(y_i)y_i\right\} - E\left\{f_j(y_j)y_j\right\}\right] dx_{ij}^2, \tag{6.52}$$

where $f_i'(y_i)$ denotes the derivative of $f_i(y_i)$ with respect to y_i. From (6.52), the stability conditions are given by

$$\chi_i + \chi_j > 0, \tag{6.53}$$

where

$$\chi_i = E\left\{f_i'(y_i)\right\} - E\left\{f_i(y_i)y_i\right\}. \tag{6.54}$$

The stability condition given in (6.53) coincides with that in [140], but we arrived at this result in the framework of the natural gradient in the Stiefel manifold. For each y_i, the condition

$$\chi_i > 0 \tag{6.55}$$

is a sufficient condition for stability.

Amari and Cardoso [23] studied this problem from the point of view of semi-parametric statistical models for information geometry [35, 36]. They consider a general form of learning equation,

$$\Delta\mathbf{W}(k) = \eta(k)\mathbf{F}\left[\mathbf{y}(k), \mathbf{W}(k)\right]\mathbf{W}(k) \tag{6.56}$$

which has an arbitrary smooth nonlinear matrix function $\mathbf{F} = [f_{ij}] \in \mathbb{R}^{n\times n}$. The results shows that (1) the diagonal entries f_{ii} of $\mathbf{F}$ can be arbitrary, and (2) general admissible (efficient) forms of $\mathbf{F}$ are spanned by $\mathbf{f}(\mathbf{y})\,\mathbf{y}^T$. This implies that another form $\mathbf{y}\,\mathbf{f}^T(\mathbf{y})$ is also a good candidate for the learning equation, because it is a linear combination of the former. Amari and Cardoso have also shown that the general form of an admissible learning function is

$$\mathbf{F}[\mathbf{y}, \mathbf{W}] = \mathbf{\Lambda} - \alpha\,\mathbf{f}(\mathbf{y})\,\mathbf{y}^T - \beta\,\mathbf{y}\,\mathbf{f}^T(\mathbf{y}), \tag{6.57}$$

where α and β are suitably chosen coefficients, and the diagonal entries of $\mathbf{F}$ have arbitrarily assigned nonnegative values.

6.3 NATURAL GRADIENT ALGORITHMS FOR BLIND EXTRACTION OF ARBITRARY GROUPS OF SOURCES

6.3.1 Stiefel and Grassmann-Stiefel Manifolds Approaches

Consider the case where we do not want to extract one single source at a time but to extract simultaneously a specified number of sources, say e, where $1 \leq e \leq n$, with $m \geq n$. (The number of sensors m is greater or equal to the number of sources n and the number of sources is generally unknown [20]). Let us also assume that the sensor signals are prewhitened, for example, by using the PCA technique described previously. Then, the transformed sensor signals satisfy the condition

$$\mathbf{R}_{\overline{\mathbf{x}}\,\overline{\mathbf{x}}} = E\{\overline{\mathbf{x}}\,\overline{\mathbf{x}}^T\} = \mathbf{I}_n, \tag{6.58}$$

where $\overline{\mathbf{x}} = \mathbf{x}_1 = \mathbf{Q}\,\mathbf{x}$ and the new global $n \times n$ mixing matrix $\mathbf{A} = \mathbf{Q}\,\mathbf{H}$ is orthogonal, that is, $\mathbf{A}\mathbf{A}^T = \mathbf{A}^T\mathbf{A} = \mathbf{I}_n$. Hence, the ideal $n \times n$ separating matrix is $\mathbf{W}_* = \mathbf{A}^{-1} = \mathbf{A}^T$ for $e = n$.

In order to solve this problem, we can formulate the appropriate cost function expressed by the Kullback-Leibler divergence

$$K_{pq} = K(p(\mathbf{y}, \mathbf{W}_e)||q(\mathbf{y})) = \int p(\mathbf{y}, \mathbf{W}_e) \log \frac{p(\mathbf{y}, \mathbf{W}_e)}{q(\mathbf{y})} d\mathbf{y}, \tag{6.59}$$

where $q(\mathbf{y}) = \prod_{i=1}^{e} q_i(y_i)$ represents an adequate independent probability distribution of the output signals. Hence, the cost function takes the form:

$$\rho(\mathbf{y}, \mathbf{W}_e) = -\sum_{i=1}^{e} \log q_i(y_i) \tag{6.60}$$

subject to constraints $\mathbf{W}_e\mathbf{W}_e^T = \mathbf{I}_e$ or equivalently $\mathbf{w}_i\mathbf{w}_j^T = \delta_{ij}$, where $\mathbf{W}_e \in \mathbb{R}^{e \times n}$ is a demixing (separating) matrix, with $e \leq n$, and $\mathbf{w}_i$ is the i-th row vector of matrix $\mathbf{W}_e$.

These constraints follow from the mixing matrix $\mathbf{A} = \mathbf{Q}\,\mathbf{H}$ being a square orthogonal matrix and the demixing matrix $\mathbf{W}_e$ should satisfy the following relationship after successful extraction of e sources (ignoring scaling and permutation for simplicity):

$$\mathbf{W}_e\mathbf{A} = [\mathbf{I}_e, \mathbf{0}_{n-e}]. \tag{6.61}$$

We say that the matrix $\mathbf{W}_e$, satisfying the above condition, forms a Stiefel manifold since its rows are mutually orthogonal ($\mathbf{w}_i\mathbf{w}_j = \delta_{ij}$). In order to satisfy the constraints during the learning process, we employ the following natural gradient formula (Amari 1998) [20]

$$\boxed{\Delta\mathbf{W}_e(k) = \mathbf{W}_e(k+1) - \mathbf{W}_e(k) = -\eta \left[\frac{\partial\rho(\mathbf{y}, \mathbf{W}_e)}{\partial\mathbf{W}_e} - \mathbf{W}_e \left(\frac{\partial\rho(\mathbf{y}, \mathbf{W}_e)}{\partial\mathbf{W}_e}\right)^T \mathbf{W}_e\right].} \tag{6.62}$$

It can be shown that the separating matrix $\mathbf{W}_e$ satisfies the relation $\mathbf{W}_e(k)\mathbf{W}_e^T(k) = \mathbf{I}_e$ in each iteration step under the condition that $\mathbf{W}_e(0)\mathbf{W}_e^T(0) = \mathbf{I}_e$.

Applying the natural gradient formula (6.62), we obtain a learning rule:

$$\boxed{\mathbf{W}_e(k+1) = \mathbf{W}_e(k) - \eta(k) \left[\mathbf{f}[\mathbf{y}(k)]\, \overline{\mathbf{x}}^T(k) - \mathbf{y}(k)\, \mathbf{f}^T[\mathbf{y}(k)]\, \mathbf{W}_e(k)\right],} \tag{6.63}$$

with initial $\mathbf{W}_e(0)$ satisfying the condition $\mathbf{W}_e(0)\mathbf{W}_e^T(0) = \mathbf{I}_e$.

It is noteworthly, that for $e = n$, the separating matrix $\mathbf{W}_e = \mathbf{W}$ is orthogonal and thus, our learning rule simplifies to the well-known algorithm proposed by Cardoso and Laheld in [148]:

$$\mathbf{W}(k+1) = \mathbf{W}(k) - \eta(k) \left[\mathbf{f}[\mathbf{y}(k)]\mathbf{y}^T(k) - \mathbf{y}(k)\mathbf{f}^T[\mathbf{y}(k)]\right] \mathbf{W}(k) \tag{6.64}$$

since $\overline{\mathbf{x}} = \mathbf{Q}\,\mathbf{x} = \mathbf{Q}\,\mathbf{H}\,\mathbf{s} = \mathbf{W}^{-1}\,\mathbf{y} = \mathbf{W}^T\,\mathbf{y}$. The above algorithm can be extended by performing prewhitening and blind separation simultaneously [148]:

$$\boxed{\Delta\mathbf{W}(k) = \eta(k) \left[\mathbf{I}_n - \mathbf{y}(k)\,\mathbf{y}^T(k) - \mathbf{f}[\mathbf{y}(k)]\,\mathbf{y}^T(k) + \mathbf{y}(k)\,\mathbf{f}^T[\mathbf{y}(k)]\right] \mathbf{W}(k).} \tag{6.65}$$

Remark 6.3 *It can be shown that the separating matrix $\mathbf{W}_e(k) \in \mathbb{R}^{e \times n}$ satisfies approximately (for sufficiently small values of the learning rate $\eta(k)$) the relation $\mathbf{W}_e(k)\mathbf{W}_e^T(k) = \mathbf{I}_e$ in each iteration step k under the condition that $\mathbf{W}_e(0)\mathbf{W}_e^T(0) = \mathbf{I}_e$ because the above learning rule satisfies approximately the relation:*

$$\Delta\mathbf{W}_e(k)\mathbf{W}_e^T(k) + \mathbf{W}_e(k)\Delta\mathbf{W}_e^T(k) = \mathbf{0}. \tag{6.66}$$

An interesting modification of the natural gradient learning algorithm with the orthogonality constraint has been proposed by Nishimori [882]. The algorithm can be represented as

$$\mathbf{W}(k+1) = \exp[-\eta_k \tilde{\nabla}_{\mathbf{W}} J]\; \mathbf{W}(k), \tag{6.67}$$

where the gradient $\tilde{\nabla}_{\mathbf{W}} J$ can take a special forms:

$$\begin{aligned} \tilde{\nabla}_{\mathbf{W}} J &= \nabla_k \mathbf{W}^T(k) - \mathbf{W}(k)\nabla_k^T \\ &= \mathbf{f}(\mathbf{y}(k))\mathbf{y}^T(k) - \mathbf{y}(k)\mathbf{f}^T(\mathbf{y}(k)), \end{aligned} \tag{6.68}$$

with the standard gradient denoted as $\nabla_k = \frac{\partial J(\mathbf{W}(k))}{\partial \mathbf{W}}$. It can be easily shown that in the special case of Grassmannn-Stiefel manifolds the learning rule (6.67) exactly satisfies semi-orthogonality constraints independent of the value of the learning rate η_k under condition that $\mathbf{W}(0)\mathbf{W}^T(0) = \mathbf{I}$.

Proof. Let us assume that $\mathbf{W}(k)\mathbf{W}^T(k) = \mathbf{I}$. Then we have $\mathbf{W}(k+1)\mathbf{W}^T(k+1) = \exp[-\eta_k \tilde{\nabla}_{\mathbf{W}} J]\; \mathbf{W}(k)\mathbf{W}^T(k) \exp[\eta_k \tilde{\nabla}_{\mathbf{W}} J] = \exp(\mathbf{0}) = \mathbf{I}$.

The above learning formula can be extended or generalized to the following forms (that not necessarily satisfy orthogonality constraints):

$$\mathbf{W}(k+1) = \exp[\eta_k \mathbf{F}(\mathbf{y}(k))]\; \mathbf{W}(k), \tag{6.69}$$

and

$$\widehat{\mathbf{H}}(k+1) = \widehat{\mathbf{H}}(k) \exp[-\eta_k \mathbf{F}(\mathbf{y}(k))], \tag{6.70}$$

where the matrix $\mathbf{F}(\mathbf{y}(k))$ can take various forms as discussed in previous sections. In fact, the learning rules (6.27) and (6.29) can be considered as special cases of the above formulas. The matrix $\exp[\eta_k \mathbf{F}(\mathbf{y}(k))]$ can be efficiently computed in MATLAB using a Padé approximation.

6.4 GENERALIZED GAUSSIAN DISTRIBUTION MODEL FOR ICA – PRACTICAL IMPLEMENTATION OF THE ALGORITHMS

The optimal nonlinear function $f_i(y_i)$ is given by (6.17). However, it requires *a priori* knowledge of the probability distributions of the sources that is not usually available. A variety of hypothesized density models have been used (see Table 6.1). For example, for super-Gaussian source signals, the hyperbolic-Cauchy distribution model leads to the nonlinear function given by

$$f_i(y_i) = \tanh(\gamma_i y_i), \tag{6.71}$$

typically, with $\gamma_i = 1/\sigma_{y_i}^2$. For sub-Gaussian source signals, the cubic nonlinear function $\mathbf{f}_i(y_i) = y_i^3$ has been a favorite choice. For mixtures of sub- and super-Gaussian source signals, according to the estimated kurtosis of the extracted signals, the nonlinear function can be selected from two different choices [391]. Several approaches [484, 270, 481] are already available. In this section, we present an adaptive nonlinear function derived using the generalized Gaussian density model [270, 269, 226]. It will be shown that the nonlinear function is self-adaptive and controlled by the Gaussian exponent.

Let us assume that the source signals have the generalized Gaussian distribution of the form [270, 269, 226]:

$$q_i(y_i) = \frac{r_i}{2\sigma_i \Gamma(1/r_i)} \exp\left(-\frac{1}{r_i} \left| \frac{y_i}{\sigma_i} \right|^{r_i} \right), \tag{6.72}$$

where $r_i > 0$ is a variable parameter, $\Gamma(r) = \int_0^\infty y^{r-1} \exp(-y) dy$ is the gamma function and $\sigma_i^r = E\{|y_i|^r\}$ is a generalized measure of variance known as the dispersion of the distribution. The parameter r_i can change from zero, through 1 (the Laplace distribution) and $r_i = 2$ (standard Gaussian distribution), to r_i going to infinity (for uniform distribution) (see Fig. 6.3 (a), (b)).

The optimal normalized nonlinear activation functions[2] can be expressed in such cases as:

$$f_i(y_i) = -\frac{d \log(q_i(y_i))}{dy_i} = |y_i|^{r_i - 1} \text{sign}(y_i), \quad r_i \geq 1. \tag{6.73}$$

[2]For adaptive algorithms with nonholonomic constraints, it is recommended to use a rescaled activation function of the form $\hat{f}_i(y_i) = (|y_i|^{r_i - 1} \, \text{sign}(y_i))/|\sigma_{y_i}|^{r_i}$, where $\sigma_{y_i} = \sqrt{E\{y_i^2\}}$ is the estimated deviation of output signal y_i.

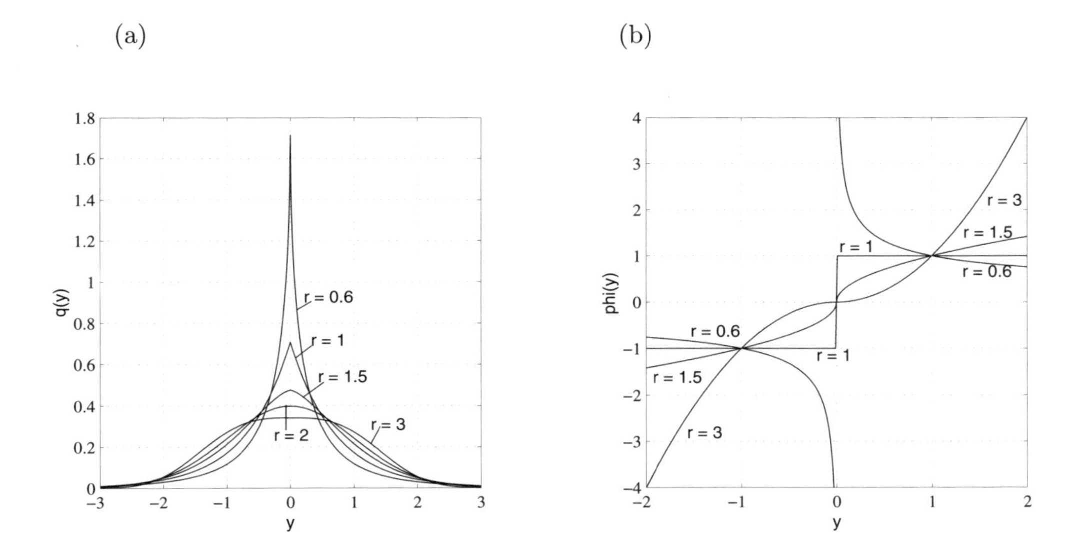

Fig. 6.3 (a) Plot of the generalized Gaussian pdf for various values of parameter r (with $\sigma^2 = 1$) and (b) corresponding nonlinear activation functions.

Taking into account that $\text{sign}(y) = y/|y|$, we obtain (see Fig. 6.3 (b))

$$f_i(y_i) = \frac{y_i}{|y_i|^{2-r_i}}. \tag{6.74}$$

In the special case of spiky or impulsive signals, the parameters r_i can take the values between zero and one. In such a case, we can use the slightly modified activation functions

$$f_i(y_i) = \frac{y_i}{[|y_i|^{2-r_i} + \epsilon]}, \qquad 0 < r_i < 1, \tag{6.75}$$

where ϵ is a very small positive parameter (typically 10^{-4}), which avoids the singularity of the function when $y_i = 0$. Alternatively, we can take the moving average of the instantaneous values of the nonlinear function as

$$f_i(y_i) = \frac{y_i}{\langle |y_i|^{2-r_i} \rangle} = \frac{y_i(k)}{\widehat{\sigma}_i^{(2-r_i)}(k)}, \quad 0 \leq r_i < \infty, \tag{6.76}$$

with estimation of $\widehat{m}_{2-r_i} = \widehat{\sigma}_i^{(2-r_i)}$ by the moving average as

$$\widehat{\sigma}_i^{(2-r_i)}(k+1) = (1-\eta_0)\widehat{\sigma}_i^{(2-r_i)}(k) + \eta_0 |y_i(k)|^{2-r_i}. \tag{6.77}$$

Such an activation function can be considered as "linear" time variable function modulated in time by the fluctuating estimated moment $m_{2-r_i} = \widehat{\sigma}_i^{(2-r_i)}$.

Table 6.1 Typical pdf $q(y)$ and the corresponding normalized activation functions $f(y) = -d \log q(y)/dy$.

Name	Density Function $q(y)$	Activation Function $f(y)$
Gaussian	$\frac{1}{\sqrt{2\pi}\sigma} \exp \; -\frac{\vert y\vert^2}{2\sigma^2}$	$\frac{y}{\sigma^2}$
Laplace	$\frac{1}{2\sigma} \exp \; -\frac{\vert y\vert}{\sigma}$	$\frac{\text{sign}(y)}{\sigma}$
Cauchy	$\frac{1}{\pi\sigma} \frac{1}{1+(y/\sigma)^2}$	$\frac{2y}{\sigma^2+y^2}$
Hyperbolic cosine	$\frac{1}{\pi \cosh(y/\sigma^2)}$	$\tanh(y/\sigma^2)$
Unimodal	$\frac{\exp(-2\sigma^{-2}y)}{(1+\exp(-2\sigma^{-2}y))^2}$	$\tanh(y/\sigma^2)$
Triangular	$\frac{1}{\sigma\,(1-\vert y\vert/\sigma)}$ $\vert y\vert < \sigma$	$\frac{\text{sign}(y)}{\sigma(1-\vert y\vert/\sigma)}$
Generalized Gaussian	$\frac{r}{2\sigma\Gamma(1/r)} \exp(-\frac{1}{r}\vert\frac{y}{\sigma}\vert^r)$ $r \geq 1$	$\frac{\vert y\vert^{r-1}}{\sigma^r} \text{sign}(y)$
Robust Generalized Gaussian	$\frac{r}{2\sigma\Gamma(1/r)} \exp(-\vert\frac{\rho(y)}{\sigma}\vert^r)$ $\rho(y)$ – robust function $r \geq 1$	$\frac{\vert\rho(y)\vert^{r-1}}{\sigma^r} \frac{\partial\rho}{\partial y}$

The parameters r_i can be fixed if some *a priori* knowledge about the statistics of the source signals is available or obtained through learning. The stochastic gradient-based rule for adjusting parameter $r_i(k)$ takes the form

$$\Delta r_i(k) = -\eta_i(k)\frac{\partial \rho}{\partial r_i} = \eta_i(k)\frac{\partial \log(q_i(y_i))}{\partial r_i} \tag{6.78}$$

$$\cong \eta_i(k)\frac{0.1 r_i(k) + |y_i(k)|^{r_i(k)}(1 - \log(|y_i(k)|^{r_i(k)})}{r_i^2(k)}. \tag{6.79}$$

It is interesting to note that, in the special case of spiky signals for $r_i = 0$, we obtain as the optimal function a "linear" time-variable function proposed by Matsuoka *et al.* [828] as

$$f_i(y_i) = \frac{y_i}{\langle |y_i|^2 \rangle} = \frac{y_i(k)}{\widehat{\sigma}_i^2(k)}, \tag{6.80}$$

Analogously, for $r_i = 1$ (Laplace distribution), we obtain

$$f_i(y_i) = \frac{y_i}{\langle |y_i| \rangle} = \frac{y_i(k)}{\widehat{\sigma}_i(k)}, \tag{6.81}$$

and for large $r_i \gg 1$, say $r_i = 10$ (approximately uniform distribution)

$$f_i(y_i) = y_i(k)\left\langle |y_i(k)|^8 \right\rangle = y_i(k)\widehat{\sigma}_i^8(k), \tag{6.82}$$

Remark 6.4 *It should be noted that such an activation function satisfies the conditions*

$$\langle f_i(y_i) y_i \rangle = 1, \quad \forall i \tag{6.83}$$

independently of the non-zero variance of the output signals. Hence, the normalization of scales of the output signals in the learning algorithm is automatically taken into account (cf. nonholonomic constraints [29]).

More generally, for a mixture of sub- and super-Gaussian signals distorted by impulsive noise (with large outliers), using this model and Amari's natural gradient approach, we may use a learning algorithm of the form

$$\Delta \mathbf{W}(k) = \eta \mathbf{F}(\mathbf{y}) \mathbf{W}(k), \tag{6.84}$$

with the elements of matrix $\mathbf{F}(\mathbf{y})$ defined by

$$f_{ij} = \begin{cases} -f_i(y_i) y_j, & \text{for } \gamma_{ij} < 1 \\ -y_i f_j(y_j), & \text{otherwise,} \end{cases} \tag{6.85}$$

where $\gamma_{ij} = \sigma_i^2 \sigma_j^2 E\{f_i'(y_i)\} E\{f_j'(y_j)\}$ and $f_i(y_i) = y_i/(|y_i|^{2-r_i} + \epsilon)$ with r_i between zero and one (see Appendix C).

Alternatively, after some simplifications, we can use the learning rule (6.84) with diagonal matrix $\mathbf{\Lambda}(k) = \mathrm{diag}\{\mathbf{f}(\mathbf{y}(k))\boldsymbol{g}^T(\mathbf{y}(k))\}$ and with adaptive time-variable activation functions which are robust respect to outliers [270, 269]

$$f_i(y_i) = \begin{cases} \frac{y_i}{\widehat{\sigma}_i^{2-r_i}} & \text{for } \kappa_4(y_i) > \delta_0, \\ y_i & \text{otherwise,} \end{cases} \tag{6.86}$$

$$g_i(y_i) = \begin{cases} y_i & \text{for } \kappa_4(y_i) > -\delta_0 \\ \frac{y_i}{\widehat{\sigma}_i^{2-r_i}} & \text{otherwise,} \end{cases} \tag{6.87}$$

where $\kappa_4(y_i) = E\{y_i^4\}/E^2\{y_i^2\} - 3$ is the normalized value of kurtosis and $\delta_0 \geq 0$ is a small threshold. The learning algorithm (6.84), with (6.86)–(6.87) monitors and estimates the statistics of each output signal, depending on the sign or value of its normalized kurtosis (which is a measure of distance from the Gaussianity). It then automatically selects (or switches) suitable nonlinear activation functions, such that successful (stable) separation of all the non-Gaussian source signals is possible. In this approach, the activation functions are adaptive time-varying nonlinearities.

Similar methods can be applied for other parameterized distributions. For example, for the generalized Cauchy distribution defined in terms of three parameters $r_i > 0$, $v_i > 0$ and σ_i^2 (see Fig. 6.4 (a) and (b))

$$q_i(y_i) = \frac{B(r_i, v_i)}{\left(1 + \frac{1}{v_i}\left[\frac{y_i}{A(r_i)}\right]^{r_i}\right)^{v_i + 1/r_i}}, \tag{6.88}$$

with $A(r_i) = [\sigma_i^2 \Gamma(1/r_i)/\Gamma(3/r_i)]^{1/2}$ and $B(r_i, v_i) = r_i v_i^{-1/r_i} \Gamma(r_i + 1/r_i)/2A(r_i)\Gamma(v_i)\Gamma(1/r_i)$, we have the following activation function (see Fig. 6.4)

$$f_i(y_i) = \frac{(v_i r_i + 1)}{(v_i |A(r_i)|^{r_i} + |y_i|^{r_i})} |y_i|^{r_i - 1} \mathrm{sign}(y_i). \tag{6.89}$$

Similarly, for the generalized Rayleigh distribution, one obtains $f_i(y_i) = |y_i|^{r_i - 2} y_i$ for complex-valued signals and coefficients.

6.4.1 Moments of the Generalized Gaussian Distribution

In order to fully understand the generalized Gaussian distribution, it is useful to look at its moments (especially the 2nd and 4th moments which give the kurtosis). The n-th moment of the generalized Gaussian distribution is given by

$$m_n = \int_{-\infty}^{\infty} y^n p(y; r) dy. \tag{6.90}$$

If n is odd, the integrand is the product of an even and an odd function over the whole real line, which integrates to zero. In particular, this implies that the mean of the distribution

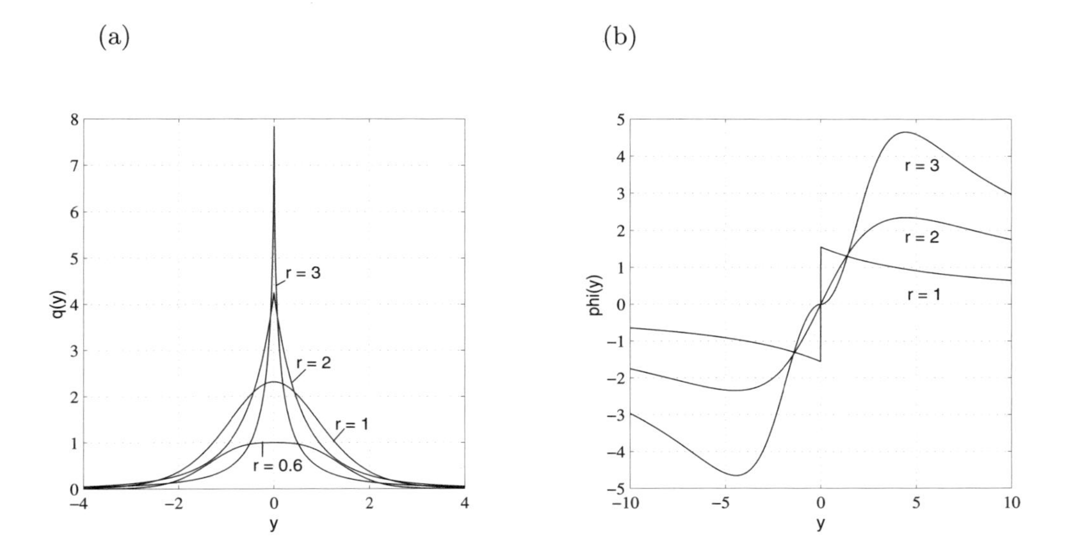

Fig. 6.4 (a) Plot of the generalized Cauchy pdf for various values of the parameter r (with $\sigma^2 = 1$) and (b) corresponding nonlinear activation functions.

given in (6.72) is zero and it is symmetric about its mean (which means its skewness is zero).

The even moments, on the other hand, completely characterize the distribution. In computing these moments, we use the following integral formula [226]

$$\int_0^{\infty} y^{\nu-1} e^{-\mu y^a} dy = \frac{1}{a} \mu^{-\frac{1}{\nu}} \Gamma\left(\frac{\nu}{a}\right). \tag{6.91}$$

The 2nd moment of the generalized Gaussian distribution is determined by

$$\begin{aligned} m_2 &= \int_{-\infty}^{\infty} y^2 p(y; r) dy \\ &= 2 \int_0^{\infty} y^2 \frac{r}{2\sigma\Gamma\left(\frac{1}{r}\right)} e^{-\left|\frac{y}{\sigma}\right|^r} dy. \end{aligned} \tag{6.92}$$

Since the integration is only over the positive values of y, we can remove the absolute value in the exponent. Thus

$$m_2 = \frac{r}{\sigma\Gamma\left(\frac{1}{r}\right)} \int_0^{\infty} y^2 e^{-\left(\frac{y}{\sigma}\right)^r} dy. \tag{6.93}$$

Making the substitution $z = \frac{y}{\sigma}$ $(dy = \sigma dz)$, we find

$$m_2 = \frac{r\sigma^2}{\Gamma\left(\frac{1}{r}\right)} \int_0^{\infty} z^2 e^{-z^r} dz. \tag{6.94}$$

Invoking the integral formula (6.91), we have

$$m_2 = \sigma^2 \frac{\Gamma\left(\frac{3}{r}\right)}{\Gamma\left(\frac{1}{r}\right)}. \tag{6.95}$$

In a similar way, we can find the 4th moment given by

$$m_4 = \sigma^4 \frac{\Gamma\left(\frac{5}{r}\right)}{\Gamma\left(\frac{1}{r}\right)}. \tag{6.96}$$

In general, the $(2k)$-th moment is given by

$$m_{2k} = \sigma^{2k} \frac{\Gamma\left(\frac{2k+1}{r}\right)}{\Gamma\left(\frac{1}{r}\right)}. \tag{6.97}$$

6.4.2 Kurtosis and Gaussian Exponent

The kurtosis is a dimensionless quantity. It measures the relative peakedness or flatness of a distribution. A distribution with positive kurtosis is termed *leptokurtic* (super-Gaussian). A distribution with negative kurtosis is termed *platykurtic* (sub-Gaussian). The normalized kurtosis of a distribution is defined in terms of the 2nd- and 4th-order moments as

$$\kappa(y) = \frac{m_4}{m_2^2} - 3, \tag{6.98}$$

where the constant term -3 makes the value zero for the standard normal distribution.

For a generalized Gaussian distribution, the kurtosis can be expressed in terms of the Gaussian exponent, given by

$$\kappa_4 = \frac{\Gamma\left(\frac{5}{r}\right)\Gamma\left(\frac{1}{r}\right)}{\Gamma^2\left(\frac{3}{r}\right)} - 3. \tag{6.99}$$

Plots of kurtosis κ_4 versus the Gaussian exponent r for leptokurtic (super-Gaussian) and platykurtic (sub-Gaussian) signals are shown in Figure 6.5.

6.4.3 The Flexible ICA Algorithm

From the parameterized generalized Gaussian density model, the nonlinear function in the algorithm (6.44) is given by [226, 270, 269]

$$f_i(y_i) = -\frac{d \log q_i(y_i)}{dy_i} = |y_i|^{r_i - 1} \operatorname{sign}(y_i), \tag{6.100}$$

where $\operatorname{sign}(y_i)$ is the sign function of y_i.

Note that for $r_i = 1$, $f_i(y_i)$ in (6.100) becomes a sign function (which can also be derived from the Laplacian density model for sources). The sign nonlinear function is favorable for the separation of speech signals since certain natural speech sources are often modelled as

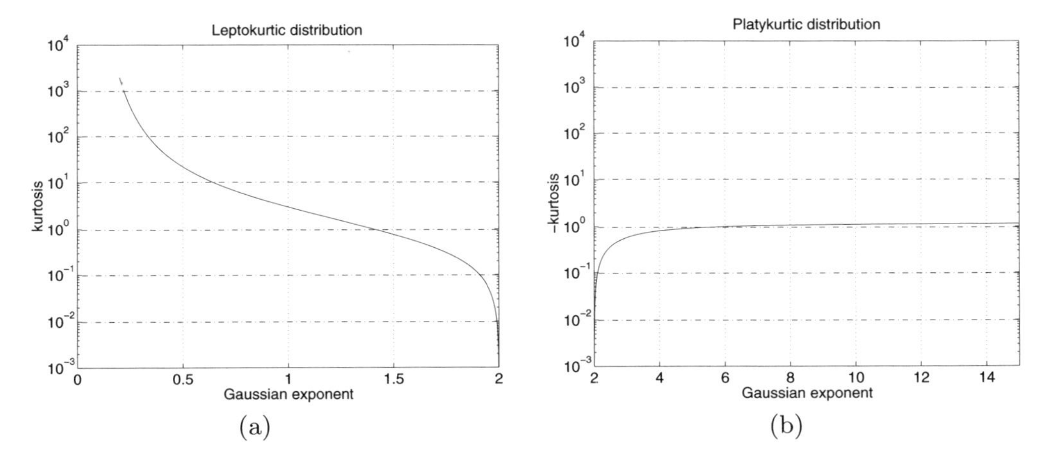

Fig. 6.5 Plots of kurtosis $\kappa_4(r)$ versus Gaussian exponent r: (a) for a leptokurtic signal; (b) for a platykurtic signal [226].

Laplacian distribution. Note also that for $r_i = 4$, $f_i(y_i)$ in (6.100) becomes a cubic function, which is known to be a good choice for sub-Gaussian sources.

In order to select a proper value for the Gaussian exponent r_i, we estimate the kurtosis of the output signal y_i and select the corresponding r_i from the relationship in Figure 6.5. The kurtosis of y_i, κ_i, can be estimated via the following iterative algorithm:

$$\kappa_i^{(k)} = \frac{m_{4i}^{(k)}}{[m_{2i}^{(k)}]^2} - 3, \tag{6.101}$$

where

$$m_{4i}^{(k)} = (1 - \eta_0) m_{4i}^{(k-1)} + \eta_0 |y_i(k)|^4, \tag{6.102}$$

$$m_{2i}^{(k)} = (1 - \eta_0) m_{2i}^{(k-1)} + \eta_0 |y_i(k)|^2, \tag{6.103}$$

and η_0 is a small constant, e.g., 0.01.

In the general case, the estimated kurtosis of the demixing filter output does not exactly match the kurtosis of the original source. However, it provides an idea whether the estimated source is a sub-Gaussian or super-Gaussian signal. Moreover, it was shown [142, 20] that the performance of source separation is not degraded even if the hypothesized density does not match the true density. From these reasons, we suggest a practical method using several different forms of nonlinear functions [226].

The kurtosis of a platykurtic source does not change much as the Gaussian exponent varies (see Figure 6.5 (b)), so we use $r_i = 4$ if the estimated kurtosis of y_i is negative. The cubic nonlinearity for the sub-Gaussian source is also involved with the kurtosis minimization method [148]. For the leptokurtic source, one can see that kurtosis varies much according to the Gaussian exponent (see Figure 6.5 (a)). Thus, we suggest using several different values of r_i, in contrast to the case of a sub-Gaussian source. From our experience, two or three different values of the Gaussian exponent are enough to handle various super-Gaussian sources.

Several different nonlinear functions were suggested in the flexible ICA algorithm [226]. Here, we discuss the stability of stationary points of the algorithm (6.44) for three different cases: (1) $r_i = 4$ for $\kappa_i < 0$; (2) $r_i = 1$; (3) $r_i = .8$ for $\kappa_i > 0$ [226].

Case 1: $r_i = 4$ - **sub-Gaussian distributions**.

The choice of $r_i = 4$ was suggested for the sub-Gaussian source ($\kappa_i < 0$). The choice of $r_i = 4$ results in the cubic nonlinear function, i.e., $f_i(y_i) = |y_i|^2 y_i$. With this selection, one can easily see that the left-hand side of (6.55) is the kurtosis of y_i multiplied by -1. Since y_i is sub-Gaussian, the condition (6.55) is satisfied.

Case 2: $r_i = 1$ - **Laplacian distribution**.

With the choice of $r_i = 1$, the generalized Gaussian density (6.72) becomes the Laplacian density, i.e.,

$$q_i(y_i) = \frac{1}{2\sigma_i} e^{-|\frac{y_i}{\sigma_i}|}. \tag{6.104}$$

The choice of $r_i = 1$ leads to the sign function (hard limiter), i.e.,

$$f_i(y_i) = \text{sign}(y_i) = \frac{y_i}{|y_i|}. \tag{6.105}$$

In order to calculate the derivative of the sign function, we model it as the sum of two unit step functions, i.e.,

$$\text{sign}(y_i) = u(y_i) - u(-y_i), \tag{6.106}$$

where $u(y_i)$ is the unit step function. Then, we can calculate the derivative, $f_i'(y_i)$

$$f_i'(y_i) = 2\delta(y_i), \tag{6.107}$$

where δ denotes Dirac's delta function. We compute $E\{f_i'(y_i)\}$

$$E\{f_i'(y_i)\} = \int_{-\infty}^{\infty} 2\delta(y_i) \frac{1}{2\sigma_i} e^{-|\frac{y_i}{\sigma_i}|} dy_i = \frac{1}{\sigma_i}. \tag{6.108}$$

We also compute $E\{f_i(y_i)y_i\}$

$$E\{f_i(y_i)y_i\} = E\{|y_i|\} = \sigma_i. \tag{6.109}$$

The normalized constraint, $E\{y_i^2\} = 1$ gives

$$E\{y_i^2\} = 2\sigma_i^2 = 1. \tag{6.110}$$

Then, we have $\sigma_i = \sqrt{\frac{1}{2}}$. Note that χ_i is given by

$$\chi_i = \frac{1 - \sigma_i^2}{\sigma_i}. \tag{6.111}$$

Since $\lambda_i = \sqrt{\frac{1}{2}}$, χ_i is positive for $\kappa_i > 0$.

Case 3: $r_i < 1$ **- sparse distribution**.

For highly peaky sources ($\kappa_i >> 1$), it might be desirable to choose the value of r_i less than 1. This gives a non-increasing nonlinear function. With this choice, the nonlinear function is singular around the origin. Thus, in practical application, for $y_i \in [-\epsilon, \epsilon]$ where ϵ is a very small positive number, the corresponding nonlinear function is restricted to have constant values.

The variance of y_i for the generalized Gaussian distribution is given by [226]

$$E\{y_i^2\} = \sigma_i^2 \frac{\Gamma\left(\frac{3}{r_i}\right)}{\Gamma\left(\frac{1}{r_i}\right)}. \tag{6.112}$$

From the normalization constraint, $E\{y_i^2\} = 1$, σ_i has the following value,

$$\sigma_i = \sqrt{\frac{\Gamma\left(\frac{1}{r_i}\right)}{\Gamma\left(\frac{3}{r_i}\right)}}. \tag{6.113}$$

Besides the region for $y_i \in [-\epsilon, \epsilon]$, we can compute $E\{f_i'(y_i)\}$ and $E\{f_i(y_i)y_i\}$ as follows

$$\begin{aligned} E\{f_i'(y_i)\} &= \int_{-\infty}^{\infty} (r_i - 2)|y_i|^{(r_i-2)} \frac{r_i}{2\sigma_i \Gamma\left(\frac{1}{r_i}\right)} e^{-\frac{|y_i|^{r_i}}{\sigma_i^{r_i}}} dy_i \\ &= \frac{(r_i - 2)\sigma_i^{r_i-2}}{\Gamma\left(\frac{1}{r_i}\right)} \Gamma\left(\frac{r_i - 1}{r_i}\right), \\ E\{f_i(y_i)y_i\} &= \int_{-\infty}^{\infty} y_i |y_i|^{(r_i-1)} \operatorname{sign}(y_i) \frac{r_i}{2\sigma_i \Gamma\left(\frac{1}{r_i}\right)} e^{-\frac{|y_i|^{r_i}}{\sigma_i^{r_i}}} dy_i \\ &= \frac{r_i \sigma_i^{r_i+1}}{\Gamma\left(\frac{1}{r_i}\right)} \frac{1}{r_i} \Gamma\left(\frac{r_i + 1}{r_i}\right). \end{aligned} \tag{6.114}$$

Note that the gamma function $\Gamma(x)$ has many singular points especially for $x < 0$. Thus, special care is required with the choice of $r_i < 1$. For instance, the choice of $r_i = 0.5$ does not satisfy the condition (6.55) because $\Gamma(-1) = \infty$. For the case of $r_i = 0.8$, one can easily see that the stability condition (6.55) is satisfied.

Summarizing, in this section we derived and analyzed unsupervised adaptive on-line and micro batch algorithms for blind separation of sources (BSS), especially when the source signals are nonstationary and have spiky (impulsive) behavior. The algorithms are applicable to mixtures of an unknown number of independent source signals with unknown statistics [270, 269]. Nonlinear activation functions are rigorously derived by assuming that the sources are modelled by generalized Gaussian distributions. As a special case, we derived and justified the time variable "linear" activation function

$$\boxed{f(y_i) = \frac{y_i(k)}{\langle y_i^2 \rangle} = \frac{y_i(k)}{\hat{\sigma}_{y_i}^2}} \tag{6.115}$$

proposed by Matsuoka *et al.* [828] for the blind separation of nonstationary signals. Applying the moving average estimation to the output variance $\widehat{\sigma}_i^2 = \langle y_i^2 \rangle$, we may use the same concept even for stationary signals since the estimation of variance continuously fluctuates in time. Extensive computer simulations have confirmed that the proposed algorithms are able to separate spiky and non-stationary sources (such as biomedical signals, especially magnetoencephalographic (MEG) signals) [270, 269, 120, 625, 29].

6.4.4 Pearson System

The Generalized Gaussian and Cauchy pdf models discussed in the previous section belong to symmetric distribution families. The natural gradient algorithm based on these models may fail to successfully separate independent sources with strongly asymmetric (skewed) distributions or non Gaussian sources with close to zero kurtosis.

A wide class of both symmetric and asymmetric distributions can be modelled by the Pearson system described by the differential equation [678]

$$q'(y) = \frac{(y-a)\, q(y)}{b_0 + b_1\, y + b_2\, y^2}, \tag{6.116}$$

where a, b_0, b_1 and b_2 are the parameters depending on the distribution of the estimated sources. The optimal activation function according to Eq. (6.17) for the Pearson system can be expressed as

$$f(y) = -\frac{q'(y)}{q(y)} = \frac{a-y}{b_0 + b_1\, y + b_2\, y^2}. \tag{6.117}$$

Many widely used distributions, including beta, gamma, normal and Student's t distributions are special forms of the Pearson system. The parameters a, b_0, b_1 and b_2 can be estimated directly by the method of moments [678] as

$$a = b_1 = -\frac{m_3(m_4 + 3m_2^2)}{d}, \tag{6.118}$$

$$b_0 = -\frac{m_2(4m_2 m_4 - 3m_3^2)}{d}, \tag{6.119}$$

$$b_2 = -\frac{(2m_2 m_4 - 3m_3^2 - 6m_2^3)}{d}, \tag{6.120}$$

where $d = 10m_2m_4 - 18m_3^3 - 12m_3^2$ and m_2, m_3, m_4 denote the second, third and fourth order sample moments of y. The advantage of using the Pearson system is that it separates sources with unsymmetrical distributions.

6.5 NATURAL GRADIENT ALGORITHMS FOR NON-STATIONARY SOURCES

In many applications, e.g., speech or biomedical applications, signals are non-stationary. The objective of this section is to derive natural gradient learning algorithms for non-stationary sources (in the sense that their variances are time varying).

The key assumption in source separation lies in the statistical independence of sources. When sources are mutually independent and are also temporally i.i.d. non-Gaussian signals, it is necessary to use higher-order statistics to achieve source separation. In such a case, source separation is related to independent component analysis (ICA), where the goal is to decompose multivariate data into a linear sum of non-orthogonal basis vectors with basis coefficients being statistically independent. Second-order statistics are sufficient for blind source separation when the sources are statistically nonstationary. Methods based on this observation include algorithms described in Chapter 4. The nonstationarity of sources was first exploited by Matsuoka *et al.* in the context of blind source separation [828].

6.5.1 Model Assumptions

Throughout this section, as in [828], the following assumptions are made:

AS1 The mixing matrix $\mathbf{H}$ has full column rank.

AS2 The source signals $\{s_i(k)\}$ are statistically independent with zero-mean. This implies that the covariance matrix of the source signal vector, $\mathbf{R_{ss}} = E\{\mathbf{s}(k)\mathbf{s}^T(k)\}$ is a diagonal matrix, i.e.,

$$\mathbf{R_{ss}} = \text{diag}\{\sigma_{s_1}^2(k), \sigma_{s_2}^2(k), \ldots, \sigma_{s_n}^2(k)\}, \tag{6.121}$$

where $\sigma_{s_i}^2(k) = E\{s_i^2(k)\}$ is the variance of the source signal s_i.

AS3 The ratio $\hat{\sigma}_{s_i}^2(k)/\hat{\sigma}_{s_j}^2(k)$ $(i, j = 1, \ldots, n$ and $i \neq j)$ are not constant with time[3].

We have to point out that the first two assumptions (AS1, AS2) are common in most existing approaches to source separation, however, the third assumption (AS3) is essential for algorithms presented in this section. In fact, the third assumption allows us to separate mixtures by using only second-order statistics.

[3]The variance of a signal $y_i(k)$ is usually estimated by taking the moving average (MA) as $\hat{\sigma}_{y_i}^2(k) = (1-\eta_0)\hat{\sigma}_{y_i}^2(k-1) + \eta_0 y_i^2(k)$.

6.5.2 Second Order Statistics Cost Function

A standard cost function for ICA is based on mutual information which requires the knowledge of the underlying distributions of sources. Since probability distributions of the sources are not known in advance, most ICA algorithms rely on hypothesized distributions. Higher-order statistics should be incorporated either explicitly or implicitly.

To eliminate all cross-correlations $E\{y_i(k)y_j(k)\}$, the following cost function was proposed by Matsuoka *et al.* in [828]

$$J(\mathbf{W}, k) = \frac{1}{2}\left(\sum_{i=1}^{n} \log(\hat{\sigma}_{y_i}^2(k)) - \log\det \widehat{\mathbf{R}}_{\mathbf{yy}}^{(k)}\right), \tag{6.122}$$

where $\hat{\sigma}_{y_i}^2(k) = \langle y_i^2(k) \rangle$ is the on-line estimate of time-varying variance, $\det(\cdot)$ denotes the determinant of a matrix and $\widehat{\mathbf{R}}_{\mathbf{yy}}^{(k)} = \langle \mathbf{y}(k)\, \mathbf{y}^T(k) \rangle$. The covariance matrix $\mathbf{R}_{\mathbf{yy}}^{(k)}$ is usually estimated on-line as $\widehat{\mathbf{R}}_{\mathbf{yy}}^{(k)} = (1-\eta_0)\widehat{\mathbf{R}}_{\mathbf{yy}}^{(k-1)} + \eta_0 \mathbf{y}(k)\mathbf{y}^T(k)$. The cost function given in (6.122) is a non-negative function which reaches minima if and only if $E\{y_i(k)y_j(k)\} = 0$, for $i, j = 1, \ldots, n,\ i \neq j$. This is a direct consequence of Hadamard's inequality (applied for the on-line estimation of the time variable covariance matrix $\widehat{\mathbf{R}}_{\mathbf{yy}}^{(k)}$) which is summarized below [228].

Theorem 6.1 (Hadamard's Inequality) *Suppose* $\mathbf{R} = [r_{ij}]$ *is a non-negative definite symmetric* $n \times n$ *matrix. Then,*

$$\det(\mathbf{R}) \leq \prod_{i=1}^{n} r_{ii}, \tag{6.123}$$

with equality if and only if $r_{ij} = 0$, *for* $i \neq j$.

6.5.3 Derivation of Natural Gradient Learning Algorithms

Feed-forward Network

We consider a simple feed-forward network as shown in Figure 6.6 (a) for the source separation task. The output of the network, $\mathbf{y}(k)$ is given by

$$\mathbf{y}(k) = \mathbf{W}\mathbf{x}(k) \tag{6.124}$$

We calculate the total differential $dJ(\mathbf{W})$,

$$\begin{aligned} dJ(\mathbf{W}) &= J(\mathbf{W} + d\mathbf{W}) - J(\mathbf{W}) \\ &= \frac{1}{2} d\left\{\sum_{i=1}^{n} \log \langle y_i^2(k) \rangle \right\} - \frac{1}{2} d\left\{\log\det \langle \mathbf{y}(k)\mathbf{y}^T(k) \rangle \right\}, \end{aligned} \tag{6.125}$$

due to the change $d\mathbf{W}$.

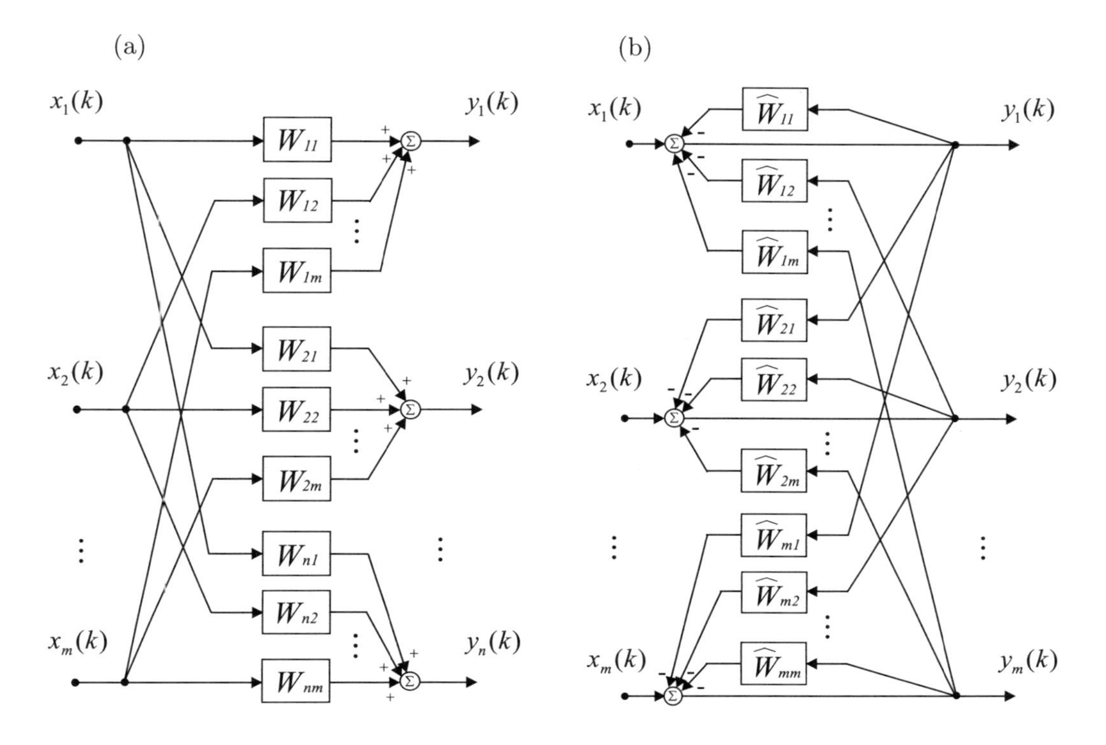

Fig. 6.6 (a) Architecture of a feed-forward neural network. (b) Architecture of a fully connected recurrent neural network.

The second term in (6.125) can be evaluated as

$$\begin{aligned} d\left\{\log\det\left\langle \mathbf{y}(k)\mathbf{y}^T(k)\right\rangle\right\} &= d\left\{\log\det(\mathbf{W}\widehat{\mathbf{R}}_{\mathbf{x}\,\mathbf{x}}\mathbf{W}^T)\right\} \\ &= 2d\left\{\log\det\left(\mathbf{W}\right)^{-1}\right\} + d\left\{\log\det\widehat{\mathbf{R}}_{\mathbf{xx}}\right\} \\ &= 2\mathrm{tr}\{d\mathbf{W}\,\mathbf{W}^{-1}\} + d\left\{\log\det\widehat{\mathbf{R}}_{\mathbf{xx}}\right\}. \end{aligned} \tag{6.126}$$

Note that the term $\widehat{\mathbf{R}}_{\mathbf{xx}}$ does not depend on the weight matrix $\mathbf{W}$, so it can be eliminated.

Define a modified differential matrix $d\mathbf{X}$, under the assumption that the separating matrix $\mathbf{W} \in \mathbb{R}^{n\times n}$ (with $m = n$) is nonsingular, as

$$d\mathbf{X} = d\mathbf{W}\,\mathbf{W}^{-1}. \tag{6.127}$$

Then,

$$d\left\{\log\det\left\langle \mathbf{y}(k)\mathbf{y}^T(k)\right\rangle\right\} = 2\mathrm{tr}\{d\mathbf{X}\}. \tag{6.128}$$

Similarly, we can evaluate the first term of (6.125) as

$$\begin{aligned} d\left\{\sum_{i=1}^{n} \log\left\langle y_i^2(k)\right\rangle\right\} &= \sum_{i=1}^{n} \frac{2\left\langle y_i(k) dy_i(k)\right\rangle}{\left\langle y_i^2(k)\right\rangle} \\ &= 2E\{\mathbf{y}^T(k)\boldsymbol{\Lambda}^{-1}(k) d\mathbf{y}(k)\} \\ &= 2E\{\mathbf{y}^T(k)\boldsymbol{\Lambda}^{-1}(k) d\mathbf{X}\mathbf{y}(k)\}, \end{aligned} \tag{6.129}$$

where $\boldsymbol{\Lambda}(k)$ is a diagonal matrix whose i-th diagonal element is $\lambda_i = \hat{\sigma}_{y_i}^2 = \left\langle y_i^2(k)\right\rangle$.

In terms of $d\mathbf{X}$, we have

$$\frac{dJ(\mathbf{W})}{d\mathbf{X}} = E\left\{\boldsymbol{\Lambda}^{-1}(k)\mathbf{y}(k)\mathbf{y}^T(k)\right\} - \mathbf{I}. \tag{6.130}$$

Taking into account that $\Delta\mathbf{W}(k) = \Delta\mathbf{X}(k)\,\mathbf{W}(k)$ and applying the stochastic gradient descent method, we obtain the on-line learning algorithm:

$$\begin{aligned} \Delta\mathbf{W}(k) &= -\eta(k)\frac{dJ(\mathbf{W})}{d\mathbf{X}}\mathbf{W}(k) \\ &= \eta(k)\left[\mathbf{I} - \boldsymbol{\Lambda}^{-1}(k)\mathbf{y}(k)\mathbf{y}^T(k)\right]\mathbf{W}(k) \end{aligned} \tag{6.131}$$

or equivalently

$$\mathbf{W}(k+1) = \mathbf{W}(k) + \tilde{\eta}(k)\left[\boldsymbol{\Lambda}(k) - \mathbf{y}(k)\mathbf{y}^T(k)\right]\mathbf{W}(k), \tag{6.132}$$

where $\tilde{\boldsymbol{\eta}}(k) = \eta(k)\boldsymbol{\Lambda}^{-1}(k)$.

It should be noted that the algorithm (6.132) derived above can be viewed as a special form of the natural gradient ICA algorithms derived in the previous sections.

We can reformulate the algorithm (6.132) using the moving average approach as

$$\boxed{\mathbf{W}(k+1) = \mathbf{W}(k) + \tilde{\eta}(k)\left[\boldsymbol{\Lambda}(k) - \widehat{\mathbf{R}}_{\mathbf{yy}}^{(k)}\right]\mathbf{W}(k),} \tag{6.133}$$

where $\boldsymbol{\Lambda}$ is the diagonal matrix whose i-th diagonal element is $\lambda_i \approx E\{y_i^2\}$ that can be estimated on-line by

$$\lambda_i(k) = (1-\eta_0)\lambda_i(k-1) + \eta_0 y_i^2(k), \tag{6.134}$$

and

$$\widehat{\mathbf{R}}_{\mathbf{yy}}^{(k)} = (1-\eta_0)\widehat{\mathbf{R}}_{\mathbf{yy}}^{(k-1)} + \eta_0\mathbf{y}(k)\mathbf{y}^T(k). \tag{6.135}$$

The algorithm (6.133) leads to a special form of nonholonomic ICA algorithm described in the previous sections with variable step sizes $\tilde{\boldsymbol{\eta}}(k) = \eta(k)\boldsymbol{\Lambda}^{-1}(k)$ for nonstationary sources [25, 228].

It should be noted that the learning algorithms presented in this section are always locally stable and do not depend on the probability distribution of sources [228].

Remark 6.5 *In order to successfully separate nonstationary sources the learning rate* $0 < \eta_0 \leq 1$ *should be suitably chosen to ensure variability of estimated variance of the output signals (typical value* $\eta_0 = 0.1$*).*

Fully-connected Recurrent Network

Let us consider now a fully-connected recurrent network as shown in Figure 6.6 (with $m = n$) for source separation task. The output of the network, $\mathbf{y}(k)$ is given by

$$\begin{aligned} \mathbf{y}(k) &= \mathbf{x}(k) - \widehat{\mathbf{W}}\mathbf{y}(k) \\ &= (\mathbf{I} + \widehat{\mathbf{W}})^{-1}\mathbf{x}(k) \end{aligned} \tag{6.136}$$

under the assumption that the matrix $(\mathbf{I} + \widehat{\mathbf{W}})$ is nonsingular.

Similarly to the feed-forward model, we can directly derive the learning algorithm for $\widehat{\mathbf{W}}$ that has the form [228]

$$\boxed{\Delta\widehat{\mathbf{W}}(k) = -\tilde{\eta}(k)\left[\mathbf{I} + \widehat{\mathbf{W}}(k)\right]\left[\mathbf{\Lambda}(k) - \widehat{\mathbf{R}}_{\mathbf{yy}}^{(k)}\right].} \tag{6.137}$$

Appendix A. Derivation of Local Stability Conditions for the Natural Gradient ICA Algorithm (6.19)

By stability conditions, we mean conditions for which a learning algorithm converges to a suitable equilibrium point corresponding to the correct separation of independent sources. In this section, we consider only local stability conditions.

The learning equation (6.19) is a stochastic difference equation depending on random inputs $\mathbf{y}(k)$. To analyze its behavior, we consider the ensemble average of the equation (see Eq.(6.27)) which is the approximation of the following differential equation (with continuous time t),

$$\frac{d\mathbf{W}}{dt} = \mu(t)E\left\{\mathbf{I} - \mathbf{f}(\mathbf{y}(t))\,\mathbf{y}^T(t)\right\}\mathbf{W}(t), \tag{A.1}$$

We will find that the true separating solution $\mathbf{W}$ with which y_i and y_j are independent is an equilibrium solution of the averaged equation, because, when y_i and y_j are independent, the off-diagonal term of the equilibrium is

$$E\{f_i(y_i)y_j\} = 0, \qquad i \neq j. \tag{A.2}$$

This condition is satisfied when y_i and y_j are independent. The diagonal term is

$$E\{f_i(y_i)y_i\} = 1, \tag{A.3}$$

which determines the scaling of the recovered signals.

The stability of the equilibrium is analyzed by the variational equation

$$\frac{d}{dt}\delta\mathbf{W}(t) = \eta(t)\frac{\partial E\left\{\mathbf{I} - \mathbf{f}\left[\mathbf{y}(t)\right]\mathbf{y}^T(t)\right\}}{\partial \mathbf{W}}\delta\mathbf{W}(t), \tag{A.4}$$

which shows the dynamic behavior of small perturbations $\delta\mathbf{W}(t)$ in the neighborhood of the true solution $\mathbf{W}$ and $(\partial/\partial\mathbf{W})\delta\mathbf{W}$ implies $\sum(\partial/\partial w_{ij})\delta w_{ij}$ in the component form. In order to show stability of the algorithm, we need to check the eigenvalues of the expectation of the extended Hessian matrix

$$-E\left\{\frac{\partial^2\rho(\mathbf{y},\mathbf{W})}{\partial\mathbf{W}\partial\mathbf{W}^T}\mathbf{W}^T\mathbf{W}\right\} \tag{A.5}$$

at the equilibrium point $\mathbf{W}_*$. When all the real parts of the eigenvalues of the above quantity are negative, the solution is stable. We present the analysis in terms of differential calculus.

In order to establish stability conditions, we need to evaluate all the eigenvalues of the operator. This can be done in terms of $d\mathbf{X}$, as follows. Since $\mathbf{I}-\mathbf{f}(\mathbf{y})\,\mathbf{y}^T$ is derived from the gradient of ρ, we shall calculate its second order differential which is the quadratic form (Hessian)

$$d^2l = \sum \frac{\partial^2\rho(\mathbf{y},\mathbf{W})}{\partial w_{ij}\partial w_{kl}} dw_{ij}dw_{kl}$$

in terms of $d\mathbf{X}$. The equilibrium is stable if and only if the expectation of the above quadratic form is positive definite. We calculate the second total differential, which is the quadratic form of the Hessian of ρ, as [24]

$$\begin{aligned} d^2l &= \mathbf{y}^T d\mathbf{X}^T\,\mathbf{f}'(\mathbf{y})\,d\mathbf{y} + \mathbf{f}^T(\mathbf{y})\,d\mathbf{X}d\mathbf{y} \\ &= \mathbf{y}^T d\mathbf{X}^T\mathbf{f}'(\mathbf{y})d\mathbf{X}\mathbf{y} + \mathbf{f}(\mathbf{y}^T)\,d\mathbf{X}d\mathbf{X}\mathbf{y}, \end{aligned} \tag{A.6}$$

where $\mathbf{f}'(\mathbf{y})$ is the diagonal matrix whose diagonal elements are $f'(y_i)$. The expectation of the first term is

$$\begin{aligned} &E\{\mathbf{y}^T d\mathbf{X}^T\mathbf{f}'(\mathbf{y})\,d\mathbf{X}\mathbf{y}\} = \sum E\{y_i dx_{ji} f_j'(y_j) dx_{jk} y_k\} \\ &= \sum_{j\neq i} E\{(y_i)^2\}E\{f_j'(y_j)\}(dx_{ji})^2 + \sum_i E\{(y_i)^2 f_i'(y_i)\}(dx_{ii})^2 \\ &= \sum_{j\neq i} \sigma_i^2\kappa_j(dx_{ji})^2 + \sum_i m_i(dx_{ii})^2, \end{aligned}$$

where $m_i = E\{y_i^2 f_i'(y_i)\}$, $\kappa_i = E\{f_i'(y_i)\}$, $\sigma_i^2 = E\{|y_i|^2\}$, y_i is the source signal extracted at the i-th output, and $f_i'(y) = df_i(y_i)/dy_i$. Here, the expectation is taken at $\mathbf{W} = \mathbf{H}^{-1}$ where the y_i's are independent.
Similarly,

$$\begin{aligned} &E\{\mathbf{f}(\mathbf{y})^T\,d\mathbf{X}d\mathbf{X}\mathbf{y}\} = \sum E\{f_i(y_i)dx_{ij}dx_{jk}y_k\} \\ &= \sum E\{y_i f_i(y_i)\}dx_{ij}dx_{ji} = \sum_{i,j} dx_{ij}dx_{ji}, \end{aligned} \tag{A.7}$$

because of $E\{y_i f_i(y_i)\} = 1$ (the normalization condition). Hence,

$$E\{d^2l\} = \sum_{j\neq i}\{\sigma_i^2\kappa_j(dx_{ji})^2 + dx_{ij}dx_{ji}\} + \sum_i (m_i+1)(dx_{ii})^2. \tag{A.8}$$

For a pair (i, j), $i \neq j$, the summand in the first term is rewritten as

$$k_{ij} = \sigma_i^2 \kappa_j (dx_{ji})^2 + \sigma_j^2 \kappa_i (dx_{ij})^2 + 2dx_{ij}dx_{ji}. \tag{A.9}$$

This k_{ij} $(i \neq j)$ is the quadratic form in (dx_{ij}, dx_{ji}), and

$$E\{d^2 l\} = \sum_{i \neq j} k_{ij} + \sum (m_i + 1)(dx_{ii})^2. \tag{A.10}$$

The matrix (k_{ij}) is positive definite, if and only if the following stability conditions hold:

$$m_i + 1 > 0, \tag{A.11}$$

$$\kappa_i > 0, \tag{A.12}$$

$$\gamma_{ij} = \sigma_i^2 \sigma_j^2 \kappa_i \kappa_j > 1, \quad \text{for all } 1 \leq i < j \leq m. \tag{A.13}$$

In other words, the true solution is a stable equilibrium of the on-line learning algorithm for assumed pdf model $q(\mathbf{y})$ and the corresponding activation functions $\mathbf{f}(\mathbf{y})$, when the above conditions are satisfied. It is easy to show that the conditions are satisfied when $q(\mathbf{y})$ is equal to the true source distribution $p_s(\mathbf{s})$ or close to it.

Appendix B. Derivation of the Learning Rule (6.32) and Stability Conditions for ICA

In order to derive the learning algorithm (6.32), we follow the notation: [24, 45, 270]

$$d\mathbf{X} \equiv d\mathbf{W}\mathbf{W}^{-1} \tag{B.1}$$

and

$$d\rho(\mathbf{y}, \mathbf{W}) = -\operatorname{tr}(d\mathbf{X}) + \boldsymbol{g}^T(\mathbf{y}) d\mathbf{X}\mathbf{y}. \tag{B.2}$$

The standard stochastic gradient method for $\mathbf{X}$ leads to the natural gradient learning algorithm for updating $\mathbf{W}$. Let us apply a different update rule as follows

$$\Delta \mathbf{X} = -\eta \left(\frac{\partial \rho}{\partial \mathbf{X}} \right)^T = \eta \left[\mathbf{I} - \mathbf{y} \boldsymbol{g}^T(\mathbf{y}) \right]. \tag{B.3}$$

On the other hand, we have

$$\frac{\partial \rho(\mathbf{y}, \mathbf{W})}{\partial \mathbf{X}} = \frac{\partial \rho(\mathbf{y}, \mathbf{W})}{\partial \mathbf{W}} \mathbf{W}^T. \tag{B.4}$$

Hence,

$$\Delta \mathbf{W} = -\eta \mathbf{W} \left[\frac{\partial \rho(\mathbf{y}, \mathbf{W})}{\partial \mathbf{W}} \right]^T \mathbf{W} \tag{B.5}$$

or explicitly

$$\Delta \mathbf{W}(k) = \eta(k) \left[\mathbf{I} - \mathbf{y}(k) \boldsymbol{g}^T[\mathbf{y}(k)] \right] \mathbf{W}(k). \tag{B.6}$$

It should be noted that in general the update rule does not lead to a gradient descent algorithm. However, we will show now that if the function $g_i(y_i)$ satisfies the following conditions [1351, 1352, 1354]

$$E\{g_i(y_i)y_i\} \approx \langle g_i(y_i)y_i\rangle > 0, \qquad (i = 1, 2, \dots, n) \tag{B.7}$$

the learning algorithm (B.6) is a stochastic gradient descent learning algorithm with the cost function $\rho(\mathbf{y}, \mathbf{W})$, which means that the algorithm converges to one of the local minima of the cost function. It should be noted that all the odd functions $g_i(y_i)$ satisfy the above conditions.

From (B.2), we have

$$\frac{\partial \rho(\mathbf{y}, \mathbf{W})}{\partial x_{ij}} = -\delta_{ij} + y_i g_j(y_j). \tag{B.8}$$

We consider the decrease of the objective function during one step of learning

$$\begin{aligned}
\Delta\rho(\mathbf{y}, \mathbf{W}) &= \rho(\mathbf{y}, \mathbf{W} + \Delta\mathbf{W}) - \rho(\mathbf{y}, \mathbf{W}) \\
&=< \frac{\partial \rho(\mathbf{y}, \mathbf{W})}{\partial \mathbf{X}}, \Delta\mathbf{X}^T > + O(|\Delta\mathbf{X}|^2) \\
&= -\eta \sum_{ij} \frac{\partial \rho}{\partial x_{ij}} \frac{\partial \rho}{\partial x_{ji}} + O(|\Delta\mathbf{X}|^2) \\
&= -\eta \left(\sum_{i=1}^{n} (1 - y_i g_i(y_i))^2 + \sum_{i \neq j} y_i g_j(y_j) y_j g_i(y_i) \right) + O(|\Delta\mathbf{X}|^2).
\end{aligned}$$

If $\langle y_i g_i(y_i)\rangle > 0$, then

$$< \frac{\partial \rho}{\partial \mathbf{X}}, \frac{\partial \rho}{\partial \mathbf{X}} >= \mathrm{tr}\left(\frac{\partial \rho}{\partial \mathbf{X}}, \frac{\partial \rho}{\partial \mathbf{X}^T} \right) > 0. \tag{B.9}$$

This means that if the learning rate η is small enough, then the cost function is decreasing during the learning process until the system achieves a minimum during the learning procedure.

Stability conditions. In order to analyze the stability condition of a separating solution, we take a variation with respect to $\mathbf{W}$ of the continuous-time learning algorithm as

$$\begin{aligned}
\frac{d\delta\mathbf{W}}{dt} &= -\eta\delta E\{\mathbf{y}(t)\boldsymbol{g}^T[\mathbf{y}(t)]\}\mathbf{W} \\
&= -\eta E\{\delta\mathbf{y}\boldsymbol{g}^T(\mathbf{y}) + \mathbf{y}\boldsymbol{g}'(\mathbf{y}^T)\delta\mathbf{y}^T)\}\mathbf{W}.
\end{aligned} \tag{B.10}$$

Substituting $d\mathbf{W} = d\mathbf{X}\mathbf{W}$ and $d\mathbf{y} = \mathbf{dXy}$, we obtain

$$\begin{aligned}
\frac{d\delta\mathbf{X}}{dt} &= -\eta E\{d\mathbf{X}\mathbf{y}\boldsymbol{g}^T(\mathbf{y}) + \mathbf{y}\boldsymbol{g}'(\mathbf{y}^T)(\mathbf{y}^T d\mathbf{X}^T)\} \\
&= -\eta \left(d\mathbf{X} + E\{\mathbf{y}\boldsymbol{g}'(\mathbf{y}^T)(\mathbf{y^T dX^T})\}\right).
\end{aligned} \tag{B.11}$$

Since the output signals y_i are mutually independent, we have at the equilibrium point

$$E\{g_i(y_i)y_iy_j\} = 0, \quad \text{for } i \neq j. \tag{B.12}$$

The components of Eq.(B.11) can be written as

$$\frac{d\delta x_{ii}}{dt} = -\eta\left(1 + E\{g_i'(y_i)y_i^2\}\right)\delta x_{ii}, \qquad (i = 1, 2, \ldots n) \tag{B.13}$$

and for $i \neq j$

$$\frac{d\delta x_{ij}}{dt} = -\eta\left(\delta x_{ij} + E\{y_i^2\}E\{g_j'(y_j)\}\delta x_{ji}\right), \tag{B.14}$$

$$\frac{d\delta x_{ji}}{dt} = -\eta\left(\delta x_{ji} + E\{y_j^2\}E\{g_i'(y_i)\}\delta x_{ji}\right). \tag{B.15}$$

It is straightforward to write the stability conditions in an explicit form:

$$1 + E\{y_i^2 g_i'(y_i)\} > 0, \qquad (i = 1, 2, \ldots, n) \tag{B.16}$$

and

$$\gamma_{ij} = E\{y_i^2\}E\{y_j^2\}E\{g_i'(y_i)\}E\{g_j'(y_j)\} < 1, \qquad (i \neq j). \tag{B.17}$$

From the above stability conditions, we see that one important advantage of the learning rule (6.32) is that this algorithm is still stable in contrast to the algorithm (6.19) even when some of the source signals become silent (decay to zero).

It is observed that the above stability conditions for learning algorithm (6.32) are mutually complementary to the one for Amari's natural gradient learning algorithm (6.19).

Appendix C. Stability of the Generalized Adaptive Learning Algorithm

Let us consider the learning algorithm proposed by Amari *et al.* [24]

$$\Delta\mathbf{W} = \eta\mathbf{F}(\mathbf{y}, \mathbf{W})\mathbf{W}, \tag{C.1}$$

with entries of matrix $\mathbf{F}$ defined by

$$f_{ii}(\mathbf{y}, \mathbf{W}) = \delta_i - y_i f_i(y_i), \qquad (i = 1, 2, \ldots, n). \tag{C.2}$$

and

$$f_{ij}(\mathbf{y}, \mathbf{W}) = \begin{cases} -f_i(y_i)y_j, & if\ \gamma_{ij} > 1, \\ -y_i f_j(y_j), & if\ \gamma_{ij} \leq 1, \end{cases} \tag{C.3}$$

where δ_i are given positive constants and $\gamma_{ij} = \sigma_i^2\sigma_j^2\kappa_i\kappa_j$. The separating solution satisfies the system of nonlinear algebraic equations

$$E\{\mathbf{F}(\mathbf{y}, \mathbf{W})\} = \mathbf{0}. \tag{C.4}$$

In order to establish the stability conditions, we write the above learning algorithm in the continuous-time form as

$$\frac{d\mathbf{W}}{dt} = \eta E\{\mathbf{F}(\mathbf{y}, \mathbf{W})\}\mathbf{W}. \tag{C.5}$$

Taking the variation of $\mathbf{W}$ at an equilibrium point, we have

$$\frac{d\delta\mathbf{W}}{dt} = \eta E\{\delta\mathbf{F}(\mathbf{y}, \mathbf{W})\}\mathbf{W}. \tag{C.6}$$

Hence, for $\gamma_{ij} > 1, i \neq j$, we have

$$\frac{d\delta x_{ij}}{dt} = -\left(E\{f_i'(y_i)\}\sigma_j^2 \delta x_{ij} + \delta x_{ji}\right) \tag{C.7}$$

and

$$\frac{d\delta x_{ji}}{dt} = -\left(\delta x_{ij} + E\{f_i'(y_i)\}\sigma_j^2 \delta x_{ij}\right). \tag{C.8}$$

Similarly, if $\gamma_{ij} < 1$, then the variations δx_{ij} and δx_{ji} satisfy the conditions derived in Appendix D. The above results can be summarized in the form of the following Theorem [24]:

Theorem C.2 *Suppose that $\langle f_i(y_i) y_i \rangle > 0$, $\gamma_{ij} = \sigma_i^2 \sigma_j^2 \kappa_i \kappa_j \neq 1$ (for $i \neq j$), $m_i + 1 > 0$ and $\kappa_i = E\{f_i'(y_i)\} > 0$ $(i = 1, 2, \ldots, n)$, then the separating solution by employing the learning algorithm (C.1) is stable.*

The form of the nonlinear functions $\mathbf{F}$ depends on the parameters $\gamma_{ij} = \sigma_i^2 \sigma_j^2 \kappa_i \kappa_j$, which cannot be explicitly determined, but we can evaluate the parameters γ_{ij} dynamically during the learning process.

As in the natural gradient algorithm (6.19) the learning algorithm (C.1) possesses two important properties. One is the equivariant property. The other one is the non-singularity of the learning matrix $\mathbf{W}(t)$, which can be observed from the following derivation [1316, 1348, 1351]. We define

$$< \mathbf{X}, \mathbf{Y} >= \operatorname{tr}(\mathbf{X}^T \mathbf{Y}) \tag{C.9}$$

and calculate

$$\begin{aligned} \frac{d \det(\mathbf{W}(t))}{dt} &= < \frac{\partial \det(\mathbf{W})}{\partial \mathbf{W}}, \frac{d\mathbf{W}}{dt} > \\ &= < \det(\mathbf{W})\mathbf{W}^{-T}, \eta \mathbf{F}(\mathbf{y}, \mathbf{W})\mathbf{W} > \\ &= \eta \ \operatorname{tr}(\mathbf{F}(\mathbf{y}, \mathbf{W}) \det(\mathbf{W})) \\ &= \eta \sum_{i=1}^{n} [\lambda_i - y_i g_i(y_i)] \det(\mathbf{W}). \end{aligned} \tag{C.10}$$

Then the determinant $\det(\mathbf{W}(t))$ is expressed by

$$\det(\mathbf{W}(t)) = \det(\mathbf{W}(0)) \exp\left(\eta \int_0^t \sum_{i=1}^{n} (\lambda_i - g_i(y_i) y_i) d\tau\right). \tag{C.11}$$

This means that if $\mathbf{W}(0)$ is nonsingular, then $\mathbf{W}(t)$ returns its non-singularity.

Appendix D. Dynamic Properties and Stability of Nonholonomic Natural Gradient Algorithms

Non-Holonomicity and Orthogonality

In order to make the scaling indeterminacies clearer, we introduce an equivalence relation in the n^2 -dimensional space of nonsingular matrices $Gl(n) = \{\mathbf{W}\}$. We define that $\mathbf{W}$ and $\mathbf{\Lambda W}$ are equivalent,

$$\mathbf{W} \sim \mathbf{\Lambda W}, \tag{D.1}$$

where $\mathbf{\Lambda}$ is an arbitrary nonsingular diagonal matrix. Given $\mathbf{W}$, its equivalence class

$$\boldsymbol{C}_W = \{\mathbf{W}' |\ \mathbf{W}' = \mathbf{\Lambda W}\} \tag{D.2}$$

is an n-dimensional subspace of $Gl(n)$ consisting of all the matrices equivalent to $\mathbf{W}$. Therefore, a learning algorithm need not search for the non-identifiable $\mathbf{W} = \mathbf{H}^{-1}$ but searches for the equivalence class $\boldsymbol{C}_W$ that contains the true $\mathbf{W} = \mathbf{H}^{-1}$ except for permutations.

The space $Gl(n) = \{\mathbf{W}'\}$ is partitioned into equivalence classes $\boldsymbol{C}_\mathbf{W}$, where any $\mathbf{W}'$ belonging to the same $\boldsymbol{C}_\mathbf{W}$ is regarded as an equivalent demixing matrix. Let $d\mathbf{W}_C$ be a direction tangent to $\boldsymbol{C}_\mathbf{W}$, that is, both $\mathbf{W} + d\mathbf{W}_C$ and $\mathbf{W}$ belong to the same equivalence class $\boldsymbol{C}_\mathbf{W}$.

The learning equation determines $\Delta\mathbf{W}(k)$ depending on the current $\mathbf{y}$ and $\mathbf{W}$. When $\Delta\mathbf{W}(k)$ includes components belonging to the tangent directions to $\boldsymbol{C}_\mathbf{W}$, such components are ineffective because they drive $\mathbf{W}$ within the equivalent class. Therefore, it is better to design a learning rule such that its trajectories $\Delta\mathbf{W}(k)$ are always orthogonal to the equivalence classes. Since $\boldsymbol{C}_\mathbf{W}$ are n-dimensional subspaces, if we could find a family of $n^2 - n$ dimensional subspaces $\mathbf{Q}$'s such that $\boldsymbol{C}_W$ and $\mathbf{Q}$ are orthogonal, we could impose the constraints that the learning trajectories would belong to $\mathbf{Q}$. Therefore, one interesting question arises: Is there a family of (n^2-n)-dimensional sub-manifolds $\mathbf{Q}$ that is orthogonal to the equivalence classes $\boldsymbol{C}_W$? The answer is no. We can prove now that there does not exist a sub-manifold that is orthogonal to the families of sub-manifolds $\boldsymbol{C}_W$s.

Theorem D.3 *The direction* $d\mathbf{W}$ *is orthogonal to* $\boldsymbol{C}_\mathbf{W}$, *if and only if*

$$dx_{ii} = 0, \qquad (i = 1, 2, \ldots, n) \tag{D.3}$$

where $d\mathbf{X} = d\mathbf{W}\mathbf{W}^{-1}$.

Proof. Since the equivalent class $\boldsymbol{C}_\mathbf{W}$ consists of matrices $\mathbf{\Lambda W}$, $\mathbf{\Lambda}$ is regarded as a coordinate system in $\boldsymbol{C}_\mathbf{W}$. A small deviation of $\mathbf{W}$ in $\boldsymbol{C}_\mathbf{W}$ is written as

$$d\mathbf{W}_C = d\mathbf{\Lambda W}, \tag{D.4}$$

where $d\mathbf{\Lambda}$ is $\mathrm{diag}(d\lambda_1, \ldots, d\lambda_n)$. The inner product of $d\mathbf{W}$ and $d\mathbf{W}_C$ is given by

$$\begin{aligned} < d\mathbf{W}, d\mathbf{W}_C >_\mathbf{W} &= < d\mathbf{W}\mathbf{W}^{-1}, d\mathbf{W}_C\mathbf{W}^{-1} >_\mathbf{I} \\ &= < d\mathbf{X}, d\mathbf{\Lambda} >_\mathbf{I} = \sum_{i,j} dx_{ij} d\lambda_{ij} \\ &= \sum dx_{ii} d\lambda_{ii}. \end{aligned} \tag{D.5}$$

Therefore, $d\mathbf{W}$ is orthogonal to $\boldsymbol{C}_{\mathbf{W}}$ when and only when $d\mathbf{W}$ satisfies $dx_{ii} = 0$ for all i.

We next show that $d\mathbf{X}$ is not integrable, that is, there are no matrix functions $\boldsymbol{G}(\mathbf{W})$ such that

$$d\mathbf{X} = \mathrm{tr}(\frac{\partial \boldsymbol{G}}{\partial \mathbf{W}} d\mathbf{W}^T), \tag{D.6}$$

where

$$\mathrm{tr}(\frac{\partial \boldsymbol{G}}{\partial \mathbf{W}} d\mathbf{W}^T) = \sum_{i,j} \frac{\partial \boldsymbol{G}}{\partial w_{ij}} dw_{ij}. \tag{D.7}$$

If such a $\boldsymbol{G}$ exists, $\mathbf{X} = \boldsymbol{G}(\mathbf{W})$ defines another coordinate system in $Gl(n)$. Even when such $\boldsymbol{G}$ does not exist, $d\mathbf{X}$ is well-defined and it forms a basis in the tangent space of $Gl(n)$ at $\mathbf{W}$. Such a basis is called a nonholonomic basis (Schouten, 1954; Frankel, 1997).

Theorem D.4 *The basis defined by $d\mathbf{X} = d\mathbf{W}\mathbf{W}^{-1}$ is nonholonomic.*

Proof. Let us assume that there exists a function $\boldsymbol{G}(\mathbf{W})$ such that

$$d\mathbf{X} = d\boldsymbol{G}(\mathbf{W}) = d\mathbf{W}\mathbf{W}^{-1}. \tag{D.8}$$

We now consider another small deviation from $\mathbf{W}$ to $\mathbf{W} + \delta\mathbf{W}$. We then have

$$\delta d\mathbf{X} = \delta d\boldsymbol{G} = d\mathbf{W}\delta(\mathbf{W}^{-1}). \tag{D.9}$$

We have

$$\begin{aligned} \delta(\mathbf{W}^{-1}) &= (\mathbf{W} + \delta\mathbf{W})^{-1} - \mathbf{W}^{-1} \\ &\simeq -\mathbf{W}^{-1}\delta\mathbf{W}\mathbf{W}^{-1}. \end{aligned} \tag{D.10}$$

Hence,

$$\delta d\mathbf{X} = -d\mathbf{W}\mathbf{W}^{-1}\delta\mathbf{W}\mathbf{W}^{-1}. \tag{D.11}$$

Since matrices are in general non-commutative, we have

$$\delta d\mathbf{X} \neq d\delta\mathbf{X}. \tag{D.12}$$

This shows that such a matrix $\mathbf{X}$ does not exist because $d\delta\boldsymbol{G} = \delta d\boldsymbol{G}$ always holds when a matrix $\boldsymbol{G}$ exists.

Our constraints

$$dx_{ii} = 0, \qquad (i = 1, 2, \ldots, n) \tag{D.13}$$

restrict the possible directions of $\Delta\mathbf{W}$, and define $(n^2 - n)$-dimensional movable directions at each point $\mathbf{W}$ of $Gl(n)$. These directions are orthogonal to $\boldsymbol{C}_{\mathbf{W}}$. However, by the same reasoning, there does not exist functions $g_i(\mathbf{W})$ such that

$$dx_{ii} = dg_i(\mathbf{W}) = \sum_{j,k} \frac{\partial g_i(\mathbf{W})}{\partial w_{jk}} dw_{jk}. \tag{D.14}$$

This implies that no subspace defined by $g_i(\mathbf{W}) = 0$ exists.

Such constraints are said to be nonholonomic. The learning equation (D.1) defines learning dynamics with nonholonomic constraints. At each point $\mathbf{W}$, $\Delta\mathbf{W}$ is constrained in (n^2-n) directions. However, the trajectories are not constrained in any (n^2-n)-dimensional subspace, and they can reach any points in $Gl(n)$.

This property is important when the amplitudes of $s_i(t)$ change over time. If the constraints are

$$E\{h_i(y_i)\} = 0, \tag{D.15}$$

for example,

$$E\{y_i^2 - 1\} = 0, \tag{D.16}$$

when $E\left[s_i^2\right]$ suddenly become 10 times smaller, the ordinary learning dynamics makes the i-th row of $\mathbf{W}$ become 10 times larger in order to compensate for this change and keep $E\left[y_i^2\right] = 1$. Therefore, even when $\mathbf{W}$ converges to the true $\boldsymbol{C}_{\mathbf{W}}$, large fluctuations emerge from the ineffective movements caused by changes in the amplitude of s_i. This sometimes causes numerical instability. On the other hand, our nonholonomic dynamics are always orthogonal to $\boldsymbol{C}_{\mathbf{W}}$ so that such ineffective fluctuations are suppressed.

Nonholonomic basis are used in classical analytical dynamics uses nonholonomic bases to analyze the spinning gyro (Whittaker, 1940). They were also used in relativity. G. Kron (1952) used nonholonomic constraints to present a general theory of electro-mechanical dynamics of generators and motors. Nonholonomic properties play a fundamental role in continuum mechanics of distributed dislocations (see, e.g., Amari, 1962). An excellent explanation is found in Brockett (1993), where the controllability of nonlinear dynamical systems is shown by using the related Lie algebra. Recently, robotics researchers have been eager to analyze dynamics with nonholonomic constraints (Suzuki and Nakamura 1995) [25].

Remark D.6 *Theorem D.4 shows that the orthogonal natural gradient descent algorithm evolves along a trajectory path which doesn't include redundant (useless) components in the directions of $\boldsymbol{C}_W$. Therefore, it seems likely that the orthogonal algorithm can be more efficient than other algorithms as has been confirmed by preliminary computer simulations.*

Stability Analysis

In this section, we discuss the local stability of the nonholonomic orthogonal natural gradient descent algorithm defined by

$$\Delta\mathbf{W}(k) = \eta(k)\,\mathbf{F}(\mathbf{y}(k))\,\mathbf{W}(k), \tag{D.17}$$

with $f_{ii} = 0$ and $f_{ij} = -f_i(y_i)\,y_j$, if $i \neq j$, in the vicinity of the desired equilibrium submanifold $\boldsymbol{C}_{\mathbf{W}}$. Similar to a theorem established in our previous publications (Amari *et al.* 1997), we formulate the following Theorem:

Theorem D.5 *When the following inequalities hold*

$$\gamma_{ij} > \delta_i\,\delta_j, \qquad (i, j = 1, 2, \ldots, n) \tag{D.18}$$

where $\gamma_{ij} = \kappa_i \kappa_j \sigma_i^2 \sigma_j^2$ *and* $\delta_i = E\{y_i f_i(y_i)\}$, *then the desired* $\mathbf{W} = \mathbf{P\Lambda H}^{-1}$ *is a stable equilibrium, where* $\mathbf{\Lambda}$ *is a diagonal matrix and* $\mathbf{P}$ *is any permutation matrix.*
If

$$\gamma_{ij} < \delta_i \, \delta_j, \qquad (i, j = 1, 2, \ldots, n) \tag{D.19}$$

then, replacing $f_{ij} = f_i(y_i) y_j$ *by*

$$f_{ij} = y_i f_j(y_j), \qquad i \neq j \tag{D.20}$$

it is guaranteed that $\mathbf{W} = \mathbf{P\Lambda H}^{-1}$ *is stable.*

Proof. Let $\mathbf{W} = \mathbf{H}^{-1}$ and $(d\mathbf{W})_i$ denote the i-th row of $d\mathbf{W}$, and $(\mathbf{W}^{-1})^j$ be the j-th column of $\mathbf{W}^{-1}$. Then taking into account that $dx_{ij} = f_i(y_i) y_j$ and $dx_{ii} = 0$, we evaluate the second differentials in the Hessian matrix as follows,

$$\begin{aligned} d^2 x_{ij} &= E\{d(f_i(y_i) y_j)\} \\ &= E\{f_i'(y_i) y_j dy_i\} + E\{f_i(y_i) dy_j\} \\ &= E\{f_i'(y_i) y_j (d\mathbf{W})_i (\mathbf{W}^{-1}\mathbf{s})\} + E\{f_i'(y_i)(d\mathbf{W})_j(\mathbf{W}^{-1}\mathbf{s})\} \\ &= E\{f_i'(y_i) y_j (d\mathbf{W})_i (\mathbf{W}^{-1})_j\} + E\{f_i(y_i) y_i (d\mathbf{W})_j (\mathbf{W}^{-1})_i\} \\ &= -E\{f_i'(s_i) s_j^2\} dx_{ij} - E\{f_i(s_i) s_i\} dx_{ji}. \end{aligned} \tag{D.21}$$

Hence,

$$d^2 x_{ij} = -E\{f_i'(s_i) s_j^2\} dx_{ij} - E\{f_i(s_i) s_i\} dx_{ji}, \tag{D.22}$$

$$d^2 x_{ji} = -E\{f_j(s_j) s_j\} dx_{ij} - E\{f_j'(s_j) s_i^2\} dx_{ji}, \tag{D.23}$$

$$d^2 x_{ii} = 0. \tag{D.24}$$

From the above, we see that the equilibrium point $\mathbf{W} = \mathbf{P\Lambda H}^{-1}$ is stable.

Appendix E. Summary of Stability Conditions

We summarize the stability conditions of the learning rule

$$\frac{d\mathbf{W}}{dt} = \eta \, \mathbf{F}(\mathbf{y}) \, \mathbf{W}, \tag{E.1}$$

where off-diagonal entries of the matrix $\mathbf{F}$ are expressed as

$$f_{ij} = -f_i(y_i) \, y(_i), \qquad i \neq j, \tag{E.2}$$

and the diagonal entries of $\mathbf{F}$ can take the following forms

$$f_{ii} = 1 - y_i \, f(y_i), \tag{E.3}$$

$$f_{ii} = 1 - y_i^2, \tag{E.4}$$

$$f_{ii} = 0. \tag{E.5}$$

The ordinary algorithm is (E.3), while (E.4) forces the power of the recovered signals to be equal to 1, and (E.5) is a nonholonomic constraint so that the scale is infinite.

Theorem E.6 *The stability condition for the ordinary case (E.3) is:*

$$1 + m_i > 0, \quad \kappa_i > 0, \quad \gamma_{ij} > 1. \tag{E.6}$$

Theorem E.7 *The stability criterion for the case (E.4), where the recovered signals are normalized (we assume $\delta_i = E\{y_i\, f_i(y_i)\} > 0$) is:*

$$\kappa_i > 0, \qquad \kappa_i \kappa_j > \delta_i\, \delta_j. \tag{E.7}$$

Theorem E.8 *The stability condition for the nonholonomic case (E.2c) is:*

$$\kappa_i > 0, \qquad \gamma_{ij} > \delta_i \delta_j. \tag{E.8}$$

Remark E.7 *If $f_i(y_i)$ is derived from a probability distribution $q(y_i)$, then*

$$f_i(y_i) = -\frac{\log q(y_i)}{dy_i}, \tag{E.9}$$

$$\delta_i = E\left\{y_i f_i(y_i)\right\} = 1 \tag{E.10}$$

holds. However, the scale of y_i is indefinite in the nonholonomic case, so that the probability distribution of $\delta_i y_i$ is different from that of y_i. Hence, δ_i depends on which point in $\boldsymbol{C}_{\mathbf{W}}$ $\mathbf{W}$ converges.

Remark E.8 *Let $\boldsymbol{C}_{\mathbf{W}}$ be the class containing the true $\mathbf{W} = \mathbf{H}^{-1}$. δ_i is a function defined in the class. Hence, it is possible that the stability condition $\gamma_{ij} > \delta_i\, \delta_j$ holds in some part of $\boldsymbol{C}_{\mathbf{W}}$ but does not in some other part.*

Appendix F. Natural Gradient for a Non-square Separating Matrix

Let us consider the case where the number of measurements (sensors) is larger than the number of sources, $(m > n)$. The recovered signals are given by

$$\mathbf{y} = \mathbf{W}\,\mathbf{x} = \mathbf{W}(\mathbf{H}\,\mathbf{s} + \boldsymbol{\nu}), \tag{F.1}$$

where $\mathbf{W} \in \mathbb{R}^{n\times m}$.

The set $\mathbb{R}^{n\times m}$ does not form a Lie group, although we show a trial to introduce a Lie group structure in the next Appendix G. Here, we introduce an inner product in the space $\mathbb{R}^{n\times m}$, and the related natural gradient. Let $\delta\mathbf{W} = (\delta\mathbf{w}_{ij})$ be a small deviation of $\mathbf{W}$. The magnitude of $\delta\mathbf{W}$ is given by the inner product $< \delta\mathbf{W}, \delta\mathbf{W} >_{\mathbf{W}}$, which introduces a Riemannian matrix in $\mathbb{R}^{n\times m}$ defined by

$$< \delta\mathbf{W}, \delta\mathbf{W} >_{\mathbf{W}} = \mathrm{tr}\,\left(\delta\mathbf{W}\,\mathbf{R}_{\mathbf{xx}}\,\delta\mathbf{W}^T\,\mathbf{R}_{\mathbf{ss}}\right), \tag{F.2}$$

where

$$\mathbf{R}_{\mathbf{x}\,\mathbf{x}} = E\{\mathbf{x}\,\mathbf{x}^T\} = \mathbf{H}\,\mathbf{H}^T + \mathbf{R}_{\boldsymbol{\nu}\boldsymbol{\nu}}, \tag{F.3}$$

and $\mathbf{R}_{\mathbf{s}\,\mathbf{s}} = E\{\mathbf{s}\,\mathbf{s}^T\} = \mathbf{I}_n$ are taken into account.

Let $\rho(\mathbf{W})$ be a loss function of $\mathbf{W}$. The natural gradient $\widetilde{\nabla}\rho$ is given by the equation

$$d\rho = \mathrm{tr}\,(\nabla\rho\, d\mathbf{W}) =< \widetilde{\nabla}\rho, d\mathbf{W} > . \tag{F.4}$$

Assuming that $\mathbf{R}_{\mathbf{xx}}$ is not singular due to nonzero noise, we have

$$\widetilde{\nabla}\rho = \nabla\rho\,\mathbf{R}^{+}_{\mathbf{xx}}. \tag{F.5}$$

When the covariance matrix $\mathbf{R}_{\boldsymbol{\nu}\boldsymbol{\nu}} = \sigma^2\mathbf{I}_m$ is negligibly small, then $\mathbf{R}_{\mathbf{xx}}$ is singular. But we have

$$\lim_{\sigma^2\to 0} \mathbf{R}^{-1}_{\mathbf{xx}} = \mathbf{R}^{+}_{\mathbf{xx}}, \tag{F.6}$$

where $\mathbf{R}^{+}_{\mathbf{xx}}$ is the generalized inverse. Taking into account that

$$\mathbf{R}^{+}_{\mathbf{x}\,\mathbf{x}} = \left(\mathbf{H}\,\mathbf{H}^T\right)^{+} = \mathbf{W}^T\mathbf{W}, \tag{F.7}$$

we can express the relationship between the natural gradient and the standard gradient as follows

$$\widetilde{\nabla}\rho = \nabla\rho\,\mathbf{W}^T\mathbf{W}. \tag{F.8}$$

Appendix G. Lie Groups and Natural Gradient for the General Case

In many cases, the number of source signals is changing and is unknown. Therefore, the mixing matrix $\mathbf{H}$ is not square and not invertible. Recently Amari (1999) [21] extended the natural gradient approach to the over and under-complete cases (i.e., the cases when the number of sensors is larger or less than the number of sources) under the condition that the sensor signals are prewhitened.

In this Appendix, we introduce Lie Group structures on the manifold of the under-complete mixture matrices, and endorse a Riemannian metric on the manifolds based on the property of the Lie groups (see Zhang *et al.* for more detail [1348, 1361, 1363]). Then, we derive the natural gradient on the manifold using the isometry of the Riemannian metric.

Denote by $Gl(n, m) = \{\mathbf{W} \in \mathbb{R}^{n\times m} | rank(\mathbf{W}) = min(n, m)\}$, the set of the $n \times m$ matrices of full rank. Assume that $m > n$. For $\mathbf{W} \in Gl(n, m)$, there exists a permutation matrix $\mathbf{P}_Q \in \mathbb{R}^{m\times m}$ such that

$$\mathbf{W}\mathbf{P}_Q = [\mathbf{W}_1, \mathbf{W}_2], \tag{G.1}$$

where $\mathbf{W}_1 \in \mathbb{R}^{n\times n}$ is a nonsingular matrix. Since the permutation of components is acceptable in the blind separation case, for simplicity, we assume that the square matrix consisting of the first n columns of $\mathbf{W}$ is always nonsingular. Therefore, every $\mathbf{W} \in Gl(n, m)$ can be decomposed into the following form $\mathbf{W} = [\mathbf{W}_1\ \mathbf{W}_2]$, where $\mathbf{W}_1 \in \mathbb{R}^{n\times n}$ is nonsingular.

G.0.1 Lie Group $Gl(n, m)$

The Lie group has played a crucial role in deriving the natural gradient in the manifold $Gl(n, m)$, whose element $\mathbf{W} \in \mathbb{R}^{n \times m}$ is a non-square full column rank matrix.

The natural gradient induced by the Lie group, proposed by Zhang *et al.* [1361] can be written in general form as

$$\boxed{\tilde{\nabla}\rho(\mathbf{y}, \mathbf{W}) = \nabla\rho(\mathbf{y}, \mathbf{W})\mathbf{W}^T\mathbf{W} + \nabla\rho(\mathbf{y}, \mathbf{W})\mathbf{N}_I,} \tag{G.2}$$

where

$$\mathbf{N}_I = \mathbf{P}_Q^T \begin{pmatrix} \mathbf{0} & \mathbf{0} \\ \mathbf{0} & \mathbf{I}_{m-n} \end{pmatrix} \mathbf{P}_Q, \tag{G.3}$$

with $\mathbf{P}_Q \in \mathbb{R}^{m \times m}$ is any permutation matrix. The result indicates that the natural gradient for under-complete mixture $(m > n)$ is not unique, which depends on the permutation matrix $\mathbf{P}_Q$.

Assume that $m > n$. For $\mathbf{W} \in Gl(n, m)$, there exists a permutation matrix $\mathbf{P}_Q \in \mathbb{R}^{m \times m}$ such that

$$\mathbf{W} = [\mathbf{W}_1, \mathbf{W}_2]\mathbf{P}_Q, \tag{G.4}$$

where $\mathbf{W}_1 \in \mathbb{R}^{n \times n}$ is a nonsingular matrix. Two basic operations for the Lie group are defined as follows,

$$\begin{aligned} \mathbf{X} \circledast \mathbf{Y} &= [\mathbf{X}_1\mathbf{Y}_1, \mathbf{X}_1\mathbf{Y}_2 + \mathbf{X}_2]\mathbf{P}_Q, && \text{(G.5)} \\ \mathbf{X}^\dagger &= [\mathbf{X}_1^{-1}, -\mathbf{X}_1^{-1}\mathbf{X}_2]\mathbf{P}_Q, && \text{(G.6)} \end{aligned}$$

where $\mathbf{X} = [\mathbf{X}_1, \mathbf{X}_2]\mathbf{P}_Q$ and $\mathbf{Y} = [\mathbf{Y}_1, \mathbf{Y}_2]\mathbf{P}_Q$ are in $Gl(n, m)$, $\circledast$ is the multiplication operator of two matrices in $Gl(n, m)$ and $\dagger$ is the inverse operator on $Gl(n, m)$. The identity is defined by $\mathbf{E} = [\mathbf{I}_n, \mathbf{0}]\mathbf{P}_Q$.

Lie Group has the special property that it allows an invariant Riemannian metric. Let $T_\mathbf{W}$ be the tangent space of $Gl(n, m)$, $\mathbf{X}$ and $\mathbf{Y} \in T_\mathbf{W}$ be the tangent vectors. The Riemannian metric can be induced easily by the following inner product

$$< \mathbf{X}, \mathbf{Y} >_\mathbf{W} = < \mathbf{X} \circledast \mathbf{W}^\dagger, \mathbf{Y} \circledast \mathbf{W}^\dagger >_\mathbf{E} . \tag{G.7}$$

The inner product at $\mathbf{E}$ is naturally defined by

$$< \mathbf{X}, \mathbf{Y} >_\mathbf{E} = \mathrm{tr}(\mathbf{X}\mathbf{Y}^T). \tag{G.8}$$

From the definition for $\mathbf{W} = [\mathbf{W}_1, \mathbf{W}_2]\mathbf{P}_Q \in Gl(n, m)$, we have

$$\mathbf{W}^\dagger = [\mathbf{W}_1^{-1}, -\mathbf{W}_1^{-1}\mathbf{W}_2]\mathbf{P}_Q \tag{G.9}$$

and

$$\begin{aligned} \mathbf{X} \circledast \mathbf{W}^\dagger &= [\mathbf{X}_1\mathbf{W}_1^{-1}, -\mathbf{X}_1\mathbf{W}_1^{-1}\mathbf{W}_2 + \mathbf{X}_2]\mathbf{P}_Q, && \text{(G.10)} \\ \mathbf{Y} \circledast \mathbf{W}^\dagger &= [\mathbf{Y}_1\mathbf{W}_1^{-1}, -\mathbf{Y}_1\mathbf{W}_1^{-1}\mathbf{W}_2 + \mathbf{Y}_2]\mathbf{P}_Q, && \text{(G.11)} \end{aligned}$$

then, we have

$$< \mathbf{X}, \mathbf{Y} >_{\mathbf{W}} = \operatorname{tr}\left([\mathbf{X}_1 \mathbf{W}_1^{-1}, -\mathbf{X}_1 \mathbf{W}_1^{-1} \mathbf{W}_2 + \mathbf{X}_2][\mathbf{Y}_1 \mathbf{W}_1^{-1}, -\mathbf{Y}_1 \mathbf{W}_1^{-1} \mathbf{W}_2 + \mathbf{Y}_2]^T \right). \tag{G.12}$$

For a function $\rho(\mathbf{y}, \mathbf{W})$ defined on the manifold $Gl(n, m)$, the natural gradient $\tilde{\nabla}\rho(\mathbf{y}, \mathbf{W})$ is the contravariant form of $\nabla\rho(y, \mathbf{W})$ denoting the steepest direction of the function $\rho(\mathbf{y}, \mathbf{W})$ as measured by the Riemannian metric of $Gl(n, m)$, which is defined by

$$< \mathbf{X}, \tilde{\nabla}\rho(\mathbf{y}, \mathbf{W}) >_{\mathbf{W}} = < \mathbf{X}, \nabla\rho(\mathbf{y}, \mathbf{W}) >_{\mathbf{E}}, \tag{G.13}$$

for any $\mathbf{X} \in Gl(n, m)$. Using definition (G.12), and comparing both sides of (G.13), we have

$$\tilde{\nabla}\rho(\mathbf{y}, \mathbf{W}) = \nabla\rho(\mathbf{y}, \mathbf{W})\mathbf{W}^T\mathbf{W} + \nabla\rho(y, \mathbf{W})\mathbf{N}_I, \tag{G.14}$$

where

$$\mathbf{N}_I = \mathbf{P}_Q^T \begin{pmatrix} \mathbf{0} & \mathbf{0} \\ \mathbf{0} & \mathbf{I}_{m-n} \end{pmatrix} \mathbf{P}_Q. \tag{G.15}$$

7

Locally Adaptive Algorithms for ICA and their Implementations

Nothing is invented and perfected at the same time.

—Latin Proverb

In the previous chapter, we rigorously derived a family of equivariant algorithms for ICA/BSS using Amari's natural gradient approach. In this chapter, the original Jutten-Hérault (J-H) algorithm [649, 306, 651] is introduced, and several quasi-heuristically derived extensions and modifications for independent component analysis are presented. In particular, we focus on simple locally adaptive Hebbian/anti-Hebbian learning algorithms and propose their implementation using multilayer neural networks.

The main purpose of this chapter is to describe and overview models and to present a family of practical and efficient adaptive or locally adaptive learning algorithms with special advantages such as efficiency, simplicity and straightforward implementation. Some of the described algorithms have special advantages in cases of noisy, badly scaled or ill-conditioned signals. The proposed algorithms are extended for the case when the number of sources and their statistics are unknown. Finally, we will address the problem of an optimum nonlinear activation function and discuss general local stability conditions.

7.1 MODIFIED JUTTEN-HÉRAULT ALGORITHMS FOR BLIND SEPARATION OF SOURCES

7.1.1 Recurrent Neural Network

In this section, we consider a recurrent neural network and an associated locally adaptive learning rule, first developed by Jutten and Hérault [649, 306, 651, 647]. We focus only on linearly mixed sensor signals, i.e.,

$$x_i(k) = \sum_{j=1}^{n} h_{ij}\, s_j(k), \qquad (i = 1, 2, \ldots, n), \tag{7.1}$$

where h_{ij} are unknown mixing parameters, $s_j(k)$ are unknown, zero-mean, independent sources and $x_i(k)$ are the observed sensor signals.

Jutten and Hérault proposed a recurrent neural network shown in Fig. 7.1 (a), described by the matrix equations

$$\mathbf{y}(k) = \mathbf{x}(k) - \widehat{\mathbf{W}}(k)\mathbf{y}(k), \tag{7.2}$$

$$\mathbf{x}(k) = \mathbf{H}\,\mathbf{s}(k). \tag{7.3}$$

Hence

$$\mathbf{y}(k) = [\mathbf{I} + \widehat{\mathbf{W}}(k)]^{-1}\mathbf{x}(k), \tag{7.4}$$

where

$$\widehat{\mathbf{W}} = \begin{bmatrix} 0 & \widehat{w}_{12} & \cdots & \widehat{w}_{1n} \\ \widehat{w}_{21} & 0 & \cdots & \widehat{w}_{2n} \\ \vdots & \vdots & \ddots & \vdots \\ \widehat{w}_{n1} & \widehat{w}_{n2} & \cdots & 0 \end{bmatrix}.$$

It is required to adapt the weights $\widehat{w}_{ij}$ (with $\widehat{w}_{ii} = 0$) of a linear system (often referred to as a single-layer recurrent neural network) to combine the observations $x_i(k)$ to form optimal estimates of the source signals $\widehat{s}_j(k) = y_j(k) = x_j(k) - \sum_{i=1}^{n} \widehat{w}_{ji}\, y_i(k)$. The optimal weights correspond to the statistical independence of the output signals $y_j(t)$. Of course, the linear superposition is the simplest case of a combination of signals; other cases discussed in the following chapters include filtering and convolution operations.

7.1.2 Statistical Independence

It is assumed that no *a priori* information about the primary sources $s_i(t)$ is available except that they are mutually independent. Independence of random variables is a more general concept than decorrelation. Roughly speaking, we say that random variables y_i and y_j are statistically independent if knowledge of the values of y_i provides no information about the values of y_j.

Mathematically, the independence of y_i and y_j can be expressed by the relationship

$$p(y_i, y_j) = p(y_i)p(y_j), \tag{7.5}$$

(a)

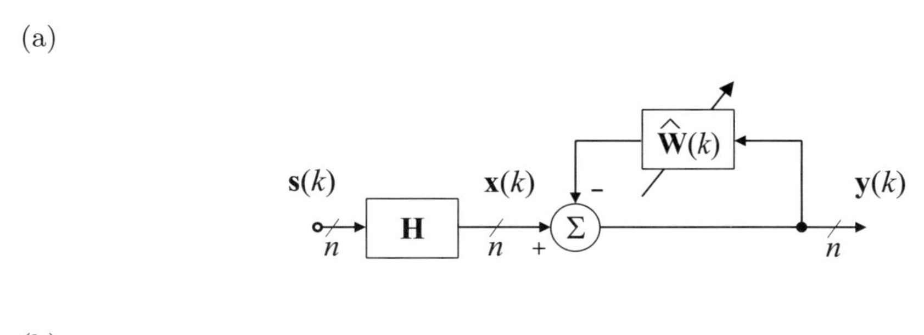

(b)

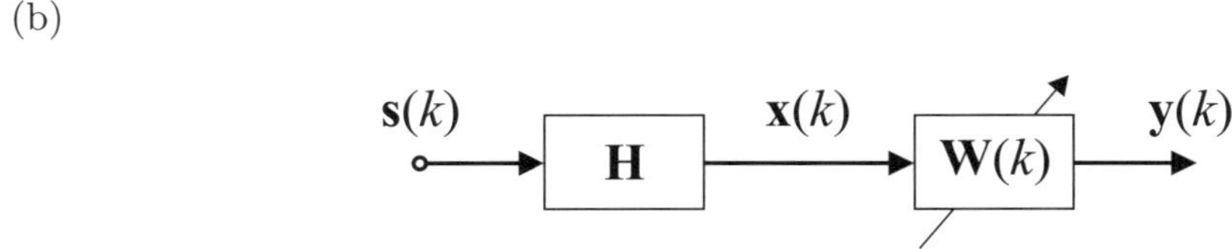

Fig. 7.1 Block diagrams: (a) Recurrent and (b) feed-forward neural network for blind source separation.

where $p(y)$ denotes the probability density function (pdf) of the random variable y. In other words, signals are independent if their joint pdf can be factorized.

If independent signals are zero-mean, then the generalized covariance matrix of $f(y_i)$ and $g(y_j)$, where $f(y)$ and $g(y)$ are different, odd nonlinear activation functions (e.g., $f(y) = y^3$ and $g(y) = \tanh(y)$) is a non-singular diagonal matrix

$$\mathbf{R}_{\mathbf{f}\,\mathbf{g}} = E\{\mathbf{f}(\mathbf{y})\mathbf{g}^T(\mathbf{y})\} - E\{\mathbf{f}(\mathbf{y})\}E\{\mathbf{g}^T(\mathbf{y})\} =$$
$$\begin{bmatrix} E\{f(y_1)g(y_1)\} - E\{f(y_1)\}E\{g(y_1)\} & & 0 \\ & \ddots & \\ 0 & & E\{f(y_n)g(y_n)\} - E\{f(y_n)\}E\{g(y_n)\} \end{bmatrix}, \tag{7.6}$$

i.e., the covariances $E\{f(y_i)g(y_j)\} - E\{f(y_i)\}E\{g(y_j)\}$ are all zero and all variances $E\{f(y_i)g(y_i)\} - E\{f(y_i)\}E\{g(y_i)\}$ are non-zero. It should be noted that for odd $f(y)$ and $g(y)$, if the probability density function of each zero-mean source signal is even, then the terms of the form $E\{f(y_i)\}E\{g(y_i)\}$ equal zero. The true general condition for statistical independence of signals is the vanishing of high-order cross-cumulants [251].

On the basis of the criterion for independence, Jutten and Hérault proposed a simple heuristic adaptive learning rule (see Fig. 7.2)which for the continuous-time model can be given by

$$\frac{d\widehat{w}_{ij}(t)}{dt} = \mu(t) f(y_i(t)) g(y_j(t)), \tag{7.7}$$

where $f(y)$ and $g(y)$ are different odd activation functions (for example $f(y) = y^3$, $g(y) = \tanh(10y)$ for sub-Gaussian sources and $f(y) = \tanh(10y)$, $g(y) = y^3$ for super-Gaussian sources, although a wide variety of functions can be used). The Jutten-Hérault learning

(a)

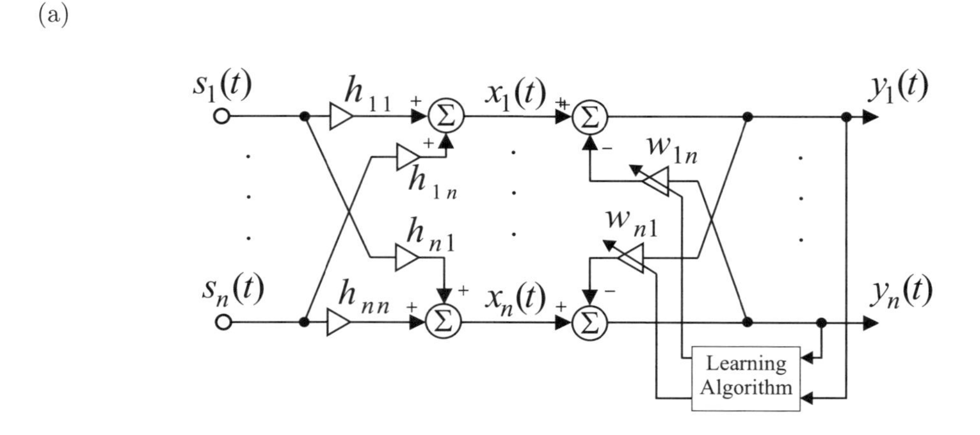

(b)

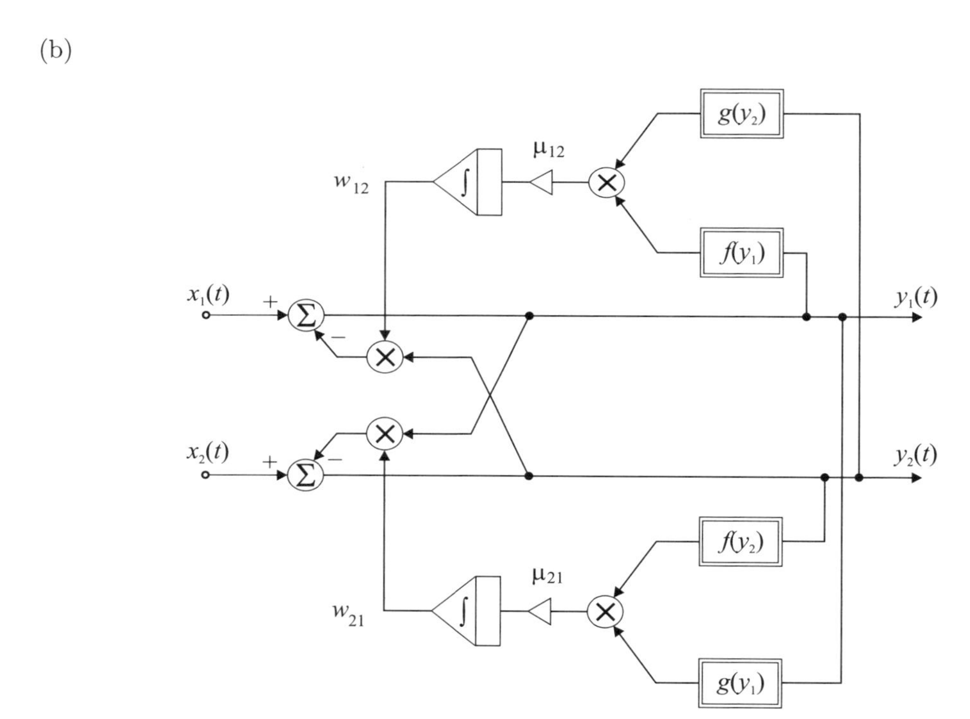

Fig. 7.2 (a) Neural network model and (b) implementation of the Jutten-Hérault basic continuous-time algorithm for two channels.

algorithm can be written in a compact matrix form as

$$\frac{d\,\widehat{\mathbf{W}}(t)}{dt} = \mu(t)\mathbf{f}[\mathbf{y}(t)]\,\mathbf{g}^T[\mathbf{y}(t)], \tag{7.8}$$

where

$$\begin{aligned}\mathbf{f}(\mathbf{y}) &= [f(y_1), f(y_2), \ldots, f(y_n)]^T, \\ \mathbf{g}(\mathbf{y}) &= [g(y_1), g(y_2), \ldots, g(y_n)]^T,\end{aligned}$$

provided that $E\{\mathbf{f}(\mathbf{y})\} = \mathbf{0}$ or $E\{\mathbf{g}(\mathbf{y})\} = \mathbf{0}$. To satisfy these conditions for arbitrary distributed sources (especially non-symmetric distributed signals around their means), we usually select nonlinearities as follows:

$$f_i(y_i) = \varphi_i(y_i) \quad \text{and} \quad g_i(y_i) = y_i \tag{7.9}$$

or

$$f_i(y_i) = y_i \quad \text{and} \quad g_i(y_i) = \varphi(y_i), \tag{7.10}$$

where $\varphi_i(y_i)$ are suitably designed nonlinear functions.

Note that in the Jutten-Hérault algorithm, the synaptic weights $\widehat{w}_{ii} = 0$ $(i = 1, 2, \ldots, n)$, i.e., the neural network in Fig. 7.2 (a) has no self-feedback loops. The Jutten-Hérault algorithm is very simple, however, it has failed in a number of cases, e.g., for weak signals or an ill-conditioned mixing matrix.

7.1.3 Self-normalization

Eq. (7.6) provides a practical criterion for the independence of the output signals. While this simple criterion has been proven to be effective in many cases, it is often very ill-conditioned for certain classes of signals, especially for badly scaled signals. To alleviate this problem, we propose an auxiliary condition that assigns to the variances $E\{f(y_i)g(y_i)\}$ specific values of λ_i (typically they are all equal to unity). This means that the output signals $y_i(t)$ should be adaptively scaled so that all the generalized variances are normalized as $E\{f(y_i)g(y_i)\} = \lambda_i$ (typically $\lambda_i = 1$). It can be shown by computer simulations that this auxiliary condition leads to considerable improvement in the performance of a neural network without any increase of the computational complexity of the learning algorithm.

Initially, we modified the Jutten-Hérault algorithm by incorporating a fully recurrent neural network in which every neuron is connected to all neurons, including itself (i.e., it contains self-loops with $\widehat{w}_{ii} \neq 0$). This network is described by the set of equations

$$y_i(t) = x_i(t) - \sum_{j=1}^{n} \widehat{w}_{ij}(t)\, y_j(t). \tag{7.11}$$

Starting from the independence and self-normalization criteria discussed above, it is straightforward to derive the associated on-line local learning algorithm as

$$\frac{d\widehat{w}_{ii}(t)}{dt} = \mu(t)\,[\lambda_i - f[y_i(t)]g[y_i(t)]] \tag{7.12}$$

and

$$\frac{d\widehat{w}_{ij}(t)}{dt} = \mu(t) f[y_i(t)] g[y_j(t)], \qquad \text{for } i \neq j, \tag{7.13}$$

which can be written in a compact matrix form as

$$\boxed{\frac{d\,\widehat{\mathbf{W}}(t)}{dt} = \mu(t)\left[\mathbf{\Lambda} - \mathbf{f}[\mathbf{y}(t)]\,\mathbf{g}^T[\mathbf{y}(t)]\right],} \tag{7.14}$$

where $\mathbf{\Lambda} = \text{diag}\{\lambda_1, \lambda_2, \ldots, \lambda_n\}$.

The main justification for employing weights $\widehat{w}_{ii}$ is to provide self-normalization of the generalized variances of the output signals. Such self-normalization improves the performance of the network, especially if the signals are badly scaled, i.e., if the mixing matrix $\mathbf{H}$ is nearly singular.

The idea behind algorithm (7.14) is the following. Suppose, we take the expectation of both sides of Eq. (7.14). If the entries of the matrix $\widehat{\mathbf{W}}(t)$ approach constants and if $\widehat{w}_{ij}$ are smooth functions of time, the left side of the equation approaches zero. Hence the matrix $[E\{\mathbf{f}[\mathbf{y}(t)]\,\mathbf{g}^T[\mathbf{y}(t)]\} - \mathbf{\Lambda}]$ tends to zero. This means that the system achieves separation if, $E\{f(y_i)g(y_j)\} = E\{f(y_i)\}E\{g(y_j)\} = 0$, i.e., the output signals are mutually independent, provided $E\{f(y_i)\} = 0$ or $E\{g(y_i)\} = 0$ and $E\{f(y_i)g(y_i)\} = \lambda_i$ represent the self-normalization conditions. In practice, the expected values are not available and according to the stochastic approximation procedure, they are replaced by instantaneous or time-averaged values.

The learning algorithms (7.12)–(7.14) are very simple and relatively efficient for well-posed problems. However, the network shown in Fig. 7.1 (a) and the associated learning algorithms have several disadvantages; they require the inversion of a matrix (Eq. (7.4)) at every iteration, which is computationally ill-conditioned and time consuming. In addition, for the algorithm to work, the number of outputs must be equal to the number of sensors and the number of sources.

7.1.4 Feed-forward Neural Network and Associated Learning Algorithms

To avoid the drawbacks mentioned above, let us consider a single-layer feed-forward neural network shown in Fig. 7.1 (b). It is required to find the weights w_{ji} to optimally estimate the source signals as

$$\widehat{s}_j = y_j(t) = \sum_{i=1}^{n} w_{ji} x_i(t), \qquad (j = 1, 2, \ldots, n). \tag{7.15}$$

Note the system shown in Fig. 7.1 (b) is described by the matrix equations

$$\mathbf{x}(t) = \mathbf{H}\,\mathbf{s}(t) \tag{7.16}$$

and

$$\mathbf{y}(t) = \mathbf{W}\,\mathbf{x}(t) = \mathbf{W}\,\mathbf{H}\,\mathbf{s}(t), \tag{7.17}$$

where $\mathbf{H} = [h_{ij}]_{n\times n} \in \mathbb{R}^{n\times n}$ is an unknown mixture matrix, $\mathbf{W} = [w_{ji}]_{n\times n} \in \mathbb{R}^{n\times n}$ is a matrix of the adaptive synaptic weights, $\mathbf{s}(t) = [s_1(t), s_2(t), \ldots, s_n(t)]^T$ is an unknown vector of the independent sources, $\mathbf{x}(t) = [x_1(t), x_2(t), \ldots, x_n(t)]^T$ is the vector of the observable (available) sensor signals and $\mathbf{y}(t) = [y_1(t), y_2(t), \ldots, y_n(t)]^T$ is the vector of desired output signals, which must be mutually independent.

The optimal value of the synaptic weights corresponds to statistical independence of the output signals $y_j(t)$ $(j = 1, 2, \ldots, n)$. It should be noted that the source separation is achieved as soon as the composite matrix $\mathbf{G}(t) = \mathbf{W}(t)\mathbf{H}$ has exactly one non-zero element in every row and every column. Such a matrix $\mathbf{G}(t)$ is called a generalized permutation matrix.

The feed-forward network of Fig. 7.1 (b) is equivalent to the recurrent network of Fig. 7.1 (a), if the following relation holds

$$\mathbf{W}(t) = [\mathbf{I} + \widehat{\mathbf{W}}(t)]^{-1} \tag{7.18}$$

or equivalently

$$\widehat{\mathbf{W}}(t) = \mathbf{W}^{-1}(t) - \mathbf{I}, \tag{7.19}$$

which requires that the matrices $\widehat{\mathbf{W}}(t) + \mathbf{I}$ and $\mathbf{W}(t)$ are nonsingular for every time instant t. Differentiating Eq. (7.19) gives

$$\frac{d\,\widehat{\mathbf{W}}}{dt} = \frac{d}{dt}(\mathbf{W}^{-1} - \mathbf{I}) = -\mathbf{W}^{-1}\frac{d\,\mathbf{W}}{dt}\mathbf{W}^{-1} \tag{7.20}$$

and hence, we have

$$\mathbf{W}^{-1}\frac{d\,\mathbf{W}}{dt}\mathbf{W}^{-1} = \mu(t)\left[\mathbf{\Lambda} - \mathbf{f}[\mathbf{y}(t)]\,\mathbf{g}^T[\mathbf{y}(t)]\right] \tag{7.21}$$

and therefore

$$\frac{d\,\mathbf{W}}{dt} = \mu(t)\mathbf{W}(t)\left[\mathbf{\Lambda} - \mathbf{f}[\mathbf{y}(t)]\,\mathbf{g}^T[\mathbf{y}(t)]\right]\mathbf{W}(t), \tag{7.22}$$

with non-zero initial conditions, typically $\mathbf{W}(0) = \mathbf{I}$.

Eq. (7.22) constitutes an equivalent learning algorithm for the feed-forward network which has the same convergence properties as the algorithm given by Eq. (7.14). Unfortunately, the resulting adaptive learning algorithm, (7.22) is relatively more complex.

A careful study of Eq. (7.22) motivated us to propose a few heuristic modifications, that we found experimentally to have advantages in implementation with respect to computational efficiency and robustness. The first of these modifications came by noting that the right hand side factors $\mathbf{W}(t)$ in Eq. (7.22) have no influence on the equilibrium point but only change the trajectory. Thus the key part of the right hand side of Eq. (7.22) is $-\mu(t)[\mathbf{f}[\mathbf{y}(t)]\,\mathbf{g}^T[\mathbf{y}(t)] - \mathbf{\Lambda}]$. In fact, the equilibrium point depends only on the term $[\mathbf{f}[\mathbf{y}(t)]\,\mathbf{g}^T[\mathbf{y}(t)] - \mathbf{\Lambda}]$, i.e., the equilibrium point is achieved if the expected value of this term is equal to zero. Thus, the algorithm in Eq. (7.22), can be considerably simplified by replacing the terms $\mathbf{W}(t)$ by an identity matrix, giving

$$\boxed{\frac{d\,\mathbf{W}(t)}{dt} = \mu(t)\left[\mathbf{\Lambda} - \mathbf{f}[\mathbf{y}(t)]\,\mathbf{g}^T[\mathbf{y}(t)]\right],} \tag{7.23}$$

which can be written in a simple scalar form as

$$\boxed{\frac{d\, w_{ij}(t)}{dt} = \mu(t)\ [\delta_{ij}\lambda_i - f[y_i(t)]g[y_i(t)]]\,,} \tag{7.24}$$

where δ_{ij} is the Kronecker delta.

We found that this algorithm successfully separates the sources for rather moderately ill-conditioned problems (if the ratio of the energies of the signals is not larger than 1:10), leading to the same equilibrium point with fewer computations and a less complex network. A functional block diagram illustrating the implementation of the simplified learning algorithm (Eq. (7.24)) is shown in Fig. 7.3. It is interesting to note that the algorithm in Eq. (7.24) for the feed-forward network is identical in form (up to the sign factor) to the algorithm (7.12)–(7.14) for the recurrent network.

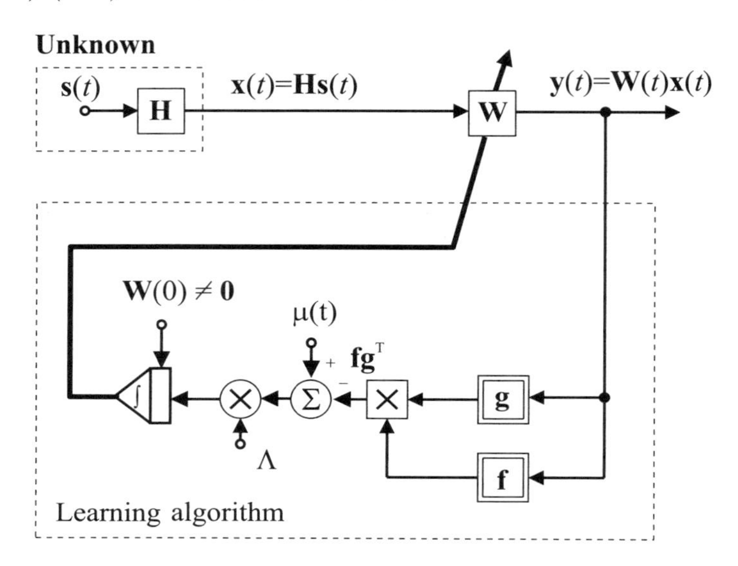

Fig. 7.3 Block diagram of the continuous-time locally adaptive learning algorithm (7.23).

For ill-conditioned problems, we can use an equivariant algorithm. One modification of such an algorithm can be obtained intuitively from the algorithm (Eq. 7.22) by assuming that $\mu(t)$ is not a scalar learning rate but a learning matrix defined as $\mu_0(t)\mathbf{W}^{-1}(t)$, where $\mu_0(t)$ is a scalar. Replacing $\boldsymbol{\mu}(t)$ in Eq. (7.22) by $\mu_0(t)\mathbf{W}^{-1}$, we obtain a learning algorithm for the feed-forward system with an updating rule given by

$$\boxed{\frac{d\,\mathbf{W}(t)}{dt} = \mu_0(t)\left[\mathbf{\Lambda} - \mathbf{f}[\mathbf{y}(t)]\,\mathbf{g}^T[\mathbf{y}(t)]\right]\mathbf{W}(t),} \tag{7.25}$$

which can be written in a scalar form as

$$\frac{d\, w_{ij}(t)}{dt} = \mu_0(t)\left[\delta_{ij}\lambda_i - f[y_i(t)]\sum_{p=1}^{m} w_{pj}(t)\, g[y_p(t)]\right]. \tag{7.26}$$

The corresponding discrete-time algorithm can take the following form

$$\boxed{w_{ij}(k+1) = w_{ij}(k) + \eta(k) \left[\delta_{ij}\lambda_i - f[y_i(k)] \sum_{p=1}^{m} w_{pj}(k)\, g[y_p(k)] \right].} \tag{7.27}$$

Figures 7.4 (a) and (b) show a block diagram illustrating the improved learning algorithm of Eq. (7.25).

In the special case, for

$$\mathbf{g}(\mathbf{y}) = \mathbf{f}(\mathbf{y}) - \mathbf{y} \quad \text{and} \quad \mathbf{\Lambda} = \mathbf{0}, \tag{7.28}$$

we have

$$\Delta\mathbf{W} = \eta\, \mathbf{f}(\mathbf{y}) \left[\mathbf{y}^T - \mathbf{f}^T(\mathbf{y})\right] \mathbf{W}. \tag{7.29}$$

Assuming further that the signals are pre-whitened, so that $\mathbf{W}^T\mathbf{W} = \mathbf{I}$, we obtain the nonlinear PCA rule developed by Oja and Karhunen [913] as:

$$\Delta\mathbf{W} = \eta\, \mathbf{f}(\mathbf{y}) \left[\mathbf{x}_1^T - \mathbf{f}^T(\mathbf{y})\mathbf{W}\right], \tag{7.30}$$

which needs a pre-whitening of the sensor signals ($\mathbf{x}_1 = \mathbf{Q}\mathbf{x}$).

Assuming that the constraint $\mathbf{W}^T\mathbf{W} = \mathbf{I}$ is satisfied during the learning process, we can easily prove that the above algorithm reduces approximately to the learning rule proposed by Cardoso and Laheld [148] as:

$$\Delta\mathbf{W} = -\eta \left[\mathbf{f}(\mathbf{y})\, \mathbf{y}^T - \mathbf{y}\, \mathbf{f}^T(\mathbf{y})\right] \mathbf{W}. \tag{7.31}$$

Connections of the nonlinear PCA rule (7.30) with other blind separation approaches are studied in [665].

7.1.5 Multilayer Neural Networks

In order to improve the flexibility and efficiency of BSS/ICA schemes, we can use multistage sequential or multilayer neural networks. Fig. 7.5 shows various possible configurations of multilayer neural networks.

In this way, we can dramatically improve the flexibility and performance of blind separation or extraction by applying different algorithms or different activation functions in each layer. Furthermore, this allows us to combine SOS and HOS algorithms to extract at various layers various sources with different statistical properties. The layers should work sequentially, one by one, i.e., after achieving convergence in the first layer, the second layer will start to work, and so on.

In general, synaptic weights in each layer can be updated, by employing any suitable algorithm described in this book. Especially, the following simple local learning rule can be applied

$$\frac{d\,\mathbf{W}^{(p)}(t)}{dt} = \mu(t) \left[\mathbf{\Lambda}^{(p)} - \left\langle \mathbf{f}_p[\mathbf{y}^{(p)}(t)]\, \mathbf{g}_p^T[\mathbf{y}^{(p)}(t)] \right\rangle \right], \quad (p = 1, 2, \ldots, K) \tag{7.32}$$

(a)

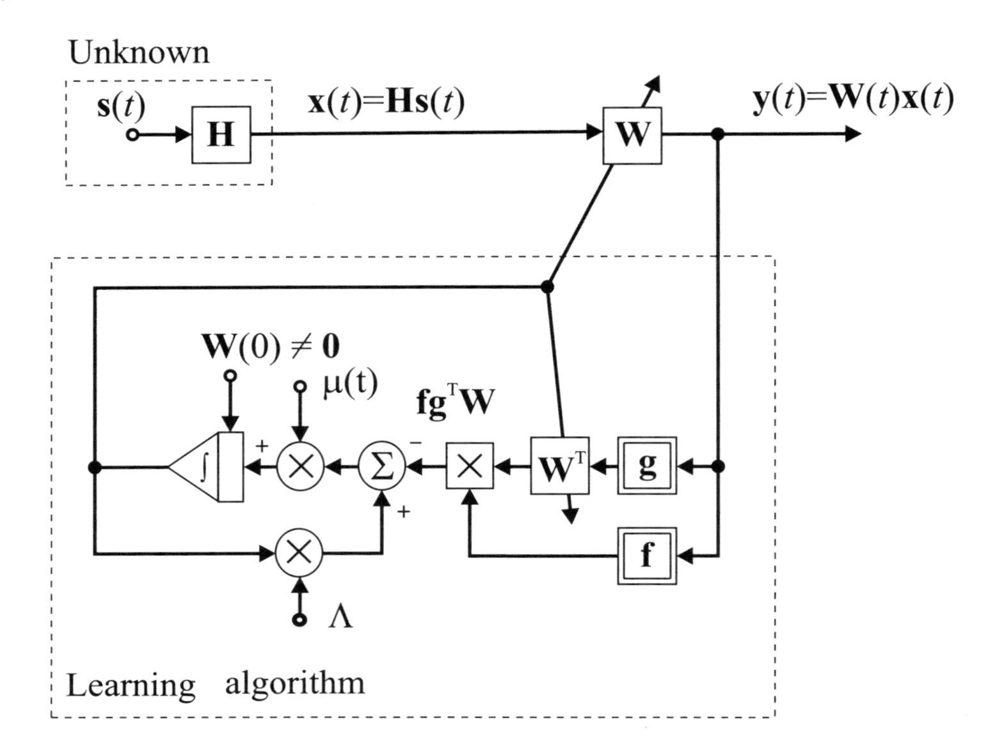

(b)

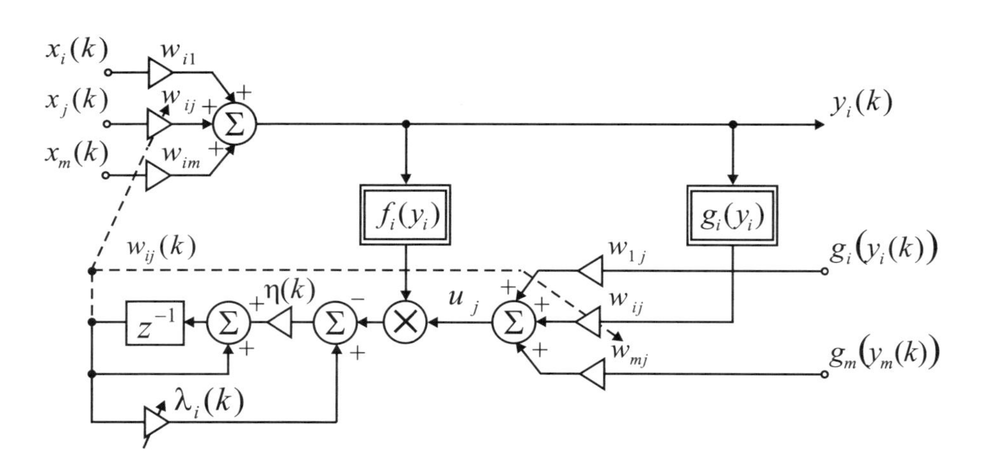

Fig. 7.4 (a) Block diagram illustrating implementation of a continuous-time robust learning algorithm, (b) illustration of implementation of a discrete-time robust learning algorithm.

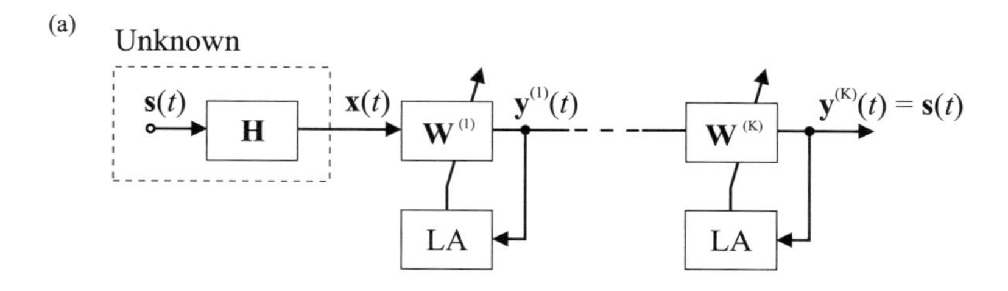

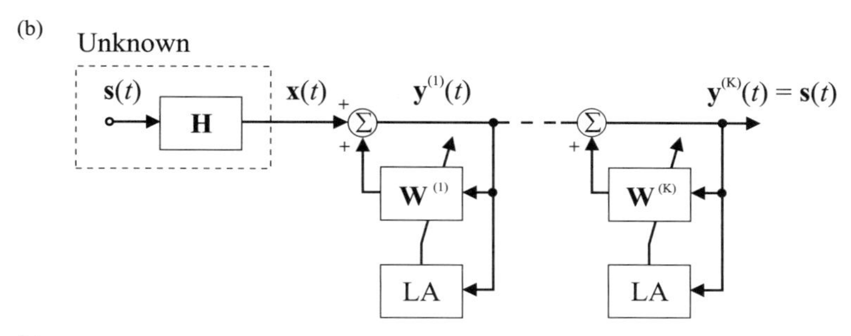

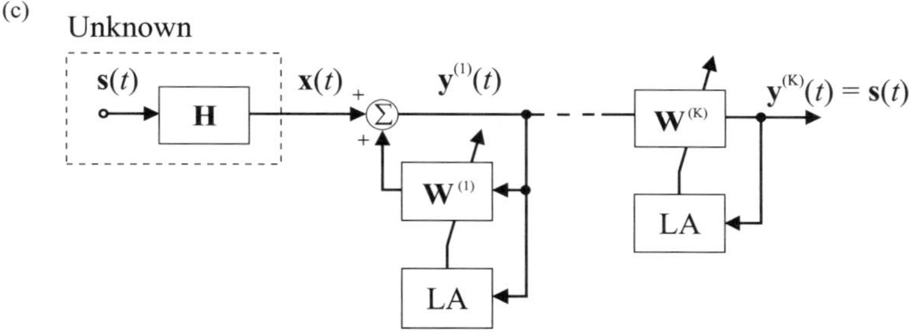

Fig. 7.5 Various configurations of multilayer neural networks for blind source separation: (a) Feed-forward model, (b) recurrent model, (c) hybrid model (LA means learning algorithm).

which can be written in a compact scalar form as

$$\frac{d\, w_{ij}^{(p)}(t)}{dt} = \mu(t) \left[\delta_{ij}\lambda_i^{(p)} - \left\langle f_p[y_i^{(p)}(t)] g_p[y_i^{(p)}(t)] \right\rangle \right], \tag{7.33}$$

where $\mathbf{\Lambda}^{(p)}$ is a nonsingular diagonal matrix (typically $\mathbf{\Lambda}^{(p)} = \mathbf{I}$), and $\langle \rangle$ denotes the statistical averaging operator.
The associated discrete-time locally adaptive learning rule takes the form

$$\mathbf{W}^{(p)}(k+1) = \mathbf{W}^{(p)}(k) + \eta^{(p)}(k) \left[\mathbf{\Lambda}^{(p)} - \left\langle \mathbf{f}_p[\mathbf{y}^{(p)}(k)] \mathbf{g}_p^T[\mathbf{y}^{(p)}(k)] \right\rangle \right], \quad (p = 1, 2, \ldots, K) \tag{7.34}$$

or in a scalar form as

$$w_{ij}^{(p)}(k+1) = w_{ij}^{(p)}(k) + \eta^{(p)}(k) \left[\delta_{ij}\lambda_i^{(p)} - \left\langle f_p[y_i^{(p)}(k)] g_p[y_i^{(p)}(k)] \right\rangle \right]. \tag{7.35}$$

Remark 7.1 *It is interesting to note that the same adaptive learning algorithm (7.35) can be used both for feed-forward and recurrent neural sub-models with either positive or negative learning rates $\eta^{(p)}$ (see Chapter 4 for more explanation).*

Extensive computer simulations have shown that this algorithm provides relatively good performance that was not possible to obtain using only a single stage (layer), especially for very ill-conditioned problems (e.g., if the signals are badly scaled or the mixing matrix $\mathbf{H}$ is extremely ill-conditioned). Moreover, due to its extremely low complexity in comparison to other algorithms, it is suitable for hardware VLSI implementations.

Of course, we are not satisfied with the present pragmatic but rather heuristic or intuitive engineering approaches presented so far in this chapter. However, the value (simplicity, biological plausibility and good performance) of the presented algorithms justifies their presentation prior to an adequate theory. In the next section, we will present a more rigorous derivation of several of these algorithms with some tips about their practical implementation.

7.2 ITERATIVE MATRIX INVERSION APPROACH TO THE DERIVATION OF A FAMILY OF ROBUST ICA ALGORITHMS

In this section, we discuss some interesting properties of the considered algorithms, and also present an alternative derivation of the robust ICA algorithm proposed by Cruces *et al.* [315, 326]. Only time-discrete learning algorithms are considered. The learning algorithm (7.25) can be derived using the basic concepts of the diagonalization of a generalized covariance matrix:

$$\mathbf{R}_{\mathbf{f}\,\mathbf{g}}(\mathbf{W}) = E\{\mathbf{f}(\mathbf{y})\mathbf{g}^T(\mathbf{y})\}. \tag{7.36}$$

Without loss of generality, we assume that this generalized covariance matrix is equal to the identity matrix at the optimum separation point $\mathbf{W}_*$, i.e.,

$$\mathbf{R}_{\mathbf{f}\,\mathbf{g}}(\mathbf{W}_*) = \mathbf{I}. \tag{7.37}$$

This diagonalization can also be viewed as an implicit matrix inversion, because if we define a new non-linear function of the outputs $\mathbf{h}(\mathbf{y}) = \mathbf{W}^{-1}\mathbf{f}(\mathbf{y})$ and a new generalized covariance matrix

$$\mathbf{R}_{\mathbf{h}\,\mathbf{g}}(\mathbf{W}) = E\{\mathbf{h}(\mathbf{y})\mathbf{g}^T(\mathbf{y})\} = \mathbf{W}^{-1}\mathbf{R}_{\mathbf{f}\,\mathbf{g}}(\mathbf{W}), \tag{7.38}$$

the diagonalization of (7.37) is equivalent to the following inversion

$$\mathbf{R}_{\mathbf{f}\,\mathbf{g}}(\mathbf{W}_*) = \mathbf{I} \Rightarrow \mathbf{W}_* = \mathbf{R}_{\mathbf{h}\,\mathbf{g}}^{-1}(\mathbf{W}_*). \tag{7.39}$$

The implicit matrix inversion (7.39) cannot be carried out directly, since we do not know $\mathbf{h}(\mathbf{y})$ and $\mathbf{g}(\mathbf{y})$ at the optimum separating matrix $\mathbf{W}_*$. Instead, we perform this in an iterative fashion. Let $\mathbf{W}(l)$ be the separating system at the l-th iteration step. If we assume that $\widehat{\mathbf{R}}_{\mathbf{h}\,\mathbf{g}}^{(l)} = \widehat{\mathbf{R}}_{\mathbf{h}\,\mathbf{g}}(\mathbf{W}(l))$, it is more convenient to estimate explicitly the mixing matrix $\widehat{\mathbf{H}}(l) = \mathbf{W}^{-1}(l)$ rather than the separating matrix $\mathbf{W}(l)$, by minimizing the following time-varying cost function

$$\mathbf{\Phi}(\widehat{\mathbf{H}}(l+1)) = \|\widehat{\mathbf{R}}_{\mathbf{h}\,\mathbf{g}}^{(l)} - \widehat{\mathbf{H}}(l)\|_F^2, \tag{7.40}$$

where $\|\cdot\|_F$ denotes the Frobenius norm of a matrix. This optimization problem is similar to those used in the Bussgang technique for blind deconvolution (see [545] for more detail) or in EM (expectation maximization).

It is important to note the alternating ("back and forth") nature of the proposed approach: The estimate $\mathbf{W}(l+1)$ resulting from the minimization of (7.40) is then used to estimate a new covariance matrix and construct the modified or updated cost function $\mathbf{\Phi}(\widehat{\mathbf{H}}(l+2))$ whose minimization yields to $\widehat{\mathbf{H}}(l+2)$ and so on.

The primary advantage of the cost function $\mathbf{\Phi}(\widehat{\mathbf{H}}(l+1))$ is that it is quadratic with respect to $\widehat{\mathbf{H}}(l) = \mathbf{W}^{-1}(l)$ and can hence be efficiently minimized using the following quasi-Newton method [315]

$$\widehat{\mathbf{H}}(l+1) = \widehat{\mathbf{H}}(l) + \eta\,(\widehat{\mathbf{R}}_{\mathbf{h}\,\mathbf{g}}^{(l)} - \widehat{\mathbf{H}}(l)). \tag{7.41}$$

We can rewrite the above formula as

$$\begin{aligned}\widehat{\mathbf{H}}(l+1) &= (1-\eta)\,\widehat{\mathbf{H}}(l) + \eta\,\widehat{\mathbf{R}}_{\mathbf{h}\,\mathbf{g}}^{(l)} &(7.42)\\ &= \widehat{\mathbf{H}}(l)\left(\mathbf{I} - \eta\,(\mathbf{I} - \widehat{\mathbf{R}}_{\mathbf{f}\,\mathbf{g}}^{(l)})\right). &(7.43)\end{aligned}$$

It is interesting to note that from equation (7.42) the matrix $\widehat{\mathbf{H}}(l+1) = \mathbf{W}^{-1}(l+1)$ can be interpreted as an estimate of $\widehat{\mathbf{R}}_{\mathbf{h}\,\mathbf{g}}$ using an exponential window. This complies with our interpretation of $\widehat{\mathbf{R}}_{\mathbf{h}\,\mathbf{g}}$ as an estimate of the mixing matrix $\mathbf{H}$.

The algorithm (7.43) can be alternatively derived by finding the zeros of the following nonlinear matrix estimation function

$$\mathcal{F}(\widehat{\mathbf{H}}) = \widehat{\mathbf{R}}_{\mathbf{h}\,\mathbf{g}} - \widehat{\mathbf{H}} = \mathbf{0}. \tag{7.44}$$

Applying the Newton-Raphson method (see S. Cruces *et al.* [315, 326] for more detail), we can find the zeros of $\mathcal{F}(\widehat{\mathbf{H}})$ by the following quasi-Newton recursion

$$\widehat{\mathbf{H}}(l+1) = \widehat{\mathbf{H}}(l) - \eta\,\mathcal{F}(\widehat{\mathbf{W}}(l))\mathbf{B}(l), \tag{7.45}$$

where $\mathbf{B}(l)$ is an approximation to the inverse of the derivative matrix $\frac{\partial \mathcal{F}(\widehat{\mathbf{H}})}{\partial \widehat{\mathbf{H}}}$ that must satisfy certain conditions described in [687]. Assuming that the previous estimation of matrix $\widehat{\mathbf{R}}_{\mathbf{h}\,\mathbf{g}}$ is independent of the mixing matrix, yields

$$\left(\frac{\partial \mathcal{F}(\widehat{\mathbf{H}}(l))}{\widehat{\mathbf{H}}(l)}\right)^{-1} \approx -\mathbf{I} = \mathbf{B}(l) \tag{7.46}$$

which boils down to the same algorithm (7.43).

Let us rewrite the algorithm (7.43) in terms of the separating matrix $\mathbf{W}(l)$ rather than $\widehat{\mathbf{H}} = \mathbf{W}^{-1}(l)$. If matrix $\left(\mathbf{I} - \eta \left(\mathbf{I} - \widehat{\mathbf{R}}_{\mathbf{f}\,\mathbf{g}}^{(l)}\right)\right)$ is not singular, the result of inverting (7.43) is

$$\boxed{\mathbf{W}(l+1) = \left[\mathbf{I} - \eta(l) \left(\mathbf{I} - \widehat{\mathbf{R}}_{\mathbf{f}\,\mathbf{g}}^{(l)}\right)\right]^{-1} \mathbf{W}(l),} \tag{7.47}$$

where $\widehat{\mathbf{R}}_{\mathbf{f}\,\mathbf{g}}^{(l)} = (\mathbf{1}/N) \sum_{k=1}^{N} \mathbf{f}(\mathbf{y}^{(l)}(k))\, \mathbf{g}^T(\mathbf{y}^{(l)}(k))$ and $\mathbf{y}^{(l)}(k) = \mathbf{W}(l)\, \mathbf{x}(k)$. It should be noted that the above algorithm is numerically stable, if the learning rate η is sufficiently small. In other words, the function $\mathcal{F}(\widehat{\mathbf{H}})$ and its derivatives should be continuous. Assuming that $\mathbf{f}(\cdot)$ and $\mathbf{g}(\cdot)$ are twice differentiable functions, discontinuities occur only at the points where the separation matrix $\mathbf{W}$ is singular. Therefore, by constraining the algorithm to be stable in a closed region that avoids the singularity of the separating matrix $\mathbf{W}$, discontinuities on the cost function will not occur. The sufficient condition to ensure numerical stability of the algorithm is [326]

$$\eta(l) < \frac{1}{\|\widehat{\mathbf{R}}_{\mathbf{f}\,\mathbf{g}}^{(l)} - \mathbf{I}\|_p}, \tag{7.48}$$

where $\|\cdot\|_p$ denotes the p-norm of a matrix. When the above constraints are satisfied, we can prevent the matrix $\mathbf{I} + \eta \left(\widehat{\mathbf{R}}_{\mathbf{f}\,\mathbf{g}}^{(l)} - \mathbf{I}\right)$ from being singular which guarantees its invertibility. As a consequence of (7.48), we can express the inverse in (7.47) as an infinite series and make the following approximation [494]

$$\begin{aligned}
\left(\mathbf{I} - \eta(l) \left(\mathbf{I} - \widehat{\mathbf{R}}_{\mathbf{f}\,\mathbf{g}}^{(l)}\right)\right)^{-1} &= \sum_{i=0}^{\infty} \left(\eta(l) \left(\mathbf{I} - \widehat{\mathbf{R}}_{\mathbf{f}\,\mathbf{g}}^{(l)}\right)\right)^{i} \\
&\approx \mathbf{I} + \eta(l) \left(\mathbf{I} - \widehat{\mathbf{R}}_{\mathbf{f}\,\mathbf{g}}^{(l)}\right).
\end{aligned} \tag{7.49}$$

Finally, substituting (7.49) into (7.47), we obtain the following simplified robust learning algorithm

$$\boxed{\mathbf{W}(l+1) = \mathbf{W}(l) + \eta(l) \left(\mathbf{I} - \widehat{\mathbf{R}}_{\mathbf{f}\,\mathbf{g}}^{(l)}\right) \mathbf{W}(l).} \tag{7.50}$$

It is interesting to note that this algorithm can be rewritten as

$$\mathbf{W}(l+1) = \mathbf{W}(l) + \eta(l) \left[\mathbf{I} - \mathbf{W}(l) \widehat{\mathbf{R}}_{\mathbf{h}\,\mathbf{g}}^{(l)}\right] \mathbf{W}(l), \tag{7.51}$$

which is an iterative version of the quasi-Newton algorithm for the inversion of $\widehat{\mathbf{R}}_{\mathbf{h}\,\mathbf{g}}^{(l)}$ (see [315] for more detail on matrix inversion based iterative algorithms). In addition, the algorithm (7.50) is also the batch version of the family of adaptive algorithms proposed by Cichocki *et al.* in [279, 277, 270]. These algorithms exhibit the equivariant property in the absence of noise.

To summarize, we presented the derivation of a family of dual robust adaptive algorithms [326]:

for estimating the mixing matrix

$$\widehat{\mathbf{H}}(k+1) = \widehat{\mathbf{H}}(k)\left[\mathbf{I} - \eta(k)\,(\mathbf{I} - \widehat{\mathbf{R}}_{\mathbf{f}\,\mathbf{g}}^{(k)})\right], \tag{7.52}$$

$$\widehat{\mathbf{H}}(k+1) = \widehat{\mathbf{H}}(k)\left[\mathbf{I} + \eta(k)\,(\mathbf{I} - \widehat{\mathbf{R}}_{\mathbf{f}\,\mathbf{g}}^{(k)})\right]^{-1} \tag{7.53}$$

and for estimating the separating matrix

$$\mathbf{W}(k+1) = \left[\mathbf{I} + \eta(k)\,(\mathbf{I} - \widehat{\mathbf{R}}_{\mathbf{f}\,\mathbf{g}}^{(k)})\right]\,\mathbf{W}(k), \tag{7.54}$$

$$\mathbf{W}(k+1) = \left[\mathbf{I} - \eta(k)\,(\mathbf{I} - \widehat{\mathbf{R}}_{\mathbf{f}\,\mathbf{g}}^{(k)})\right]^{-1}\mathbf{W}(k), \tag{7.55}$$

where

$$\widehat{\mathbf{R}}_{\mathbf{f}\,\mathbf{g}}^{(k)} = (1-\eta_0)\widehat{\mathbf{R}}_{\mathbf{f}\,\mathbf{g}}^{(k-1)} + \eta_0 \mathbf{f}(\mathbf{y}(k))\mathbf{g}^T(\mathbf{y}(k)).$$

It should be noted that the algorithms (7.52) and (7.54) have slightly different convergence speed, since the product $\mathbf{W}(k+1)\widehat{\mathbf{H}}(k+1)$ is not exactly equal to the identity matrix, even if we start from the same initial conditions (say $\mathbf{W}(0) = \widehat{\mathbf{H}}(0) = \mathbf{I}$). This feature can be explained by the approximation in (7.49). However, it is straightforward to show that the following equality holds

$$\|\mathbf{W}(k+1)\widehat{\mathbf{H}}(k+1) - \mathbf{I}\|_p = \eta^2(k)\,\|\widehat{\mathbf{R}}_{\mathbf{f}\,\mathbf{g}}^{(k)} - \mathbf{I}\|_p^2. \tag{7.56}$$

This means that both algorithms have almost the same convergence properties if $\eta(k)$ is small and/or $\mathbf{R}_{\mathbf{f}\,\mathbf{g}}^{(k)}$ is very close to the identity matrix. Moreover, assuming that $\eta(k)$ satisfies (7.48) we have the following constraint

$$\|\mathbf{W}(k+1)\widehat{\mathbf{H}}(k+1) - \mathbf{I}\|_p < 1. \tag{7.57}$$

It is interesting to note that the learning rules (7.52)-(7.55) can be generalized to the following forms

$$\widehat{\mathbf{H}}(k+1) = \widehat{\mathbf{H}}(k)\ \exp[-\eta_k \mathbf{F}(\mathbf{y}(k))], \tag{7.58}$$

and

$$\mathbf{W}(k+1) = \exp[\eta_k \mathbf{F}(\mathbf{y}(k))]\ \mathbf{W}(k), \tag{7.59}$$

where the matrix $\mathbf{F}(\mathbf{y}(k))$ can take various forms as discussed in previous sections (see also Chapter 6 for details).

7.2.1 Derivation of Robust ICA Algorithm Using Generalized Natural Gradient Approach

The family of ICA algorithms discussed in the previous section can be derived in slightly less general form using the generalized natural gradient formula [453]

$$\Delta\mathbf{W} = -\eta\,\frac{\partial J(\mathbf{y},\mathbf{W})}{\partial \mathbf{W}}\mathbf{W}^T\mathbf{D}_1(\mathbf{y})\mathbf{W}, \tag{7.60}$$

where $J(\mathbf{y}, \mathbf{W})$ is a suitably selected cost function and $\mathbf{D}_1(\mathbf{y})$ is a scaling positive definite diagonal matrix. It should be noted that similarly to the standard natural gradient, the formula (7.60) ensures stable gradient descent search of a local minimum of the cost function because the term $\mathbf{W}^T \mathbf{D}_1(\mathbf{y}) \mathbf{W}$ is a symmetric positive definite matrix for a nonsingular separating matrix.

Let us consider an example of the cost function given by

$$J(\mathbf{y}, \mathbf{W}) = -\log|\det(\mathbf{W})| - \sum_{i=1}^{n} E\{\log(q_i(y_i))\}. \tag{7.61}$$

The gradient of the cost function (7.61) can be expressed as

$$\frac{\partial J(\mathbf{y}, \mathbf{W})}{\partial \mathbf{W}} = -\mathbf{W}^{-T} + E\{\mathbf{f}(\mathbf{y})\mathbf{x}^T\}, \tag{7.62}$$

where $\mathbf{f}(\mathbf{y}) = [f_1(y_1), f_2(y_2), \ldots, f_n(y_n)]^T$, with $f_i(y_i) = -d\log(q_i(y_i)/dy_i$. Hence, applying the generalized natural gradient formula (7.60), we obtain the known robust learning rule

$$\Delta\mathbf{W}(l) = \boldsymbol{\eta}(l)\ \left[\boldsymbol{\Lambda}(l) - \left\langle \mathbf{f}(\mathbf{y})\mathbf{g}^T(\mathbf{y})\right\rangle\right] \mathbf{W}(l), \tag{7.63}$$

where $\boldsymbol{\Lambda} = \mathbf{D}_1(\mathbf{y})$ and $\mathbf{g}(\mathbf{y}) = \mathbf{y}^T \mathbf{D}_1(\mathbf{y}) = [g_1(y_1), g_2(y_2), \ldots, g_n(y_n)]^T$.

In a special case, for symmetric pdf distributions of sources and odd activation functions $f_i(y_i)$ and

$$\mathbf{D}_1 = \text{diag}\{\langle |y_1|^p\rangle, \langle |y_2|^p\rangle, \ldots, \langle |y_n|^p\rangle\},$$

we obtain the median learning rule for $p = -1$

$$\Delta\mathbf{W}(l) = \boldsymbol{\eta}(l)\ \left[\boldsymbol{\Lambda}(l) - \left\langle \mathbf{f}(\mathbf{y})[\text{sign}(\mathbf{y})]^T\right\rangle\right] \mathbf{W}(l), \tag{7.64}$$

where $\text{sign}(\mathbf{y}) = [\text{sign}(y_1), \text{sign}(y_2), \ldots \text{sign}(y_n)]^T$. Simulation results show that such a median learning rule with the sign activation function is robust to additive noise.

7.2.2 Practical Implementation of the Algorithms

The generalized covariance matrix $\widehat{\mathbf{R}}_{\mathbf{f}\,\mathbf{g}}^{(k)}$ employed in the robust learning algorithms (7.52) and (7.54) can be typically estimated as a statistical average on the basis of available (incoming) output data. There are two possibilities: On-line adaptations in which the generalized covariance matrix $\widehat{\mathbf{R}}_{\mathbf{f}\,\mathbf{g}}^{(l)}$ is replaced by its on-line sample estimate for each discrete time instant k, i.e.,

$$\widehat{\mathbf{R}}_{\mathbf{f}\,\mathbf{g}}^{(k)} \approx \mathbf{f}(\mathbf{y}(k))\, \mathbf{g}^T(\mathbf{y}(k)) \tag{7.65}$$

and batch adaptations where, if we assume that the stationarity and ergodicity properties hold on a block of observations of L samples, we can replace the statistical average by the moving average

$$\widehat{\mathbf{R}}_{\mathbf{f}\,\mathbf{g}}^{(k)} = (1 - \eta_0)\, \widehat{\mathbf{R}}_{\mathbf{f}\,\mathbf{g}}^{(k-1)} + \eta_0\, \mathbf{f}(\mathbf{y}(k))\, \mathbf{g}^T(\mathbf{y}(k)). \tag{7.66}$$

Another important practical issue is the estimation of the learning rate in order to ensure both numerical stability and possibly high convergence speed. Let us start by considering the on-line version of the algorithm (7.43) with one sample estimate of the covariance matrix

$$\mathbf{W}(k+1) = \left[\mathbf{I} \;-\; \eta\, \left(\mathbf{I} - \mathbf{f}(\mathbf{y}(k))\mathbf{g}^T(\mathbf{y}(k))\right)\right]^{-1} \mathbf{W}(k). \tag{7.67}$$

To ensure numerical stability of the algorithm, the constraint $\eta < \frac{1}{1+|\mathbf{g}^T(\mathbf{y})\mathbf{f}(\mathbf{y})|}$ must be satisfied whenever $\mathbf{g}^T(\mathbf{y})\,\mathbf{f}(\mathbf{y}) < 0$. Let us rewrite (7.67) in terms of $\eta_0 = \frac{\eta}{1-\eta}$ as

$$\mathbf{W}(k+1) = (1+\eta_0)\, \left[\mathbf{I} + \eta_0\, \mathbf{f}(\mathbf{y}(k))\mathbf{g}^T(\mathbf{y}(k))\right]^{-1} \mathbf{W}(k). \tag{7.68}$$

Applying the Sherman-Morrison inversion formula (see Appendix), we obtain

$$\mathbf{W}(k+1) = (1+\eta_0)\left(\mathbf{I} \;-\; \eta_0\, \frac{\mathbf{f}(\mathbf{y})\mathbf{g}^T(\mathbf{y})}{1+\eta_0\, \mathbf{g}^T(\mathbf{y})\mathbf{f}(\mathbf{y})}\right) \mathbf{W}(k).$$

For small values of η_0, it is possible to neglect the higher order terms with $\eta_0^i \approx 0$ for $i \geq 2$, which yields a normalized stochastic version of the algorithm (7.54)

$$\boxed{\mathbf{W}(k+1) = \left(\mathbf{I} + \eta_0\, \frac{\mathbf{I} - \mathbf{f}(\mathbf{y})\mathbf{g}^T(\mathbf{y})}{1+\eta_0\, \mathbf{g}^T(\mathbf{y})\mathbf{f}(\mathbf{y})}\right) \mathbf{W}(k).} \tag{7.69}$$

It is interesting to note that equation (7.69) can be interpreted as the algorithm (7.54) with the self-adaptive learning rate

$$\eta(k) = \frac{\eta_0}{1+\eta_0\, \mathbf{g}^T(\mathbf{y}(k))\mathbf{f}(\mathbf{y}(k))}. \tag{7.70}$$

To prevent the algorithm from escaping from the continuous region, where it should be stable, the variable learning rate should satisfy condition (7.48). One way to ensure this is to choose $\eta < 1$ and setting $\eta < \frac{1}{|\mathbf{g}^T(\mathbf{y})\mathbf{f}(\mathbf{y})|}$ whenever $\mathbf{g}^T(\mathbf{y})\mathbf{f}(\mathbf{y}) < 0$.

An alternative way to estimate a simple learning rate that satisfies (7.48) is by employing the following condition: Impose the constraint $\eta < 1$, by replacing the denominator of (7.70) by its positive upper bound value $1 + \eta_0\, |\mathbf{g}^T(\mathbf{y})\mathbf{f}(\mathbf{y})|$, i.e.,:

$$\boxed{\eta(k) = \frac{\eta_0}{1+\eta_0\, |\mathbf{g}^T(\mathbf{y}(k))\mathbf{f}(\mathbf{y}(k))|},} \tag{7.71}$$

where $0 < \eta_0 < 1$ is a fixed constant. This normalization is similar to that used in the normalized or posterior LMS [390] and is equivalent to that proposed in [148] for the EASI algorithm. In practical implementations of the on-line algorithm, we will typically choose $\eta << 1$ because the single noisy sample estimate of $\mathbf{R}_{\mathbf{f}\,\mathbf{g}}^{(l)}$ introduces considerable bias into the algorithm.

For the batch adaptation, we can use the following step size which is a generalization of Eq. (7.71)

$$\boxed{\eta(k) = \frac{\eta_0}{1+\eta_0\, ||\mathbf{R}_{\mathbf{f}\,\mathbf{g}}^{(k)}||_p},} \tag{7.72}$$

where $0 < \eta_0 < 1$. Note that when $p = 2$ and $\mathbf{R}_{\mathbf{f}\,\mathbf{g}}^{(k)} \approx \mathbf{f}(\mathbf{y}(k))\mathbf{g}^T(\mathbf{y}(k))$, Eq. (7.72) reduces to (7.71). Taking into account that $\|\mathbf{R}_{\mathbf{f}\,\mathbf{g}}^{(k)} - \mathbf{I}\|_p \leq 1 + \|\mathbf{R}_{\mathbf{f}\,\mathbf{g}}^{(k)}\|_p$, it is easy to show that the batch step size (7.72) always satisfies condition (7.48) for any given p-norm. The selection of the norm in (7.72) depends on practical considerations, i.e., acceptable complexity of the algorithm. Some norms such as the 1-norm or the ∞-norm are easy to compute and therefore lead to a relatively simple evaluation of step-sizes. On the other hand, the 2-norm of matrices provides the fastest convergence for a fixed η, but its exact evaluation is computationally more demanding than the 1-norm. In practical implementations of the batch algorithm, we will typically choose η close to unity, since we are interested in an algorithm that converges as fast as possible.

7.2.3 Special Forms of the Flexible Robust Algorithm

In this section, we show how several well known learning algorithms for BSS discussed previously can be derived as special cases of the algorithm (7.54). This enables us to interpret them as quasi-Newton algorithms that iteratively invert a generalized covariance matrix. We focus here on adaptive on-line algorithms only, although the extension to batch algorithms is straightforward.

7.2.4 Decorrelation Algorithm

Choosing $\mathbf{f}(\mathbf{y}) = \mathbf{g}(\mathbf{y}) = \mathbf{y}$, $\mathbf{\Lambda} = \mathbf{I}$ and the normalized step given by (7.71) the adaptation rule (7.54) reduces to the decorrelation algorithm proposed by Almeida *et al.* [13] (see also [390])

$$\boxed{\mathbf{W}(k+1) = \mathbf{W}(k) + \eta_0 \frac{\mathbf{I} - \mathbf{y}\,\mathbf{y}^T}{1 + \eta_0\,\mathbf{y}^T\mathbf{y}} \mathbf{W}(k).} \tag{7.73}$$

7.2.5 Natural Gradient Algorithms

Selecting $\mathbf{g}(\mathbf{y}) = \mathbf{y}$ and using the normalized learning rate given by (7.71), the algorithm (7.54) simplifies to

$$\boxed{\mathbf{W}(k+1) = \mathbf{W}(k) + \eta_0 \frac{\mathbf{I} - \mathbf{f}(\mathbf{y})\,\mathbf{y}^T}{1 + \eta_0\,|\mathbf{y}^T\,\mathbf{f}(\mathbf{y})|} \mathbf{W}(k),} \tag{7.74}$$

which is a normalized version of the Natural Gradient algorithm developed rigorously in Chapter 6 [20, 29, 326].

7.2.6 Generalized EASI Algorithm

The family of EASI algorithms was proposed by Cardoso and Laheld [148] and extended by Karhunen *et. al.* in [671, 672]. To derive this algorithm from the learning rule (7.54), we employ an approach similar to that described in [148]. Let us start decomposing the separating stage into two stages, i.e., $\mathbf{W} = \mathbf{W}_b \mathbf{W}_a$. The first matrix $\mathbf{W}_a$ will be selected to decorrelate the input, i.e., to diagonalize the symmetric matrix $\mathbf{R}_{\mathbf{y}\mathbf{y}}$. This can be done with the algorithm (7.73). On the other hand, the second matrix $\mathbf{W}_b$ will be selected to diagonalize the matrix $\mathbf{R}_{\mathbf{f}\,\mathbf{g}}$. This objective can be achieved using the following on-line version of (7.50)

$$\mathbf{W}_b(k+1) = \mathbf{W}_b(k) + \eta_0 \frac{\mathbf{I} - \mathbf{f}(\mathbf{y})\, \mathbf{g}^T(\mathbf{y})}{1 + \eta_0\, |\mathbf{g}^T(\mathbf{y})\mathbf{f}(\mathbf{y})|}\, \mathbf{W}_b(k). \tag{7.75}$$

However, in order to integrate (7.75) and (7.73) into a single recursion for the overall separating matrix $\mathbf{W}$, it is necessary that both adaptations be approximately orthogonal (at the first order in η) [148, 326]. This can be accomplished if we replace $\mathbf{I} - \mathbf{f}(\mathbf{y})\mathbf{g}^T(\mathbf{y})$ by its projection onto the space of skew-symmetric matrices. Then, the resulting algorithm becomes

$$\mathbf{W}_b(k+1) = \left(\mathbf{I} - \frac{\eta_0}{2} \frac{\mathbf{f}(\mathbf{y})\mathbf{g}^T(\mathbf{y}) - \mathbf{g}(\mathbf{y})\mathbf{f}^T(\mathbf{y})}{1 + \eta_0\, |\mathbf{g}^T(\mathbf{y})\mathbf{f}(\mathbf{y})|}\right) \mathbf{W}_b(k).$$

Combining (7.73) and (7.75) so that $\mathbf{W}(k+1) = \mathbf{W}_b(k+1)\, \mathbf{W}_a(k+1)$ and neglecting the higher order terms with η (which is a valid approximation if η is small enough), we arrive at

$$\boxed{\mathbf{W}(k+1) = \left(\mathbf{I} + \eta_0 \frac{\mathbf{I} - \mathbf{y}\mathbf{y}^T}{1 + \eta_0\, |\mathbf{y}^T\mathbf{y}|} - \frac{\eta_0}{2} \frac{\mathbf{f}(\mathbf{y})\mathbf{g}^T(\mathbf{y}) - \mathbf{g}(\mathbf{y})\mathbf{f}^T(\mathbf{y})}{1 + \eta_0\, |\mathbf{g}^T(\mathbf{y})\mathbf{f}(\mathbf{y})|}\right) \mathbf{W}(k),} \tag{7.76}$$

which is a normalized version of the generalized EASI algorithm proposed by Karhunen and Pajunen in [671, 148].

7.2.7 Non-linear PCA Algorithm

When the separation matrix $\mathbf{W}(k)$ is restricted to be orthogonal, it is possible to relate the algorithm (7.54) with the non-linear PCA algorithm developed by Oja and Karhunen in [908, 670]. First, let us redefine the generalized (non-linear) covariance matrix as

$$\bar{\mathbf{R}}^{(l)}_{\mathbf{f}\,\mathbf{g}} = \langle \mathbf{f}(\mathbf{y})\, \mathbf{g}^T(\mathbf{y}) \rangle + \mathbf{I},$$

where $\mathbf{f}(\mathbf{y})$ is an odd function and $\mathbf{g}(\mathbf{y}) = \mathbf{f}(\mathbf{y}) - \mathbf{y}$. Therefore, the on-line implementation of the algorithm (7.54) reduces to

$$\begin{aligned} \mathbf{W}(k+1) &= \mathbf{W}(k) - \eta\, \mathbf{f}(\mathbf{y})\mathbf{g}^T(\mathbf{y})\mathbf{W}(k) \\ &= \mathbf{W}(k) - \eta\, \mathbf{f}(\mathbf{y}) \left(\mathbf{f}^T(\mathbf{y}) - \mathbf{y}^T\right) \mathbf{W}(k). \end{aligned} \tag{7.77}$$

Because we are assuming that $\mathbf{W}(k)$ is an orthogonal matrix, i.e., $\mathbf{W}^T(k)\mathbf{W}(k) = \mathbf{I}$, we can rewrite (7.77) as

$$\boxed{\mathbf{W}(k+1) = \mathbf{W}(k) + \eta\, \mathbf{f}(\mathbf{y}) \left(\mathbf{x}^T - \mathbf{f}^T(\mathbf{y})\mathbf{W}(k)\right),} \tag{7.78}$$

which is the non-linear PCA algorithm proposed in [908, 670]. However, it should be noted that the dynamics of the algorithm (7.54) and the non-linear **PCA** algorithm are different in general, because none of the algorithms force the orthogonality of $\mathbf{W}(k)$ at each iteration.

7.2.8 Flexible ICA Algorithm for Unknown Number of Sources and their Statistics

In practice, the number of sources is generally unknown and may change over time. Moreover, their statistics are also usually unknown. When the number of sensors is larger or equal to the number of sources, there are at least three different approaches to solve this problem. The first possible approach is to apply PCA or robust orthogonalization based on SVD (discussed in Chapters 3 and 4) in order to determine the number of sources and simultaneously reduce the total mixing matrix to a square $n \times n$ nonsingular matrix. The second approach is based on the extraction of the sources sequentially one by one, using the methods described in Chapter 5. The third efficient approach, discussed in this section, is to apply directly the separating network with the number of outputs equal to the number of sensors, so a separating matrix is a square matrix.

In this case, we propose to use the following nonholonomic algorithm for $\mathbf{W} \in \mathbb{R}^{m \times m}$ with $m \geq n$

$$\Delta\mathbf{W}(k) = \eta(k) \left[\mathbf{\Lambda}^{(k)} - \mathbf{R}_{\mathbf{f}\,\mathbf{g}}^{(k)}\right] \mathbf{W}(k), \tag{7.79}$$

where

$$\mathbf{\Lambda}^{(k)} = (1 - \eta_0)\, \mathbf{\Lambda}^{(k-1)} + \eta_0 \; \mathrm{diag}\{\mathbf{f}(\mathbf{y}(k))\, \mathbf{g}^T(\mathbf{y}(k))\}, \tag{7.80}$$

$$\mathbf{R}_{\mathbf{f}\,\mathbf{g}}^{(k)} = (1 - \eta_0)\, \mathbf{R}_{\mathbf{f}\,\mathbf{g}}^{(k-1)} + \eta_0\, \mathbf{f}(\mathbf{y}(k))\, \mathbf{g}^T(\mathbf{y}(k)), \tag{7.81}$$

whereas diag$\{\mathbf{R}\}$ denotes a diagonal matrix which is the main diagonal of $\mathbf{R}$.

Usually for a noise-free case when the number of sensors is larger than the number of sources, some outputs should be automatically set to zero if they are below some threshold value. This way, for the ideal noiseless case, the redundant $(m-n)$ output signals y_i should decay to zero during the adaptive learning process. Then, only n outputs will correspond to the recovered sources. It should be noted that the matrix $\mathbf{W}$ is assumed to be a square matrix, i.e., the number of outputs of the separating system is equal to the number of sensors, although in practice the number of sources can be less than the number of sensors $(m \geq n)$. Such a model is justified by the fact that the number of sources may change over time. Furthermore, in practice, we have additive noise which can be considered as auxiliary unknown sources; so it is reasonable to extract these noise signals, too.

For small magnitudes of the additive noise and $m > n$, the above algorithm may be unstable in the sense that the Frobenius norm of the separating matrix $\mathbf{W}(k)$ slowly tends to infinity for a large number of iterations. In order to stabilize the algorithm, we can use the modified algorithm

$$\boxed{\Delta\mathbf{W}(k) = \eta(k) \left[\mathbf{\Lambda}^{(k)} - \mathbf{R}_{\mathbf{f}\,\mathbf{g}}^{(k)} - \beta\gamma(k)\, \mathbf{I}_m\right] \mathbf{W}(k),} \tag{7.82}$$

with the forgetting factor:

$$\gamma(k) = \mathrm{tr}\left(\mathbf{W}^T(k)(\mathbf{\Lambda}^{(k)} - \mathbf{R}_{\mathbf{f}\,\mathbf{g}}^{(k)})\mathbf{W}(k)\right). \tag{7.83}$$

and sacling factor $\beta > 0$. It has been shown in Chapter 4 that such an algorithm stabilizes the Frobenius norm of $\mathbf{W}(k)$ to value β^{-1}.

Nonlinear functions of vectors $\mathbf{f}[\mathbf{y}(k)]$ and $\mathbf{g}[\mathbf{y}(k)]$ should be suitably designed as explained in the previous chapter. In order to satisfy stability conditions if the measured signals $x_i(k)$ contain mixtures of both sub-Gaussian and super-Gaussian sources (see Appendix A), we can use the switching activation functions:

$$f_i(y_i) = \begin{cases} \varphi_i(y_i) & \text{for } \kappa_4(y_i) > \delta \\ y_i & \text{otherwise} \end{cases} \tag{7.84}$$

$$g_i(y_i) = \begin{cases} y_i & \text{for } \kappa_4(y_i) > -\delta \\ \varphi_(y_i) & \text{otherwise} \end{cases} \tag{7.85}$$

$\kappa_4(y_i) = E\{y_i^4\}/E^2\{y_i^2\} - 3$ is the normalized value of kurtosis, $\delta \geq 0$ is a small threshold and $\varphi_i(y_i)$ are suitably designed nonlinear functions depending on the distribution of source signals. There are many possible choices of activation functions such as $\varphi_i(y_i) = \tanh(\theta_i y_i)$ or $\varphi_i(y_i) = \text{sign}(y_i)\exp(-\theta_i|y_i|)$, with $\theta_i \geq 0$ [1]. It should be noted that such nonlinearities provide a degree of robustness to outliers that is not shared by nonlinearities of the form $f_i(y_i) = \text{sign}(y_i)|y_i|^{r_i - 1}$ (for $r_i \geq 3$).

The above learning algorithms (7.79)-(7.85) monitor and estimate the statistics of each output signal and depending on the sign or value of its normalized kurtosis (which is a measure of the distance from Gaussianity) automatically select (or switch) a suitable nonlinear activation functions in order to successfully separate source signals. In this approach, activation functions are chosen to be adaptive time-varying nonlinearities. For this choice, the parameters $\theta_i \geq 2$ can be either fixed in value or adapted during the learning process [269, 1302].

7.3 BLIND SOURCE SEPARATION WITH NON-NEGATIVITY CONSTRAINTS

In many applications such as computer tomography and biomedical image processing non-negative constraints are imposed for entries ($h_{ij} \geq 0$) of the mixing matrix $\mathbf{H}$ and/or estimated source signals ($s_j(k) \geq 0$) [114, 874, 948, 973, 568]. Moreover, recently several authors suggested that a decomposition of a observation $\mathbf{X} = [\mathbf{x}(1), \mathbf{x}(2), \ldots \mathbf{x}(N)] = \mathbf{HS}$ into non-negative factors or Non-Negative Matrix Factorization (NMF), is able to produce useful and meaningful representation of real- world data, especially in image analysis, hyperspectral data processing, in biological modeling and sparse coding. [948, 731, 972, 568].

In this section, we present very simple and practical technique for estimation of nonnegative sources and entries of the mixing matrix using standard ICA approach and suitable postprocessing. In other words, we will show that by simple modifications of existing ICA or BSS algorithms we are able to satisfy non-negativity constraints of sources and simultaneously impose they are sparse or independent as possible. Without loss of generality, we assume that matrix and all sources are non-negative, i.e., $s_j(k) = \tilde{s}_j(k) + c_j \geq 0 \quad \forall j, k$.

[1]Since the activation function $\varphi_i(y_i) = \text{sign}(y_i)\exp(-\theta_i|y_i|)$ is not a differentiable function at zero it can be closely approximated by $\varphi_i(y_i) = \tanh(\beta y_i)\exp(-\theta_i|y_i|)$, with $\beta >> 1$.

Moreover, we assume that the zero mean subcomponents $\widetilde{s}_j(k)$ are mutually statistically independent[2].

Furthermore, we may assume if necessary that entries of nonsingular mixing matrix $\mathbf{H}$ are also non-negative i.e, $h_{ij} \geq 0 \;\; \forall i, j$ and optionally that columns of the mixing matrix are normalized vectors with 1-norm equal to unity [948, 874, 114].

We propose a two stage procedure. In the first stage, we can apply any standard ICA or BSS algorithm for zero-mean (pre-processed) sensor signals without any constraints in order to estimate the separating matrix $\mathbf{W}$ up to arbitrary scaling and permutation and estimate the waveforms of the original sources by projecting the (non-zero mean) sensor signals $s_j(k)$ via the estimated separating matrix ($\widehat{\mathbf{s}}(k) = \mathbf{W}_* \mathbf{x}(k)$). It should be noted that since the global mixing-unmixing matrix defined as $\mathbf{G} = \mathbf{W}_* \mathbf{H}$ after successful extraction of the sources is a generalized permutation matrix containing only one nonzero (negative or positive) element in each row and each column, thus each estimated source in the first stage will be either non-negative or non-positive for every time instant.

In the second stage in order to recover the original waveforms of the sources with correct sign all the estimated non positive sources should be inverted, i.e. multiplied by -1. It should be noted that this procedure is valid for an arbitrary nonsingular mixing matrix with both positive and negative elements.

If the original mixing matrix $\mathbf{H}$ has non-negative entries then in order to identify it the corresponding vectors of the estimating matrix $\widehat{\mathbf{H}} = \mathbf{W}^{-1}$ should be multiplied by the factor -1. In this way, we can estimate the original sources and blindly identify the mixing matrix satisfying non-negativity constraints. Furthermore, if necessary, we can redefine $\widehat{\mathbf{H}}$ and $\hat{\mathbf{s}}$ as follows: $\widehat{\hat{h}}_{kj} = \hat{h}_{kj} / \sum_{i=1}^{n} \hat{h}_{ij}$ and $\widehat{\hat{s}}_j = \hat{s}_j (\sum_{i=1}^{n} \hat{h}_{ij})$. After such transformation, the new estimated mixing matrix $\widehat{\widehat{\mathbf{H}}}$ has column sums equal to one and the vector $\mathbf{x} = \widehat{\widehat{\mathbf{H}}} \, \widehat{\widehat{\mathbf{s}}}$ is unchanged.

Summarizing, from this simple explanation it follows that it is not necessary to develop any special kind of algorithms for BSS with non-negativity constraints (see for example [731, 973, 948]). Any standard ICA algorithm (batch or on-line) can be applied first for zero-mean signals, and the waveforms of the original sources and optionally the desired mixing matrix with non-negativity constraints can be estimated exploiting basic properties of the assumed model.

7.4 COMPUTER SIMULATIONS

The validity and performance of the adaptive learning algorithms presented in this chapter have been extensively simulated for a large variety of difficult separation problems. Very promising results have been obtained. We shall present here only several illustrative examples. In all examples, we used $f(y) = y^3$ and $g(y) = \tanh(y)$, although we explored

[2]It should be noted that the non negative sources $s_j(k) = \tilde{s}_j(k) + c_j$ are non independent since dc (constant) sub-components c_j are dependent. Due to this reason we refer to the problem as non-negative blind source separation rather than non-negative ICA [972, 973].

many other functions with similar results. For all of these examples, the H–J algorithm, (Eq. (7.7)) failed or performed very poorly.

Example 7.1 (Blind separation for ill-conditioned problem) Five badly scaled signals are shown in Fig. 7.6 (a):

$$\begin{aligned}
s_1(t) &= 10^{-6} \times [\sin(350t)][\sin(60t)], \\
s_2(t) &= 10^{-5} \times \text{triangular}\,(70t) \qquad \text{triangular waveform}, \\
s_3(t) &= 10^{-4} \times [\sin(800t)][\sin(80t)], \\
s_4(t) &= 10^{-3} \times [\cos(400t + 4\cos(60t))], \\
s_5(t) &= 1.0 \times n(t) \qquad \text{Gaussian noise in the range: -1 to 1.}
\end{aligned}$$

These signals are mixed together by the 5×5 Hilbert mixing matrix $\mathbf{H}$, which is extremely ill-conditioned. We employed a single-layer feed-forward neural network with the learning algorithm (7.25). The separation process after 2000 iterations (time of 500 ms) is shown in Fig. 7.6 (a) and (b). Note that the amplitude ratios of the source signals are $1 : 10 : 10^2 : 10^3 : 10^6$. Of course, in this example the very weak and badly scaled signals are absolutely not visible from the observed sensor signals which all appear identical to the noise signal, within a scale factor (see Fig. 7.6 (a)). However, the source signals are successfully and completely retrieved by using for example, the learning algorithm, (7.25), as shown in Fig. 7.6 (c). The other local algorithms had difficulties in separating such weak and badly scaled signals.

Example 7.2 (Extraction of Fetal ECG sources) The ECG data of a pregnant woman shown in Fig. 7.7 (a) are the electric voltage potential recordings during an 8-channel experiment. Only 2.5 seconds of the recordings (resampled at 200 Hz) are displayed. In this experiment, the electrodes were placed on the abdomen and the cervix of the mother. Abdominal signals measured near the fetus are shown in channels 1 to 5. The weak fetal contributions are contained in x_1 to x_5, although they are not clearly visible. The ECG raw data measured through 8 channels are dominated by the mother's ECG (MECG). The flexible ICA algorithm (7.79) was applied to process the ECG raw data, and the result is shown in Figure 7.7. The 3rd and 5th independent components (output signals y_3, y_5) correspond to the FECG signal. The 2nd and 7th independent components contain the MECG. The rest of the extracted signals might contain noise contributions. The weak FECG signal was well extracted by the ICA algorithm (7.79), whereas it was not well extracted with PCA.

Example 7.3 (Estimation of sources for noisy data) Five independent sources mixed by a randomly generated full column rank mixing matrix $\mathbf{H} \in \mathbb{R}^{9\times 5}$ and additive uniform noise was added with SNR 15dB. The plots of original sources and their noisy mixtures are shown respectively in Fig 7.8 (a) and (b). In order to reconstruct original sources the learning rule (7.82) has been applied for square separating matrix $\mathbf{W} \in \mathbb{R}^{9\times 9}$ with $\mathbf{F}(y) = \mathbf{I}_5 - \langle \mathbf{f}(y)\mathbf{y}^T \rangle$, $\beta = 0.01$ and $\mathbf{f}(y) = \tanh(2\mathbf{y})$. After 100 iterations the algorithm was able to estimate the sources (see Fig. 7.8 (c) by keeping the Frobenius norm of the separating matrix in the range $||\mathbf{W}(k)||_F^2 = \mathrm{tr}(\mathbf{W}^T(k)\mathbf{W}(k)) \approx 100$. Without a stabilizing factor the norm of $\mathbf{W}$ grows to a very large value as the number of iteration increases.

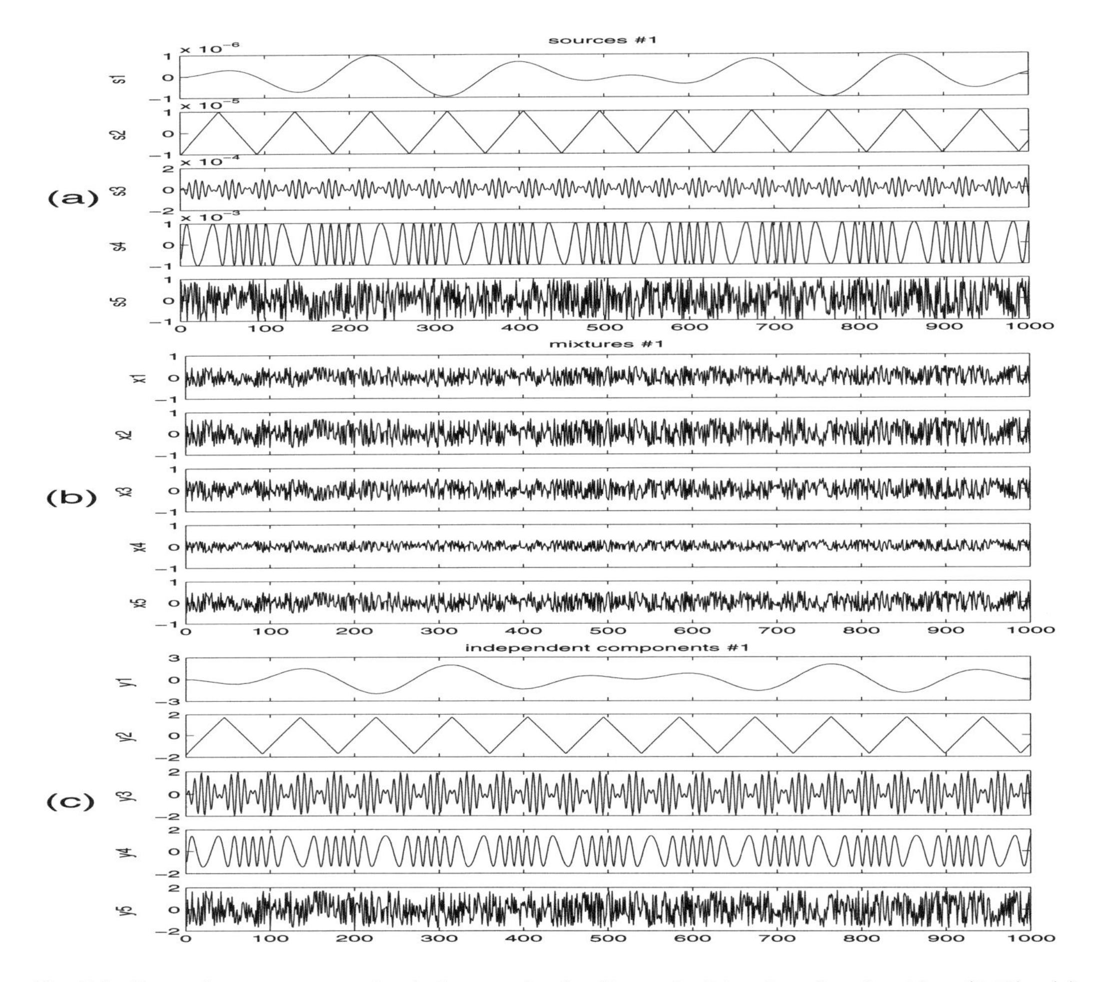

Fig. 7.6 Exemplary computer simulation results for Example 7.1 using the algorithm (7.25): (a) Waveforms of primary sources, (b) noisy sensor signals and (c) reconstructed source signals.

Example 7.4 (Estimation of sparse sources with non-negativity constraints) In this example sparse non-negative source signals are mixed by a normalized sparse matrix with non-negative entries (assumed to be unknown).

$$\mathbf{H} = \begin{bmatrix} 0 & 0.7 & 0.2 & 0 \\ 0.4 & 0 & 0 & 0.3 \\ 0.6 & 0.3 & 0 & 0.5 \\ 0 & 0 & 0.8 & 0.2 \end{bmatrix} \tag{7.86}$$

The mixed (observed sensor) signals are shown in Fig. 7.9 (a). After standard prewhitening the natural gradient algorithm (7.54) with hyperbolic tangent nonlinearity has been applied. The estimated sources are shown in Fig 7.9 (b) and the estimating mixing matrix has the

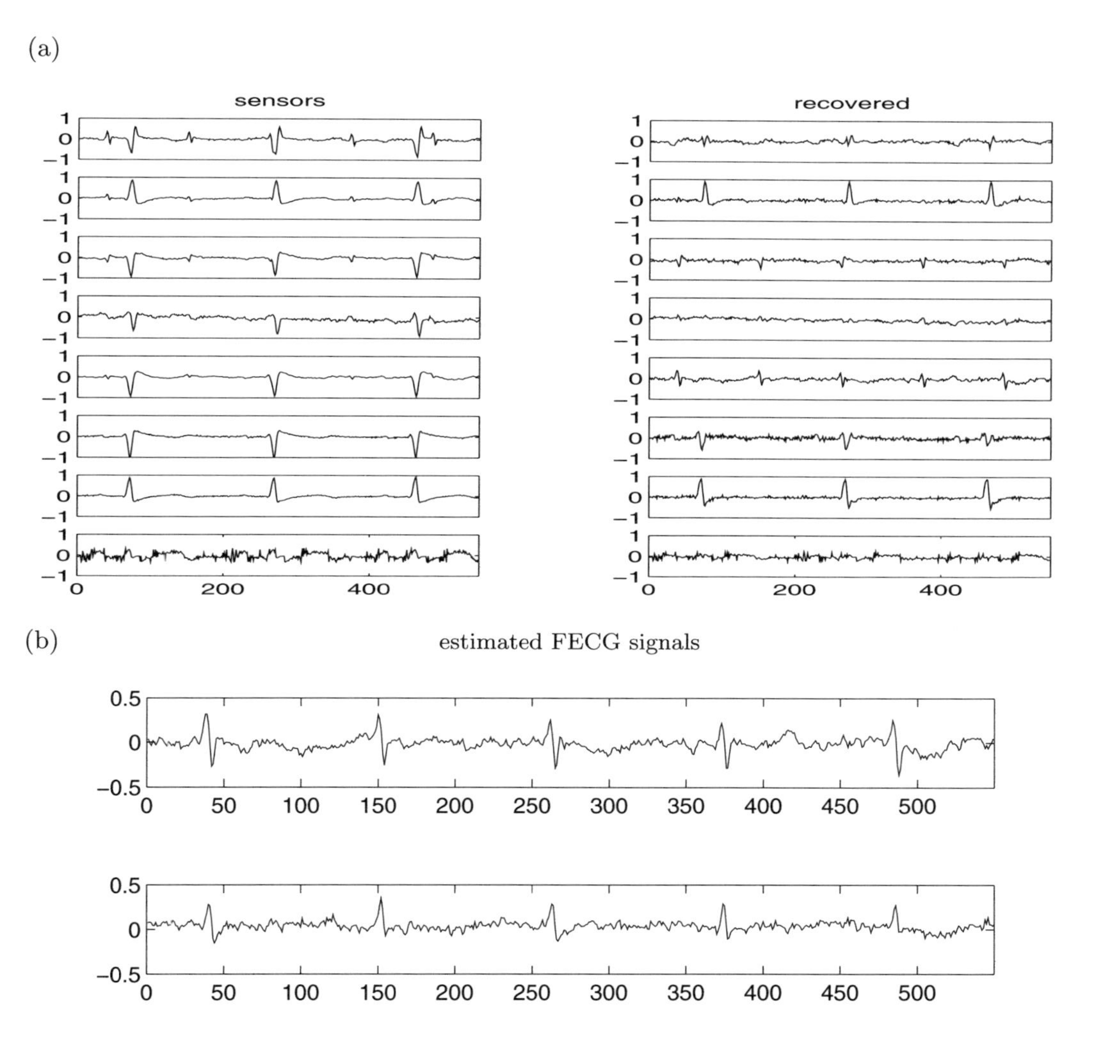

Fig. 7.7 (a) Eight ECG signals are separated into: Four maternal signals, two fetal signals and two noise signals. (b) Detailed plots of extracted fetal ECG signals. The mixed signals were obtained from 8 electrodes located on the abdomen of a pregnant woman. The signals are 2.5 seconds long and sampled at 200 Hz.

form

$$\widehat{\mathbf{H}} = \mathbf{W}^{-1} = \begin{bmatrix} 0.000 & 3.549 & 0.000 & 0.888 \\ -1.861 & 0.000 & -1.245 & 0.000 \\ -2.783 & 1.521 & -2.095 & 0.001 \\ -0.005 & 0.003 & -0.909 & 3.563 \end{bmatrix}$$

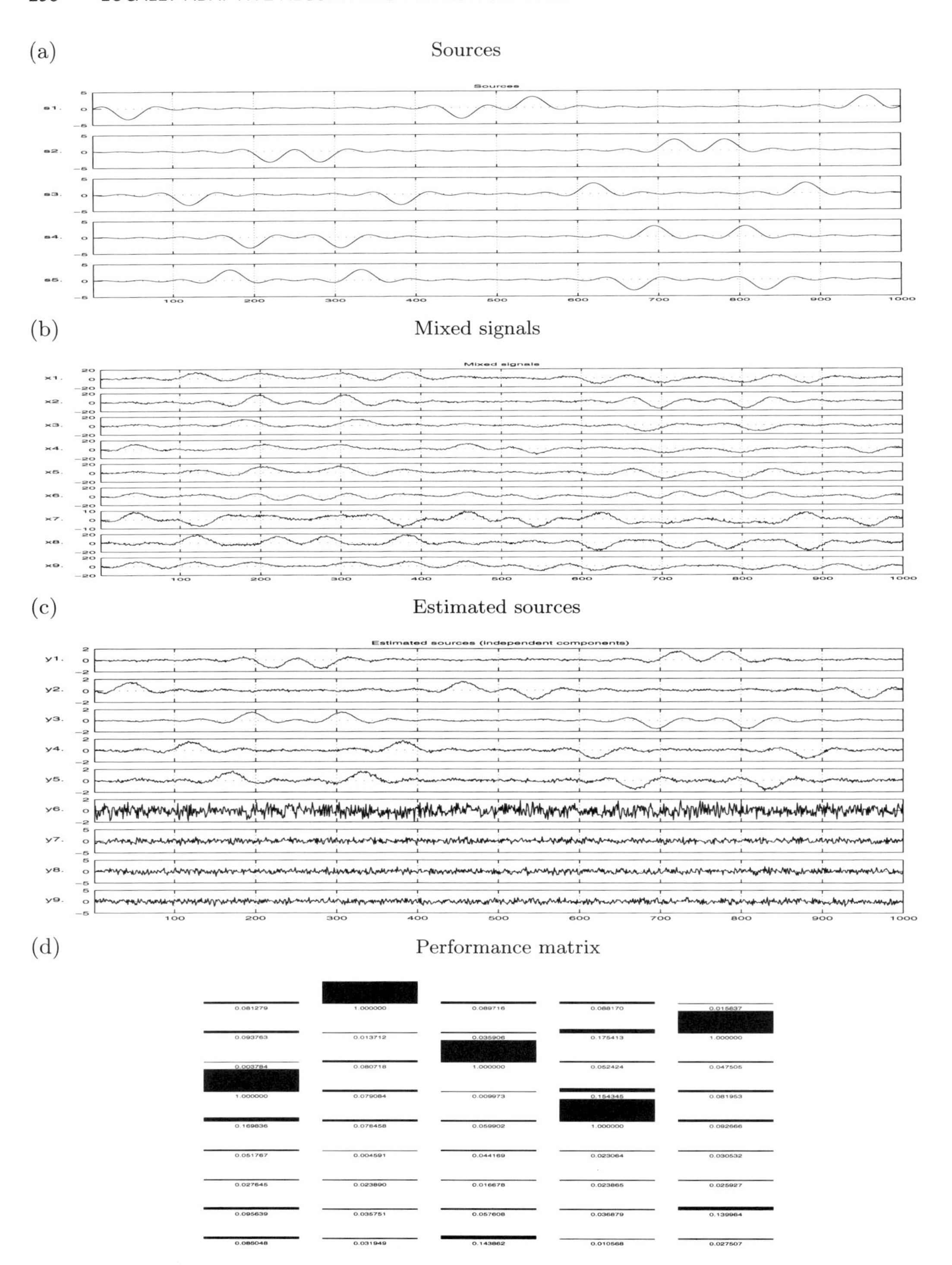

Fig. 7.8 Plots illustrating Example 7.3: (a) Original sources, (b) mixed (observed) noisy signals, (c) estimated source signals, (d) performance matrix $\mathbf{G} = \mathbf{W}\mathbf{H}$.

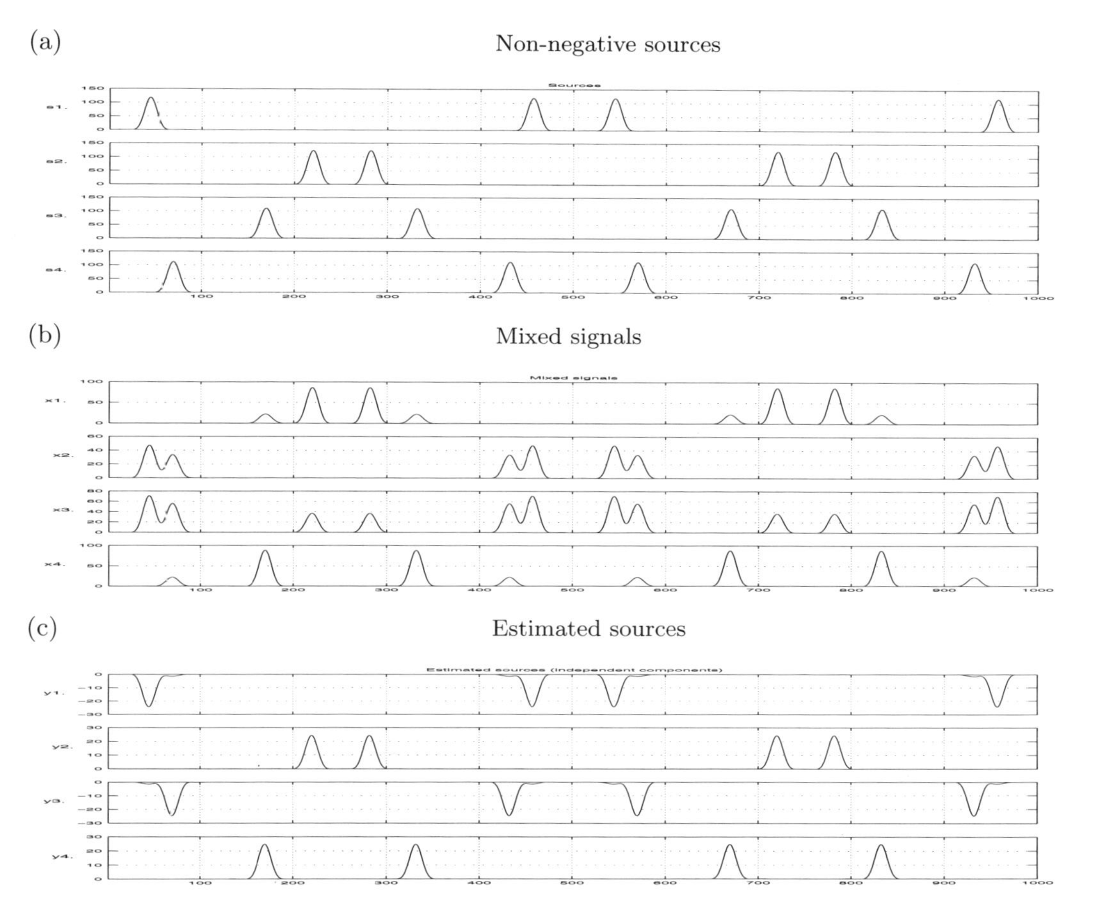

Fig. 7.9 Plots illustrating Example 7.4: (a) Original non-negative sources, (b) mixing signals, (c) estimated sources before post-processing.

It is seen that source $\hat{s}_1(k)$ and $\hat{s}_3(k)$ are non positive so they should be inverted and the corresponding columns (first and third one) of the matrix $\widehat{\mathbf{H}}$ should be multiplied by -1. After normalizing the mixing matrix we obtained

$$\widehat{\widehat{\mathbf{H}}} = \begin{bmatrix} 0.000 & 0.700 & 0.000 & 0.199 \\ 0.395 & 0.000 & 0.293 & 0.000 \\ 0.605 & 0.300 & 0.493 & 0.000 \\ 0.000 & 0.000 & 0.214 & 0.801 \end{bmatrix} \tag{7.87}$$

which is a very close approximation of the original mixing matrix $\mathbf{H}$ neglecting permutation ambiguity. In this way, we were able to reconstruct the original sources and estimate the mixing matrix with only permutation ambiguity. We reduced the scaling ambiguity with the assumption that the mixing matrix $\mathbf{H}$ is normalized and non-negative.

Example 7.5 (Blind separation of highly correlated images) In this experiment 6 human faces are mixed with the Hilbert (very ill-conditioned) mixing matrix $\mathbf{H}$. The mixing

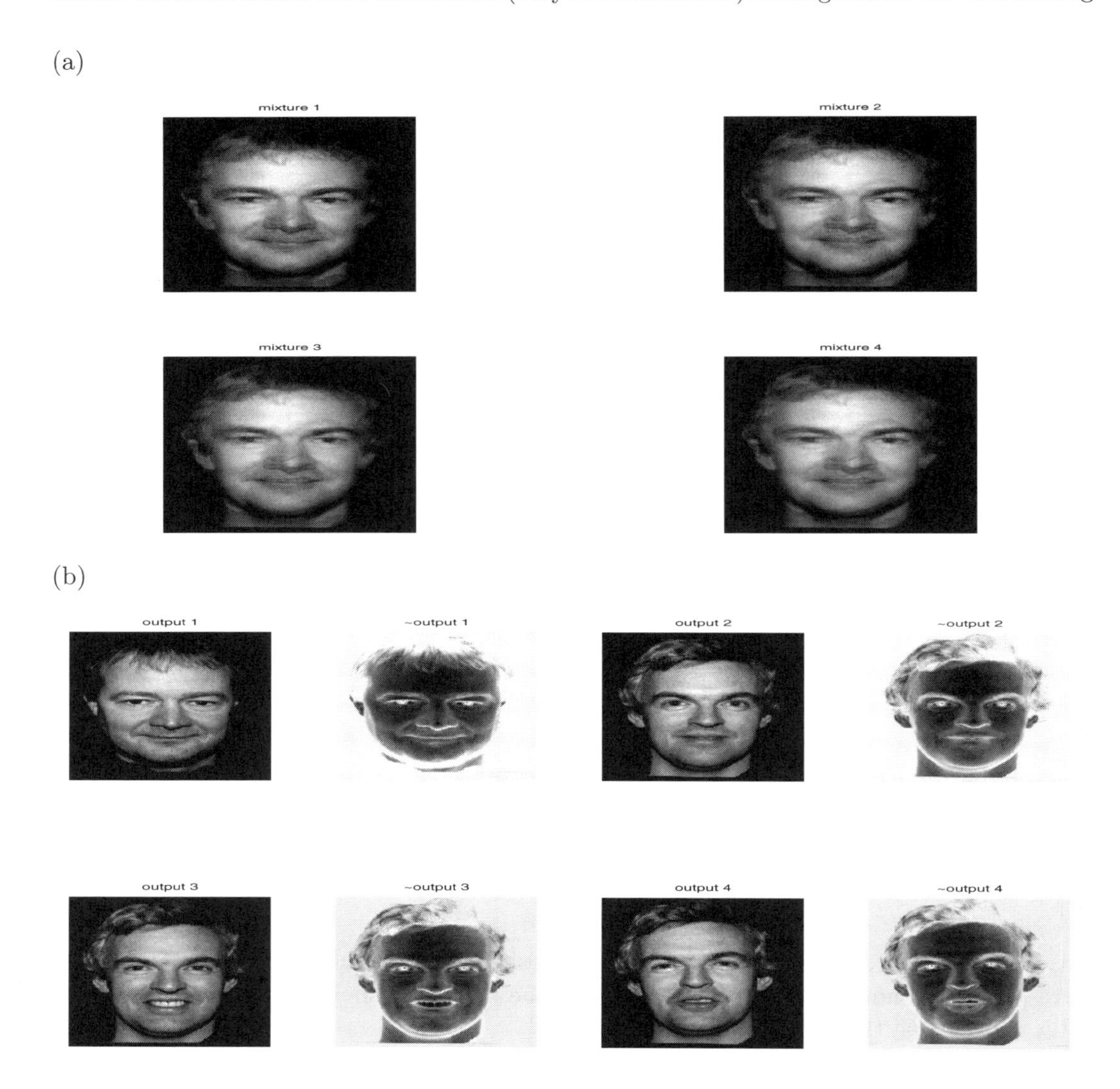

Fig. 7.10 Example 7.5: (a) Mixed (superimposed) images, (b) reconstructed original images.

images shown in Fig. 7.10 (a) are strongly correlated, thus any classical ICA algorithm would fail. In order to reconstruct the original images we applied the preprocessing stage first by high pass filtering the observed images in order to enhance edges (which appear to be independent for the original images). For such preprocessed mixed images we have applied the learning rule (7.54). The estimated original images are shown in Fig. 7.10 (b).

It should be noted that original images are reconstructed almost perfectly. Similar results have been obtained for other natural images and randomly generated matrices.

Appendix A. Stability Conditions for the Robust ICA Algorithm (7.50) [326]

In this Appendix, we present the local stability conditions of the algorithm (7.50) derived in the general form by Cruces *et al.* [326]. Let us adapt the following short notation $f_i = f_i(s_i)$, $f_i' = \frac{\partial f_i(s_i)}{\partial s_i}$, $g_i = g_i(s_i)$, $g_i' = \frac{\partial g_i(s_i)}{\partial s_i}$.

Theorem A.1 *Let us assume that*

1. *The vector* $\mathbf{s} = [s_1, s_2, \ldots, s_n]^T$ *consists of n real and zero-mean sources satisfying the following scaling conditions*

$$E\{f_i g_i\} = 1. \tag{A.1}$$

2. *The nonlinear functions* $f_i(y_i)$ *and* $g_i(y_i)$, $\forall i$ *are at least twice differentiable in the domain of support.*

3. *The learning rate* η *is sufficiently small to ensure numerical stability of the algorithm (7.50).*

Then, the algorithm (7.50) with a constant step-size which is repeated here for convenience

$$\mathbf{W}(l+1) \quad = \quad \mathbf{W}(l) + \eta \left(\mathbf{I} - \mathbf{R}_{\mathbf{f}\,\mathbf{g}}^{(l)}\right) \mathbf{W}(l) \tag{A.2}$$

has an asymptotically stable point at the separation solution if and only if the following conditions are satisfied

$$E\{f_i' s_i g_i\} + E\{f_i s_i g_i'\} \quad > \quad 0, \tag{A.3}$$
$$E\{f_i'\}E\{s_j g_j\} + E\{f_j'\}E\{s_i g_i\} \quad > \quad 0, \tag{A.4}$$
$$E\{f_i'\}E\{f_j'\}E\{s_i g_i\}E\{s_j g_j\} \quad > \quad E\{g_i'\}E\{g_j'\}E\{s_i f_i\}E\{s_j f_j\} \tag{A.5}$$

for all $i, j|_{i \neq j} = 1, 2, \ldots, n$.

The above stability conditions can be considered as an extension or generalization of the stability conditions derived by Amari *et al.* [24]. From this Theorem, we can obtain the following Corollary, first presented by Amari in [20].

Corollary 1 *If* $\mathbf{g}(\mathbf{y}) = \mathbf{y}$ *and the sources are properly scaled such that*

$$E\{f_i s_i\} = 1, \;\; \forall i, \tag{A.6}$$

then the asymptotic stability conditions reduce to

$$E\{f_i' s_i^2\} + 1 \quad > \quad 0, \tag{A.7}$$
$$E\{f_i'\} \quad > \quad 0, \tag{A.8}$$
$$E\{f_i'\}E\{f_j'\}E\{s_i^2\}E\{s_j^2\} \quad > \quad 1 \tag{A.9}$$

for all $i, j|_{i \neq j} = 1, 2, \ldots, n$.

Note that when $\mathbf{g}(\mathbf{y}) = \mathbf{y}$, the constraint $E\{f_i g_j\} = 0, \ \forall i \neq j$ is always true at the correct equilibrium point due to the independence and zero-mean assumptions for the sources, i.e., $E\{f_i g_j\} = E\{f_i\}E\{g_j\} = E\{f_i\}E\{y_j\} = 0, \ \forall i \neq j$. The conditions (A.7)-(A.9) are obtained directly from (A.3)-(A.5) taking into account (A.6) and that $g_i = s_i$ and $g_i' = 1$ for all $i = 1, 2, \ldots, N$.

If the sources have even pdfs and the non-linearities are strictly monotonically increasing odd functions, (A.3)-(A.4) are always satisfied. These assumptions however, do not guarantee that the condition (A.5) is satisfied. In this case, the following Corollary shows how to stabilize the algorithm.

Corollary 2 *If* $\mathbf{f}(\mathbf{y}) = \mathbf{y}$, *then the asymptotic stability conditions reduce to:*

$$\begin{aligned} E\{g_i' s_i^2\} + 1 &> 0, && \text{(A.10)} \\ E\{g_i'\}E\{g_j'\}E\{s_i^2\}E\{s_j^2\} &< 1 && \text{(A.11)} \end{aligned}$$

for all $i, j|_{i \neq j} = 1, 2, \ldots, n$.

Corollary 3 *If*

$$E\{f_i'\}E\{f_j'\}E\{s_i g_i\}E\{s_j g_j\} < E\{g_i'\}E\{[g_j'\}E\{s_i f_i\}E\{s_j f_j\} \qquad \text{(A.12)}$$

then we can interchange $f_i = f_i(s_i)$, *with* $g_i = g_i(s_i)$ *which guarantees the asymptotical stability of the algorithm (7.50).*

Proof. If the pdfs of the sources are even and the nonlinearities are strictly odd functions, then $E\{\mathbf{f}(\mathbf{s})\} = \mathbf{0}$ and $E\{\mathbf{g}(\mathbf{s})\} = \mathbf{0}$. On the other hand, when the nonlinearities are strictly monotonically increasing odd functions, $E\{f_i' s_i g_i\} > 0$ and $E\{f_i s_i g_i'\} > 0, \ \forall i$ and condition (A.3) is true. For the same reasons, $E\{f_i'\} > 0$ and $E\{s_i g_i\} > 0, \ \forall i$ and (A.4) is also true. Finally, the condition (A.5), when the equality does not hold, is not critical because it can be forced by just permuting the order of the functions $\mathbf{f}(\cdot)$ and $\mathbf{g}(\cdot)$.

These facts have been observed by Sorouchyari in [1098] for the local Jutten-Hérault algorithm and later by Amari for the natural gradient algorithm in [24] (for $\mathbf{g}(\mathbf{y}) = \mathbf{y}$) and Zhang for convolutive models [1367]. However, the stability conditions presented in this chapter are valid for the more general case of two non-linear activation functions and for the flexible robust algorithm (7.50) [326, 277].

Corollary 4 *The algorithm (7.43)*

$$\widehat{\mathbf{H}}(l+1) = \widehat{\mathbf{H}}(l)\left(\mathbf{I} - \eta\,(\mathbf{I} - \widehat{\mathbf{R}}_{\mathbf{f}\mathbf{g}}^{(l)})\right) \qquad \text{(A.13)}$$

has the same asymptotical stability conditions given by Eq. (A.3)-(A.5) as the algorithm (7.50).

The intuitive justification of this Corollary is based on the fact demonstrated with equation (7.56), that both algorithms have almost the same convergence properties and the same equilibrium points corresponding to the true estimation of the source signals.

Proof of Theorem A.1. We will now prove that the conditions (A.3)-(A.5) ensure the asymptotic stability of the algorithm (7.50) under the assumptions of Theorem A.1 [326]. Let us start multiplying both sides of Eq. (A.2) by the mixing matrix $\mathbf{H}$ to express the algorithm (7.50) in terms of the global transfer matrix $\mathbf{G}$ as

$$\mathbf{G}(l+1) \quad = \quad \mathbf{G}(l) + \eta \left(\mathbf{I} - \mathbf{R}_{\mathbf{f}\,\mathbf{g}}^{(l)} \right) \mathbf{G}(l). \tag{A.14}$$

Let $C(\mathbf{Gs}) = \mathbf{f}(\mathbf{Gs})\mathbf{g}^T(\mathbf{Gs}) - \mathbf{I}$. Condition (A.1) is necessary to allow the separation solution $\mathbf{G}_* = \mathbf{I}$ to be an equilibrium point of the algorithm, i.e., $E\{C(\mathbf{G}_*\mathbf{s})\} = \mathbf{0}$.

The next step is to study the asymptotic stability of the algorithm at this point. For this purpose, we will study the behavior of the Ordinary Differential Equation (ODE) associated with the algorithm at the separation solution [88, 89]. Let us define the mean field of the algorithm as $M(\mathbf{G}) = E\{C(\mathbf{Gs})\mathbf{G}\}$ and an arbitrary small matrix perturbation $\boldsymbol{\epsilon}$ of the global transfer matrix at the separation solution $\mathbf{G}_*$. Then,

$$M(\mathbf{I}+\boldsymbol{\epsilon}) = E\{C(\mathbf{s}+\boldsymbol{\epsilon}\mathbf{s})(\mathbf{I}+\boldsymbol{\epsilon})\} = E\{C(\mathbf{s}+\boldsymbol{\epsilon}\mathbf{s})\} + \mathcal{O}(\boldsymbol{\epsilon}), \tag{A.15}$$

where $\mathcal{O}(\boldsymbol{\epsilon}) = E\{C(\mathbf{s}+\boldsymbol{\epsilon})\boldsymbol{\epsilon}\}$.

To find a linear approximation to the mean field in terms of $\boldsymbol{\epsilon}$, we will replace both functions $\mathbf{f}(\mathbf{s}+\boldsymbol{\epsilon}\mathbf{s})$ and $\mathbf{g}(\mathbf{s}+\boldsymbol{\epsilon}\mathbf{s})$ by first-order Taylor expansions at $\mathbf{G}_* = \mathbf{I}$ as

$$\begin{aligned} \mathbf{f}(\mathbf{s}+\boldsymbol{\epsilon}\mathbf{s}) &= \mathbf{f}(\mathbf{s}) + \mathbf{f}'(\mathbf{s})\boldsymbol{\epsilon}\mathbf{s} + o(\boldsymbol{\epsilon}), \\ \mathbf{g}(\mathbf{s}+\boldsymbol{\epsilon}\mathbf{s}) &= \mathbf{g}(\mathbf{s}) + \mathbf{g}'(\mathbf{s})\boldsymbol{\epsilon}\mathbf{s} + o(\boldsymbol{\epsilon}), \end{aligned} \tag{A.16}$$

where $o(\boldsymbol{\epsilon})/||\boldsymbol{\epsilon}||$ tends to zero as $\|\boldsymbol{\epsilon}\|$ tends to zero and $\mathbf{f}'(\mathbf{s}) = \frac{\partial \mathbf{f}(\mathbf{s})}{\partial \mathbf{s}}$ and $\mathbf{g}'(\mathbf{s}) = \frac{\partial \mathbf{g}(\mathbf{s})}{\partial \mathbf{s}}$ denote two diagonal matrices since $\mathbf{f}(\cdot)$ and $\mathbf{g}(\cdot)$ act component-wise. Substituting (A.16) into (A.15) and having in mind that $E\{C(\mathbf{G}_*\mathbf{s})\} = \mathbf{0}$, we can express the mean field of the algorithm as

$$M(\mathbf{I}+\boldsymbol{\epsilon}) = E\{\mathbf{f}'(\mathbf{s})\boldsymbol{\epsilon}\mathbf{s}\mathbf{g}^T(\mathbf{s})\} + E\{\mathbf{f}(\mathbf{s})\mathbf{s}^T\boldsymbol{\epsilon}^T(\mathbf{g}'(\mathbf{s}))^T\} + \mathcal{O}(\boldsymbol{\epsilon}) + o(\boldsymbol{\epsilon}). \tag{A.17}$$

Denoting $f_i = \{\mathbf{f}(\mathbf{s})\}_i$, $f_i' = \{\mathbf{f}'(\mathbf{s})\}_{ii}$, the terms of the mean field can be written as

$$M_{ij}(I+\boldsymbol{\epsilon}) = \sum_k E\{f_i' s_k g_j\}\epsilon_{ik} + \sum_k E\{f_i s_k g_j'\}\epsilon_{jk} + \mathcal{O}(\boldsymbol{\epsilon}) + o(\boldsymbol{\epsilon}). \tag{A.18}$$

Taking into account the independence assumptions of the sources at the separation solution, we obtain

$$\begin{aligned} M_{ii}(I+\boldsymbol{\epsilon}) &= E\{f_i' s_i g_i\}\epsilon_{ii} + E\{f_i s_i g_i'\})\epsilon_{ii} + \textstyle\sum_{k\neq i} E\{f_i' g_i\}E\{s_k\}\epsilon_{ik} \\ &\quad + \textstyle\sum_{k\neq i} E\{f_i g_i'\}E\{s_k\}\epsilon_{ik} + \mathcal{O}(\boldsymbol{\epsilon}) + o(\boldsymbol{\epsilon}), \\ M_{ij}(I+\boldsymbol{\epsilon}) &= E\{f_i'\}E\{s_j g_j\}\epsilon_{ij} + E\{f_i' s_i\}E\{g_j\}\epsilon_{ii} + \textstyle\sum_{k\neq i,j} E\{f_i'\}E\{s_k\}E\{g_j\} \\ &+ E\{f_i s_i\}E\{g_j'\}\epsilon_{ji} + E\{f_i\}E\{s_j g_j'\}\epsilon_{jj} \\ &+ \textstyle\sum_{k\neq(i,j)} E\{f_i\}E\{s_k\}E\{g_j'\} + \mathcal{O}(\boldsymbol{\epsilon}) + o(\boldsymbol{\epsilon}). \end{aligned}$$

Using the zero-mean assumption of the sources ($E\{s_i\} = 0 \ \forall i$), we obtain

$$\begin{aligned} M_{ii}(I+\boldsymbol{\epsilon}) &= E\{f_i' s_i g_i\}\epsilon_{ii} + E\{f_i s_i g_i'\})\epsilon_{ii} + \mathcal{O}(\boldsymbol{\epsilon} + o(\boldsymbol{\epsilon}), \\ M_{ij}(I+\boldsymbol{\epsilon}) &= E\{f_i'\}E\{s_j g_j\}\epsilon_{ij} + E\{f_i' s_i\}E\{g_j\}\epsilon_{ii} \\ &\quad + E\{f_i s_i\}E\{g_j'\}\epsilon_{ji} + E\{f_i\}E\{s_j g_j'\}\epsilon_{jj} + \mathcal{O}(\boldsymbol{\epsilon}) + o(\boldsymbol{\epsilon}). \end{aligned}$$

We next vectorize the average Eq. (A.14) and consider the corresponding differential equation

$$\frac{d\mathbf{\Theta}}{dt} = \mathcal{M}(\mathbf{\Theta}), \tag{A.19}$$

where $\mathbf{\Theta} = vec\mathbf{G}$.

Using the above relationships, it is found that the gradient of $\mathcal{M}(\mathbf{\Theta})$ at the separation takes a simple form: It is a block-diagonal matrix with the diagonal block elements of size one and two [24, 315, 148, 908].

One of the conditions for asymptotic stability is extracted from the 1×1 diagonal blocks

$$\left.\frac{\partial M_{ii}(I+\boldsymbol{\epsilon})}{\partial \epsilon_{ii}}\right|_{\epsilon=0} = E\{f_i' s_i g_i\} + E\{f_i s_i g_i'\} > 0. \tag{A.20}$$

The 2×2 diagonal blocks are of the form

$$\begin{pmatrix} \frac{\partial M_{ij}(I+\boldsymbol{\epsilon})}{\partial \epsilon_{ij}} & \frac{\partial M_{ij}(I+\boldsymbol{\epsilon})}{\partial \epsilon_{ji}} \\ \frac{\partial M_{ji}(I+\boldsymbol{\epsilon})}{\partial \epsilon_{ij}} & \frac{\partial M_{ji}(I+\boldsymbol{\epsilon})}{\partial \epsilon_{ji}} \end{pmatrix} = \begin{pmatrix} E\{f_i'\}E\{s_j g_j\} & E\{g_j'\}E\{s_i f_i\} \\ E\{g_i'\}E\{s_j f_j\} & E\{f_j'\}E\{s_i g_i\} \end{pmatrix} = \begin{pmatrix} a & b \\ c & d \end{pmatrix} \tag{A.21}$$

for all $i \neq j$. The eigenvalues of these block matrices are the roots of the characteristic polynomial equation $P(\lambda) = \lambda^2 - (a+d)\lambda + (ad - bc)$. We check directly that the real parts of the eigenvalues are positive if $a + d > 0$ and $ad - bc > 0$. Then, substituting a, b, c, d into these expressions, we arrive at the sufficient conditions for the asymptotic stability of the algorithm (7.50)

$$\begin{aligned} E\{f_i'\}E\{s_j g_j\} + E\{f_j'\}E\{s_i g_i\} &> 0, && \text{(A.22)} \\ E\{f_i'\}E\{[f_j'\}E\{s_i g_i\}E\{s_j g_j\} &> E\{g_i'\}E\{g_j'\}E\{s_i f_i\}E\{s_j f_j\} && \text{(A.23)} \end{aligned}$$

for all $i \neq j$. □

8

Robust Techniques for BSS and ICA with Noisy Data

All progress is based upon a universal innate desire of every organism to live beyond its means.
—(Samuel Butler)

8.1 INTRODUCTION

In this chapter, we focus mainly on approaches to blind separation of sources when the measured signals are contaminated by a large quantity of additive noise. We extend existing adaptive algorithms with equivariant properties in order to considerably reduce the bias caused by measurement noise for the estimation of mixing and separating matrices. Moreover, we propose dynamical recurrent neural networks for simultaneous estimation of the unknown mixing matrix, source signals and reduction of noise in the extracted output signals. The optimal choice of nonlinear activation functions for various noise distributions assuming a generalized-Gaussian-distributed noise model is also discussed. Computer simulations of selected techniques which confirm their usefulness and satisfactory performance are also provided.

As the estimation of a separating (demixing) matrix $\mathbf{W}$ and a mixing matrix $\widehat{\mathbf{H}}$ in the presence of noise is rather difficult, the majority of past research efforts have been devoted to the noiseless case or assumed that noise has a negligible effect on the performance of the algorithms. The objective of this chapter is to present several approaches and learning algorithms that are more robust with respect to noise than the techniques described in the previous chapters or can reduce the noise in the estimated output vector $\mathbf{y}(k)$. In this chapter, we assume that the source signals and additive noise components are statistically independent.

In general, the problem of noise cancellation is difficult or even impossible to handle, because we have $(m+n)$ unknown source signals (n sources and m noise signals), but only m available or measured sensor signals. However, in many practical situations, we can estimate unbiased separating matrices and reduce or cancel the noise if some information about the noise is available. For example, in some situations, the environmental noise can be measured or modelled.

8.2 BIAS REMOVAL TECHNIQUES FOR PREWHITENING AND ICA ALGORITHMS

8.2.1 Bias Removal for Whitening Algorithms

Let us consider at first a simpler problem related to ICA, that of standard decorrelation or prewhitening algorithm for $\boldsymbol{x}(k)$ given by [13, 390]

$$\mathbf{W}(k+1) = \mathbf{W}(k) + \eta(k)[\mathbf{I} - \mathbf{y}(k)\,\mathbf{y}^T(k)]\mathbf{W}(k) \tag{8.1}$$

or its averaged version given by

$$\Delta\mathbf{W}(k) = \eta(k)\ \left[\mathbf{I} - \langle \mathbf{y}(k)\,\mathbf{y}^T(k)\rangle\right]\mathbf{W}(k), \tag{8.2}$$

where $\mathbf{y}(k) = \mathbf{W}(k)\,\mathbf{x}(k)$ and $\langle\cdot\rangle$ denotes statistical average. When $\mathbf{x}(k)$ is noisy such that $\mathbf{x}(k) = \hat{\mathbf{x}}(k) + \boldsymbol{\nu}(k)$, where $\hat{\mathbf{x}}(k)$ and $\hat{\mathbf{y}}(k) = \mathbf{W}(k)\,\hat{\mathbf{x}}(k)$ are the noiseless estimates of the input and output vectors respectively. It is easy to show that the additive noise $\boldsymbol{\nu}(k)$ within $\mathbf{x}(k)$ introduces a bias in the estimated decorrelation matrix $\mathbf{W}$. The covariance matrix of the output can be evaluated as

$$\widehat{\mathbf{R}}_{\mathbf{y}\,\mathbf{y}} = \langle \mathbf{y}(k)\,\mathbf{y}^T(k)\rangle = \mathbf{W}\widehat{\mathbf{R}}_{\hat{\mathbf{x}}\hat{\mathbf{x}}}\mathbf{W}^T + \mathbf{W}\widehat{\mathbf{R}}_{\boldsymbol{\nu}\boldsymbol{\nu}}\mathbf{W}^T, \tag{8.3}$$

where $\widehat{\mathbf{R}}_{\hat{\mathbf{x}}\hat{\mathbf{x}}} = \langle \hat{\mathbf{x}}(k)\hat{\mathbf{x}}^T(k)\rangle = \frac{1}{N}\sum_{k=1}^{N}\hat{\mathbf{x}}(k)\hat{\mathbf{x}}^T(k)$ and $\widehat{\mathbf{R}}_{\boldsymbol{\nu}\boldsymbol{\nu}} = \langle \boldsymbol{\nu}(k)\boldsymbol{\nu}^T(k)\rangle$.

Assuming that the covariance matrix of the noise is known (e.g., $\widehat{\mathbf{R}}_{\boldsymbol{\nu}\boldsymbol{\nu}} = \hat{\sigma}_\nu^2\mathbf{I}$) or can be estimated, a proposed modified algorithm employing bias removal is given by

$$\boxed{\Delta\mathbf{W}(k) = \eta(k)\left[\mathbf{I} - \mathbf{R}_{\mathbf{y}\,\mathbf{y}}^{(k)} + \mathbf{W}(k)\widehat{\mathbf{R}}_{\boldsymbol{\nu}\boldsymbol{\nu}}\mathbf{W}^T(k)\right]\mathbf{W}(k).} \tag{8.4}$$

The stochastic gradient version of this algorithm for $\widehat{\mathbf{R}}_{\boldsymbol{\nu}\boldsymbol{\nu}} = \hat{\sigma}_\nu^2\mathbf{I}$ is

$$\Delta\mathbf{W}(k) = \eta(k)\left[\mathbf{I} - \mathbf{y}(k)\,\mathbf{y}^T(k) + \hat{\sigma}_\nu^2\mathbf{W}(k)\mathbf{W}^T(k)\right]\mathbf{W}(k). \tag{8.5}$$

Alternatively, to reduce the influence of the white additive noise for colored signals, we can apply the modified learning rule given by

$$\Delta\mathbf{W}(k) = \eta(k)\left[\mathbf{I} - \frac{1}{2}[\langle \mathbf{y}(k)\,\mathbf{y}^T(k-p)\rangle + \langle \mathbf{y}(k-p)\,\mathbf{y}^T(k)\rangle]\right]\mathbf{W}(k), \tag{8.6}$$

where p is a small integer time delay. It should be noted that the above algorithm is theoretically insensitive to the additive white noise. In fact, instead of diagonalizing the zero-lag covariance matrix $\mathbf{R}_{\mathbf{y}\,\mathbf{y}} = \langle \mathbf{y}(k)\,\mathbf{y}^T(k)\rangle$, we diagonalize the time-delayed covariance matrix

$$\tilde{\mathbf{R}}_{\mathbf{y}}(p) = \frac{1}{2}[\langle \mathbf{y}(k)\,\mathbf{y}^T(k-p)\rangle + \langle \mathbf{y}(k-p)\,\mathbf{y}^T(k)\rangle] \tag{8.7}$$

which is insensitive to the additive white noise on the condition that the noise-free sensor signals $\hat{\mathbf{x}}(k)$ are colored signals and the number of observations is sufficiently large (typically more than 10^4 samples).

8.2.2 Bias Removal for Adaptive ICA Algorithms

A technique similar to the one described above can be applied to remove the bias of coefficients for a class of natural gradient algorithms for ICA [396]. We illustrate the technique based on the ICA algorithm of the following form (see Chapters 6 and 7 for more explanation):

$$\Delta\mathbf{W}(k) = \eta(k)\left[\mathbf{I} - \mathbf{R}_{\mathbf{f}\,\mathbf{g}}^{(k)}\right]\mathbf{W}(k). \tag{8.8}$$

Nonlinear functions of vectors $\mathbf{f}[\mathbf{y}(k)]$ and $\mathbf{g}[\mathbf{y}(k)]$ should be suitably designed as explained in the previous chapters.

Remark 8.1 *It should be noted that the algorithm works on the condition that $E\{f_i(y_i)\} = 0$ or $E\{g_i(y_i)\} = 0$. In order to satisfy these conditions for non-symmetrically distributed sources for each index i, we use only one nonlinear function and the second one is linear. In other words, for any i we employ $f_i(y_i)$ and $g_i(y_i) = y_i$ or $f_i(y_i) = y_i$ and $g_i(y_i)$.*

The above learning algorithm has been shown to possess excellent performance when applied to noiseless signal mixtures; however, its performance deteriorates with noisy measurements due to undesirable coefficient biases and the existence of noise in the separated signals. To estimate the coefficient biases, we determine the Taylor series expansions of the nonlinearities $f_i(y_i)$ and $g_j(y_j)$ about the estimated noiseless values $\hat{y}_i$. The generalized covariance matrix $\mathbf{R}_{\mathbf{f}\,\mathbf{g}}$ can be approximately evaluated as [396]

$$\mathbf{R}_{\mathbf{f}\,\mathbf{g}} = E\{\mathbf{f}[\mathbf{y}(k)]\,\mathbf{g}[\boldsymbol{y}^T(k)]\} \cong E\{\mathbf{f}[\hat{\mathbf{y}}(k)]\,\mathbf{g}[\hat{\mathbf{y}}^T(k)]\} + \boldsymbol{k}_f\mathbf{W}\widehat{\mathbf{R}}_{\boldsymbol{\nu}\boldsymbol{\nu}}\mathbf{W}^T\boldsymbol{k}_g, \tag{8.9}$$

where $\boldsymbol{k}_f$ and $\boldsymbol{k}_g$ are diagonal matrices with entries

$$k_{fi} = E\{df_i(y_i(k))/dy_i\}, \qquad k_{gi} = E\{dg_i(y_i(k))/dy_i\},$$

respectively. Thus, a modified adaptive learning algorithm with reduced coefficient bias has the form

$$\boxed{\Delta\mathbf{W}(k) = \eta(k)\left[\mathbf{I} - \mathbf{R}_{\mathbf{f}\,\mathbf{g}}^{(k)} + \boldsymbol{k}_f\mathbf{W}(k)\widehat{\mathbf{R}}_{\boldsymbol{\nu}\boldsymbol{\nu}}\mathbf{W}^T(k)\boldsymbol{k}_g\right]\mathbf{W}(k),} \tag{8.10}$$

which can be written in a more compact form as

$$\Delta \mathbf{W}(k) = \eta(k) \left[\mathbf{I} - \mathbf{R}_{\mathbf{fg}}^{(k)} + \mathbf{C} \circ \mathbf{W}(k) \widehat{\mathbf{R}}_{\boldsymbol{\nu}\boldsymbol{\nu}} \mathbf{W}^T(k) \right] \mathbf{W}(k), \tag{8.11}$$

where $\mathbf{C} = [c_{ij}]$ is an $n \times n$ scaling matrix with entries $c_{ij} = k_{fi} k_{gj}$ and the operator $\circ$ denotes the Hadamard product.

In the special case when all of the source distributions are identical, $f_i(y_i) = f(y_i)\ \forall i$, $g_i(y_i) = g(y_i)\ \forall i$, and $\widehat{\mathbf{R}}_{\boldsymbol{\nu}\boldsymbol{\nu}} = \hat{\sigma}_\nu^2 \mathbf{I}$, the bias correction term simplifies to $\mathbf{B} = \hat{\sigma}_\nu^2 k_f k_g \mathbf{W}\mathbf{W}^T$. It is interesting to note that we can almost always select nonlinearities such that the global scaling coefficient $c = k_f k_g$ can be close to zero for a wide class of signals. For example, when $f(y_i) = |y_i|^r \mathrm{sign}(y_i)$ and $g(y_i) = \tanh(\gamma y_i)$ are chosen, or when $f(y_i) = \tanh(\gamma y_i)$ and $g(y_i) = |y_i|^r \mathrm{sign}(y_i)$, the scaling coefficient is equal to $c = k_f k_g = r\gamma E\{|y_i|^{r-1}(k)\}[1 - E\{\tanh^2(\gamma y_i(k))\}]$ for $r \geq 1$, which is smaller over the range $|y_i| \leq 1$ in contrast to the case when $g(y_i) = y_i$ is chosen. Moreover, we can optimally design the parameters r and γ so that within a specified range of y_i, the absolute value of the scaling coefficient $c = k_f k_g$ is minimal.

Another possible solution to mitigate coefficient bias is to employ nonlinearities in the form $\tilde{f}(y_i) = f(y_i) - \alpha_i y_i$ and $g(y_i) = y_i$ with $\alpha_i \geq 0$. The motivation behind the use of linear terms $-\alpha_i y_i$ is to reduce the values of the scaling coefficients $c_{ij} = k_{fi} - \alpha_i$ as well the influence of large outliers. Alternatively, the generalized Fahlman functions given by $\tanh(\gamma_i\, y_i) - \alpha_i y_i$ can be used for either $f_i(y_i)$ or $g_i(y_i)$, where appropriate [482, 481].

One disadvantage of the proposed techniques for bias removal is that any equivariant properties for the resulting algorithm are lost when a bias compensating term is added, and thus the algorithm may perform poorly or even fail to separate sources if the mixing matrix is very ill-conditioned. For this reason, it is necessary to design nonlinearities which correspond as closely as possible to those produced from the true pdf's of the source signals while also maximally reducing the bias caused by noise.

Example 8.1 We now illustrate the behavior of the bias removal algorithm in (8.10) via simulation [396]. In this example, a 3×3 mixing matrix given by

$$\mathbf{H} = \begin{bmatrix} 0.6 & 0.7 & 0.7 \\ 0.9 & 0.1 & 0.5 \\ 0.1 & 0.5 & 0.8 \end{bmatrix} \tag{8.12}$$

is employed. Three independent random sources-one uniform-$[-1, 1]$-distributed and two binary-$\{\pm 1\}$-distributed were generated. The matrix equation $\mathbf{x}(k) = \mathbf{H}\mathbf{s}(k) + \boldsymbol{\nu}(k)$ is used to create $\mathbf{x}(k)$, where each $\boldsymbol{\nu}(k)$ is a Gaussian random noise with the covariance matrix $\widehat{\mathbf{R}}_{\boldsymbol{\nu}\boldsymbol{\nu}} = \hat{\sigma}_\nu^2 \mathbf{I}$ with $\hat{\sigma}_\nu^2 = 0.01$. The condition number of $\mathbf{H}E\{\boldsymbol{s}(k)\boldsymbol{s}^T(k)\}\mathbf{H}^T$ is 51.5. Here, $f_i(y) = y^3$ and $g_i(y) = \tanh(10y)$ for all $1 \leq i \leq 3$ and $\eta(k) = 0.001$. Twenty trials were run, in which $\mathbf{W}(0)$ were different random orthogonal matrices such that $\mathbf{W}(0)\mathbf{W}^T(0) = 0.25\mathbf{I}$, and ensemble averages were taken in each case. Figure 8.1 shows the evolution of the performance factor $\zeta(k)$ defined as

$$\zeta(k) = \frac{1}{n} \sum_{i=1}^{n} \left\{ \left(\sum_{k=1}^{n} \frac{|g_{ik}|}{\max_j |g_{ij}|} - 1 \right) + \left(\sum_{k=1}^{n} \frac{|g_{ki}|}{\max_j |g_{ji}|} - 1 \right) \right\}, \tag{8.13}$$

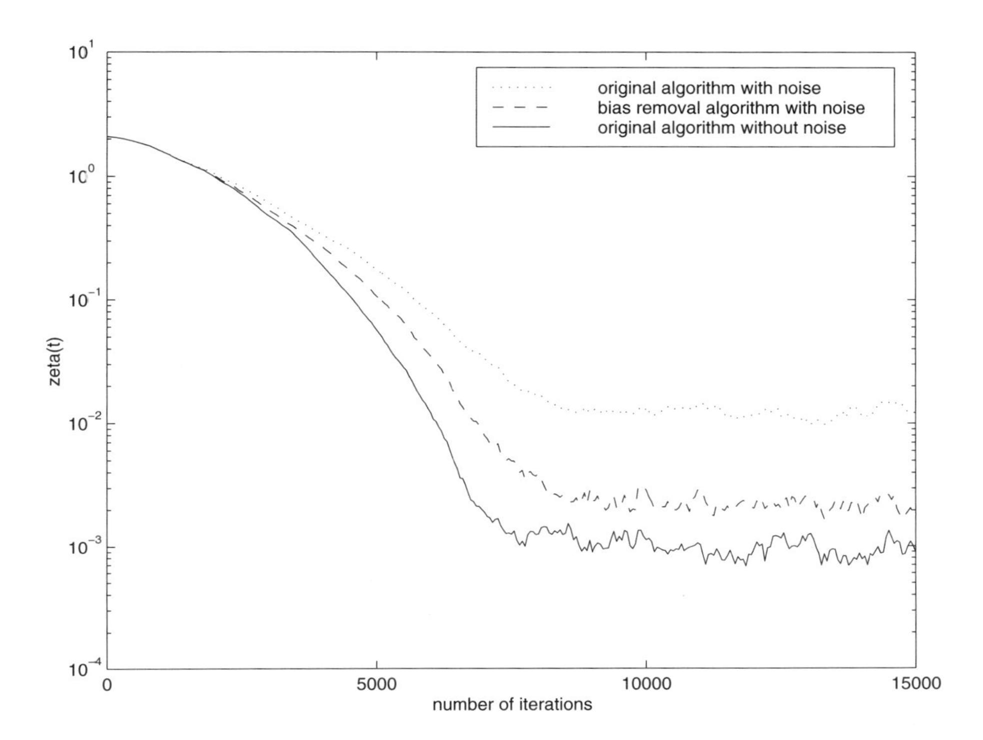

Fig. 8.1 Ensemble-averaged value of the performance index for uncorrelated measurement noise in the first example: dotted line represents the original algorithm (8.8) with noise, dashed line represents the bias removal algorithm (8.10) with noise, solid line represents the original algorithm (8.8) without noise [396].

where $n = 3$. g_{ij} is the (i, j)-element of the global system matrix $\mathbf{G} = \mathbf{W}\,\mathbf{H}$ and $\max_j g_{ij}$ represents the maximum value among the elements in the ith row vector of $\mathbf{G}$. $\max_j g_{ji}$ is the maximum value among the elements in the ith column vector of $\mathbf{G}$. The value of $\zeta(k)$ measures the average source signal crosstalk in the output signals $\{y_i(k)\}$ if no noise is present. As can be seen, the original algorithm yields a biased estimate of $\mathbf{W}(k)$, whereas the bias removal algorithm achieves a crosstalk level that is about 7 dB lower. Also shown for comparison is the original algorithm with no measurement noise, showing that the performance of the described algorithm approaches this idealized case for small learning rates.

Remark 8.2 *In the special case of Gaussian additive noise, we can design special forms of nonlinearities such that entries of the generalized covariance matrix* $\mathbf{R_{f\,g}} = E\{\mathbf{y}(k)\,\mathbf{g}^T(\mathbf{y}(k))\}$ *are expressed by higher order cross-cumulants that are theoretically non-sensitive to Gaussian noise (see Section 8.5 for details)*

8.3 BLIND SEPARATION OF SIGNALS BURIED IN ADDITIVE CONVOLUTIVE REFERENCE NOISE

We next consider the following problem: *How to efficiently separate the sources if additive colored noise can no longer be ignored ?* This can be stated alternatively as *how to cancel or suppress additive noise.* In general, the problem is rather difficult, because we have $(m+n)$ unknown signals (where m is the number of sensors). Hence the problem is highly under-determined, and without any *a priori* information about the mixture model and/or noise, it is very difficult or even impossible to solve [1227].

However, in many practical situations, we can measure or model the environmental noise. In the sequel, we shall refer to such a noise as *reference noise* $\nu_R(k)$ or a vector of reference noises (for each individual sensor $\nu_{Ri}(k)$) (see Fig. 8.2). For example, in the acoustic *cocktail party* problem, we can measure such noise during a short silence period (when all persons do not speak) or we can measure and record it on-line, continuously in time, by an additional isolated microphone. Similarly, one can measure noise in biomedical applications like EEG or ECG by additional electrodes, placed appropriately.

Due to environmental conditions the noise $\nu_R(k)$ may influence each sensor in some unknown manner. Parasitic effects such as delays, reverberation, echo, nonlinear distortion etc. may occur. It may be assumed that the reference noise is processed by some unknown linear or nonlinear dynamical system before it reaches each sensor (see Fig. 8.2). ARMA, NARMA or FIR filters are usually used for modeling convolutive colored noise. In the simplest case, a convolutive FIR model of noise can be assumed, i.e., the reference noise is processed by some finite impulse response filters, whose parameters need to be estimated. Hence, the additive noise in the i–th sensor is modelled as (see Fig.8.2 and Fig. 8.3 (a)) [261, 1227, 1279]

$$\nu_i(k) = \sum_{p=0}^{L} [h_{ip} z^{-p}]\, \nu_R(k) = \sum_{p=0}^{L} h_{ip}\, \nu_R(k-p), \tag{8.14}$$

where z^{-1} is the unit delay operator. Such a model is generally regarded as a realistic (real–world) model in both signal and image processing [1227, 1279]. In this model, we assume that a known reference noise is added to each sensor (mixture of sources) with different unknown time delays and various unknown coefficients $h_{ip}(k)$ representing attenuation coefficients. The unknown mixing and convolutive processes can be described in matrix form as

$$\mathbf{x}(k) = \mathbf{H}\,\mathbf{s}(k) + \mathbf{h}(z)\,\nu_R(k), \tag{8.15}$$

where $\mathbf{h}(z) = [H_1(z), H_2(z), ..., H_n(z)]^T$ with

$$H_i(z) = h_{i0} + h_{i1} z^{-1} + ... + h_{iL} z^{-L}. \tag{8.16}$$

Analogously, the separating and noise deconvolutive process can be described as (see Fig. 8.2)

$$\begin{aligned} \mathbf{y}(k) &= \mathbf{W}\,\mathbf{x}(k) - \tilde{\mathbf{w}}(z)\,\nu_R \\ &= \mathbf{W}\,\mathbf{H}\,\mathbf{s}(k) + \mathbf{W}\,\mathbf{h}(z)\,\nu_R - \tilde{\mathbf{w}}(z)\,\nu_R, \end{aligned} \tag{8.17}$$

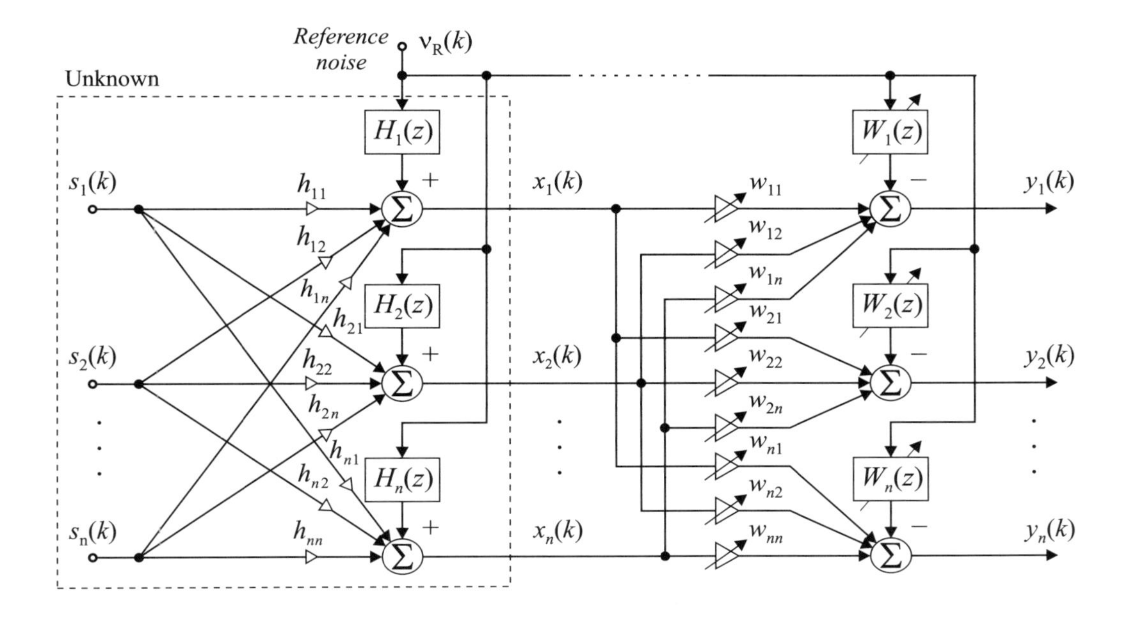

Fig. 8.2 Conceptual block diagram of mixing and demixing systems with noise cancellation. It is assumed that the reference noise is available.

where $\tilde{\mathbf{w}}(z) = [W_1(z), ..., W_n(z)]^T$ with

$$W_i(z) = \sum_{p=0}^{M} \tilde{w}_{ip} z^{-p}. \tag{8.18}$$

We can achieve the best noise reduction by minimizing the generalized energy of all its output signals under some constraints and simultaneously to enforce their mutual independence.

8.3.1 Learning Algorithms for Noise Cancellation

In the basic approach, two learning procedures are simultaneously performed: The signals are separated from their linear mixture by using any algorithm for BSS and the additive noise is estimated and subtracted by minimizing the output energy of the output signals. In the simplest case, we can formulate the following energy (cost) function

$$J(\tilde{\mathbf{w}}) = \frac{1}{2} \sum_{i=1}^{n} E\{|y_i|^2\}. \tag{8.19}$$

Minimizing the above cost function with respect to coefficients $\tilde{w}_{ip}$, we obtain the standard LMS (Least Mean Square) algorithm [1280]

$$\boxed{\tilde{w}_{ip}(k+1) = \tilde{w}_{ip}(k) - \tilde{\eta}(k)\frac{\partial J(\tilde{\mathbf{w}})}{\partial \tilde{w}_{ip}} = \tilde{w}_{ip}(k) + \tilde{\eta}(k) \left\langle y_i(k)\nu_R(k-p)\right\rangle .} \tag{8.20}$$

The number of time delay units M in the separating/deconvolutive model should be usually much larger than the corresponding number L in the mixing model ($M >> L$).

Alternatively, a multistage linear model for noise cancellation is shown in Fig. 8.3 (a). In such a model, the noise cancellation and source separation are performed sequentially in two or more stages. We first attempt to cancel the noise contained in the mixtures and then to separate the sources (see Fig. 8.3 (a)). In order to cancel additive noise and to develop an adaptive learning algorithm for unknown coefficients $\tilde{w}_{ip}(k)$, we can apply the concept of minimization of the generalized output energy of output signals $\tilde{\mathbf{x}}(k) = [\tilde{x}_1(k), \tilde{x}_2(k), \ldots, \tilde{x}_n(k)]^T$. In other words, we can formulate the following cost function (generalized energy)

$$J(\tilde{\mathbf{w}}_1) = \sum_{i=1}^{n} E\{\rho_i(\tilde{x}_i)\}, \tag{8.21}$$

where $\rho_i(\tilde{x}_i)$, $(i = 1, 2, \ldots, n)$ are suitably chosen loss functions, typically

$$\rho_i(\tilde{x}_i) = \frac{1}{\gamma}\log cosh(\gamma \tilde{x}_i) \quad \text{or} \quad \rho_i(\tilde{x}_i) = \frac{1}{2}|\tilde{x}_i|^2 \tag{8.22}$$

and

$$\tilde{x}_i(k) = x_i(k) - \sum_{p=1}^{M} \tilde{w}_{1ip}\, \nu_R(k-p), \qquad \forall i. \tag{8.23}$$

Minimization of this cost function according to the standard gradient descent leads to a simple learning algorithm given by

$$\boxed{\tilde{w}_{1ip}(k+1) = \tilde{w}_{1ip}(k) - \tilde{\eta}(k)\frac{\partial J(\tilde{\mathbf{w}}_1)}{\partial \tilde{w}_{1ip}} \approx \tilde{w}_{1ip}(k) + \tilde{\eta}(k)\, f_R[\tilde{x}_i(k)]\, \nu_R(k-p),} \tag{8.24}$$

where $f_R(\tilde{x}_i(k))$ is a suitably chosen nonlinear function

$$\boxed{f_R(\tilde{x}_i(k)) = \frac{\partial \rho_i(\tilde{x}_i)}{\partial \tilde{x}_i}.} \tag{8.25}$$

Typically, $f_R(\tilde{x}_i) = \tilde{x}_i$ or $f_R(\tilde{x}_i) = \tanh(\gamma \tilde{x}_i)$.

In linear Finite Impulse Response (FIR) adaptive noise cancellation models described above, the noise is estimated as a weighted sum of delayed samples of reference interference. However, linear adaptive noise cancellation systems may not achieve an acceptable level of noise cancellation for some real world problems when interference signals are related to the measured reference signals in a complex, dynamic and nonlinear way.

(a)

(b)

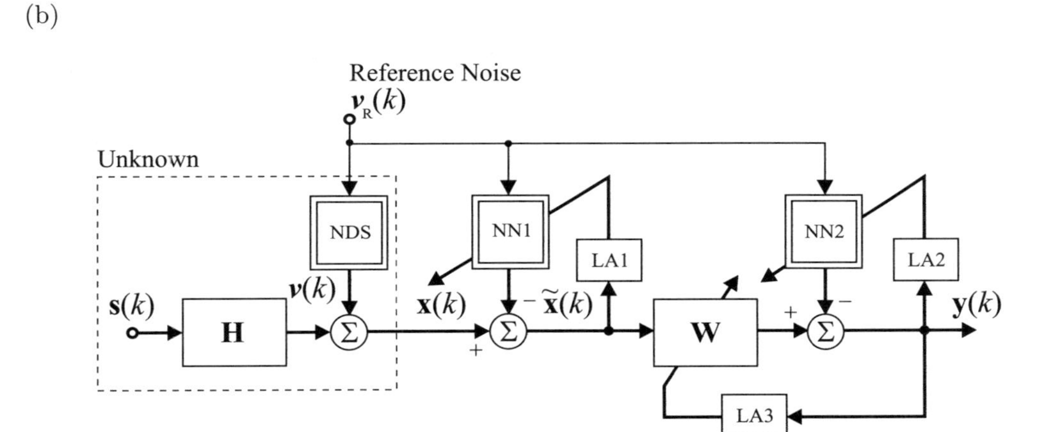

Fig. 8.3 Block diagrams illustrating multistage noise cancellation and blind source separation: (a) Linear model of convolutive noise, (b) more general model of additive noise modelled by nonlinear dynamical systems (NDS) and adaptive neural networks (NN); LA1 and LA2 denote learning algorithms performing the LMS or back-propagation supervising learning rules, whereas LA3 denotes a learning algorithm for BSS.

In such applications especially in biomedical signal processing, the optimum interference and noise cancellation usually requires nonlinear adaptive processing of recorded and measured on-line signals. In such cases, instead of linear filters, we can use standard neural network models and train them by back-propagation algorithms (see Fig. 8.3 (b)) [276, 259].

8.4 CUMULANT-BASED ADAPTIVE ICA ALGORITHMS

The main objective of this section is to present some ICA algorithms which are expressed by higher order cumulants. The main advantage of such algorithms is that they are theoretically robust to Gaussian noise (white or colored) on the condition that the number of available samples is sufficiently large to precisely estimate the cumulants. This important feature follows from the fact that all cumulants of the order higher than two are equal to zero for Gaussian noise. This property can be particularly useful in the analysis of biomedical signals buried in large Gaussian noise. However, since third-order cumulants are also equal to zero for any symmetrically distributed process (not just for Gaussian processes), then fourth-order cumulants are usually used for most applications when some source signals may have symmetrical distributions.

8.4.1 Cumulant-Based Cost Functions

In Chapter 6 as a measure of independence, we have used the Kullback-Leibler divergence which leads to the cost (risk) function

$$\mathcal{R}(\mathbf{y}, \mathbf{W}) = -\frac{1}{2}\log(|\det(\mathbf{W}\mathbf{W}^T)|) - \sum_{i=1}^{n} E\{\log(p_i(y_i))\} \tag{8.26}$$

or equivalently

$$\mathcal{R}(\mathbf{y}, \mathbf{G}) = -\log|\det(\mathbf{G})| - \sum_{i=1}^{n} E\{\log(p_i(y_i))\}, \tag{8.27}$$

where $\mathbf{G} = \mathbf{W}\mathbf{H} \in \mathbb{R}^{n \times n}$ is a nonsingular global mixing-separating matrix. In many practical applications, however, the pdf of the sources is not known *a priori* and in the derivation of practical algorithms we have to use a set of nonlinearities that in some special cases may not exactly match the true pdf of primary sources. In this section, we use an alternative cost function which is based on cumulants.

In this chapter, we will use the following notation: $C_q(y_1)$ denotes the q-order cumulants of the signal y_i and $\mathbf{C}_{p,q}(\mathbf{y}, \mathbf{y})$ denotes the cross-cumulant matrix whose elements are $[\mathbf{C}_{pq}(\mathbf{y}, \mathbf{y})]_{ij} = Cum(\underbrace{y_i, y_i, \ldots, y_i}_{p}, \underbrace{y_j, y_j, \ldots, y_j}_{q})$ (see Appendix A for properties of matrix cumulants and their relation to higher order moments).

Let us consider a particular case of nonlinearity misadjustment that results from replacing the pseudo-entropy terms $-E\{\log(p_i(y_i))\}$ in (8.27) by a function of the cumulants of the outputs $|C_{1+q}(y_i)|/(1+q)$. In other words, let us consider the following cost function:

$$J(\mathbf{y}, \mathbf{W}) = -\frac{1}{2}\log|\det(\mathbf{W}\mathbf{W}^T)| - \frac{1}{1+q}\sum_{i=1}^{n} |C_{1+q}(y_i)|. \tag{8.28}$$

The first term assures that the determinant of the global matrix will not approach zero. By including this term, we avoid the trivial solution $y_i = 0 \ \forall i$. The second terms force the

output signals to be as far as possible from Gaussianity, because the higher order cumulants are a natural measure of non-Gaussianity and they will vanish for Gaussian signals.
Taking into account the definition and properties of the cumulants, it is easy to show that

$$\begin{aligned}\frac{\partial Cum(\overbrace{y_i, \dots, y_i}^{1+q})}{\partial w_{kj}} &= (1+q)Cum(\overbrace{y_i, \dots, y_i}^{q}, \frac{\partial y_i}{\partial w_{kj}}) \\ &= (1+q)\delta_{ik}Cum(\overbrace{y_i, \dots, y_i}^{q}, x_j). \qquad (8.29)\end{aligned}$$

Using this result, we obtain [320, 321]

$$\frac{\partial \sum_{i=1}^{n} |C_{1+q}(y_i)|}{\partial \mathbf{W}} = (1+q)\mathbf{S}_{q+1}(\mathbf{y})\,\mathbf{C}_{q,1}(\mathbf{y}, \mathbf{x}), \qquad (8.30)$$

where $\mathbf{S}_{q+1}(\mathbf{y}) = \mathrm{sign}(\mathrm{diag}(\mathbf{C}_{1,q}(\mathbf{y}, \mathbf{y})))$. In addition, since

$$\frac{\partial \log|\det(\mathbf{W}\mathbf{W}^T)|}{\partial \mathbf{W}} = 2(\mathbf{W}\mathbf{W}^T)^{-1}\mathbf{W},$$

we obtain

$$\frac{\partial J(\mathbf{y}, \mathbf{W})}{\partial \mathbf{W}} = -(\mathbf{W}\mathbf{W}^T)^{-1}\mathbf{W} + \mathbf{S}_{q+1}(\mathbf{y})\,\mathbf{C}_{q,1}(\mathbf{y}, \mathbf{x}). \qquad (8.31)$$

8.4.2 Family of Equivariant Algorithms Employing Higher Order Cumulants

In order to improve the convergence and simplify the algorithm, instead of the standard gradient technique, we can use the Atick-Redlich formula which has the following form:

$$\Delta \mathbf{W} = -\eta \mathbf{W}\left[\frac{\partial J(\mathbf{y}, \mathbf{W})}{\partial \mathbf{W}}\right]^T \mathbf{W} = \eta \mathbf{W}\left[(\mathbf{W}\mathbf{W}^T)^{-1}\mathbf{W} - \mathbf{S}_{q+1}(\mathbf{y})\,\mathbf{C}_{q,1}(\mathbf{y}, \mathbf{x})\right]^T \mathbf{W}. \qquad (8.32)$$

Hence, after some mathematical manipulations, we obtain [320, 321]

$$\Delta \mathbf{W}(l) = \mathbf{W}(l+1) - \mathbf{W}(l) = \eta(l)\left[\mathbf{I} - \mathbf{C}_{1,q}(\mathbf{y}, \mathbf{y})\,\mathbf{S}_{q+1}(\mathbf{y})\right]\mathbf{W}(l). \qquad (8.33)$$

Multiplying both sides of Eq.(8.33) from the right side by the mixing matrix $\mathbf{H}$, we obtain the following algorithm to update the global matrix $\mathbf{G}$

$$\Delta \mathbf{G}(l) = \eta(l)\left[\mathbf{I} - \mathbf{C}_{1,q}(\mathbf{y}, \mathbf{y})\,\mathbf{S}_{q+1}(\mathbf{y})\right]\mathbf{G}(l). \qquad (8.34)$$

Taking into account the triangular inequality [315, 321]

$$\|\mathbf{C}_{1,q}(\mathbf{y}, \mathbf{y})\,\mathbf{S}_{q+1}(\mathbf{y}) - \mathbf{I}\|_p \le 1 + \|\mathbf{C}_{1,q}(\mathbf{y}, \mathbf{y})\,\mathbf{S}_{q+1}(\mathbf{y})\|_p = 1 + \|\mathbf{C}_{1,q}(\mathbf{y}, \mathbf{y})\|_p, \qquad (8.35)$$

it is sufficient to apply the following constraints to ensure the numerical stability of the algorithm

$$\boxed{\eta(l) = \min\left(\frac{2\eta_0}{1+\eta_0\, q}, \frac{\eta_0}{1+\eta_0\,\|\mathbf{C}_{1,q}(\mathbf{y}, \mathbf{y})\|_p}\right) < \frac{1}{1+\|\mathbf{C}_{1,q}(\mathbf{y}, \mathbf{y})\|_p},} \qquad (8.36)$$

where $0 < \eta_0 \leq 1$. It is interesting that for some special cases, the algorithm (8.33) simplifies to some well known algorithms: For example for $q = 1$, we obtain the globally stable algorithm for blind decorrelation proposed first by Almeida and Silva [13]

$$\mathbf{W}(l+1) = \mathbf{W}(l) + \eta(l) \left[\mathbf{I} - \langle \mathbf{y}\mathbf{y}^T \rangle\right] \mathbf{W}(l). \tag{8.37}$$

Analogously, for $q = 2$, we obtain a special form of the equivariant algorithm which is useful if all source signals have non-symmetric distributions, [277, 315] as

$$\boxed{\mathbf{W}(l+1) = \mathbf{W}(l) + \eta(l) \left[\mathbf{I} - \langle \mathbf{y}\mathbf{g}^T(\mathbf{y}) \rangle\right] \mathbf{W}(l),} \tag{8.38}$$

where components of vector $\mathbf{g}(\mathbf{y})$ have the following nonlinear functions $g(y_i) = \text{sign}(y_i)|y_i|^2$. Such an algorithm is insensitive to any symmetrically distributed noise and interference.

A more general algorithm (which is suitable for both symmetrically and non-symmetrically distributed sources) can be easily derived for $q = 3$, as (see Appendix A for more detail)

$$\begin{aligned}
\mathbf{W}(l+1) &= \mathbf{W}(l) + \eta(l) \left[\mathbf{I} - \mathbf{C}_{1,3}(\mathbf{y}, \mathbf{y})\, \mathbf{S}_4(\mathbf{y})\right] \mathbf{W}(l) \\
&= \mathbf{W}(l) + \eta(l) \left[\mathbf{I} - \langle \mathbf{y}\, \mathbf{g}^T(\mathbf{y}) \rangle\right] \mathbf{W}(l),
\end{aligned} \tag{8.39}$$

where $\mathbf{C}_{1,3}(\mathbf{y}, \mathbf{y}) = \langle \mathbf{y}(\mathbf{y} \circ \mathbf{y} \circ \mathbf{y})^T \rangle - 3 \langle \mathbf{y}\, \mathbf{y}^T \rangle \, \text{diag}\{\langle \mathbf{y} \circ \mathbf{y} \rangle\}$ and the diagonal elements of the matrix $\mathbf{S}_4(\mathbf{y})$ are defined by

$$[\mathbf{S}_4(\mathbf{y})]_{jj} = \text{sign}\left(E\{y_j^4\} - 3(E\{y_j^2\})^2\right) = \text{sign}\, \kappa_4(y_j).$$

It should be noted that we can write

$$\mathbf{C}_{1,3}(\mathbf{y}, \mathbf{y})\, \mathbf{S}_4(\mathbf{y}) = \langle \mathbf{y}\, \mathbf{g}^T(\mathbf{y}) \rangle, \tag{8.40}$$

where entries of the vector $\mathbf{g}(y) = [g_1(y_1), g_2(y_2), \ldots, g_n(y_n)]^T$ have the form

$$g_j(y_j) = (y_j^3 - 3 \langle y_j^2 \rangle y_j) \, \text{sign}(\kappa(y_j)). \tag{8.41}$$

In other words, the estimated cross-cumulants can be expressed as

$$C_{1,3}(y_i, y_j) \, \text{sign}(\kappa_4(y_j)) = \langle y_i g_j(y_j) \rangle = \left\langle y_i (y_j^3 - 3\hat{\sigma}_{y_j}^2 y_j) \, \text{sign}(\kappa(y_j)) \right\rangle. \tag{8.42}$$

Such cross-cumulants are insensitive to Gaussian noise. Moreover, the local stability conditions (derived in Chapter 7) are given by:

$$E\{g_i' s_i^2\} + 1 \;>\; 0, \tag{8.43}$$

$$E\{g_i'\}E\{g_j'\}E\{s_i^2\}E\{s_j^2\} \;<\; 1, \quad \forall i \neq j \tag{8.44}$$

and are always satisfied for any sources with non-zero kurtosis if the above nonlinear activation function is employed in the algorithm (8.41).

The on-line version of the above algorithm can be formulated as

$$\mathbf{W}(k+1) = \mathbf{W}(k) + \eta(k) \left[\mathbf{I} - \mathbf{R}_{\mathbf{y}\,\mathbf{g}}^{(k)}\right] \mathbf{W}(k), \tag{8.45}$$

where

$$\mathbf{R}_{\mathbf{y}\mathbf{g}}^{(k)} = (1 - \eta_0)\, \mathbf{R}_{\mathbf{y}\mathbf{g}}^{(k-1)} + \eta_0\, \mathbf{y}(k)\, \mathbf{g}^T(\mathbf{y}(k)), \tag{8.46}$$

with $g_j(y_j) = (y_j^3 - 3 y_j \hat{\sigma}_{y_j}^2)\, \mathrm{sign}(\kappa_4(y_j))$ and $\hat{\sigma}_{y_j}^2(k) = (1-\eta_0)\, \hat{\sigma}_{y_j}^2(k-1) + \eta_0\, y_j^2(k)$.

It should be noted that the learning algorithm (8.33) achieves its equilibrium point when [315, 321]

$$\mathbf{C}_{1,q}(\mathbf{y}, \mathbf{y})\, \mathbf{S}_{1+q}(\mathbf{y}) = \mathbf{G}\, \mathbf{C}_{1,q}(\mathbf{s}, \mathbf{s})\, (\mathbf{G}^{\circ q})^T\, \mathbf{S}_{1+q}(\mathbf{y}) = \mathbf{I}_n, \tag{8.47}$$

where $\mathbf{G}^{\circ q}$ denotes the q-th order Hadamard product of the matrix $\mathbf{G}$ with itself, i.e., $\mathbf{G}^{\circ q} = \mathbf{G} \circ \cdots \circ \mathbf{G}$. Hence, assuming that the primary independent sources are properly scaled such that $\mathbf{C}_{q,1}(\mathbf{s}, \mathbf{s}) = \mathbf{S}_{q+1}(\mathbf{s}) = \mathbf{I}_n$, (i.e., they all have the same value and sign of cumulants $C_{1+q}(s_i)$) the above nonlinear algebraic matrix equation simplifies to

$$\mathbf{G}\, (\mathbf{G}^{\circ q})^T = \mathbf{I}_n. \tag{8.48}$$

Possible solutions of this equation can be expressed as $\mathbf{G} = \mathbf{P}$, where $\mathbf{P}$ is any permutation matrix.

8.4.3 Possible Extensions

One of the disadvantages of the cost function (8.28) and the associated algorithm (8.33) is that they cannot be used when the specific $(q+1)$ order cumulant $\mathbf{C}_{1+q}(s_i)$ is zero or very close to zero for any of the source signals s_i. To overcome this limitation, we can use several matrix cumulants of different orders. In other words, we can attempt to find a set of indices $\Omega = \{q_1, \ldots, q_{N_\Omega} : q_i \in \mathcal{N}^+, q_i \neq 1\}$ such that the following sum of cumulants $\sum_{q \in \Omega} |C_{1+q}(y_i)|$ does not vanish for any of the sources. The existence of at least one possible set Ω is ensured due to the assumption of non-Gaussianity for the sources. Then, we can replace the terms $\frac{\mathbf{C}_{1+q}(y)}{1+q}$ in (8.28) by using a weighted sum of several cumulants of the outputs signals whose index q belongs to set Ω, i.e.,

$$J(\mathbf{y}, \mathbf{W}) = \sum_{q \in \Omega} \alpha_q\, \frac{|C_{1+q}(y_i)|}{1+q}, \tag{8.49}$$

where the positive weighting terms α_q are chosen such that $\sum_{q \in \Omega} \alpha_q = 1$. With $1 \notin \Omega$, we will assume that the second order statistics information is excluded from this weighted sum.

Following the same steps as before, it is straightforward to see that the learning algorithm takes the form [320, 321]:

$$\boxed{\Delta \mathbf{W}(l) = \eta(l)[\mathbf{I} - \sum_{q \in \Omega} \alpha_q\, \mathbf{C}_{1,q}(\mathbf{y}, \mathbf{y})\, \mathbf{S}_{q+1}(\mathbf{y})]\mathbf{W}(l),} \tag{8.50}$$

with a learning rate satisfying the constraints

$$\eta(l) \;=\; \min \left\{ \frac{2\eta_0}{1 + \eta_0 \sum_{q \in \Omega} \alpha_q\, q}\, ,\, \frac{\eta_0}{1 + \eta_0\, \| \sum_{q \in \Omega} \alpha_q\, \mathbf{C}_{1,q}(\mathbf{y}, \mathbf{y})\, \mathbf{S}_{q+1}(\mathbf{y}) \|_p} \right\}, \tag{8.51}$$

where $\eta_0 \leq 1$ and $\sum_{q \in \Omega} \alpha_q = 1$.

8.4.4 Cumulants for Complex Valued Signals

The extension of the previous algorithms to the case of complex sources and mixtures can be obtained by replacing the transpose operator $(\cdot)^T$ with the Hermitian operator $(\cdot)^H$ and by changing the definition of cumulants and cross-cumulants.
Whenever $(1+q)$ is an even number, it is possible to define self-cumulants

$$C_{1+q}(y_i) = Cum(\underbrace{y_i, \ldots, y_i}_{\frac{1+q}{2}}, \underbrace{y_i^*, \ldots, y_i^*}_{\frac{1+q}{2}}) \tag{8.52}$$

and the cross-cumulants

$$C_{1,q}(y_i, y_j) = Cum(y_i, \underbrace{y_j, \ldots, y_j}_{\frac{1+q}{2}-1}, \underbrace{y_j^*, y_j^*, \ldots, y_j^*}_{\frac{1+q}{2}}) \tag{8.53}$$

for complex-valued signals as [315, 321]. It is easy to show that cumulants are always real. With the above mentioned modification and changing the transpose operator to the Hermitian operator, it is possible to use the same algorithms for complex-valued signals [104, 321].

8.4.5 Blind Separation with More Sensors than Sources

The derivations of the presented class of robust algorithms (8.33) with respect to the Gaussian noise and its stability study can be performed in terms of the global transfer matrix $\mathbf{G}$ [321]. As long as this matrix remains square and non-singular these derivations in terms of $\mathbf{G}$ still fully apply. This also includes the case where the number of sensors m is greater than the number of sources $(m > n)$.
Taking the equivariant algorithm (8.34) expressed in terms of $\mathbf{G}$

$$\mathbf{G}(l+1) = [\mathbf{I} + \eta(l)(\mathbf{I} - \mathbf{C}_{1,q}(\mathbf{y}, \mathbf{y})\, \mathbf{S}_{q+1}(\mathbf{y}))]\, \mathbf{G}(l), \tag{8.54}$$

we can post-multiply it by the pseudo-inverse of the mixing matrix $\mathbf{H}^+ = (\mathbf{H}^T\mathbf{H})^{-1}\mathbf{H}^T$ to obtain

$$\mathbf{W}(l+1)\mathbf{P_H} = [\mathbf{I} + \eta(l)(\mathbf{I} - \mathbf{C}_{1,q}(\mathbf{y}, \mathbf{y})\, \mathbf{S}_{q+1}(\mathbf{y}))]\, \mathbf{W}(l)\mathbf{P_H}, \tag{8.55}$$

where $\mathbf{P_H} = \mathbf{H}\mathbf{H}^+$ is the projection matrix onto the space spanned by the columns of $\mathbf{H}$. Thus the algorithm is defined in terms of the separation matrix $\mathbf{W}(l+1)\mathbf{P_H}$ instead of $\mathbf{W}(l+1)$. If we omit[1] the projection $\mathbf{P_H}$, we obtain the same form of algorithm for the non-square separating matrix $\mathbf{W} \in \mathbb{R}^{n \times m}$ with $m \geq n$ as

$$\boxed{\mathbf{W}(l+1) = [\mathbf{I} + \eta(l)(\mathbf{I} - \mathbf{C}_{1,q}(\mathbf{y}, \mathbf{y})\, \mathbf{S}_{q+1}(\mathbf{y}))]\, \mathbf{W}(l).} \tag{8.56}$$

[1] With this omission, we will not affect the signal component of the outputs because $\mathbf{W}\mathbf{P_H}\mathbf{H} = \mathbf{I}_n$ implies $\mathbf{W}\mathbf{H} = \mathbf{I}_n$ and therefore, the separation can still be performed.

Table 8.1 Basic cost functions for ICA/BSS algorithms without prewhitening.

No.	Cost function $J(\mathbf{y}, \mathbf{W})$
1.	$-\frac{1}{2}\log\det(\mathbf{W}\mathbf{W}^T) - \sum_{i=1}^n E\{\log(p_i(y_i))\}$
2.	$-\frac{1}{2}\log\det(\mathbf{W}\mathbf{W}^T) - \frac{1}{1+q}\sum_{i=1}^n \lvert C_{1+q}(y_i)\rvert$ For $q=3, \quad C_4(y_i) = \kappa_4(y_i) = \frac{E\{y_i^4\}}{E^2\{y_i^2\}} - 3$
3.	$-\frac{1}{2}\log\det(\mathbf{W}\mathbf{W}^T) - \frac{1}{q}\sum_{i=1}^n E\{\lvert y_i\rvert^q)\}$ $q = 2$ for nonstationary sources, $q > 2$ for sub-Gaussian sources, $q < 2$ for super-Gaussian sources
4.	$\frac{1}{2}[-\log\det E\{\mathbf{y}\,\mathbf{y}^T\} + \sum_{i=1}^n \log E\{y_i^2\}]$ Sources are assumed to be nonstationary

Analogously, we can derive a similar algorithm for the estimation of the mixing matrix $\mathbf{H} \in \mathbb{R}^{m\times n}$ as

$$\boxed{\widehat{\mathbf{H}}(l+1) = \widehat{\mathbf{H}}(l) - \eta(l)\left[\widehat{\mathbf{H}}(l) - \mathbf{C}_{1,q}(\mathbf{x},\mathbf{y})\,\mathbf{S}_{1+q}(\mathbf{y})\right]} \tag{8.57}$$

or equivalently

$$\boxed{\widehat{\mathbf{H}}(l+1) = \widehat{\mathbf{H}}(l) - \eta(l)\widehat{\mathbf{H}}(l)\left[\mathbf{I} - \mathbf{C}_{1,q}(\mathbf{y},\mathbf{y})\,\mathbf{S}_{q+1}(\mathbf{y})\right].} \tag{8.58}$$

The equivariant algorithms presented in this section have several interesting properties which can be formulated in the form of the following theorems (see S. Cruces *et al.* for more details) [315, 320, 321].

Theorem 8.1 *The local convergence of the cumulant based equivariant algorithm (8.34) is isotropic, i.e., it does not depend on the source distribution, as long as their $(1+q)$-order cumulants do not vanish.*

Theorem 8.2 *The presence of additive Gaussian noise in the mixture does not change (for a sufficiently large number of samples) the convergence properties of the algorithm asymptotically, i.e., the estimated separating matrix is not biased theoretically by the additive Gaussian noise.*

This property is a consequence of using higher order cumulants instead of nonlinear activation functions. It should also be noted that the higher order cumulants of Gaussian distributed signals are always zero. However, we should point out that in practice, the cumulants of the outputs should be estimated from a finite set of data. This theoretical robustness to Gaussian noise only occurs in cases where there are sufficient numbers of samples (typically more than 5000), which allow reliable estimation of the cumulants.

Tables 8.1 and 8.2 summarize the typical cost functions and fundamental equivariant ICA algorithms presented and analyzed in this and the previous chapters.

8.5 ROBUST EXTRACTION OF A GROUP OF SOURCE SIGNALS ON THE BASIS OF CUMULANT COST FUNCTIONS

The approach discussed in the previous section can be relatively easily extended to the case when it is desired to extract a certain number of source signals e, with $1 \leq e \leq n$, defined by the user [21, 29, 322, 323, 325].

8.5.1 Blind Extraction of Sparse Sources with Largest Positive Kurtosis Using Prewhitening and Semi-Orthogonality Constraint

Let us assume that source signals are ordered according to the decreasing absolute value of their kurtosis as

$$|\kappa_4(s_1)| \geq |\kappa_4(s_2)| \geq \cdots |\kappa_4(s_e)| > |\kappa_4(s_{e+1})| \geq \cdots \geq |\kappa_4(s_n)| \,, \tag{8.59}$$

with $\kappa_4(s_j) = E\{s_j^4\} - 3E^2\{s_j^2\} \neq 0$, then we can formulate the following constrained optimization problem for extraction of e sources

$$\text{minimize} \qquad J(\mathbf{y}) = -\sum_{j=1}^{e} |\kappa_4(y_j)| \tag{8.60}$$

$$\text{subject to the constraint } \mathbf{W}_e \mathbf{W}_e^T = \mathbf{I}_e,$$

where $\mathbf{W} \in \mathbb{R}^{e \times n}$, $\mathbf{y}_e(k) = \mathbf{W}_e \overline{\mathbf{x}}(k)$ under the assumption that the sensor signals $\overline{\mathbf{x}} = \mathbf{Q}\,\mathbf{x}$ are prewhitened or the mixing matrix is orthogonalized in such way that $\mathbf{A} = \mathbf{Q}\,\mathbf{H}$ is orthogonal. The global minimization of such a constrained cost function leads to extraction of the e sources with largest absolute values of kurtosis.

To derive the learning algorithm, we can apply the natural gradient approach. A particularly simple and useful method to minimize the proposed cost function under semi-orthogonality constraints is to use the natural Riemannian gradient descent in the Stiefel

Table 8.2 Family of equivariant adaptive learning algorithms for ICA for complex-valued signals.

No.	Learning Algorithm	References
1.	$\Delta \mathbf{W} = \eta \left[\mathbf{\Lambda} - \langle \mathbf{f}(\mathbf{y})\, \mathbf{g}^H(\mathbf{y}) \rangle \right] \mathbf{W}$	Cichocki, *et al.* (1994) [279, 278, 277]
	$\mathbf{\Lambda}$ is a diagonal matrix with nonnegative elements λ_{ii}	
	$\mathbf{W}(l+1) = \left[\mathbf{I} \mp \eta \left[\mathbf{I} - \langle \mathbf{f}(\mathbf{y})\, \mathbf{g}^H(\mathbf{y}) \rangle \right] \right]^{\mp 1} \mathbf{W}(l)$	Cruces *et al.* (1999) [315, 326]
2.	$\Delta \mathbf{W} = \eta \left[\mathbf{\Lambda} - \langle \mathbf{f}(\mathbf{y})\, \mathbf{y}^H \rangle \right] \mathbf{W}$	Amari, Cichocki, Yang (1996) [28]
	$\lambda_{ii} = \langle f(y_i(k)) y_i^*(k) \rangle$ or $\lambda_{ii} = 1, \;\forall i$	Amari, Chen, Cichocki (1998) [25]
3.	$\Delta \mathbf{W} = \eta \left[\mathbf{I} - \langle \mathbf{y}\, \mathbf{y}^H \rangle - \langle \mathbf{f}(\mathbf{y})\, \mathbf{y}^H \rangle + \langle \mathbf{y}\, \mathbf{f}^H(\mathbf{y}) \rangle \right] \mathbf{W}$	Cardoso, Laheld, (1996) [148]
4.	$\Delta \mathbf{W} = \eta \left[\mathbf{I} - \langle \mathbf{y}\, \mathbf{y}^H \rangle - \langle \mathbf{f}(\mathbf{y})\, \mathbf{y}^H \rangle + \langle \mathbf{f}(\mathbf{y})\, \mathbf{f}^H(\mathbf{y}) \rangle \right] \mathbf{W}$	Karhunen, Pajunen (1997) [673]
5.	$\Delta \mathbf{W} = -\eta \left[\mathbf{W}\mathbf{W}^H \langle \mathbf{f}(\mathbf{y})\, \mathbf{y}^H \rangle - \langle \mathbf{y}\, \mathbf{f}^H(\mathbf{y}) \rangle \right] \mathbf{W}$	Douglas (1999) [381]
6.	$\tilde{\mathbf{W}} = \mathbf{W} + \eta \left[\mathbf{\Lambda} - \langle \mathbf{f}(\mathbf{y})\, \mathbf{y}^H \rangle \right] \mathbf{W}, \;\; \lambda_{ii} = \langle f(y_i)\; y_i^* \rangle$	Hyvärinen, Oja (1999) [589]
	$\eta_{ii} = [\lambda_{ii} + \langle f'(y_i) \rangle]^{-1}; \;\; \mathbf{W} = (\tilde{\mathbf{W}} \tilde{\mathbf{W}}^H)^{-1/2}\, \tilde{\mathbf{W}}$	
7.	$\Delta \mathbf{W} = \eta \left[\mathbf{I} - \mathbf{\Lambda}^{-1} \langle \mathbf{y}\, \mathbf{y}^H \rangle \right] \mathbf{W}$	Amari, Cichocki (1998) [26]
	$\lambda_{ii}(k) = \langle y_i(k)) y_i^*(k) \rangle$	Choi, Cichocki, Amari (2000) [228]
8.	$\Delta \mathbf{W} = \eta \left[\mathbf{I} - \mathbf{C}_{1,q}(\mathbf{y}, \mathbf{y})\, \mathbf{S}_q(\mathbf{y}) \right] \mathbf{W}$	Cruces *et al.* (1999) [320]
	$C_{1,q}(y_i, y_j) = Cum(y_i, \underbrace{y_j, \ldots, y_j}_{\frac{1+q}{2}-1}, \underbrace{y_j^*, y_j^*, \ldots, y_j^*}_{\frac{1+q}{2}})$	
9.	$\mathbf{W}(l+1) = \exp(\eta\, \mathbf{F}[\mathbf{y}])\, \mathbf{W}(l)$	Nishimori, (1999) [882]
	$\mathbf{F}(\mathbf{y}) = \mathbf{\Lambda} - \langle \mathbf{y}\, \mathbf{y}^H \rangle - \langle \mathbf{f}(\mathbf{y})\, \mathbf{y}^H \rangle + \langle \mathbf{y}\, \mathbf{f}^H(\mathbf{y}) \rangle$	Cichocki, Georgiev (2002) [286]
10.	$\Delta \mathbf{W} = \eta \mathbf{F}[\mathbf{y}] \mathbf{W}$	Amari, (1997) [23]
	$f_{ij} = \left[\lambda_{ii} \delta_{ij} - \alpha_{1i} \left\langle y_i y_j^* \right\rangle - \alpha_{2i} \left\langle f(y_i) y_j^* \right\rangle + \alpha_{3i} \left\langle y_i f(y_j^*) \right\rangle \right]$	L. Zhang *et al.* (2000) [1350]

manifold of the semi-orthogonal matrix [29] which is given by (see Chapter 6 for more detail)

$$\Delta \mathbf{W}_e(l) = -\eta \left[\nabla_{\mathbf{W}_e} J - \mathbf{W}_e (\nabla_{\mathbf{W}_e} J)^T \mathbf{W}_e\right]. \tag{8.61}$$

In our case this leads to the following gradient algorithm [323]

$$\boxed{\mathbf{W}_e(l+1) = \mathbf{W}_e(l) + \eta \left[\mathbf{S}_4(\mathbf{y})\, \mathbf{C}_{3,1}(\mathbf{y}, \overline{\mathbf{x}}) - \mathbf{C}_{1,3}(\mathbf{y}, \mathbf{y})\, \mathbf{S}_4(\mathbf{y}) \mathbf{W}_e(l)\right],} \tag{8.62}$$

where $\mathbf{S}_4(\mathbf{y})$ is a diagonal matrix with entries $s_{ii} = [\text{sign}(\kappa_4(y_i))]_{ii}$ (the actual kurtosis signs of the output signals) and $\mathbf{C}_{1,3}(\mathbf{y}, \mathbf{y})$ is the fourth order cross-cumulant matrix with elements $[\mathbf{C}_{1,3}(\mathbf{y}, \mathbf{y})]_{ij} = Cum(y_i(k), y_j(k), y_j(k), y_j(k))$; analogously the cross-cumulant matrix is defined by $\mathbf{C}_{3,1}(\mathbf{y}, \overline{\mathbf{x}}) = (\mathbf{C}_{1,3}(\overline{\mathbf{x}}, \mathbf{y}))^T$.
For $e = n$, ($\mathbf{W}_e = \mathbf{W}$) the algorithm takes the form

$$\mathbf{W}(l+1) = \mathbf{W}(l) + \eta \left[\mathbf{S}_4(\mathbf{y})\, \mathbf{C}_{3,1}(\mathbf{y}, \mathbf{y}) - \mathbf{C}_{1,3}(\mathbf{y}, \mathbf{y})\, \mathbf{S}_4(\mathbf{y})\right] \mathbf{W}(l). \tag{8.63}$$

The algorithm (8.62) can be implemented as an on-line (moving average) algorithm:

$$\boxed{\mathbf{W}_e(k+1) = \mathbf{W}_e(k) + \eta(k) \left[\mathbf{R}^{(k)}_{\mathbf{g}\overline{\mathbf{x}}} - \mathbf{R}^{(k)}_{\mathbf{y}\mathbf{g}}\, \mathbf{W}_e(k)\right],} \tag{8.64}$$

where

$$\mathbf{R}^{(k)}_{\mathbf{g}\overline{\mathbf{x}}} = (1 - \eta_0)\, \mathbf{R}^{(k-1)}_{\mathbf{g}\overline{\mathbf{x}}} + \eta_0\, \mathbf{g}(\mathbf{y}(k))\, \overline{\mathbf{x}}^T(k), \tag{8.65}$$

$$\mathbf{R}^{(k)}_{\mathbf{y}\mathbf{g}} = (1 - \eta_0)\, \mathbf{R}^{(k-1)}_{\mathbf{y}\mathbf{g}} + \eta_0\, \mathbf{y}(k)\, \mathbf{g}^T(\mathbf{y}(k)), \tag{8.66}$$

and $\mathbf{g}(\mathbf{y}) = [g(y_1), g(y_2), \ldots, g(y_e)]^T$, with

$$g(y_i) = \left(y_i^3 - 3 y_i E\{y_i^2\}\right)\, \text{sign}(\kappa_4(y_i)). \tag{8.67}$$

It is important to note that such a cost function has no spurious minima in the sense that all local minima correspond to separating equilibria of true sources $s_1, s_2, \ldots, s_n$, with nonzero kurtosis.

Note that because the mixing matrix is orthogonal and the demixing matrix is semi-orthogonal, the global matrix $\mathbf{G} = \mathbf{W}\mathbf{A} \in \mathbb{R}^{e \times n}$ is also semi-orthogonal, thus the constraints $\mathbf{W}_e \mathbf{W}_e^T = \mathbf{I}_e$ can be replaced equivalently by $\mathbf{G}\mathbf{G}^T = \mathbf{I}_e$ or equivalently $\mathbf{g}_i^T \mathbf{g}_j = \delta_{ij}, \forall i$, where $\mathbf{g}_i$ is the i-th row of $\mathbf{G}$. Taking into account this fact, the optimization problem (8.60) can be converted into an equivalent set of e simultaneous maximization problems:

$$\text{maximize} \qquad J_i(\mathbf{g}_i) = \beta_i\, \kappa_4(y_i) = \beta_i \sum_{j=1}^{n} g_{ij}^4 \kappa_4(s_j), \tag{8.68}$$

subject to the constraints $\mathbf{g}_i^T \mathbf{g}_j = \delta_{ij}$ for $i, j = 1, 2, \ldots, e$, where $\beta_i = \text{sign}(\kappa_4(y_i))$.

Because the vectors $\mathbf{g}_i$ are orthogonal and of unit length, the above cost functions satisfy the conditions of Lemma A.2, and Theorem A.1 presented in Chapter 5; thus the local minimum is achieved if and only if $\mathbf{g}_i = \pm \mathbf{e}_i$. Also due to orthogonality constraints the above cost functions ensure the extraction of e different sources.

In the special case, when all sources $\{s_i(k)\}$ have the same kurtosis sign, such that $\kappa_4(s_i) > 0$ or $\kappa_4(s_i) < 0$ for all $1 \le i \le n$, we can formulate a somewhat simpler optimization problem to (8.60) as

$$\text{minimize} \qquad \tilde{J}(\mathbf{y}) = -\frac{\beta}{4} \sum_{i=1}^{e} E\{y_i^4\} \;\; \text{subject to} \;\; \mathbf{W}_e \mathbf{W}_e^T = \mathbf{I}_e, \tag{8.69}$$

where β satisfies $\beta \kappa_4[s_i] > 0, \quad \forall i$. This leads to the following simplified batch learning algorithm proposed by Amari and Cruces [21, 322]:

$$\boxed{\mathbf{W}_e(l+1) = \mathbf{W}_e(l) + \eta \, \beta \, \left[\langle \mathbf{y}^3 \overline{\mathbf{x}}^T \rangle - \langle \mathbf{y} (\mathbf{y}^3)^T \rangle \, \mathbf{W}_e(l) \right],} \tag{8.70}$$

where $\mathbf{g}(\mathbf{y}) = \mathbf{y}^3 = \mathbf{y} \circ \mathbf{y} \circ \mathbf{y} = [y_1^3, y_2^3, \ldots, y_e^3]^T$.

8.5.2 Blind Extraction of an Arbitrary Group of Sources without Prewhitening

In the section 8.4.5, we have shown that the mixing matrix $\mathbf{H} \in \mathbb{R}n \times n$ can be estimated using the matrix cumulants $\mathbf{C}_{1,q}(\mathbf{y}, \mathbf{y})$ by applying the following learning rule

$$\widehat{\mathbf{H}}(l+1) \;\; = \;\; \widehat{\mathbf{H}}(l) - \eta(l) \left[\widehat{\mathbf{H}}(l) - \mathbf{C}_{1,q}(\mathbf{x}, \mathbf{y}) \, \mathbf{S}_{1+q}(\mathbf{y}) \right]. \tag{8.71}$$

Cruces *et al.* note that the above learning form can be written in a more general form as [325]

$$\boxed{\widehat{\mathbf{H}}_e(l+1) = \widehat{\mathbf{H}}_e(l) - \eta(l) \left[\widehat{\mathbf{H}}_e(l) - \mathbf{C}_{1,q}(\mathbf{x}, \mathbf{y}) \, \mathbf{S}_{1+q}(\mathbf{y}) \right],} \tag{8.72}$$

where $\mathbf{y} = \mathbf{W}_e \, \mathbf{x}$ and $\widehat{\mathbf{H}}_e = [\widehat{\mathbf{h}}_1, \widehat{\mathbf{h}}_2, \ldots, \widehat{\mathbf{h}}_e]$ are the first e vectors of the estimating mixing matrix $\widehat{\mathbf{H}}$. This means, we can estimate an arbitrary number of the columns of the mixing matrix without knowledge of other columns, on the condition that the whole separating matrix can also be estimated.

In order to extract e sources, we can minimize the Mean Square Error (MSE) between the estimated sources $\widehat{\mathbf{s}}_e$ and the outputs $\mathbf{y} = \mathbf{W}_e \, \mathbf{x}$. This is equivalent to the minimization of the output power of estimated output signals subject to some signal preserving constraints, i.e. [325],

$$\min_{\mathbf{W}_e} \operatorname{tr}\{\mathbf{R}_{\mathbf{y}\,\mathbf{y}}\} \;\; \text{subject to} \;\; \mathbf{W}_e(l) \, \mathbf{H}_e(l) = \mathbf{I}_e. \tag{8.73}$$

We can solve this constrained minimization problem by means of Lagrange multipliers, which after some simple mathematical operations yields [325]

$$\boxed{\mathbf{y}(k) = \mathbf{W}_e(l) \, \mathbf{x}(k) = \left[\widehat{\mathbf{H}}_e^T(l) \, \mathbf{R}_{\mathbf{x}\,\mathbf{x}}^{+} \, \widehat{\mathbf{H}}_e(l) \right]^{-1} \widehat{\mathbf{H}}_e^T(l) \, \mathbf{R}_{\mathbf{x}\,\mathbf{x}}^{+} \, \mathbf{x}(k)} \tag{8.74}$$

Table 8.3 Typical cost functions for blind signal extraction of a group of e-sources $(1 \leq e \leq n)$ with prewhitening of sensor signals, i.e., $\mathbf{A}\mathbf{A}^T = \mathbf{I}$.

No.	Cost function $J(\mathbf{y}, \mathbf{W})$	Remarks
1.	$\sum_{i=1}^{e} h(y_i) = -\sum_{i=1}^{e} E\{\log p_i(y_i)\}$ s.t. $\mathbf{W}_e \mathbf{W}_e^T = \mathbf{I}_e$	Minimization of entropy Amari [21, 29]
2.	$\frac{1}{1+q} \sum_{i=1}^{e} \lvert C_{1+q}(y_i) \rvert$ s.t. $\mathbf{W}_e \mathbf{W}_e^T = \mathbf{I}_e$ Cumulants $C_{1+q}(y_i)$ must not vanish for the extracted source signals. For $q = 3$, $\quad C_4(y_i) = \kappa_4(y_i) = E\{y_i^4\} - 3E^2\{y_i^2\}$	Maximization of distance from Gaussianity Cruces *et al.* [322, 323]
3.	$-\frac{1}{q} \sum_{i=1}^{e} E\{\lvert y_i \rvert^q\}$ s.t. $\mathbf{W}_e \mathbf{W}_e^T = \mathbf{I}_e$	Minimization of generalized energy [228, 277]
4.	$\frac{1}{2} \sum_{i=1}^{e} E\{\log(y_i^2)\}$ s.t. $\mathbf{W}_e \mathbf{W}_e^T = \mathbf{I}_e$	Maximization of negentropy Matsuoka *et al.* [827, 828] Choi *et al.* [228]

and

$$\boxed{\widehat{\mathbf{H}}_e(l+1) = \widehat{\mathbf{H}}_e(l) - \eta(l) \left[\widehat{\mathbf{H}}_e(l) - \mathbf{C}_{1,q}(\mathbf{x}, \mathbf{y})\, \mathbf{S}_{1+q}(\mathbf{y}) \right],} \tag{8.75}$$

where $\mathbf{R}_{\mathbf{xx}}^{+} \in \mathbb{R}^{m \times m}$ (with $m \geq n$ and $1 \leq e \leq n$) is the pseudo-inverse of the estimated covariance matrix of the observed sensor signals. Then, in order to extract the sources, we only have to alternatively iterate equations (8.74) and (8.75), which constitute the kernel of the implementation of the robust BSE algorithm. The algorithm is able to recover an arbitrary number $e < n$ of sources without the prewhitening step. However, this is done at the extra cost of having to compute the pseudo-inverse of the correlation matrix of the

Table 8.4 BSE algorithm based on cumulants without prewhitening [325].

1. Initialization: $e \leq rank(\mathbf{R_{xx}})$, $\eta_0 < 1$, δ, c, $\mathbf{H}_e(0)$.

   Computing: $\mathbf{R}^+_{\mathbf{xx}}$, $\mathbf{W}_e(0) = [\mathbf{H}^H_e(0)\mathbf{R}^+_{\mathbf{xx}}\mathbf{H}_e(0)]^{-1}\mathbf{H}^H_e(0)\mathbf{R}^+_{\mathbf{xx}}$ and $\mathbf{y} = \mathbf{W}_e(0)\mathbf{x}$.

2. Estimation of the optimal learning rate:

$$\eta(l) = min\left\{\frac{2\eta_0}{1+3\eta_0}, \frac{\eta_0}{1+\eta_0\|\mathbf{C}_{1,3}(\mathbf{y},\mathbf{y})\|_c}\right\}$$

3. Estimation of a column of the mixing matrix and extraction of a group of sources:

$$\begin{aligned}
\widehat{\mathbf{H}}_e(l+1) &= \widehat{\mathbf{H}}_e(l) - \eta(l)[\widehat{\mathbf{H}}_e(l) - \mathbf{C}_{1,3}(\mathbf{x},\mathbf{y})\,\mathbf{S}_4(\mathbf{y})], \\
\mathbf{W}_e(l+1) &= [\widehat{\mathbf{H}}^H_e(l+1)\,\mathbf{R}^+_{\mathbf{xx}}\,\widehat{\mathbf{H}}_e(l+1)]^{-1}\widehat{\mathbf{H}}^H_e(l+1)\,\mathbf{R}^+_{\mathbf{xx}}, \\
\mathbf{y}(k) &= \mathbf{W}_e(l+1)\,\mathbf{x}(k).
\end{aligned}$$

4. $n = n + 1$,
   UNTIL ($\|\mathbf{C}_{1,3}(\mathbf{y},\mathbf{y})\,\mathbf{S}_4(\mathbf{y}) - \mathbf{I}_e\|_c < \delta$) RETURN TO 2.

5. IF deflation
   STORE $\mathbf{y}$,
   $\mathbf{x} = (\mathbf{I}_n - \widehat{\mathbf{H}}_e(l)[\widehat{\mathbf{H}}_e(l)]^+)\,\mathbf{x}$,
   RETURN TO 1.
   ELSE END.

sensor signals. The advantage of this algorithm is that it is insensitive to Gaussian noise on the condition that the covariance of the noise is known or can be reliably estimated. Practical implementation of the algorithm for the complex-valued signals using fourth order cumulants is summarized in Table 8.4 [325].

8.6 RECURRENT NEURAL NETWORK APPROACH FOR NOISE CANCELLATION

8.6.1 Basic Concept and Algorithm Derivation

Assume that we have successfully estimated an unbiased estimate of the separating matrix $\mathbf{W}$ via one of the previously described approaches. Then, we can estimate a mixing matrix

$\widehat{\mathbf{H}} = \mathbf{W}^+ = \mathbf{H}\,\mathbf{P}\,\mathbf{D}$, where $\mathbf{W}^+$ is the pseudo-inverse of $\mathbf{W}$, $\mathbf{P}$ is any $n \times n$ permutation matrix, and $\mathbf{D}$ is an $n \times n$ non-singular diagonal scaling matrix. We now propose approaches for cancelling the effects of noise in the estimated source signals.

In order to develop a viable neural network approach for noise cancellation, we define the error vector

$$\mathbf{e}(t) = \mathbf{x}(t) - \widehat{\mathbf{H}}\,\widehat{\mathbf{y}}(t), \tag{8.76}$$

where $\mathbf{e}(t) = [e_1(t), e_2(t), \ldots, e_m(t)]^T$ and $\widehat{\mathbf{y}}(t)$ is an estimate of the source $\mathbf{s}(t)$. To compute $\widehat{\mathbf{y}}(t)$, consider the minimum entropy (ME) cost function

$$E\{J(\mathbf{e}(t))\} = -\sum_{i=1}^{m} E\{\log[p_i(e_i(t))]\}, \tag{8.77}$$

where $p_i(e_i)$ is the true pdf of the additive noise $\nu_i(t)$. It should be noted that we have assumed the noise sources to be i.i.d.; thus, stochastic gradient descent of the ME function yields stochastic independence of the error components as well as the minimization of their magnitude in an optimal way. The resulting system of differential equations is

$$\frac{d\widehat{\mathbf{y}}(t)}{dt} = \mu(t)\widehat{\mathbf{H}}^T\,\boldsymbol{\Psi}[\mathbf{e}(t)], \tag{8.78}$$

where $\boldsymbol{\Psi}[\mathbf{e}(t)] = [\Psi_1[e_1(t)], \ldots, \Psi_m[e_m(t)]]^T$ with nonlinearities

$$\Psi_i(e_i) = -\frac{\partial \log p_i(e_i)}{\partial e_i}. \tag{8.79}$$

Remark 8.3 *In a more general case, functions $\Psi_i(e_i)$ $(i = 1, 2, \ldots, m)$ can be implemented via nonlinear filters that perform filtering and nonlinear noise shaping in every channel. For example, we can use the FIR filter of the form*

$$\Psi_i(e_i(k)) = \psi_i\left(\sum_{p=0}^{L} b_{ip}\, e_i(k-p)\right), \tag{8.80}$$

where the parameters of filters $\{b_{ip}\}$ and nonlinearities $\psi_i(e_i)$ are suitably chosen.

A block diagram illustrating the implementation of the above algorithm is shown in Fig. 8.4, where Learning Algorithm denotes an appropriate bias removal learning rule (8.10).

In the proposed algorithm, the optimal choice for nonlinearities $\Psi_i(e_i)$ depends on the noise distributions. Assume that all of the noise signals are drawn from a generalized Gaussian distribution of the form [509]

$$p_i(e_i) = \frac{r_i}{2\sigma_i\Gamma(1/r_i)} \exp\left(-\frac{1}{r_i}\left|\frac{e_i}{\sigma_i}\right|^{r_i}\right), \tag{8.81}$$

where $r_i > 0$ is a variable parameter, $\Gamma(r) = \int_0^\infty u^{r-1}\exp(-u)du$ is the Gamma function and $\sigma^r = E\{|e|^r\}$ is a generalized measure of the noise variance known as dispersion. Note that a unity value of r_i yields a Laplacian distribution, $r_i = 2$ yields the standard Gaussian distribution, and $r_i \to \infty$ yields a uniform distribution. In general, we can select any value of $r_i \geq 1$, in which case the locally-optimal nonlinear activation functions are of the form

$$\Psi_i(e_i) = -\frac{\partial \log(p_i(e_i))}{\partial e_i} = \frac{1}{|\sigma_i|^{r_i}} |e_i|^{r_i - 1} \text{sign}(e_i), \quad r_i \geq 1. \tag{8.82}$$

Remark 8.4 *For the standard Gaussian noise with known variances σ_i^2, $i = 1, 2, \ldots, m$ with $m > n$, the optimal nonlinearities simplify to linear functions of the form $\Psi_i(e_i) = e_i/\sigma_i^2 \ \forall i$. The Amari-Hopfield neural network realizes on-line implementation of the BLUE (Best Linear Unbiased Estimator) algorithm for which the numerical form has been discussed in Chapter 2. On the other hand, for very impulsive (spiky) sources with high value of kurtosis, the optimal parameter r_i typically takes a value between zero and one. In such cases, we can use the modified activation functions $\Psi_i(e_i) = e_i/[|\sigma_i|^{r_i} |e_i|^{2-r_i} + \epsilon]$, where ϵ is a small positive constant to avoid the singularity of the function at $e_i = 0$.*

If the distribution of noise is not known *a priori*, we can attempt to adapt the value of $r_i(k)$ for each error signal $e_i(k)$ according to its estimated distance from Gaussianity. A simple gradient-based rule for adjusting each parameter $r_i(k)$ is

$$\Delta r_i(k) = -\eta_i(k) \frac{\partial \phi}{\partial r_i} = \eta_i(k) \frac{\partial \log(p_i(e_i))}{\partial r_i} \tag{8.83}$$

$$\cong \eta_i(k) |e_i(k)|^{r_i(k)} \log(|e_i(k)|). \tag{8.84}$$

or

$$\Delta r_i(k) = -\eta_i \frac{\partial \phi}{\partial r_i} = \eta_i \frac{\partial \log(p_i(e_i))}{\partial r_i} \tag{8.85}$$

$$\cong \eta_i \frac{0.1 r_i(k) + |e_i(k)|^{r_i(k)} (1 - \log(|e_i(k)|^{r_i(k)})}{r_i^2(k)}. \tag{8.86}$$

Similar methods can be applied for other parameterized noise distributions. For example, when $p_i(e_i)$ is a generalized Cauchy distribution, then $\Psi_i(e_i) = [(v r_i + 1)/(v|A(r_i)|^{r_i} + |e_i|^{r_i})] |e_i|^{r_i - 1} \text{sign}(e_i)$. Similarly, for the generalized Rayleigh distribution, one obtains $\Psi_i(e_i) = |e_i|^{r_i - 2} e_i$ for complex-valued signals and coefficients.

It should be noted that the continuous–time algorithm in (8.78) can be easily converted to a discrete-time algorithm as

$$\widehat{\mathbf{y}}(k+1) = \widehat{\mathbf{y}}(k) + \eta(k) \widehat{\mathbf{H}}^T(k) \ \mathbf{\Psi}[\mathbf{e}(k)]. \tag{8.87}$$

The proposed system in Fig. 8.5 can be considered as a form of nonlinear post-processing that effectively reduces the additive noise component in the estimated source signals. In the next subsection, we propose a more efficient architecture that simultaneously estimates the mixing matrix $\mathbf{H}$ while reducing the amount of noise in the separated sources.

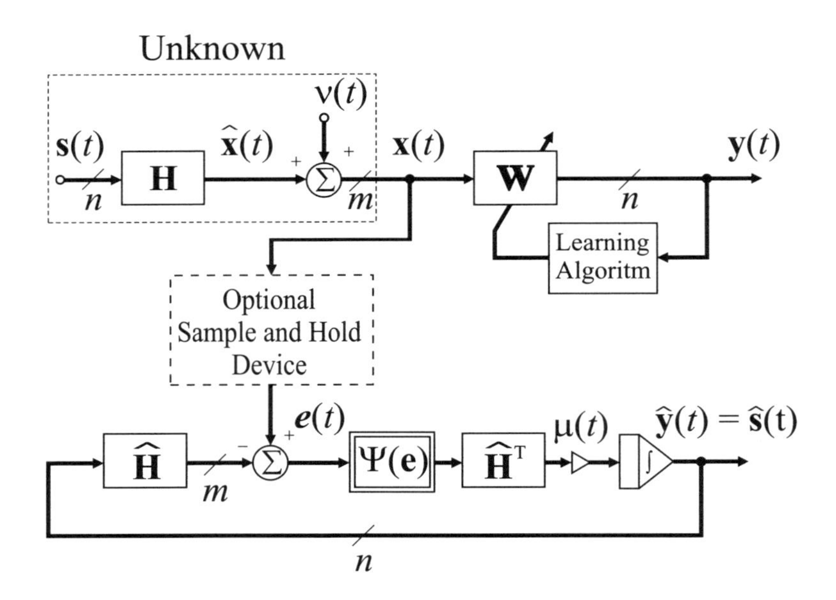

Fig. 8.4 Analog Amari-Hopfield neural network architecture for estimating the separating matrix and reducing the noise in the separated sources.

8.6.2 Simultaneous Estimation of a Mixing Matrix and Noise Reduction

Let us next consider on-line estimation of the mixing matrix $\mathbf{H}$ rather than estimation of the separating matrix $\mathbf{W}$. Based upon the analysis from previous sections, it is easy to derive such a learning algorithm. Taking into account that $\widehat{\mathbf{H}} = \mathbf{W}^{+}$ and, for $m \geq n$, from $\mathbf{W}\widehat{\mathbf{H}} = \mathbf{W}\mathbf{W}^{+} = \mathbf{I}_n$, we have a simple relation [277]

$$\frac{d\mathbf{W}}{dt}\widehat{\mathbf{H}} + \mathbf{W}\frac{d\widehat{\mathbf{H}}}{dt} = \mathbf{0}. \tag{8.88}$$

Hence, we obtain the learning algorithm (see Chapters 6 and 7 for derivation)

$$\frac{d\widehat{\mathbf{H}}(t)}{dt} = -\widehat{\mathbf{H}}\frac{d\mathbf{W}}{dt}\widehat{\mathbf{H}} = -\mu_1(t)\widehat{\mathbf{H}}(t)[\mathbf{\Lambda}(t) - \mathbf{f}[\mathbf{y}(t)]\,\mathbf{g}^T[\mathbf{y}(t)]], \tag{8.89}$$

where $\mathbf{\Lambda} = \mathrm{diag}\{\lambda_1, \lambda_2, \ldots, \lambda_n\}$ is a diagonal positive definite matrix (typically, $\mathbf{\Lambda} = \mathbf{I}_n$ or $\mathbf{\Lambda} = \mathrm{diag}\{\mathbf{f}[\mathbf{y}(t)]\,\mathbf{y}^T(t)\}$) and $\mathbf{f}[\mathbf{y}(t)] = [f_1(y_1), \ldots, f_n(y_n)]^T$, $\mathbf{g}(\mathbf{y}) = [g_1(y_1), \ldots, g_n(y_n)]^T$ are vectors of suitably chosen nonlinear functions. We can replace the output vector $\mathbf{y}(t)$ by an improved estimate $\widehat{\mathbf{y}}(t)$ to derive a learning algorithm as

$$\frac{d\widehat{\mathbf{H}}(t)}{dt} = -\mu_1(t)\widehat{\mathbf{H}}(t)\left[\mathbf{\Lambda}(t) - \mathbf{f}[\widehat{\mathbf{y}}(t)]\,\mathbf{g}^T[\widehat{\mathbf{y}}(t)]\right] \tag{8.90}$$

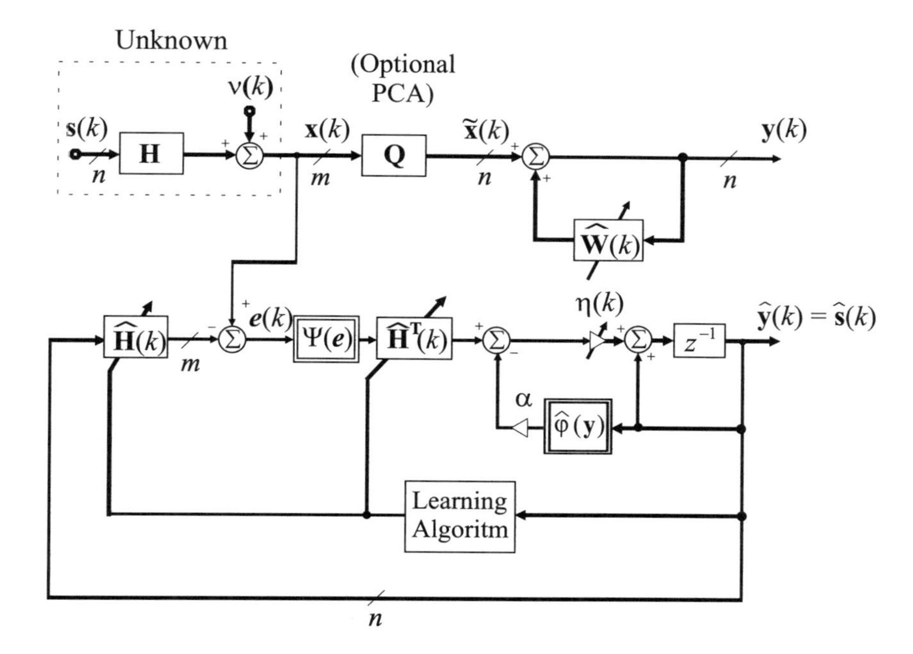

Fig. 8.5 Architecture of the Amari-Hopfield recurrent neural network for simultaneous noise reduction and mixing matrix estimation: Conceptual discrete-time model with optional PCA.

and

$$\frac{d\widehat{\mathbf{y}}(t)}{dt} = \mu(t)\widehat{\mathbf{H}}^T(t)\ \boldsymbol{\Psi}[\mathbf{e}(t)], \tag{8.91}$$

or in the discrete-time,

$$\begin{aligned} \Delta\widehat{\mathbf{H}}(k) &= \widehat{\mathbf{H}}(k+1) - \widehat{\mathbf{H}}(k) \\ &= \eta_1(k)\widehat{\mathbf{H}}(k)\left[\boldsymbol{\Lambda}(k) - \mathbf{f}[\widehat{\mathbf{y}}(k)]\,\mathbf{g}^T[\widehat{\mathbf{y}}^T(k)]\right] \end{aligned} \tag{8.92}$$

and

$$\widehat{\mathbf{y}}(k+1) = \widehat{\mathbf{y}}(k) + \eta(k)\widehat{\mathbf{H}}^T(k)\ \boldsymbol{\Psi}[\mathbf{e}(k)], \tag{8.93}$$

where $\mathbf{e}(k) = \mathbf{x}(k) - \widehat{\mathbf{H}}(k)\widehat{\mathbf{s}}(k)$ and $\mathbf{x}(k) = \widehat{\mathbf{x}}(k) + \boldsymbol{\nu}(k)$. A functional block diagram illustrating the implementation of the algorithm using Amari-Hopfield neural network is shown in Fig. 8.5.

8.6.2.1 Regularization For some systems, the mixing matrix $\mathbf{H}$ may be highly ill-conditioned. Thus, in order to estimate reliable and stable solutions, it is necessary to apply optional prewhitening (e.g., PCA) and/or regularization methods. For ill-conditioned cases, we can

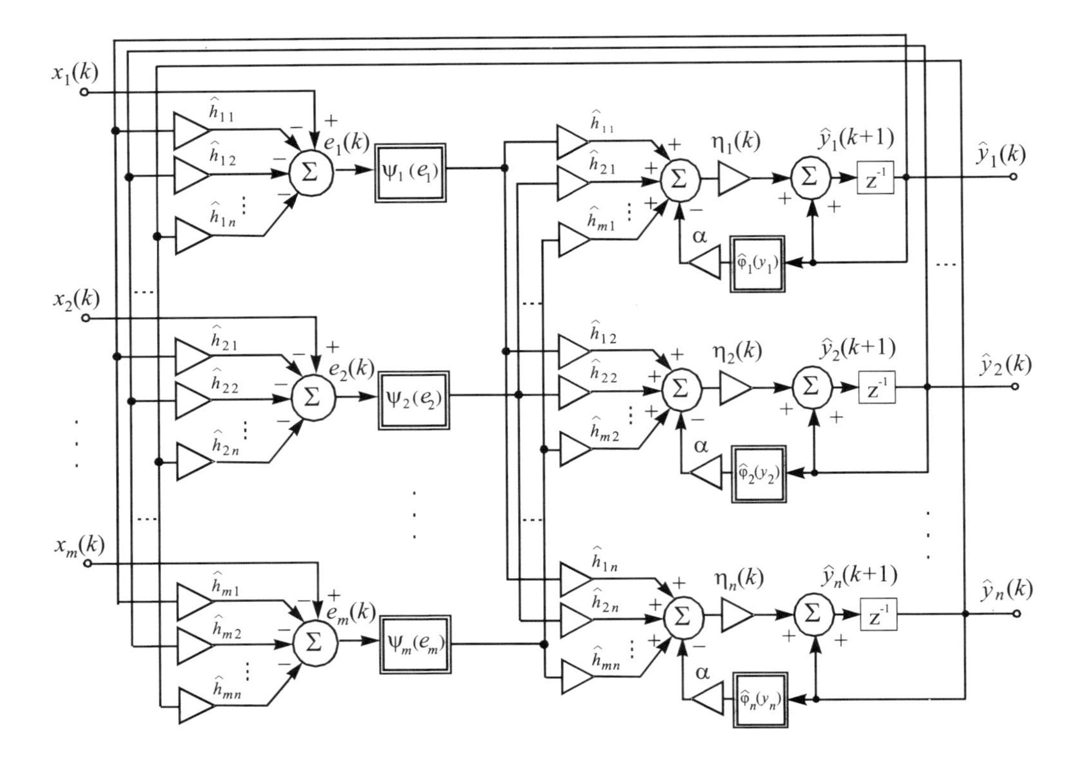

Fig. 8.6 Detailed architecture of the discrete-time Amari-Hopfield recurrent neural network with regularization.

use the contrast function with a regularization term

$$J_\rho(\mathbf{e}) = \rho(\mathbf{x} - \widehat{\mathbf{H}}\widehat{\mathbf{y}}) + \frac{\alpha}{p} \, \|\mathbf{L}\,\widehat{\mathbf{y}}\|_p^p, \tag{8.94}$$

where $\mathbf{e} = \mathbf{x} - \widehat{\mathbf{H}}\widehat{\mathbf{y}}$, $\alpha > 0$ is a regularization constant chosen to control the size of the solution and $\mathbf{L}$ is a regularization matrix that defines a (semi)norm of the solution. Typically, matrix $\mathbf{L}$ is equal to the identity matrix $\mathbf{I}_n$, and sometimes it represents a first- or second-order operator. If $\mathbf{L}$ is the identity matrix, $\rho(\mathbf{e}) = \frac{1}{2}||\mathbf{e}||_2^2$ and $p = 2$, then the problem reduces to the least squares problem with the standard Tikhonov regularization term [276]

$$J_2(\mathbf{e}) = \frac{1}{2}||\mathbf{x} - \widehat{\mathbf{H}}\widehat{\mathbf{y}})||_2^2 + \frac{\alpha}{2} \, \|\widehat{\mathbf{y}}\|_2^2. \tag{8.95}$$

Minimization of the cost function (8.94), with $\mathbf{L} = \mathbf{I}_n$, leads to the learning rule

$$\widehat{\mathbf{y}}(k+1) = \widehat{\mathbf{y}}(k) + \eta(k) \left[\widehat{\mathbf{H}}^T(k)\; \mathbf{\Psi}(\mathbf{x}(k) - \widehat{\mathbf{H}}(k)\,\widehat{\mathbf{y}}(k)) - \alpha\; \widehat{\boldsymbol{\varphi}}(\widehat{\mathbf{y}}(k))\right], \tag{8.96}$$

where $\mathbf{\Psi}(\mathbf{e}) = [\Psi_1(e_1), \Psi_2(e_2), \ldots, \Psi_m(e_m)]^T$ with entries $\Psi_i(e_i) = \partial\rho/\partial e_i$ and $\widehat{\boldsymbol{\varphi}}(\widehat{\mathbf{y}}) = [\widehat{\varphi}_1(\widehat{y}_1), \widehat{\varphi}_2(\widehat{y}_2), \ldots, \widehat{\varphi}_n(\widehat{y}_n)]^T$ with nonlinearities $\widehat{\varphi}_i(\widehat{y}_i) = |\widehat{y}_i|^{p-1}\text{sign}(\widehat{y}_i)$.

By combining this with the learning equation (8.90), we obtain an algorithm for simultaneous estimation of the mixing matrix and source signals for ill-conditioned problems. Figure 8.6 illustrates the detailed architecture of the recurrent Amari-Hopfield neural network according to Eq. (8.96). It can be proved that the above learning algorithm is stable if nonlinear activation functions $\psi_i(y_i)$ are monotonically increasing odd functions.

8.6.3 Robust Prewhitening and Principal Component Analysis (PCA)

As an improvement to the above approaches and to reduce the effects of data conditioning or to reduce the effects of noise when $m > n$, we can perform either robust prewhitening or principal component analysis of the measured sensor signals in the preprocessing stage. This preprocessing step is represented in Fig. 8.5 by the $n \times m$ matrix $\mathbf{Q}$. Prewhitening for noisy data can be performed by using the learning algorithm (8.5) (see Fig. 8.5)

$$\Delta\mathbf{Q}(t) = \eta(k)\left[\mathbf{I} - \langle \tilde{\mathbf{x}}(k)\,\tilde{\mathbf{x}}^T(k)\rangle + \hat{\sigma}_\nu^2 \mathbf{Q}(k)\,\mathbf{Q}^T(k)\right]\mathbf{Q}(k) \tag{8.97}$$

where $\tilde{\mathbf{x}}(k) = \mathbf{Q}\,\mathbf{x}(k) = \mathbf{Q}\,[\mathbf{H}\,\mathbf{s}(k) + \boldsymbol{\nu}(k)]$. Alternatively, for a nonsingular covariance matrix $\mathbf{R}_{\mathbf{xx}} = E\{\mathbf{x}(k)\mathbf{x}^T(k)\} = \mathbf{V}\boldsymbol{\Lambda}\mathbf{V}^T$ with $m = n$, we can use the following algorithm:

$$\mathbf{Q} = \left[\mathrm{diag}\left\{\frac{\lambda_1}{\lambda_1^2 + \hat{\sigma}_\nu^2} \quad \cdots \quad \frac{\lambda_n}{\lambda_n^2 + \hat{\sigma}_\nu^2}\right\}\right]^{1/2} \mathbf{V}^T, \tag{8.98}$$

where $\hat{\sigma}_\nu^2$ is the estimated variance of the noise and λ_i are eigenvalues and $\mathbf{V} = [\mathbf{v}_1, \mathbf{v}_2, \ldots, \mathbf{v}_n]$ is an orthogonal matrix of the corresponding eigenvectors of $\mathbf{R}_{\mathbf{xx}}$.

8.6.4 Computer Simulation Experiments for the Amari-Hopfield Network

Example 8.2 The performance of the proposed neural network model will be illustrated by two examples. The three sub-Gaussian source signals shown in Fig. 8.2 have been mixed using the mixing matrix whose rows are $\mathbf{h}_1 = [0.8 \ -0.4 \ 0.9]$, $\mathbf{h}_2 = [0.7 \ 0.3 \ 0.5]$, and $\mathbf{h}_3 = [-0.6 \ 0.8 \ 0.7]$. Uncorrelated Gaussian noise signals with variance 1.6 were added to each of the elements of $\mathbf{x}(k)$. The neural network model depicted in Fig. 8.4 with the associated learning rules in (8.92) – (8.93) and nonlinearities $f(y_i) = y_i$, $g(y_i) = \tanh(10y_i)$ and $\Psi(e_i) = \alpha_i e_i$ with $\alpha_i = 1/\sigma_{\nu_i}^2$, where $\sigma_{\nu_i}^2$ is the variance of noise was used to separate these signals, where $\widehat{\mathbf{H}}(0) = \mathbf{I}$. Shown in Fig.8.2 are the resulting separated signals, in which the source signals are accurately estimated. The resulting three rows of the combined system matrix $\widehat{\mathbf{H}}^{-1}\mathbf{H}$ after 400 milliseconds (with sampling period 0.0001) are $[0.0034 \ -0.0240 \ 0.8541]$, $[-0.0671 \ 0.6251 \ -0.0142]$ and $[-0.2975 \ -0.0061 \ -0.0683]$ respectively, indicating that separation has been achieved. Note that standard algorithms that assume noiseless measurements fail to separate such noisy signals.

Example 8.3 In the second illustrative example, the sensor signals were contaminated by additive impulsive (spiky) noise as shown in Fig. 8.8. The same learning rule was employed but with nonlinear functions $\Psi(e_i) = \tanh(10e_i)$. The neural network of Fig.8.4 was able to reduce the influence of the noise in separating signals considerably.

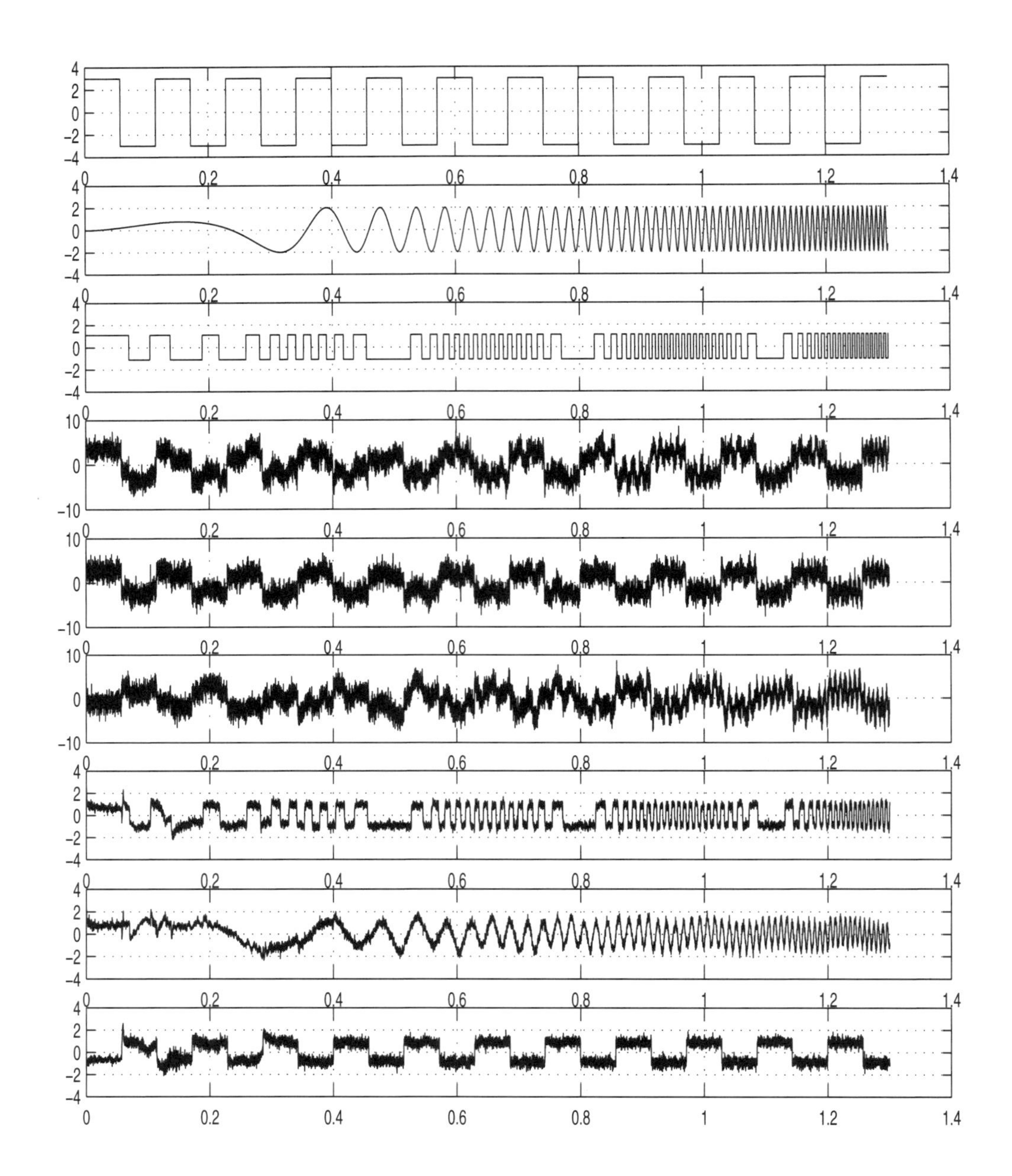

Fig. 8.7 Exemplary simulation results for the neural network in Fig.8.4 for signals corrupted by Gaussian noise. The first three signals are the original sources, the next three signals are the noisy sensor signals, and the last three signals are the on-line estimated source signals using the learning rule given in (8.92)-(8.93). The horizontal axis represents time in seconds.

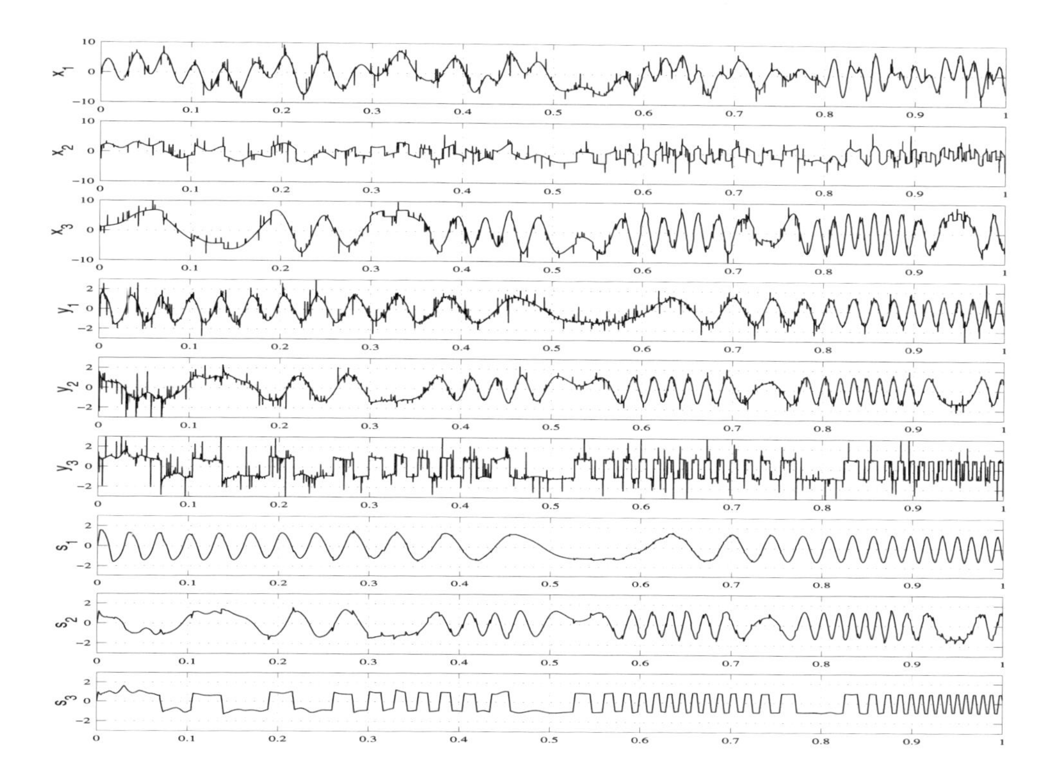

Fig. 8.8 Exemplary simulation results for the neural network in Fig. 8.4 for impulsive noise. The first three signals are the mixed sensors signals contaminated by the impulsive (Laplacian) noise, the next three signals are the source estimated on-line signals using the learning rule (8.8) and the last three signals are the on-line estimated source signals using the learning rule (8.92)-(8.93).

Appendix A. Cumulants in Terms of Moments

For cumulant evaluation in terms of the moments of signals, we can use the following formula (see [888, 889, 992])

$$Cum(y_1, y_2, \ldots, y_n) = \sum_{(p_1, 2, \ldots, p_m)} (-1)^{m-1}(m-1)! \cdot$$

$$E\{\prod_{i \in p_1} y_i\} E\{\prod_{i \in p_2} y_i\} \ldots E\{\prod_{i \in p_m} y_i\} \tag{A.1}$$

where the sum is extended to all the possible partitions $(p_1, 2, \ldots, p_m)$, $m = 1, 2, \ldots, n$, of the set of natural numbers $(1, 2, \ldots, n)$.

This calculus results in low complexity for lower orders but its complexity rapidly increases for higher orders. In our case, the fact that the cross-cumulants of our interest take the form $\mathbf{C}_{1,q}(\mathbf{y},\mathbf{y}) = E\{\mathbf{y}\,\mathbf{g}^T(\mathbf{y})\}$ simplifies this task for real and zero-mean sources, because many partitions disappear or give rise to the same kind of sets.

We define the moment and cross-moment matrices of the outputs as

$$\begin{aligned} \mathbf{M}_q(\mathbf{y}) &= E\{\underbrace{\mathbf{y}\circ\ldots\circ\mathbf{y}}_{q}\} = E\{\mathbf{g}(\mathbf{y})\}, \\ \mathbf{M}_{1,q}(\mathbf{y},\mathbf{y}) &= E\{\mathbf{y}(\underbrace{\mathbf{y}\circ\ldots\circ\mathbf{y}}_{q})^T\} = E\{\mathbf{y}\,\mathbf{g}^T(\mathbf{y})\}, \end{aligned} \tag{A.2}$$

where $\circ$ denotes the Hadamard product of vectors and $\mathbf{g}(\mathbf{y}) = [g_1(y_1), g_2(y_2), \ldots, g_n(y_n)]^T = \underbrace{\mathbf{y}\circ\ldots\circ\mathbf{y}}_{q} = \mathbf{y}^{\circ q}$, with entries $g_j(y_j) = y_j^q$.

Below are presented expressions of the cross-cumulant matrices $\mathbf{C}_{1,q}(\mathbf{y},\mathbf{y})$ in terms of the moment matrices for $q = 1, 2, \ldots, 7$.

$$\begin{aligned}
&\mathbf{C}_{1,1}(\mathbf{y},\mathbf{y}) = \mathbf{M}_{1,1}(\mathbf{y},\mathbf{y}) = E\{\mathbf{y}\mathbf{y}^T\},\\
&\mathbf{C}_{1,2}(\mathbf{y},\mathbf{y}) = \mathbf{M}_{1,2}(\mathbf{y},\mathbf{y}) = E\{\mathbf{y}(\mathbf{y}\circ\mathbf{y})^T\} = E\{\mathbf{y}\;[\mathbf{y}^{\circ\,2}]^T\},\\
&\mathbf{C}_{1,3}(\mathbf{y},\mathbf{y}) = \mathbf{M}_{1,3}(\mathbf{y},\mathbf{y}) - 3\mathbf{M}_{1,1}(\mathbf{y},\mathbf{y})\ \mathrm{diag}(\mathbf{M}_2(\mathbf{y})) = E\{\mathbf{y}\;[\mathbf{y}^{\circ\,3} - 3\ \mathrm{diag}(\mathbf{M}_2(\mathbf{y}))\,\mathbf{y}]^T\},\\
&\mathbf{C}_{1,4} = \mathbf{M}_{1,4} - 4\mathbf{M}_{1,1}\ \mathrm{diag}(\mathbf{M}_3) - 6\mathbf{M}_{1,2}\ \mathrm{diag}(\mathbf{M}_2),\\
&\mathbf{C}_{1,5} = \mathbf{M}_{1,5} - 5\mathbf{M}_{1,1}\ \mathrm{diag}(\mathbf{M}_4) - 10\mathbf{M}_{1,2}\ \mathrm{diag}(\mathbf{M}_3) - 10\mathbf{M}_{1,3}\ \mathrm{diag}(\mathbf{M}_2)\\
&\qquad + 30\mathbf{M}_{1,1}\ \mathrm{diag}(\mathbf{M}_2)^2,\\
&\mathbf{C}_{1,6} = \mathbf{M}_{1,6} - 6\mathbf{M}_{1,1}\ \mathrm{diag}(\mathbf{M}_5) - 15\mathbf{M}_{1,2}\ \mathrm{diag}(\mathbf{M}_4) - 20\mathbf{M}_{1,3}\ \mathrm{diag}(\mathbf{M}_3)\\
&\qquad - 15\mathbf{M}_{1,4}\ \mathrm{diag}(\mathbf{M}_2) + 120\mathbf{M}_{11}\ \mathrm{diag}(\mathbf{M}_2)\ \mathrm{diag}(\mathbf{M}_3) + 90\mathbf{M}_{1,2}\ \mathrm{diag}(\mathbf{M}_2)^2,\\
&\mathbf{C}_{1,7} = \mathbf{M}_{1,7} - 7\mathbf{M}_{1,1}\ \mathrm{diag}(\mathbf{M}_6) - 21\mathbf{M}_{1,2}\ \mathrm{diag}(\mathbf{M}_5) - 35\mathbf{M}_{1,3}\ \mathrm{diag}(\mathbf{M}_4)\\
&\qquad - 35\mathbf{M}_{1,4}\ \mathrm{diag}(\mathbf{M}_3) - 21\mathbf{M}_{1,5}\ \mathrm{diag}(\mathbf{M}_2) + 210\mathbf{M}_{1,1}\ \mathrm{diag}(\mathbf{M}_2)\ \mathrm{diag}(\mathbf{M}_4)\\
&\qquad + 140\mathbf{M}_{1,1}\ \mathrm{diag}(\mathbf{M}_3)^2 + 420\mathbf{M}_{1,2}\ \mathrm{diag}(\mathbf{M}_2)\ \mathrm{diag}(\mathbf{M}_3) + 210\mathbf{M}_{1,3}\ \mathrm{diag}(\mathbf{M}_2)^2\\
&\qquad - 630\mathbf{M}_{1,1}\ \mathrm{diag}(\mathbf{M}_2)^3,
\end{aligned}$$

where $\mathrm{diag}(\mathbf{M}_2(\mathbf{y})) = \mathrm{diag}\{E\{y_1^2\}, E\{y_2^2\}, \ldots, E\{y_n^2\}\}$.

For the case of complex sources, the expressions are much more complicated. As an example, we can see that for $q = 3$, the cumulant matrix is

$$\mathbf{C}_{1,3}(\mathbf{y},\mathbf{y}) = E\{\mathbf{y}(\mathbf{y}\circ\mathbf{y}^*\circ\mathbf{y}^*)^T\} - E\{\mathbf{y}\mathbf{y}^T\}\ \mathrm{diag}(E\{\mathbf{y}^*\mathbf{y}^H\}) - 2E\{\mathbf{y}\mathbf{y}^H\}\ \mathrm{diag}(E\{\mathbf{y}\mathbf{y}^H\}) \tag{A.3}$$

which can be compared with the real case to note the rise of computational complexity.

9

Multichannel Blind Deconvolution: Natural Gradient Approach

The truth is rarely pure, and never simple.

—(Oscar Wilde)

Blind separation/deconvolution of source signals has been a subject under consideration for more than two decades [545, 1041]. There are significant potential applications of blind separation/deconvolution in various fields, for example, wireless telecommunication systems, sonar and radar systems, audio and acoustics, image enhancement and biomedical signal processing (EEG/MEG signals) [797, 103, 371, 472]. In these applications, single or multiple unknown but independent temporal signals propagate through a mixing and filtering medium. The blind source separation/deconvolution problem is concerned with recovering independent sources from sensor outputs without assuming any *a priori* knowledge of the original signals, except for certain statistical features [29, 589].

In this chapter, using various models and assumptions, we present relatively simple and efficient, adaptive and batch algorithms for blind deconvolution and equalization for single-input/multiple-output (SIMO) and multiple-input/multiple-output (MIMO) dynamical minimum phase and non-minimum phase systems. The basic relationships between the standard ICA/BSS (Independent Component Analysis and Blind Source Separation) and multichannel blind deconvolution are discussed in detail. They enable us to extend algorithms such as the one used for the natural gradient ICA approach for instantaneous mixture to convolutive dynamical models. We also derive a family of equivariant algorithms and analyze their stability and convergence properties. Furthermore, a Lie group and Riemannian metric are introduced on the manifold of FIR filters Using the isometry of the Riemannian metric, we discuss the natural gradient on the FIR manifold [1363, 1367]. Based on the minimization of mutual information, we then present a natural gradient algorithm for the causal minimum phase finite impulse response (FIR) multichannel filter. Using information back-propagation, we also discuss an efficient implementation of the learning algorithm for non-causal FIR filters. Computer simulations are presented to illustrate the validity and satisfactory performance of the proposed algorithms.

Existing adaptive algorithms for blind deconvolution and equalization can generally be classified into two categories: The mutual information minimization/entropy maximization and the cumulant-based algorithms [301, 549, 539, 496, 1198]. The Bussgang algorithms [472, 88, 75] and the natural gradient algorithm [33, 103, 385] are two typical examples of the first category. The Bussgang techniques are iterative equalization schemes that employ stochastic gradient descent procedures to minimize non-convex cost functions depending on the equalizer output signals. The Bussgang algorithms are simple and easy to implement, but, may converge to wrong solutions resulting in poor performance of the equalizer. In the Cumulant Fitting Procedure (CFP) [1185, 1186, 1196], the channel identification process directly employs the minimization of higher-order cumulant-based nonlinear cost functions. The underlying cost functions in the CFP are multi-modal, as in the case of Bussgang algorithms. The natural gradient approach was developed by Amari *et al.* to overcome the drawback of the Bussgang algorithm [33, 34]. It has been shown that the natural gradient algorithm can considerably improve efficiency in blind separation and blind deconvolution [103, 397]. The main objective of this chapter is to review and extend existing adaptive natural gradient algorithms for various multichannel blind deconvolution models.

9.1 SIMO CONVOLUTIVE MODELS AND LEARNING ALGORITHMS FOR ESTIMATION OF A SOURCE SIGNAL

In many applications it is necessary to reconstruct an unknown single source signal which is transmitted through several channels with unknown convolutive characteristics. In other words, only distorted (convolutive or filtered) versions of the source are available at the outputs of the channels. It is thus necessary to reconstruct the original source and/or identify the unknown channels with some, usually nonessential, intrinsic ambiguities (such as arbitrary scaling and time delay). This problem is referred to as blind equalization of SIMO (single-input/multiple-output) channels, and it has found numerous applications in digital communication, cable HDTV, the global positioning system, and some biomedical applications [549, 472, 371].

Typical scenarios are illustrated in Fig.9.1 (a), (b) and (c). For example, in the acoustic speech reconstruction problem an unknown speech signal recorded by an array of microphones is distorted by reverberation and echo (see Fig.9.1 (a)). Similarly, in neuroscience (e.g., EEG recordings) a brain source is recorded by several electrodes or sensors which measure convolutive (distorted, low-pass filtered) versions of the source produced by propagation effects.

In a wireless communication scenario, the transmitted signals are received by several antennas (Fig.9.1 (b)). The transmitted signals interact with various objects in the physical environment before reaching the antenna array. Each path follows a different direction,

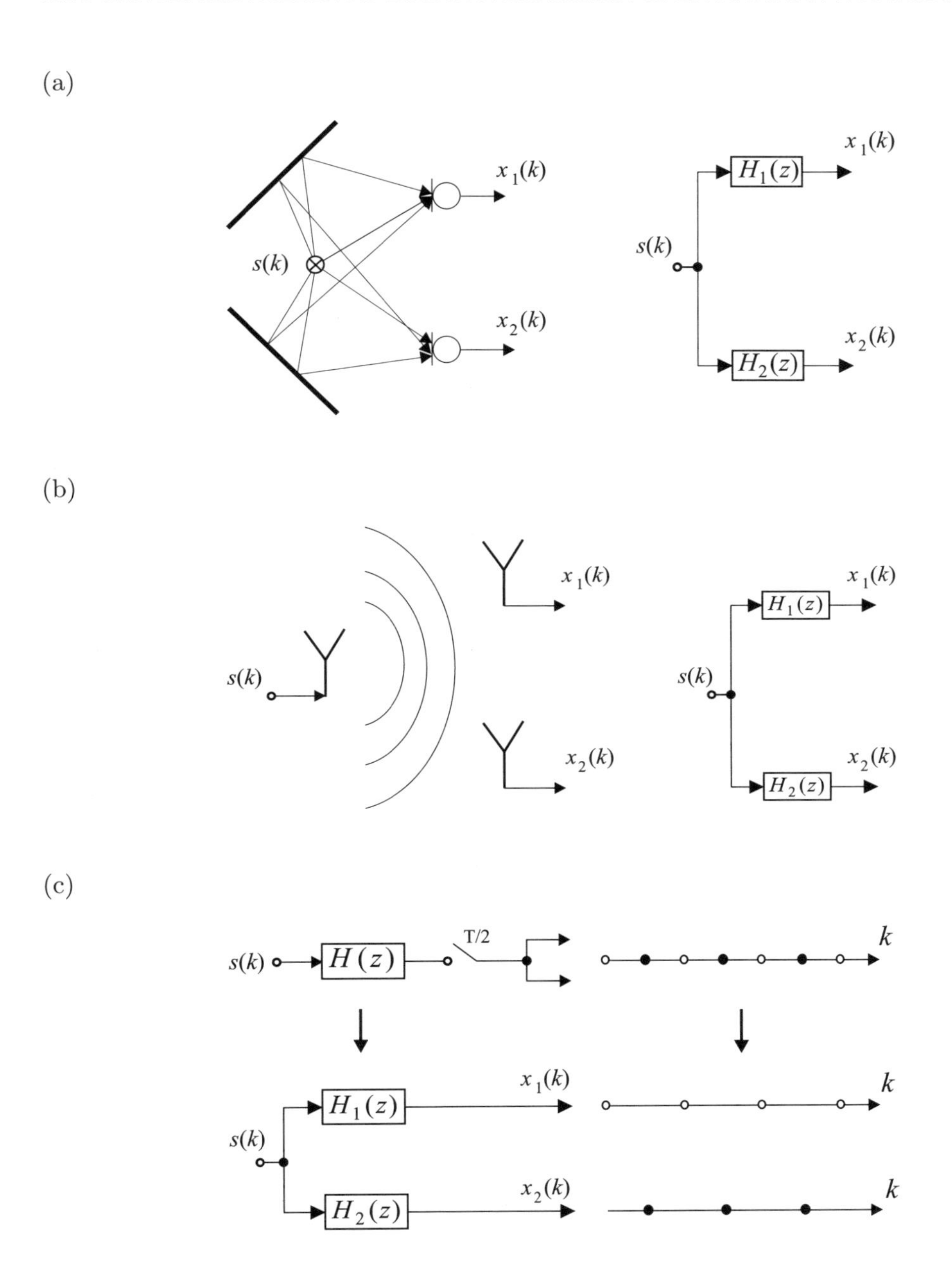

Fig. 9.1 Conceptual models of single-input/multiple-output (SIMO) dynamical system: (a) Recording an unknown acoustic signal distorted by reverberation by an array of microphones, (b) array of antenna receiving distorted version of transmitted signal, (c) illustration of oversampling principle for two channels.

with some unknown propagation delay and attenuation. This phenomenon of receiving a superposition of many delayed signals is called multi-path fading [472]. Furthermore, a sensor signal can be oversampled or fractionally sampled as illustrated in Fig.9.1 (c), leading several distorted observations of the same source signal [371].

In the blind equalization of SIMO channels (Fig.9.2), it is assumed that the ith sensor output $x_i(k)$ is generated from a linear time-invariant filter as

$$\begin{aligned} x_i(k) &= \sum_{p=0}^{M} h_{ip}\, s(k-p) + \nu_i(k), \\ &= [H_i(z)]\, s(k) + \nu_i(k), \qquad (i = 1, 2, \ldots, m) \end{aligned} \tag{9.1}$$

where $s(k)$ is an unknown (zero-mean uncorrelated with noise) source signal, and,

$$H_i(z) = \sum_{p=0}^{M} h_{ip}\, z^{-p}$$

is the transfer function of the ith channel, (z^{-p} is the delay operator such that $z^{-p}s(k) = s(k-p)$), and $\nu_i(k)$ is an additive noise (typically assumed to be white Gaussian noise). In this model, we assume that lengths of the channels are equal to or less than M and the number of channels m (equals the number of sensors) is greater than or equal to two.

The task of blind equalization of SIMO channels is to design a multiple-input/single-output (MISO) system (Fig.9.2 (a)), so that its composite output:

$$y(k) = y_1(k) + \cdots + y_m(k), \tag{9.2}$$

where

$$y_i(k) = [W_i(z)]\, x_i(k) = \sum_{p=0}^{L} w_{ip}\, x_i(k-p) \tag{9.3}$$

is possibly a scaled and time-delayed estimate of the original source signal $s(k)$, i.e., $y(k) = g\, s(k-\Delta)$, where g is some scaling factor and Δ is a time-delay. The term "blind" implies that the problem should be solved with no knowledge about channels $\{H_i(z)\}$ and the source signal $s(k)$.

9.1.1 Equalization Criteria for SIMO Systems

In the blind equalization problem certain conditions about the channels and the source signal must be satisfied to ensure identifiability [1160]. One fundamental condition is the channel diversity. By channel diversity, we mean usually different characteristic or modes of the channels.

Throughout this section, the following assumptions are made:

AS1: The source signal $s(k)$ is a zero-mean random sequence and temporally uncorrelated (white) with finite variance, i.e.,

$$\begin{aligned} E\{s(k)\} &= 0, \\ E\{s(k)\, s(k-p)\} &= \sigma_s^2(k)\, \delta_{p0}, \end{aligned} \tag{9.4}$$

where δ_{p0} is the Kronecker delta equal to 1 if $p = 0$, and 0 otherwise.

AS2: The channels $\{H_i(z)\}$ are coprime, that is, that finite impulse response channels have different zeros.

Under the assumptions (AS1) and (AS2), it has been shown [221] that if the composite output $y(k) = y_1(k) + \cdots + y_m(k)$ (the sum of the outputs of the equalizer-bank) is a white signal,[1], then the composite output $y(k)$ is a scaled and time-delayed version of the original source signal $s(k)$. Based on this property, a causal FIR filter equalizer-bank can be constructed and the temporal decorrelation-based blind equalization (TDBE) algorithm can be used [221, 222]. The equalization criterion for the i.i.d. (white) sources is summarized in the following Theorem [221].

Theorem 9.1 *Suppose that the assumptions (AS1) and (AS2) are satisfied. Then, the composite output $y(k) = y_1(k) + \cdots + y_m(k)$ $(m \geq 2)$ is equal to a scaled and delayed version of the original source signal, i.e., $y(k) = g\, s(k - \Delta)$, if all $y_i(k)$ satisfy*

$$E\{y_i(k)\, y_j(k-p)\} = 0, \qquad \forall\, i \neq j, \text{ and } p \neq 0, \tag{9.5}$$

$$E\{y_i^2(k)\} \neq 0, \qquad \forall i \quad (i = 1, 2, \ldots, m), \tag{9.6}$$

Proof. Note that the output signal $y(k)$ is determined as $y(k) = y_1(k) + \cdots + y_m(k)$. Then one can easily show that the conditions (9.5) and (9.6) imply that

$$E\{y(k)\, y(k-p)\} = 0, \quad \forall\, p \neq 0, \tag{9.7}$$

$$E\{y^2(k)\} \neq 0. \tag{9.8}$$

If the assumption (AS2) is satisfied, then the inverse of $\{H_i(z\})$ is also causal and stable. Thus, the global system (combining a channel and an equalizer) is causal and stable, and this can be represented as an infinite FIR [371]. The composite output $y(k)$ can be written as

$$y(k) = [G(z)]\, s(k) = \sum_{p=0}^{\infty} g_p\, s(k-p), \tag{9.9}$$

where $G(z) = \sum_{i=1}^{m} W_i(z)\, H_i(z)$ is the global transfer function. Suppose g_d is the leading non-zero coefficient (i.e., $g_0 = g_1 = \cdots = g_{d-1} = 0$). Then (9.9) becomes

$$y(k) = \sum_{p=d}^{\infty} g_p\, s(k-p). \tag{9.10}$$

[1]The white signal satisfies the relationship $E\{y(k)y(k+\tau)\} = 0$ for all $\tau \neq 0$.

Using (9.10), the correlation between $y(k)$ and $y(k+q)$ is given by

$$E\{y(k)\,y(k+q)\} = \sum_{p=d}^{\infty} g_p\, g_{p+q} E\{s^2(k-p)\}. \tag{9.11}$$

It follows from (9.7) and $E\{s^2(k-p)\} \neq 0$, $\forall p$ that $g_p = 0$ for $p = d+1, \ldots, \infty$. Therefore, the composite output $y(k)$ is equal to $g_d\, s(k-d)$, if the conditions (9.7) and (9.8) are satisfied. □

9.1.2 SIMO Blind Identification and Equalization via Robust ICA/BSS

In this section, we present a simple batch method based on the approximate maximum likelihood source separation (AMLSS) and the natural gradient approach proposed by Choi and Cichocki [220]. The blind equalization of SIMO with FIR channels described by Equation (9.1) can be reformulated as the instantaneous blind source separation problem or independent component analysis problem. Equation (9.1) can be reformulated as follows: An m-dimensional observation vector $\mathbf{x}(k)$ is assumed to be generated from an unknown source signal $s(k)$ through m different FIR filters, i.e.,

$$\mathbf{x}(k) = \sum_{p=0}^{M} \mathbf{h}(p)\, s(k-p) + \boldsymbol{\nu}(k), \tag{9.12}$$

where $\{\mathbf{h}(p)\}$ $(\mathbf{h}(p) = [h_{1p}, h_{2p}, \ldots, h_{mp}]^T)$ are the impulse responses of channels with length M and $\boldsymbol{\nu}(k)$ is the additive white Gaussian noise that is assumed to be statistically independent of the source signal $s(k)$.

Stacking N successive samples of the observation vector, i.e.,

$$\boldsymbol{\mathcal{X}}(k) = [\mathbf{x}^T(k), \mathbf{x}^T(k-1), \ldots, \mathbf{x}^T(k-N+1)]^T \in \mathbb{R}^N, \tag{9.13}$$

the model (9.12) can be reformulated as a system of linear algebraic equations

$$\boldsymbol{\mathcal{X}}(k) = \boldsymbol{\mathcal{H}}_N \boldsymbol{\mathcal{S}}(k) + \boldsymbol{\mathcal{V}}(k), \tag{9.14}$$

where

$$\boldsymbol{\mathcal{H}}_N = \begin{bmatrix} \mathbf{h}(0) & \mathbf{h}(1) & \cdots & \mathbf{h}(M) & 0 & \cdots & \cdots & 0 \\ 0 & \mathbf{h}(0) & \mathbf{h}(1) & \cdots & \mathbf{h}(M) & 0 & \cdots & 0 \\ \vdots & & & & & & & \vdots \\ 0 & \cdots & \cdots & 0 & \mathbf{h}(0) & \mathbf{h}(1) & \cdots & \mathbf{h}(M) \end{bmatrix}, \tag{9.15}$$

and

$$\begin{aligned} \boldsymbol{\mathcal{S}}(k) &= [s(k), s(k-1), \ldots, s(k-N-M+1)]^T \in \mathbb{R}^{N+M} && (9.16) \\ \boldsymbol{\mathcal{V}}(k) &= [\boldsymbol{\nu}^T(k), \boldsymbol{\nu}^T(k-1), \ldots, \boldsymbol{\nu}^T(k-N+1)]^T \in \mathbb{R}^{Nm}. && (9.17) \end{aligned}$$

Here, we assume that the channels are coprime. Another necessary condition for almost all batch algorithms for blind equalization is that $\boldsymbol{\mathcal{H}}_N \in \mathbb{R}^{Nm \times (N+M)}$ be full rank. Full rank requires that the number of rows be greater than the number of columns, i.e., $N \geq \frac{M}{m-1}$. In the model described by Eq. (9.14), we can see some similarity with ICA models for instantaneous mixtures discussed in the previous chapters. The main difference is that in blind equalization, the matrix $\mathcal{H}_N$ has block Toeplitz structure and entries of the vector $\mathcal{S}(k)$ are time-delayed versions of the same signal. However, for i.i.d sources we can treat $\mathcal{S}(k)$ as a signal vector formed by $M+N$ "independent" signals, although they build up the same sequence at different time-lags. We can also treat $\mathcal{H}_N$ as an unstructured mixing matrix. Such treatments are justified due to the assumption that the source signal is i.i.d., and some ambiguities are acceptable. Such an assumption motivates the use of an ICA/BSS approach to the blind equalization problem.

In the preprocessing stage, we need to first estimate the rank of $\mathcal{H}_N$. This can be done by computing the sampled covariance matrix $\mathbf{R}_{\mathcal{X}}(0)$ of the observed vector $\mathcal{X}$ and the computing its eigenvalue decomposition. In fact, by taking into account the relation

$$\begin{aligned}
\mathbf{R}_{\mathcal{X}}(0) &= \mathcal{H}_N \mathcal{H}_N^T + \sigma_\nu^2 \mathbf{I}_N \\
&= [\mathbf{V}_{\mathcal{S}}, \mathbf{V}_{\mathcal{N}}] \begin{bmatrix} \mathbf{\Lambda}_{\mathcal{S}} & \mathbf{0} \\ \mathbf{0} & \mathbf{\Lambda}_{\mathcal{N}} \end{bmatrix} [\mathbf{V}_{\mathcal{S}}, \mathbf{V}_{\mathcal{N}}]^T \\
&= \mathbf{V}_{\mathcal{S}} \mathbf{\Lambda}_{\mathcal{S}} \mathbf{V}_{\mathcal{S}}^T + \mathbf{V}_{\mathcal{N}} \mathbf{\Lambda}_{\mathcal{N}} \mathbf{V}_{\mathcal{N}}^T,
\end{aligned} \tag{9.18}$$

which holds under the assumption of an i.i.d. source signal and white noise, we can estimate the rank of $\mathcal{H}_N$ and variance of the noise $\sigma_{\boldsymbol{\nu}}^2$.

Our objective is to estimate source signal $\mathcal{S}(k)$ and/or identify the "mixing" matrix $\widehat{\mathcal{H}}_N$ or separating matrix $\mathcal{W} = \widehat{\mathcal{H}}_N^{+}$, especially for the ill-conditioned channels in the presence of Gaussian noise. On the basis of methods presented in Chapters 6 and 8, we have developed the following iterative algorithm [220]:

Algorithm Outline: Robust ICA/BLUE Algorithm for Blind SIMO Equalization

1. Given the current estimate of the mixing matrix, $\widehat{\mathcal{H}}_N(k)$, we estimate the source vector by using the BLUE (Best Linear Unbiased Estimator)

$$\boxed{\widehat{\mathcal{S}}(k) = \left[\widehat{\mathcal{H}}_N^T(k) \widehat{\mathcal{R}}_{\boldsymbol{\nu\nu}}^{-1} \widehat{\mathcal{H}}_N(k)\right]^{-1} \widehat{\mathcal{H}}_N^T(k) \widehat{\mathcal{R}}_{\boldsymbol{\nu\nu}}^{-1} \mathcal{X}(k),} \tag{9.19}$$

where $\widehat{\mathcal{R}}_{\boldsymbol{\nu\nu}}$ is the estimated covariance matrix of the noise. If the noise covariance matrix $\widehat{\mathcal{R}}_{\boldsymbol{\nu\nu}}$ is not available, then we assume that $\widehat{\mathcal{R}}_{\boldsymbol{\nu\nu}} = \hat{\sigma}_\nu^2 \mathbf{I}$ and the above formula simplifies to

$$\boxed{\widehat{\mathcal{S}}(k) = \left(\widehat{\mathcal{H}}_N^T(k) \widehat{\mathcal{H}}_N(k)\right)^{-1} \widehat{\mathcal{H}}_N^T(k) \mathcal{X}(k).} \tag{9.20}$$

2. Using $\widehat{\mathcal{S}}(k)$ and $\widehat{\mathcal{H}}_N(k)$, we can find a new estimate of the mixing matrix, $\widehat{\mathcal{H}}_N(k+1)$ by applying the standard NG ICA algorithm (see Chapter 6)

$$\boxed{\widehat{\mathcal{H}}_N(k+1) = \widehat{\mathcal{H}}_N(k) - \eta \widehat{\mathcal{H}}_N(k) \left[\mathbf{I} - \mathbf{R}_{\mathbf{fg}}^{(k)}\right],} \tag{9.21}$$

where

$$\mathbf{R}_{\mathbf{f}\,\mathbf{g}}^{(k)} = (1-\eta_0)\,\mathbf{R}_{\mathbf{f}\,\mathbf{g}}^{(k-1)} + \eta_0\,\mathbf{f}(\widehat{\boldsymbol{\mathcal{S}}}(k))\widehat{\boldsymbol{\mathcal{S}}}^T(k). \tag{9.22}$$

Alternatively, for a large Gaussian noise (see Chapter 8 for theoretical justification)

$$\mathbf{R}_{\mathbf{f}\,\mathbf{g}}^{(k)} = (1-\eta_0)\,\mathbf{R}_{\mathbf{f}\,\mathbf{g}}^{(k-1)} + \eta_0\,\widehat{\boldsymbol{\mathcal{S}}}(k)\mathbf{g}^T(\widehat{\boldsymbol{\mathcal{S}}}(k)), \tag{9.23}$$

where entries of the vector $\mathbf{g}$ have the form $g_i(\boldsymbol{\mathcal{S}}_i) = (\boldsymbol{\mathcal{S}}_i^3 - 3\boldsymbol{\mathcal{S}}_i E\{\boldsymbol{\mathcal{S}}_i^2\})\,\mathrm{sign}(\kappa_4(\boldsymbol{\mathcal{S}}_i)$.

These two steps are repeated until $\widehat{\boldsymbol{\mathcal{H}}}_N$ converges.

Compared with other batch algorithms that exploit structural properties of the channels or data, the ICA approach is rather insensitive to channel order estimation and robust to ill-conditioned channels or additive Gaussian noise [220, 1008, 1378]. Moreover, it was confirmed [220] by computer simulation experiments that the natural gradient based BSS methods could recover the source signal for the blind SIMO equalization successfully, even when some zeros of channels $H_i(z)$ are close to each other, whereas second-order based blind identification methods may fail [371, 1160, 865]. The main disadvantage of the approach is that the matrix $\widehat{\boldsymbol{\mathcal{H}}}_N \in \mathbb{R}^{Nm\times(N+M)}$ is large if the length M of the filters and/or the number of samples N are large, which require large memory and extensive computations. In such a case, it would be difficult to implement the above algorithm in practice. An alternative solution is to apply an adaptive on-line algorithm with good convergence rate and good tracking ability, as discussed in the next section.

9.1.3 Feed-forward Deconvolution Model and Natural Gradient Learning Algorithm

Let us consider the feed-forward model shown in Fig.9.2 (a), described as

$$y(k) = \sum_{i=1}^{m} y_i(k), \tag{9.24}$$

with

$$y_i(k) = \sum_{p=0}^{L} w_{ip}(k)\,x_i(k-p) = \mathbf{w}_i^T\mathbf{x}_i(k), \qquad (i=1,2,\ldots,m) \tag{9.25}$$

where $\mathbf{w}_i(k) = [w_{i0}(k), w_{i1}(k), \ldots, w_{iL}(k)]^T$ and $\mathbf{x}_i(k) = [x_i(k), x_i(k-1), \ldots, x_i(k-L)]^T$. Theorem 9.1 allows us to use the following cost functions

$$J_i(y, \mathbf{w}_i) = \|\boldsymbol{\Lambda}_i - \widehat{\mathbf{R}}_{\mathbf{y}_i\,\mathbf{y}}\|_F, \tag{9.26}$$

where $\boldsymbol{\Lambda}_i = \mathrm{diag}\{\lambda_{i0}, \ldots, \lambda_{iL}\}$, $\widehat{\mathbf{R}}_{\mathbf{y}_i\,\mathbf{y}} = \langle \mathbf{y}_i(k)\,\mathbf{y}^T(k)\rangle$ is the cross correlation matrix with $\mathbf{y}_i(k) = [y_i(k), y_i(k-1), \ldots, y_i(k-L)]^T$ and $\mathbf{y}(k) = [y(k), y(k-1), \ldots, y(k-L)]^T$. Applying the method presented previously (Chapters 4 and 7), we can derive the following nonholonomic algorithm

$$\boxed{\Delta\mathbf{w}_i(k) = \eta(k)\left[\boldsymbol{\Lambda}_i^{(k)} - \langle \mathbf{y}_i(k)\,\mathbf{y}^T(k)\rangle\right]\,\mathbf{w}_i(k),} \tag{9.27}$$

where the entries of the diagonal matrix $\boldsymbol{\Lambda}_i^{(k)} = \text{diag}\{\lambda_{i0}, \lambda_{i1}, \ldots, \lambda_{iL}\}$ are chosen as $\lambda_{ip} = \langle y_i(k-p)\, y(k-p)\rangle$ with $p = 0, 1, \ldots, L$.

For the sensor signals corrupted by additive Gaussian noise, instead of the cross-correlation matrix $\widehat{\mathbf{R}}_{\mathbf{y}_i\,\mathbf{y}}$, we can alternatively use the matrix cumulant of the form $\mathbf{C}_{1,q}(\mathbf{y}_i, \mathbf{y})$ (see Chapter 8). In such a case, the algorithm can be generalized to the following form:

$$\Delta \mathbf{w}_i(k) = \eta(k) \left[\boldsymbol{\Lambda}_i^{(k)} - \langle \mathbf{y}_i(k)\, \mathbf{g}^T(\mathbf{y}(k)) \rangle \right] \mathbf{w}_i(k), \tag{9.28}$$

where $\mathbf{g}(\mathbf{y}) = [g_0(y(k)), g_1(y(k-1)), \ldots, g_L(y(k-L))]^T$. The nonlinear functions can take various forms, depending on the distribution of source signal. For example, for $q = 1$ the algorithm (9.28) simplifies to the algorithm (9.27) with $g_p(y(k-p)) = y(k-p)$; for $q = 3$ the nonlinear functions take the form $g_p(y_p) = -(y_p^3 - 3\, y_p\, \hat{\sigma}_{y_p}^2)\, \kappa_4(y_p)$ for $p = 0, 1, \ldots, L$.

The learning rule (9.28) can be formulated in a slightly modified form as

$$\boxed{\Delta \mathbf{w}_i(k) = \eta(k) \left[\boldsymbol{\Lambda}_i^{(k)} - \mathbf{R}_{\mathbf{y}_i\,\mathbf{g}}^{(k)} \right] \mathbf{w}_i(k), \qquad (i = 1, 2, \ldots, m)} \tag{9.29}$$

where

$$\begin{aligned} \boldsymbol{\Lambda}_i^{(k)} &= (1-\eta_0)\, \boldsymbol{\Lambda}_i^{(k-1)} + \eta_0\, \text{diag}\{\mathbf{y}_i(k)\, \mathbf{g}^T(\mathbf{y}(k))\}, & (9.30) \\ \mathbf{R}_{\mathbf{y}_i\,\mathbf{g}}^{(k)} &= (1-\eta_0)\, \mathbf{R}_{\mathbf{y}_i\,\mathbf{g}}^{(k-1)} + \eta_0\, \mathbf{y}_i(k)\, \mathbf{g}^T(\mathbf{y}(k)). & (9.31) \end{aligned}$$

It is interesting to note that the above algorithm has a form similar to the natural gradient algorithms derived in Chapters 6 and 8.

9.1.4 Recurrent Neural Network Model and Hebbian Learning Algorithm

According to Theorem 9.1, it is desirable to build a neural network which can perform a spatio-temporal decorrelation. In this section, we extend Földiák's idea [435] on anti-Hebbian learning to the spatio-temporal domain. It has been shown that the anti-Hebbian learning rule can be used to decorrelate signals in the spatial domain without instability, provided that the learning rate is small enough [221, 225].

Let us consider a linear feedback (recurrent) neural network with FIR synapses (see Fig.9.2 (b) and (c)) whose outputs $y_i(k)$ are described as

$$\begin{aligned} y_i(k) &= x_i(k) - \sum_{j=1}^{m} \sum_{p=0}^{L} \hat{w}_{ij\,p}(k)\, y_j(k-p) \\ &= x_i(k) - \sum_{j=1}^{m} [\widehat{W}_{ij}(z,k)]\, y_j(k) & (9.32) \end{aligned}$$

or in the compact matrix form

$$\begin{aligned} \mathbf{y}(k) &= \mathbf{x}(k) - \sum_{p=0}^{L} \widehat{\mathbf{W}}_p(k)\, \mathbf{y}(k-p) \\ &= \mathbf{x}(k) - [\widehat{\mathbf{W}}(z)]\, \mathbf{y}(k), & (9.33) \end{aligned}$$

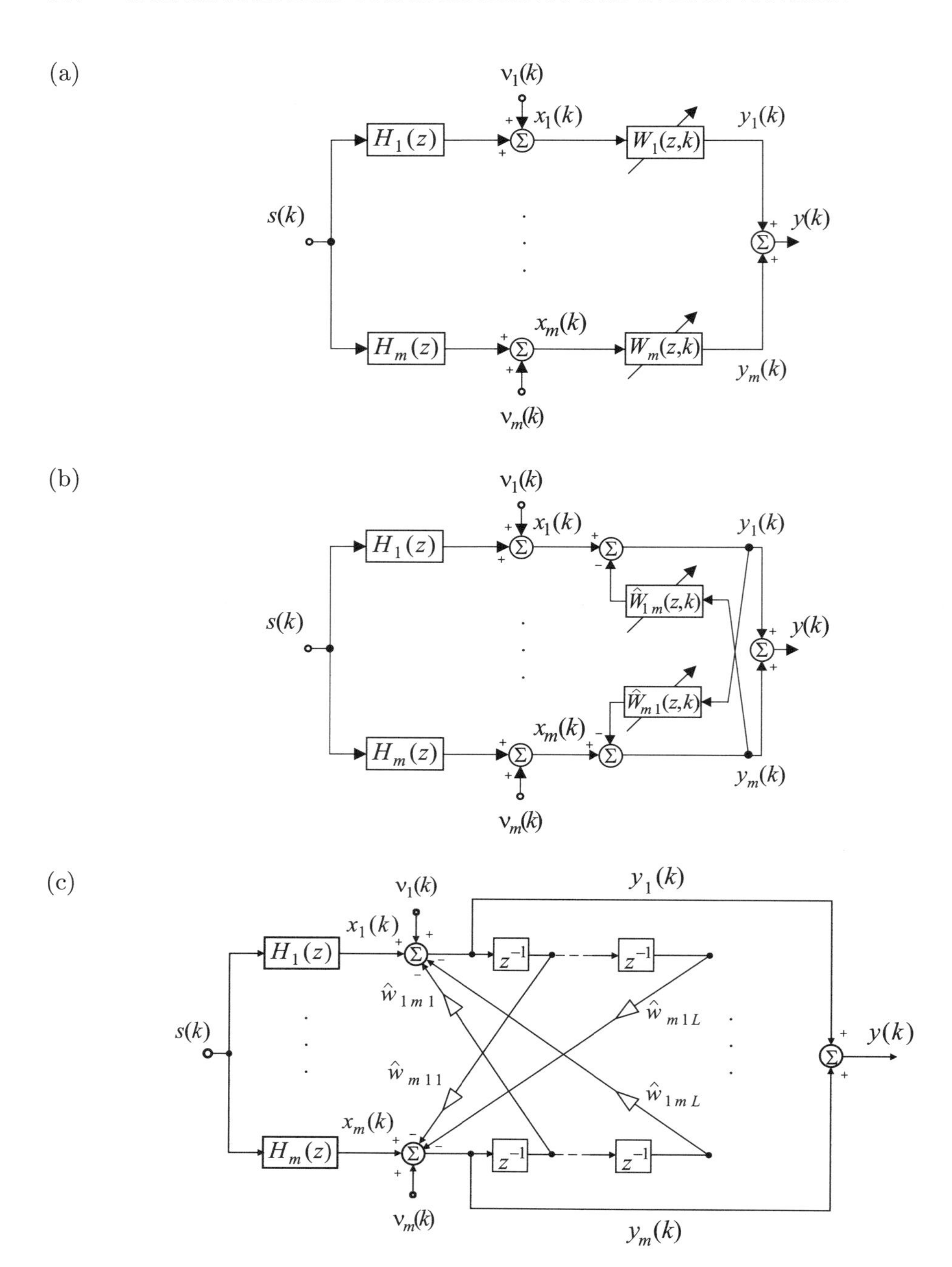

Fig. 9.2 Functional diagrams illustrating SIMO blind equalization models: (a) Feed-forward model, (b) recurrent model, (c) detailed structure of the recurrent model.

where $\widehat{\mathbf{W}}_p(k) \in \mathbb{R}^{m\times m}$ is the synaptic weight matrix whose (i,j)th element is $\hat{w}_{ij\,p}(k)$ ($\widehat{\mathbf{W}}(z,k) = \sum_{p=0}^{L} \widehat{\mathbf{W}}_p(k)\, z^{-p}$, typically, with $\widehat{\mathbf{W}}_0 = \mathbf{0}$). Theoretically, L should approach infinity, but in practice it can assume a finite number. The synaptic weight $w_{ij\,p}(k)$ represents the connection strength between $y_i(k)$ and $y_j(k-p)$.

The neural network should be trained in such a way that the network output signals are spatially and temporally uncorrelated. The composite output $y(k) = \sum_{i=1}^{m} y_i(k)$ then becomes the scaled and/or delayed estimate of the source signal $s(k)$ according to Theorem 9.1. We want to minimize the statistical correlations of the entries of the output vector $\mathbf{y}(k)$ in the spatio-temporal domain. To minimize the statistical correlations, we choose the Kullback-Leibler divergence, which is an asymmetric measure of the distance between two probability distributions. It is straightforward to show that we can achieve spatio-temporal decorrelation of the output signals by minimizing the following cost (risk) function

$$\mathcal{R}(\{\widehat{\mathbf{W}}_p\}) = -\sum_{i=1}^{m} E\{\log q_i(y_i)\}, \tag{9.34}$$

where $q_i(y_i)$ are the pdfs of the signals y_i. The above cost function can be approximated as

$$\hat{\mathcal{R}}(\{\widehat{\mathbf{W}}_p\}) = -\frac{1}{N}\sum_{i=1}^{m}\sum_{k=1}^{N} \log q_i(y_i(k)). \tag{9.35}$$

In order to derive the spatio-temporal anti-Hebbian learning rule, one can try then to minimize the instantaneous realization of the sum of marginal entropies of $y_i(k)$ in an N-point time block. This can be viewed as an extension of minimum entropy coding (or factorial coding) in the spatio-temporal domain [873].

Applying the standard gradient descent method, we obtain a simple learning rule

$$\boxed{\widehat{\mathbf{W}}_p(k+1) = \widehat{\mathbf{W}}_p(k) + \eta(k)\,\left\langle \mathbf{f}(\mathbf{y}(k))\,\mathbf{y}^T(k-p)\right\rangle,} \tag{9.36}$$

where $\mathbf{f}(\mathbf{y}(k)) = [f_1(y_1(k)), \ldots, f_m(y_m(k))]^T$ with entries

$$f_i(y_i(k)) = -\frac{\partial \log q_i(y_i(k))}{\partial y_i(k)}. \tag{9.37}$$

In general, $\widehat{\mathbf{W}}_p(k)$ is updated in such a way that the higher-order cross-correlation between $\mathbf{y}(k)$ and $\mathbf{y}(k-p)$ is minimized. As a special case of this, with hypothesized Gaussian model for $\mathbf{y}(k)$, one can obtain the linear learning rule, i.e., $\mathbf{f}(\mathbf{y}(k)) = \mathbf{y}(k)$.

Note that $\left\langle \mathbf{f}(\mathbf{y}(k))\,\mathbf{y}^T(k-p)\right\rangle = \frac{1}{N}\sum_{k=1}^{N} \mathbf{f}(\mathbf{y}(k))\,\mathbf{y}^T(k-p)$ is the time-average, which can very roughly be approximated by a single sample. In this way, we obtain the simple on-line learning rule

$$\boxed{\widehat{\mathbf{W}}_p(k+1) = \widehat{\mathbf{W}}_p(k) + \eta(k)\mathbf{f}(\mathbf{y}(k))\,\mathbf{y}^T(k-p)} \tag{9.38}$$

or in scalar form

$$\boxed{\hat{w}_{ij\,p}(k+1) = \hat{w}_{ij\,p}(k) + \eta(k) f(y_i(k))\, y_j(k-p).} \tag{9.39}$$

This is a spatio-temporal anti-Hebbian local learning rule which performs the spatio-temporal decorrelation [220, 222, 225].

(a)

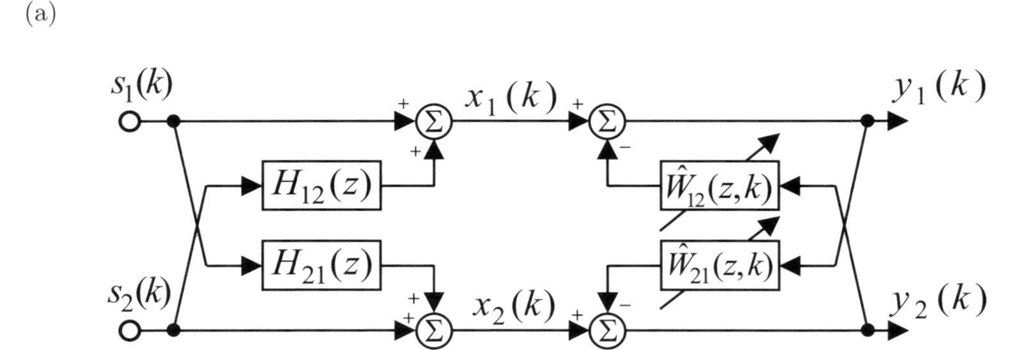

(b)

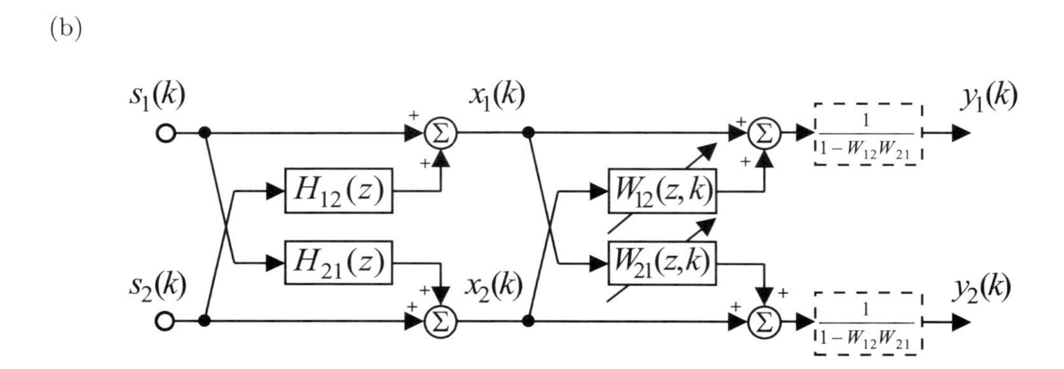

Fig. 9.3 Block diagrams illustrating the multichannel blind deconvolution problem: (a) Recurrent neural network, (b) feed-forward neural network (for simplicity, models for only two channels are shown).

9.2 MULTICHANNEL BLIND DECONVOLUTION WITH CONSTRAINTS IMPOSED ON FIR FILTERS

In many practical applications (e.g., the cocktail party problem) the multichannel convolutive and deconvolutive models take special forms with some constraints imposed on synaptic weights (parameters) of convolutive FIR filters. For example, if each individual sensor is located close to a specific source and far away from other sources, then it is possible to apply the convolutive model shown in Fig.9.3 (a). To be more specific, in a typical cocktail party scenario with two speakers, one microphone can be located close to one speaker and another one close to the second speaker (or loudspeaker). Let us consider the mixing and convolutive model described by the feed-forward model shown in Fig.9.3:

$$x_i(k) = s_i(k) + \sum_{p=0}^{L} \sum_{j \neq i} h_{ij\,p}\, s_j(k-p) = s_i(k) + \sum_{j \neq i} [H_j(z)]\, s_j(k). \tag{9.40}$$

Alternatively, in some applications, we can employ a stable multichannel autoregressive model described by the set of equations

$$x_i(k) = s_i(k) - \sum_{p=0}^{L} \sum_{j \neq i} \hat{h}_{ij\,p}\, x_j(k-p) = s_i(k) - \sum_{j \neq i} [\widehat{H}_j(z)]\, x_j(k). \tag{9.41}$$

In order to estimate the original source signals (which can be white or colored signals, similar to speech signals) in the feed-forward mixing/convolutive model, we can use the recurrent deconvolution model shown in Fig.9.3 (a) or the feed-forward model with post-processing shown in Fig.9.3 (b). For brevity, we will discuss only the derivation of the learning algorithm for the recurrent deconvolution model of Fig.9.3 (a), described by the set of equations:

$$y_i(k) = x_i(k) - \sum_{p=0}^{L} \sum_{j \neq i} \hat{w}_{ij\,p}\, y_j(k-p) = x_i(k) - \sum_{j \neq i} [\widehat{W}_j(z)]\, y_j(k) \tag{9.42}$$

or in a compact matrix form

$$\mathbf{y}(k) = \mathbf{x}(k) - \sum_{p=0}^{L} \widehat{\mathbf{W}}_p\, \mathbf{y}(k-p) = (\mathbf{I} + \widehat{\mathbf{W}}_0)^{-1} \left[\mathbf{x}(k) - \sum_{p=1}^{L} \widehat{\mathbf{W}}_p\, \mathbf{y}(k-p) \right], \tag{9.43}$$

where all diagonal elements of the synaptic weight matrices $\widehat{\mathbf{W}}_p$ are zero and $\widehat{\mathbf{W}}_0 \neq \mathbf{0}$. In other words, the model imposes restrictions on some synaptic weights and it does not allow self-feedback connections [209, 233, 235].

In order to derive the learning algorithm, let us consider the minimization of the following cost function [233]

$$J(\widehat{\mathbf{W}}, \mathbf{y}) = -\sum_{i=1}^{n} E\{\log(q_i(y_i))\}, \tag{9.44}$$

where $q_i(y_i)$ $(i = 1, 2, \ldots, n)$ denote the true probability density functions[2] of the source signals. In fact, minimization of the above cost function provides the minimization of mutual information (see Chapter 6 for more explanation). The differential (infinitesimal increment) of the cost function can be evaluated as follows

$$dJ = \left\langle \mathbf{f}^T(\mathbf{y}(k))\, d\mathbf{y}(k) \right\rangle, \tag{9.45}$$

[2] In practice, approximate (hypothesized) pdfs of the source signals are used.

where $\mathbf{f}(\mathbf{y}) = [f_1(y_1), \ldots, f_n(y_n)]^T$, with $f_i(y_i) = -d \log(q_i(y_i))/dy_i$ and

$$d\mathbf{y}(k) = -(\mathbf{I} + \widehat{\mathbf{W}}_0)^{-1} \left[d\widehat{\mathbf{W}}_0\, \mathbf{y}(k) + \sum_{p=1}^{L} (d\widehat{\mathbf{W}}_p\, \mathbf{y}(k-p) + \widehat{\mathbf{W}}_p\, d\mathbf{y}(k-p) \right]. \tag{9.46}$$

Assuming that small changes of the output signals are affected only by the small variation of the synaptic weights $\{\hat{w}_{ij\,p}\}$, we can approximate Eq. (9.46) by

$$d\mathbf{y}(k) = -(\mathbf{I} + \widehat{\mathbf{W}}_0)^{-1} \sum_{p=1}^{L} d\widehat{\mathbf{W}}_p\, \mathbf{y}(k-p). \tag{9.47}$$

Hence, on the basis of the standard gradient descent method, we obtain the approximate learning rule

$$\Delta \widehat{\mathbf{W}}_p = -\eta \frac{d\,J}{d\,\widehat{\mathbf{W}}_p} = \eta(k) \left[\mathbf{I} + \widehat{\mathbf{W}}_0 \right]^{-T} \left\langle \mathbf{f}(\mathbf{y}(k))\, \mathbf{y}^T(k-p) \right\rangle. \tag{9.48}$$

In order to avoid the inversion of the matrix $[\mathbf{I} + \widehat{\mathbf{W}}_0]$ in each iteration step and to increase the convergence speed, we can apply the natural gradient concept by introducing a new differential, defined as

$$d\,\mathbf{X}_p = \left[\mathbf{I} + \widehat{\mathbf{W}}_0 \right]^{-1} d\,\widehat{\mathbf{W}}_p \tag{9.49}$$

and

$$d\,\mathbf{y}(k) = -\sum_{p=1}^{L} d\,\mathbf{X}_p\, \mathbf{y}(k-p). \tag{9.50}$$

Hence, we obtain the batch algorithm

$$\Delta \widehat{\mathbf{W}}_p(l) = -\eta(l) \left[\mathbf{I} + \widehat{\mathbf{W}}_0(l) \right] \frac{d\,J}{d\,\mathbf{X}_p} = \eta(l) \left[\mathbf{I} + \widehat{\mathbf{W}}_0(l) \right] \left\langle \mathbf{f}(\mathbf{y}(k))\, \mathbf{y}^T(k-p) \right\rangle. \tag{9.51}$$

Using the MA (moving average) approach, we obtain the on-line algorithm

$$\boxed{\widehat{\mathbf{W}}(k+1) = \widehat{\mathbf{W}}(k) + \eta(k)\, [\mathbf{I} + \widehat{\mathbf{W}}_0]\, \hat{\mathbf{R}}_{\mathbf{f}\,\mathbf{y}}^{(k)}(p),} \tag{9.52}$$

where $\hat{\mathbf{R}}_{\mathbf{f}\,\mathbf{y}}^{(k)}(p) = (1 - \eta_0)\, \hat{\mathbf{R}}_{\mathbf{f}\,\mathbf{y}}^{(k-1)}(p) + \eta_0\, \mathbf{f}(\mathbf{y}(k))\, \mathbf{y}^T(k-p)$. The above algorithm provides improved performance and convergence speed over the local algorithm [1230, 1231, 1145, 1146]

$$\Delta \widehat{\mathbf{W}}_p(k) = \eta(k) \left\langle \mathbf{f}(\mathbf{y}(k))\, \mathbf{y}^T(k-p) \right\rangle. \tag{9.53}$$

9.3 GENERAL MODELS FOR MULTIPLE-INPUT MULTIPLE-OUTPUT BLIND DECONVOLUTION

9.3.1 Fundamental Models and Assumptions

Let us consider a multichannel, linear time-invariant (LTI), discrete-time dynamical system described in the most general form as

$$\mathbf{x}(k) = \sum_{p=-\infty}^{\infty} \mathbf{H}_p \, \mathbf{s}(k-p), \tag{9.54}$$

where $\mathbf{H}_p$ is an $m \times n$-dimensional matrix of mixing coefficients at time-lag p (called the impulse response at time-lag p) and $\mathbf{s}(k)$ is an n-dimensional vector of source signals with mutually independent components. It should be noted that the causality in the time domain is satisfied only when $\mathbf{H}_p = \mathbf{0}$ for all $p < 0$ [3].

The goal of multichannel blind deconvolution is to estimate the source signals using only sensor signals $\mathbf{x}(k)$ with some knowledge of the source signal distributions and statistics. In the most general case, we attempt to estimate the sources by employing another multichannel, LTI, discrete-time, stable dynamical system (Fig. 9.4 (a) and (b)) described as

$$\mathbf{y}(k) = \sum_{p=-\infty}^{\infty} \mathbf{W}_p \, \mathbf{x}(k-p), \tag{9.55}$$

where $\mathbf{y}(k) = [y_1(k), y_2(k), \ldots, y_n(k)]^T$ is an n-dimensional vector of the outputs and $\mathbf{W}_p$ is an $n \times m$-dimensional coefficient matrix at time lag p. We use the operator form notation

$$\mathbf{H}(z) = \sum_{p=-\infty}^{\infty} \mathbf{H}_p \, z^{-p}, \tag{9.56}$$

$$\mathbf{W}(z) = \sum_{p=-\infty}^{\infty} \mathbf{W}_p \, z^{-p}. \tag{9.57}$$

In practical applications, we need to implement the blind deconvolution problem with a finite impulse response (FIR) multichannel filter with matrix transfer function:

$$\mathbf{W}(z) = \sum_{p=0}^{L} \mathbf{W}_p \, z^{-p}, \tag{9.58}$$

or apply a non-causal (doubly-finite) feed-forward multichannel filter

$$\mathbf{W}(z) = \sum_{p=-K}^{L} \mathbf{W}_p \, z^{-p}, \tag{9.59}$$

[3]For images or two dimensional signals, where time is replaced by spatial coordinates, there is no problem with causality if the whole image is available.

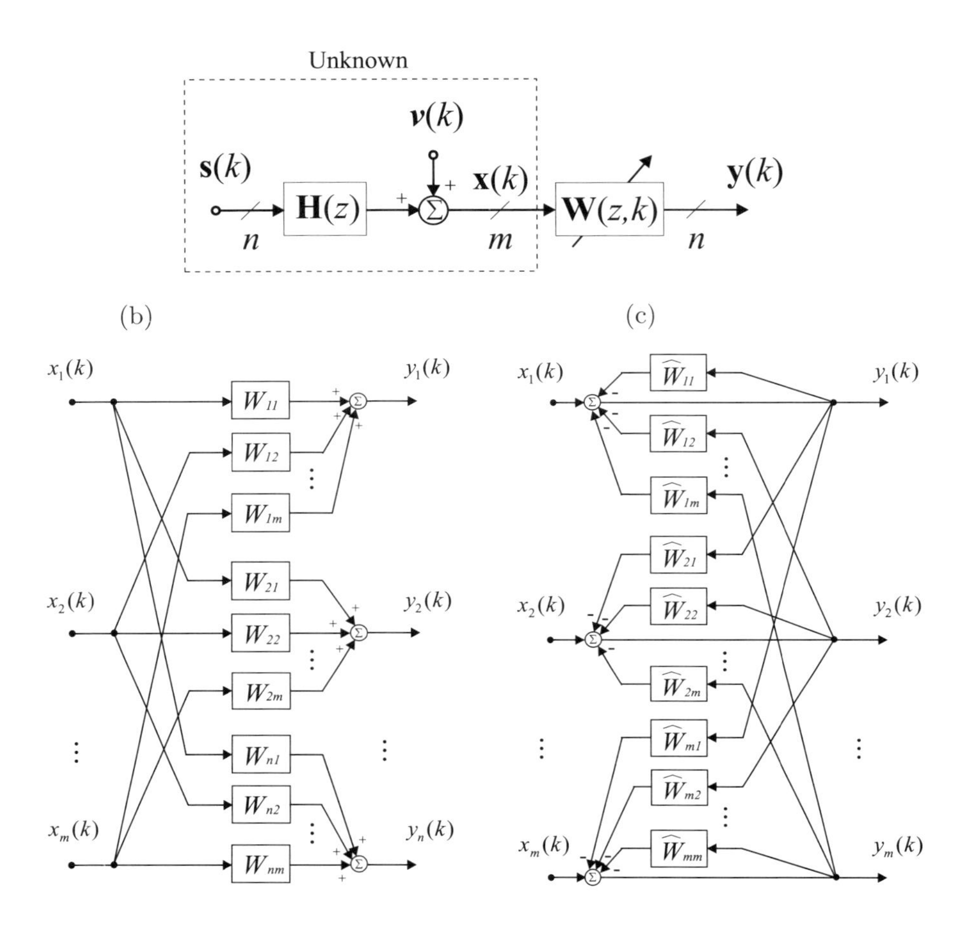

Fig. 9.4 Illustration of the multichannel deconvolution models: (a) Functional block diagram of the feed-forward model, (b) architecture of the feed-forward neural network (each synaptic weight $W_{ij}(z,k)$ is an FIR or stable IIR filter, (c) architecture of the fully connected recurrent neural network.

where K and L are two given positive integers.

The global transfer function is defined by

$$\mathbf{G}(z) = \mathbf{W}(z)\,\mathbf{H}(z). \tag{9.60}$$

In order to ensure that the mixing system is recoverable, we put the following constraints on the convolutive/mixing systems.

1. The filter $\mathbf{H}(z)$ is stable, i.e., its impulse response satisfies the absolute summability condition

$$\sum_{p=-\infty}^{\infty} \|\mathbf{H}_p\|_2 < \infty, \tag{9.61}$$

where $\|\cdot\|_2$ denotes the Euclidean norm.

2. The filter matrix transfer function $\mathbf{H}(z)$ is full rank on the unit circle ($|z| = 1$), that is, it has no zeros on the unit circle.

9.3.2 Separation-Deconvolution Criteria

The blind deconvolution task is to find a matrix transfer function $\mathbf{W}(z)$ such that

$$\mathbf{G}(z) = \mathbf{W}(z)\,\mathbf{H}(z) = \mathbf{P}\,\mathbf{\Lambda}\,\mathbf{D}(z), \tag{9.62}$$

where $\mathbf{P} \in \mathbb{R}^{n\times n}$ is a permutation matrix, $\mathbf{\Lambda} \in \mathbb{R}^{n\times n}$ is a nonsingular diagonal scaling matrix, and a diagonal matrix $\mathbf{D}(z) = \text{diag}\{D_1(z), \ldots, D_n(z)\}$ represents a bank of arbitrary stable filters with transfer functions $D_i(z) = \sum_p d_{ip} z^{-p}$. In other words, the objective of multichannel blind deconvolution, in the most general case, is to recover the source vector $\mathbf{s}(k)$ from the observation vector $\mathbf{x}(k)$, up to possibly scaled, reordered, and filtered estimates. We can only recover a filtered version of each source signal $s_i(k)$ because we assume nothing about the temporal structure of each source. However, if we assume that sources are i.i.d., then we can relax the conditions to the form:

$$\mathbf{G}(z) = \mathbf{W}(z)\,\mathbf{H}(z) = \mathbf{P}\,\mathbf{\Lambda}\,\mathbf{D}_0(z), \tag{9.63}$$

where $\mathbf{D}_0(z) = \text{diag}\{z^{-\Delta_1}, \ldots, z^{-\Delta_n}\}$. In such a case, the original source signals can be reconstructed up to arbitrary scaled, reordered, and delayed estimates. In other words, we can preserve their waveforms exactly.

For some models it is difficult or even impossible to find an exact inverse of the channels in the sense described above, since no knowledge of the channel and source signals is available in advance. Hence, instead of finding an inverse decomposition (9.62) or (9.63) in one step, we often attempt to find a matrix $\mathbf{W}(z)$ that satisfies the generalized zero-forcing (ZF) condition, given by

$$\mathbf{G}(z) = \mathbf{W}(z)\,\mathbf{H}(z) = \mathbf{\Gamma}\,\mathbf{D}_0(z), \tag{9.64}$$

where $\mathbf{\Gamma}$ is an $n \times n$ nonsingular memoryless (constant) mixture matrix and $\mathbf{D}_0(z) = \text{diag}\{z^{-\Delta_1}, \ldots, z^{-\Delta_n}\}$.

Let us denote the output signal of the system $\mathbf{W}(z)$ by $\mathbf{y}(k)$, i.e., $\mathbf{y}(k) = [\mathbf{W}(z)]\,\mathbf{x}(k)$. It will be shown that the generalized zero-forcing condition (9.64) is achieved if $\{y_i(k)\}$ are uncorrelated in the temporal domain as well as in the spatial domain. Suppose that the system $\mathbf{W}(z)$ satisfies the generalized zero-forcing condition (9.64), then one can easily see that the signals $\{y_i(k)\}$ are instantaneous mixtures of source signals $\{s_i(k)\}$. Instantaneous mixtures can be separated in the second stage by blind source separation or independent component analysis.

Theorem 9.2 (Zero-Forcing Conditions) *Let the channel $\mathbf{H}(z)$ satisfy the following assumptions:*
(AS1) There are more sensors than sources, i.e., $m > n$.
(AS2) $\mathbf{H}(z)$ is causal and rational.
(AS3) $\mathbf{H}(z)$ is full rank for all z.
Furthermore, suppose that source signals $\{s_i(k)\}$ are spatially independent and temporally i.i.d. sequences. Then the generalized zero-forcing condition (9.64) is satisfied, if the following relation holds

$$E\{\mathbf{y}(k)\,\mathbf{y}^T(q)\} = \mathbf{\Gamma}\,\delta_{kq}, \tag{9.65}$$

where $\mathbf{\Gamma} \in \mathbb{R}^{n\times n}$ is the constant matrix and δ_{kq} is the Kronecker delta equal to 1 for $k = q$, and 0 otherwise.

Proof. Since both the channel and its inverse are causal and stable, the global system $\mathbf{G}(z)$ is also causal and stable. Then

$$\mathbf{y}(k) = \sum_{p=0}^{\infty} \mathbf{G}_p\,\mathbf{s}(k-p). \tag{9.66}$$

Suppose that $\mathbf{G}_d$ is the leading nonsingular coefficient matrix, i.e., $\mathbf{G}_0 = \cdots = \mathbf{G}_{d-1} = \mathbf{0}$. Then,

$$\mathbf{y}(k) = \sum_{p=d}^{\infty} \mathbf{G}_p\,\mathbf{s}(k-p). \tag{9.67}$$

Invoking $E\{\mathbf{y}(k)\,\mathbf{y}^T(k+q)\} = \mathbf{0}$ for $\forall q \neq 0$, we have

$$\sum_{p=d}^{\infty} \mathbf{G}_p E\{\mathbf{s}(k-p)\,\mathbf{s}^T(k-p)\}\mathbf{G}_{p+q}^T = \mathbf{0}, \qquad \forall q \neq 0. \tag{9.68}$$

Since $E\{\mathbf{s}(k-p)\,\mathbf{s}^T(k-p)\}$ and $\mathbf{G}_d$ are nonsingular matrices, the condition (9.65) implies that $\mathbf{G}_p = \mathbf{0}$ for $i = d+1, \ldots, \infty$. Therefore, $\mathbf{y}(k) = \mathbf{G}_d\,\mathbf{s}(k-d)$ if the condition (9.65) is satisfied. □

From the above criteria and discussion it follows that blind multichannel deconvolution is a fairly complex process which can be performed in two or even more stages or by neural systems containing multi-layer structures. Fig.9.5 illustrates typical architectures for two stage procedures. In the schema shown in Fig.9.5 (a), blind spatio-temporal decorrelation of sensor signals is performed initially by using a recurrent neural network (see Chapter 4). The output signals represented by the vector $\widehat{\mathbf{y}}$ after convergence should be a linear mixture of a time-delayed version of the original sources $\mathbf{s}(k)$ satisfying the zero-forcing conditions, i.e.,

$$\widehat{\mathbf{y}}(k) = [(\mathbf{I} + \widehat{\mathbf{W}}(z)]^{-1}\,\mathbf{x}(k) = \mathbf{\Gamma}\,\mathbf{D}_0(z)\,\mathbf{s}(k), \tag{9.69}$$

(a)

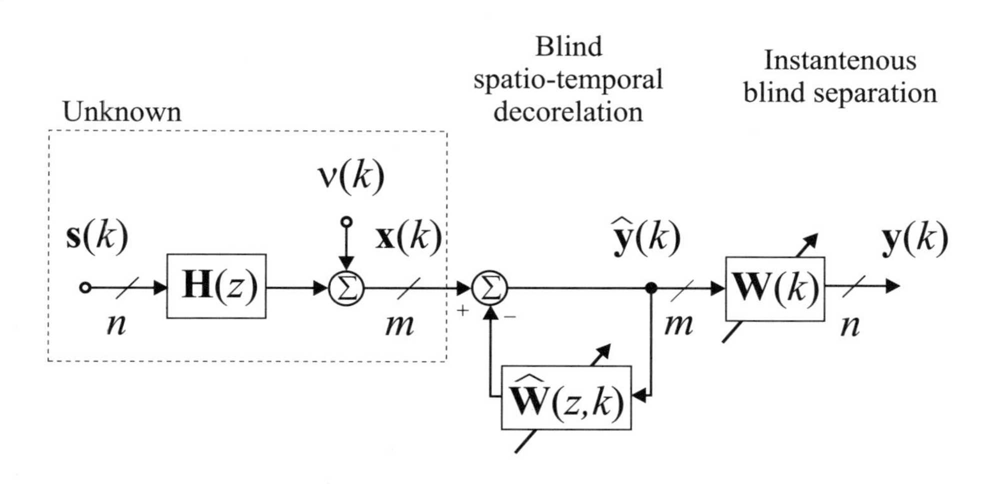

(b)

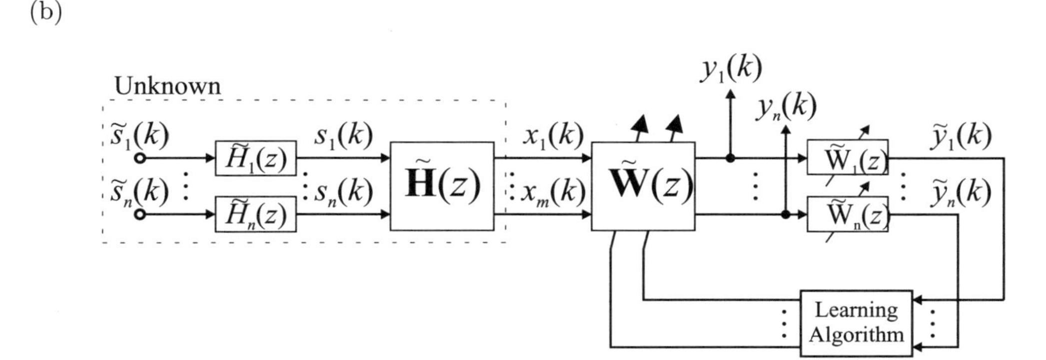

Fig. 9.5 Exemplary architectures for two stage multichannel deconvolution.

where $\mathbf{\Gamma} \in \mathbb{R}^{m \times m}$ is an arbitrary nonsingular constant (memoryless) matrix and $\mathbf{D}_0(z) = \text{diag}\{z^{-\Delta_1}, \ldots, z^{-\Delta_m}\}$. In order to reconstruct original signals (up to arbitrary, permutation, scaling and time delay) in the second stage, we can apply ICA or BSS for an instantaneous mixture (since the matrix $\mathbf{\Gamma}$ is memoryless).

Fig.9.5 (b) shows an alternative model for blind deconvolution, assuming that colored source signals are modelled as

$$s_i(k) = [\widetilde{H}_i(z)]\, \widetilde{s}_i(k) \tag{9.70}$$

where $\widetilde{H}_i(z)$ describe an unknown AR or ARMA process and $\widetilde{s}_i(k)$ is innovation i.i.d. signals. Assuming that only sensor signals ($\mathbf{x}(k) = [\widetilde{\mathbf{H}}(z)]\,\mathbf{s}(k)$) are available in the first stage, we perform multichannel deconvolution in such a way that the outputs $y_i(k)$ are filtered versions of the original sources $s_j(k)$. In the second stage, we perform SISO or

SIMO blind equalization using one of the available methods, described in the previous sections of this chapter.

The multichannel blind deconvolution task will be largely driven by the assumption that sources are mutually independent. In some cases, however, we make additional assumption that sources are white signals. In order to separate independent sources by a deconvolution model, we can reformulate the blind deconvolution problem as an optimization problem, so that selection of a suitable cost function will be necessary.

9.4 RELATIONSHIPS BETWEEN BSS/ICA AND MBD

The algorithms described in the previous chapters (especially, algorithms for ICA/BSS described in Chapters 4 and 8) are very efficient. However, a key problem arises when one extends or generalizes these algorithms (and others algorithms for ICA and BSS/BES) for real-world problems concerned with robust multichannel blind deconvolution/separation (MBD). In other words, the question arises: Is it possible to establish a direct relationship between ICA/BSS and MBD and to convert directly a learning algorithm for BSS to an "equivalent" algorithm for MBD? This problem is the subject of this section. We will show that the learning algorithms for blind source separation (BSS) can be generalized or extended to achieve multichannel blind deconvolution (MBD) [33, 34, 393, 397]. We use relationships between convolution in the time domain and multiplication in the frequency domain and more generally exploit abstract algebra with their useful and powerful properties to obtain new results [1026, 398, 399, 549, 715].

9.4.1 Multichannel Blind Deconvolution in the Frequency Domain

The simplest idea of extending the blind source separation and ICA algorithms to multichannel blind deconvolution is to use the frequency domain techniques. A convolutive mixture in the time domain corresponds to an instantaneous mixture of complex-valued signals and parameters in the frequency domain [41, 867, 809]. An n-point windowed DFT (Discrete Fourier Transform) is used to convert time domain signals $y_i(k)$ into frequency domain complex-valued time-series signals:

$$Y_i(\omega, b) = \sum_{k=0}^{N-1} e^{-j\omega k}\, y_i(k)\, win(k - b\Delta), \tag{9.71}$$

$$for \quad \omega = 0, \frac{1}{N}2\pi, \ldots, \frac{N-1}{N}2\pi, \tag{9.72}$$

where win denotes a window function and Δ is the shifting interval of the window. The number of frequency bins is equal to the frame length N and it corresponds to the length of FIR filters of the deconvolution system.

By using the Fourier transform, the multichannel deconvolution model is represented by

$$\mathbf{Y}(\omega, b) = \mathbf{W}(\omega)\,\mathbf{X}(\omega, b), \tag{9.73}$$

where $\mathbf{Y}(\omega, b) = [Y_1(\omega, b), \ldots, Y_n(\omega, b)]^T$ and $\mathbf{X}(\omega, b) = [X_1(\omega, b), \ldots, X_m(\omega, b)]^T$.

For each frequency bin ω we can apply any ICA algorithm discussed in the previous chapters in order to obtain matrix $\mathbf{W}(\omega) \in \mathbb{R}^{n \times m}$ and n output signals $\mathbf{Y}(\omega, b)$ which are estimates of the source signals in the frequency domain. For example, the natural gradient ICA algorithms can be adopted as

$$\Delta \mathbf{W} \;\; = \;\; \eta \left[\mathbf{\Lambda} - \mathbf{f}(\mathbf{Y}) \, \mathbf{Y}^H \right] \mathbf{W}, \tag{9.74}$$

where superscript H denotes the Hermitian conjugate, $\mathbf{\Lambda}$ is a diagonal positive definite matrix (typically, $\mathbf{\Lambda} = \mathbf{I}$ or $\mathbf{\Lambda} = \text{diag}\{\mathbf{f}(\mathbf{Y}) \, \mathbf{Y}^H\}$) and $\mathbf{f}(\mathbf{Y}) = [f(Y_1), \ldots, f(Y_n)]^T$ is the vector of suitably chosen nonlinear functions of complex-valued signals:

$$Y_i = Y_i^{(R)} + j Y_i^{(I)} = |Y_i|^{j\varphi_i}, \tag{9.75}$$

with $\varphi_i = \tan^{-1}(Y_i^{(I)} / Y_i^{(R)})$.

A good choice of nonlinear function for super-Gaussian source signals is

$$f(Y_i) = \frac{1}{\hat{\sigma}_i^2} \tanh(\gamma_i |Y_i|) \, e^{j\varphi_i}, \tag{9.76}$$

where $\hat{\sigma}_i^2$ is the estimated variance of signal Y_i and $\gamma_i > 0$ is a parameter which controls the steepness of the nonlinearity. It should be noted that for such a class of nonlinear functions the diagonal terms in Eq. (9.74) can be expressed as

$$f(Y_i) \, Y_i^* = \frac{1}{\sigma_i^2} \tanh(\gamma_i |Y_i|) \, e^{j\varphi_i} \, |Y_i| \, e^{-j\varphi_i} = \frac{1}{\sigma_i^2} \tanh(\gamma_i |Y_i|) \, |Y_i|, \tag{9.77}$$

which are always real-valued.

The matrix $\mathbf{W}(\omega)$ converges for each frequency bin ω to an equilibrium point that satisfies

$$f(Y_i) Y_j^* = \lambda_i \delta_{ij}. \tag{9.78}$$

After convergence, we can obtain the coefficient of the FIR filters of length $L = N$ by applying the inverse DFT to all matrices $\mathbf{W}(\omega)$.

Unfortunately, the algorithms in the frequency domain are batch or block algorithms and are computationally expensive [42, 124, 516, 867, 1079, 1160]. Moreover, they are not responsive to changing environments. In the following sections of this chapter, we will discuss more direct algorithms in the time domain that are computationally less expensive i.e. they have lower computational complexity.

9.4.2 Algebraic Equivalence of Various Approaches

Notational conventions used in this section are as follows. The discrete-time signals and parameters are assumed to be real-valued. An infinite time series (on both sides of the time axis) is denoted as $\underline{x}(k) = \{\ldots, x(k-1), x(k), x(k+1), \ldots\}$. A reversed-order time series is defined as $\underline{x}(-k) = \{\ldots, x(k+1), x(k), x(k-1), \ldots\}$. A multi-variable time series, that is, a vector time series, is denoted as $\underline{\mathbf{x}}(k) = \{\ldots, \mathbf{x}(k-1), \mathbf{x}(k), \mathbf{x}(k+1), \ldots\}$, where $\mathbf{x}(k) =$

$[x_1(k), x_2(k), \ldots, x_m(k)]^T$. Similarly, an $m \times n$ time-domain impulse response is denoted as $\underline{\mathbf{H}} = \underline{\mathbf{H}}_p = \{\mathbf{H}_p\} = \{\ldots, \mathbf{H}_{p-1}, \mathbf{H}_p, \mathbf{H}_{p+1}, \ldots\}$ and its reversed-order is $\underline{\mathbf{H}}_{(-p)} = \{\mathbf{H}_{-p}\} = \{\ldots, \mathbf{H}_{p+1}, \mathbf{H}_p, \mathbf{H}_{p-1}, \ldots\}$. Equivalently, by using the $\mathcal{Z}$-transform transfer function, we may write $\mathbf{H}(z) = \sum_{p=-\infty}^{\infty} \mathbf{H}_p z^{-p}$, where $\mathbf{H}_p$ is a matrix of mixing coefficients at lag p and $\mathbf{H}(z^{-1}) = \sum_{p=-\infty}^{\infty} \mathbf{H}_p z^p$ [1026].

We aim to recover the original signal $\underline{\mathbf{s}}(k)$ by applying separating/deconvolution filters characterized by an impulse response $\underline{\mathbf{W}} = \{\mathbf{W}_p\} = \{\ldots, \mathbf{W}_{p-1}, \mathbf{W}_p, \mathbf{W}_{p+1}, \ldots\}$. The recovered signals can be described as $\underline{\mathbf{y}}(k) = \underline{\mathbf{W}} * \underline{\mathbf{x}}(k)$, where the k-th element of the resulting series $\underline{\mathbf{y}}(k)$ is computed as $\mathbf{y}(k) = \sum_{p=-\infty}^{\infty} \mathbf{W}_p \, \mathbf{x}(k-p), \quad (k = -\infty, \ldots, \infty)$. In the scalar form the i-th output can be expressed as $y_i(k) = \sum_{j=1}^{n} \sum_{p=-\infty}^{\infty} w_{ijp} x_j(k-p)$.

A global system that describes the convolution-deconvolution process with input $\underline{\mathbf{s}}(k)$ and output $\underline{\mathbf{y}}(k)$ is

$$\underline{\mathbf{y}}(k) = \underline{\mathbf{G}} * \underline{\mathbf{s}}(k) = \underline{\mathbf{W}} * \underline{\mathbf{H}} * \underline{\mathbf{s}}(k). \tag{9.79}$$

Notice that $\underline{\mathbf{H}} = \{\mathbf{H}_p\}$, $\underline{\mathbf{W}} = \{\mathbf{W}_p\}$ and $\underline{\mathbf{G}} = \{\mathbf{G}_p\}$ are double-infinite non-causal linear filters, which, in implementation applications, are replaced by their truncated finite-duration impulse response (FIR) versions. An equivalent description in the $\mathcal{Z}$-transform domain is as follows

$$\mathbf{Y}(z) = \mathbf{W}(z)\,\mathbf{X}(z) = \mathbf{W}(z)\,\mathbf{H}(z)\,\mathbf{S}(z) = \mathbf{G}(z)\,\mathbf{S}(z), \tag{9.80}$$

where $\mathbf{Y}(z) = \sum_{k=-\infty}^{\infty} \mathbf{y}(k)\, z^{-k}$ and $\mathbf{W}(z) = \sum_{p=-\infty}^{\infty} \mathbf{W}_p\, z^{-p}$ are the $\mathcal{Z}$-transforms of the time domain infinite series $\underline{\mathbf{y}}(k)$ and $\underline{\mathbf{W}}(k)$, respectively. $\mathbf{H}(z), \mathbf{G}(z), \mathbf{X}(z), \mathbf{S}(z)$ are defined accordingly. If $\underline{\mathbf{H}} = \{\ldots, \mathbf{0}, \mathbf{H}_0, \mathbf{0}, \ldots\}$ and $\underline{\mathbf{W}} = \{\ldots, \mathbf{0}, \mathbf{W}_0, \mathbf{0}, \ldots\}$, we obtain a much simpler task of blind source separation of an instantaneous mixture.

Different ways of describing physical phenomena of signal propagation (IIR filter, FIR filter, DFT, $\mathcal{Z}$-transform, wavelets, other transforms) result in different but equivalent mathematical models. In blind deconvolution, the key operation is a linear convolution (FIR and/or IIR filtering) of a time series. The usual notations in the discrete time domain involve the $\mathcal{Z}$-transform, z^{-1}-delay operator, DFT (discrete Fourier Transform) or convolution (others are also possible). Transformations create suitable relationships for the available data set and algebraic operations [1026].

The basic data set in our case consists of infinite series of elements such as w_p, $x(k)$ and $y(k)$, that is, $\underline{w}_p = \{\ldots, w_{p-1}, w_p, w_{p+1}, \ldots\}$, $\underline{x}(k) = \{\ldots, x(k-1), x(k), x(k+1), \ldots\}$ and $\underline{y}(k) = \{\ldots, y(k-1), y(k), y(k+1), \ldots\}$, where p and k are indices. In the case of time-domain filtering, $\underline{x}(k)$ and $\underline{y}(k)$ are time series of samples and k is a discrete time index. The data can be rearranged to other forms which best suit solving a specific problem, provided that the new data set with the new operators defined fall into one of the algebraic categories. In the case of the $\mathcal{Z}$-transform, the data set is formed in polynomials with polynomial multiplication and addition [715]. In the case of the DFT, the data set is transformed into a complex number series (the frequency domain) with point-by-point multiplication and addition of complex series. The case of using a time series in its time-domain format $\underline{y}(k)$ and a convolution as the multiplicative operator is described next.

9.4.3 Convolution as a Multiplicative Operator

Let us consider the inner product for a complete signal space (Hilbert space). For a finite energy signal space, the inner product is (e.g., for time series $\underline{x}(k)$ and $\underline{y}(k)$)[4]

$$(\underline{x}, \underline{y}) = \sum_{k=-\infty}^{\infty} x(k)y(k). \tag{9.81}$$

For a finite power signal space, the inner product is

$$(\underline{x}, \underline{y}) = \lim_{N\to\infty} \frac{1}{2N+1} \sum_{k=-N}^{N} x(k)y(k). \tag{9.82}$$

Let the symbols $\underline{0}$ (zero element) and $\underline{1}$ (one element) denote the time series $\{\ldots, 0, 0, 0, \ldots\}$, $\{\ldots, 0, 1, 0, \ldots\}$, respectively with 1 for $k = 0$. Let symbols + (additive operation) denote point-by-point series addition. Let * (multiplicative operation) denote linear convolution. The convolution of two time series $\underline{x}(k)$ and $\underline{y}(k)$ results in a series $\underline{u}(k) = \underline{x}(k) * \underline{y}(k)$, where the k-th entry ($k = -\infty, \ldots, \infty$) of the series $\underline{u}(k)$ is

$$u(k) = \sum_{p=-\infty}^{\infty} x(p)y(k-p) = \sum_{p=-\infty}^{\infty} x(k-p)y(p).$$

For finite power signals, it may be infinite, and a more informative quantity is the average $\langle \underline{x}(k) * \underline{y}(k) \rangle$ defined by analogy to (9.82) with the k-th entry equal to

$$\langle \underline{x}(k) * \underline{y}(k) \rangle = \lim_{N\to\infty} \frac{1}{2N+1} \sum_{p=-N}^{N} x(p)y(k-p). \tag{9.83}$$

A set of time series together with the zero element, one element, additive operation, and convolution operation creates a *general algebra* that fulfills the standard conditions of a commutative ring. By adding some assumptions it becomes a field. We define symbol / (division operator) by $\underline{x}/\underline{y} = \underline{x}(k) * \underline{y}^{-1}(k), \underline{y} \neq \underline{0}$. Since $\underline{y}(k) * \underline{y}^{-1}(k) = \underline{1}$, a problem may arise concerning the existence of any bounded inverse element $\underline{y}^{-1}$. The problem is easily solved by a suitable choice of the origin $k = 0$ of the series $\underline{y}$.

All the rules of differential and integral calculus hold since these are linear operations. For example, expressions for the differentiation of filtering

$$y(k) = \underline{w} * \underline{x}(k) = \sum_{p=-\infty}^{\infty} w_p x(k-p) \quad k = -\infty, \ldots, \infty, \tag{9.84}$$

is as follows,

$$\frac{\partial}{\partial \underline{w}} [y(k)] = \frac{\partial}{\partial \underline{w}} [\underline{w} * \underline{x}(k)] = \underline{x}(-k) \tag{9.85}$$

[4]For complex-valued data instead of $y(k)$, we will use the complex conjugate $y^*(k)$.

The reversed order of the resulting time series is a consequence of the multiplicative operator definition.

The time series field reduces to the real number field for the special case of the time series $\underline{y}(k) = \{\ldots, 0, y, 0, \ldots\}$ with only one non-zero entry for $k = 0$. This is the notation of a real number y in terms of time series, which is analogous to writing a real number y by using the complex notation $y + j0$. The probability density function of a series at the point y may be described in this way $\{\ldots, 0, f_y(y), 0, \ldots\}$. When one of the operands of a multiplicative operation is of this form, the symbol $*$ will be omitted.

The above single-channel approach easily extends to a multichannel approach. The resulting algebra is a non-commutative ring with division, exactly as the standard matrix algebra of the instantaneous BSS case. It is easily seen that all the algebraic properties of the instantaneous and convolutive mixing models are equivalent [715, 1026, 549].

Remark 9.1 *Inevitable ambiguities in blind deconvolution are those of permutation, scaling and delay. In instantaneous blind source separation, only those of permutation and scaling are encountered. However, from an algebraic point of view, there is a precise equivalence between instantaneous and convolutive cases, and the ambiguities should also be equivalent [1026]. The arbitrary time-delay ambiguity arises as a consequence of index ordering k regarded as time instants and due to causality of the convolving filters. However, from the algebraic point of view, there is no restriction on incorporating the time-delay with the convolving filter, and regarding its output as non-causal, thus making it time-delayed.*

Basic analogies and relationships between instantaneous BSS and MBD (for m=n and complex-valued signals and parameters) are collected in Table 9.1.

9.4.4 Natural Gradient Learning Rules for Multichannel Blind Deconvolution (MBD)

The natural gradient algorithm [28, 33, 34, 26] can be generalized for MBD as (see also Table 9.1)

$$\Delta\underline{\mathbf{W}}(l) = \underline{\mathbf{W}}(l+1) - \underline{\mathbf{W}}(l) = -\eta \frac{\partial K_{fq}}{\partial \underline{\mathbf{W}}} * \underline{\mathbf{W}}^T_{(-p)}(l) * \underline{\mathbf{W}}(l), \tag{9.86}$$

where K_{fq} is the Kullback - Leibler objective function and $\mathbf{W}(l)$ is an estimate at time instant n of the inverse system for recovering the source signals. Alternatively, we can employ the natural gradient by using the $\mathcal{Z}$-transform as

$$\Delta\mathbf{W}(z, l) = -\eta(l) \frac{\partial K_{fq}}{\partial \mathbf{W}(z)} \mathbf{W}^T(z^{-1}, l)\, \mathbf{W}(z, l). \tag{9.87}$$

The Atick-Redlich learning rule discussed in Chapter 4 can be generalized as follows

$$\Delta\underline{\mathbf{W}}(l) = -\eta(l)\underline{\mathbf{W}}(l) * \left[\frac{\partial K_{fq}}{\partial \underline{\mathbf{W}}}\right]^T * \underline{\mathbf{W}}(l) \tag{9.88}$$

or using the $\mathcal{Z}$-transform

$$\Delta\mathbf{W}(z, l) = -\eta(l)\mathbf{W}(z, l) \left[\frac{\partial K_{fq}}{\partial \mathbf{W}(z)}\right]^T \mathbf{W}(z, l). \tag{9.89}$$

9.4.5 NG Algorithms for Double Infinite Filters

Since the convolutive case is of interest here, all the symbols follow from the properties of time series with the convolution operator $*$. We can derive a family of algorithms by minimization of the Kullback-Leibler divergence between the actual distribution $p_y(\mathbf{y})$ and the factorizable model distribution $q_s(\mathbf{y})$, which leads to the objective function (see Chapter 6)

$$\rho(\underline{\mathbf{W}}) = -\frac{1}{2}\log|\det(\underline{\mathbf{W}}_{(-p)} * \underline{\mathbf{W}}^T)| - \sum_{i=1}^{n} \log q_i(y_i). \tag{9.90}$$

Applying the natural gradient rule of the form (9.86), we obtain the learning algorithm first derived rigorously by Amari *et al.* [33, 34]

$$\underline{\mathbf{u}}(k) = \underline{\mathbf{W}}^T_{(-p)}(l) * \underline{\mathbf{y}}(k) \tag{9.91}$$

$$\underline{\mathbf{R}}^{(l)}_{\mathbf{f}\,\mathbf{u}} = \langle \mathbf{f}[\underline{\mathbf{y}}(k)] * \underline{\mathbf{u}}^T(-k)\rangle \tag{9.92}$$

$$\underline{\mathbf{W}}(l+1) = \underline{\mathbf{W}}(l) + \eta(l)\left(\underline{\mathbf{W}}(l) - \underline{\mathbf{R}}^{(l)}_{\mathbf{f}\,\mathbf{u}}\right) \tag{9.93}$$

with nonlinearities $f_i(y_i) = -q_i'(y_i)/q_i(y_i)$, where samples of $\underline{\mathbf{u}}(k)$ can be computed as

$$\mathbf{u}(k) = \sum_{p=-\infty}^{\infty} \mathbf{W}_p^T\, \mathbf{y}(k+p), \quad (k = -\infty, \ldots, \infty). \tag{9.94}$$

The averaging operator $\langle\cdot\rangle$ is used here in the same sense as in (9.83). Equation (9.92) defines the multichannel cross-correlation function. The Atick-Redlich gradient rule (9.88) leads to a learning algorithm

$$\underline{\mathbf{v}}(k) = \underline{\mathbf{W}}^T_{(-p)}(l) * \mathbf{g}[\underline{\mathbf{y}}(k)], \tag{9.95}$$

$$\underline{\mathbf{R}}^{(l)}_{\mathbf{v}\,\mathbf{g}} = \langle \underline{\mathbf{y}}(k) * \underline{\mathbf{v}}^T(-k)\rangle \tag{9.96}$$

$$\underline{\mathbf{W}}(l+1) = \underline{\mathbf{W}}(l) + \eta(l)\left(\underline{\mathbf{W}}(l) - \underline{\mathbf{R}}^{(l)}_{\mathbf{v}\,\mathbf{g}}\right), \tag{9.97}$$

where samples of $\underline{\mathbf{v}}(k)$ can be computed as

$$\mathbf{v}(k) = \sum_{p=-\infty}^{\infty} \mathbf{W}_p^T\, \mathbf{g}[\mathbf{y}(k+p)], \quad (k = -\infty, \ldots, \infty). \tag{9.98}$$

Detailed learning rules for the calculation of the equalizer output, filtered output, the cross-correlation matrix and separating/deconvolution matrices $\mathbf{W}_p$ are given in the next section. Nonlinearities $g_i(y_i)$ are now inverse (dual) functions to the functions $f_i(y_i) = -q'_{s_i}(y_i)/q_{s_i}(y_i)$. For example, instead of $f(y_i) = y_i^{1/3}$, we use the cubic function $g_i(y_i) = y_i^3$, or instead $f_i(y_i) = \tanh(y_i)$ we use the inverse function $g_i(y_i) = \tanh^{-1}(y_i) = \frac{1}{2}\log(\frac{1+y_i}{1-y_i})$.

9.4.6 Implementation of Algorithms for a Minimum Phase Non-causal System

9.4.6.1 Batch Update Rules The general formulas (9.91)-(9.94) can be implemented as a generalized batch learning rule. The batch of training data $\mathbf{x}(1), \mathbf{x}(2), \ldots, \mathbf{x}(N)$ is used, with zero values assumed for: $k < 1$ and $k > N$. A biased covariance estimator is used as

$$\begin{aligned}
\text{For each} \quad k &= 1, \ldots, N, \\
\mathbf{y}(k) &= \sum_{p=-L}^{L} \mathbf{W}_p(l)\, \mathbf{x}(k-p) \qquad (9.99) \\
\mathbf{v}(k) &= \sum_{p=-L}^{L} \mathbf{W}_p^T(l)\, \mathbf{g}[\mathbf{y}(k+p)] \qquad (9.100) \\
\text{For each} \quad p &\in -L, \ldots, L: \\
\widehat{\mathbf{R}}_{\mathbf{f}\,\mathbf{v}}^{(l)}(p) &= \frac{1}{N-p} \sum_{k=p+1}^{N} \mathbf{f}[\mathbf{y}(k)]\, \mathbf{v}^T(k-p) \qquad (9.101) \\
\mathbf{W}_p(l+1) &= \mathbf{W}_p(l) + \eta(l)[\mathbf{W}_p(l) - \widehat{\mathbf{R}}_{\mathbf{f}\,\mathbf{v}}^{(l)}(p)] \qquad (9.102)
\end{aligned}$$

9.4.6.2 On-line Update Rule Since the filtering operation leading to $\mathbf{u}(k)$ is non-causal, the most recent time instant of $\mathbf{u}(k)$ that can be calculated by using samples of $\mathbf{x}(k)$ and $\mathbf{y}(k)$ up to time instant k is $\mathbf{u}(k-L)$. The generalized covariance matrix is replaced by its instantaneous value:

$$\begin{aligned}
\mathbf{y}(l) = \mathbf{y}(k) &= \sum_{p=0}^{L} \mathbf{W}_p(k)\, \mathbf{x}(k-p) \qquad (9.103) \\
\mathbf{u}(l) = \mathbf{u}(k-L) &= \sum_{p=0}^{L} \mathbf{W}_p^T\, \mathbf{g}[\mathbf{y}(k-L+p)] \qquad (9.104) \\
\text{For each} \quad p &= 0, \ldots, L: \\
\widehat{\mathbf{R}}_{\mathbf{f}\,\mathbf{u}}^{(k)}(L+p) &= \mathbf{f}(\mathbf{y}(k-L))\, \mathbf{u}^T(k-L-p) \qquad (9.105) \\
\mathbf{W}_p(k+1) &= \mathbf{W}_p(k) + \eta(k)(\mathbf{W}_p(k) - \widehat{\mathbf{R}}_{\mathbf{f}\,\mathbf{u}}^{(k)}(L+p)) \qquad (9.106)
\end{aligned}$$

The formulas (9.103-9.106) describe the $(l+1)$-th step of the on-line update rule by using samples up to the k-th sample.

9.4.6.3 Block On-line Update Rule Let us assume that data samples arrive continuously and are gathered into blocks: $\{\ldots, [x((l-1)N+1), \ldots, x(lN)], [x(lN+1), \ldots, x((l+1)N)], \ldots\}$. The processing of the data can be done in the time-domain (or by using the DFT in

Table 9.1 Relationships between instantaneous blind source separation and multichannel blind deconvolution for complex-valued signals and parameters.

Blind Source Separation	Multichannel Blind Deconvolution - Time Domain	Multichannel Blind Deconvolution - z-Transform Domain
	Mixing-Unmixing Model	
$\mathbf{x}(k) = \mathbf{H}\,\mathbf{s}(k)$	$\underline{\mathbf{x}}(k) = \underline{\mathbf{H}} * \underline{\mathbf{s}}(k)$	$\mathbf{X}(z) = \mathbf{H}(z)\mathbf{S}(z)$
$\mathbf{y}(k) = \mathbf{W}\,\mathbf{x}(k)$	$\underline{\mathbf{y}}(k) = \underline{\mathbf{W}} * \underline{\mathbf{x}}(k)$	$\mathbf{Y}(z) = \mathbf{W}(z)\mathbf{X}(z)$
$x_i(k) = \sum_{j=1}^{n} h_{ij} s_j(k)$	$\underline{x}_i(k) = \sum_{j=1}^{n} \underline{h}_{ij} * \underline{s}_j(k)$	$X_i(z) = \sum_{j=1}^{n} H_{ij}(z) S_j(z)$
$y_i(k) = \sum_{j=1}^{n} w_{ij}(l) x_j(k)$	$\underline{y}_i(k) = \sum_{j=1}^{n} \underline{w}_{ij}(l) * \underline{x}_j(k)$	$Y_i(z) = \sum_{j=1}^{n} W_{ij}(z) X_j(z)$
	Contrast Functions: $\phi(\mathbf{y}, \mathbf{W})$ or $\phi(\mathbf{y}, \underline{\mathbf{W}})$ or $\phi(\mathbf{W}(z))$	
$-\log \lvert \det(\mathbf{W}) \rvert$	$-\log \lvert \det(\underline{\mathbf{W}}) \rvert$	$-\frac{1}{2\pi j} \oint \log \lvert \det \mathbf{W}(z) \rvert z^{-1} dz$
$-\sum_{i=1}^{n} \log(q_i(y_i))$	$-\sum_{i=1}^{n} \log(q_i(y_i))$	$-\sum_{i=1}^{n} \log(q(y_i))$
	Natural Gradient Rules: $\Delta\mathbf{W}(l)$ or $\Delta\underline{\mathbf{W}}(l)$ or $\Delta\mathbf{W}(z)$	
$-\eta \frac{\partial \phi}{\partial \mathbf{W}} \mathbf{W}^H(l) \mathbf{W}(l)$	$-\eta \frac{\partial \phi}{\partial \underline{\mathbf{W}}} * \underline{\mathbf{W}}^H_{(-p)}(l) * \underline{\mathbf{W}}(l)$	$-\eta \frac{\partial \phi}{\partial \mathbf{W}(z)} \mathbf{W}^H(z^{-1}) \mathbf{W}(z)$
$-\eta \mathbf{W}(l) \left[\frac{\partial \phi}{\partial \mathbf{W}}\right]^H \mathbf{W}(l)$	$-\eta \underline{\mathbf{W}}(l) * \left[\frac{\partial \phi}{\partial \underline{\mathbf{W}}}\right]^H * \underline{\mathbf{W}}(l)$	$-\eta \mathbf{W}(z) \; \frac{\partial \phi}{\partial \mathbf{W}(z)}^{\;H} \mathbf{W}(z)$
	Batch Learning Algorithms: $\Delta\mathbf{W}(l)$ or $\Delta\underline{\mathbf{W}}(l)$ or $\Delta\mathbf{W}(z,l)$	
$\eta\Big[\mathbf{W}(l) - \langle \mathbf{f}[\mathbf{y}(k)]\, \mathbf{u}^H(k) \rangle\Big]$ where $\mathbf{u}(k) = \mathbf{W}^H(l)\, \mathbf{y}(k)$	$\eta\Big[\underline{\mathbf{W}}(l) - \langle \mathbf{f}[\underline{\mathbf{y}}(k)] * \underline{\mathbf{u}}^H(-k) \rangle\Big]$ where $\underline{\mathbf{u}}(k) = \underline{\mathbf{W}}^H_{(-p)}(l) * \underline{\mathbf{y}}(k)$	$\eta\Big[\mathbf{W}(z,l) - \langle Z\{\mathbf{f}[\mathbf{y}(k)]\}\mathbf{U}^H(z^{-1}) \rangle\Big]$ where $\mathbf{U}(z) = \mathbf{W}^H(z^{-1}, l)\mathbf{Y}(z)$
$\eta\Big[\mathbf{W}(l) - \langle \mathbf{y}(k)\, \mathbf{v}^H(k) \rangle\Big]$ where $\mathbf{v}(k) = \mathbf{W}^H(l)\, \mathbf{g}[\mathbf{y}(k)]$	$\eta\Big[\underline{\mathbf{W}}(l) - \langle \underline{\mathbf{y}}(k) * \underline{\mathbf{v}}^H(-k) \rangle\Big]$ where $\underline{\mathbf{v}}(k) = \underline{\mathbf{W}}^H_{(-p)}(l) * \mathbf{g}[\underline{\mathbf{y}}(k)]$	$\eta\Big[\mathbf{W}(z,l) - \langle \mathbf{Y}(z)\mathbf{V}^H(z^{-1}) \rangle\Big]$ where $\mathbf{V}(z) = \mathbf{W}^H(z^{-1}, l)\, Z\{\mathbf{g}[\mathbf{y}(k)]\}$

the frequency-domain) as

$$\begin{aligned}
\text{For each} \quad k &= (l-1)N+1, \ldots, lN : \\
\mathbf{y}(l) = \mathbf{y}(k) &= \sum_{p=0}^{L} \mathbf{W}_p(l)\, \mathbf{x}(k-p) \qquad (9.107) \\
\mathbf{v}(l) = \mathbf{v}(k-L) &= \sum_{p=0}^{L} \mathbf{W}_p^T\, \mathbf{g}[\mathbf{y}(k-L+p)] \qquad (9.108) \\
\text{For each} \quad p &= 0, \ldots, L :
\end{aligned}$$

$$\widehat{\mathbf{R}}_{\mathbf{f}\,\mathbf{u}}^{(l)}(L+p) = \frac{1}{N} \sum_{k=(l-1)N+1}^{lN} \mathbf{f}[\mathbf{y}(k-L)]\, \mathbf{u}^T(k-L-p) \qquad (9.109)$$

$$\mathbf{W}_p(l+1) = \mathbf{W}_p(l) + \eta(l) \left[\mathbf{W}_p(l) - \widehat{\mathbf{R}}_{\mathbf{f}\,\mathbf{u}}^{(l)}(L+p) \right] \qquad (9.110)$$

Summarizing this section, relationships and equivalences between instantaneous blind source separation (BSS) and multichannel blind deconvolution (MBD) are given in terms of algebraic properties. These relationships have been developed by using the notion of the abstract algebra of formal series. The basic input data set consists of a series of time-ordered sampled data which can be transformed into another set, more suitable for a given problem. Algebraic operations are suitably modified to preserve algebraic field properties. The algorithms may be implemented by using on-line data (sampled signals, convolution operator and $\mathcal{Z}$-transform) or pre-processed data (DFT). The practical implementation of the algorithms has been discussed in terms of instantaneous mixing and linear convolution. Using the established analogies and relationships, the learning rules can be automatically generated.

In the next sections, we derive directly and rigorously some improved algorithms based on the natural gradient approach.

9.5 NATURAL GRADIENT ALGORITHMS WITH NONHOLONOMIC CONSTRAINTS

The natural gradient approach described in Chapter 6 can be extended to MBD using an approximation of an infinite impulse response system by a causal finite impulse response system. In practical situations, it is reasonable to assume that there are always more sensors than sources. Throughout this section, we will consider the case where the number of sensors is strictly greater than the number of sources. The overdetermined restriction is also one of the sufficient conditions for the existence of a finite length FIR equalizer [1196].

9.5.1 Equivariant Learning Algorithm for Causal FIR Filters in the Lie Group Sense

In the framework of ICA, it has been shown in Chapter 6 that a nonholonomic constraint [25] improves the convergence and performance of the natural gradient learning algorithms for the overdetermined case or for nonstationary source signals. Since the condition $m > n$ is one of the sufficient conditions for signal-separability for multichannel blind deconvolution with causal FIR filters, we derive the natural gradient learning algorithm [33, 34] by incorporating a nonholonomic constraint into the algorithm.

Let us consider a finite length MIMO FIR equalizer (see Fig.9.4(a)) whose m dimensional output $\mathbf{y}(k)$ is described by

$$\mathbf{y}(k) = \sum_{p=0}^{L} \mathbf{W}_p(l)\,\mathbf{x}(k-p), \tag{9.111}$$

where $\mathbf{W}_p(l)$ is the synaptic weight matrix at the l-th iteration and represents the connection strength between $\mathbf{y}(k)$ and $\mathbf{x}(k-p)$. Note that for on-line learning, the index l can be replaced by the time index k. We define $\mathbf{W}(z,l)$ as

$$\mathbf{W}(z,l) = \sum_{p=0}^{L} \mathbf{W}_p(l) z^{-p}. \tag{9.112}$$

We consider n observations $\{x_i(k)\}$ and n output signals $\{y_i(k)\}$ over an N-point time block. Let us define the following vectors:

$$\begin{aligned} \underline{\mathbf{x}} &= [x_1(1), \dots, x_n(1), \dots, x_1(N), \dots, x_n(N)]^T, \\ \underline{\mathbf{y}} &= [y_1(1), \dots, y_n(1), \dots, y_1(N), \dots, y_n(N)]^T. \end{aligned}$$

The MIMO FIR equalizer should be trained such that the joint probability density of $\underline{\mathbf{y}}$ can be factorized as follows:

$$p(\underline{\mathbf{y}}) = \prod_{i=1}^{n} \prod_{k=1}^{N} r_i(y_i(k)), \tag{9.113}$$

where $\{r_i(\cdot)\}$ are probability densities of the source signals. As a cost function, we choose the Kullback-Leibler divergence which is an asymmetric measure of distance between two different probability distributions. The risk function $\mathcal{R}(\mathbf{W}(z,l))$ is formulated as [1349, 1351, 1358]

$$\begin{aligned} \mathcal{R}(\mathbf{W}(z,l)) &= E\{\rho(\mathbf{W}(z,l))\} \\ &= \frac{1}{N} \int p(\underline{\mathbf{y}}) \log \frac{p(\underline{\mathbf{y}})}{\prod_{i=1}^{n} \prod_{k=1}^{N} q_i(y_i(k))} d\underline{\mathbf{y}}, \end{aligned} \tag{9.114}$$

where we replace $r_i(\cdot)$ by a hypothesized density model for sources, $q_i(\cdot)$, since we do not know the true probability distribution of sources, $r_i(\cdot)$.

To derive the relation between $p(\underline{\mathbf{x}})$ and $p(\underline{\mathbf{y}})$, we write (9.111) in matrix form,

$$\underline{\mathbf{y}} = \boldsymbol{\mathcal{W}}_N \; \underline{\mathbf{x}}, \tag{9.115}$$

where the $\boldsymbol{\mathcal{W}}_N$ is given by

$$\boldsymbol{\mathcal{W}}_N = \begin{bmatrix} \mathbf{W}_0 & \mathbf{0} & \cdots & \mathbf{0} \\ \mathbf{W}_1 & \mathbf{W}_0 & \cdots & \mathbf{0} \\ \vdots & & & \vdots \\ \mathbf{W}_{N-1} & \mathbf{W}_{N-2} & \cdots & \mathbf{W}_0 \end{bmatrix} \tag{9.116}$$

The length of the time delay L in the equalizer is much smaller than N, i.e., $\mathbf{W}_{L+1} = \cdots = \mathbf{W}_{N-1} = \mathbf{0}$. The input-output equation (9.115) written in matrix form leads to the following relation between $p(\underline{\mathbf{x}})$ and $p(\underline{\mathbf{y}})$:

$$p(\underline{\mathbf{y}}) = \frac{p(\underline{\mathbf{x}})}{|\det \mathbf{W}_0^N|}. \tag{9.117}$$

Invoking the relation (9.117), our loss function $\rho(\mathbf{W}(z,l))$ is given by

$$\rho(\mathbf{W}(z,l)) = -\log|\det \mathbf{W}_0| - \sum_{i=1}^{n} \langle \log q_i(y_i(k)) \rangle, \tag{9.118}$$

where $\langle \cdot \rangle$ represents the time-average, i.e.,

$$\langle \log q_i(y_i(k)) \rangle = \frac{1}{N} \sum_{k=1}^{N} \log q_i(y_i(k)). \tag{9.119}$$

Note that $p(\underline{\mathbf{x}})$ was not included in (9.118) because it does not depend on the parameters of the matrix $\{\mathbf{W}_p(l)\}$.

Here, we follow the derivation of the algorithm that was presented by Amari *et al.* [33, 34]. To determine a learning algorithm which minimizes the loss function (9.118), we calculate an infinitesimal increment

$$d\rho((\mathbf{W}(z,l)) = \rho(\mathbf{W}(z,l) + d\mathbf{W}(z,l)) - \rho(\mathbf{W}(z,l)), \tag{9.120}$$

corresponding to an increment $d\mathbf{W}(z,l)$. Simple algebra and differential calculus yields

$$d\rho((\mathbf{W}(z,l)) = \left\langle \mathbf{f}^T(\mathbf{y}(k))\, d\mathbf{X}(z,l)\, \mathbf{y}(k) \right\rangle - \operatorname{tr}\{d\mathbf{X}_0(l)\}, \tag{9.121}$$

where $\mathbf{dX}_0(l) = d\mathbf{W}_0(l)\mathbf{W}_0^{-1}(l)$ and $\mathbf{f}(\mathbf{y}(k))$ is a column vector whose components are defined as

$$f_i(y_i(k)) = -\frac{d \log q_i(y_i(k))}{dy_i(k)}. \tag{9.122}$$

Let us introduce the following notation

$$d\mathbf{X}(z,l) = d\mathbf{W}(z,l) \circledast \mathbf{W}^{\dagger}(z) = \left[d\mathbf{W}(z,l)\mathbf{W}^{-1}(z,l) \right]_L, \tag{9.123}$$

where $[\mathbf{W}(z)]_L$ denotes a truncation operator that omits or sets to zero all the terms with a length greater than L in the polynomial matrix $\mathbf{W}(z)$ (see Appendix A). We can rewrite $d\rho(\mathbf{W}(z,l))$ in the following form

$$d\rho(\mathbf{W}(z,l)) = -\operatorname{tr}(d\mathbf{X}_0) + \langle \mathbf{f}(\mathbf{y})^T \rangle \, d\mathbf{W}(z)\mathbf{W}^{-1}(z)\, \mathbf{y}. \tag{9.124}$$

From this equation, we obtain the partial derivatives of $\rho(\mathbf{y}, \mathbf{W}(z,l))$ with respect to $d\mathbf{X}(z)$

$$\frac{\partial \rho(\mathbf{W}(z,l))}{\partial \mathbf{X}_p} = -\delta_{0,p}\, \mathbf{I} + \langle \mathbf{f}(\mathbf{y})\, \mathbf{y}^T(k-p) \rangle, \qquad (p = 0, 1, \ldots, L). \tag{9.125}$$

Using the natural gradient descent learning rule, Zhang *et al.* derived the simple natural gradient learning rule as follows [1358, 1361, 1362]

$$\begin{aligned} \Delta \mathbf{W}_p &= -\eta \sum_{q=0}^{p} \frac{\partial \rho(\mathbf{W}(z))}{\partial \mathbf{X}_q} \mathbf{W}_{p-q} \\ &= \eta \sum_{q=0}^{p} \left(\delta_{0q} \mathbf{I} - \langle \mathbf{f}(\mathbf{y}(k))\, \mathbf{y}^T(k-q) \rangle \right) \mathbf{W}_{p-q}, \end{aligned} \tag{9.126}$$

for $p = 0, 1, \ldots, L$, where η is a learning rate. In particular, the learning algorithm for $p = 0, 1$ is described by

$$\Delta \mathbf{W}_0 = \eta \left(\mathbf{I} - \langle \mathbf{f}(\mathbf{y}(k))\, \mathbf{y}^T(k) \rangle \right) \mathbf{W}_0, \tag{9.127}$$

$$\Delta \mathbf{W}_1 = \eta [(\mathbf{I} - \langle \mathbf{f}(\mathbf{y}(k))\, \mathbf{y}^T(k) \rangle) \mathbf{W}_1 - \langle \mathbf{f}(\mathbf{y}(k))\, \mathbf{y}^T(k-1) \rangle \, \mathbf{W}_0]. \tag{9.128}$$

The stationary points of (9.126) satisfy

$$E\{\langle f_i(y_i(k))\, y_i(k) \rangle\} = 1. \tag{9.129}$$

In other words, the learning algorithm (9.126) forces $\{y_i(k)\}$ to have constant magnitude. Such constraints for the output signals are undesirable and they lead to the case that the extracted signals have nearly flat frequency spectra [103]. Moreover, this may cause a problem for $m > n$ if we do not know the number of source signals. To avoid this drawback, we follow the proposal of the nonholonomic constraint that was applied to ICA [25]. We replace the identity matrix by a diagonal matrix $\mathbf{\Lambda}^{(k)}$ which is given by

$$\mathbf{\Lambda}^{(k)} = \operatorname{diag}\{f_1(y_1(k))\, y_1(k), f_2(y_2(k))\, y_2(k), \ldots, f_n(y_n(k))\, y_n(k)\}. \tag{9.130}$$

Then, the natural gradient algorithm with nonholonomic constraints takes the form:

$$\boxed{\Delta \mathbf{W}_p(l) = \eta(l) \sum_{q=0}^{p} \left(\left\langle \mathbf{\Lambda}^{(k)} \right\rangle \delta_{0q} - \langle \mathbf{f}(\mathbf{y}(k))\, \mathbf{y}^T(k-q) \rangle \right) \mathbf{W}_{p-q}(l).} \tag{9.131}$$

Applying the moving average approach, we can easily transform it to an on-line version:

$$\boxed{\mathbf{W}_p(k+1) = \mathbf{W}_p(k) + \eta(k) \sum_{q=0}^{p} \left(\mathbf{\Lambda}^{(k)}\, \delta_{0q} - \mathbf{R}_{\mathbf{f}\,\mathbf{y}}^{(k)}(q) \right) \mathbf{W}_{p-q}(k),} \tag{9.132}$$

where

$$\boxed{\mathbf{R}_{\mathbf{f}\,\mathbf{y}}^{(k)}(q) = (1-\eta_0)\, \mathbf{R}_{\mathbf{f}\,\mathbf{y}}^{(k-1)}(q) + \eta_0\, \mathbf{f}(\mathbf{y}(k))\, \mathbf{y}^T(k-q), \qquad (q = 0, 1, \ldots, L)} \tag{9.133}$$

and

$$\boxed{\mathbf{\Lambda}^{(k)} = (1-\eta_0)\, \mathbf{\Lambda}^{(k-1)} + \eta_0\, \mathrm{diag}\{\mathbf{f}(\mathbf{y}(k))\, \mathbf{y}^T(k)\}.} \tag{9.134}$$

It should be noted that the algorithm looks similar, but in fact it is not identical to the algorithms presented in the previous sections. The essential difference is that the update rule for $\Delta \mathbf{W}_p$ in this section depends only on $\mathbf{W}_q$, in the range $q = 0, 1, \ldots, p$, while in other algorithms it depends on all parameters $\mathbf{W}_q$, in the range $q = 0, 1, \ldots, L$. The algorithm (9.126) has two important properties, the uniform performance (the equivariant property [148]) and the nonsingularity of $\mathbf{W}_0$. In the multichannel deconvolution problem, an algorithm is equivariant if its dynamical behavior depends on the global transfer function $\mathbf{G}(z)$ but not on the specific mixing filter $\mathbf{H}(z)$. In fact, the learning algorithm (9.126) satisfies the equivariant property in the Lie group sense (see Appendix A).

It should be noted that the learning algorithm (9.132) keeps the filter $\mathbf{W}(z)$ on the manifold $\mathcal{M}(L)$ if the initial filter is on the manifold. The equilibrium points of the learning algorithm satisfy the following equations

$$E\{\mathbf{f}(\mathbf{y}(k))\, \mathbf{y}^T(k-p)\} = \mathbf{0}, \qquad (p = 1, 2, \ldots, L), \tag{9.135}$$

$$E\{\mathbf{f}(\mathbf{y}(k))\, \mathbf{y}^T(k)\} = \mathbf{\Lambda}, \tag{9.136}$$

where $\mathbf{\Lambda}$ is a diagonal positive definite matrix [5] which is automatically adjusted during the learning process. The nonholonomic constraints imply no particular temporal structure on the estimated output signals.

Example 9.1 The geometrical interpretation of the Lie group's inverse operation is illustrated in Fig.9.6, where $\mathbf{H}(z)$ is a two channel filter of length $L = 50$, $\mathbf{W}(z) = \mathbf{H}^{\dagger}(z)$ is the Lie group inverse filter of length 50 and the composite transfer function $\mathbf{G}(z) = \sum_{p=0}^{2L} \mathbf{G}_p z^{-p} = \mathbf{W}(z)\, \mathbf{H}(z)$ is a filter of length $2L$. In this figure, the label $H(z)_{1,1}$ denotes the sub-channel transfer function $H_{11}(z) = \sum_{p=0}^{L} h_{11\,p} z^{-p}$, where the horizontal axis indicates time delays ($p = 0, 1, \ldots, L$), and the vertical axis indicates the magnitudes $h_{11\,p}$. From this illustration, we see that the composite transfer function $\mathbf{G}(z)$ is not the exact identity matrix because there still exist small fluctuations in entries of the matrix $\mathbf{G}_p$, for $p > L$. The fluctuations can be negligibly small if we make the length L of $\mathbf{W}(z)$ sufficiently large. However, considering the multiplication in the Lie group sense, we have $\mathbf{G}(z) = \mathbf{W}(z) \circledast \mathbf{H}(z) = \mathbf{I}$.

[5] For more sensors than sources ($m > n$), the matrix $\mathbf{\Lambda}$ is semi-positive definite.

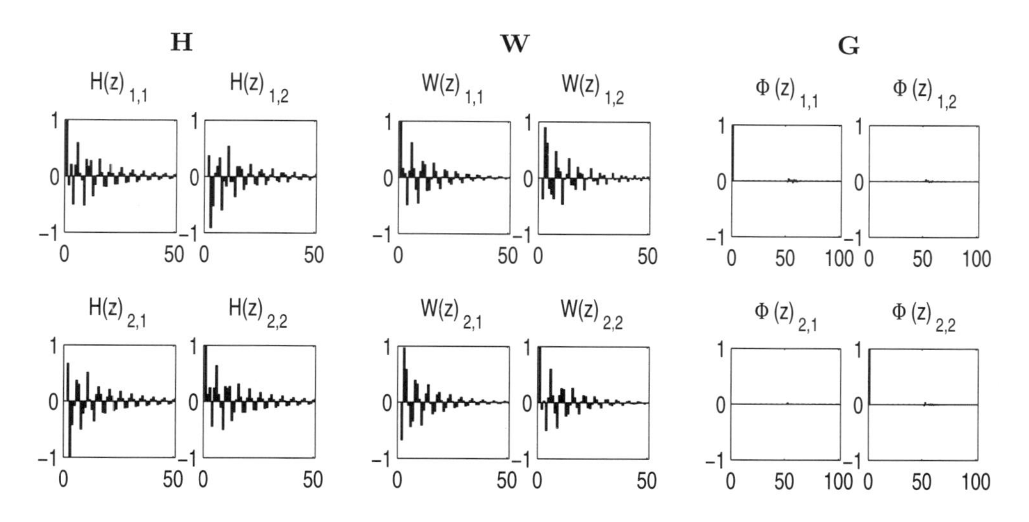

Fig. 9.6 Illustration of the Lie group's inverse of an FIR filter, where $\mathbf{H}(z)$ is an FIR filter of length $L = 50$, $\mathbf{W}(z)$ is the Lie group's inverse of $\mathbf{H}(z)$, and $\mathbf{G}(z) = \mathbf{W}(z)\mathbf{H}(z)$ is the composite transfer function.

9.5.2 Natural Gradient Algorithm for a Fully Recurrent Network

Let us consider a linear feedback network with FIR weights (see Fig.9.4 (c)). The network output $\mathbf{y}(k)$ is given by

$$\begin{aligned} \mathbf{y}(k) &= \mathbf{x}(k) - \sum_{p=0}^{L} \widehat{\mathbf{W}}_p(l)\, \mathbf{y}(k-p) \\ &= [\mathbf{I} + \widehat{\mathbf{W}}(z,l)]^{-1} \mathbf{x}(k), \end{aligned} \tag{9.137}$$

where $\widehat{\mathbf{W}}(z,l) = \sum_{p=0}^{L} \widehat{\mathbf{W}}_p(l) z^{-p}$. We write (9.137) in matrix form as [206, 225]

$$\underline{\mathbf{x}} = \widehat{\mathcal{W}}_N\, \underline{\mathbf{y}}, \tag{9.138}$$

where $\widehat{\mathcal{W}}$ is the corresponding $(Lm) \times (Lm)$ matrix, defined as

$$\widehat{\mathcal{W}}_N = \begin{bmatrix} \mathbf{I} + \widehat{\mathbf{W}}_0 & \mathbf{0} & \cdots & \mathbf{0} \\ \widehat{\mathbf{W}}_1 & \mathbf{I} + \widehat{\mathbf{W}}_0 & \cdots & \mathbf{0} \\ \vdots & & & \vdots \\ \widehat{\mathbf{W}}_L & \widehat{\mathbf{W}}_{L-1} & \cdots & \mathbf{I} + \widehat{\mathbf{W}}_0 \end{bmatrix}. \tag{9.139}$$

Then, to minimize the spatio-temporal statistical dependence, our loss function is described as

$$\rho(\widehat{\mathbf{W}}(z,l)) = -\sum_{i=1}^{m} \langle \log q_i(y_i(k)) \rangle - \log|\det(\mathbf{I} + \widehat{\mathbf{W}}_0(l))^{-1}|. \tag{9.140}$$

Simple calculation yields

$$\widehat{\mathbf{W}}(z,l)) = \left\langle \mathbf{f}^T(\mathbf{y}(k))\, d\widehat{\mathbf{X}}(z,l)\, \mathbf{y}(k) \right\rangle - \mathrm{tr}\{d\widehat{\mathbf{X}}_0(l)\}, \tag{9.141}$$

where $d\widehat{\mathbf{X}}(z,l)$ is defined as

$$d\widehat{\mathbf{X}}(z,l) = [\mathbf{I} + \widehat{\mathbf{W}}(z,l)]^{-1} d\widehat{\mathbf{W}}(z,l). \tag{9.142}$$

The learning algorithm that minimizes (9.140) is therefore given by, in terms of $d\widehat{\mathbf{X}}(z,l)$

$$\Delta\widehat{\mathbf{X}}_p(l) = -\eta(l)\frac{d\rho(\widehat{\mathbf{W}}(z,l))}{d\widehat{\mathbf{X}}_p(l)} = -\eta(l)\{\mathbf{I}\delta_{p0} - \langle \mathbf{f}(\mathbf{y}(k))\, \mathbf{y}^T(k-p)\rangle\}. \tag{9.143}$$

We replace the identity matrix in (9.143) by the diagonal matrix $\mathbf{\Lambda}^{(k)}$ defined in (9.130). Taking into account that

$$\Delta\widehat{\mathbf{W}}_p(l) = \Delta\widehat{\mathbf{X}}_p(l) + \sum_{q=0}^{L} \widehat{\mathbf{W}}_q(l)\Delta\widehat{\mathbf{X}}_{p-q}(l) \tag{9.144}$$

the learning algorithm in terms of $d\widehat{\mathbf{W}}(z,l)$ is given by

$$\Delta\widehat{\mathbf{W}}_p(l) = -\eta(l)\left[\left\langle \mathbf{\Lambda}^{(k)} \right\rangle \delta_{p0} - \langle \mathbf{f}(\mathbf{y}(k))\, \mathbf{y}^T(k-p)\rangle - \widehat{\mathbf{W}}_p(l)\left\langle \mathbf{\Lambda}^{(k)} \right\rangle + \left\langle \widehat{\mathbf{Y}}_p(k) \right\rangle\right], \tag{9.145}$$

where $\widehat{\mathbf{Y}}_p(k)$ is defined as

$$\widehat{\mathbf{Y}}_p(k) = \sum_{q=0}^{L} \widehat{\mathbf{W}}_q(k)\, \mathbf{f}(\mathbf{y}(k))\, \mathbf{y}^T(k-p+q). \tag{9.146}$$

For on-line adaptation, we have [206]

$$\boxed{\Delta\widehat{\mathbf{W}}_p(k) = -\eta(k)\left\{\mathbf{\Lambda}^{(k)}\,\delta_{p0} - \mathbf{f}(\mathbf{y}(k))\,\mathbf{y}^T(k-p) - \widehat{\mathbf{W}}_p(k)\,\mathbf{\Lambda}^{(k)} + \widehat{\mathbf{Y}}_p(k-L)\right\},} \tag{9.147}$$

where

$$\boxed{\widehat{\mathbf{Y}}_p(k-L) = \sum_{q=0}^{L} \widehat{\mathbf{W}}_{L-q}(k)\, \mathbf{f}(\mathbf{y}(k-L+p))\, \mathbf{y}^T(k-q).} \tag{9.148}$$

9.6 BLIND DECONVOLUTION OF A NON-MINIMUM PHASE SYSTEM USING FILTER DECOMPOSITION APPROACH

In this section, we present a filter decomposition approach for multichannel blind deconvolution of non-minimum phase systems proposed by Zhang, Amari and Cichocki [1351, 1358].

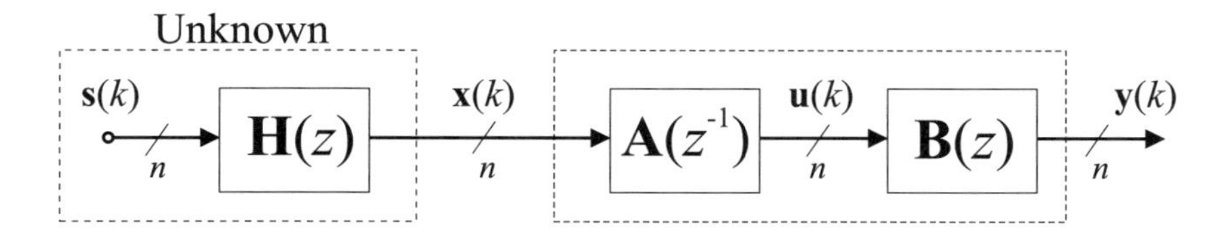

Fig. 9.7 Cascade of two FIR filters (non-causal and causal) for blind deconvolution of non-minimum phase system.

Let us denote the set $\mathcal{M}(L, K)$ of filters (9.59) as

$$\mathcal{M}(L, K) = \left\{ \mathbf{W}(z) | \mathbf{W}(z) = \sum_{p=-K}^{L} \mathbf{W}_p z^{-p} \right\}. \tag{9.149}$$

For simplicity in the following discussion, we assume that $K = L$. In general, the multiplication of two transfer functions belonging to $\mathcal{M}(L, L)$ of filters connected in a cascade will increase the order of the global transfer function, but this does not belong to $\mathcal{M}(L, L)$. This makes it difficult to introduce the natural gradient on the manifold of doubly-finite multichannel non-causal filters. In order to explore the geometric structure of $\mathcal{M}(L, L)$ and develop an efficient learning algorithm for $\mathbf{W}(z)$, we present the following filter decomposition approach and present the basic algebraic operations of filters in the Lie group framework (see Appendix A for theoretical background). It is plausible to decompose a doubly-finite non-causal filter into a cascade form of a causal FIR filter and a non-causal FIR filter, since then it becomes easy to study the invertibility of the filter and to develop efficient learning algorithms.

Let us decompose a non-causal filter $\mathbf{W}(z)$ in $\mathcal{M}(L, L)$ into a cascade form of two FIR filters (see Fig.9.7) as

$$\mathbf{W}(z) = \mathbf{B}(z)\, \mathbf{A}(z^{-1}), \tag{9.150}$$

where $\mathbf{A}(z^{-1}) = \sum_{p=0}^{L} \mathbf{A}_p z^p$, with the constraint $\mathbf{A}_0 = \mathbf{I}$, and $\mathbf{B}(z) = \sum_{p=0}^{L} \mathbf{B}_p z^{-p}$, and both are one-sided finite multichannel FIR filters. The first one is a forward filter (with non-causal components) and the second one is the standard FIR filter (with causal components). The relation between the coefficients of three filters is as follows

$$\mathbf{W}_p = \sum_{r-q=p,\ \ 0 \leq r,q \leq L} \mathbf{B}_r\, \mathbf{A}_q, \qquad (p = -L, \ldots, L). \tag{9.151}$$

By applying this decomposition, it becomes possible to achieve invertibility of double-finite multichannel non-causal filters in the Lie group sense [1362]. An important theoretical question arises: What conditions guarantee that a non-causal filter $\mathbf{W}(z)$ in $\mathcal{M}(L, L)$ can be decomposed as defined by (9.150)? In other words, the problem can be formulated as follows: Given a double-finite filter $\mathbf{W}(z)$, it is desired to find an FIR filter $\mathbf{A}(z^{-1})$ such that the multiplication of two filters in the Lie Group sense satisfies

$$\left[\mathbf{W}(z)\, \mathbf{C}(z^{-1})\right]_L = \mathbf{W}(z) \circledast \mathbf{C}(z^{-1}) = \mathbf{B}(z), \tag{9.152}$$

where matrix $\mathbf{C}(z^{-1}) = \sum_{p=0}^{L} \mathbf{C}_p z^p = \mathbf{A}^{\dagger}(z^{-1})$, with $\mathbf{C}_0 = \mathbf{I}$, $\mathbf{C}_p = \sum_{q=1}^{p} \mathbf{C}_{p-q}^{p} \mathbf{A}_q$ is the generalized inverse of $\mathbf{A}(z^{-1})$ in the Lie Group sense and $[\mathbf{W}(z)]_L$ denotes a truncated operator by which all terms with a length greater than L in the polynomial matrix $\mathbf{W}(z)$ are omitted.

Unlike doubly-infinite filters, the doubly-finite filters do not have the self-closed multiplication and inverse operations in the manifold of fixed length filters. In general, the multiplication of two filters with a given length makes a filter with an extended length, so does the inverse operation [1349, 1362] (see Appendix A).

For the estimation of the parameters of the causal filter, we can use the on-line learning algorithm derived in the Section 9.5

$$\boxed{\Delta \mathbf{B}_p = \eta \sum_{q=0}^{p} \left(\delta_{0q} \mathbf{I} - \mathbf{f}(\mathbf{y}(k))\, \mathbf{y}^T(k-q)\right) \mathbf{B}_{p-q}} \tag{9.153}$$

9.6.1 Information Back-propagation

The capability of achieving learning in a cascade of two adaptive filters is of fundamental importance to the blind deconvolution of non-minimum phase systems. To explore this fact, we begin with the filter decomposition and show how to estimate parameters (weights) of the non-causal filter by using the mutual information back-propagation approach [1362].

According to the filter decomposition (see Fig.9.7), we denote

$$\mathbf{u}(k) = [\mathbf{A}(z^{-1})]\, \mathbf{x}(k), \tag{9.154}$$

$$\mathbf{y}(k) = [\mathbf{B}(z)]\, \mathbf{u}(k). \tag{9.155}$$

If we consider $\mathbf{u}(k)$ as the observed signals, we can apply the natural gradient learning rule (9.126) to update parameters in the filter $\mathbf{B}(z)$. In order to develop an efficient learning algorithm for the non-causal filter $\mathbf{A}(z^{-1})$, we use the information back-propagation technique. The back-propagation rule is described as follows

$$\begin{aligned} \frac{\partial \rho(\mathbf{W}(z))}{\partial \mathbf{u}(k)} &= \sum_{q=0}^{L} \sum_{i=1}^{n} \frac{\partial \rho(\mathbf{W}(z))}{\partial y_i(k+q)} \frac{\partial y_i(k+q)}{\partial \mathbf{u}(k)} \\ &= \sum_{q=0}^{L} \mathbf{B}_q^T \mathbf{f}(\mathbf{y}(k+q)) = [\mathbf{B}^H(z)]\, \mathbf{f}(\mathbf{y}(k)), \end{aligned} \tag{9.156}$$

where $\mathbf{B}^H(z) = \sum_{q=0}^{L} \mathbf{B}_q^T z^q$ is the conjugate operator of $\mathbf{B}(z)$, and

$$\begin{aligned} \frac{\partial \rho(\mathbf{W}(z))}{\partial \mathbf{A}_p} &= \sum_{i=1}^{n} \frac{\partial \rho(\mathbf{W}(z))}{\partial u_i(k)} \frac{\partial u_i(k)}{\partial \mathbf{A}_p}, \\ &= \frac{\partial \rho(\mathbf{W}(z))}{\partial \mathbf{u}(k)} \mathbf{x}^T(k+p), \end{aligned} \tag{9.157}$$

for $p = 1, 2, \ldots, L$. The structure of the information back-propagation process is illustrated in Figure 9.8.

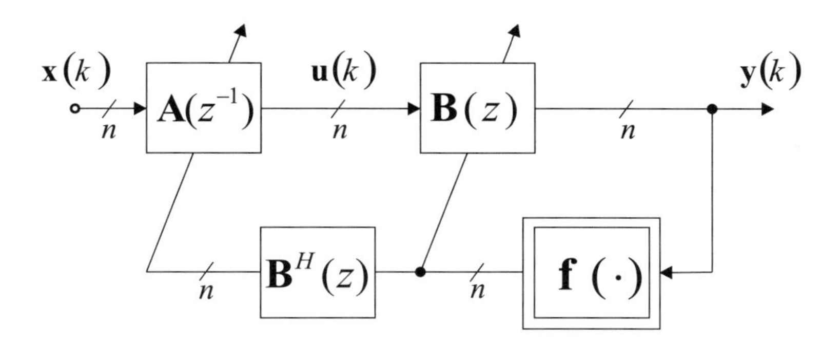

Fig. 9.8 Illustration of the information back-propagation learning.

In this structure, the blind statistical error $\frac{\partial \rho(\mathbf{W}(z))}{\partial \mathbf{y}(k)}$ is backward propagated through channel $\mathbf{B}^H(z)$ to form the blind error $\frac{\partial \rho(\mathbf{W}(z))}{\partial \mathbf{u}(k)}$, which is used to estimate the parameters of the non-causal filter $\mathbf{A}(z^{-1})$. There are several schemes for learning the parameters of $\mathbf{A}(z^{-1})$. One typical scheme is the ordinary gradient descent learning algorithm, which is described as follows

$$\Delta \mathbf{A}_p = -\eta \left(\frac{\partial \rho(\mathbf{W}(z))}{\partial \mathbf{u}(k))} \mathbf{x}^T(k+p) \right). \tag{9.158}$$

Using the natural gradient approach and information back-propagation, we can derive the following learning rule [1349, 1362]

$$\Delta \mathbf{A}_p = -\eta \sum_{q=0}^{p} (\sum_{r=0}^{L} \mathbf{B}_r^T \mathbf{f}(\mathbf{y}(k)) \mathbf{u}^T(k-r+q)) \mathbf{A}_{p-q}. \tag{9.159}$$

9.6.2 Batch Natural Gradient Learning Algorithm

In this section, we consider the efficient implementation of the learning algorithms for blind deconvolution of non-minimum phase systems. We can employ directly the learning algorithm (9.153) to estimate the parameters of the causal filter $\mathbf{B}(z)$. However, using the batch approach, we can improve the convergence speed. Stacking $(L+1)$ samples of the vectors $\mathbf{x}(k)$ and $\mathbf{u}(k)$

$$\begin{aligned} \mathcal{X}(k) &= [\mathbf{x}^T(k), \mathbf{x}^T(k+1), \ldots, \mathbf{x}^T(k+L)]^T, &(9.160)\\ \mathcal{U}(k) &= [\mathbf{u}^T(k), \mathbf{u}^T(k-1), \ldots, \mathbf{u}^T(k-L)]^T, &(9.161)\end{aligned}$$

and using the notation

$$\overline{\mathbf{A}} = [\mathbf{A}_0, \mathbf{A}_1, \ldots, \mathbf{A}_L], \qquad \overline{\mathbf{B}} = [\mathbf{B}_0, \mathbf{B}_1, \ldots, \mathbf{B}_L], \tag{9.162}$$

we can describe the demixing system shown in Fig. 9.7 in the following matrix form

$$\mathbf{u}(k) = \sum_{p=0}^{L} \mathbf{A}_p \mathbf{x}(k+p) = \overline{\mathbf{A}}\, \boldsymbol{\mathcal{X}}(k), \tag{9.163}$$

$$\mathbf{y}(k) = \sum_{p=1}^{L} \mathbf{B}_p \mathbf{u}(k-p) = \overline{\mathbf{B}} \boldsymbol{\mathcal{U}}(k). \tag{9.164}$$

The natural gradient algorithm (9.153) can be reformulated in the batch matrix form as [1349, 1362]

$$\Delta \overline{\mathbf{B}}(l+1) = \eta\, \boldsymbol{\mathcal{F}}(l)\, \overline{\mathbf{B}}(l), \tag{9.165}$$

where the matrix $\boldsymbol{\mathcal{F}}(l)$ is defined as

$$\boldsymbol{\mathcal{F}}(l) = \begin{bmatrix} \boldsymbol{F}_0 & \mathbf{0} & \mathbf{0} & \cdots & \mathbf{0} \\ \boldsymbol{F}_1 & \boldsymbol{F}_0 & \mathbf{0} & \cdots & \mathbf{0} \\ \boldsymbol{F}_2 & \boldsymbol{F}_1 & \boldsymbol{F}_0 & \cdots & \mathbf{0} \\ \vdots & \vdots & \vdots & \ddots & \vdots \\ \boldsymbol{F}_L & \boldsymbol{F}_{L-1} & \boldsymbol{F}_{L-2} & \cdots & \boldsymbol{F}_0 \end{bmatrix}. \tag{9.166}$$

and the sub-matrices $\boldsymbol{F}_p$ are estimated by

$$\boldsymbol{F}_p(l) = \delta_{0p}\, \mathbf{I} - \left\langle \mathbf{f}(\mathbf{y}(k))\, \mathbf{y}^T(k-p) \right\rangle, \qquad (p = 0, 1, \ldots, L). \tag{9.167}$$

Similarly, we can formulate the batch natural gradient algorithm for the non causal filter parameters as [1349, 1362]

$$\Delta \overline{\mathbf{A}}(l+1) = \eta\, \boldsymbol{\mathcal{L}}(l)\, \overline{\mathbf{A}}(l), \tag{9.168}$$

where the matrix $\boldsymbol{\mathcal{C}}(l)$ is defined as

$$\boldsymbol{\mathcal{C}}(l) = \begin{bmatrix} \mathbf{0} & \mathbf{0} & \mathbf{0} & \cdots & \mathbf{0} \\ \boldsymbol{C}_1 & \mathbf{0} & \mathbf{0} & \cdots & \mathbf{0} \\ \boldsymbol{C}_2 & \boldsymbol{C}_1 & \mathbf{0} & \cdots & \mathbf{0} \\ \vdots & \vdots & \vdots & \ddots & \vdots \\ \boldsymbol{C}_l & \boldsymbol{C}_{L-1} & \boldsymbol{C}_{L-2} & \cdots & \mathbf{0} \end{bmatrix}. \tag{9.169}$$

with the sub-matrices $\boldsymbol{C}_p$ evaluated as

$$\boldsymbol{C}_p(l) = -\sum_{q=0}^{L} \left\langle \mathbf{B}_q^T(l)\, \mathbf{f}(\mathbf{y}(k))\, \mathbf{u}^T(k-q+p) \right\rangle, \qquad (p = 1, 2, \ldots, L) \tag{9.170}$$

and $\mathbf{A}_0 = \mathbf{I}$.
Computer simulations show that the natural gradient algorithm has much better convergence properties and performance than the ordinary gradient algorithm (9.158).

9.7 COMPUTER SIMULATION EXPERIMENTS

In this section, we present some illustrative examples of computer simulations to demonstrate the validity and effectiveness of the natural gradient algorithms for SIMO and MIMO blind deconvolution/equalization problems.

Example 9.2 Let us consider the simple SIMO equalization model of Fig.9.2 (a) with $m = 2$ and non-minimum phase transfer functions of order 7 which are assumed to be completely unknown (see Fig.9.9 (a)). Applying the NG algorithm (9.28) with nonlinearities $g(y) = \text{sign}(y)\, y^2$, after 1000 iterations, we obtain the filter coefficients illustrated in Fig.9.9 (a). It is clearly seen that the global system achieves equalization, although each single channel is not able to do this (see Fig.9.9 (b) and (c)).

9.7.1 The Natural Gradient Algorithm vs. the Ordinary Gradient Algorithm

Several computer simulations have been performed to compare the performance of the natural (9.132) with the ordinary gradient algorithm for the multichannel blind deconvolution. To evaluate the performance of the proposed learning algorithms, we employ the criterion of multichannel inter-symbol interference (M_{ISI}), defined as [611]

$$M_{ISI} = \sum_{i=1}^{n} \frac{|\sum_j \sum_p |g_{ij\,p}| - \max_{p,j} |g_{ij\,p}|}{\max_{p,j} |g_{ij\,p}|} + \sum_{j=1}^{n} \frac{|\sum_i \sum_p |g_{ij\,p}| - \max_{p,i} |g_{ij\,p}|}{\max_{p,i} |g_{ij\,p}|} \tag{9.171}$$

It is straightforward to show that $M_{ISI} = 0$ if and only if $\mathbf{G}(z) = \mathbf{W}(z)\,\mathbf{H}(z)$ is of the form $G(z) = \mathbf{P}\mathbf{D}_0(z)$ where $\mathbf{P}$ is any permutation matrix and $\mathbf{D}_0(z) = \text{diag}\{z^{-\Delta_1}, z^{-\Delta_2}, \ldots, z^{-\Delta_n}\}$. We evaluate the performance of the algorithms by using the ensemble average approach, where in each trial a time sequence of M_{ISI} is obtained in order to compute the average ISI performance.

The mixing model used for computer simulations is the multichannel ARMA model

$$\mathbf{x}(k) = -\sum_{p=1}^{M} \widehat{\mathbf{H}}_p\, \mathbf{x}(k-p) + \sum_{p=0}^{L} \mathbf{H}_p\, \mathbf{s}(k-p) + \boldsymbol{\nu}(k) \tag{9.172}$$

where $\mathbf{x}, \mathbf{s}$ and $\boldsymbol{\nu} \in \mathbb{R}^3$.

Example 9.3 For simplicity, we assume here that $m = n = 3$, and thus the matrices $\widehat{\mathbf{H}}_i \in \mathbb{R}^{3\times 3}$ and $\mathbf{H}_i \in \mathbb{R}^{3\times 3}$ are randomly chosen such that the mixing system is stable and of minimum phase. The source signals $\mathbf{s}$ are randomly generated as i.i.d. signals, uniformly distributed in the range (-1,1), and $\boldsymbol{\nu}$ are the Gaussian noises with zero-mean and a covariance matrix of $0.1\mathbf{I}$. The nonlinear activation function is chosen to be $f_i(y_i) = y_i^3$ for any i. We employ an AR model of order $L = 20$ as a mixing system, which can be exactly inverted by an FIR filter. A large number of simulations show that the natural

(a)

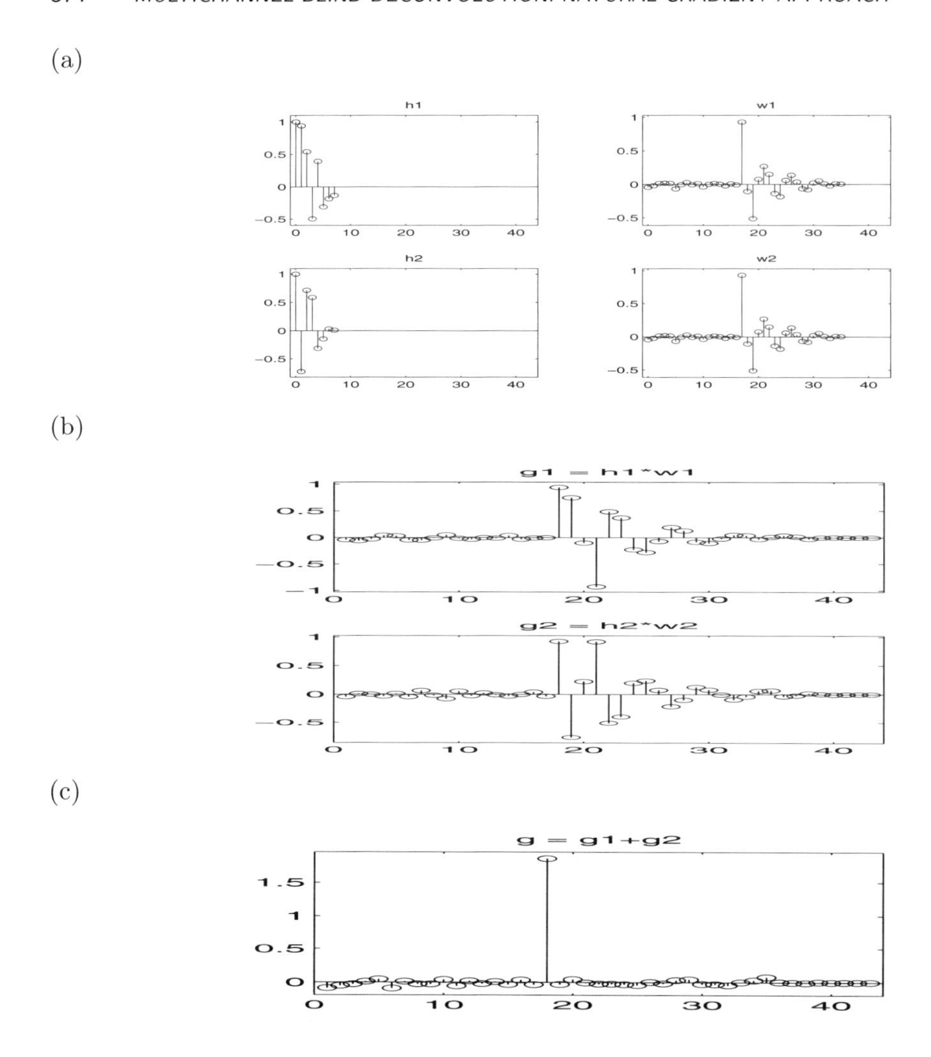

Fig. 9.9 Simulation results of two channel blind deconvolution for the SIMO system in Example 9.2: (a) Parameters of mixing filters $(H_1(z), H_2(z))$ and estimated parameters of adaptive deconvoluting filters $(W_1(z), W_2(z))$, (b) coefficients of global sub-channels $(G_1(z) = W_1(z)H_1(z), G_2(z) = W_2(z)H_2(z))$, (c) parameters of global system $(G(z) = G_1(z) + G_2(z))$.

gradient learning algorithm can easily and relatively quickly recover source signals in the sense of $\mathbf{W}(z)\mathbf{H}(z) = \mathbf{P}\,\mathbf{D}_0(z)$. Fig.9.10 illustrates typical (100 trial ensemble average) M_{ISI} performances of the natural and the ordinary gradient learning algorithms. It is observed that the natural gradient algorithm usually needs less than 3000 iterations to obtain satisfactory results, while the ordinary gradient algorithm needs more than 20000 iterations for the same problem. Fig.9.11 illustrates the distribution of the coefficients of

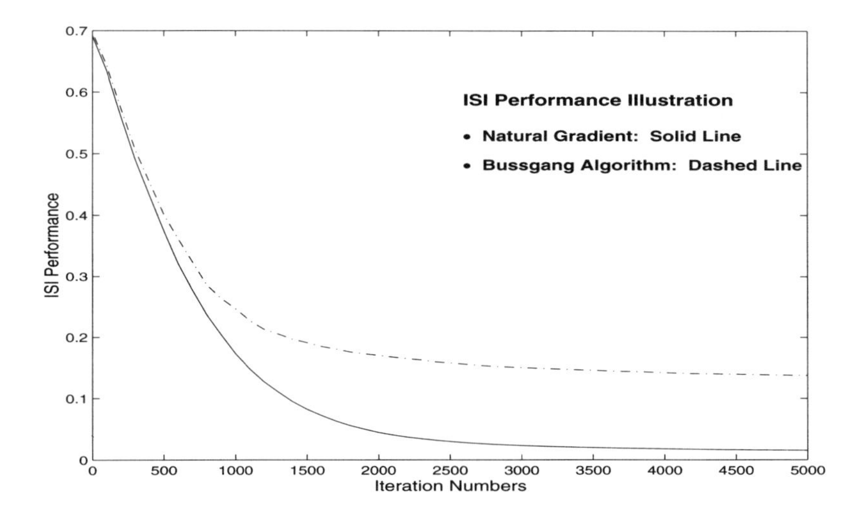

Fig. 9.10 Typical performance index M_{ISI} of the natural gradient algorithm for multichannel blind deconvolution in comparison with the standard gradient algorithm [1362].

the global transfer function $\mathbf{G}(z) = \mathbf{W}(z)\,\mathbf{H}(z)$ at the initial state (Fig.9.11 (a)) and after 3000 iterations (Fig.9.11 (b)) respectively, where the (i,j)th sub-figure plots the coefficients of the transfer function $G_{ij}(z) = \sum_{p=0}^{\infty} g_{ijp} z^{-p}$ up to the length of 80. The corresponding M_{ISI} performance is similar to the one given in the previous example.

9.7.2 Information Back-propagation Example

In the previous example, we assumed a minimum phase mixing system (i.e., all zeros and poles of the transfer functions are located inside the unit circle). However, in practice many systems are non-minimum phase dynamical systems whose transfer functions have zeros outside and inside the unit circle. In such a case, we can use a cascade of two FIR filters (see Fig.9.12). In this example sensor signals are produced by the multichannel ARMA model (9.172), of which the matrices are chosen such that the mixing system is stable and non-minimum phase.

It is straightforward to prove that the system is stable and non-minimum phase. The location of zeros and poles of the mixing system are plotted in Fig. 9.12. In order to estimate source signals, the learning algorithms (9.132) and (9.168) have been employed to estimate parameters of the demixing model. Fig.9.13 (a) and (b) illustrate the distribution of coefficients of the global transfer function $\mathbf{G}(z) = \mathbf{W}(z)\,\mathbf{H}(z)$ at the initial state and after 3000 iterations respectively, where the (i,j)th sub-figure plots the coefficients of the transfer function $G_{ij}(z) = \sum_{p=0}^{\infty} g_{ijp} z^{-p}$ up to the length of 80.

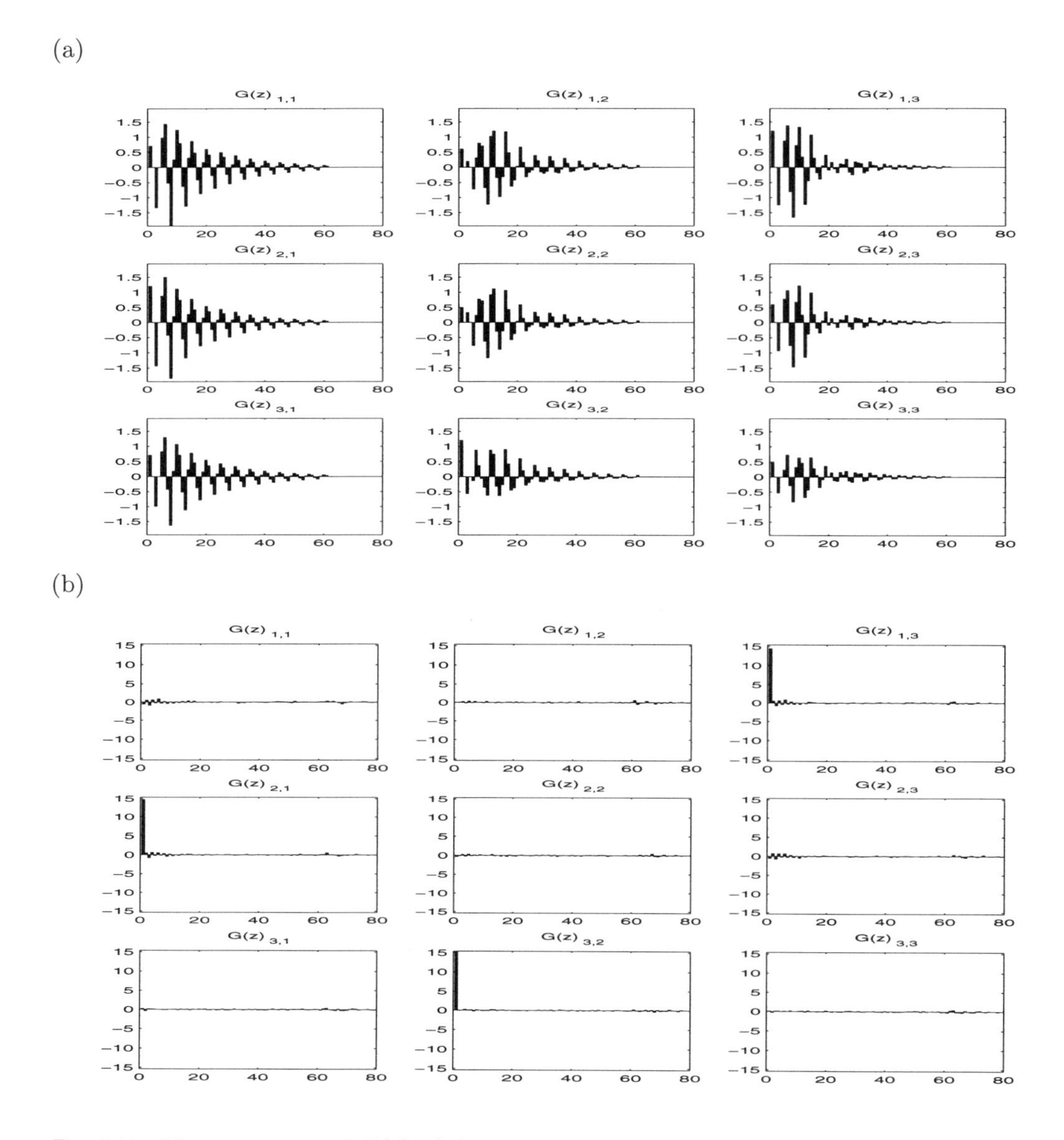

Fig. 9.11 The parameters of $\mathbf{G}(z)$ of the causal system in Example 9.3: (a) The initial state, (b) after 3000 iterations [1361, 1367].

Appendix A. Lie Group and Riemannian Metric on FIR Manifold

In this Appendix, we discuss geometrical structures on the manifold of FIR filters $\mathbf{B}(z)$ [1362, 1368]. In the following discussion, we denote by $\mathcal{M}(L)$, the subset of $\mathcal{M}(L, 0)$ having the constraint that $\mathbf{B}_0$ is nonsingular. Firstly, we introduce the Lie group and the Rieman-

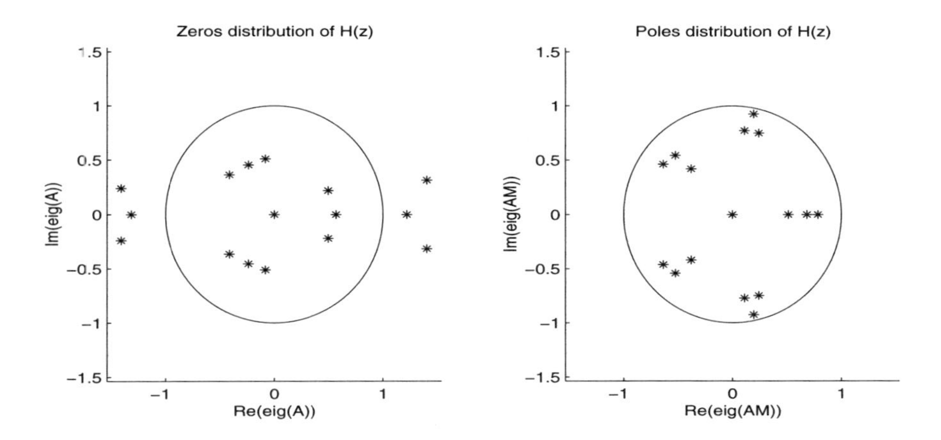

Fig. 9.12 Distributions of the zeros and poles of the mixing ARMA model in Example 9.4.

nian structure on the manifold of FIR filters. Using the isometric property of Lie groups, we derive the natural gradient for any cost function defined on the manifold $\mathcal{M}(L)$.

A.0.1 Lie Group

In the manifold $\mathcal{M}(L)$, the Lie group has two basic operations: Multiplication $\circledast$ and inverse $\dagger$, which are defined as follows. For $\mathbf{B}(z),\ \mathbf{C}(z) \in \mathcal{M}(L)$:

$$\mathbf{B}(z) \circledast \mathbf{C}(z) = \sum_{p=0}^{L} \sum_{q=0}^{p} \mathbf{B}_q \mathbf{C}_{(p-q)} z^{-p}, \tag{A.1}$$

$$\mathbf{B}^{\dagger}(z) = \sum_{p=0}^{L} \mathbf{B}_p^{\dagger} z^{-p}, \tag{A.2}$$

where $\mathbf{B}_p^{\dagger}$ is recursively defined by

$$\mathbf{B}_0^{\dagger} = \mathbf{B}_0^{-1}, \quad \mathbf{B}_1^{\dagger} = -\mathbf{B}_0^{\dagger}\mathbf{B}_1, \tag{A.3}$$

$$\mathbf{B}_p^{\dagger} = -\sum_{q=1}^{p} \mathbf{B}_{p-q}^{\dagger} \mathbf{B}_q, \qquad (p = 1, 2, \ldots, L). \tag{A.4}$$

With these operations, both $\mathbf{B}(z) \circledast \mathbf{C}(z)$ and $\mathbf{B}^{\dagger}$ remain on the manifold $\mathcal{M}(L)$. It is straightforward to verify that the manifold $\mathcal{M}(L)$ with the above operations forms a Lie Group. The identity element is $\boldsymbol{E}(z) = \mathbf{I}$.

In the following discussion, we consider the global transfer function in the Lie group sense $\mathbf{G}(z) = \mathbf{W}(z) \circledast \mathbf{H}(z)$. Moreover, the Lie group possesses the following properties

$$\begin{array}{ll} \text{1) Associative Law :} & \mathbf{A}(z) \circledast (\mathbf{B}(z) \circledast \mathbf{C}(z)) = (\mathbf{A}(z) \circledast \mathbf{B}(z)) \circledast \mathbf{C}(z), \\ \text{2) Inverse Property :} & \mathbf{B}(z) \circledast \mathbf{B}^{\dagger}(z) = \mathbf{B}^{\dagger}(z) \circledast \mathbf{B}(z) = \mathbf{I}. \end{array} \tag{A.5}$$

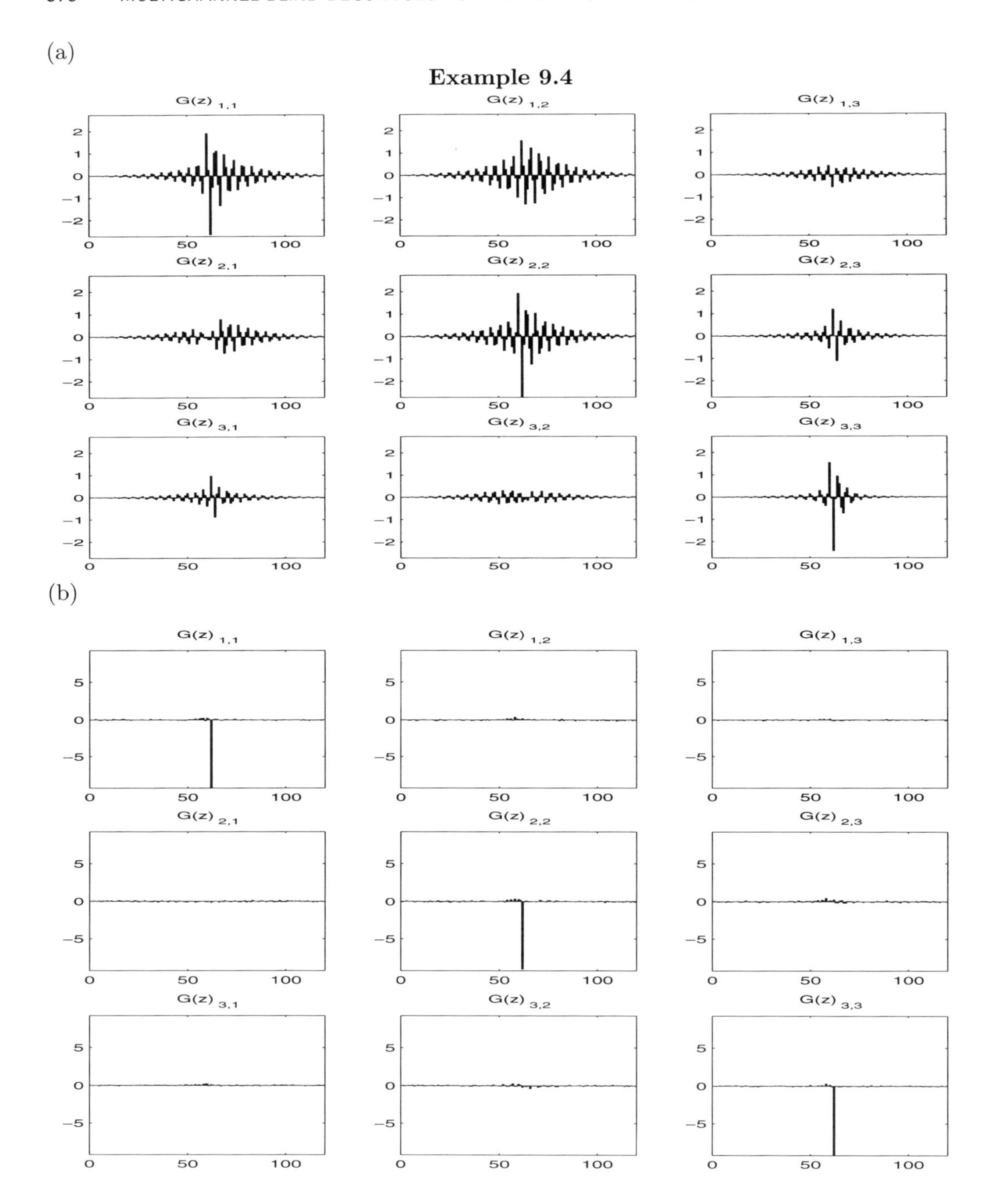

Fig. 9.13 The distribution of the parameters of the global transfer function $\mathbf{G}(z)$ of the non-causal system in Example 9.4: (a) The initial state, (b) after convergence [1362].

In fact, the Lie multiplication of $\mathbf{B}(z)$, $\mathbf{C}(z) \in \mathcal{M}(L)$ is the truncated form of ordinary multiplication up to the length L, that is

$$\mathbf{B}(z) \circledast \mathbf{C}(z) = [\mathbf{B}(z)\mathbf{C}(z)]_L , \tag{A.6}$$

where $[\mathbf{B}(z)]_L$ is such a truncation operator that any terms with orders higher than L in the polynomial matrix $\mathbf{B}(z)$ are omitted.

A.0.2 Riemannian Metric and Natural Gradient in the Lie Group Sense

The Lie Group has an important property that allows an invariant Riemannian metric (see Zhang *et al.* [1348, 1358] for more detail). Let $\mathcal{T}_{\mathbf{W}(z)}$ and $\mathbf{X}(z)$, $\mathbf{Y}(z) \in \mathcal{T}_{\mathbf{W}(z)}$ be tangent vectors of the Lie group at $\mathbf{W}(z)$. In other words, for small $\varepsilon > 0$, the terms $\mathbf{W}(z) + \varepsilon\mathbf{X}(z)$ and $\mathbf{W}(z) + \varepsilon\mathbf{Y}(z)$ represent deviations of $\mathbf{W}(z)$ in two different directions $\mathbf{X}(z)$ and $\mathbf{Y}(z)$. We introduce the inner product with respect to $\mathbf{W}(z)$ as $< \mathbf{X}(z), \mathbf{Y}(z) >_{\mathbf{W}(z)}$. Since $\mathcal{M}(L)$ is a Lie group, any $\mathbf{B}(z) \in \mathcal{M}(L)$ defines an onto-mapping: $\mathbf{W}(z) \rightarrow \mathbf{W}(z) \circledast \mathbf{B}(z)$. The multiplication transformation maps a tangent vector $\mathbf{X}(z)$ at $\mathbf{W}(z)$ to a tangent vector $\mathbf{X}(z) \circledast \mathbf{B}(z)$ at $\mathbf{W}(z) \circledast \mathbf{B}(z)$. Therefore, we can define a Riemannian metric on $\mathcal{M}(L)$, such that the right multiplication transformation is isometric, that is, it preserves the Riemannian metric on $\mathcal{M}(L)$. Explicitly, we write it as

$$< \mathbf{X}(z), \mathbf{Y}(z) >_{\mathbf{W}(z)} = < \mathbf{X}(z) \circledast \mathbf{B}(z), \mathbf{Y}(z) \circledast \mathbf{B}(z) >_{\mathbf{W}(z) \circledast \mathbf{B}(z)} . \tag{A.7}$$

If we define the inner product at the identity $\boldsymbol{E}(z) = 1$ by

$$\begin{aligned} < \mathbf{X}(z), \mathbf{Y}(z) >_{\mathbf{E}(z)} &= \sum_{p=0}^{L} \operatorname{tr}(\mathbf{X}_p \mathbf{Y}_p^T) \\ &= \frac{1}{2\pi j} \oint \operatorname{tr}\left[\mathbf{X}(z^{-1})\mathbf{Y}^T(z)\right] z^{-1} dz, \end{aligned} \tag{A.8}$$

then $< \mathbf{X}(z), \mathbf{Y}(z) >_{\mathbf{W}(z)}$ is automatically induced by

$$< \mathbf{X}(z), \mathbf{Y}(z) >_{\mathbf{W}(z)} = < \mathbf{X}(z) \circledast \mathbf{W}(z)^{\dagger}, \mathbf{Y}(z) \circledast \mathbf{W}(z)^{\dagger} >_{\mathbf{E}(z)} . \tag{A.9}$$

For a loss function $\rho(\mathbf{W}(z))$ defined on the Riemannian manifold $\mathcal{M}(L)$, the natural gradient $\tilde{\nabla}\rho(\mathbf{W}(z))$ is the steepest descent direction of the cost function $\rho(\mathbf{W}(z))$ as measured by the Riemannian metric on $\mathcal{M}(L)$, which is the contravariant form of partial derivatives (ordinary gradient)

$$\frac{\partial \rho(\mathbf{W}(z))}{\partial \mathbf{W}(z)} = \sum_{p=0}^{L} \frac{\partial \rho(\mathbf{W}(z))}{\partial \mathbf{W}_p} z^{-p}, \tag{A.10}$$

where $\frac{\partial \rho(\mathbf{W}(z))}{\partial \mathbf{W}_p} = \left[\frac{\partial \rho(\mathbf{W})}{\partial w_{pij}}\right]_{n \times n}$. We introduce the following notation for the natural gradient of the cost function $\rho(\mathbf{W}(z))$

$$\tilde{\nabla}\rho(\mathbf{W}(z)) = \sum_{p=0}^{L} \tilde{\partial}_p \rho(\mathbf{W}(z)) z^{-p}. \tag{A.11}$$

From definitions (A.8)-(A.9), we have

$$\begin{aligned}\langle \mathbf{X}(z), \nabla\rho(\mathbf{W}(z))\rangle_{\boldsymbol{E}} &= \langle \mathbf{X}(z) \circledast \mathbf{W}(z)^{\dagger}, \tilde{\nabla}\rho(\mathbf{W}(z)) \circledast \mathbf{W}(z)^{\dagger}\rangle_{\boldsymbol{E}(z)} \\ &= \sum_{k=0}^{L} \mathrm{tr}\left(\sum_{p=0}^{k} \mathbf{X}_p \mathbf{W}_{k-p}^{\dagger}\left(\sum_{q=0}^{k} \tilde{\partial}_q \rho \mathbf{W}_{k-q}^{\dagger}\right)^T\right) \\ &= \sum_{p=0}^{L} \mathrm{tr}\left(\mathbf{X}_p \sum_{k=p}^{L} \mathbf{W}_{k-p}^{\dagger}\left(\sum_{q=0}^{k} \tilde{\partial}_q \rho \mathbf{W}_{k-q}^{\dagger}\right)^T\right) \end{aligned} \tag{A.12}$$

for any $\mathbf{X}(z)$ in $T\mathcal{M}_{\mathbf{W}(z)}$. This induces the following relation

$$\partial_p \rho = \sum_{k=p}^{L} \sum_{q=0}^{k} \tilde{\partial}_q l \mathbf{W}_{k-q}^{\dagger} \mathbf{W}_{k-p}^{\dagger T}. \tag{A.13}$$

On the other hand, if we take the new differential variables

$$d\mathbf{X}(z) = d\mathbf{W}(z) \circledast \mathbf{W}^{\dagger}(z), \tag{A.14}$$

which are nonholonomic, we have

$$\partial_p \rho = \sum_{k=p}^{L} \frac{\partial \rho(\mathbf{W}(z))}{\partial \mathbf{X}_k} \mathbf{W}_{k-p}^{\dagger T}. \tag{A.15}$$

Comparing equations (A.13) and (A.15) leads to the following relationship

$$\frac{\partial \rho(\mathbf{W}(z))}{\partial \mathbf{X}_k} = \sum_{q=0}^{k} \tilde{\partial}_q l \mathbf{W}_{k-q}^{\dagger}, \tag{A.16}$$

or writing it in operator form, we have

$$\frac{\partial \rho(\mathbf{W}(z))}{\partial \mathbf{X}(z)} = \tilde{\nabla}\rho(\mathbf{W}(z)) \circledast \mathbf{W}^{\dagger}(z). \tag{A.17}$$

According to the definition of the inverse operation in the Lie group, we have

$$\tilde{\nabla}\rho(\mathbf{W}(z)) = \frac{\partial \rho(\mathbf{W}(z))}{\partial \mathbf{X}(z)} \circledast \mathbf{W}(z), \tag{A.18}$$

where $\mathbf{X}(z)$ is a nonholonomic base, which is defined by

$$d\mathbf{X}(z) = d\mathbf{W}(z) \circledast \mathbf{W}^{\dagger}(z) = \left[d\mathbf{W}(z)\mathbf{W}^{-1}(z)\right]_L. \tag{A.19}$$

It should be noted that $d\mathbf{X}(z) = \left[d\mathbf{W}(z)\mathbf{W}^{-1}(z)\right]_L$ is a nonholonomic basis, which has a definite geometrical meaning and proves to be useful in blind separation algorithms [25].

Appendix B. Properties and Stability Conditions for the Equivariant Algorithm

B.0.1 Proof of Fundamental Properties and Stability Analysis of Equivariant NG Algorithm (9.126)

The learning algorithm (9.126) in the Lie group form may be written as

$$\Delta \mathbf{W}(z) = -\eta \frac{\partial \rho(\mathbf{W}(z))}{\partial \mathbf{X}(z)} \circledast \mathbf{W}(z). \tag{B.1}$$

Multiplying both sides by the mixing filter $\mathbf{H}(z)$ in the Lie group sense and using the associative law of the Lie group, we have

$$\Delta \mathbf{G}(z) = -\eta \frac{\partial \rho(\mathbf{W}(z))}{\partial \mathbf{X}(z)} \circledast \mathbf{G}(z), \tag{B.2}$$

where $\mathbf{G}(z) = \mathbf{W}(z) \circledast \mathbf{H}(z)$. From equation (9.125), we know that $\frac{\partial \rho(\mathbf{W}(z))}{\partial \mathbf{X}(z)}$ is formally independent of the mixing channel $\mathbf{H}(z)$. This infers that the learning algorithm (9.126) is equivariant.

Another important property of the learning algorithm (9.128) is that it keeps the nonsingularity of $\mathbf{W}_0$ if the initial $\mathbf{W}_0$ is nonsingular [1361]. In fact, if we denote the inner product of two matrices by $< \mathbf{A}, \mathbf{B} >= \operatorname{tr}(\mathbf{A}^T \mathbf{B})$, we can easily calculate the derivative of the determinant $|\mathbf{W}_0|$ in the following way

$$\begin{aligned} \frac{d|\mathbf{W}_0|}{dt} &= \langle \frac{\partial |\mathbf{W}_0|}{\partial \mathbf{W}_0}, \frac{d\mathbf{W}_0}{dt} \rangle = \langle |\mathbf{W}_0| \mathbf{W}_0^{-T}, \frac{d\mathbf{W}_0}{dt} \rangle & \text{(B.3)} \\ &= \operatorname{tr}\left(|\mathbf{W}_0| \mathbf{W}_0^{-1} (\mathbf{I} - \mathbf{f}(\mathbf{y})\, \mathbf{y}^T(k)) \mathbf{W}_0 \right) = \operatorname{tr}(\mathbf{I} - \mathbf{f}(\mathbf{y})\, \mathbf{y}^T(k)) |\mathbf{W}_0|) . & \text{(B.4)} \end{aligned}$$

This equation results in

$$|\mathbf{W}_0(t)| = |\mathbf{W}_0(0)| \exp\left(\int_0^t \operatorname{tr}(\mathbf{I} - \mathbf{f}(\mathbf{y}(\tau))\, \mathbf{y}^T(\tau)) d\tau \right). \tag{B.5}$$

Therefore, the matrix $\mathbf{W}_0(t)$ is nonsingular if the initial matrix $\mathbf{W}_0(0)$ is nonsingular. This means that the learning algorithm (9.126) keeps the filter $\mathbf{W}(z)$ on the manifold $\mathcal{M}(L)$ if the initial filter is on the manifold.

It is easily observed that separated signals $\mathbf{y}(k)$ can achieve the highest possible mutual independence if the nonlinear activation functions $\mathbf{f}(\mathbf{y})$ are suitably chosen. If the mixing model is simplified into the instantaneous mixture, the learning algorithm (9.126) is the same as the one discussed in Chapter 6 [28].

B.0.2 Stability Analysis of the Learning Algorithm

Since the learning algorithm for updating $\mathbf{W}_p$, $(p = 0, 1, \ldots, L)$, is a linear combination of $\mathbf{X}_p$, the stability of the learning algorithm for $\mathbf{X}_p$ implies the stability of the learning algorithm (9.126). In order to analyze the stability of the learning algorithm, we suppose that the separating signals $\mathbf{y} = [y_1, y_2, \dots, y_n]^T$ are not only spatially mutually independent but temporally identically and independently distributed for each component (i.e., independent for different components). Now consider the continuous-time learning algorithm for updating $\mathbf{X}_p$:

$$\frac{d\mathbf{X}_p}{dt} = \eta(\delta_{0,p}\mathbf{I} - \langle \mathbf{f}(\mathbf{y}(t)\, \mathbf{y}^T(t-p)) \rangle\,, \qquad (p = 0, 1, \dots, L). \tag{B.6}$$

To analyze the asymptotic properties of the learning algorithm, we take the expectation in equation (B.6)

$$\frac{d\mathbf{X}_p}{dt} = \eta\left(\delta_{0,p}\mathbf{I} - E\{\mathbf{f}(\mathbf{y})\, \mathbf{y}^T(t-p)\}\right), \qquad (p = 0, 1, \dots, L). \tag{B.7}$$

Taking a variation $\delta\mathbf{X}_p$ on $\mathbf{X}_p$, we have

$$\frac{d\delta\mathbf{X}_p}{dt} = -\eta E\{\mathbf{f}'(\mathbf{y})\delta\mathbf{y}\mathbf{y}^T(t-p) + \mathbf{f}(\mathbf{y}(t))\, \delta\mathbf{y}^T(t-p)\}, \qquad (p = 0, 1, 2, \dots, L) \tag{B.8}$$

where $\delta\mathbf{y}(k-p) = [\delta\mathbf{W}(z)]\mathbf{u}(k-p) = [\delta\mathbf{X}(z)]\, \mathbf{y}(k-p) = \sum_{j=0}^{\infty} \delta\mathbf{X}_p\, \mathbf{y}(k-p-j)$. Using the mutual independence and i.i.d. properties of signals $\mathbf{y}_i$ and the normalized condition (9.136), we deduce that

$$\begin{aligned} \frac{d\delta\mathbf{f}X_0}{dt} &= -\eta\left(E\{(\mathbf{f}'(\mathbf{y})\delta\mathbf{X}_0\mathbf{y})\, \mathbf{y}^T\} + \delta\mathbf{X}_0^T\right), && \text{(B.9)} \\ \frac{d\delta\mathbf{X}_p}{dt} &= -\eta\left(E\{(\mathbf{f}'(\mathbf{y})\delta\mathbf{X}_p\, \mathbf{y}(t-p)\, \mathbf{y}(t-p)^T\}\right), \qquad p = 1, 2, \dots, L. && \text{(B.10)} \end{aligned}$$

Then the stability conditions for (B.9)-(B.10) are

$$\begin{aligned} m_i + 1 &> 0, \qquad (i = 1, 2, \dots, n), && \text{(B.11)} \\ \kappa_i &> 0, \qquad (i = 1, 2, \dots, n), && \text{(B.12)} \\ \kappa_i\kappa_j\sigma_i^2\sigma_j^2 &> 1, \qquad (i, j = 1, 2, \dots, n), && \text{(B.13)} \end{aligned}$$

where $m_i = E\{f'(y_i)\, y_i^2\}$, $\kappa_i = E\{f_i'(y_i)\}$, $\sigma_i^2 = E\{|y_i|^2\}$, $i = 1, 2, \dots, n$. The conditions are identical to the ones derived by Amari *et al.* [24] for instantaneous blind source separation. It can be easily shown that the stability conditions are satisfied when one chooses, for example, the nonlinear activation function

$$f(y) = (|y|^{r-1} + \alpha|y|)sign(y), \quad r > 2, \ 0 < \alpha \ll 1, \tag{B.14}$$

for sub-Gaussian signals (with negative kurtosis) and

$$f(y) = tanh(\gamma y), \ \gamma > 2 \tag{B.15}$$

for super-Gaussian signals (with positive kurtosis).

10

Estimating Functions and Superefficiency for ICA and Deconvolution

Climb mountains to see lowlands.

—-Old Asian Proverb

We have so far studied a great number of algorithms for BSS, ICA and MBD. Some are on-line and some are in the batch mode, although most online algorithms can be easily converted to the batch mode by taking the average. Some algorithms extract all the components in parallel, while some do one by one sequentially. Intermediate algorithms extract several components in parallel. The present chapter searches for a unified viewpoint to explain most of these algorithms from the statistical point of view.

The present chapter introduces the method of estimating functions to elucidate the common structures in most of the existing algorithms. We use information geometry for this purpose, and define estimating functions in semiparametric statistical models which include unknown functions as parameters [19, 1351]. Differences in most existing algorithms are only in the choices of estimating functions.

We then give error analysis and stability analysis in terms of estimating functions. This makes it possible to design various adaptive methods for choosing unknown parameters included in estimating functions, which control accuracy and stability. The Newton method is automatically derived by the standardized estimating functions.

First the standard BSS/ICA problem is formulated in the framework of the semiparametric model and a family of estimating functions.

Furthermore, the present chapter will discuss issues related to convergence and efficiency of the batch estimator and natural gradient learning via the semiparametric statistical model, estimating functions, standardized estimating functions derived by using efficient score functions elucidated recently by Amari *et al.* [19, 23, 35, 36] and Zhang *et al.* [1351, 1361, 1365]. We present the geometrical properties of the manifold of FIR filters based on the Lie group structure and formulate the multichannel blind deconvolution problem within the framework of the semiparametric model, deriving a family of estimating functions for blind deconvolution. We then analyze the efficiency of the batch estimator (its convergence rate) based on the estimating function. Finally, we show that both batch learning and natural gradient learning are superefficient under given nonsingular conditions.

10.1 ESTIMATING FUNCTIONS FOR STANDARD ICA

10.1.1 What is an Estimating Function?

We begin with the simplest mixture model

$$\boldsymbol{x}(k) = \boldsymbol{H}\boldsymbol{s}(k) + \boldsymbol{\nu}(k), \tag{10.1}$$

where $\boldsymbol{H}$ is an $m \times n$ unknown full rank mixing matrix. Here, s_i and s_j are independent, but each signal may have temporal correlations. We assume, for the moment $m = n$, and that the noise vector $\boldsymbol{\nu}$ is negligibly small. Let $\boldsymbol{W}$ be an $n \times n$ matrix, which is a candidate for the separating matrix,

$$\boldsymbol{y}(k) = \boldsymbol{W}\boldsymbol{x}(k) \tag{10.2}$$

such that $y_i(k)$, $i = 1, 2, \ldots, n$, are original independent random source signals, or their scaled and permuted version. An on-line learning method uses a candidate matrix $\boldsymbol{W}(k)$ at discrete-time k, and calculates $\boldsymbol{y}(k) = \boldsymbol{W}(k)\boldsymbol{x}(k)$, which is an approximation of the original independent source signals. The candidate is then updated by

$$\boldsymbol{W}(k+1) = \boldsymbol{W}(k) + \eta\, \boldsymbol{F}\left(\boldsymbol{x}(k), \boldsymbol{W}(k)\right), \tag{10.3}$$

where η is a learning constant (which may depend on k) and $\boldsymbol{F}(\boldsymbol{x}, \boldsymbol{W}) \in \mathbb{R}^{n\times n}$ is a matrix-valued function, such that $\boldsymbol{W}(k)$ converges to the true solution. Usually, $\boldsymbol{F}$ depends on $\boldsymbol{x}$ through $\boldsymbol{y} = \boldsymbol{W}\boldsymbol{x}$, and is in the form of $\tilde{\boldsymbol{F}}(\boldsymbol{y})\boldsymbol{W}$, as is the case with the natural gradient.

There have been proposed various functions $\boldsymbol{F}$, which are derived in many cases (but not in all cases) as the gradients of cost functions to be minimized. The cost functions are, for example, higher-order cumulants, entropy, negative log likelihood and others. In many cases, algorithms include free parameters, sometimes free functions, to be chosen adequately or to be determined adaptively. Since the probability density functions of the source signals are usually unknown, there is no way to avoid such parameters.

For an algorithm to converge to the true solution, the function $\boldsymbol{F}$ should satisfy some conditions. The true $\boldsymbol{W}$ should be an equilibrium of dynamic equation (10.3). Because this is a stochastic difference equation, it is more convenient to use its continuous time version for mathematical analysis,

$$\frac{d}{dt}\boldsymbol{W}(t) = \mu\, \boldsymbol{F}\left[\boldsymbol{x}(t), \boldsymbol{W}(t)\right]. \tag{10.4}$$

Because $\boldsymbol{x}(t)$ is a stochastic process, its expected version is written as

$$\frac{d}{dt}\boldsymbol{W}(t) = \mu\, E\left\{\boldsymbol{F}\left[\boldsymbol{x}(t), \boldsymbol{W}(t)\right]\right\}. \tag{10.5}$$

The condition that the true solution $\boldsymbol{W}$ is an equilibrium of (10.5) is given by

$$E\left\{\boldsymbol{F}(\boldsymbol{x}, \boldsymbol{W})\right\} = \boldsymbol{0}, \tag{10.6}$$

where the expectation is taken over $\boldsymbol{x} = \boldsymbol{H}\boldsymbol{s}$, except for indeterminacy due to permutations and scaling.

A function $\boldsymbol{F}(\boldsymbol{x}, \boldsymbol{W})$ satisfying (10.6) for the true (desired) $\boldsymbol{W}$, and $E\left\{\boldsymbol{F}(\boldsymbol{x}, \boldsymbol{W}')\right\} \neq \boldsymbol{0}$ for false $\boldsymbol{W}'$, is called the estimating function. This is defined in relation to a semiparametric statistical model, as is shown in the next section.

10.1.2 Semiparametric Statistical Model

We formulate the problem in the statistical framework. Let $r_i(s_i)$ be the true probability density function of s_i. The joint probability density of $\boldsymbol{s}$ is written as

$$r(\boldsymbol{s}) = \prod_{i=1}^{n} r_i(s_i) \tag{10.7}$$

since they are independent. The observation vector $\boldsymbol{x}$ is a linear function of $\boldsymbol{s}$, so that its probability density function is given in terms of $\boldsymbol{W} = \boldsymbol{H}^{-1}$ by

$$p_X(\boldsymbol{x}; \boldsymbol{W}, r) = \det|\boldsymbol{W}|\, r(\boldsymbol{W}\boldsymbol{x}). \tag{10.8}$$

Because we do not know r except that it is in a product from (10.7), the probability model (10.8) of $\boldsymbol{x}$ includes two parameters, $\boldsymbol{W}$ called the "parameter of interest", which we want to estimate, and an unknown function $\mathbf{r} = r_1 \cdots r_n$ called the "nuisance parameter (function)" which is unimportant. Such a statistical model including an infinite or functional degree of freedom of nuisance parameters is called a semiparametric model. In general due to the existence of unknown functions, it is difficult to estimate the parameter of interest.

A method of estimation in semiparametric statistical models has been developed in the framework of information geometry (Amari and Kawanabe [35, 36]; see also Amari [16], Amari and Nagaoka [37] as for information geometry).

The advantage of using the semiparametric approach is that we do not need to estimate the nuisance parameters: the probability density functions of source signals in blind separation and deconvolution problems. It is inferred from the theory of estimating functions that the batch estimator of the estimating equation converges to the true solution as the number of observed data tends to infinity.

An estimating function in the present case is a matrix-valued function $\boldsymbol{F}(\boldsymbol{x},\boldsymbol{W}) = [F_{ab}(\boldsymbol{x},\boldsymbol{W})]$ of $\boldsymbol{x}$ and $\boldsymbol{W}$ not including the nuisance parameter r, that satisfies

$$E_{\mathbf{W},r}\{\boldsymbol{F}(\boldsymbol{x},\boldsymbol{W}')\} = \mathbf{0}, \quad \text{when } \boldsymbol{W}' = \boldsymbol{W} \tag{10.9}$$

$$E_{\mathbf{W},r}\{\boldsymbol{F}(\boldsymbol{x},\boldsymbol{W}')\} \neq \mathbf{0}, \quad \text{when} \boldsymbol{W}' \neq \boldsymbol{W}, \tag{10.10}$$

where $E_{\mathbf{W},r}$ denotes expectation with respect to the probability distribution given by (10.8). It is required that (10.9) holds for all r of the form (10.7). Here, suffixes $a, b, c, \cdots$ represent components of the original source signals or recovered signals, $\boldsymbol{s}$ or $\boldsymbol{y}$. Sometimes, we require a milder condition that

$$\boldsymbol{\mathcal{K}} = E_{\mathbf{W},r}\left\{\frac{\partial}{\partial \boldsymbol{W}}\boldsymbol{F}(\boldsymbol{x},\boldsymbol{W})\right\} \tag{10.11}$$

be non-degenerate. That is, condition (10.11) holds only locally. It should be noted that $\boldsymbol{\mathcal{K}}$ is a matrix-by-matrix linear operator that maps a matrix to a matrix. The components of $\boldsymbol{\mathcal{K}}$ are

$$\mathcal{K}_{ab,ij} = E_{\mathbf{W},r}\left\{\frac{\partial}{\partial W_{ij}}F_{ab}(\boldsymbol{x},\boldsymbol{W})\right\}, \tag{10.12}$$

where W_{ij} denotes elements of $\boldsymbol{W}$, and suffixes i, j, a, b, etc. represent components of the observed signals $\boldsymbol{x}$. It is convenient to use capital indices A, $B, \cdots$ to represent a pair (a, b), (i, j) and so. Then, for $A = (a, b)$, $B = (c, i)$, $\boldsymbol{\mathcal{K}}$ has a matrix representation $\boldsymbol{\mathcal{K}} = [\mathcal{K}_{AB}]$ that operates on $(W_B) = (W_{ij})$ as

$$\boldsymbol{\mathcal{K}}\boldsymbol{W} = \sum_B \mathcal{K}_{AB}W_B = \sum_{i,j}\mathcal{K}_{ab,ij}W_{ij}. \tag{10.13}$$

The inverse of $\boldsymbol{\mathcal{K}}$ is defined by the inverse matrix of $\boldsymbol{\mathcal{K}} = [\mathcal{K}_{AB}]$.

Given an estimating function $\boldsymbol{F}(\boldsymbol{x},\boldsymbol{W})$, a batch estimator $\tilde{\boldsymbol{W}}$ for observed data $\boldsymbol{x}(1), \ldots,$ $\boldsymbol{x}(N)$, is given by the estimating equation

$$\sum_{k=1}^{N}\boldsymbol{F}\{\boldsymbol{x}(k),\boldsymbol{W}\} = \mathbf{0}. \tag{10.14}$$

This is derived by replacing the expectation in (10.9) by the empirical sum of observations and the on-line learning algorithm is given by (10.3), so the batch estimating equation works without using the unknown r. The problem is 1) if there exists an estimating function which works without knowing r, and 2) how to find the best estimating function $\boldsymbol{F}$ when there are many.

10.1.3 Admissible Class of Estimating Functions

Algorithms proposed by Jutten and Herault [650]; Cichocki *et al.* [278, 279, 277], Bell and Sejnowski [73]; Amari *et al.* [28]; Cardoso and Laheld [148]; Oja and Kahrunen [912] etc. use various estimating functions found heuristically. There are good ones and bad ones. Estimating function $\boldsymbol{F}$ is better than $\boldsymbol{F}'$, when the expected error of estimator $\hat{\boldsymbol{W}}$ derived by $\boldsymbol{F}$ is smaller than that by $\boldsymbol{F}'$. However, it may happen that $\boldsymbol{F}$ is better than $\boldsymbol{F}'$ when the true (unknown) distribution is $r(\boldsymbol{s})$ but $\boldsymbol{F}'$ is better when it is $r'(\boldsymbol{s})$. Hence, they are in general not comparable. A family of estimating functions is said to be admissible, when, given any estimating function, an equivalent or better estimating function can be found in the family. We may focus only on an admissible class of estimating functions. Moreover, this class includes the best estimator in the sense that it satisfies the extended Cramér-Rao bound asymptotically (that is, the Fisher efficient estimator).

Amari and Cardoso [23] applied the information geometrical theory to the estimating functions of Amari and Kawanabe [35, 36], and proved that estimating functions of the form

$$\boldsymbol{F}(\boldsymbol{x}, \boldsymbol{W}) = \mathbf{I} - \boldsymbol{\varphi}(\boldsymbol{y})\boldsymbol{y}^T, \tag{10.15}$$

or

$$F_{ij}(\boldsymbol{x}, \boldsymbol{W}) = \delta_{ij} - \varphi_i(y_i)y_j$$

in component form, give a set of admissible estimating functions, where

$$\boldsymbol{\varphi}(\boldsymbol{y}) = [\varphi_1(y_1), \varphi_2(y_2), \ldots, \varphi_n(y_n)]^T \tag{10.16}$$

are arbitrary non-trivial functions φ_i. This is indeed an estimating function as is easily shown. When $\boldsymbol{W}$ is the true solution, y_j and y_j are independent. Therefore, whatever r is,

$$E_{r,\mathbf{W}}\left\{\varphi_i\left(y_i\right)y_j\right\} = E\left\{\varphi_i\left(y_i\right)\right\}E\left\{y_j\right\} = 0, \quad i \neq j. \tag{10.17}$$

However, when $\boldsymbol{W}$ is not the true solution, the above equation does not hold in general. For the diagonal terms, $i = j$, we have

$$E\left\{\varphi_i\left(y_i\right)y_i\right\} = 1, \tag{10.18}$$

which specifies the magnitude of the recovered signal y_i. Since the magnitude may be arbitrary, we may set the diagonal terms F_{ii} arbitrarily, including the nonholonomic one where $F_{ii} = 0$.

We give some typical examples of estimating functions. Let

$$q(\boldsymbol{s}) = \prod_{i=1}^{n} q_i\left(s_i\right) \tag{10.19}$$

be a (misspecified) joint probability density function of $\boldsymbol{s}$, which might be different from the true one

$$r(\boldsymbol{s}) = \prod_{i=1}^{n} r_i\left(s_i\right).$$

The negative log likelihood of $\boldsymbol{x}$ derived therefrom is

$$\rho(\boldsymbol{x}, \boldsymbol{W}) = -\det|\boldsymbol{W}| - \sum_{i=1}^{n} \log q_i\left(y_i\right), \tag{10.20}$$

where y_i is the i-th component of $\boldsymbol{y} = \boldsymbol{W}\boldsymbol{x}$, depending on both $\boldsymbol{x}$ and $\boldsymbol{W}$. The criterion of minimizing ρ is interpreted as maximization of the entropy, or maximization of the likelihood. Let us put

$$\varphi_i\left(y_i\right) = -\frac{d}{dy_i}\log q_i\left(y_i\right). \tag{10.21}$$

The gradient of ρ gives an estimating function

$$\tilde{\boldsymbol{F}}(\boldsymbol{x}, \boldsymbol{W}) = -\frac{\partial \rho(\boldsymbol{x}, \boldsymbol{W})}{\partial \boldsymbol{W}} = \boldsymbol{W}^{-T} - \varphi(\boldsymbol{y})\boldsymbol{x}^T. \tag{10.22}$$

We can prove that $\tilde{\boldsymbol{F}}$ is an estimating function. However, when $\tilde{\boldsymbol{F}}$ is an estimating function,

$$\boldsymbol{F}(\boldsymbol{y}) = \tilde{\boldsymbol{F}}(\boldsymbol{x}, \boldsymbol{W})\boldsymbol{W}^T\boldsymbol{W} = \left[\mathbf{I} - \varphi(\boldsymbol{y})\boldsymbol{y}^T\right]\boldsymbol{W} \tag{10.23}$$

is also an estimating function. It is easy to prove that

$$E\{\boldsymbol{F}(\boldsymbol{y})\} = \mathbf{0} \tag{10.24}$$

and

$$E\left\{\tilde{\boldsymbol{F}}(\boldsymbol{x}, \boldsymbol{W})\right\} = \mathbf{0} \tag{10.25}$$

are equivalent.

When the true distributions are r_i, the best choice of φ_i is

$$\varphi_a(i) = -\frac{d}{ds}\log r_i(s).$$

This gives the maximum likelihood estimator (Pham [963, 967]). However, even when we use a different φ_a, the estimating equation (10.14) gives a $\sqrt{N}$-consistent estimator, that is, the estimation error converges to 0 in probability in the order of $1/\sqrt{N}$ as N goes to infinity, when N is the number of observations. It is easy to show that similar estimating functions are derived from the criterion of maximizing higher-order cumulants and others. The algorithms given by Cardoso, Jutten-Herault, Karhunen-Oja etc. use respective estimating functions [138, 277, 650, 670, 589].

We have shown that $\tilde{\boldsymbol{F}}(\boldsymbol{x}, \boldsymbol{W})$ and $\boldsymbol{F}(\boldsymbol{y})$ are equivalent estimating functions, because they are linearly related and their estimating equations give the same solution. More generally, let $\boldsymbol{\mathcal{R}}(\boldsymbol{W})$ be an arbitrary nonsingular linear operator acting on matrices. When $\boldsymbol{F}(\boldsymbol{x}, \boldsymbol{W})$ is an estimating function matrix, $\boldsymbol{\mathcal{R}}(\boldsymbol{W})\,\boldsymbol{F}(\boldsymbol{x}, \boldsymbol{W})$ is also an estimating function matrix, because

$$E_{\mathbf{W},r}\left\{\boldsymbol{\mathcal{R}}(\boldsymbol{W})\,\boldsymbol{F}(\boldsymbol{x}, \boldsymbol{W})\right\} = \boldsymbol{\mathcal{R}}(\boldsymbol{W})\,E_{\mathbf{W},r}\{\boldsymbol{F}(\boldsymbol{x}, \boldsymbol{W})\} = \mathbf{0}. \tag{10.26}$$

Moreover, $\boldsymbol{F}(\boldsymbol{x}, \boldsymbol{W})$ and $\boldsymbol{\mathcal{R}}(\boldsymbol{W})\,\boldsymbol{F}(\boldsymbol{x}, \boldsymbol{W})$ are equivalent in the sense that the derived batch estimators are exactly the same, because the two estimating equations

$$\sum_{k=1}^{N} \boldsymbol{F}[\boldsymbol{x}(k), \boldsymbol{W}] = \mathbf{0}, \tag{10.27}$$

$$\sum_{k=1}^{N} \boldsymbol{\mathcal{R}}(\boldsymbol{W})\,\boldsymbol{F}[\boldsymbol{x}(k), \boldsymbol{W}] = \mathbf{0} \tag{10.28}$$

give the same solution $\hat{\boldsymbol{W}}_*$ (ignoring the arbitrary scaling and permutation). This defines an equivalent class of estimating functions which are essentially the same for batch estimation.

However, two equivalent estimating functions $\boldsymbol{F}(\boldsymbol{x}, \boldsymbol{W})$ and $\boldsymbol{\mathcal{R}}(\boldsymbol{W})\,\boldsymbol{F}(\boldsymbol{x}, \boldsymbol{W})$ give different dynamical properties in on-line learning. That is, the dynamical properties of on-line learning algorithms

$$\begin{aligned} \boldsymbol{W}(k+1) &= \boldsymbol{W}(k) + \eta\, \boldsymbol{F}\left(\boldsymbol{x}(k), \boldsymbol{W}(k)\right), & (10.29) \\ \boldsymbol{W}(k+1) &= \boldsymbol{W}(k) + \eta\, \boldsymbol{\mathcal{R}}\left(\boldsymbol{W}(k)\right)\, \boldsymbol{F}\left(\boldsymbol{x}(k), \boldsymbol{W}(k)\right) & (10.30) \end{aligned}$$

are completely different. Therefore, instead of the form (10.15), we need to consider an enlarged type of estimating function of the form $\boldsymbol{\mathcal{R}}(\boldsymbol{W})\,\boldsymbol{F}(\boldsymbol{x}, \boldsymbol{W})$ to derive a good on-line estimator.

10.1.4 Stability of Estimating Functions

One of the important dynamical properties of on-line learning is the stability of the algorithm at the true solution $\boldsymbol{W}$, which is guaranteed to be an equilibrium of the dynamic equation by using $\boldsymbol{F}$. We begin with the averaged dynamic equation of natural gradient learning

$$\frac{d}{dt}\boldsymbol{W}(t) = \mu E\{\boldsymbol{F}(\boldsymbol{x}, \boldsymbol{W}(t)\}\boldsymbol{W}(t), \tag{10.31}$$

with $\boldsymbol{F}(\boldsymbol{x}, \boldsymbol{W}(t))$. The stability of dynamic equation (10.31) at the equilibrium is given by studying the eigenvalues of its Hessian. For the stability analysis, let us put

$$\boldsymbol{W}(t) = \boldsymbol{W} + \delta\boldsymbol{W}(t), \tag{10.32}$$

where $\delta\boldsymbol{W}(t)$ is a small deviation from the true $\boldsymbol{W}$. Then, (10.31) is rewritten as

$$\frac{d}{dt}\delta\boldsymbol{W}(t) = \mu\, E\left\{\boldsymbol{F}\left[\boldsymbol{x}, \boldsymbol{W} + \delta\boldsymbol{W}(t)\right]\right\}(\boldsymbol{W}(t) + \delta\boldsymbol{W}(t)). \tag{10.33}$$

It is convenient to use the nonholonomic variables

$$\delta\boldsymbol{X} = \delta\boldsymbol{W}\boldsymbol{W}^{-1}, \tag{10.34}$$

and rewrite the dynamic equation in the neighborhood of the true solution as

$$\frac{d}{dt}\delta\boldsymbol{X}(t) = \mu\, E\left\{\boldsymbol{F}(\boldsymbol{x}, \boldsymbol{W} + \delta\boldsymbol{X}\boldsymbol{W})\right\}. \tag{10.35}$$

By Taylor expansion, we have

$$\frac{d}{dt}\delta\boldsymbol{X}(t) = \mu\,\boldsymbol{\mathcal{K}}(\boldsymbol{W})\delta\boldsymbol{X}(t), \tag{10.36}$$

where

$$\boldsymbol{\mathcal{K}}(\boldsymbol{W}) = \frac{\partial E\left\{\boldsymbol{F}(\boldsymbol{x}, \boldsymbol{W})\right\}}{\partial \boldsymbol{X}} = \frac{\partial E\left\{\boldsymbol{F}(\boldsymbol{x}, \boldsymbol{W})\right\}}{\partial \boldsymbol{W}} \circ \boldsymbol{W} \tag{10.37}$$

is a linear operator which maps a matrix to another matrix. Since both $\boldsymbol{F} = [F_{ab}]$ and $\boldsymbol{X} = [X_{cd}]$ are matrices, $\boldsymbol{\mathcal{K}}$ will have four indices a, b, c, d and is given by

$$\mathcal{K}_{ab,cd} = \frac{\partial E\{F_{ab}\}}{\partial X_{cd}} \tag{10.38}$$

in the component form. At the true value $\boldsymbol{W}$, where $y_a = s_a$ and $\boldsymbol{F}$ is given by (10.15), $\boldsymbol{\mathcal{K}}$ is calculated as

$$\mathcal{K}_{ab,cd} = E\{\varphi_a'(s_a)\, s_b^2\}\, \delta_{bd}\, \delta_{ac} + \delta_{ad}\, \delta_{bc}, \tag{10.39}$$

where φ' denotes the derivative of φ. We derive the above result in the following.

In order to calculate the gradient of $\boldsymbol{F}$ with respect to $\boldsymbol{X}$, we put

$$\begin{aligned} d\boldsymbol{F}(\boldsymbol{x}, \boldsymbol{W}) &= \boldsymbol{F}(\boldsymbol{x}, \boldsymbol{W} + d\boldsymbol{W}) - \boldsymbol{F}(\boldsymbol{x}, \boldsymbol{W}) \\ &= \boldsymbol{F}(\boldsymbol{x}, \boldsymbol{W} + d\boldsymbol{X}\boldsymbol{W}) - \boldsymbol{F}(\boldsymbol{x}, \boldsymbol{W}), \end{aligned}$$

where $d\boldsymbol{F}$ denotes the increment of $\boldsymbol{F}$ due to change $d\boldsymbol{W}$ of $\boldsymbol{W}$, and expand it in the form

$$dF_{ab}(\boldsymbol{x}, \boldsymbol{W}) = \sum M_{ab,cd}(\boldsymbol{x}, \boldsymbol{W})\, dX_{cd}, \qquad \textit{with}$$

$$M_{ab,cd} = \frac{\partial F_{ab}}{\partial X_{cd}}$$

and its expectation gives $\mathcal{K}_{ab,cd}$.

For $\boldsymbol{F} = (F_{ab})$ given by

$$F_{ab} = \delta_{ab} - \varphi(y_a) y_b,$$

we have

$$\begin{aligned} dF_{ab} &= d\varphi(y_a) y_b + \varphi(y_a) dy_b \\ &= \varphi'(y_a) dy_a y_b + \varphi(y_a) dy_b. \end{aligned}$$

From

$$d\boldsymbol{y} = d\boldsymbol{W}\boldsymbol{x} = d\boldsymbol{W}\boldsymbol{W}^{-1}\boldsymbol{W}\boldsymbol{x} = d\boldsymbol{X}\boldsymbol{y},$$

we have

$$dy_a = \sum_{d=1}^{n} dX_{ad} y_d = \sum_{c,d=1}^{n} y_d \delta_{ac} dX_{cd}.$$

Therefore,

$$M_{ab,cd} = \varphi'(y_a) y_b y_d \delta_{ac} + \varphi(y_a) y_b \delta_{bc}.$$

At the true $\boldsymbol{W}$, y_a and y_b are independent for $a \neq b$. Hence,

$$\begin{aligned} &E\{\varphi'(y_a) y_b y_d\}\delta_{ac} = E\{\varphi'(s_a) y_b^2\}\delta_{ac}\delta_{bd}, \\ &E\{\varphi(y_a) y_d\}\delta_{bc} = \delta_{ad}. \end{aligned}$$

The diagonal term F_{aa} may be disregarded, because it can be arbitrary.

Many components of $\boldsymbol{\mathcal{K}}$ may vanish. For $a \neq b$,

$$\frac{\partial F_{ab}}{\partial X_{cd}} = 0, \tag{10.40}$$

except for the cases where $(a, b) = (c, d)$ or $(a, b) = (d, c)$. When the pairs (a, b) and (c, d) are equal, (4.5) gives

$$\begin{aligned} \mathcal{K}_{ab,ab} &= \kappa_a \sigma_b^2, \\ \mathcal{K}_{ab,ba} &= 1, \mathcal{K}_{aa,bb} = \frac{\partial E\{F_{aa}}{\partial X_{aa}} = E\{\varphi'(s_a)s_a^2\} + E\{\varphi(s_a)s_a\} = \kappa_a \sigma_a^2 + 1, \end{aligned}$$

where

$$\kappa_a = E\{\varphi'(s_a)\}. \tag{10.41}$$

and

$$\sigma_a^2 = E\{y_a^2\}. \tag{10.42}$$

Let us summarize the above results. For the pairwise components of the enlarged matrix $\boldsymbol{\mathcal{K}} = (\mathcal{K}_{AB})$, $\mathcal{K}_{AB} = 0$ except for $A = (a, b)$, $a \neq b$, and $A' = (b, a)$. This shows that $\boldsymbol{\mathcal{K}} = (\mathcal{K}_{AB})$ is decomposed into the two-by-two minor matrices of $\partial F_{ab}/\partial X_{ab}$, $\partial F_{ab}/\partial X_{ba}$, $\partial F_{ba}/\partial X_{ab}$ and $\partial F_{ba}/\partial X_{ba}$,

$$\begin{bmatrix} \mathcal{K}_{AA} & \mathcal{K}_{AA'} \\ \mathcal{K}_{A'A} & \mathcal{K}_{A'A'} \end{bmatrix} = \begin{bmatrix} \kappa_a \sigma_b^2 & 1 \\ 1 & \kappa_b \sigma_a^2 \end{bmatrix}, \tag{10.43}$$

where $A = (a, b)$ and $A' = (b, a)$ (see also [24, 148, 23]).

The inverse of $\boldsymbol{\mathcal{K}}$ has also the same diagonalized form, for (A, A')-part,

$$\begin{bmatrix} \kappa_a \sigma_b^2 & 1 \\ 1 & \kappa_b \sigma_a^2 \end{bmatrix}^{-1} = c_{ab} \begin{bmatrix} \kappa_b \sigma_a^2 & -1 \\ -1 & \kappa_a \sigma_b^2 \end{bmatrix}, \tag{10.44}$$

where

$$c_{ab} = \frac{1}{\kappa_a \kappa_b \sigma_a^2 \sigma_b^2 - 1}. \tag{10.45}$$

The on-line dynamic equation is stable at the true solution, when $\boldsymbol{\mathcal{K}} = (\mathcal{K}_{A,B})$ is positive definite. Since it is decomposed in the two-by-two sub-matrices, it is positive definite when all the sub-matrices $\mathcal{K}_{AA'}$ are positive definite. Hence, we have the following stability Theorem.

Theorem 10.1 (Stability Theorem) *Assume that $E\{\varphi_a s_a\} = 1$. Then, the learning dynamic equation is stable when*

$$\kappa_i \kappa_j \sigma_j^2 \sigma_j^2 \;>\; 1 \tag{10.46}$$

$$\kappa_i > 0. \tag{10.47}$$

The stability depends on the parameters κ_i and σ_i^2, which are related to $\boldsymbol{\varphi}$ and r.

Remark 10.1 *We may choose the diagonal terms F_{aa} arbitrarily. In order to arrange y_a in the same scale, we may choose $F_{aa} = y_a^2 - 1$. Then, $\sigma_a^2 = E\{y_a^2\} = 1$ holds. The stability conditions are: $\kappa_i \kappa_j > \delta_i \, \delta_j$, $\kappa_i > 0$, where $\delta_i = E\{\varphi_i(y_i)\, y_i\}$.*

10.1.5 Standardized Estimating Function and Adaptive Newton Method

The learning dynamic equation

$$\Delta \boldsymbol{W}(k) = \boldsymbol{W}(k+1) - \boldsymbol{W}(k) = \eta\, \boldsymbol{F}\left[\boldsymbol{x}(k), \boldsymbol{W}(k)\right] \boldsymbol{W}(k) \tag{10.48}$$

can be accelerated by the Newton method, given by

$$\Delta \boldsymbol{X}(k) = \eta\, \mathcal{K}^{-1}\left[\boldsymbol{W}(k)\right] \boldsymbol{F}\left[\boldsymbol{x}(k), \boldsymbol{W}(k)\right]. \tag{10.49}$$

Note that $\mathcal{K}^{-1}\boldsymbol{F}$ is an estimating function equivalent to $\boldsymbol{F}$. That is, the Newton method is derived by the following estimating function with superlinear convergence,

$$\boldsymbol{F}^*(\boldsymbol{x}, \boldsymbol{W}) = \mathcal{K}^{-1}(\boldsymbol{W})\boldsymbol{F}(\boldsymbol{x}, \boldsymbol{W}). \tag{10.50}$$

Its convergence is superlinear. Moreover, the true solution $\boldsymbol{W}$ is always stable, because the Hessian of $\boldsymbol{F}^*$ is the identity matrix. This is easily shown from

$$\mathcal{K}^* = E\left\{\frac{\partial \boldsymbol{F}^*}{\partial \boldsymbol{X}}\right\} = \frac{\partial \mathcal{K}^{-1}}{\partial \boldsymbol{X}} E\left\{\boldsymbol{F}\right\} + \mathcal{K}^{-1} \circ \mathcal{K} = \mathbf{I}. \tag{10.51}$$

We call $\boldsymbol{F}^*$ the standardized estimating function, for which $\mathcal{K}^*$ is the identity operator.

By using (10.43) or (10.44), the standardized estimating function matrix $\boldsymbol{F}^*$ is derived as

$$F^*_{ab} = c_{ab}\{\kappa_b \sigma_a^2 \varphi_a(y_a) y_b - \varphi_b(y_b) y_a\}, \quad a \neq b. \tag{10.52}$$

The standardized estimating function $\boldsymbol{F}^*$ includes the parameters σ_a^2 and κ_a, which are usually known. They depend on the statistical properties of the source signal s_a. Therefore, an adaptive method is necessary to implement the Newton method, which estimates the parameters. This not only accelerates the convergence, but automatically stabilizes the separating solution.

Let $\kappa_a(k)$ and $\sigma_a^2(k)$ be their estimates at discrete-time k. Then, we can use the following adaptive rules to update them:

$$\kappa_a(k+1) = (1-\eta_0)\, \kappa_a(k) + \eta_0\, \varphi_a'\left(y_a(k)\right), \tag{10.53}$$

$$\sigma_a^2(k+1) = (1-\eta_0)\, \sigma_a^2(k) + \eta_0\, y_a^2(k), \tag{10.54}$$

where η_0 is the learning rate.

We may require the diagonal term of $\boldsymbol{F}$ to be equal to

$$F_{aa} = 1 - y_a^2. \tag{10.55}$$

Then, the recovered signal is normalized to $\sigma_a^2 = 1$, so that $\boldsymbol{F}^*$ is simplified to

$$F^*_{ab} = -\frac{1}{\kappa_a \kappa_b}\{\kappa_b \varphi_a(y_a) y_b - \varphi_b(y_b) y_a\}, \quad a \neq b. \tag{10.56}$$

10.1.6 Analysis of Estimation Error and Superefficiency

Let us consider the estimation error in the case of batch estimator $\hat{\mathbf{W}}$, which is the solution of the estimating equation

$$\sum_{k=1}^{N} \mathbf{F}\left(\mathbf{x}(k), \mathbf{W}\right) = \mathbf{0}. \tag{10.57}$$

The error depends on $\mathbf{F}$ and the number N of observations. By using the standard method of statistical analysis, we can calculate the covariance of estimator $\hat{\mathbf{W}} = \mathbf{W} + \Delta\mathbf{W}$, where $\Delta\mathbf{W}$ is the error. It is easier to calculate $E\left\{\Delta\mathbf{X}\,\Delta\mathbf{X}\right\}$ in terms of $\Delta\mathbf{X} = \Delta\mathbf{W}\,\mathbf{W}^{-1}$.

It should be noted that $\mathbf{F}$ and $\boldsymbol{\mathcal{R}}\,\mathbf{F}$ give the same error, since the estimating equations are equivalent. This is a big difference in comparison with online learning, where $\mathbf{F}$ and $\boldsymbol{\mathcal{R}}\,\mathbf{F}$ are different in convergence speed and stability. The covariance of $\Delta\mathbf{X}$ is now calculated explicitly. To this end, we put

$$l_a = E\{\varphi_a(s_a)\}, \tag{10.58}$$
$$\mathcal{G}^*_{ab,cd} = E\{F^*_{ab}(\mathbf{x}, \mathbf{W}) F^*_{cd}(\mathbf{x}, \mathbf{W})\} \tag{10.59}$$

by using the standardized estimating function $\mathbf{F}^*$.

Lemma 10.1 *The covariances of* $\Delta X^{(N)}_{ab}$ *are given as*

$$E\{\Delta X^{(N)}_{ab}\,\Delta X^{(N)}_{cd}\} = \frac{1}{N}\mathcal{G}^*_{ab,cd} + \mathcal{O}\left(\frac{1}{N^2}\right), \tag{10.60}$$

$$\mathcal{G}^*_{ac,bc} = c_{ac}c_{bc}\sigma_a^2\sigma_b^2\sigma_c^2\kappa_c^2 l_a l_b, \qquad a \neq b,\; c \neq a,\; c \neq b. \tag{10.61}$$

It is possible to evaluate the error by the covariance matrix of the error $\Delta\mathbf{y}$ in the recovered signals, where we assume the magnitudes are adjusted, that is $\mathbf{y} = \mathbf{s}$ and

$$\mathbf{y} = (\mathbf{W} + \Delta\mathbf{X}\,\mathbf{W})\,\mathbf{x} = \mathbf{s} + \Delta\mathbf{s}, \tag{10.62}$$
$$\Delta\mathbf{s} = \Delta\mathbf{X}\,\mathbf{s}. \tag{10.63}$$

Let us put

$$V^{(N)}_{ab} = E\left\{\Delta s_a\,\Delta s_b\right\} \tag{10.64}$$

We calculate, for $a \neq b$,

$$\begin{aligned} E\{y_a(k)\,y_b(k)\} &= E\{[s_a(k) + \sum_c \Delta X_{ac}\,s_c(k)]\,[s_b(k) + \sum_d \Delta X_{bd}\,s_d(k)]\} \\ &= E\{\Delta s_a\,\Delta s_b\} = \sum_{c,d} E\{\Delta X_{ac}\,\Delta X_{bd}\,s_c\,s_d\} \\ &= \sum_{c,d} E\{\Delta X_{ac}\,\Delta X_{bd}\}\,E\{s_c\,s_d\} = \sum_c E\{\Delta X_{ac}\,\Delta X_{bc}\}\,\sigma_c^2. \end{aligned}$$

Hence, we have

$$\begin{aligned} V_{ab}^{(N)} &= E\{\Delta s_a\, \Delta s_b\} = E\{y_a(k)\, y_b(k)\} \\ &= \sum_c E\{\Delta X_{ac}^{(N)}\, \Delta X_{bc}^{(N)}\}\, \sigma_c^2. \end{aligned} \tag{10.65}$$

Lemma 10.2 *The covariance matrix $\boldsymbol{V}_N$ of $\Delta \boldsymbol{s}$ is given by*

$$V_{ab}^{(N)} = \frac{1}{N} \sum_c \mathcal{G}_{ac,bc}^{*}\, \sigma_c^2 + \mathcal{O}\left(\frac{1}{N^2}\right), \qquad (a \neq b). \tag{10.66}$$

The lemma shows that the covariances

$$V_{ab}^{(N)} = E\{\Delta s_a\, \Delta s_b\} = E\{y_a\, y_b\} = (1/N) \sum_{k=1}^{N} y_a(k)\, y_b(k)$$

of the recovered signals $y_a(k)$ and $y_b(k)$ $(a \neq b)$ decrease in the order of $1/N$. This fact agrees with the ordinary asymptotic statistical analysis, as is expected. However, it happens that the covariance of any two recovered signals decreases in the order of $1/N^2$ under a certain condition. This is much smaller than (10.66) of order $1/N$. We call this property superefficiency.

Theorem 10.2 *A batch estimator is superefficient, $V_{ab}^{(N)} = \mathcal{O}\left(\frac{1}{N^2}\right)$ when*

$$l_a = E\left\{\varphi_a(y)\right\} = 0 \tag{10.67}$$

is satisfied, because $\mathcal{G}_{ac,bc}^{*} = 0$.

The condition (10.67) holds when

$$\varphi_a = -\frac{d}{dy} \log(r_a(y)). \tag{10.68}$$

It also holds when $r_a(y)$ is an even function, that is, the distribution is symmetric.

The superefficiency holds in the case of on-line learning

$$\Delta \boldsymbol{W}(k) = \eta\, \boldsymbol{F}(\boldsymbol{x}(k), \boldsymbol{W}(k)) \boldsymbol{W}(k). \tag{10.69}$$

When learning rate η is a small positive constant, $\boldsymbol{W}$ converges to the true solution $\boldsymbol{W}_*$, but finally fluctuates in its neighborhood. The magnitude of the fluctuation is

$$\lim_{k \to \infty} E\{||\boldsymbol{W}(k) - \boldsymbol{W}_*||_2^2\} = \mathcal{O}(\eta), \tag{10.70}$$

as has been proved in the general case of the stochastic gradient dynamic equation [14]. However, when (10.67) holds, we have superefficiency of on-line learning

$$\lim_{k \to \infty} E\{||\boldsymbol{W}(k)\boldsymbol{x}(k) - \boldsymbol{s}||_2^2\} = \mathcal{O}(\eta). \tag{10.71}$$

10.1.7 Adaptive Choice of φ Function

The estimation error depends on the choice of $\boldsymbol{F}(\boldsymbol{x}, \boldsymbol{W})$ or $\boldsymbol{F}^*(\boldsymbol{x}, \boldsymbol{W})$, that is, the functions $\boldsymbol{\varphi}$. Note that the standardized $\boldsymbol{F}^*$ improves the stability and convergence, but the asymptotic error for the batch mode and the on-line learning depend on $\boldsymbol{\varphi}$.

In order to improve the error, an adaptive choice of $\boldsymbol{\varphi}$ is useful. An adaptive choice of $\boldsymbol{\varphi}$ is also useful for guaranteeing stability. When $\boldsymbol{\varphi}$ is derived from the true probability distributions of the sources, the estimated $\hat{\boldsymbol{W}}$ is MLE, and is efficient in the sense that the asymptotic error is minimal and equal to the inverse of the Fisher information matrix. However, it is highly computationally expensive to estimate the probability density functions of the sources. Instead, we use a parametric family of $\boldsymbol{\varphi}$,

$$\varphi_a = \varphi_a\left(y; \boldsymbol{\Theta}_a\right) \tag{10.72}$$

for each source s_a and update the parameter $\boldsymbol{\Theta}_a$, which specifies φ_a by

$$\Delta \boldsymbol{\Theta}_a = -\eta_{\Theta} \frac{\partial \rho}{\partial \boldsymbol{\Theta}_a}. \tag{10.73}$$

There are several models to specify φ_a. The Gaussian mixture is one method for approximating the source probability density. It is the parametric family

$$q(y; \boldsymbol{\Theta}) = \sum_{i=1}^{u} v_i \exp\left\{-\frac{\left(x-\mu_i\right)^2}{2\sigma_i^2}\right\}, \tag{10.74}$$

where $\boldsymbol{\Theta}$ consists of a number of v_i, μ_i and σ_i^2. The corresponding parametric $\varphi(y; \boldsymbol{\Theta})$ is derived therefrom. This covers both sub-Gaussian and super-Gaussian distributions. However, this family is computationally expensive .

A simpler method is to use the generalized Gaussian family

$$q(y, \Theta) = c \exp\left\{-|y|^{\Theta}\right\}, \tag{10.75}$$

where Θ is the only parameter to be adjusted. This family covers both super-Gaussian and sub-Gaussian cases. The adaptive nonlinear activation function commonly used in an ICA algorithm has in this case the following form (see Chapter 6 for detailed explanation)

$$\varphi(y, \Theta) = \tilde{c}\, \mathrm{sign}(y)\, |y|^{\Theta-1} \tag{10.76}$$

where $\tilde{c}$ is some positive scaling constant.

Zhang *et al.* [1350] proposed an exponential family connecting three typical distributions; Gaussian, super-Gaussian and sub-Gaussian. It is the following exponential family of distributions [1350]

$$q_a\left(s, \boldsymbol{\theta}_a\right) = \exp\left\{\boldsymbol{\theta}_a^T \boldsymbol{g}(s) - \psi\left(\boldsymbol{\theta}_a\right)\right\}, \tag{10.77}$$

where $\boldsymbol{\theta}_a$ is a vector of canonical parameters, $\boldsymbol{g}(s)$ is an adequate vector function and ψ is a normalization factor. The function φ_a is derived as

$$\varphi_a(y) = -\frac{d}{dy} \log q_a\left(y, \boldsymbol{\theta}_a\right) = \boldsymbol{\theta}_a^T \boldsymbol{g}'(y). \tag{10.78}$$

Zhang [1350] proposed to use the three-dimensional model,

$$\boldsymbol{g}(y) = \left[\log \operatorname{sech}(y), -y^4, -y^2\right]^T \tag{10.79}$$

or

$$\boldsymbol{g}'(y) = \left[\tanh(y), y^3, y\right]^T, \tag{10.80}$$

of which components correspond to the typical φ proposed so far. They are responsible for the super-Gaussian, sub-Gaussian and linear cases, respectively. The $\varphi_a(y)$ is their linear combination, covering all the cases. The parameter $\boldsymbol{\theta}_a$ is adaptively determined as

$$\boldsymbol{\theta}_a(k+1) = \boldsymbol{\theta}_a(k) - \eta(k)\left[\boldsymbol{g}\left(y_a(k)\right) + E\left\{\boldsymbol{g}\left(y_a\right)\right\}\right], \tag{10.81}$$

where $E\left\{\boldsymbol{g}\left(y_a\right)\right\}$ may be adaptively estimated.

10.2 ESTIMATING FUNCTIONS IN NOISY CASES

Let us analyze the noisy case

$$\boldsymbol{x} = \boldsymbol{H}\boldsymbol{s} + \boldsymbol{\nu}, \tag{10.82}$$

where $\boldsymbol{\nu}$ is a noise vector in the measurement. We assume $\boldsymbol{\nu}$ is Gaussian with uncorrelated components.

$$\mathbf{R}_{\nu\nu} = E\{\boldsymbol{\nu}\boldsymbol{\nu}^T\} = \operatorname{diag}\{\sigma_1^2, \sigma_2^2, \ldots, \sigma_n^2\} \tag{10.83}$$

be its covariance matrix. In order to fix the scale, we also assume

$$E\{s_i^2\} = 1. \tag{10.84}$$

Let $\boldsymbol{W} = \boldsymbol{H}^{-1}$ be the true separating matrix, and put

$$\boldsymbol{y} = \boldsymbol{W}\boldsymbol{x}. \tag{10.85}$$

Then, we have

$$\boldsymbol{y} = \boldsymbol{s} + \boldsymbol{W}\boldsymbol{\nu} = \boldsymbol{s} + \tilde{\boldsymbol{\nu}}, \tag{10.86}$$

where $\tilde{\boldsymbol{\nu}} = \boldsymbol{W}\boldsymbol{\nu}$ is a noise vector whose components are correlated.

In the noisy case, functions of the type $\boldsymbol{F} = \mathbf{I} - \varphi(\boldsymbol{y})\boldsymbol{y}^T$ are not in general estimating functions. Indeed,

$$E\left\{\mathbf{I} - \varphi(\boldsymbol{y})\boldsymbol{y}^T\right\} \neq \mathbf{0} \tag{10.87}$$

even when $\boldsymbol{y}$ is derived from the true $\boldsymbol{W}$, because y_i and y_j are no longer independent even when $\boldsymbol{W} = \boldsymbol{H}^{-1}$. However, estimating functions exist even in the noisy case.

For the true $\boldsymbol{W} = \boldsymbol{H}^{-1}$, the noise term is

$$\tilde{\boldsymbol{\nu}} = \boldsymbol{W}\boldsymbol{\nu}, \tag{10.88}$$

which is Gaussian. Let its covariance matrix be

$$\boldsymbol{V} = E\left\{\tilde{\boldsymbol{\nu}}\tilde{\boldsymbol{\nu}}^T\right\} = E\left\{\boldsymbol{W}\boldsymbol{\nu}\boldsymbol{\nu}^T\boldsymbol{W}^T\right\} = \boldsymbol{W}\mathbf{R}_{\nu\nu}\boldsymbol{W}^T. \tag{10.89}$$

Kawanabe and Murata [682] studied all possible estimating functions. The following is the simplest estimating matrix function $\boldsymbol{F}(\boldsymbol{y}, \boldsymbol{W})$ with entries

$$F_{ab}(\boldsymbol{y}, \boldsymbol{W}) = y_a^3 y_b - 3v_{aa} y_a y_b - 3v_{ab} y_a^2 + 3v_{aa} v_{ab}, \tag{10.90}$$

where v_{ab} are elements of $\boldsymbol{V}$. We can easily prove that

$$E\{\boldsymbol{F}(\boldsymbol{y}, \boldsymbol{W})\} = \boldsymbol{0}, \tag{10.91}$$

when $\boldsymbol{W} = \boldsymbol{H}^{-1}$. Hence, the adaptive learning algorithm

$$\boldsymbol{W}(k+1) = \boldsymbol{W}(k) + \eta(k)\, \boldsymbol{F}\{\boldsymbol{y}(k), \boldsymbol{W}(k)\}\, \boldsymbol{W}(k) \tag{10.92}$$

is effective even under large Gaussian noise.

When the covariance matrix $\mathbf{R}_{\nu\nu}$ of the measurement noise is unknown, we need to estimate it. Factor analysis provides a method of estimating it (Ikeda and Toyama [598]). The off-diagonal term can be adaptively estimated from

$$v_{ab}(k+1) = (1-\eta_0)\, v_{ab}(k) + \eta_0\, y_a(k)\, y_b(k), \tag{10.93}$$

where η_0 is a learning rate.

The learning algorithm (10.92) is not necessarily stable. A stable algorithm is given by the standardized estimating function $\boldsymbol{F}^*$, which is an adaptive Newton method. We can obtain $\boldsymbol{F}^*$ explicitly by a method similar to the one used in the noiseless case.

10.3 ESTIMATING FUNCTIONS FOR TEMPORALLY CORRELATED SOURCE SIGNALS

10.3.1 Source Model

Independent source signals $s_i(k)$ are temporally correlated in many cases. If we use this fact, separation can be done much easier, even if we do not know the exact temporal correlation coefficients. Moreover, using the second order correlations is sufficient for separation of sources with different spectra (see Chapter 4). We begin with the description of the temporally correlated source models.

Let us consider a stationary stochastic model described by a linear model

$$s_i(k) = \sum_{p=1}^{L_i} a_{ip} s_i(k-p) + \varepsilon_i(k), \tag{10.94}$$

where L_i is either finite or infinite and $\varepsilon_i(k)$ is a zero mean independent and identically distributed (i.e., white) time series called innovation. The present section will involve such source models.

The innovation may be Gaussian or non-Gaussian. We assume that they satisfy

$$\begin{aligned} &E\{\varepsilon_i(k)\} = 0, \\ &E\{\varepsilon_i(k)\varepsilon_j(k')\} = 0, \quad (i \neq j \ \text{ or } \ k \neq k'). \end{aligned} \tag{10.95}$$

When L_i is finite, this is an AR model of degree L_i. By introducing the time shift operator z^{-1} such that $z^{-1}s_i(k) = s_i(k-1)$, we can rewrite (10.94) as

$$\left[A_i\left(z^{-1}\right)\right] s_i(k) = \varepsilon_i(k), \tag{10.96}$$

where

$$A_i\left(z^{-1}\right) = 1 - \sum_{p=1}^{L_i} a_{ip} z^{-p}. \tag{10.97}$$

By using the inverse of the polynomial A_i, the source signal is written as

$$s_i(k) = \left[A_i^{-1}\left(z^{-1}\right)\right] \varepsilon_i(k), \tag{10.98}$$

where $A_i^{-1}\left(z^{-1}\right)$ is a formal infinite power series of z^{-1}

$$A_i^{-1}\left(z^{-1}\right) = \sum_{p=0}^{\infty} \bar{a}_{ip} z^{-p}. \tag{10.99}$$

Function $A_i^{-1}\left(z^{-1}\right)$ represents the impulse response of the i-th source, by which $\{s_i(k)\}$ is generated from white signals $\{\varepsilon_i(k)\}$. Let $r_i(\varepsilon_i)$ be the probability density function of $\varepsilon_i(k)$. Then, the conditional probability density function of $s_i(k)$ conditioned on the past signals can be written as

$$p_i\left\{s_i(k)|s_i(k-1), s_i(k-2), \ldots\right\} = r_i\left\{s_i(k) - \sum_p a_{ip} s_i(k-p)\right\} = r_i\left\{A_i\left(z^{-1}\right) s_i(k)\right\}. \tag{10.100}$$

Therefore, for vector source signals $\boldsymbol{s}(k) = [s_1(k), \ldots, s_n(k)]^T$ at time k, the conditional probability density is

$$p\left\{\boldsymbol{s}(k)|\boldsymbol{s}(k-1), \boldsymbol{s}(k-2), \ldots\right\} = \prod_{i=1}^{n} r_i\left\{[A_i(z^{-1})]s_i(k)\right\}. \tag{10.101}$$

We introduce the following notations:

$$\boldsymbol{\varepsilon} = \left[\varepsilon_1, \varepsilon_2, \ldots, \varepsilon_n\right]^T, \tag{10.102}$$

$$\boldsymbol{A}\left(z^{-1}\right) = \operatorname{diag}\{A_1(z^{-1}), \ldots, A_n(z^{-1})\}, \tag{10.103}$$

$$r(\boldsymbol{\varepsilon}) = \prod_{i=1}^{n} r_i\left(\varepsilon_i\right), \tag{10.104}$$

and use the following abbreviation

$$\boldsymbol{s}_k = \boldsymbol{s}(k), \quad \boldsymbol{x}_k = \boldsymbol{x}(k) \quad \boldsymbol{y}_k = \boldsymbol{y}(k) = \boldsymbol{W}\boldsymbol{x}(k), \tag{10.105}$$

when there is no confusion. We also denote the past signals by

$$\boldsymbol{s}(k, \text{past}) = \{\boldsymbol{s}(k-1), \ \boldsymbol{s}(k-2), \cdots\}. \tag{10.106}$$

Then, eq. (10.101) is rewritten as

$$p\left\{\boldsymbol{s}(k)|\boldsymbol{s}(k,\text{past})\right\}=r\left\{\boldsymbol{A}\left(z^{-1}\right)\boldsymbol{s}(k)\right\}. \tag{10.107}$$

The joint probability density function of $\{\boldsymbol{s}(1),\boldsymbol{s}(2),\ldots,\boldsymbol{s}(N)\}$ is written as

$$\begin{aligned} p\left(\boldsymbol{s}(1),\boldsymbol{s}(2),\ldots,\boldsymbol{s}(N)\right) &= \prod_{k=1}^{N} p\left\{\boldsymbol{s}(k)|\boldsymbol{s}(k,\text{past})\right\} \\ &= \prod_{k=1}^{N} r\left\{\boldsymbol{A}\left(z^{-1}\right)\boldsymbol{s}(k)\right\}, \end{aligned} \tag{10.108}$$

where $\boldsymbol{s}(k)$ $(k\leq 0)$ are put equal to 0. Practically $\boldsymbol{s}(k)$ $(k<0)$ are not equal to 0 so that (10.108) is an approximation which holds asymptotically, that is, for large N.

The source models are specified by n functions $r_i(\varepsilon_i)$ and n inverse impulse response functions $A_i\left(z^{-1}\right)$. Blind source separation should extract the independent signals from their instantaneous linear mixtures $\boldsymbol{x}(k)$ without knowing the exact forms of $r_i(\varepsilon_i)$ and $A_i\left(z^{-1}\right)$. In other words, they are treated as unknown nuisance parameters.

Given N observations $\{\boldsymbol{x}(1),\boldsymbol{x}(2),\ldots,\boldsymbol{x}(N)\}$, their joint probability density function is easily derived from (10.108) and $\boldsymbol{s}_k=\boldsymbol{W}\boldsymbol{x}_k$, where $\boldsymbol{W}=\boldsymbol{H}^{-1}$. It is written as

$$p\left\{\boldsymbol{x}_1,\ldots,\boldsymbol{x}_N,\boldsymbol{W};\boldsymbol{A},r\right\}=\det|\boldsymbol{W}|^N\prod_{k=1}^{N} r_k\left\{\boldsymbol{A}\left(z^{-1}\right)\boldsymbol{W}\boldsymbol{x}_k\right\}, \tag{10.109}$$

which is specified by the unmixing parameter $\boldsymbol{W}=\boldsymbol{H}^{-1}$ and the nuisance parameters $\boldsymbol{A}$ and r's of the source models.

10.3.2 Likelihood and Score Functions

For the moment, we assume that r and $\boldsymbol{A}$ are known. We are then able to use the maximum likelihood (ML) method to estimate $\boldsymbol{W}$. The log likelihood is derived from (10.109) as

$$\begin{aligned} \rho^{(N)}\left(\boldsymbol{x}_1,\ldots,\boldsymbol{x}_N;\boldsymbol{W},\boldsymbol{A},r\right) &= -\log p\left\{\boldsymbol{x}_1,\ldots,\boldsymbol{x}_N;\boldsymbol{W},\boldsymbol{A},r\right\} \\ &= -N\log|\boldsymbol{W}|-\sum_{k=1}^{N}\log r\left\{\boldsymbol{A}\left(z^{-1}\right)\boldsymbol{W}\boldsymbol{x}_k\right\} \\ &= -N\log|\boldsymbol{W}|-\sum_{k=1}^{N}\log r\left\{\boldsymbol{A}\left(z^{-1}\right)\boldsymbol{y}_k\right\}, \end{aligned} \tag{10.110}$$

where we put $\boldsymbol{y}_k=\boldsymbol{W}\boldsymbol{x}_k$. The MLE (Maximum Likelihood Estimator) is the one that maximizes the above likelihood for given N observations $\boldsymbol{x}_1,\ldots,\boldsymbol{x}_N$.

We put

$$\rho\left(\boldsymbol{y}_k,\boldsymbol{W}\right)=-\log r\left(\boldsymbol{A}\left(z^{-1}\right)\boldsymbol{y}_k\right). \tag{10.111}$$

Note that ρ depends not only on $\boldsymbol{y}_k$ but also on the past $\boldsymbol{y}_{k-1}, \boldsymbol{y}_{k-2}, \ldots$, due to the operator $\boldsymbol{A}\left(z^{-1}\right)$. Note also that ρ is a function of $\boldsymbol{W}$ only through $\boldsymbol{y}_k$'s. We then have

$$\rho^{(N)} = -N \log |\boldsymbol{W}| - \sum_{k=1}^{N} \rho\left(\boldsymbol{y}_k, \boldsymbol{W}\right). \tag{10.112}$$

The small change $d\rho$ of ρ due to a small change of $\boldsymbol{W}$ to $\boldsymbol{W} + d\boldsymbol{W}$ is

$$d\rho = -\boldsymbol{\varphi}_r\left(\boldsymbol{A}\boldsymbol{y}_k\right)^T d\left(\boldsymbol{A}\boldsymbol{y}_k\right), \tag{10.113}$$

where $\boldsymbol{\varphi}_r(\boldsymbol{y}) = -\frac{\partial}{\partial \boldsymbol{y}} \log r(\boldsymbol{y})$ is a vector. Noting that $d\left(\boldsymbol{A}\boldsymbol{y}_k\right) = \boldsymbol{A} d\boldsymbol{y}_k$ and

$$d\boldsymbol{y}_k = d\boldsymbol{W}\boldsymbol{x}_k = d\boldsymbol{X}\boldsymbol{y}_k, \tag{10.114}$$

we have

$$d\rho = \boldsymbol{\varphi}_r\left(\boldsymbol{A}\boldsymbol{y}_k\right)^T \boldsymbol{A}\, d\boldsymbol{X}\boldsymbol{y}_k. \tag{10.115}$$

We finally have the score function in terms of $d\boldsymbol{X}$,

$$\frac{\partial \rho^{(N)}}{d\boldsymbol{X}} = \frac{\partial \rho^{(N)}}{\partial \boldsymbol{W}}\boldsymbol{W} = \sum_{k=1}^{N} \left[\mathbf{I} - \left\{\boldsymbol{\varphi}_r\left(\boldsymbol{A}\boldsymbol{y}_k\right)\boldsymbol{A}\right\}\boldsymbol{y}_k^T\right], \tag{10.116}$$

where $\boldsymbol{\varphi}_r\boldsymbol{A}$ is a column vector whose components are $\varphi_j A_j\left(z^{-1}\right)$. Note that, when $d\,\rho$ is written in component form as $d\rho = \sum c_{ij} dX_{ij}$, the derivative $\partial\rho/d\boldsymbol{X}$ is a matrix whose elements are c_{ij}. Hence, $\boldsymbol{\varphi}_r\left(\boldsymbol{A}\boldsymbol{y}\right)\boldsymbol{A}\boldsymbol{y}^T$ is represented in component form as $\varphi_i\left(A_i\left(z^{-1}\right)y_i\right)$ $A_i\left(z^{-1}\right)y_j$.

By putting

$$\frac{\partial \rho}{d\boldsymbol{X}} = \boldsymbol{F}\left(\boldsymbol{y}, \boldsymbol{W}; r, \boldsymbol{A}\right) = \mathbf{I} - \left\{\boldsymbol{\varphi}_r\left(\boldsymbol{A}\boldsymbol{y}\right) \circ \boldsymbol{A}\right\}\boldsymbol{y}^T, \tag{10.117}$$

the likelihood equation is given by

$$\sum_{k=1}^{N} \boldsymbol{F}\left(\boldsymbol{y}_k, \boldsymbol{W}; r, \boldsymbol{A}\right) = \mathbf{0}, \tag{10.118}$$

whose solution $\hat{\boldsymbol{W}}$ gives the maximum likelihood estimator.

10.3.3 Estimating Functions

Since we do not know the true distributions of sources $\{r_i\}$ and filters $\{A_i(z^{-1})\}$, we cannot use the above considered estimation function $\boldsymbol{F}$, which depends on r and $\boldsymbol{A}$. We search for estimating functions in the following class:

$$\boldsymbol{F}(\boldsymbol{y}, \boldsymbol{W}, q, \boldsymbol{B}) = \frac{d\rho(\boldsymbol{y}, \boldsymbol{W}, q, \boldsymbol{B})}{d\boldsymbol{X}} = \mathbf{I} - \boldsymbol{\varphi}_q\left\{\left(\boldsymbol{B}\left(z^{-1}\right)\boldsymbol{y}\right)\boldsymbol{B}\left(z^{-1}\right)\right\}\boldsymbol{y}^T, \tag{10.119}$$

for any fixed independent distribution q and matrix $\boldsymbol{B}\left(z^{-1}\right) = \mathrm{diag}\{B_1(z^{-1}), \ldots, B_n(z^{-1})\}$, with fixed filters $B_i(z^{-1}) = \sum_{p=0}^{L_i} b_{ip} z^{-p}$. This is an estimating function whatever q and $\boldsymbol{B}$ are, because it satisfies

$$\boldsymbol{E}_{\mathbf{W},r,\mathbf{A}}\left[\boldsymbol{F}(\boldsymbol{y}, \boldsymbol{W}, q, \boldsymbol{B})\right] = \mathbf{0} \tag{10.120}$$

for any sources having true independent distributions r and filters $\boldsymbol{A}\left(z^{-1}\right)$. It should be noted that $\boldsymbol{F}(\boldsymbol{y}, \boldsymbol{W}, q, \boldsymbol{B})$ is the true score function when the true nuisance parameters happen to be $r = q$ and $\boldsymbol{A} = \boldsymbol{B}\left(z^{-1}\right)$. However, even when q and $\boldsymbol{B}$ are misspecified, it works as an estimating function.

To be more precise, we state the identifiability conditions (Tong et al. 1991 [1157, 1158]; Comon, 1994 [299]).

Identifiability Condition:

1) All the independent sources have different spectra, that is, all $A_i\left(z^{-1}\right)$ are different, or

2) when some sources have the same spectra, the distributions r_i of these sources are non-Gaussian except for one source.

Summarizing these, we obtain the following Theorem from the general theory (Amari [16]).

Theorem 10.3 *When the identifiability condition is satisfied, the smallest admissible class of estimating functions is spanned by the non-diagonal elements of $\boldsymbol{F}(\boldsymbol{y}, \boldsymbol{W}, q, \boldsymbol{B})$, where q and $\boldsymbol{B}$ are arbitrary.*

The estimating equation is

$$\sum_{k=1}^{N} \boldsymbol{F}\left(\boldsymbol{y}_k, \boldsymbol{W}, q, \boldsymbol{B}\right) = \mathbf{0}. \tag{10.121}$$

An adaptive learning algorithm on the basis of such estimation function can take the form

$$\Delta \boldsymbol{W}(k) = \eta(k)\, \boldsymbol{F}\left[\boldsymbol{y}(k), \boldsymbol{W}(k)\right] \boldsymbol{W}(k) \tag{10.122}$$

or more generally by using the standardized estimating function,

$$\Delta \boldsymbol{W}(k) = \eta(k)\, \boldsymbol{F}^*\left[\boldsymbol{y}(k), \boldsymbol{W}(k)\right], \tag{10.123}$$

which will be presented in the following sections.

10.3.4 Simultaneous and Joint Diagonalization of Covariance Matrices and Estimating Functions

Joint diagonalization of the covariance matrices

$$\boldsymbol{R}_x(\tau) = E\{\boldsymbol{x}(k)\boldsymbol{x}^T(k-\tau)\} \tag{10.124}$$

for various τ is a standard method for blind separation of temporally correlated sources (Tong *et al.*, 1991; Molgedey and Schuster, 1994; Ikeda and Murata, 1999; Belouchrani *et al.*, 1997 [1159, 82, 84, 850, 866, 867]). This type of estimator looks quite different from those derived from estimating functions. To our surprise, this method is also given by an estimating function when the source signals are colored Gaussian. Although joint diagonalization is important for practical applications and interesting from a computational viewpoint, it is not admissible, and there always exist better estimating functions.

Another example of non-admissible estimating functions is the following,

$$\tilde{\boldsymbol{F}}(\boldsymbol{y}) = \varphi\left(\boldsymbol{B}\left(z^{-1}\right)\boldsymbol{y}\right)\left[\boldsymbol{C}\left(z^{-1}\right)\boldsymbol{y}\right]^{T}, \tag{10.125}$$

where $\boldsymbol{B}$ and $\boldsymbol{C}$ are arbitrary filters, which may be equal to each other.

Let $\boldsymbol{R}_s(\tau), \boldsymbol{R}_x(\tau)$ and $\boldsymbol{R}_y(\tau)$ be the cross-correlation matrices of $\boldsymbol{s}(t), \boldsymbol{x}(t)$ and $\boldsymbol{y}(t)$, respectively, defined by

$$\begin{aligned} \boldsymbol{R}_s(\tau) &= E\{\boldsymbol{s}(k)\boldsymbol{s}^T(k-\tau)\}, & (10.126)\\ \boldsymbol{R}_x(\tau) &= E\{\boldsymbol{x}(k)\boldsymbol{x}^T(k-\tau)\}, & (10.127)\\ \boldsymbol{R}_y(\tau) &= E\{\boldsymbol{y}(k)\boldsymbol{y}^T(k-\tau)\}. & (10.128) \end{aligned}$$

Because different sources are independent, $\boldsymbol{R}_s(\tau)$ is a diagonal matrix for any time delay τ. These matrices are connected by the relations

$$\begin{aligned} \boldsymbol{R}_x(\tau) &= \boldsymbol{H}\boldsymbol{R}_s(\tau)\boldsymbol{H}^T, & (10.129)\\ \boldsymbol{R}_x(\tau) &= \boldsymbol{W}\boldsymbol{R}_x(\tau)\boldsymbol{W}^T & (10.130) \end{aligned}$$

so that the true $\boldsymbol{W}$ is the one that diagonalizes $\boldsymbol{R}_x(\tau)$ for all τ simultaneously.

This leads us to the following batch algorithm for estimating $\boldsymbol{W}$ (see Chapter 4 for more details and explanation).

1. From the observed signals $\{\boldsymbol{x}_1, \ldots, \boldsymbol{x}_N\}$, calculate the empirical cross-correlation

$$\hat{\boldsymbol{R}}_x(\tau) = \frac{1}{N}\sum_{k=1}^{N}\boldsymbol{x}(k)\boldsymbol{x}^T(k-\tau). \tag{10.131}$$

2. Prewhiten the signals by $\bar{\boldsymbol{x}} = \boldsymbol{\Lambda}^{1/2}\boldsymbol{Q}\boldsymbol{x}$ such that the correlation matrix with zero delay $\hat{\boldsymbol{R}}_{\bar{x}}(0) = \left\langle\bar{\boldsymbol{x}}(k)\bar{\boldsymbol{x}}^T(k)\right\rangle$ becomes equal to the identity matrix, where $\boldsymbol{Q}$ is the orthogonal matrix composed of the eigenvectors of $\hat{\boldsymbol{R}}_x(0) = \mathbf{V}\boldsymbol{\Lambda}\mathbf{V}^T$ and $\boldsymbol{\Lambda}^{1/2}$ is a diagonal matrix composed of the inverse square roots of eigenvalues. We then have the resultant cross-correlation matrix $\hat{\boldsymbol{R}}_{\bar{x}}(\tau)$ of $\boldsymbol{x}_1$, where $\hat{\boldsymbol{R}}_{\bar{x}}(0)$ is the identity matrix.

3. Let $\boldsymbol{W} = \boldsymbol{U}\boldsymbol{\Lambda}^{1/2}\boldsymbol{Q}$ be the singular value decomposition of $\boldsymbol{W}$. We have fixed $\boldsymbol{\Lambda}^{1/2}$ and $\boldsymbol{Q}$ by using $\hat{\boldsymbol{R}}_x(0)$. The remaining task is to diagonalize $\hat{\boldsymbol{R}}_{\bar{x}}(\tau), \tau = 1, 2, \ldots$, by finding a suitable orthogonal matrix $\boldsymbol{U}$ such that $\hat{\boldsymbol{R}}_y(\tau)$ are diagonal where $\boldsymbol{y} = \boldsymbol{U}\bar{\boldsymbol{x}}$. A typical algorithm is to find $\boldsymbol{U}$ that minimizes the weighted sum of the squares of the off-diagonal elements of $\boldsymbol{R}_y(\tau)$,

$$\hat{J}(\boldsymbol{U}) = \sum_{i\neq j}\sum_{\tau\neq 0} c(\tau)\left\{\hat{R}_{ij}(\tau)\right\}^2, \tag{10.132}$$

where $R_{ij}(\tau) = E\{y_i(k)y_j(\tau)\}$ and $c(\tau)$ are suitably chosen non-negative weights.

We show how the above method is related to estimating functions. We have, for $i \neq j$, $\tau > 0$, and in the Gaussian case,

$$\begin{aligned} C_{ij}(\tau) &= E\{(y_i(k)y_j(k-\tau))^2\} \\ &= E\left\{\{y_i(k)\}^2\right\} E\left\{\{y_j(k-\tau)\}^2\right\} + 2E^2\left\{y_i(k)y_j(k-\tau)\right\} \\ &= 1 + 2\left\{R_{ij}(\tau)\right\}^2, \end{aligned} \tag{10.133}$$

because $E\{\bar{x}_i(k)^2\} = E\{y_i(k)^2\} = 1$ holds when $\boldsymbol{U}$ is an orthogonal matrix. Therefore, the cost function (10.132) to be minimized is written as

$$J(\boldsymbol{U}) = \frac{1}{2} \sum_{i \neq j} \sum_{\tau} c_\tau C_{ij}(\tau). \tag{10.134}$$

Given a set of observations $\boldsymbol{y}(1), \ldots, \boldsymbol{y}(N)$, where $\boldsymbol{y}(k) = \boldsymbol{U}\bar{\boldsymbol{x}}(k)$, $C_{ij}(\tau)$ is replaced by its empirical estimate

$$\hat{C}_{ij}(\tau) = \frac{1}{N} \sum_{k=1}^{N} \left[y_i(k)y_j(k-\tau)\right]^2, \tag{10.135}$$

where $y_j(k)$, $k \leq 0$, are put equal to 0. Therefore, the estimator $\hat{\boldsymbol{U}}$ minimizes

$$\hat{C}(\boldsymbol{U}) = \frac{1}{2} \sum_{i \neq j} \sum_{\tau} c_\tau \hat{C}_{ij}(\tau). \tag{10.136}$$

In order to calculate the derivative of $\hat{C}_{ij}(\tau)$, we use

$$\begin{aligned} d\left[y_i^2(k)y_j^2(k-\tau)\right] &= 2\Big[y_i(k)dy_i(k)y_j^2(k-\tau) + y_i(k)^2 y_j(k-\tau)dy_j(k-\tau)\Big] \\ &= 2\sum_l \Big[y_i(k)dX_{il}y_l(k)y_j^2(k-\tau) + y_i^2(k)y_j(k-\tau)dX_{jl}y_l(k-\tau)\Big], \end{aligned} \tag{10.137}$$

where dX_{il} are entries of the matrix

$$d\boldsymbol{X} = d\boldsymbol{U}\boldsymbol{U}^{-1}, \tag{10.138}$$

which is an antisymmetric matrix, because $\boldsymbol{U}$ is an orthogonal matrix.

We note that

$$\begin{aligned} &\frac{N}{2} d\left[\sum_{i \neq j} \hat{C}_{ij}(\tau)\right] \\ &= \sum_{i \neq j} \sum_l \sum_{k=1}^{N} \left[y_i(k)dX_{ik}y_l(k)y_j^2(k-\tau) + y_i^2(k)y_j(k-\tau)dX_{jl}y_l(k-\tau)\right] \\ &= \sum_{k=1}^{N} \sum_l \sum_{i \neq j} dX_{il}\left[y_i(k)y_l(k)y_j^2(k-\tau) + y_j^2(k+\tau)y_i(k)y_l(k)\right], \end{aligned} \tag{10.139}$$

where $\boldsymbol{y}(k)$, $k > N$ is put equal to 0. We take up the coefficient of dX_{il} in the first term on the right-hand side of (10.139),

$$\sum_{j \neq i} y_i(k) y_l(k) y_j^2(k-\tau). \tag{10.140}$$

Because $dX_{il} = -dX_{li}$ holds, the coefficients of dX_{il} are summarized into

$$\begin{aligned} & y_i(k) y_l(k) \left[\sum_{j \neq i} y_j^2(k-\tau) - \sum_{j \neq l} y_j^2(k-\tau) \right] \\ = \quad & y_i(k) y_l(k) \left[y_l^2(k-\tau) - y_i^2(k-\tau) \right]. \end{aligned} \tag{10.141}$$

A similar relation holds for the second terms. Therefore, the derivatives are given by

$$\begin{aligned} & \frac{\partial}{\partial X_{il}} \frac{N}{2} \sum_{p \neq q} \hat{C}_{pq}(\tau) \\ & = \sum_k y_i(k) y_l(k) \left[y_l^2(k-\tau) - y_i^2(k-\tau) + y_l^2(k+\tau) - y_i^2(k+\tau) \right]. \end{aligned} \tag{10.142}$$

Summarizing these results, we have the following Theorem.

Theorem 10.4 *When the source signals are colored Gaussian, the method of simultaneous diagonalization is equivalent to the estimating function method with the entries of matrix estimating function*

$$F_{ij}(\boldsymbol{y}, \boldsymbol{U}) = y_i y_j \sum_{\tau} c(\tau) \left(z^{-\tau} + z^{\tau} \right) \left(y_j^2 - y_i^2 \right). \tag{10.143}$$

Note that the estimating function (10.143) does not belong to the admissible class. Hence, one can always find a better estimating function.

10.3.5 Standardized Estimating Function and Newton Method

Let $\boldsymbol{\mathcal{R}}(\boldsymbol{W}) = (\mathcal{R}_{AB})$ be a nonsingular matrix operator which may depend on $\boldsymbol{W}$. Then, $\boldsymbol{F}$ and $\tilde{\boldsymbol{F}} = \boldsymbol{\mathcal{R}} \boldsymbol{F}$ are equivalent estimating functions.

Among a class of equivalent estimating functions, the one $\boldsymbol{F}^*$ that satisfies

$$\boldsymbol{\mathcal{K}}^* = E \left\{ \frac{\partial \boldsymbol{F}^*}{\partial \boldsymbol{X}} \right\} = \text{identity operator} \tag{10.144}$$

is called the standardized estimating function (Amari, 1999 [22]). Given an estimating function $\boldsymbol{F}$, its standardized form is given by

$$\boldsymbol{F}^* = \boldsymbol{\mathcal{K}}^{-1} \boldsymbol{F}, \tag{10.145}$$

where

$$\boldsymbol{\mathcal{K}} = E \left\{ \frac{\partial \boldsymbol{F}}{\partial \boldsymbol{X}} \right\}. \tag{10.146}$$

We now calculate

$$\mathcal{K} = E\left\{\frac{\partial \boldsymbol{F}(\boldsymbol{y}, \boldsymbol{W}, q, \boldsymbol{B})}{\partial \boldsymbol{X}}\right\} \tag{10.147}$$

at the true solution $\boldsymbol{W} = \boldsymbol{H}^{-1}$. Rewriting $\boldsymbol{F} = \partial \rho / d\boldsymbol{X}$ or

$$d\rho = -\mathrm{tr}\, d\boldsymbol{X} + \left[\boldsymbol{\varphi}\left(\tilde{\boldsymbol{y}}\right)^T \boldsymbol{B}\left(z^{-1}\right)\right] d\boldsymbol{X}\boldsymbol{y} \tag{10.148}$$

in component form, where we put

$$\tilde{\boldsymbol{y}} = [\boldsymbol{B}\left(z^{-1}\right)]\boldsymbol{y}, \tag{10.149}$$

we calculate the second-order differential in the component form as follows:

$$\begin{aligned} d^2\rho &= d[\sum_{i,j,p} \varphi_i(\tilde{y}_i) b_{ip} y_j(k-p)] dX_{ij} \\ &= \sum_{i,j,p} \{\varphi_i'\left(\tilde{y}_i\right) d\tilde{y}_i b_{ip} y_j(k-p) + \varphi_i\left(\tilde{y}_i\right) b_{ip} dy_j(k-p)\} \, dX_{ij} \\ &= \sum_{i,j,m,q,k} \varphi_i'\left(\tilde{y}_i\right) b_{iq} y_m(k-q) b_{ip} y_j(k-p) dX_{im} dX_{ij} \\ &\quad + \sum_{i,j,m,k} \varphi\left(\tilde{y}_i\right) b_{ip} y_m(k-p) dX_{jm} dX_{ij}, \end{aligned} \tag{10.150}$$

where $\varphi_i'(y) = d\,\varphi_i(y)/dy$.

At the true solution, we have

$$E\{\varphi'\left(\tilde{y}_i\right) y_j y_m\} = E\{\varphi\left(\tilde{y}_i\right) y_j\} = 0, \quad \text{unless } \ i = j = m \tag{10.151}$$

and

$$\begin{aligned} d^2\rho &= \sum_i E\left\{\varphi_i'\left(\tilde{y}_i\right) \tilde{y}_i^2\right\} \left(dX_{ii}\right)^2 + \sum_{i,j} E\left\{\varphi\left(\tilde{y}_i\right) \tilde{y}_i\right\} dX_{ij} dX_{ji} \\ &\quad + \sum_{i \neq j} E\{\varphi_i'(\tilde{y}_i)[\sum_{p=0} b_{ip} y_j(k-p)]^2\}(dX_{ij})^2. \end{aligned} \tag{10.152}$$

Hence, the quadratic form $d^2\rho$ in terms of dX_{ij}'s splits into the diagonal terms

$$\sum \left(\tilde{m}_i + 1\right) \left(dX_{ii}\right)^2 \tag{10.153}$$

and 2×2 minor matrices consisting of dX_{ij} and dX_{ji} $(i \neq j)$,

$$\sum_{i \neq j} \left\{\tilde{\kappa}_i \tilde{\sigma}_{ij}^2 \left(dX_{ij}\right)^2 + dX_{ij} dX_{ji}\right\}, \tag{10.154}$$

where we put

$$\begin{aligned} \tilde{m}_i &= E\{\varphi_i'(\tilde{y}_i)\tilde{y}_i^2\}, & (10.155) \\ \tilde{\kappa}_i &= E\{\varphi_i'\left(\tilde{y}_i\right)\}, & (10.156) \\ \tilde{\sigma}_{ij}^2 &= E\{[\sum_{p=0} b_{ip} y_j(k-p)]^2\}. & (10.157) \end{aligned}$$

When we put

$$F_{ii}(\boldsymbol{y}) = 1 - y_i^2, \tag{10.158}$$

the recovered signals satisfy

$$E\{y_i^2\} = 1. \tag{10.159}$$

In this case, the diagonal term is

$$2\sum dX_{ii}^2, \tag{10.160}$$

and the 2×2 diagonal terms are

$$\tilde{\kappa}_i \tilde{\sigma}_{ij}^2 \left(dX_{ij}\right)^2 + \tilde{h}_i dX_{ij} dX_{ji}, \tag{10.161}$$

where

$$\tilde{h}_i = E\{\varphi\left(\tilde{y}_i\right) \tilde{y}_i\}. \tag{10.162}$$

From this analysis, we have the stability condition of the algorithm

$$\Delta \boldsymbol{W} = \eta\, \boldsymbol{F}(\boldsymbol{y}, \boldsymbol{W})\, \boldsymbol{W}. \tag{10.163}$$

Theorem 10.5 *The separating solution is asymptotically stable, when and only when*

$$1) \qquad \tilde{m}_i + 1 > 0, \tag{10.164}$$

$$2) \qquad \tilde{\kappa}_i > 0, \tag{10.165}$$

$$3) \qquad \tilde{\kappa}_i \tilde{\kappa}_j \tilde{\sigma}_{ij}^2 \tilde{\sigma}_{ji}^2 > 1. \tag{10.166}$$

The inverse of $\boldsymbol{\mathcal{K}}$ has the same block structure as $\boldsymbol{\mathcal{K}}$. Its diagonal $\mathcal{K}_{AA}$ parts for $A = (i,i)$ are

$$k_{ii,ii} = \frac{1}{1 + \tilde{m}_i} \tag{10.167}$$

and its 2×2 diagonal parts $\mathcal{K}_{AA'}$ for $A = (i,j)$ and $A' = (j,i)$, $i \neq j$ are

$$\mathcal{K}_{AA'} = c_{ij} \begin{bmatrix} \tilde{\kappa}_j \tilde{\sigma}_{ji}^2 & -1 \\ -1 & \tilde{\kappa}_i \tilde{\sigma}_{ij}^2 \end{bmatrix} \tag{10.168}$$

where

$$c_{ij} = \frac{1}{\tilde{\kappa}_i \tilde{\kappa}_j \tilde{\sigma}_{ij}^2 \tilde{\sigma}_{ji}^2 - 1}. \tag{10.169}$$

We have similar expressions for $F_{ii} = 1 - y_i^2$.

We thus have the standardized estimating function.

Theorem 10.6 *The standardized estimating function $\boldsymbol{F}^*(\boldsymbol{y}, \boldsymbol{W})$ has entries given by*

$$F_{ij}^* = c_{ij} \left[-\tilde{\kappa}_j \tilde{\sigma}_{ji}^2 \varphi_i\left(\tilde{y}_i\right) \tilde{y}_j^{(i)} + \varphi_j\left(\tilde{y}_j\right) \tilde{y}_i^{(j)} \right], \tag{10.170}$$

$$F_{ii}^* = \frac{1}{\tilde{m}_i + 1} \left\{1 - \varphi_i\left(\tilde{y}_i\right) \tilde{y}_i\right\}, \tag{10.171}$$

where

$$\tilde{y}_j^{(i)} = B_i\left(z^{-1}\right) y_j. \tag{10.172}$$

The associated adaptive learning algorithm

$$\Delta \boldsymbol{W} = \eta\, \boldsymbol{F}^*\left(\boldsymbol{y}, \boldsymbol{W}\right) \tag{10.173}$$

is the Newton method.

10.3.6 Asymptotic Errors

The asymptotic estimation error is easily obtained from

$$\mathcal{G}^*_{AB} = E\left\{F^*_A F^*_B\right\}. \tag{10.174}$$

The results in this section are derived in a similar way to that used in Amari (1999) [21].

Theorem 10.7

$$\mathcal{G}^*_{ik,jk} = c_{ik}\, c_{jk}\, \tilde{\sigma}^2_{ik}\, \tilde{\sigma}^2_{jk}\, \tilde{\sigma}^2_{kk}\, \tilde{\kappa}^2_j\, \tilde{l}_i\, \tilde{l}_j, \tag{10.175}$$

$$\mathcal{G}^*_{ii,ij} = \frac{1}{\tilde{m}_i + 1}\, c_{ij}\, \tilde{\kappa}_j\, \tilde{\sigma}^2_{ii}\, \tilde{\sigma}^2_{ij}\, \tilde{l}_j\, E\{\tilde{y}^2_i\, \varphi_i\left(\tilde{y}_i\right)\}, \tag{10.176}$$

where

$$\tilde{l}_i = E\{\varphi_i\left(\tilde{y}_i\right)\}, \qquad \tilde{m}_i = E\{\varphi'_i(\tilde{y}_i)\tilde{y}^2_i\}, \qquad \tilde{\kappa}_i = E\{\varphi'_i\left(\tilde{y}_i\right)\}, \tag{10.177}$$

$$\tilde{\sigma}^2_{ij} = E\{[\sum_{p=0} b_{ip} y_j(k-p)]^2\}, \qquad c_{ij} = [\tilde{\kappa}_i \tilde{\kappa}_j \tilde{\sigma}^2_{ij} \tilde{\sigma}^2_{ji} - 1]^{-1}. \tag{10.178}$$

The error covariances of recovered signals are given by

$$E\{\Delta y_i\, \Delta y_j\} = \sum E\{\Delta X_{ik}\, \Delta X_{jk}\}\, \sigma^2_k. \tag{10.179}$$

It is remarkable that "superefficiency" of

$$E\{\Delta y_i\, \Delta y_j\} = \mathcal{O}\left(\frac{1}{N^2}\right), \quad (i \neq j) \tag{10.180}$$

holds, when the condition

$$\tilde{l}_i = E\{\varphi_i\left(\tilde{y}_i\right)\} = 0 \tag{10.181}$$

holds. The proof is similar to the non-correlated case (Amari, 1999 [22]).

10.4 SEMIPARAMETRIC MODELS FOR MULTICHANNEL BLIND DECONVOLUTION

Most theories treat only blind source separation of instantaneous mixtures and it is only recently that the natural gradient approach has been proposed for multichannel blind deconvolution [34, 1351, 1361]. Amari *et al.* [33, 34] discussed the geometric structures of the

IIR (Infinite Impulse Response) filter manifold, to develop an efficient learning algorithm for blind deconvolution. However, in most practical implementations, it is necessary to employ a filter of finite length as a deconvolution model. Zhang *et al.* [1351, 1361, 1365] directly investigated the geometric structures of the FIR filter manifold and derived the natural gradient algorithm for training FIR filters. Local stability condition for natural gradient learning is also extended for the blind deconvolution case.

The present section will examine further convergence and efficiency of the batch estimator and natural gradient learning for blind deconvolution via the semiparametric statistical model and estimating functions [96]. First we introduce the geometrical properties of the manifold of FIR filters based on the Lie group structure and formulate the blind deconvolution problem within the framework of the semiparametric model, deriving a family of estimating functions for blind deconvolution. We then analyze the efficiency of the batch estimator based on estimating function. Finally, we prove that batch and natural gradient learning are superefficient under given nonsingular conditions [1351, 1361].

10.4.1 Notation and Problem Statement

As a convolutive mixing model, we consider a multichannel *linear time-invariant* (LTI) system of the form (see Chapter 9 for more detail and learning algorithms):

$$\mathbf{x}(k) = \sum_{p=0}^{\infty} \mathbf{H}_p \mathbf{s}(k-p), \tag{10.182}$$

where $\mathbf{H}_p$ is an $n \times n$-dimensional matrix of mixing coefficients at time-lag p, called the impulse response at time p, $\mathbf{s}(k) = [s_1(k), s_2(k) \ldots, s_n(k)]^T$ is an n-dimensional vector of source signals, zero-mean and *independent and identically distributed (i.i.d.)*, and $\mathbf{x}(k) = [x_1(k), \ldots, x_n(k)]^T$ is an n-dimensional vector of sensor signals. For simplicity, we use the notation

$$\mathbf{H}(z) = \sum_{p=0}^{\infty} \mathbf{H}_p z^{-p}, \tag{10.183}$$

where z is the z-transform variable. $\mathbf{H}(z)$ is usually called the mixing filter, which is unknown in blind deconvolution.

The goal of multichannel blind deconvolution is to retrieve source signals only using sensor signals $\mathbf{x}(k)$ and some knowledge of source signal distributions. Generally, we carry out the blind deconvolution with another multichannel LTI and non-causal system of the form

$$\mathbf{y}(k) = \sum_{p=-\infty}^{\infty} \mathbf{W}_p \mathbf{x}(k-p), \tag{10.184}$$

where $\mathbf{y}(k) = [y_1(k), y_2(k), \ldots, y_n(k)]^T$ is an n-dimensional vector of the outputs and $\mathbf{W}_p$ is an $n \times n$-dimensional coefficient matrix at time lag p, whose components are the parameters to be determined during training. The matrix transfer function of deconvolutive filters can be expressed as

$$\mathbf{W}(z) = \sum_{p=-\infty}^{\infty} \mathbf{W}_p z^{-p}, \tag{10.185}$$

The objective of blind deconvolution is to make the output signals $\mathbf{y}(k)$ of the separating model maximally spatially mutually independent and temporarily i.i.d. In this section, we employ a semiparametric model to derive a family of estimating functions and develop efficient learning algorithms for training the separating filter $\mathbf{W}(z)$. Finally, we analyze the convergence and efficiency of the learning algorithms.

In practice, we can easily implement the blind deconvolution problem with a *finite impulse response* (FIR) filter

$$\mathbf{W}(z) = \sum_{p=0}^{L} \mathbf{W}_p z^{-p}, \tag{10.186}$$

where L is the maximum order (length) of the deconvolutive/separating filters. Alternatively we can employ non-causal filters of the symmetrical form $\mathbf{W}(z) = \sum_{p=-L/2}^{L/2} \mathbf{W}_p z^{-p}$. In general, the multiplication of two filters of form (10.186) will enlarge the filter length. Below, we will discuss briefly some geometrical structures of the FIR manifold (see Chapter 9 for more details).

10.4.2 Geometrical Structures on FIR Manifold

Geometrical structures, such as the Riemannian metric on the parameter space, can help us develop efficient learning algorithms for training parameters. The commonly used gradient descent learning is not optimal in minimizing a cost function defined on a Riemannian space. The steepest search direction is given by the natural gradient. It has been demonstrated that the natural gradient search scheme is an efficient approach for solving iterative parameter estimation problems [19]. In order to develop an efficient learning algorithm for blind deconvolution, we first explore some geometrical properties of the manifold of FIR filters.

The set of all FIR filters $\mathbf{W}(z)$ of length L, having the constraint $\mathbf{W}_0$ being nonsingular, is denoted by $\mathcal{M}(L)$,

$$\mathcal{M}(L) = \left\{ \mathbf{W}(z) \mid \mathbf{W}(z) = \sum_{p=0}^{L} \mathbf{W}_p z^{-p}, \ \det(\mathbf{W}_0) \neq 0 \right\}. \tag{10.187}$$

$\mathcal{M}(L)$ is a manifold of dimension $n^2(L+1)$. In general, multiplication of two filters in $\mathcal{M}(L)$ will enlarge the filter length. This makes it difficult to introduce the Riemannian structure to the manifold of multichannel FIR filters. In order to explore possible geometrical structures of $\mathcal{M}(L)$, which will lead to effective learning algorithms for $\mathbf{W}(z)$, we define the algebraic operations of filters in the Lie group framework.

10.4.3 Lie Group

In the manifold $\mathcal{M}(L)$, Lie operations, *multiplication* $\circledast$ and *inverse* $\dagger$, are defined as follows: For $\mathbf{W}(z),\ \mathbf{H}(z) \in \mathcal{M}(L)$,

$$\mathbf{W}(z) \circledast \mathbf{H}(z) = \sum_{p=0}^{L} \sum_{q=0}^{p} \mathbf{W}_q \mathbf{H}_{(p-q)} z^{-p}, \tag{10.188}$$

$$\mathbf{W}^{\dagger}(z) = \sum_{p=0}^{L} \mathbf{W}_p^{\dagger} z^{-p}, \tag{10.189}$$

where $\mathbf{W}_p^{\dagger}$ are recurrently defined by $\mathbf{W}_0^{\dagger} = \mathbf{W}_0^{-1},\ \mathbf{W}_p^{\dagger} = -\sum_{q=1}^{p} \mathbf{W}_{p-q}^{\dagger} \mathbf{B}_q \mathbf{W}_0^{-1},\ p = 1, 2, \ldots, L$. With these operations, both $\mathbf{W}(z) \circledast \mathbf{H}(z)$ and $\mathbf{W}^{\dagger}(z)$ still remain on the manifold $\mathcal{M}(L)$. It is easy to verify that the manifold $\mathcal{M}(L)$ with the above operations forms a Lie Group [1351]. The identity element is $\mathbf{E}(z) = \mathbf{I}$, where $\mathbf{I}$ is the identity matrix. In fact the Lie multiplication of two $\mathbf{W}(z),\ \mathbf{H}(z) \in \mathcal{M}(L)$ is the truncated form of the ordinary multiplication up to order L, that is

$$\mathbf{W}(z) \circledast \mathbf{H}(z) = [\mathbf{W}(z)\mathbf{H}(z)]_L \tag{10.190}$$

where $[\mathbf{W}(z)]_L$ is a truncating operator such that any terms with orders higher than L in the matrix polynomial $\mathbf{W}(z)$ are omitted.

The fluctuations will be negligible if we make the length L of $\mathbf{W}(z)$ sufficiently large. However, considering the multiplication in the Lie group sense, we have $\mathbf{G}(z) = \mathbf{W}(z) \circledast \mathbf{H}(z) = \mathbf{I}$. In the following discussion, we consider the global transfer function in the Lie group sense $\mathbf{G}(z) = \mathbf{W}(z) \circledast \mathbf{H}(z)$.

10.4.4 Natural Gradient Approach for Multichannel Blind Deconvolution

The Lie group has an important property that admits an invariant Riemannian metric [1351]. Using the Lie group structure, we derive the natural gradient of a cost function $\rho(\mathbf{W}(z))$ defined on the manifold $\mathcal{M}(L)$

$$\tilde{\nabla}\rho(\mathbf{W}(z)) = \frac{\partial \rho(\mathbf{W}(z))}{\partial \mathbf{X}(z)} \circledast \mathbf{W}(z) = \nabla\rho(\mathbf{W}(z)) \circledast \mathbf{W}(z), \tag{10.191}$$

where $d\mathbf{X}(z)$ is a nonholonomic variable [25], defined by the following equation

$$d\mathbf{X}(z) = d\mathbf{W}(z) \circledast \mathbf{W}^{\dagger}(z) = \left[d\mathbf{W}(z)\mathbf{W}^{-1}(z)\right]_L. \tag{10.192}$$

Alternatively, the natural gradient can be expressed as

$$\tilde{\nabla}\rho(\mathbf{W}(z)) = \nabla\rho(\mathbf{W}(z)) \circledast \mathbf{W}^T(z^{-1}) \circledast \mathbf{W}(z). \tag{10.193}$$

However, it is much easier to evaluate the natural gradient, if we introduce the nonholonomic differential variable $d\mathbf{X}(z)$ defined by (10.192). There are two ways to calculate the $\frac{\partial \rho(\mathbf{W}(z))}{\partial \mathbf{X}(z)}$.

One is to evaluate it by the following relation

$$\frac{\partial \rho(\mathbf{W}(z))}{\partial \mathbf{X}(z)} = \frac{\partial \rho(\mathbf{W}(z))}{\partial \mathbf{W}(z)} \circledast \mathbf{W}^T(z^{-1}). \tag{10.194}$$

The other way is to directly calculate it by using the following property,

$$d\mathbf{y}(k) = d\mathbf{W}(z)\mathbf{x}(k) = d\mathbf{X}(z)\mathbf{y}(k). \tag{10.195}$$

From the above equation, we see that the differential $d\mathbf{X}(z)$ defines a channel variation with respect to variation of output of the separating model. This property is critical for the derivation of learning algorithms with equivariance property.

Assuming that $d\mathbf{X}(z) = \sum_{p=0}^{L} d\mathbf{X}_p z^{-p}$ is in $\mathcal{M}(L)$, and $\rho(\mathbf{X}(z))$ is a cost function defined on $\mathcal{M}(L)$, we can define

$$\frac{\partial \rho(\mathbf{X}(z))}{\partial \mathbf{X}_p} = \left(\frac{\partial \rho(\mathbf{X}(z))}{\partial X_{p,ij}} \right)_{n \times n}. \tag{10.196}$$

Hence, we can write

$$\frac{\partial \rho(\mathbf{X}(z))}{\partial \mathbf{X}(z)} = \sum_{p=0}^{L} \frac{\partial \rho(\mathbf{X}(z))}{\partial \mathbf{X}_p} z^{-p}. \tag{10.197}$$

The estimating function for blind deconvolution is denoted by

$$\mathbf{F}(\mathbf{y}, \mathbf{X}(z)) = \sum_{p=0}^{L} \mathbf{F}_p(\mathbf{y}, \mathbf{X}(z)) z^{-p} \tag{10.198}$$

where $\mathbf{F}_p \in \mathbb{R}^{n \times n}$, $p = 0, 1, \ldots, L$ are matrix functions on $\mathcal{M}(L)$. Given p, q, the derivative $\frac{\partial \mathbf{F}_p}{\partial \mathbf{X}_q}$ is a 4-dimensional tensor, defined by $\frac{\partial \mathbf{F}_p}{\partial \mathbf{X}_q} = \left(\frac{\partial F_{p,ij}}{\partial X_{q,lk}} \right)_{n \times n \times n \times n}$. For any matrix $\mathbf{P} \in \mathbb{R}^{n \times n}$, the operation $\frac{\partial \mathbf{F}_p}{\partial \mathbf{X}_q} \mathbf{P}$ is defined by $\frac{\partial \mathbf{F}_p}{\partial \mathbf{X}_q} \mathbf{P} = \sum_{l,k} \frac{\partial \mathbf{F}_p}{\partial X_{q,lk}} P_{lk}$. Therefore, the derivative $\frac{\partial \mathbf{F}(\mathbf{y}, \mathbf{X}(z))}{\partial \mathbf{X}(z)}$ is an operator mapping $\mathcal{M}(L)$ to $\mathcal{M}(L)$, defined by

$$\frac{\partial \mathbf{F}(\mathbf{y}, \mathbf{X}(z))}{\partial \mathbf{X}(z)} \mathbf{P}(z) = \sum_{p=0}^{L} \sum_{q=0}^{L} \frac{\partial \mathbf{F}_p}{\partial \mathbf{X}_q} \mathbf{P}_q z^{-p} \tag{10.199}$$

for any filter $\mathbf{P}(z) \in \mathcal{M}(L)$.

Using the above properties, we derive the natural gradient learning algorithm for multichannel deconvolution.

In order to implement the adaptive on-line learning, we formulate the standard cost function as

$$J(\mathbf{y}, \mathbf{W}(z)) = E\{\rho(\mathbf{y}, \mathbf{W}(z))\} = -\log|\det(\mathbf{W}_0)| - \sum_{i=1}^{n} E\{\log q(y_i)\}, \tag{10.200}$$

where $q(y_i)$ is an estimator of the true probability density function of source signals.

We evaluate the total differential $d\rho(\mathbf{y}, \mathbf{W}(z))$

$$\begin{aligned} d\rho(\mathbf{y}, \mathbf{W}(z)) &= d(-\log|\det(\mathbf{W}_0)| - \sum_{i=1}^{n} \log q(y_i)) \\ &= -tr(d\mathbf{W}_0 \mathbf{W}_0^{-1}) + \boldsymbol{\varphi}^T(\mathbf{y})(\mathbf{y})^T d\mathbf{y}, \end{aligned} \tag{10.201}$$

where tr represents the trace of a matrix and $\boldsymbol{\varphi}(\mathbf{y})$ is a vector of nonlinear activation functions,

$$\varphi_i(y_i) = -\frac{d \log q_i(y_i)}{dy_i} = -\frac{q_i'(y_i)}{q_i(y_i)}. \tag{10.202}$$

By introducing the nonholonomic differential base (10.192), we rewrite (10.201) as

$$d\rho(\mathbf{y}, \mathbf{W}(z)) = -\operatorname{tr}(d\mathbf{X}_0) + \boldsymbol{\varphi}^T(\mathbf{y}) d\mathbf{W}(z) \mathbf{W}^{-1}(z)\mathbf{y}. \tag{10.203}$$

Hence, we obtain

$$\frac{\partial \rho(\mathbf{y}, \mathbf{W}(z))}{\partial \mathbf{X}_p} = -\delta_{0,p}\, \mathbf{I} + \boldsymbol{\varphi}^T(\mathbf{y})\, \mathbf{y}^T(k-p), \;\; p = 0, 1, \ldots, L \tag{10.204}$$

Using the natural gradient descent learning rule, we obtain an efficient on-line learning algorithm as follows

$$\begin{aligned} \Delta \mathbf{W}_p(k) &= -\eta(k) \sum_{q=0}^{p} \frac{\partial \rho(\mathbf{W}(z))}{\partial \mathbf{X}_q} \mathbf{W}_{p-q} \\ &= \eta(k) \sum_{q=0}^{p} \left[\delta_{0,q}\, \mathbf{I} - \boldsymbol{\varphi}(\mathbf{y}(k)) \mathbf{y}^T(k-q)\right] \mathbf{W}_{p-q}(k), \end{aligned} \tag{10.205}$$

for $p = 0, 1, \ldots, L$, where η is the learning rate. In particular, the learning algorithm for $\mathbf{W}_0$ is described by

$$\Delta \mathbf{W}_0(k) = \eta(k) \left[\mathbf{I} - \boldsymbol{\varphi}(\mathbf{y}(k))\, \mathbf{y}^T(k)\right] \mathbf{W}_0(k). \tag{10.206}$$

Alternatively, we can use an adaptive batch version of the algorithm

$$\boxed{\Delta \mathbf{W}_p(k) = \eta \sum_{q=0}^{p} \left[\delta_{0,q}\, \mathbf{I} - \mathbf{R}_{\boldsymbol{\varphi}\mathbf{y}}^{(k)}(q)\right] \mathbf{W}_{p-q}(k),} \tag{10.207}$$

where

$$\boxed{\mathbf{R}_{\boldsymbol{\varphi}\mathbf{y}}^{(k)}(q) = (1-\eta_0)\mathbf{R}_{\boldsymbol{\varphi}\mathbf{y}}^{(k-1)}(q) + \eta_0 \boldsymbol{\varphi}(\mathbf{y}(k))\mathbf{y}^T(k-q).} \tag{10.208}$$

The NG algorithms (10.205) and (10.207) have two important properties, uniform performance (the equivariant property) and invariance of nonsingularity of $\mathbf{W}_0$.

Remark 10.2 *In multichannel blind deconvolution, an algorithm is equivariant if its dynamical behavior depends on the global transfer function $\mathbf{G}(z) = \mathbf{W}(z) \circledast \mathbf{H}(z)$, but not on the specific mixing filter $\mathbf{H}(z)$. In fact the learning algorithm (10.205) has the equivariant property in the Lie group sense. Writing the learning algorithm in the Lie group form and multiplying both sides by the mixing filter $\mathbf{H}(z)$ in the Lie group sense, we obtain*

$$\Delta \mathbf{G}(z) = -\eta \frac{\partial \rho(\mathbf{W}(z))}{\partial \mathbf{X}(z)} \circledast \mathbf{G}(z). \tag{10.209}$$

where $\mathbf{G}(z) = \mathbf{W}(z) \circledast \mathbf{H}(z)$. From equation (10.204), we know $\frac{\partial \rho(\mathbf{W}(z))}{\partial \mathbf{X}(z)}$ is formally independent of the mixing channel $\mathbf{H}(z)$. This means that the algorithm (10.205) is equivariant.

Another important property of the learning algorithm (10.206) is that it keeps the nonsingularity of $\mathbf{W}_0$ provided the initial $\mathbf{W}_0$ is nonsingular [1315]. In fact if we denote the inner product of two matrices by $\langle \mathbf{A}, \mathbf{B} \rangle = \mathrm{tr}(\mathbf{A}^T \mathbf{B})$, we can easily calculate the derivative of the determinant $|\mathbf{W}_0| = \det \mathbf{W}_0$ in the following way

$$\frac{d|\mathbf{W}_0|}{dt} = \langle \frac{\partial |\mathbf{W}_0|}{\partial \mathbf{W}_0}, \frac{d\mathbf{W}_0}{dt} \rangle = \langle |\mathbf{W}_0| \mathbf{W}_0^{-T}, \frac{d\mathbf{W}_0}{dt} \rangle \tag{10.210}$$

$$= \mathrm{tr}\left(|\mathbf{W}_0| \mathbf{W}_0^{-1} (\mathbf{I} - \varphi(\mathbf{y}) \mathbf{y}^T(k)) \mathbf{W}_0 \right) = tr\left(\mathbf{I} - \varphi(\mathbf{y}) \mathbf{y}^T(k)) |\mathbf{W}_0| \right). \tag{10.211}$$

This equation results in

$$|\mathbf{W}_0(t)| = |\mathbf{W}_0(0)| \exp(\int_0^t \mathrm{tr}(\mathbf{I} - \varphi(\mathbf{y}(\tau)) \mathbf{y}^T(\tau)) d\tau). \tag{10.212}$$

Therefore, the matrix $\mathbf{W}_0$ is nonsingular whenever the initial matrix $\mathbf{W}_0(0)$ is nonsingular. This means that the learning algorithm (10.205) keeps the filter $\mathbf{W}(z)$ on the manifold $\mathcal{M}(N)$ if the initial filter is on the manifold. The condition implies that the equilibrium points of the learning algorithm satisfy the following equations

$$E\left\{ \varphi(\mathbf{y}(k)) \, \mathbf{y}^T(k-p) \right\} = \mathbf{0}, \qquad \text{for } p = 1, 2, \ldots, L, \tag{10.213}$$

$$E\left\{ \mathbf{I} - \varphi(\mathbf{y}(k)) \, \mathbf{y}^T(k) \right\} = \mathbf{0}. \tag{10.214}$$

The nonlinear activation function $\varphi(\mathbf{y})$ originally is defined by the score function of the logarithm of source distribution functions. The choice of $\varphi(\mathbf{y})$ depends on both the statistics of the source signals and the stability conditions of the learning algorithm.

10.4.5 Efficient Score Matrix Function and its Representation

In this section, we give an explicit form of the score function, using a local nonholonomic reparameterization. We then derive the efficient score by projecting the score function onto the subspace orthogonal to the nuisance tangent space.

Assume that the mixing filter $\mathbf{H}(z)$ is in $\mathcal{M}(L)$. The blind deconvolution problem looks for a separating FIR filter $\mathbf{W}(z)$ such that the output $\mathbf{y}(k)$ of the separating model is maximally spatially mutually independent and temporarily i.i.d. To this end, we first define the score functions of log-likelihood with respect to $\mathbf{W}(z)$. Since the mixing model is a matrix FIR filter, we write an estimating function in the same matrix filter format

$$\mathbf{F}(\mathbf{x}, \mathbf{W}(z)) = \sum_{p=0}^{L} \mathbf{F}_p(\mathbf{x}, \bar{\mathbf{W}}) z^{-p}, \tag{10.215}$$

where $\mathbf{F}_p(\mathbf{x}, \bar{\mathbf{W}})$ are matrix functions of $\mathbf{x}$ and $\bar{\mathbf{W}} = [\mathbf{W}_0, \mathbf{W}_1, \ldots, \mathbf{W}_L]$.

Now let us consider the $\bar{\mathbf{W}}$-score function, which is a filter in $\mathcal{TM}(L)$, defined by [1351]

$$\frac{\partial \log p(\mathbf{y}; \bar{\mathbf{W}}, r)}{\partial \mathbf{W}(z)} = \sum_{p=0}^{L} \frac{\partial \log p(\mathbf{y}; \bar{\mathbf{W}}, r)}{\partial \mathbf{W}_p} z^{-p}, \tag{10.216}$$

where $p(\mathbf{y}; \bar{\mathbf{W}}, r)$ is the probability density function of $\mathbf{y}$, and $\frac{\partial \log p(\mathbf{y}; \bar{\mathbf{W}}, r)}{\partial \mathbf{W}_p}$ denotes the gradient in matrix form, whose (i,j)-element is defined by $\frac{\partial \log p(\mathbf{y}; \bar{\mathbf{W}}, r)}{\partial W_{pij}}$.

Using Amari's natural gradient approach [19, 148], we define

$$d\mathbf{X}(z) = d\mathbf{W}(z) \circledast \mathbf{W}^{\dagger}(z), \tag{10.217}$$

a nonholonomic differential variable which is not integrable. The variation $d\mathbf{H}(z)$ of $\mathbf{H}(z)$ is represented as $d\mathbf{H}(z) = -\mathbf{H}(z) \circledast d\mathbf{X}(z)$ in terms of $d\mathbf{X}(z)$ assuming $\mathbf{W}(z) = \mathbf{H}^{\dagger}(z)$.

Denote the inner product of any two filters $\mathbf{W}(z)$ and $\mathbf{H}(z)$ in tangent space $\mathcal{TM}_{\mathbf{W}(z)}$ by [1351]

$$< \mathbf{W}(z), \mathbf{H}(z) >= \sum_{p=0}^{L} tr(\mathbf{W}_p^T \mathbf{H}_p).$$

Consider the differential $d \log p(\mathbf{y}; \bar{\mathbf{W}}, r)$ with respect to the new variables,

$$d \log p(\mathbf{y}; \bar{\mathbf{W}}, r) =< \frac{\partial \log p(\mathbf{y}; \bar{\mathbf{W}}, r)}{\partial \mathbf{X}(z)}, d\mathbf{X}(z) > . \tag{10.218}$$

On the other hand, using the relation (10.217), we have

$$\begin{aligned} d \log p(\mathbf{y}; \bar{\mathbf{W}}, r) &= < \frac{\partial \log p(\mathbf{y}; \bar{\mathbf{W}}, r)}{\partial \mathbf{W}(z)}, d\mathbf{W}(z) > \\ &= < \frac{\partial \log p(\mathbf{y}; \bar{\mathbf{W}}, r)}{\partial \mathbf{W}(z)} \circledast \mathbf{W}^T(z^{-1}), d\mathbf{X}(z) > . \end{aligned} \tag{10.219}$$

Comparing the two equations (10.218) and (10.219), and using the invariant property of the differential expression, we deduce

$$\frac{\partial \log p(\mathbf{y}; \bar{\mathbf{W}}, r)}{\partial \mathbf{X}(z)} = \frac{\partial \log p(\mathbf{y}; \bar{\mathbf{W}}, r)}{\partial \mathbf{W}(z)} \circledast \mathbf{W}^T(z^{-1}). \tag{10.220}$$

Using the relation (10.195), we evaluate the score function at $\mathbf{X}(z) = \mathbf{0}$

$$\frac{\partial \log p(\mathbf{y}; \bar{\mathbf{W}}, r)}{\partial X_{p,ij}}|_{\mathbf{X}(z)=\mathbf{0}} = \varphi_i(y_i(k))y_j(k-p), \tag{10.221}$$

where $\varphi_i(y_i) = -\frac{d \log(r_i(y_i))}{dy_i}$, $i = 1, 2, \ldots, n$. This can also be re-written in the compact form

$$\mathbf{U}(\mathbf{x}, \mathbf{W}(z), r) = \sum_{p=0}^{L} \mathbf{U}_p z^{-p} = \sum_{p=0}^{L} \boldsymbol{\varphi}(\mathbf{y}(k))\, \mathbf{y}^T(k-p) z^{-p}, \tag{10.222}$$

where $\boldsymbol{\varphi}(\mathbf{y}) = [\varphi_1(y_1), \ldots, \varphi_n(y_n)]^T$, and $\mathbf{y}$ is the vector of estimated source signals. It should be noted that the score function $\mathbf{U}(\mathbf{x}, \mathbf{W}(z), r)$ generally depends on the sensor signals $\mathbf{x}(k)$ and the separating filter $\mathbf{W}(z)$. However, by introducing the nonholonomic reparameterization, we derive a score function that only depends on the output of the separating model or the global transfer function $\mathbf{G}(z)$. This property is called equivariance in blind separation of instantaneous mixtures [148]. The relative or the natural gradient of a cost function on the Riemannian manifold can be automatically derived from this nonholonomic representation [21, 1361]

The efficient scores, denoted by $\mathbf{U}^E(\mathbf{x}; \mathbf{W}(z), r)$, can be obtained by projecting the score function onto the subspace orthogonal to the nuisance tangent space $\mathcal{T}^{\mathcal{N}}_{\mathbf{W}(z),r}$. (See Zhang *et al.* [1351] for more details).

In summary we have the following theorem [1351]

Theorem 10.8 *The efficient score, $\mathbf{U}^E(\mathbf{x}; \mathbf{W}(z), r)$ is expressed by*

$$\mathbf{U}^E(\mathbf{x}; \mathbf{W}(z), r) = \sum_{p=0}^{L} \mathbf{U}^E_p z^{-p}, \tag{10.223}$$

where

$$\mathbf{U}^E_p = \boldsymbol{\varphi}(\mathbf{y})(k)\mathbf{y}^T(k-p), \quad \text{for } p \geq 1; \tag{10.224}$$

$$\mathbf{U}^E_0 = \begin{cases} \boldsymbol{\varphi}(\mathbf{y}(k))\, \mathbf{y}^T(k), & \text{for off} - \text{diagonal elements}, \\ c_2(\varphi_i(y_i(k))\, y_i(k) - 1), & \text{for diagonal elements}. \end{cases} \tag{10.225}$$

10.5 ESTIMATING FUNCTION AND STANDARDIZED ESTIMATING FUNCTION FOR MULTICHANNEL BLIND DECONVOLUTION

In this section, we discuss a family of estimating functions and standardized estimating functions for blind deconvolution.

It has been shown by Zhang *et al.* [1351, 1361, 1365] that the efficient score function is an estimating function which can be expressed as

$$\mathbf{F}(\mathbf{x}(k), \mathbf{W}(z)) = \sum_{p=0}^{L} \boldsymbol{\varphi}(\mathbf{y}(k))\mathbf{y}(k-p)^T z^{-p} - \mathbf{I}, \tag{10.226}$$

where $\mathbf{y}(k) = \sum_{p=0}^{L} \mathbf{W}_p \mathbf{x}(k-p)$, and φ is a vector of given activation functions, provided that the derivative operator $\boldsymbol{\mathcal{K}}(z) = E\left\{\frac{\partial \mathbf{F}(\mathbf{x}, \mathbf{W}(z))}{\partial \mathbf{X}(z)}\right\}$ is invertible. The estimating function is the efficient score function, when and $F_{ii}(y_i) = \varphi_i(y_i) y_i - 1$.

The derivative operator $\boldsymbol{\mathcal{K}}(z) = E\left\{\frac{\partial \mathbf{F}(\mathbf{x}, \mathbf{W}(z))}{\partial \mathbf{X}(z)}\right\}$ is a tensor filter, represented by

$$\boldsymbol{\mathcal{K}}(z) = \sum_{p=0}^{L} \boldsymbol{\mathcal{K}}_p z^{-p}. \tag{10.227}$$

(See Appendix A for detailed derivation.)

We take the following notations

$$m_i = E\{y_i^2 \varphi_i'(y_i)\}, \quad \kappa_i = E\{\varphi_i'(y_i)\}, \quad \sigma_i^2 = E\{y_i^2\}, \tag{10.228}$$

$$c_{ij} = [\kappa_i \kappa_j \sigma_i^2 \sigma_j^2 - 1]^{-1}, \quad l_i = E\{\varphi(y_i)\}. \tag{10.229}$$

Lemma 10.3 *The coefficients of the operator* $\boldsymbol{\mathcal{K}}(z) = \sum_{p=0}^{L} \boldsymbol{\mathcal{K}}_p z^{-p}$ *can be expressed by*

$$\mathcal{K}_{p,ij,lm} = E\{\varphi'(y_i(k)) s_j^2(k-p)\} \delta_{il}\, \delta_{jm} + \delta_{im}\, \delta_{jl}\, \delta_{0p}. \tag{10.230}$$

Furthermore, if the following conditions are satisfied

$$\kappa_i \neq 0, \quad \kappa_i \kappa_j \sigma_i^2 \sigma_j^2 - 1 \neq 0, \quad m_i + 1 \neq 0, \tag{10.231}$$

then the derivative operator $\boldsymbol{\mathcal{K}}(z)$ *is invertible.*

The semiparametric approach suggests the use of the following estimating equation [23, 138] for the parameters of interest,

$$\sum_{k=1}^{N} \mathbf{F}(\mathbf{x}(k), \mathbf{W}(z)) = \mathbf{0}. \tag{10.232}$$

The estimator obtained from (10.232) is called an M-estimator. An M-estimator is consistent, that is, the estimator $\mathbf{W}(z,k)$ converges to the true value as N tends to infinity without reference to the true distribution of the sources $r(\mathbf{s})$. The estimating function is not unique, as for any nonsingular linear operator $\boldsymbol{\mathcal{R}}(z)$ mapping from $\mathcal{M}(L)$ to $\mathcal{M}(L)$, $\boldsymbol{\mathcal{R}}(z)\, \mathbf{F}(\mathbf{x}, \mathbf{W}(z))$ is also an estimating function. It has already been established that the two estimating functions are equivalent in the sense that the derived batch estimators give exactly the same solution. This defines an equivalent class of estimating functions that are essentially the same in batch estimation. However, when we consider online learning, the learning dynamics is not equivalent and this necessitates introduction of an estimating function that will make the learning algorithm more stable and efficient. To this end, we introduce the concept of standardized estimating function. The standardized estimating function [19] is defined as follows: If the derivative operator $\boldsymbol{\mathcal{K}}(z) = E\left\{\frac{\partial \mathbf{F}(\mathbf{x}, \mathbf{W}(z))}{\partial \mathbf{X}(z)}\right\}$ is an identity operator, the estimating function is called the standardized estimating function [1351, 1361, 1365].

Lemma 10.4 *Given any estimating function* $\mathbf{F}(\mathbf{x}, \mathbf{W}(z))$*, if the operator* $\boldsymbol{\mathcal{K}}(z)$ *is invertible, then*

$$\boldsymbol{\mathcal{K}}^{-1}(z)\mathbf{F}(\mathbf{x}, \mathbf{W}(z)) \tag{10.233}$$

is a standardized estimating function.

Using lemma (10.4) we can formulate a family of standardized estimating functions for the blind deconvolution problem.

Theorem 10.9 ([1351, 1361]) *Given an estimating function of form (10.226), the standardized estimating function is expressed by*

$$\mathbf{F}^*(\mathbf{x}, \mathbf{W}(z)) = \sum_{p=0}^{L} \mathbf{F}_p^*(\mathbf{x}, \mathbf{W}(z)) z^{-p}, \tag{10.234}$$

where

$$F_{0,ii}^* = \frac{1}{m_i + 1} \{\varphi_i(y_i) y_i - 1\}, \qquad \text{for } i = 1, 2, \ldots, n \tag{10.235}$$

$$F_{0,ij}^* = c_{ij} \{\kappa_j \sigma_i^2 \varphi_i(y_i) y_j - \varphi_j(y_j) y_i\}, \qquad \text{for } i \neq j \tag{10.236}$$

$$F_{p,ij}^* = \varphi_i(y_i) y_j (k - p)/(\kappa_i \sigma_j^2), \qquad \text{for } p \geq 1. \tag{10.237}$$

Proof In order to compute the inverse of the operator $\boldsymbol{\mathcal{K}}(z)$, we consider the following equation

$$\boldsymbol{\mathcal{K}}(z)\mathbf{F}^*(\mathbf{x}, \mathbf{W}(z)) = \mathbf{F}(\mathbf{x}, \mathbf{W}(z)) \tag{10.238}$$

Using expression (10.230), we can rewrite (10.238) in the following component form

$$(m_i + 1) F_{0,ii}^* = F_{0,ii}, \quad \text{for } i = 1, 2, \ldots, n, \tag{10.239}$$

$$\kappa_i \sigma_j^2 F_{0,ij}^* + F_{0,ji}^* = F_{0,ij}, \quad \text{for } i, j = 1, 2, \ldots, n, \ i \neq j. \tag{10.240}$$

$$\kappa_i \sigma_j^2 F_{p,ij}^* = F_{p,ij}, \quad \text{for } p \geq 1, \ i, j = 1, 2, \ldots, n. \tag{10.241}$$

Solving the above equations, we obtain the results. □

There are some advantages of using the standardized estimating function in on-line learning. The natural gradient learning is given by

$$\Delta \mathbf{W}(z) = -\eta\, \mathbf{F}^*(\mathbf{x}, \mathbf{W}(z)) \circledast \mathbf{W}(z). \tag{10.242}$$

It can be proved that the true solution $\mathbf{W}(z) = \mathbf{H}^{\dagger}(z)$ is always the stable equilibrium of the natural gradient learning above, provided conditions (10.231) are satisfied. This property is called universal convergence. See [19] for further information. The statistics in (10.228) and (10.229) require an on-line estimate in order to implement learning algorithm (10.242). In particular, if the source signals are binary, taking values $1, -1$, we can calculate the statistics for the standardized estimating function. if we choose the cubic function $\varphi_i(y_i) = y_i^3$ as activation function, the statistics are evaluated by

$$m_i = 3, \ \kappa_i = 3, \ \sigma_i^2 = 1, \ \tilde{\gamma}_{ij} = c_{ij}^{-1} = 8. \tag{10.243}$$

Therefore, the standardized estimating function can be given explicitly.

10.5.1 Superefficiency of Batch Estimator

Amari [19] proves that in the instantaneous case, the covariance $V_{ij}(N) = E\{y_i y_j\}$ $(i \neq j)$ vanishes at a rate $\frac{1}{N^2}$ under certain simple conditions. This property is called superefficiency. Zhang *et al.* proved that superefficiency remains valid in blind deconvolution [1351, 1361, 1365].

Suppose that $\mathbf{F}^*(\mathbf{x}, \mathbf{W}(z))$ is a standardized estimating function:

$$E\{\Delta \mathbf{X}_N(z,k) \otimes \Delta \mathbf{X}_N^T(z,k)\} = \frac{1}{N}\mathcal{G}^*(z) + O(\frac{1}{N^2}), \tag{10.244}$$

where $\mathcal{G}^*(z) = \mathcal{K}^{-1}(z)\mathcal{G}(z)\mathcal{K}^{-T}(z) = E\{\mathbf{F}^*(\mathbf{x}, \mathbf{W}(z)) \otimes \mathbf{F}^{*T}(\mathbf{x}, \mathbf{W}(z))\}$.

Lemma 10.5 *The coefficients of $\mathcal{G}^*(z)$ are expressed by*

$$G^*_{0,il,jl} = c_{il}c_{jl}\sigma_i^2\sigma_j^2\sigma_l^2 k_l^2 l_i l_j, \text{ for } i \neq j, j \neq l, l \neq i, \tag{10.245}$$

$$G^*_{0,ii,ji} = \frac{1}{m_i+1} c_{ji}\kappa_i\sigma_j^2 l_j E\{s_i^2\varphi_i(s_i)\}, \text{ for } i \neq j, \tag{10.246}$$

$$G^*_{p,il,jl} = \frac{l_i l_j}{\kappa_i\kappa_j}, \text{ for } p \geq 1, \ i,j = 1,2,\ldots,n. \tag{10.247}$$

Using the expression for $\mathbf{F}^*(\mathbf{x}, \mathbf{W}(z))$ in Theorem 10.9, we can derive the result by direct calculation [1351].

Theorem 10.10 *A batch estimator is superefficient when the following condition is satisfied*

$$l_i = E\{\varphi_i(s_i)\} = 0, \text{ for } i = 1,2,\ldots,n. \tag{10.248}$$

Proof Using Lemma 10.5 and (10.248), we have

$$G^*_{p,il,jl} = 0, \text{ for } i \neq j, \ p = 0,1,\ldots,L, \ l = 1,2,\ldots,n. \tag{10.249}$$

Writing the estimate (10.244) in component form,

$$E\{\Delta X_{p,il}^{(N)} \, \Delta X_{p,jl}^{(N)}\} = \frac{1}{N} G^*_{p,il,jl} + \mathcal{O}(\frac{1}{N^2}) = \mathcal{O}(\frac{1}{N^2}), \tag{10.250}$$

for $i \neq j$, leads to the following estimation

$$V_{ij}^{(N)} = \sum_{l=1}^{n}\sum_{p=0}^{L} E\{\Delta X_{p,il}^{(N)} \, \Delta X_{p,jl}^{(N)}\}\sigma_l^2 = \mathcal{O}(\frac{1}{N^2}). \tag{10.251}$$

This proves our result. □

From the arguments above we can see that superefficiency of both batch estimator and natural gradient algorithm require that

$$l_i = E\{\varphi_i(s_i)\} = 0, \qquad \text{for } i = 1, 2, \ldots, n. \tag{10.252}$$

Fortunately, the commonly used activation functions, such as the cubic function and the hyperbolic tangent function satisfy these conditions.

In this chapter, we have discussed estimating functions and the semiparametric approach to blind separation/deconvolution problem. We also tackled issues of convergence and efficiency of the batch estimator and natural gradient learning. Firstly, blind separation and multichannel deconvolution is formulated in the framework of the semiparametric model and a family of estimating functions and standardized estimating functions are derived by using efficient score functions. The advantage of using the semiparametric approach is that we do not need to estimate the nuisance parameters- the probability density functions of source signals in blind deconvolution. It is inferred from the theory of estimating functions that the batch estimator of the estimating equation converges to the true solution as the number of observed data tends to infinity. If stability conditions are satisfied, the natural gradient learning also converges to the true solution irrespective of the probability density function of the source signals. The superefficiency of both the batch estimator and natural gradient learning is proven when certain local conditions are satisfied.

Appendix A. Representation of Operator $\mathcal{K}(z)$

In this appendix, we present the explicit form of operator $\mathcal{K}(z)$ and its inverse $\mathcal{K}^{-1}(z)$ and give a definition of the transpose $\mathcal{K}^T(z)$ of $\mathcal{K}(z)$ [1351]. Assume that the recovered signal $\mathbf{y}(k)$ is spatially mutually independent and temporally i.i.d.

Lemma A.6 *For any $p \neq q$,*

$$E\left\{\frac{\partial \mathbf{F}_p}{\partial \mathbf{X}_q}\right\} = \mathbf{0}. \tag{A.1}$$

Proof By definition, $F_{p,ij} = \varphi(y_i)\, y_j(k-p) - \delta_{0,p}$. Using the i.i.d. properties of $\mathbf{y}(k)$ and the relation (10.195), we have , for $p \neq q$,

$$E\left\{\frac{\partial F_{p,ij}}{\partial X_{q,lm}}\right\} = E\left\{\varphi'(y_i)\frac{\partial y_i(k)}{\partial X_{q,lm}} y_j(k-p) + \varphi(y_i)\frac{\partial y_j(k-p)}{\partial X_{q,lm}}\right\} = 0. \tag{A.2}$$

Proposition A.1 *The derivative operator $\mathcal{K}(z)$ can be represented as*

$$\mathcal{K}(z) = \sum_{p=0}^{L} \mathcal{K}_p z^{-p} = \sum_{p=0}^{L} E\left\{\frac{\partial \mathbf{F}_p}{\partial \mathbf{X}_p}\right\} z^{-p}, \tag{A.3}$$

which maps $\mathbf{P}(z) \in \mathcal{M}(L)$ to $\mathcal{K}(z)\mathbf{P}(z) = \sum_{p=0}^{L} \mathcal{K}_p \mathbf{P}_p z^{-p}$. Furthermore, the coefficients of $\mathcal{K}(z)$ are given by

$$\mathcal{K}_{p,ij,lm} = E\{\varphi'(y_i(k))\, y_j^2(k-p)\}\delta_{il}\,\delta_{jm} + \delta_{im}\,\delta_{jl}\,\delta_{0p}. \tag{A.4}$$

Proof From definition (10.199) and using (A.1) we have

$$\mathcal{K}(z)\mathbf{P}(z) = \sum_{p=0}^{L}\sum_{q=0}^{L} E\left\{\frac{\partial \mathbf{F}_p}{\partial \mathbf{X}_q}\right\}\mathbf{P}_q z^{-p} = \sum_{p=0}^{L} E\left\{\frac{\partial \mathbf{F}_p}{\partial \mathbf{X}_p}\right\}\mathbf{P}_p z^{-p}. \tag{A.5}$$

Using the i.i.d. properties of $\mathbf{y}(k)$ and (10.195), we have

$$\begin{aligned} E\left\{\frac{\partial F_{p,ij}}{\partial X_{p,lm}}\right\} &= E\left\{\varphi'(y_i)\frac{\partial y_i(k)}{\partial X_{p,lm}} y_j(k-p) + \varphi(y_i)\frac{\partial y_j(k-p)}{\partial X_{p,lm}}\right\} \\ &= E\{\varphi'(y_i(k))\, y_j^2(k-p)\}\,\delta_{il}\,\delta_{jm} + \delta_{im}\,\delta_{jl}\,\delta_{0p}. \end{aligned} \tag{A.6}$$

The result follows.

□

In order to calculate the inverse of $\mathcal{K}(z)$, consider the following equation

$$\mathcal{K}(z)\mathbf{X}(z) = \mathbf{P}(z), \tag{A.7}$$

where $\mathbf{X}(z)$ and $\mathbf{P}(z) \in \mathcal{M}(L)$. Substitute (A.4) into (A.7), and write it in component form

$$(m_i+1)X_{0,ii} = P_{0,ii}, \quad \text{for } i = 1, 2, \ldots, n, \tag{A.8}$$

$$\kappa_i \sigma_j^2 X_{0,ij} + X_{0,ji} = P_{0,ij}, \quad \text{for } i,j = 1, 2, \ldots, n,\ i \neq j, \tag{A.9}$$

$$\kappa_i \sigma_j^2 X_{p,ij} = P_{p,ij}, \quad \text{for } p \geq 1,\ i,j = 1, 2, \ldots, n. \tag{A.10}$$

We can directly solve $X_{0,ii}$ and $X_{p,ij}$ from (A.8) and (A.10). For $X_{0,ij}, i \neq j$, we can write (A.9) in the following 2×2 self-closed subsystem

$$\begin{bmatrix} \kappa_i \sigma_j^2 & 1 \\ 1 & \kappa_j \sigma_i^2 \end{bmatrix}\begin{bmatrix} X_{0,ij} \\ X_{0,ji} \end{bmatrix} = \begin{bmatrix} P_{0,ij} \\ P_{0,ji} \end{bmatrix}. \tag{A.11}$$

If $\tilde{\gamma}_{ij} = \kappa_i \kappa_j \sigma_i^2 \sigma_j^2 - 1 \neq 0$, we can uniquely solve the above equations. Therefore, we have the following result.

Proposition A.2 *If* $m_i + 1 \neq 0,\ \kappa_i \neq 0,\ \ \tilde{\gamma}_{ij} = c_{ij}^{-1} = \kappa_i\kappa_j\sigma_i^2\sigma_j^2 - 1 \neq 0$, *then the operator* $\mathcal{K}(z)$ *is invertible and the inverse* $\mathcal{K}^{-1}(z) = \sum_{p=0}^{L} \mathcal{R}_p z^{-p}$ *is expressed by*

$$\mathcal{R}_{0,ii,lm} = \frac{1}{m_i+1}\delta_{il}\,\delta_{im}, \quad \mathcal{R}_{p,ij,lm} = \frac{1}{\kappa_i\,\sigma_j^2}\delta_{il}\delta_{jm} \tag{A.12}$$

$$\mathcal{R}_{0,ij,lm} = c_{ij}(\kappa_j\,\sigma_i^2\,\delta_{il}\,\delta_{jm} - \delta_{im}\,\delta_{jl}). \tag{A.13}$$

Now we give a definition of the transpose operation of tensor filters. The transpose of a tensor filter $\boldsymbol{\mathcal{K}}(z)$ is given by

$$\boldsymbol{\mathcal{K}}^T(z) = \sum_{p=0}^{L} \boldsymbol{\mathcal{K}}_p^T z^{-p}, \tag{A.14}$$

where $\boldsymbol{\mathcal{K}}_p^T = [\mathcal{K}_{p,lm,ij}]$, given $\boldsymbol{\mathcal{K}}_p = [\mathcal{K}_{p,ij,lm}]$.

11

Linear Blind Filtering and Separation Using a State-Space Approach

Every tool carries with it the spirit by which it has been created.

—(Werner Karl Heisenberg; 1901-1976)

In this chapter, we present a flexible and universal framework using the state-space approach to blind separation and filtering. As a special case, we consider the standard multichannel blind deconvolution problem with causal FIR filters.

The state-space description of dynamical systems [653, 893] is a powerful and flexible generalized model for blind separation and deconvolution or more generally for filtering and separation. There are several reasons why the state-space models are advantageous for blind separation and filtering. Transfer function models in the z-domain or the frequency domain are equivalent to the state-space models in the time domain for any linear, stable time-invariant dynamical system. However, using transfer functions directly, it is difficult to exploit the internal representation of real dynamical systems. The main advantage of the state-space description is two fold. It gives the internal description of a system, but there are various equivalent canonical types of state-space realizations for a system, such as balanced realization and observable canonical forms [653, 893]. In particular, it is possible to parameterize some specific classes of models which are of interest in applications. In addition, it is relatively easy to tackle the stability problem of state-space systems using the Kalman filter. Moreover, the state-space model enables a much more general description than the standard finite impulse response (FIR) convolutive filtering models discussed in Chapter 9. In fact, all the known filtering models, such as the AR, MA, ARMA, ARMAX and Gamma filtering, could also be considered as special cases of flexible state-space models [989, 1358].

The state-space approach to blind source separation/deconvolution problems has been developed by Zhang, Cichocki and Amari [1348]-[1368], [281]-[284] and independently by Salam *et al.* [1032, 1033, 1034, 1251]. Efficient natural gradient learning algorithms have been derived by Zhang and Cichocki [1358] to adaptively estimate the state matrices by minimizing the mutual information. In order to compensate for the model bias and to reduce the effect of noise, a state estimator approach [1366] has been recently proposed using the Kalman filter. We also extend the state-space approach to nonlinear systems [283], and two-stage learning algorithms [281] for estimating parameters in nonlinear demixing models (see Chapter also 12).

In this chapter, we briefly review adaptive learning algorithms based on the natural gradient approach and give some new insight into multiple-input multiple-output blind separation and filtering in the state-space framework.

11.1 PROBLEM FORMULATION AND BASIC MODELS

Suppose that the unknown source signals $\mathbf{s}(k)$ are mixed by a stable but unknown causal, linear time invariant dynamical system described by the following set of matrix difference equations (see Fig. 11.1)

$$\begin{aligned} \overline{\boldsymbol{\xi}}(k+1) &= \overline{\mathbf{A}}\ \overline{\boldsymbol{\xi}}(k) + \overline{\mathbf{B}}\,\mathbf{s}(k) + \overline{\mathbf{N}}\ \boldsymbol{\nu}_P(k), & (11.1)\\ \mathbf{x}(k) &= \overline{\mathbf{C}}\ \overline{\boldsymbol{\xi}}(k) + \overline{\mathbf{D}}\,\mathbf{s}(k) + \boldsymbol{\nu}(k), & (11.2) \end{aligned}$$

where $\overline{\boldsymbol{\xi}}(k) \in \mathbb{R}^d$ is the state vector of the system, $\mathbf{s}(k) \in \mathbb{R}^n$ is a vector of unknown source signals (assuming that they are zero-mean, i.i.d. and spatially independent), $\mathbf{x}(k) \in \mathbb{R}^m$ is an available vector of sensor signals, $\overline{\mathbf{A}} \in \mathbb{R}^{d\times d}$ is a state matrix, $\overline{\mathbf{B}} \in \mathbb{R}^{d\times n}$ is an input mixing matrix, $\overline{\mathbf{C}} \in \mathbb{R}^{m\times d}$ is an output mixing matrix, $\overline{\mathbf{D}} \in \mathbb{R}^{m\times n}$ is an input-output mixing matrix and $\overline{\mathbf{N}} \in \mathbb{R}^{d\times d}$ is a noise matrix. The integer d is called the state dimension or system order. In principle, there exists an infinite number of state space realizations for a given system. For example, the state vector $\overline{\boldsymbol{\xi}}(k)$ might contain some states that are not excited by the input or are not observed in the output. In practice, we consider models for which the order d is minimal. Even for the minimal order (called canonical form), the state representation is not unique. An equivalent system representation giving the same transfer function is obtained by applying a state transformation nonsingular matrix $\overline{\mathbf{T}} \in \mathbb{R}^{d\times d}$ to define a new state vector $\overline{\boldsymbol{\xi}}'(k) = \overline{\mathbf{T}}\overline{\boldsymbol{\xi}}(k)$. The eigenvalues of the state matrix $\overline{\mathbf{A}}$ are invariant under this transformation. In order to ensure the stability of the system, the eigenvalues of $\overline{\mathbf{A}}$ must be smaller than 1 in absolute value, i.e., the eigenvalues must be located inside the unit circle. In fact, the eigenvalues of $\overline{\mathbf{A}}$ are directly related to the poles of matrix transfer function.

Let us assume that there exists a stable inverse system (in some sense discussed later) called a separating or demixing-filtering system. In other words, we assume that the demixing/filtering model consists of another linear state-space system, described as (see Fig. 11.1)

$$\begin{aligned} \boldsymbol{\xi}(k+1) &= \mathbf{A}\,\boldsymbol{\xi}(k) + \mathbf{B}\,\mathbf{x}(k) + \mathbf{L}\boldsymbol{\nu}_R(k), & (11.3)\\ \mathbf{y}(k) &= \mathbf{C}\,\boldsymbol{\xi}(k) + \mathbf{D}\,\mathbf{x}(k), & (11.4) \end{aligned}$$

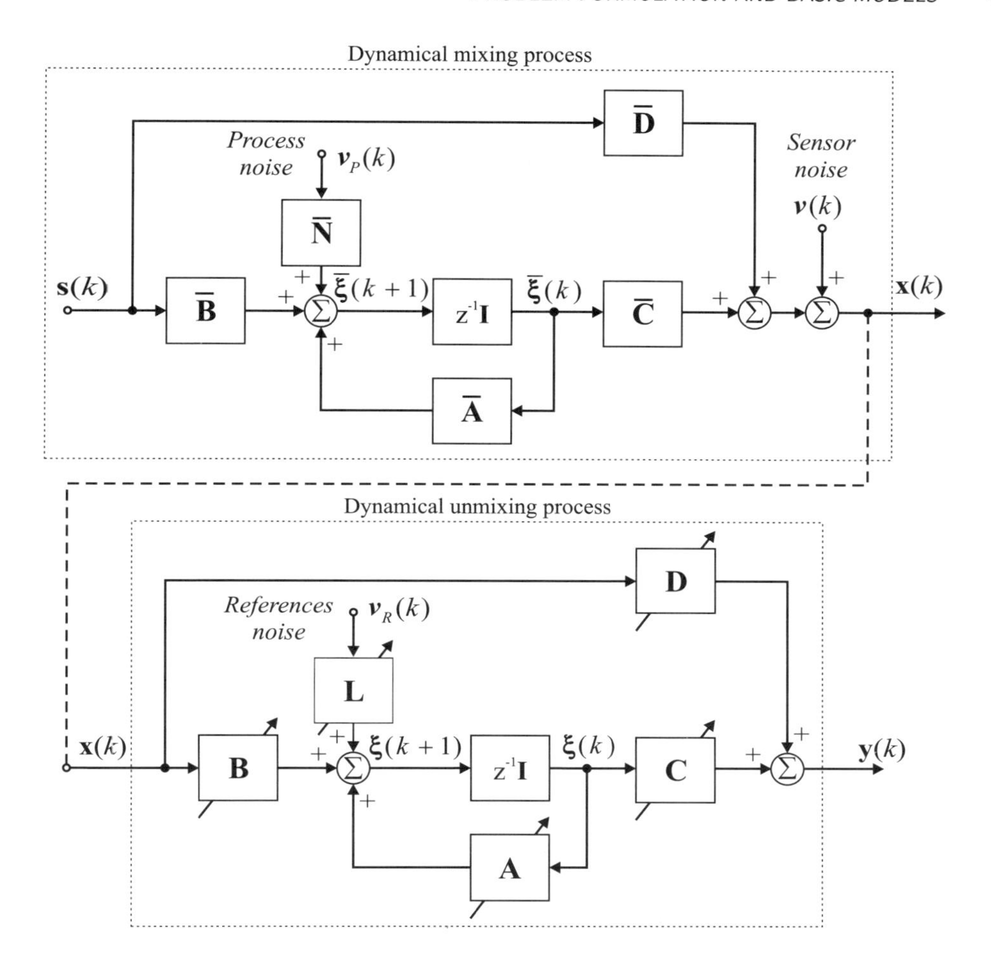

Fig. 11.1 Conceptual block diagram illustrating the general linear state-space mixing and self-adaptive demixing model for blind separation and filtering. The objective of learning algorithms is the estimation of a set of matrices $\{\mathbf{A}, \mathbf{B}, \mathbf{C}, \mathbf{D}, \mathbf{L}\}$ [281, 283, 284, 1352, 1353, 1354, 1361].

where $\boldsymbol{\xi}(k) \in \mathbb{R}^M$ is the state vector of the separating system and the unknown state-space matrices have dimensions: $\mathbf{A} \in \mathbb{R}^{M \times M}$, $\mathbf{B} \in \mathbb{R}^{M \times m}$, $\mathbf{C} \in \mathbb{R}^{n \times M}$, $\mathbf{D} \in \mathbb{R}^{n \times m}$, with $M \geq d$ (i.e., the order of the demixing system should be at least of the order of the mixing system).

Since the mixing system is completely unknown (we do not know its parameters or even its order), our objective is to identify this system or to estimate a demixing/filtering system with some kind of ambiguities. In the blind deconvolution problem, the estimation of order (dimension M) is a difficult task, and therefore we usually overestimate it, i.e., $M >> d$. The overestimation of the order M may produce auxiliary delays of output signals with respect to the original sources, but this is usually acceptable in the blind deconvolution because it is most important to recover the waveforms of the original signals.

The state-space description [653, 893] allows us to divide the variables into two types: The internal state variable $\boldsymbol{\xi}(k)$, which produces the dynamics of the system, and the external variables $\mathbf{x}(k)$ and $\mathbf{y}(k)$, which represent the input and output of the demixing/filtering system, respectively. The vector $\boldsymbol{\xi}(k)$ is known as the state of the dynamical system, which summarizes all the information about the past behavior of the system that is needed to uniquely predict its future behavior. The linear state-space model plays a critical role in the mathematical formulation of a dynamical system. It also allows us to realize the internal structure of the system and to define the controllability and observability of the system [653]. The parameters in the state equation of the separating/filtering system are referred to as internal representation parameters (or simply internal parameters), and the parameters in the output equation as external ones. Such a partition enables us to estimate the demixing model in two stages: Estimation of the internal dynamical representation and the output (memoryless) demixing system. The first stage involves a batch or adaptive on-line estimation of the state space and input parameters represented by the set of matrices $\mathbf{A}, \mathbf{B}$, such that the systems would be stable and have possibly sparse, and lowest dimensions. The second stage involves fixing the internal parameters and estimating the output (external) parameters represented by the set of matrices $\mathbf{C}, \mathbf{D}$ by employing suitable batch or adaptive algorithms. In general, the set of matrices $\boldsymbol{\Theta} = \{\mathbf{A}, \mathbf{B}, \mathbf{C}, \mathbf{D}, \mathbf{L}\}$ contains parameters, to be determined in a learning process on the basis of the sequence of available sensor signals $\mathbf{x}(k)$ and some *a priori* knowledge about the system and noise.

If we ignore the noise terms in the mixing model, its transfer function is an $m \times m$ matrix of the form

$$\mathbf{H}(z) = \overline{\mathbf{C}}\,(z\,\mathbf{I} - \overline{\mathbf{A}})^{-1}\,\overline{\mathbf{B}} + \overline{\mathbf{D}}, \tag{11.5}$$

where z^{-1} is a time delay operator.

We formulate the blind separation problem as a task to recover the original signals from observations $\mathbf{x}(k) = [\mathbf{H}(z)]\,\mathbf{s}(k)$, without *a priori* knowledge of the source signals $\mathbf{s}(k)$ or state-space matrices $[\overline{\mathbf{A}},\ \overline{\mathbf{B}},\ \overline{\mathbf{C}},\ \overline{\mathbf{D}}]$ except for certain statistical features of the source signals.

For simplicity, we assume that the noise terms both in the mixing and demixing models are negligibly small. The transfer function of the demixing model can be expressed as

$$\mathbf{W}(z) = \mathbf{C}\,(z\,\mathbf{I} - \mathbf{A})^{-1}\,\mathbf{B} + \mathbf{D}. \tag{11.6}$$

The output signals $\mathbf{y}(k)$ should estimate the source signals in the following sense

$$\mathbf{y}(k) = [\mathbf{W}(z)\,\mathbf{H}(z)]\,\mathbf{s}(k) = \mathbf{P}\,[\boldsymbol{\Lambda}(z)]\,\mathbf{s}(k), \tag{11.7}$$

where $\mathbf{P}$ is any permutation (memoryless) matrix and $\boldsymbol{\Lambda}(z)$ is a diagonal matrix with entries $\lambda_i z^{-\Delta_i}$; here λ_i is a nonzero scaling constant and Δ_i is any nonnegative integer. In other words, we assume that the independent sources are fully recoverable from the demixing model (11.3) and (11.4) if the global matrix transfer function satisfies the following relationship

$$\mathbf{G}(z) = \mathbf{W}(z)\,\mathbf{H}(z) = \mathbf{P}\,\boldsymbol{\Lambda}(z). \tag{11.8}$$

11.1.1 Invertibility by State Space Model

The fundamental question arises as to whether a set of matrices $[\mathbf{A}, \mathbf{B}, \mathbf{C}, \mathbf{D}]$ exists in the demixing model (11.3) and (11.4), such that its transfer function $\mathbf{W}(z)$ satisfies (11.8). The answer is affirmative under some weak conditions. We will show that if the matrix $\overline{\mathbf{D}}$ in the mixing model satisfies $rank(\overline{\mathbf{D}}) = n$, and $\mathbf{W}_*(z)$ is the inverse of $\mathbf{H}(z)$, then any state-space realization $[\mathbf{A}, \mathbf{B}, \mathbf{C}, \mathbf{D}]$ of a new transfer function $\mathbf{W}(z) = \mathbf{P}\,\boldsymbol{\Lambda}(z)\,\mathbf{W}_*(z)$ satisfies equation (11.4). Suppose that $\overline{\mathbf{D}}$ satisfies $rank(\overline{\mathbf{D}}) = n$, and $\overline{\mathbf{D}}^+$ is the generalized inverse of $\overline{\mathbf{D}}$ in the sense of the Penrose generalized pseudo-inverse[1].

Let

$$\mathbf{D} = \overline{\mathbf{D}}^+, \qquad \mathbf{A} - \overline{\mathbf{A}} = \mathbf{B}\,\overline{\mathbf{C}},$$

$$\mathbf{B} = \overline{\mathbf{B}}\,\mathbf{D}, \qquad \mathbf{C} = -\mathbf{D}\overline{\mathbf{C}},$$

and thus the global system can be described as

$$\mathbf{G}(z) = \mathbf{W}(z)\,\mathbf{H}(z) = \mathbf{I}_n. \tag{11.9}$$

Any nonsingular transform matrix $\mathbf{T}$ does not change the transfer functions if the following relations hold

$$\begin{aligned}
\mathbf{A} &= \mathbf{T}\,(\overline{\mathbf{A}} - \overline{\mathbf{B}}\,\overline{\mathbf{D}}^+\,\overline{\mathbf{C}})\,\mathbf{T}^{-1}, \\
\mathbf{B} &= \mathbf{T}\,\overline{\mathbf{B}}\,\overline{\mathbf{D}}^+, \\
\mathbf{C} &= -\overline{\mathbf{D}}^+\,\overline{\mathbf{C}}\,\mathbf{T}^{-1}, \\
\mathbf{D} &= \overline{\mathbf{D}}^+.
\end{aligned}$$

Therefore, source signals can, in principle, be recovered by the linear state space demixing/filtering model (11.3) and (11.4). We summarize this feature in the form of the following Theorem [1358]:

Theorem 11.1 *If the matrix $\overline{\mathbf{D}}$ in the mixing model is full column rank, i.e., $rank(\overline{\mathbf{D}}) = n$, then there exists a set of matrices $\mathbf{W} = \{\mathbf{A}, \mathbf{B}, \mathbf{C}, \mathbf{D}\}$, such that the output signals $\mathbf{y}$ of the state-space system (11.3) and (11.4) recover the independent source signals in the sense of (11.8).*

Remark 11.1 *In the blind deconvolution/separation problem, we do not know in advance neither the set of matrices $[\mathbf{A}, \mathbf{B}, \mathbf{C}, \mathbf{D}]$ nor the state-space dimension M. Before we begin to estimate the matrices $[\mathbf{A}, \mathbf{B}, \mathbf{C}, \mathbf{D}]$, we may attempt to estimate the dimension M of the system if one needs to obtain a canonical solution. There are several criteria for estimating the dimension of a system in system identification, such as the AIC and MDL criteria (see Chapter 3). However, the order estimation problem in blind demixing and filtering is a quite difficult and challenging problem. It remains an open problem that is not discussed in this book. Fortunately, if the dimension of the state vector in the demixing model can be simply overestimated, i.e., if we take it larger than necessary, we can still successfully recover the original source signals.*

[1]In practice, for $m = n$ the matrix $\mathbf{D} \in \mathbb{R}^{n \times n}$ is a square nonsingular matrix, and thus $\mathbf{D}^+ = \mathbf{D}^{-1}$.

11.1.2 Controller Canonical Form

If the transfer function of the demixing dynamical system is given by

$$\mathbf{W}(z) = \mathbf{P}(z)\,\mathbf{Q}^{-1}(z), \tag{11.10}$$

where $\mathbf{P}(z) = \sum_{i=0}^{L} \mathbf{P}_i z^{-i}$ and $\mathbf{Q}(z) = \sum_{i=0}^{L} \mathbf{Q}_i z^{-i}$, with $\mathbf{Q}_0 = \mathbf{I}$, then the matrices $\mathbf{A}, \mathbf{B}, \mathbf{C}$ and $\mathbf{D}$ required for the canonical controller form can be represented as follows

$$\mathbf{A} = \begin{bmatrix} -\boldsymbol{\mathcal{Q}} & -\mathbf{Q}_L \\ \mathbf{I}_{m(L-1)} & \boldsymbol{\mathcal{O}} \end{bmatrix}, \qquad \mathbf{B} = \begin{bmatrix} \mathbf{I}_m \\ \boldsymbol{\mathcal{O}} \end{bmatrix} \tag{11.11}$$

$$\mathbf{C} = (\mathbf{P}_1,\ \mathbf{P}_2,\ \ldots,\ \mathbf{P}_L), \qquad \mathbf{D} = \mathbf{P}_0, \tag{11.12}$$

where $\boldsymbol{\mathcal{Q}} = (\mathbf{Q}_1\ \mathbf{Q}_2,\ \ldots,\ \mathbf{Q}_{L-1})$ is an $m \times m(L-1)$ matrix, $\boldsymbol{\mathcal{O}}$ is an $m(L-1) \times m$ null matrix, $\mathbf{I}_m$ and $\mathbf{I}_{m(L-1)}$ are the $m \times m$ and $m(L-1) \times m(L-1)$ identity matrices, respectively. It should be noted that in the special case when synaptic weights are FIR filters, $\mathbf{W}(z) = \mathbf{P}(z)$, and both internal space matrices $\mathbf{A}$ and $\mathbf{B}$ are constant matrices and they can be determined explicitly.

11.2 DERIVATION OF BASIC LEARNING ALGORITHMS

The objective of this section is to derive basic adaptive learning algorithms that perform an update of the external (output) parameters $\overline{\mathbf{W}} = [\mathbf{C}, \mathbf{D}]$ in the demixing model. It should be noted that for any feed-forward dynamical model, the internal parameters represented by the set of matrices $[\mathbf{A}, \mathbf{B}]$ are fixed and they can be explicitly determined on the basis of the assumed model. In such a case, the problem is reduced to the estimation of external (output) parameters but the dynamical (internal) part of the system can be established and fixed in advance.

In fact, the separating matrix $\overline{\mathbf{W}} = [\mathbf{C}, \mathbf{D}] \in \mathbb{R}^{n \times (m+M)}$ can be estimated by any learning algorithm for ICA, BSS or BSE discussed in the previous chapters, assuming that the sensor vector has the form $\overline{\mathbf{x}} = [\boldsymbol{\xi}^T,\ \mathbf{x}^T]^T = [\xi_1, \xi_2, \ldots, \xi_M, x_1, x_2,\ \ldots, x_m]^T \in \mathbb{R}^{M+m}$. In other words, our objective in this section is to estimate the non-square full rank separating matrix $\overline{\mathbf{W}}$ for a memoryless system with $(M+m)$ inputs and n outputs.

The basic idea is to use the gradient descent approach to minimize a suitably designed cost function. We can use several criteria and cost functions discussed in the previous chapters. To illustrate the approach, we use the minimization of the Kullback-Leibler divergence as a measure of independence of the output signals. In order to obtain an improved learning performance, we define a new search direction, which is related to the natural gradient, developed by Amari [16] and Amari and Nagaoka [37].

11.2.1 Gradient Descent Algorithms for Estimation of Output Matrices $\overline{\mathbf{W}} = [\mathbf{C}, \mathbf{D}]$

Let us consider a basic cost (risk) function derived from the Kullback-Leibler divergence and mutual information as [281, 1358]

$$\mathcal{R}(\mathbf{y}, \overline{\mathbf{W}}) = -\frac{1}{2} \log \left|\det(\mathbf{D}\mathbf{D}^T)\right| - \sum_{i=1}^{n} E\{\log q_i(y_i)\}. \tag{11.13}$$

Its instantaneous representation - the loss function can be written as

$$\rho(\mathbf{y}, \overline{\mathbf{W}}) = -\frac{1}{2} \log \left|\det(\mathbf{D}\mathbf{D}^T)\right| - \sum_{i=1}^{n} \log q_i(y_i), \tag{11.14}$$

where $\det(\mathbf{D})$ is the determinant of the matrix $\mathbf{D}$. Each pdf $q_i(y_i)$ is an approximation of the true pdf $r_i(y_i)$ of an estimated source signal.

The first term of the loss function prevents all the outputs signals from decaying to zero and the second term provides the output signals to be maximally statistically independent. It should be noted that by minimization of such a loss function, we are not explicitly aiming to recover the source signals, but we only attempt to make the output signals mutually independent. The true source signals may differ from these recovered independent output signals by an arbitrary permutation and convolution.

In order to evaluate the gradient of the loss function $\rho(\mathbf{y}, \overline{\mathbf{W}})$ with respect to $\overline{\mathbf{W}}$, we calculate the total differential $d\rho(\mathbf{y}, \overline{\mathbf{W}})$ of $\rho(\mathbf{y}, \overline{\mathbf{W}})$, where we take a differential $d\overline{\mathbf{W}}$ on $\overline{\mathbf{W}}$,

$$d\rho(\mathbf{y}, \overline{\mathbf{W}}) = \rho(\mathbf{y}, \overline{\mathbf{W}} + d\overline{\mathbf{W}}) - \rho(\mathbf{y}, \overline{\mathbf{W}}). \tag{11.15}$$

which can be expressed as

$$d\rho(\mathbf{y}, \overline{\mathbf{W}}) = -tr(d\mathbf{D}\,\mathbf{D}^{-1}) + \mathbf{f}^T(\mathbf{y})d\mathbf{y}, \tag{11.16}$$

where $\mathrm{tr}(\cdot)$ is the trace of a matrix and $\mathbf{f}(\mathbf{y}) = [f_1(y_1), f_2(y_2), \ldots, f_n(y_n)]^T$ is a vector of nonlinear activation functions

$$f_i(y_i) = -\frac{d \log q_i(y_i)}{dy_i} = -\frac{q_i'(y_i)}{q_i(y_i)} \qquad (i = 1, 2, \ldots, n). \tag{11.17}$$

Taking a differential of $\mathbf{y}$ in equation (11.4), we have the following relation

$$d\mathbf{y}(k) = d\mathbf{C}\; \boldsymbol{\xi}(k) + d\mathbf{D}\; \mathbf{x}(k) + \mathbf{C}\; d\boldsymbol{\xi}(k). \tag{11.18}$$

Hence, we obtain the gradient components

$$\frac{\partial \rho(\mathbf{y}, \overline{\mathbf{W}})}{\partial \mathbf{C}} = \mathbf{f}(\mathbf{y})\; \boldsymbol{\xi}^T, \tag{11.19}$$

$$\frac{\partial \rho(\mathbf{y}, \overline{\mathbf{W}})}{\partial \mathbf{D}} = -\mathbf{D}^{-T} + \mathbf{f}(\mathbf{y})\mathbf{x}^T. \tag{11.20}$$

Let us apply now the standard gradient descent method for updating the matrix $\mathbf{C}$ and the natural gradient rule for updating the matrix $\mathbf{D}$. Then, we obtain on the basis of (11.19) and (11.20) the adaptive learning algorithm

$$\Delta \mathbf{C}(l) = \mathbf{C}(l+1) - \mathbf{C}(l) = -\eta \frac{\partial \mathcal{R}}{\partial \mathbf{C}} = -\eta \left\langle \mathbf{f}[\mathbf{y}(k)]\, \boldsymbol{\xi}^T(k) \right\rangle, \tag{11.21}$$

$$\begin{aligned} \Delta \mathbf{D}(l) &= \mathbf{D}(l+1) - \mathbf{D}(l) = -\eta \frac{\partial \mathcal{R}}{\partial \mathbf{D}} \mathbf{D}^T \mathbf{D} \\ &= \eta \left[\mathbf{I} - \langle \mathbf{f}[\mathbf{y}(k)]\, \mathbf{x}^T(k) \mathbf{D}^T(l) \rangle\right] \mathbf{D}(l) \\ &= \eta \left[\mathbf{I} - \langle \mathbf{f}[\mathbf{y}(k)]\, \mathbf{y}_{\mathbf{x}}^T(k) \rangle\right] \mathbf{D}(l), \end{aligned} \tag{11.22}$$

where $\mathbf{y}_{\mathbf{x}}(k) = \mathbf{D}(l)\, \mathbf{x}(k)$ [281].
The on-line version of the algorithm can take the following form

$$\boxed{\mathbf{C}(k+1) = \mathbf{C}(k) - \eta_C(k)\, \mathbf{R}_{\mathbf{f}\boldsymbol{\xi}}^{(k)}} \tag{11.23}$$

and

$$\boxed{\mathbf{D}(k+1) = \mathbf{D}(k) - \eta_D(k) \left[\mathbf{I} - \mathbf{R}_{\mathbf{f}\,\mathbf{y_x}}^{(k)}\right] \mathbf{D}(k),} \tag{11.24}$$

where

$$\boxed{\mathbf{R}_{\mathbf{f}\boldsymbol{\xi}}^{(k)} = (1-\eta_0)\, \mathbf{R}_{\mathbf{f}\boldsymbol{\xi}}^{(k-1)} + \eta_0\, \mathbf{f}[\mathbf{y}(k)]\, \boldsymbol{\xi}^T(k)} \tag{11.25}$$

and

$$\boxed{\mathbf{R}_{\mathbf{f}\,\mathbf{y_x}}^{(k)} = (1-\eta_0)\, \mathbf{R}_{\mathbf{f}\,\mathbf{y_x}}^{(k-1)} + \eta_0\, \mathbf{f}[\mathbf{y}(k)]\, \mathbf{x}^T(k) \mathbf{D}^T(k).} \tag{11.26}$$

The above algorithm is computationally efficient because the matrix $\mathbf{D}$ needs to be inverted at each iteration step. However, it does not dramatically improve the convergence speed in comparison to the standard gradient descent method because the modified search direction is not exactly the natural gradient one [282].

Remark 11.2 *The above algorithm uses the vector of activation functions $\mathbf{f}(\mathbf{y})$. The optimal choice of the activation function is given by equation (11.17) with $q_i(y_i) = r_i(y_i)$, if we can adaptively estimate the true source probability distribution $r_i(y_i)$. Another solution is to use a score function according to the statistics of source signals. Typically, if a source signal y_i is a super-Gaussian, one can choose $f_i(y_i) = tanh(y_i)$, and if it is a sub-Gaussian, one can choose $f_i(y_i) = y_i^3$ [24, 25, 394]. A question can be raised whether the learning algorithm will converge to a true solution if the approximated activation functions are used. The theory of the semi-parametric model for blind separation/deconvolution [23, 35, 36, 1349, 1351] shows that even if a mis-specified pdf is used in the learning algorithm, it can still converge to the true solution if certain stability conditions are satisfied (see Chapter 10) [24].*

The equilibrium points of the above learning algorithm satisfy the following equations

$$E\{\mathbf{f}(\mathbf{y}(k))\, \boldsymbol{\xi}^T(k)\} = \mathbf{0}, \tag{11.27}$$

$$E\{\mathbf{I} - \mathbf{f}(\mathbf{y}(k))\, \mathbf{y}_{\mathbf{x}}^T(k)\} = \mathbf{0}. \tag{11.28}$$

By pre-multiplying equation (11.27) by $\mathbf{C}^T$ from the right hand, and adding it to equation (11.28), we obtain

$$E\left\{\mathbf{I} - \mathbf{f}(\mathbf{y}(k))\, \mathbf{y}^T(k)\right\} = \mathbf{0}. \tag{11.29}$$

This means that the output signals $\mathbf{y}$ converge when the generalized covariance matrix

$$\mathbf{R}_{\mathbf{f}\,\mathbf{y}} = E\{\mathbf{f}(\mathbf{y}(k))\, \mathbf{y}^T(k)\} = \mathbf{I} \tag{11.30}$$

is the identity matrix (or more generally any positive definite diagonal matrix).

For the special case when matrix $\mathbf{D} = \mathbf{0}$, we can formulate the following alternative cost function [282]

$$\mathcal{R}_2(\mathbf{y}, \mathbf{C}) = -\frac{1}{2} \log(\det(\mathbf{C}_1\, \mathbf{C}_1^T)) - \sum_{i=1}^{n} E\{\log q_i(y_i)\},$$

where $\mathbf{C}_1$ is an $n \times n$ nonsingular matrix which is the block matrix of the output matrix $\mathbf{C} = [\mathbf{C}_1, \mathbf{C}_2] \in \mathbb{R}^{n \times M}$. The first term in the cost function ensures that the trivial solution $\mathbf{y} = \mathbf{0}$ is avoided, while the second term ensures mutual independence of outputs.

By using the combined natural-standard gradient approach, we obtain simple learning rules for $\mathbf{C}_1$ and $\mathbf{C}_2$ [1353, 1354]

$$\Delta \mathbf{C}_1 = -\eta \frac{\partial \mathcal{R}_2}{\partial \mathbf{C}_1} \mathbf{C}_1^T\, \boldsymbol{\Lambda}\, \mathbf{C}_1 \tag{11.31}$$

$$= \eta \left[\boldsymbol{\Lambda} - \langle \mathbf{f}[\mathbf{y}(k)]\, \mathbf{y}_1^T(k) \rangle\right] \mathbf{C}_1, \tag{11.32}$$

$$\Delta \mathbf{C}_2 = -\eta \frac{\partial \mathcal{R}_2}{\partial \mathbf{C}_2} = -\eta \left\langle \mathbf{f}[\mathbf{y}(k)]\, \boldsymbol{\xi}_2^T(k) \right\rangle, \tag{11.33}$$

where $\mathbf{y}_1 = \boldsymbol{\Lambda}\, \mathbf{C}_1\, \boldsymbol{\xi}_1$, $\boldsymbol{\xi}_1 = [\xi_1, \xi_2, \ldots, \xi_n]^T$, and $\boldsymbol{\xi}_2 = [\xi_{n+1}, \xi_{n+2}, \ldots, \xi_M]^T$.

A slightly different algorithm can be derived by applying the extended natural gradient rule, described in Chapter 6, given in general form as [1354, 1356, 1357]

$$\Delta \overline{\mathbf{W}} = -\eta \frac{\partial J}{\partial \overline{\mathbf{W}}} \left[\overline{\mathbf{W}}^T \overline{\mathbf{W}} + \mathbf{Q}^T\, \boldsymbol{\Lambda}_Q\, \mathbf{Q}\right], \tag{11.34}$$

where $\overline{\mathbf{W}} = [\mathbf{C}, \mathbf{D}] \in \mathbb{R}^{n \times (m+M)}$ represents output matrices, $\mathbf{Q} \in \mathbb{R}^{(m+M) \times (m+M)}$ is any orthogonal matrix and $\boldsymbol{\Lambda}_Q \in \mathbb{R}^{(m+M) \times (m+M)}$ is a quasi-diagonal matrix with nearly all the elements zero except the first M diagonal elements which have positive values. The optimal choice of the matrix $\mathbf{Q}$ depends on the character of noise [1355].

After simple mathematical manipulation, we can derive the on-line learning algorithm for the state-space model, and this can be considered as an extension to previously described algorithms [1358]:

$$\Delta \mathbf{C} = \eta \left(\mathbf{I} - \langle \mathbf{f}[\mathbf{y}(k)]\, \mathbf{y}^T(k) \rangle\right) \mathbf{C} - \eta \left\langle \mathbf{f}[\mathbf{y}(k)]\, \boldsymbol{\xi}^T(k) \right\rangle \boldsymbol{\Lambda}_M, \tag{11.35}$$

$$\Delta \mathbf{D} = \eta \left[\mathbf{I} - \langle \mathbf{f}[\mathbf{y}(k)]\, \mathbf{y}^T(k) \rangle\right] \mathbf{D}, \tag{11.36}$$

with $\mathbf{Q} = \mathbf{I}$, where $\mathbf{\Lambda}_M \in \mathbb{R}^{M \times M}$ is a positive definite diagonal matrix.

11.2.2 Special Case - Multichannel Blind Deconvolution with Causal FIR Filters

Let us consider the special case of a simplified, moving average (convolutive) model described as [281, 282]

$$\begin{aligned} \mathbf{y}(k) &= \mathbf{C}\,\boldsymbol{\xi}(k) + \mathbf{D}\,\mathbf{x}(k) = [\mathbf{W}(z)]\,\mathbf{x}(k) \\ &= \sum_{p=0}^{L} \mathbf{W}_p z^{-p}\,\mathbf{x}(k) = \sum_{p=0}^{L} \mathbf{W}_p\,\mathbf{x}(k-p), \end{aligned} \tag{11.37}$$

where $\mathbf{C} = [\mathbf{W}_1, \mathbf{W}_2, \ldots, \mathbf{W}_L] \in \mathbb{R}^{n \times M}$, $\mathbf{D} = \mathbf{W}_0 \in \mathbb{R}^{n \times m}$ and $\boldsymbol{\xi}(k) = \underline{\mathbf{x}}(k) = [\mathbf{x}^T(k-1), \ldots, \mathbf{x}^T(k-L)]^T$, with $M = mL$. It should be noted that for a such model, the state space vector $\boldsymbol{\xi}(k) \in \mathbb{R}^M$ is determined explicitly by the sensor vector $\mathbf{x}(k)$, so the estimation of matrices $\mathbf{A}$ and $\mathbf{B}$ is not required in this case. For this model, on the basis of the above derived generalized learning algorithms (11.21)-(11.22), we can easily obtain a learning rule for the standard blind deconvolution problem:

$$\Delta \mathbf{W}_0 = \eta\,\left[(\mathbf{I} - \langle \mathbf{f}[\mathbf{y}(k)]\,\mathbf{y}_{\mathbf{x}}^T(k)\rangle\right]\,\mathbf{W}_0, \tag{11.38}$$

$$\Delta \mathbf{W}_p = -\eta\,\langle \mathbf{f}[\mathbf{y}(k)]\,\mathbf{x}^T(k-p)\rangle, \qquad (p = 1, 2, \ldots, L). \tag{11.39}$$

Alternatively, we can use the NG learning rule (in the Lie Group sense) proposed by Zhang *et al.* [1356, 1358] (see Chapters 6 and 9 for more details):

$$\Delta \mathbf{W}_p = \eta[(\mathbf{I} - \langle \mathbf{f}[\mathbf{y}(k)]\,\mathbf{y}^T(k)\rangle)\,\mathbf{W}_p - (1 - \delta_{p0})\,\langle \mathbf{f}[\mathbf{y}(k)]\,\mathbf{x}^T(k-p)\rangle]\,\mathbf{\Lambda}_M, \tag{11.40}$$

where $p = 0, 1, \ldots, L$, δ_{p0} is the Kronecker delta and $\mathbf{\Lambda}_M$ is a diagonal positive definite matrix.

11.2.3 Derivation of the Natural Gradient Algorithm for the State Space Model

The learning algorithms presented in the previous section, although of practical value, have been derived in a heuristic or intuitive way. In this section, we derive the efficient natural gradient learning algorithm in a mathematically rigorous way [1358].

Let us assume without loss of generality that the number of outputs is equal to the number of sensors and the matrix $\mathbf{D} \in \mathbb{R}^{n \times n}$ is nonsingular. From the linear output equation (11.4), we have

$$\mathbf{x}(k) = \mathbf{D}^{-1}\,(\mathbf{y}(k) - \mathbf{C}\,\boldsymbol{\xi}(k)). \tag{11.41}$$

Substituting (11.41) into (11.18), we obtain

$$d\mathbf{y} = (d\mathbf{C} - d\mathbf{D}\,\mathbf{D}^{-1}\mathbf{C})\,\boldsymbol{\xi} + d\mathbf{D}\,\mathbf{D}^{-1}\,\mathbf{y} + \mathbf{C}\,d\boldsymbol{\xi}. \tag{11.42}$$

In order to improve the computational efficiency of learning algorithms, we introduce a new search direction defined as

$$d\mathbf{X}_1 = d\mathbf{C} - d\mathbf{D}\,\mathbf{D}^{-1}\,\mathbf{C}, \tag{11.43}$$

$$d\mathbf{X}_2 = d\mathbf{D}\,\mathbf{D}^{-1}. \tag{11.44}$$

Straightforward calculation leads to the following relationships

$$\frac{\partial \rho(\mathbf{y}, \mathbf{W})}{\partial \mathbf{X}_1} = \mathbf{f}(\mathbf{y}(k))\, \boldsymbol{\xi}^T(k), \tag{11.45}$$

$$\frac{\partial \rho(\mathbf{y}, \mathbf{W})}{\partial \mathbf{X}_2} = \mathbf{f}(\mathbf{y}(k))\, \mathbf{y}^T(k) - \mathbf{I}. \tag{11.46}$$

Using the standard gradient descent approach, we obtain $\mathbf{X}_1$ and $\mathbf{X}_2$

$$\Delta \mathbf{X}_1(k) = -\eta \frac{\partial \mathcal{R}(\mathbf{y}, \mathbf{W})}{\partial \mathbf{X}_1} = -\eta \left\langle \mathbf{f}(\mathbf{y}(k))\, \boldsymbol{\xi}^T(k) \right\rangle, \tag{11.47}$$

$$\Delta \mathbf{X}_2(k) = -\eta \frac{\partial \mathcal{R}(\mathbf{y}, \mathbf{W})}{\partial \mathbf{X}_2} = -\eta \left(\langle \mathbf{f}(\mathbf{y}(k))\, \mathbf{y}^T(k) \rangle - \mathbf{I} \right), \tag{11.48}$$

Taking into account the relationships (11.43) and (11.44), we obtain a batch version of the natural gradient learning algorithm to update the matrices $\mathbf{C}$ and $\mathbf{D}$ as

$$\boxed{\Delta \mathbf{C}(l) = \eta \left[\left(\mathbf{I} - \langle \mathbf{f}(\mathbf{y}(k))\, \mathbf{y}^T(k) \rangle \right) \mathbf{C}(l) - \left\langle \mathbf{f}(\mathbf{y}(k))\, \boldsymbol{\xi}^T(k) \right\rangle \right],} \tag{11.49}$$

$$\boxed{\Delta \mathbf{D}(l) = \eta \left[\mathbf{I} - \langle \mathbf{f}(\mathbf{y}(k))\, \mathbf{y}^T(k) \rangle \right] \mathbf{D}(l).} \tag{11.50}$$

On the basis of the above considerations we can establish the relationship between the natural gradient $\tilde{\nabla}\mathcal{R}$ and the standard gradient $\nabla\mathcal{R}$ for the state-space models as follows

$$\boxed{\tilde{\nabla}\mathcal{R} = \nabla\mathcal{R} \begin{bmatrix} \mathbf{I} + \mathbf{C}^T\mathbf{C} & \mathbf{C}^T\mathbf{D} \\ \mathbf{D}^T\mathbf{C} & \mathbf{D}^T\mathbf{D} \end{bmatrix} = \nabla\mathcal{R}\ \mathbf{M},} \tag{11.51}$$

where $\nabla\mathcal{R} = \left[\frac{\partial \mathcal{R}(\mathbf{y},\mathbf{W})}{\partial \mathbf{C}}\ \frac{\partial \mathcal{R}(\mathbf{y},\mathbf{W})}{\partial \mathbf{D}} \right]$ is the standard gradient and $\mathbf{M}$ is a symmetric positive definite pre-conditioning matrix

$$\mathbf{M} = \begin{bmatrix} \mathbf{I} + \mathbf{C}^T\mathbf{C} & \mathbf{C}^T\mathbf{D} \\ \mathbf{D}^T\mathbf{C} & \mathbf{D}^T\mathbf{D}. \end{bmatrix}$$

The natural gradient learning algorithm can be rewritten equivalently in the following form

$$\boxed{\Delta \overline{\mathbf{W}} = [\Delta \mathbf{C}\ \Delta \mathbf{D}] = -\eta\, \tilde{\nabla}\mathcal{R}(\mathbf{y}, \mathbf{W}) = -\eta\, \nabla\mathcal{R}\ \mathbf{M}.} \tag{11.52}$$

Using the moving-average method, the on-line version of the natural gradient algorithm with nonholonomic constraints (see Chapter 6) can be expressed as

$$\boxed{\mathbf{C}(k+1) = \mathbf{C}(k) + \eta_C(k) \left[\left(\boldsymbol{\Lambda}^{(k)} - \mathbf{R}_{\mathbf{f}\mathbf{y}}^{(k)} \right) \mathbf{C}(k) - \mathbf{R}_{\mathbf{f}\boldsymbol{\xi}}^{(k)} \right]} \tag{11.53}$$

and

$$\boxed{\mathbf{D}(k+1) = \mathbf{D}(k) + \eta_D(k) \left[\boldsymbol{\Lambda}^{(k)} - \mathbf{R}_{\mathbf{f}\mathbf{y}}^{(k)} \right] \mathbf{D}(k),} \tag{11.54}$$

where

$$\boxed{\mathbf{R}_{\mathbf{f}\boldsymbol{\xi}}^{(k)} = (1-\eta_0)\mathbf{R}_{\mathbf{f}\boldsymbol{\xi}}^{(k-1)} + \eta_0\, \mathbf{f}[\mathbf{y}(k)]\, \boldsymbol{\xi}^T(k),} \tag{11.55}$$

$$\boxed{\mathbf{R}_{\mathbf{f}\,\mathbf{y}}^{(k)} = (1-\eta_0)\mathbf{R}_{\mathbf{f}\,\mathbf{y}}^{(k-1)} + \eta_0\, \mathbf{f}[\mathbf{y}(k)]\, \mathbf{y}^T(k)} \tag{11.56}$$

and elements of the diagonal matrix $\boldsymbol{\Lambda}^{(k)} = \operatorname{diag}\{\lambda_1^{(k)}, \lambda_2^{(k)}, \ldots, \lambda_n^{(k)}\}$ are updated as

$$\boxed{\lambda_i^{(k)} = (1-\eta_0)\,\lambda_i^{(k-1)} + \eta_0\, f_i(y_i(k))y_i(k).} \tag{11.57}$$

It is straightforward to check that the above algorithm is an extension or generalization of the natural gradient algorithm discussed in Chapter 6 for the dynamical state-space model.

The natural gradient not only ensures better convergence properties but also provides a form by which the analysis of stability becomes much easier. The equilibrium points of the learning algorithm satisfy the following equations

$$E\{\mathbf{f}(\mathbf{y}(k))\, \boldsymbol{\xi}^T(k)\} = \mathbf{0}, \tag{11.58}$$
$$E\{\mathbf{f}(\mathbf{y}(k))\, \mathbf{y}^T(k)\} = \boldsymbol{\Lambda}, \tag{11.59}$$

where $\boldsymbol{\Lambda}$ is a positive definite matrix for $m = n$ and a semi-positive definite diagonal matrix for $m > n$. This means that separated signals in vector $\mathbf{y}$ can achieve mutual independence, if the nonlinear activation function $\mathbf{f}(\mathbf{y})$ is suitably chosen.

On the other hand, if the output signals $\mathbf{y}$ of (11.4) are spatially mutually independent and temporary i.i.d. signals, it is easy to verify that they satisfy (11.58) and (11.59). In fact, from (11.3) and (11.4), we have

$$\boldsymbol{\xi}(k+1) \;=\; \sum_{p=0}^{k-1} \tilde{\mathbf{A}}^p\, \tilde{\mathbf{B}}\, \mathbf{y}(k-p), \tag{11.60}$$

where $\tilde{\mathbf{A}} = \mathbf{A} - \mathbf{B}\,\mathbf{D}^{-1}\,\mathbf{C}$, $\tilde{\mathbf{B}} = \mathbf{B}\,\mathbf{D}^{-1}$ and we have assumed that $\boldsymbol{\xi}(0) = \mathbf{0}$. Substituting (11.60) into (11.58), we deduce that (11.58) is satisfied for i.i.d. signals.

11.3 ESTIMATION OF MATRICES $[\mathrm{A}, \mathrm{B}]$ BY INFORMATION BACK–PROPAGATION

Until now we have assumed that the matrices $[\mathbf{A}, \mathbf{B}]$ are fixed and are determined explicitly by the assumed structure of the dynamical system. However, for recurrent dynamical systems, these matrices are unknown and must be estimated by a suitably designed learning algorithm. In order to develop a learning algorithm for matrices $\mathbf{A}$ and $\mathbf{B}$, we use in this

Table 11.1 Family of adaptive learning algorithms for state-space models.

Reference	Model	Algorithm
Zhang and Cichocki (1998) [1352]	linear	$\Delta\mathbf{C} = -\eta\ \mathbf{f}(\mathbf{y})\ ^T$ $\Delta\mathbf{D} = \eta\,[\mathbf{I} - \mathbf{f}(\mathbf{y})\,\mathbf{x}^T\,\mathbf{D}^T\ ]\,\mathbf{D}$
Zhang and Cichocki (1998) [1353]	linear	$\Delta\mathbf{C} = \eta\,[(\mathbf{I} - \mathbf{f}(\mathbf{y})\,\mathbf{y}^T\ )\,\mathbf{C} - \mathbf{f}(\mathbf{y})\ ^T\ ]$ $\Delta\mathbf{D} = \eta\,[\mathbf{I} - \mathbf{f}(\mathbf{y})\,\mathbf{y}^T\ ]\,\mathbf{D}$ *with Kalman filter*
Cichocki and Zhang (1998) [1354]	nonlinear	$\Delta\mathbf{C} = \eta\,[(\mathbf{\Lambda} - \mathbf{f}(\mathbf{y})\,\mathbf{y}^T\ )\,\mathbf{C} - \mathbf{f}(\mathbf{y})\ ^T\ ]$ $\Delta\mathbf{D} = \eta\,[\mathbf{\Lambda} - \mathbf{f}(\mathbf{y})\,\mathbf{y}^T\ ]\,\mathbf{D}$
Cichocki and Zhang (1998) [281]	nonlinear	Two-stage approach
Cichocki and Zhang (1999) [282]	nonlinear	$\Delta\mathbf{C} = \eta\,[(\mathbf{I} - \mathbf{f}(\mathbf{y})\,\mathbf{y}^T\ )\,\mathbf{C} - \mathbf{f}(\mathbf{y})\ ^T\ \mathbf{\Lambda}]$ $\Delta\mathbf{D} = \eta\,[\mathbf{I} - \mathbf{f}(\mathbf{y})\,\mathbf{y}^T\ ]\,\mathbf{D}$
Salam and Waheed (2001) [1034]	linear	Recurrent network
Salam and Erten (1999) [1033]	nonlinear	Lagrange multiplier approach
Zhang and Cichocki [1358, 1359]	nonholonomic	$\Delta[\mathbf{C}\ \mathbf{D}] = -\eta\,\nabla\mathcal{R} \begin{bmatrix} \mathbf{I} + \mathbf{C}^T\mathbf{C} & \mathbf{C}^T\mathbf{D} \\ \mathbf{D}^T\mathbf{C} & \mathbf{D}^T\mathbf{D} \end{bmatrix}$
	NG Algorithm	$\Delta\mathbf{C}(k) = \eta(k)\left[\ \mathbf{\Lambda}^{(k)} - \mathbf{R}_{\mathbf{f}\,\mathbf{y}}^{(k)}\ \ \mathbf{C}(k) - \mathbf{R}_{\mathbf{f}}^{(k)}\right]$ $\Delta\mathbf{D}(k) = \eta(k)\left[\mathbf{\Lambda}^{(k)} - \mathbf{R}_{\mathbf{f}\,\mathbf{y}}^{(k)}\right]\mathbf{D}(k)$

section the information back-propagation approach discussed in Chapter 9. Combining (11.16) and (11.18), we express the gradient of $\rho(\mathbf{y}, \mathbf{W})$, with respect to $\boldsymbol{\xi}(k)$ as

$$\frac{\partial \rho(\mathbf{y})}{\partial \boldsymbol{\xi}} = \mathbf{C}^T\,\mathbf{f}(\mathbf{y}(k)). \tag{11.61}$$

Therefore, we can calculate the gradient of the loss function $\rho(\mathbf{y})$, with respect to $\mathbf{A} \in \mathbb{R}^{M\times M}$ and $\mathbf{B} \in \mathbb{R}^{M\times m}$ as follows

$$\nabla_{\mathbf{A}}\rho(\mathbf{y}) = \frac{\partial \rho(\mathbf{y})}{\partial \mathbf{A}} = \sum_{h=1}^{M} \frac{\partial \rho(\mathbf{y})}{\partial \xi_h(k)} \frac{\partial \xi_h(k)}{\partial \mathbf{A}}, \tag{11.62}$$

$$\nabla_{\mathbf{B}}\rho(\mathbf{y}) = \frac{\partial \rho(\mathbf{y})}{\partial \mathbf{B}} = \sum_{h=1}^{M} \frac{\partial \rho(\mathbf{y})}{\partial \xi_h(k)} \frac{\partial \xi_h(k)}{\partial \mathbf{B}}, \tag{11.63}$$

where entries of matrices $\frac{\partial \xi_h(k)}{\partial \mathbf{A}}$ and $\frac{\partial \xi_h(k)}{\partial \mathbf{B}}$ are obtained by the following on-line iterations

$$\frac{\partial \xi_h(k+1)}{\partial a_{ij}} = \sum_{p=1}^{M} a_{hp} \frac{\partial \xi_p(k)}{\partial a_{ij}} + \delta_{hi}\, \xi_j(k), \tag{11.64}$$

$$\frac{\partial \xi_h(k+1)}{\partial b_{iq}} = \sum_{p=1}^{M} a_{hp} \frac{\partial \xi_p(k)}{\partial b_{iq}} + \delta_{hi}\, x_q(k), \tag{11.65}$$

for $h, i, j = 1, 2, \ldots, M$ and $q = 1, 2, \ldots, m$, where δ_{hi} is the Kronecker delta.

The minimization of the loss function (11.14) by the gradient descent method leads to a mutual information back-propagation learning algorithm as follows

$$\Delta a_{ij}(k) = -\eta(k)\, \mathbf{f}^T(\mathbf{y}(k)) \sum_{h=1}^{M} \mathbf{c}_h \frac{\partial \xi_h(k)}{\partial a_{ij}}, \tag{11.66}$$

$$\Delta b_{iq}(k) = -\eta(k)\, \mathbf{f}^T(\mathbf{y}(k)) \sum_{h=1}^{M} \mathbf{c}_h \frac{\partial \xi_h(k)}{\partial b_{iq}}, \tag{11.67}$$

where $\mathbf{c}_h$ is the h-th column vector of the matrix $\mathbf{C}$.

Since the matrices $\mathbf{A}$ and $\mathbf{B}$ are sparse in the canonical forms, we need not update all elements of these matrices. Here, we elaborate the learning algorithm for the controller canonical form. In the controller canonical form, the matrix $\mathbf{B}$ is a constant matrix, and only the first n rows of matrix $\mathbf{A}$ are variable parameters. Denote the vector of the h-th row of the matrix $\mathbf{A}$ by $\mathbf{a}_h$, $(h = 1, 2, \ldots, M)$, and define a matrix

$$\frac{\partial \boldsymbol{\xi}(k)}{\partial \mathbf{a}_h} = \left[\frac{\partial \xi_i(k)}{\partial a_{hj}} \right]_{M\times M}. \tag{11.68}$$

The derivative matrix $\frac{\partial \boldsymbol{\xi}(k)}{\partial \mathbf{a}_h}$ can be calculated by the following iteration

$$\frac{\partial \boldsymbol{\xi}(k+1)}{\partial \mathbf{a}_h} = \mathbf{A} \frac{\partial \boldsymbol{\xi}(k)}{\partial \mathbf{a}_h} + \boldsymbol{\Phi}_h(k), \tag{11.69}$$

where $\boldsymbol{\Phi}_h(k) = [\delta_{hi}\, \xi_j(k)]_{M\times M}$. Substituting the above representation into (11.66) and (11.67), we get the following learning rule for $\mathbf{a}_h$,

$$\Delta \mathbf{a}_h = -\eta(k)\, \mathbf{f}^T(\mathbf{y}(k))\, \mathbf{C} \frac{\partial \boldsymbol{\xi}(k)}{\partial \mathbf{a}_h}, \qquad (h = 1, 2, \ldots, M). \tag{11.70}$$

The above learning algorithm updates on-line the internal parameters of the dynamical system. The dynamical system (11.64) and (11.65) is the variational system of the demixing model with respect to $\mathbf{A}$ and $\mathbf{B}$. The purpose of the learning algorithm is to estimate on-line the derivatives of $\boldsymbol{\xi}(k)$ with respect to $\mathbf{A}$ and $\mathbf{B}$. It should be noted that we should choose initial values of the matrices $\mathbf{A}$ and $\mathbf{B}$ very carefully to ensure the stability of the system during the learning process[2]. However, there is no guarantee that the system will remain stable during the learning process even if it were stable in initial iterations. Stability is a common problem in dynamical system identification. One possible solution is to formulate the demixing model in the Lyapunov balanced canonical form [893].

11.4 STATE ESTIMATOR – THE KALMAN FILTER

In order to overcome the above mentioned problem, an alternative approach is to employ the Kalman filter to estimate the state of the system. From output equation (11.3), it is observed that if we can accurately estimate the state vector $\boldsymbol{\xi}(k)$ of the system, then we can separate mixed signals using the learning algorithm (11.49) and (11.50).

11.4.1 Kalman Filter

The Kalman filter is a powerful approach for estimating the state vector in state-space models [111, 510, 1366]. The function of the Kalman filter is to generate on-line the state estimate of the state $\boldsymbol{\xi}(k)$. The Kalman filter dynamics are given as follows (see Fig. 11.2)

$$\hat{\boldsymbol{\xi}}(k+1|k) = \mathbf{A}\,\hat{\boldsymbol{\xi}}(k|k) + \mathbf{B}\,\mathbf{x}(k), \tag{11.71}$$

$$\hat{\boldsymbol{\xi}}(k|k) = \hat{\boldsymbol{\xi}}(k|k-1) + \mathbf{L}\,\mathbf{e}(k), \tag{11.72}$$

$$\mathbf{e}(k) = \mathbf{y}(k) - (\mathbf{C}\,\hat{\boldsymbol{\xi}}(k|k-1) + \mathbf{D}\,\mathbf{x}(k)), \tag{11.73}$$

where $\mathbf{L}$ is the Kalman filter gain matrix, and $\mathbf{e}(k)$ is called the innovation or residual which measures the error between the measured (or expected) output $\mathbf{y}(k)$ and the predicted output $\hat{\mathbf{y}}(k|k-1)$. There exists a variety of algorithms to update the Kalman filter gain matrix $\mathbf{L}$ as well as the state $\hat{\boldsymbol{\xi}}(k|k-1)$; refer to [510] for more details. In this section, we introduce a concept called hidden innovation in order to implement the Kalman filter for the blind separation and filtering problem [1366]. Since the updating matrices $\mathbf{C}$ and $\mathbf{D}$ will produce an innovation in each learning step, we can define a hidden innovation as follows

$$\mathbf{e}(k) = \Delta\mathbf{y}(k) = \mathbf{y}(k) - \hat{\mathbf{y}}(k\,|\,k-1) = \Delta\mathbf{C}(k)\,\boldsymbol{\xi}(k) + \Delta\mathbf{D}(k)\,\mathbf{x}(k), \tag{11.74}$$

[2]The system is stable if all eigenvalues of the state matrix $\mathbf{A}$ are located inside the unit circle in the complex plane.

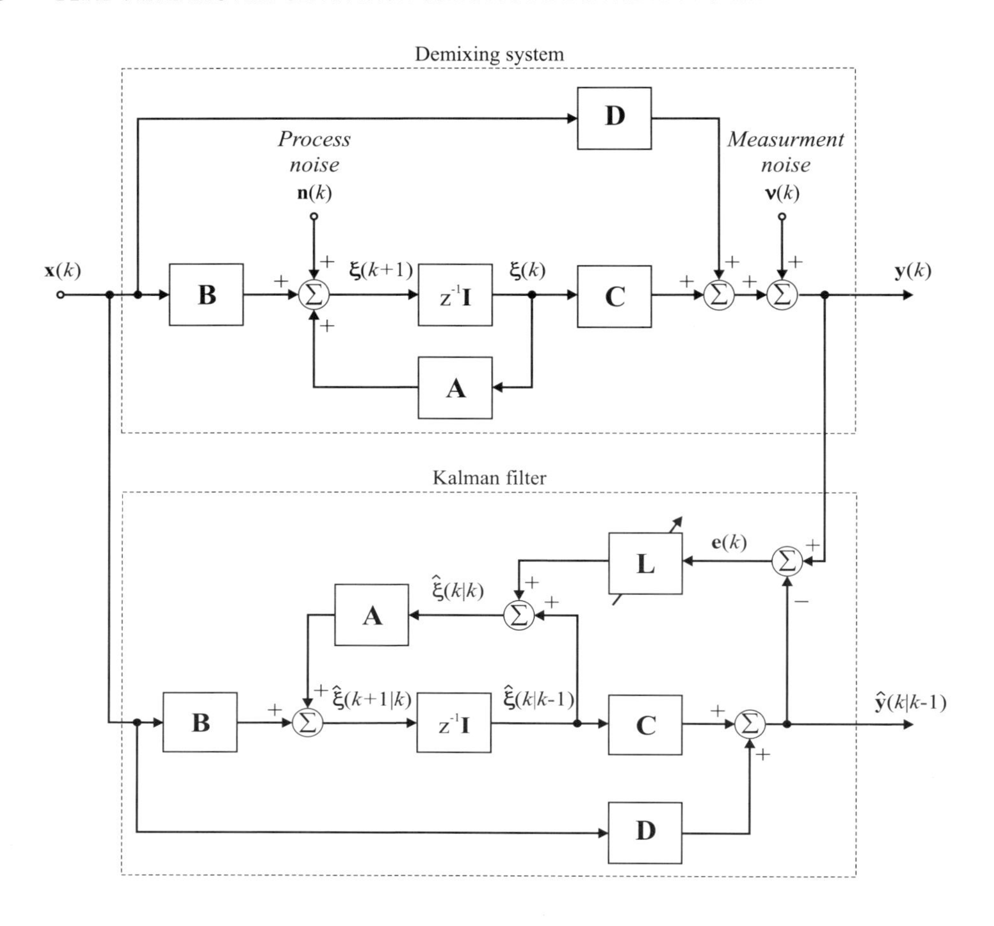

Fig. 11.2 Kalman filter for noise reduction.

where $\Delta\mathbf{C}(k) = \mathbf{C}(k+1) - \mathbf{C}(k)$ and $\Delta\mathbf{D}(k) = \mathbf{D}(k+1) - \mathbf{D}(k)$. The hidden innovation presents the adjusting direction of the output of the demixing system and is used to generate an *a posteriori* state estimate. Once we define the hidden innovation, we can employ the commonly used Kalman filter to estimate the state vector $\boldsymbol{\xi}(k)$ and to update the Kalman gain matrix $\mathbf{L}$ (see Fig. 11.2). The updating rules are described as follows [510, 1366, 1358]

Algorithm Outline: Kalman filter for robust state estimation

1. Kalman gain

$$\mathbf{L}(k) = \mathbf{R}_{\mathbf{ee}}^{(k)}\mathbf{C}^T(k)[\mathbf{C}(k)\mathbf{R}_{\mathbf{ee}}^{(k)}\mathbf{C}^T(k) + \mathbf{R}_{\boldsymbol{\nu}\boldsymbol{\nu}}^{(k)}]^{-1}.$$

2. Update the estimate with hidden innovation

$$\hat{\hat{\xi}}(k|k) = \hat{\boldsymbol{\xi}}(k|k+1) + \mathbf{L}(k)\,\mathbf{e}(k).$$

3. Update the error covariance

$$\widehat{\mathbf{R}}_{\mathbf{ee}}^{(k)} = (\mathbf{I} - \mathbf{L}(k)\,\mathbf{C}(k))\,\mathbf{R}_{\mathbf{ee}}^{(k)}.$$

4. Predict the state vector ahead

$$\hat{\boldsymbol{\xi}}(k+1|k) = \mathbf{A}(k)\,\hat{\boldsymbol{\xi}}(k|k) + \mathbf{B}(k)\,\mathbf{x}(k).$$

5. Predict the error covariance ahead

$$\mathbf{R}_{\mathbf{ee}}^{(k+1)} = \mathbf{A}(k)\,\widehat{\mathbf{R}}_{\mathbf{ee}}^{(k)}\,\mathbf{A}^T(k) + \mathbf{R}_{\mathbf{nn}}^{k)},$$

where $\mathbf{R}_{\mathbf{nn}}^{(k)}$ and $\mathbf{R}_{\boldsymbol{\nu\nu}}^{(k)}$ are the covariance matrices of the noise vector $\mathbf{n}(k)$ and output measurement noise $\boldsymbol{\nu}(k)$, respectively.

One disadvantage of the above algorithm is that it requires knowledge of the covariance matrices of the noise. Theoretical problems such as convergence analysis and stability of the above procedure remain an open problem. However, extensive simulation experiments show that this algorithm, based on the Kalman filter, can separate convolved signals efficiently [1366, 1358].

11.5 TWO–STAGE SEPARATION ALGORITHM

In this section, we present a two-stage separation algorithm for state-space models. In this approach, we decompose the separation problem into the following two stages. First, we separate the mixed signals in the following sense [281, 1358, 1367]

$$\mathbf{G}(z) = \mathbf{W}(z)\,\mathbf{H}(z) = \mathbf{P}\,\mathbf{D}(z), \tag{11.75}$$

where $\mathbf{P}$ is a permutation matrix and $\mathbf{D}(z) = \text{diag}\{D_1(z), D_2(z), \ldots, D_n(z)\}$ is a diagonal matrix in polynomials of z^{-1}. At this stage the output signals will be mutually independent and each individual output signal will be a filtered version of one source signal. So, to recover the original sources, a blind equalization of each channel is necessary. In other words, we need to apply only single channel equalization methods, such as the natural gradient approach or Bussgang methods, to obtain the temporarily i.i.d. recovered signals.

The question here is whether a set of matrices $\boldsymbol{\Theta} = \{\mathbf{A}, \mathbf{B}, \mathbf{C}, \mathbf{D}\}$ exists in the demixing model (11.3) and (11.4), such that its transfer function $\mathbf{W}(z)$ satisfies (11.75). The answer is affirmative under some weak conditions. Suppose that there is a stable inverse filter $\mathbf{W}_*(z)$ of $\mathbf{H}(z)$ in the sense of (11.75). Since $\mathbf{W}_*(z)$ is a rational polynomial of z^{-1}, we know that there is a state-space realization $[\mathbf{A}_*, \mathbf{B}_*, \mathbf{C}_*, \mathbf{D}_*]$ of $\mathbf{W}_*(z)$. Then, we rewrite $\mathbf{W}_*(z)$ as

$$\begin{aligned} \mathbf{W}_*(z) &= \mathbf{D}_* + \mathbf{C}_*(z\,\mathbf{I} - \mathbf{A}_*)^{-1}\mathbf{B}_* \\ &= \sum_{i=0}^{L} \mathbf{P}_i z^{-i} / Q(z^{-1}) \end{aligned} \tag{11.76}$$

We can construct a linear system with transfer function $\sum_{i=0}^{M} \mathbf{P}_i z^{-i}$ as follows

$$\mathbf{A} = \begin{bmatrix} \boldsymbol{\mathcal{O}}^T & \mathbf{0}_m \\ \mathbf{I}_{m(M-1)} & \boldsymbol{\mathcal{O}} \end{bmatrix}, \quad \mathbf{B} = \begin{bmatrix} \mathbf{I}_m \\ \boldsymbol{\mathcal{O}} \end{bmatrix} \tag{11.77}$$

$$\mathbf{C} = (\mathbf{P}_1, \mathbf{P}_2, \ldots, \mathbf{P}_L), \quad \mathbf{D} = \mathbf{P}_0, \tag{11.78}$$

where $\mathbf{I}_{m(M-1)}$ is an $m(M-1) \times m(M-1)$ identity matrix, $\mathbf{0}_m$ is an $m \times m$ zero matrix, and $\boldsymbol{\mathcal{O}}$ is an $m(M-1) \times m$ zero matrix, respectively. Then, we deduce that

$$\mathbf{W}(z) = \mathbf{D} + \mathbf{C}(z\,\mathbf{I} - \mathbf{A})^{-1}\mathbf{B} = \mathbf{W}_*(z)Q(z^{-1}). \tag{11.79}$$

Thus, we have

$$\mathbf{G}(z) = \mathbf{W}(z)\,\mathbf{H}(z) = \mathbf{P}\,\boldsymbol{\Lambda}(z)\,Q(z^{-1}) = \mathbf{P}\,\mathbf{D}(z), \tag{11.80}$$

where $\mathbf{D}(z) = \boldsymbol{\Lambda}(z)\,Q(z^{-1})$ is a diagonal matrix in polynomials of z^{-1} in its diagonal entities. It is easily seen that both $\mathbf{A}$ and $\mathbf{B}$ are constant matrices. Therefore, we need to develop a learning algorithm to update $\mathbf{C}$ and $\mathbf{D}$ so as to obtain the separated signals in the sense of (11.75).

On the other hand, we know that if the matrix $\overline{\mathbf{D}}$ in the mixing model satisfies $rank(\overline{\mathbf{D}}) = n$, then there exists a set of matrices $\{\mathbf{A}, \mathbf{B}, \mathbf{C}, \mathbf{D}\}$, such that the output signal $\mathbf{y}$ of state-space system (11.3) and (11.4) recovers the independent source signals in the sense of (11.75). Therefore, we have the following Theorem [1358]:

Theorem 11.2 *If the matrix $\overline{\mathbf{D}}$ in the mixing model satisfies $rank(\overline{\mathbf{D}}) = n$, then for given specific matrices $\mathbf{A}$ and $\mathbf{B}$ as (11.77), there exist matrices $[\mathbf{C}, \mathbf{D}]$, such that the transfer matrix $\mathbf{W}(z)$ of the system (11.3) and (11.4) satisfies equation (11.75).*

The two-stage blind deconvolution can be realized in the following way: First, we construct the matrices $\mathbf{A}$ and $\mathbf{B}$ of the state equation in the form (11.77), and then we employ the natural gradient algorithm to update $\mathbf{C}$ and $\mathbf{D}$. After the first stage, the outcome signals can be represented in the following form

$$\hat{y}_i(k) = Q(z)s_i(k), \qquad (i = 1, 2, \ldots, n). \tag{11.81}$$

Then we can employ the blind equalization approach discussed in Chapter 9 for double finite FIR filter to remove distortion caused by filtering or convolution of the signals. It should be noted that the two-stage approach enables us also to recover the source signals mixed by a non-minimum phase dynamical system [1360].

Appendix A. Derivation of the Cost Function

We consider n observations $\{x_i(k)\}$ and n output signals $\{y_i(k)\}$, with length N.

$$\begin{aligned} \underline{\mathbf{x}}(k) &= [\mathbf{x}^T(1), \mathbf{x}^T(2), \ldots, \mathbf{x}^T(N)]^T, \\ \underline{\mathbf{y}}(k) &= [\mathbf{y}^T(1), \mathbf{y}^T(2), \ldots, \mathbf{y}^T(N)]^T, \end{aligned}$$

where $\mathbf{x}(k) = [x_1(k), x_2(k), \ldots, x_m(k)]^T$ and $\mathbf{y}(k) = [y_1(k), y_2(k), \ldots, y_n(k)]$. The task of blind deconvolution is to estimate a state-space demixing model, such that output signals achieve independence, i.e., when the joint probability density of $\underline{\mathbf{y}}$ is factorized as follows:

$$p(\underline{\mathbf{y}}) = \prod_{i=1}^{n} \prod_{k=1}^{N} p_i(y_i(k)), \tag{A.1}$$

where $\{p_i(\cdot)\}$ is the probability density of the source signals. In order to measure the mutual independence of output signals, we employ the Kullback-Leibler divergence as a criterion, which is an asymmetric measure of the distance between two different probability distributions,

$$K_{pq}(\mathbf{W}(z)) = \frac{1}{N} \int p(\underline{\mathbf{y}}) \log \frac{p(\underline{\mathbf{y}})}{\prod_{i=1}^{n} \prod_{k=1}^{N} q_i(y_i(k))} d\underline{\mathbf{y}}, \tag{A.2}$$

where we replace $p_i(\cdot)$ by certain approximate density functions $q_i(\cdot)$ for estimated sources, since we do not know the true probability distributions $r_i(\cdot)$ of the original source signals.

Provided that initial conditions are set to $\boldsymbol{\xi}(1) = \mathbf{0}$, we have the following relation [1358]

$$\underline{\mathbf{y}} = \boldsymbol{\mathcal{W}} \underline{\mathbf{x}}, \tag{A.3}$$

where $\boldsymbol{\mathcal{W}}$ is given by

$$\boldsymbol{\mathcal{W}} = \begin{bmatrix} \mathbf{H}_0 & \mathbf{0} & \cdots & \mathbf{0} & \mathbf{0} \\ \mathbf{H}_1 & \mathbf{H}_0 & \cdots & \mathbf{0} & \mathbf{0} \\ \vdots & \vdots & \ddots & \vdots & \vdots \\ \mathbf{H}_{N-2} & \mathbf{H}_{N-3} & \ddots & \mathbf{H}_0 & \mathbf{0} \\ \mathbf{H}_{N-1} & \mathbf{H}_{N-2} & \cdots & \mathbf{H}_1 & \mathbf{H}_0 \end{bmatrix}, \tag{A.4}$$

and $\mathbf{H}_i$ are the Markov parameters defined by $\mathbf{H}_0 = \mathbf{D}$, $\mathbf{H}_i = \mathbf{C}\mathbf{A}^{i-1}\mathbf{B}$, ($i = 0, 1, \ldots, N-1$). According to the property of the probability density function, we derive the following relation between $p(\underline{\mathbf{x}})$ and $p(\underline{\mathbf{y}})$:

$$p(\underline{\mathbf{y}}) = \frac{p(\underline{\mathbf{x}})}{|\det \mathbf{H}_0^N|}. \tag{A.5}$$

Using the relation (A.2), we derive the loss function $\rho(\mathbf{W}(z))$ as follows

$$\rho(\mathbf{W}(z)) = -\log|\det \mathbf{H}_0| - \sum_{i=1}^{n} \frac{1}{N} \sum_{k=1}^{N} \log q_i(y_i(k)). \tag{A.6}$$

Note that $p(\underline{\mathbf{x}})$ was not included in (A.6) because it does not depend on the set of parameters $\{\mathbf{H}_i\}$.

12

Nonlinear State Space Models – Semi-Blind Signal Processing

We must dare to think unthinkable thoughts. We must learn to explore all the options and possibilities that confront us in a complex and rapidly changing world. We must learn to welcome and not to fear the voices of dissent. We must dare to think about unthinkable things because when things become unthinkable, thinking stops and action becomes mindless.

—(J. William Fulbright)

Beyond each corner new directions lie in wait.

—(Stanislaw Lec)

12.1 GENERAL FORMULATION OF THE PROBLEM

In this chapter we attempt to extend and generalize the results discussed in the previous chapters to nonlinear dynamical models. However, the problem is not only very challenging but intractable in the general case without *a priori* knowledge about the mixing and filtering nonlinear process. Therefore, in this chapter we consider very briefly only some simplified nonlinear models. In addition, we assume that some information about the mixing and separating system and source signals is available.

In practice, special nonlinear dynamical models are often considered in order to simplify the problem and solve it efficiently for specific applications. Specific examples include the Wiener model, the Hammerstein model, bilinear models, Volterra models, and NARMA (Nonlinear Autoregressive Moving Average) models [52, 283, 684, 1129, 1247, 1281, 1317].

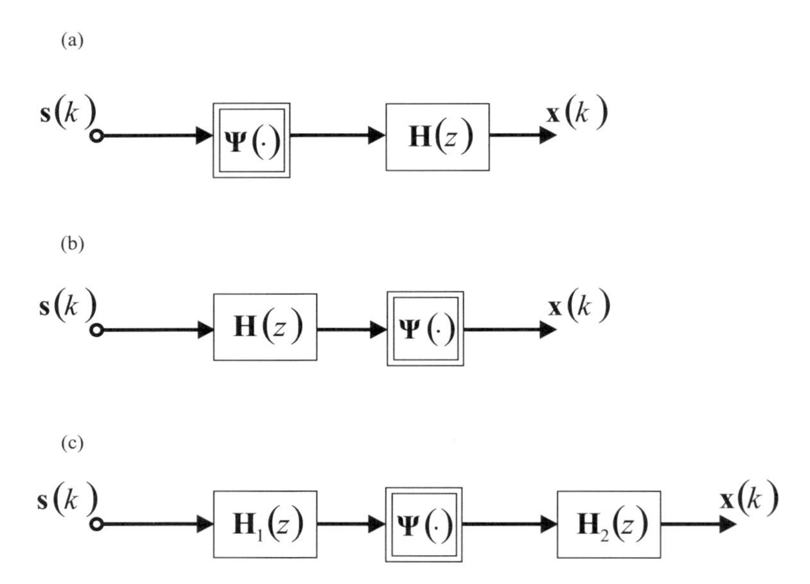

Fig. 12.1 Typical nonlinear dynamical models: (a) The Hammerstein system, (b) the Wiener system and (c)the "Sandwich" system.

The Hammerstein and Wiener systems consist of linear dynamical systems in cascade (series) with static (memoryless) nonlinearities (see Fig. 12.1). In the Wiener system, linear filters precede memoryless nonlinearities, while in the Hammerstein system a nonlinear memoryless system precedes linear (dynamical) filters. Both systems are special cases of an important class of block oriented sandwich systems with static nonlinearities sandwiched between two linear dynamical subsystems [1129]. Sandwich systems constitute a relatively simple but important class of nonlinear systems since linear combination of such systems can approximate a wide class of nonlinear dynamical systems [1126, 1128, 1129]. Such models arise in practice whenever measurement (sensor) devices have nonlinear characteristics (see Fig. 12.2). It should be noted that, in the special case when the nonlinear function is known or can be estimated and the inverse function exists, the Wiener model can be simplified to linear problems described in last three chapters, as illustrated in Fig. 12.2. However in such models, usually some typical nonlinearities such as hard limiters, dead-zone limiters, quantizers and hysteresis are excluded and for such nonlinearities it is impossible to estimate exact inverse systems.

The problems illustrated in Fig. 12.1 (a) and (c) are difficult, because of the permutation and scaling ambiguities even if the nonlinearities are known and invertible. However, in the special case when the set of all nonlinear functions $\{\boldsymbol{\Psi}\}$ are identical and are known, we can easily convert the problem to linear multichannel blind deconvolution which was discussed in the previous chapters. It should be noted, that all these specific nonlinear dynamical models can be described by a general and flexible nonlinear state-space model or NARMA model discussed below.

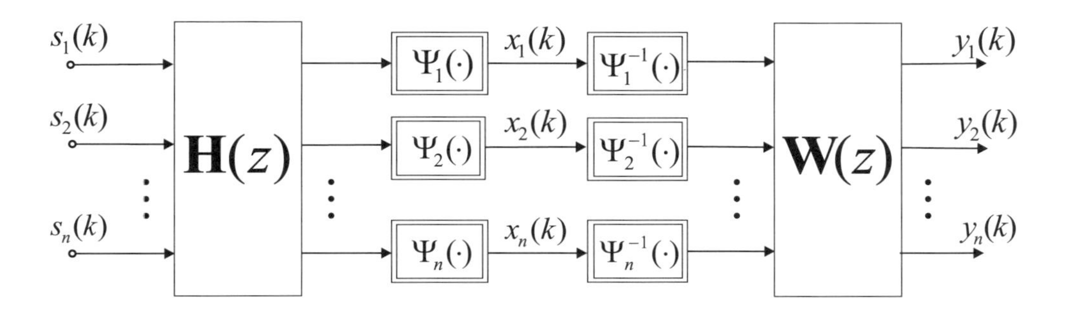

Fig. 12.2 The simple nonlinear dynamical model which leads to the standard linear filtering and separation problem if the nonlinear functions Ψ_i can be estimated and their inverses Ψ_i^{-1} exist.

Assume that unknown source signals $\mathbf{s}(k) = [s_1(k), s_2(k), \ldots, s_n(k)]^T \in \mathbb{R}^n$ are zero-mean i.i.d. and mutually statistically independent. Suppose that the unknown source signals $\mathbf{s}(k)$ are mixed by a stable unknown nonlinear dynamical system

$$\bar{\boldsymbol{\xi}}(k+1) = \overline{\boldsymbol{\mathcal{F}}}\left[\bar{\boldsymbol{\xi}}(k),\, \mathbf{s}(k),\, \boldsymbol{\nu}_P(k)\right], \tag{12.1}$$

$$\mathbf{x}(k) = \overline{\mathbf{G}}\left[\bar{\boldsymbol{\xi}}(k),\, \mathbf{s}(k),\, \boldsymbol{\nu}(k),\right], \tag{12.2}$$

where $\overline{\boldsymbol{\mathcal{F}}}$ and $\overline{\mathbf{G}}$ are two unknown nonlinear mappings, $\bar{\boldsymbol{\xi}}(k) \in \mathbb{R}^d$ is the state vector of the system, and $\mathbf{x}(k) \in \mathbb{R}^m$ (with $m \geq n$) is a vector of available sensor signals, $\boldsymbol{\nu}_P(k)$ and $\boldsymbol{\nu}(k)$ are the process noises and sensor noises of the mixing dynamical system, respectively.

Let us consider another adaptive dynamical system as a demixing model (see Fig. 12.3)

$$\boldsymbol{\xi}(k+1) = \mathbf{F}\left[\mathbf{x}(k),\, \boldsymbol{\xi}(k),\, \boldsymbol{\Theta}\right] \tag{12.3}$$

$$\mathbf{y}(k) = \mathbf{G}\left[\mathbf{x}(k),\, \boldsymbol{\xi}(\boldsymbol{k}),\, \boldsymbol{\Theta}_G\right], \tag{12.4}$$

where $\mathbf{x}(k) \in \mathbb{R}^m$ is the vector of sensor signals, $\boldsymbol{\xi}(k) \in \mathbb{R}^M$ is the state vector of the system, $\mathbf{y}(k) \in \mathbb{R}^n$ is designated to recover the source signals in a certain sense, $\mathbf{F}$ is a nonlinear mapping described by a general nonlinear capability neural network, $\boldsymbol{\Theta}$ is the set of parameters (synaptic weights and nonlinear basis functions) of the neural network, $\mathbf{G}$ is a nonlinear mapping with non-singularity of the derivative $\frac{\partial \mathbf{G}}{\partial \mathbf{x}}$, and $\boldsymbol{\Theta}_G$ are the weights of $\mathbf{G}$. The dimension M of the state vector is the order of the demixing system.

If both of the mappings, $\mathbf{F}$ and $\mathbf{G}$ are linear, the nonlinear state-space model will reduce to the standard multichannel blind filtering and separation discussed in the previous chapter. Since the problem is still intractable in the general case, we consider a slightly simplified model:

$$\boldsymbol{\xi}(k+1) = \mathbf{F}\left[\mathbf{x}(k),\, \boldsymbol{\xi}(k),\, \boldsymbol{\Theta}\right], \tag{12.5}$$

$$\mathbf{y}(k) = \mathbf{C}\,\boldsymbol{\xi}(k) + \mathbf{D}\,\mathbf{x}(k). \tag{12.6}$$

In this demixing model, the output equation is assumed to be linear. The restriction is reasonable since in many practical problems, the measurement is a linear combination of

(a)

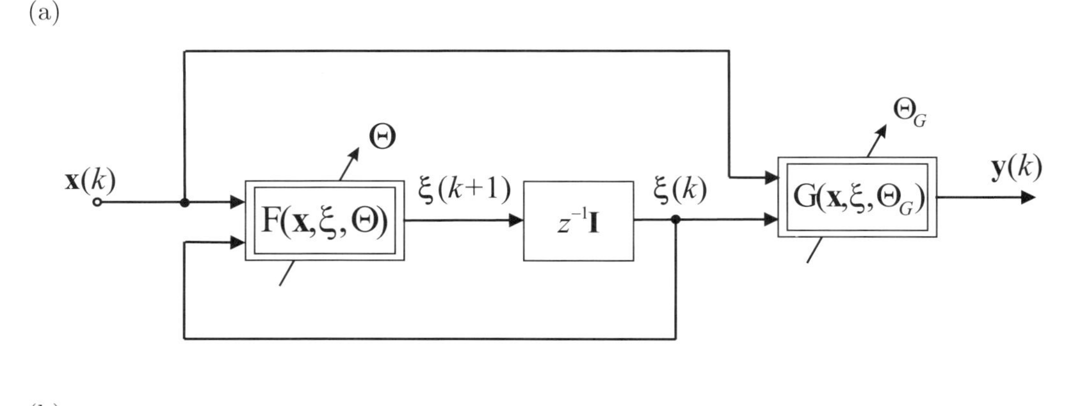

(b)

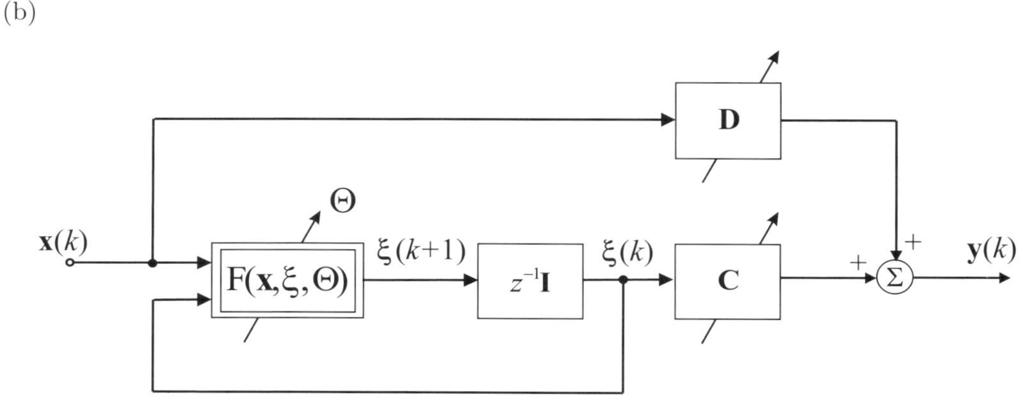

Fig. 12.3 Nonlinear state-space models for multichannel semi-blind separation and filtering: (a) Generalized nonlinear model, (b) simplified nonlinear model.

certain variables. Since the mixing system is unknown (we neither know the nonlinear mappings $\overline{\mathcal{F}}$ and $\overline{\mathbf{G}}$, nor the dimension d of the state vector $\bar{\xi}(k)$), we may attempt to estimate the order and approximate nonlinear mappings of the demixing system in order to estimate original source signals. However, in blind deconvolution, it is very difficult to determine the dimensionality M of the demixing system; therefore, usually an overestimate, i.e., $M > d$ is made. The overestimation of the order M may produce auxiliary delays in the output signals, but this is acceptable in blind deconvolution. There are several neural networks such as Radial Basis Function (RBF), Support Vector Machine (SVM) and multilayer perceptron (MLP), which can be used as demixing models. In this chapter, we employ the RBF to model and identify the nonlinear mapping $\mathbf{F}$ in the demixing model and to estimate the set of output matrices $\overline{\mathbf{W}} = [\mathbf{C}, \mathbf{D}]$, we will employ learning algorithms described in Chapter 11.

12.1.1 Invertibility by State Space Model

Assume that the number of sensor signals equals the number of source signals, i.e., $m = n$. In the following discussion, we restrict the mixing model to the following form,

$$\begin{aligned} \boldsymbol{\xi}(k+1) &= \overline{\mathcal{F}}\left[\bar{\boldsymbol{\xi}}(k),\, \mathbf{s}(k)\right], && (12.7) \\ \mathbf{x}(k) &= \overline{\mathbf{C}}\,\bar{\boldsymbol{\xi}}(k) + \overline{\mathbf{D}}\,\mathbf{s}(k), && (12.8) \end{aligned}$$

where the state equation is a nonlinear dynamical system, and the output equation is a linear one. From a theoretical point of view, we can easily find the inverse of the state-space models in the same form, if the matrix $\overline{\mathbf{D}}$ is invertible. In fact, the inverse system is expressed by

$$\bar{\boldsymbol{\xi}}(k+1) = \overline{\mathcal{F}}\left[\bar{\boldsymbol{\xi}}(k),\, \overline{\mathbf{D}}^{-1}\left(\mathbf{x}(k) - \overline{\mathbf{C}}\,\bar{\boldsymbol{\xi}}(k)\right)\right], \tag{12.9}$$

$$\mathbf{s}(k) = \overline{\mathbf{D}}^{-1}\left(\mathbf{x}(k) - \overline{\mathbf{C}}\,\bar{\boldsymbol{\xi}}(k)\right). \tag{12.10}$$

This means that if the mixing model is expressed by (12.7) and (12.8), we can recover the source signals using the inverse system (12.9) and (12.10). This is an advantage of the state-space model that we do not need to invert any non-linear function explicitly.

12.1.2 Internal Representation

As we have already discussed in Chapter 11, the state-space description allows us to divide the variables into two types: The internal state variable $\boldsymbol{\xi}(k)$, which produces the dynamics of the system, and the external variables $\mathbf{x}(k)$ and $\mathbf{y}(k)$, which represent the input and output of the system, respectively. The vector $\boldsymbol{\xi}(k)$ is known as the state of the dynamical system, which represents all the information about the past behavior of the system. Using this vector one can predict the system's behavior, except for the purely external input $\mathbf{x}(k)$. The state-space description plays a critical role in mathematical formulation of a dynamical system. It allows us to realize the internal structure of the system and to define the controllability and observability of the system as well. In the state-space framework, it becomes much easier to discuss the stability, controllability and observability of nonlinear dynamical systems.

We formulate the separating model in the framework of the state-space models for blind separation and filtering. The parameters in the state equation of the separation are referred to as internal representation parameters (or simply internal parameters), and the parameters in the output equation as external ones. Such an approach enables us to estimate the demixing nonlinear dynamical model in two stages: First, an estimation on some *a priori* knowledge of the internal nonlinear dynamical representation and second, to fix the internal set of parameters and estimate the output linear demixing subsystem. In the estimation of the internal representation stage, we will make the state-space matrix as sparse as possible such that the output signals can be represented as a sparse linear combination of the state vector $\boldsymbol{\xi}(k)$ and sensor vector $\mathbf{x}(k)$.

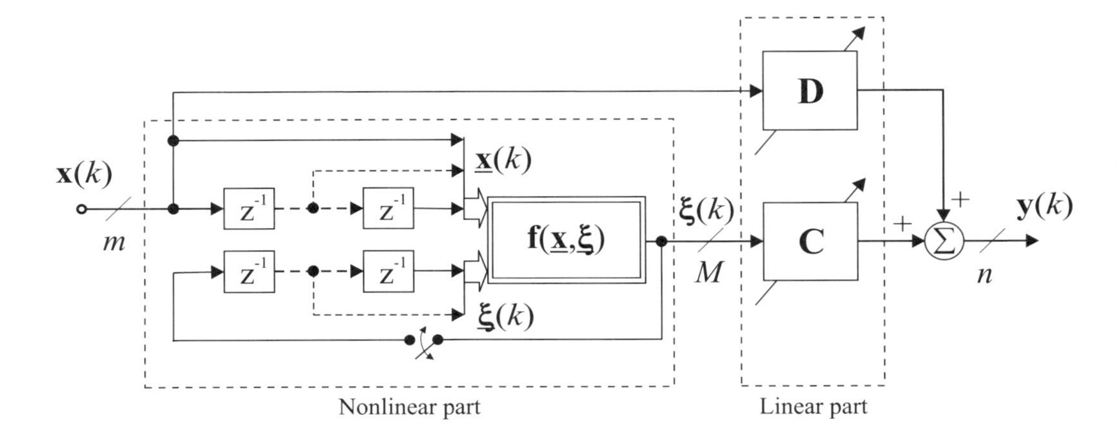

Fig. 12.4 Block diagram of a simplified nonlinear demixing NARMA model. For the switch open, we have a feed-forward nonlinear MA model, and for the switch closed we have a recurrent nonlinear ARMA model.

12.2 SUPERVISED-UNSUPERVISED LEARNING APPROACH

12.2.1 Nonlinear Autoregressive Moving Average Model

The linear state-space separating and filtering model described in the previous chapters can be extended and generalized to a flexible nonlinear model as (see Fig. 12.4) [283, 284]

$$\boldsymbol{\xi}(k) = \mathbf{f}[\underline{\mathbf{x}}(k),\ \underline{\boldsymbol{\xi}}(k)], \tag{12.11}$$

$$\mathbf{y}(k) = \mathbf{C}(k)\,\boldsymbol{\xi}(k) + \mathbf{D}(k)\,\mathbf{x}(k), \tag{12.12}$$

where $\boldsymbol{\xi}(k) = [\xi_1(k), \ldots, \xi_M(k)]^T$ is the state vector, $\mathbf{x}(k) = [x_1(k), \ldots, x_m(k)]^T$ is an available vector of sensor signals, $\mathbf{f}[\underline{\mathbf{x}}(k), \underline{\boldsymbol{\xi}}(k)]$ is an M-dimensional vector of nonlinear functions (with $\underline{\mathbf{x}}(k) = [\mathbf{x}^T(k), \ldots, \mathbf{x}^T(k - L_x)]^T$ and $\underline{\boldsymbol{\xi}}(k) = [\boldsymbol{\xi}^T(k-1), \ldots, \boldsymbol{\xi}^T(k - L_\xi)]^T)$, $\mathbf{y}(k) = [y_1(k), \ldots, y_n(k)]^T$ is a vector of output signals, and $\mathbf{C} \in \mathbb{R}^{n \times M}$ and $\mathbf{D} \in \mathbb{R}^{n \times m}$ are output matrices.

It should be noted that equation (12.11) describes the NARMA model while the output memoryless model (12.12) is linear. Our objective is to estimate or identify in the first stage the NARMA model by using a neural network approach. In the general case, we will be able to estimate the parameters if both sensor signals $\mathbf{x}(k)$ and source (desired) signals $\mathbf{s}(k)$ are available for at least short time windows to perform standard input output identification.

In other words, in order to solve this challenging and difficult problem, we attempt to apply a semi-blind approach, i.e., we combine supervised and un-supervised learning algorithms. Such an approach is justified in many practical applications, especially for time-variable models. For example, for MEG or EEG, we can use a phantom of the human head with known artificial source excitations located in specific places inside the phantom. Similarly, for the cocktail party problem we can record, for short-time windows, test speech sources. These short-time window training sources enable us to determine, a suitable nonlinear demixing model with their associated nonlinear basis functions and their parameters. However, in practice such a complex nonlinear dynamical mixing system is usually slowly time-varying, i.e., some of its parameters may fluctuate slightly with time, e.g., due to movement of source signals in space. After the access to the training signals is lost, we can apply an unsupervised learning approach and apply leaning algorithms described in previous chapters. In this way, we will be able to perform fine adjustment of the output matrices $\mathbf{C}$ and $\mathbf{D}$ (by keeping the nonlinear model fixed). The on-line update of the matrices $\mathbf{C}$ and $\mathbf{D}$ can be performed on the basis of several criteria, e.g., minimizing the mutual information.

12.2.2 Hyper Radial Basis Function Neural Network Model (HRBFN)

We assume that a small amount of training (desired) signals $\mathbf{d}(k) = \alpha \mathbf{s}(k - \Delta)$ is available. So the NARMA model can be estimated by using standard neural models such as multilayer perceptron (MLP), radial basis function (RBF), wavelets, Volterra or sigma-pi neural networks [276]. Furthermore, we assume that the output model can be adjusted to compensate for small fluctuations or slow drifts of the mixing system, and that its parameters can be estimated in the time windows by unsupervised ICA algorithm when the training (desired) signals are not available. In this section, we describe briefly a hyper radial basis function (HRBF) network introduced first by Poggio and Girosi (see Fig. 12.5) [276, 1247] to estimate a NARMA model because of its flexibility as a universal approximator of multidimensional nonlinear mapping.

HRBFN can be considered as a two-layer neural network in which the hidden layer performs an adaptive nonlinear transformation with adjustable parameters, in such a way that the $L = (L_\xi + L_x + 1)$ dimensional input space

$$\widetilde{\mathbf{x}}(k) = [\mathbf{x}^T(k), \ldots, \mathbf{x}^T(k - L_x), \boldsymbol{\xi}^T(k-1), \ldots, \boldsymbol{\xi}^T(k - L_\xi)]^T$$

is mapped to the M- dimensional output space

$$\boldsymbol{\xi}(k) = [\xi_1(k), \xi_2(k), \ldots, \xi_M(k)]^T$$

which is described by a set of nonlinear equations

$$\xi_i = w_{i0} + \sum_{j=1}^{h} w_{ij}\, \Phi_j(r_j), \qquad (i = 1, 2, \ldots, M). \tag{12.13}$$

The above nonlinear mapping can be written in a compact matrix form as[1]

$$\boldsymbol{\xi} = \mathbf{f}(\underline{\mathbf{x}}) = \mathbf{w}_0 + \mathbf{W}\, \boldsymbol{\Phi}(\mathbf{r}), \tag{12.14}$$

[1] For simplicity of our considerations, we consider here a simplified feed-forward model (nonlinear moving average (MA) model with the switch in the off position in Fig. 12.4, i.e., with the vector $\_(k) = [\ ^T(k - 1), \ldots\ ^T(k - L_\xi]^T = \mathbf{0}$, i.e., $L_\xi = 0$.

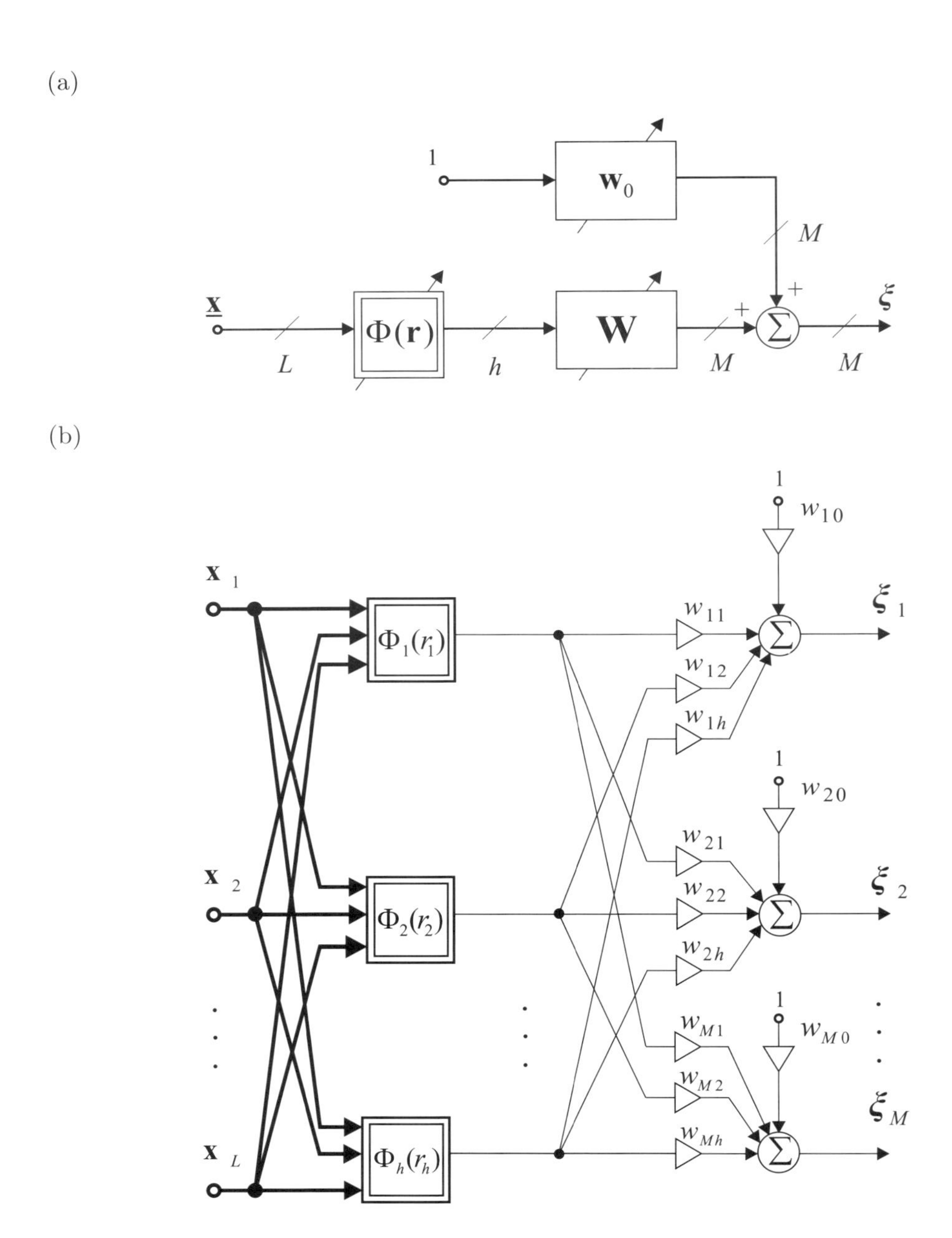

Fig. 12.5 Conceptual block diagram illustrating HRBF neural network model employed for nonlinear semi-blind separation and filtering: (a) Block diagram, (b) detailed neural network model.

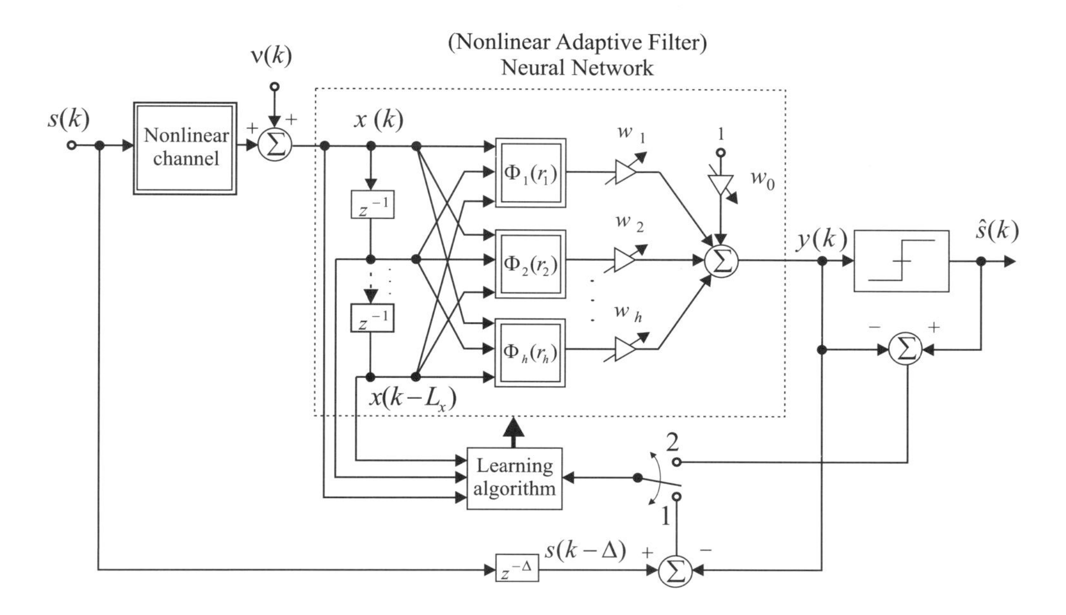

Fig. 12.6 Simplified model of an HRBF neural network for nonlinear semi-blind single channel equalization; if the switch is in position 1, we have supervised learning, if it is in position 2, we have unsupervised learning assuming binary sources.

where $\underline{\mathbf{x}} = [\mathbf{x}_1^T, \mathbf{x}_2^T, \dots, \mathbf{x}_L^T]^T$, (with $\mathbf{x}_i = \mathbf{x}(k-i+1)$, for $i = 1, 2, \dots, L$), $\mathbf{W} = [w_{ij}] \in \mathbb{R}^{M \times h}$, $\mathbf{r} = (r_1, r_2, \dots, r_h)$ and $\mathbf{\Phi}(\mathbf{r}) = [\Phi_1(r_1), \Phi_2(r_2), \dots, \Phi_h(r_h)]^T$.
The nonlinear activation functions $\Phi_i(r_i)$ are defined as generalized (hyper) multidimensional Gaussian functions

$$\Phi_j(r_j) = \frac{1}{2} \exp(-r_j^2/2) \tag{12.15}$$

and

$$r_j^2 = (\underline{\mathbf{x}} - \mathbf{c}_j)^T \, \mathbf{Q}_j^T \, \mathbf{Q}_j \, (\underline{\mathbf{x}} - \mathbf{c}_j), \quad (j = 1, 2, \dots, h), \tag{12.16}$$

with adaptive centers $\mathbf{c}_j = [c_{j1}, c_{j2}, \dots, c_{jL}]^T$.

It should be noted that, in the special case when the symmetric positive definite $L \times L$ matrix $\mathbf{Q}_j^T \, \mathbf{Q}_j$ reduces to a diagonal matrix $\mathbf{Q}_j^T \, \mathbf{Q}_j = \mathrm{diag}\{\sigma_{j1}^{-2}, \sigma_{j2}^{-2}, \dots, \sigma_{jL}^{-2}\}$, the HRBF network is simplified to the standard RBF neural network.

12.2.3 Estimation of Parameters of HRBF Networks Using Gradient Approach

Our objective is to estimate the set of parameters $\mathbf{\Theta} = \{\mathbf{C}, \, \mathbf{D}, \, \mathbf{w}_0, \, \mathbf{W}, \, \{\mathbf{Q}_j\}, \, \{\mathbf{c}_j\}\}$ of the demixing system using the standard cost function

$$J(\mathbf{\Theta}) = \frac{1}{2} \sum_k ||\mathbf{e}(k)||^2, \tag{12.17}$$

where the error vector is defined as $\mathbf{e}(k) = \mathbf{d}(k) - \mathbf{y}(k)$.
In order to avoid getting stuck in the local minima of the above cost function, we apply the Manhattan learning formula of the form [276]

$$\Delta \boldsymbol{\Theta} = -\boldsymbol{\eta}(k)\, \text{sign} \frac{\partial J}{\partial \boldsymbol{\Theta}}, \tag{12.18}$$

with self-adaptive learning step matrix $\boldsymbol{\eta}$ [276].
It can be shown by direct simple calculations that gradient components for the model shown in Fig.12.4 with the HRBF neural network depicted in Fig.12.5 can be evaluated as follows:

$$\frac{\partial J}{\partial \mathbf{C}} = -\mathbf{e}\, \boldsymbol{\xi}^T, \qquad \frac{\partial J}{\partial \mathbf{D}} = -\mathbf{e}\, \underline{\mathbf{x}}^T, \tag{12.19}$$

$$\frac{\partial J}{\partial \mathbf{w}_0} = -\mathbf{C}^T\, \mathbf{e}, \qquad \frac{\partial J}{\partial \mathbf{W}} = -\mathbf{C}^T\, \mathbf{e}\, \boldsymbol{\Phi}^T(\mathbf{r}), \tag{12.20}$$

$$\frac{\partial J}{\partial \mathbf{Q}_j} = \delta_j\, \mathbf{Q}_j\, (\underline{\mathbf{x}} - \mathbf{c}_j)\, (\underline{\mathbf{x}} - \mathbf{c}_j)^T, \tag{12.21}$$

$$\frac{\partial J}{\partial \mathbf{c}_j} = -\delta_j\, \mathbf{Q}_j^T\, \mathbf{Q}_j (\underline{\mathbf{x}} - \mathbf{c}_j), \tag{12.22}$$

where $\delta_j = \Phi_j(r_j)\, [\mathbf{W}^T\, \mathbf{C}^T\, \mathbf{e}]_j$ and $[\mathbf{e}]_j$ means j-th element of the vector $\mathbf{e}$. On the basis of these formulas, we can formulate the supervised learning rules. The unsupervised learning algorithms can be alternatively used for updating the output matrices $\mathbf{C}$, $\mathbf{D}$ as described in the Chapter 11.

It is interesting to note that in the special case of binary signals, we can estimate the desired signals by taking $\widehat{\mathbf{d}}(k) = \text{sign}(\mathbf{y}(k))$, as illustrated for the simplified network shown in Fig. 12.6. In this case, instead of back propagating the true error, we can propagate the "blind" error. Instead of propagating the error $\mathbf{e}(k)$, we can back propagate mutual information $\widehat{\mathbf{e}}(k) = \mathbf{f}(\mathbf{y}(k)) - \mathbf{y}(k)$ [1317, 1318].

The algorithms discussed in this chapter can be considered as extensions and generalizations of some existing algorithms proposed for the linear multichannel blind separation and/or deconvolution problems. Computer simulation experiments have confirmed the validity and good performance of the developed algorithms. Such algorithms can recover source signals even if the mixing dynamical model is nonlinear and structurally unknown.

References

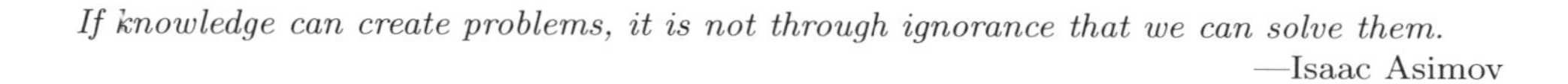

If knowledge can create problems, it is not through ignorance that we can solve them.
—Isaac Asimov

Knowledge is of no value unless you put it into practice.
—Anton Chekhov

1. K. Abed-Meraim, S. Attallah, A. Chkeif, and Y. Hua. Orthogonal Oja algorithm. *IEEE Signal Processing Letters*, 7(5):116–119, May 2000.

2. K. Abed-Meraim, J.-F. Cardoso, A. Gorokhov, P. Loubaton, and É. Moulines. On subspace methods for blind identification of single-input multiple-output FIR systems. *IEEE Trans. on Signal Processing*, 45(1):42– 55, January 1997.

3. K. Abed-Meraim, A. Chkief, and Y. Hua. Fast orthonormal PAST algorithm. *IEEE Signal Processing Letters*, 7(3):60–62, March 2000.

4. K. Abed-Meraim and Y. Hua. A least-squares approach to joint Schur decomposition. In *Proc. IEEE ICASSP*, pages 2541–2544, 1998.

5. K. Abed-Meraim, P. Loubaton, and É. Moulines. A subspace algorithm for certain blind identification problems. *IEEE Trans. Information Theory*, 43(2):499–511, February 1997.

6. K. Abed-Meraim, É. Moulines, and P. Loubaton. Prediction error method for second-order blind identification. *IEEE Trans. Signal Processing*, 45:694–705, March 1997.

7. M. Adachi, K. Aihara, and A. Cichocki. Separation of mixed patterns by a chaotic neural network. In *International Symposium on Nonlinear Theory and its Applications - NOLTA'96, Proceedings,Research Society on NTA, IEICE, Japan*, pages 93–96, Oct. 1996.

8. T. Adali, M.K. Sonmez, and K. Patel. On the dynamics of the LRE algorithm: a distribution learning approach to adaptive equalization. In *Proc. IEEE ICASSP*, pages 929–932 vol.2, Detroit, MI, 1995.

9. S. Affes, S. Gazor, and Y. Grenier. A subarray manifold revealing projection for partially blind identification and beamforming. *IEEE Signal Processing Letters*, 3(6):187–189, June 1996.

10. B.G. Agee, S.V. Schell, and W.A. Gardner. Spectral self-coherence restoral: A new approach to blind adaptive signal extraction using antenna arrays. *Proc. IEEE*, 78(4):753–767, April 1990.

11. T. Akuzawa and N. Murata. Multiplicative nonholonomic Newton -like algorithm. *Chaos, Solitons and Fractals,*, 12(2):785–781, 2001.

12. M.J. Al-Kindi and J. Dunlop. Improved adaptive noise cancellation in the presence of signal leakage on the noise reference channel. *Signal Processing*, 17:241–250, 1989.

13. L. B. Almeida and F. M. Silva. Adaptive decorrelation. *Artificial Neural Networks (Elsevier)*, 2:149–156, 1992.

14. S. Amari. Theory of adaptive pattern classifiers. *IEEE Trans. on Electrical Comput.*, 16(3):299–307, 1967.

15. S. Amari. Neural theory of association and concept formation. *Biological Cybernetics*, 26:175–185, 1977.

16. S. Amari. *Differential Geometrical Methods of Statistics, Springer Lectures Notes in Statistics*. Heidelberg, Springer Verlag., 1985.

17. S. Amari. Differential geometry of a parametric family of invertible linear systems Riemannian metric, dual affine connections and divergence. *Mathematical Systems Theory*, 20:53–82, 1987.

18. S. Amari. Mathematical theory of neural learning. *New Generation of Computing*, 8:135–143, 1991.

19. S. Amari. Super-efficiency in blind source separation. *IEEE Trans. on Signal Processing*, 1997.

20. S. Amari. Natural gradient works efficiently in learning. *Neural Computation*, 10:271–276, 1998.

21. S. Amari. Natural gradient learning for over- and under-complete bases in ICA. *Neural Computation*, 11(8):1875–1883, November 1999.

22. S. Amari. Estimating function of independent component analysis for temporally correlated signals. *Neural Computation*, 12(9):2083–2107, September 2000.

23. S. Amari and J.-F. Cardoso. Blind source separation — semi-parametric statistical approach. *IEEE Trans. on Signal Processing*, 45(11):2692–2700, Dec. 1997.

24. S. Amari, T.-P. Chen, and A. Cichocki. Stability analysis of adaptive blind source separation. *Neural Networks*, 10(8):1345–1351, 1997.

25. S. Amari, T.-P. Chen, and A. Cichocki. Non-holonomic constraints in learning algorithms for blind source separation. *Neural Computation*, 12:1463–1484, 2000.

26. S. Amari and A. Cichocki. Adaptive blind signal processing - neural network approaches. *Proceedings IEEE*, 86:1186–1187, 1998.

27. S. Amari, A. Cichocki, and H.H. Yang. Recurrent neural networks for blind separation of sources. In *in Proc. Int. Symposium Nonlinear Theory and its Applications NOLTA-95 Las Vegas*, volume 1, pages 37–42, Dec. 1995.

28. S. Amari, A. Cichocki, and H.H. Yang. A new learning algorithm for blind signal separation. In Michakel C. Mozer David S. Touretzky and Michael E. Hasselmo, editors, *Advances in Neural Information Processing Systems 1995*, volume 8, pages 757–763. MIT Press: Cambridge, MA, 1996.

29. S. Amari, A. Cichocki, and H.H. Yang. *Unsupervised Adaptive Filtering*, chapter Blind Signal Separation and Extraction - Neural and Information Theoretic Approaches. John Wiley, 1999.

30. S. Amari and S.C. Douglas. Why natural gradient. In *Proc. IEEE International Conference Acoustics, Speech, Signal Processing*, volume II, pages 1213–1216, Seattle, WA, May 1998.

31. S. Amari, S.C. Douglas, and A. Cichocki. Multichannel blind deconvolution and source separation using the natural gradient. In *submitted to IEEE Trans. Signal Processing*, September 1997.

32. S. Amari, S.C. Douglas, and A. Cichocki. Information geometry of blind source deconvolution. In *presented at Mathematical Theory of Networks and Systems*, Padova, Italy, July 1998.

33. S. Amari, S.C. Douglas, A. Cichocki, and H.H. Yang. Multichannel blind deconvolution and equalization using the natural gradient. In *Proc. IEEE Workshop on Signal Processing Advances in Wireless Communications*, pages 101–104, Paris, France, April 1997.

34. S. Amari, S.C. Douglas, A. Cichocki, and H.H. Yang. Novel on-line adaptive learning algorithms for blind deconvolution using the natural gradient approach. In *Proc. 11th IFAC Symposium on System Identification*, volume 3, pages 1057–1062, Kitakyushu City, Japan, July 1997.

35. S. Amari and M. Kawanabe. Information geometry of estimating functions in semiparametric statistical models. *Bernoulli*, 3(1):29–54, 1997.

36. S. Amari and M. Kawanabe. Estimating functions in semiparametric statistical models. In I. V. Basawa, V. Godambe, and R. Taylor, editors, *Estimating Functions*, volume 32 of *Monograph Series*, pages 65–81. IMS, 1998.

37. S. Amari and H. Nagaoka. *Methods of Information Geometry.* AMS and Oxford University Press, 1999.

38. K. Anand, G. Mathew, and V.U. Reddy. Blind separation of multiple co-channel BPSK signals arriving at an antenna array. *IEEE Signal Processing Letters*, 2(9):176–178, September 1995.

39. S. Andersson, M. Millnert, M. Viberg, and B. Wahlberg. An adaptive array for mobile communication systems. *IEEE Trans. on Veh. Tec.*, 40(1):230–236, 1991.

40. C. Anton-Haro, J.A.R. Fonollosa, and J.R. Fonollosa. Blind channel estimation and data detection using hidden Markov models. *IEEE Trans. Signal Processing*, 45(1):241–247, January 1997.

41. S. Araki, S. Makino, R. Mukai, Y. Hinamoto, T. Nishikawa, and H. Saruwatari. Equivalence between frequency domain blind source separation and frequency domain adaptive beamforming. In *ICASSP 2002*, pages 1789–1902, USA, May. 9-13 2002.

42. S. Araki, S. Makino, R. Mukai, T. Nishikawa, and H. Saruwatari. Fundamental limitation of frequency domain blind source separation for convolved mixture of speech. In *Third International Conference on Independent Component Analysis and Signal Separation (ICA-2001)*, pages 132–137, San Diego, USA, Dec. 9-13 2001.

43. G. Archer and K. Chan. Bootstrapping uncertainty in image analysis. In A. Prat, editor, *COMPSTAT. Proceedings in Computational Statistics. 12th Symposium*, pages 193–198, Barcelona, Spain, 1996. Physica-Verlag.

44. D. Asztély, B. Ottersten, and A.L. Swindlehurst. A generalized array manifold model for local scattering in wireless communications. In *Proc. IEEE ICASSP*, pages 4021–4024, Munich (Germany), 1997.

45. J. J. Atick and A. N. Redlich. Convergent algorithm for sensory receptive field development. *Neural Computation*, 5(1):45–60, 1993.

46. H. Attias and C.E. Schreiner. Blind source separation and deconvolution: the dynamic component analysis algorithm. *Neural Computation*, 10:1373–1424, 1998.

47. R.A. Axford, Jr., L.B. Milstein, and J.R. Zeidler. A dual-mode algorithm for blind equalization of QAM signals: CADAMA. In *29-th Asilomar Conference Signals, Systems, Comp.*, pages 172–176 vol.1. IEEE, 1996.

48. G.R. Ayers and J.C. Dainty. Iterative blind deconvolution method and its applications. *Optics Letters*, 13(7):547–549, July 1988.

49. L.A. Baccala and S. Roy. A new blind time-domain channel identification method based on cyclostationarity. *IEEE Signal Processing Letters*, 1(6):89–91, June 1994.

50. A.D. Back and A. Cichocki. Input variable selection using independent component analysis and higher order statistics. In *Proc. of the First International Workshop on Independent Component Analysis and Signal Separation - ICA'99*, pages 203–208, Aussois, France, 1999.

51. A.D. Back and A.C. Tsoi. Blind deconvolution of signals using a complex recurrent network. In J. Vlontzos, J. Hwang, and E. Wilson, editors, *Proc. of the 1994 IEEE Workshop Neural Networks for Signal Processing 4 (NNSP94)*, Ermioni, Greece, 1994. IEEE Press.

52. A.D. Back and A.C. Tsoi. A comparison of discrete-time operator models for nonlinear system identification. In G. Tesauro, D.S. Touretzky, and T.K. Leen, editors, *Advances in Neural Information Processing Systems*, volume 7, pages 883–890, Cambridge, MA, 1995. The MIT Press.

53. A.D. Back and A.S. Weigend. A first application of independent component analysis to extracting structure from stock returns. *Int. Journal of Neural Systems*, 8:473–484, October 1997.

54. U.-M. Bae, T.-W. Lee, and S.-Y. Lee. Blind signal separation in teleconferencing using the ICA mixture model. *Electronic Letters*, 37(7):680–682, 2000.

55. Z. Bai, J. Demmel, J. Dongarra, A. Ruhe, and H. van der Vorst (editors). *Templates for the Solution of Algebraic Eigenvalue Problems: A Practical Guide*. SIAM, Philadelphia, 2000.

56. P. Baldi and K. Hornik. Neural networks and principal component analysis: Learning from examples without local minima. *Neural Networks*, 2:53–58, 1989.

57. S. Bannour and M.R. Azimi-Sadjadi. Principal component extraction using recursive least squares learning. *IEEE Tran. on Neural Networks*, 6:456–469, 1995.

58. Y. Bar-Ness and N. Sezgin. Adaptive multiuser bootstrapped decorrelating CDMA detector for one-shot asynchronous unknown channels. In *Proc. of the 1995 IEEE Int. Conference on Acoustics, Speech, and Signal Processing*, volume 3, pages 1733–1736, Detroit, Michigan USA, May 9–12 1995.

59. S. Barnett. *Matrices - Methods and Applications*. Clarendon Press, Oxford, 1990.

60. A. K. Barros and A. Cichocki. Robust batch algorithm for sequential blind extraction of noisy biomedical signals. In *Proc.Symposium on Signal Processing and its Applications (ISSPA'99)*, pages 363–366, Brisbane, Australia, 1999.

61. A. K. Barros and A. Cichocki. Extraction of specific signals with temporal structure. *Neural Computation*, 13(9):1995–2000, September 2001.

62. A.K. Barros, H. Kawahara, A. Cichocki, S. Kojita, T. Rutkowski, M. Kawamoto, and N. Ohnishi. Enhancement of a speech signal embedded in noisy environment using two microphones. In *Proceedings of the Second International Workshop on ICA and BSS, ICA'2000*, pages 423–428, Helsinki, Finland, 19-22 June 2000.

63. M. Bartlett. *Face Image Analysis by Unsupervised Learning*, volume 612. Kluwer International Series of Engineering and Computer Science, Boston, 2001.

64. R.H.T. Bates, Hong Jiang, and B.L.K. Davey. Multidimensional system identification through blind deconvolution. *Multidimensional Systems and Signal Processing*, 1(2):127–142, June 1990.

65. R.H.T. Bates and R.G. Lane. Automatic deconvolution and phase retrieval. *Proc. SPIE*, 828:158–164, 1987.

66. R.H.T. Bates and R.G. Lane. Deblurring should now be automatic. *Scanning Microscopy*, suppl.(2):149–156, 1988.

67. R.H.T. Bates, B.K. Quek, and C.R. Parker. Some implications of zero sheets for blind deconvolution and phase retrieval. *Journal of the Optical Society of America A (Optics and Image Science)*, 7(3):468–479, March 1990.

68. S. Becker. Unsupervised learning procedures for neural networks. *Int. Journal of Neural Systems*, 2:17–33, 1991.

69. Th. Beelen and P. Van Dooren. An improved algorithm for the computation of Kronecker's canonical form of a singular pencil. *Lin. Alg. Appl.*, 105:9–65, 1988.

70. A.J. Bell and T.J. Sejnowski. Blind separation and blind deconvolution: an information-theoretic approach. In *Proc. IEEE ICASSP*, pages 3415–3418 vol.5, Detroit, MI, 1995. IEEE.

71. A.J. Bell and T.J. Sejnowski. An information maximization approach to blind separation and blind deconvolution. *Neural Computation*, 7, no. 6:1129–1159, Nov 1995.

72. A.J. Bell and T.J. Sejnowski. A non-linear information maximization approach that performs blind separation. In *Advances in Neural Information Processing Systems 7*, pages 467–474. MIT Press, Cambridge, Mass, 1995.

73. A.J. Bell and T.J. Sejnowski. Learning the higher-order structure of a natural sound. *Network: Computation in Neural Systems*, 7:261–266, 1996.

74. S. Bellini. Bussgang techniques for blind equalization. In *Proc. of IEEE Global Telecommunications Conference*, pages 1634–1640, Houston, TX, 1986.

75. S. Bellini. Blind equalization. *Alta Frequenza*, 57(7):445–450, September 1988.

76. S. Bellini. Blind equalization and deconvolution. *Proc. SPIE*, 1565:88–101, 1991.

77. S. Bellini and F. Rocca. Asymptotically efficient blind deconvolution. *Signal Processing*, 20(3):193–209, July 1990.

78. A. Belouchrani, K. Abed-Meraim, J.-F. Cardoso, and É. Moulines. A blind source separation technique using second-order statistics. *IEEE Trans. Signal Processing*, 45(2):434–444, February 1997.

79. A. Belouchrani and M.G. Amin. A new approach for blind source separation using time-frequency distributions. *Proc. SPIE*, 2846:193–203, 1996.

80. A. Belouchrani, M.G. Amin, and K. Abed-Meraim. Direction finding in correlated noise fields based on joint block-diagonalization of spatio-temporal correlation matrices. *IEEE Signal Processing Letters*, 4(9), September 1997.

81. A. Belouchrani and J.-F. Cardoso. Maximum likelihood source separation for discrete sources. In *Signal Processing VII: Theories and Applications (Proc. of the EUSIPCO-94)*, pages 768–771, Edinburgh, Scotland, Sept. 13-16 1994. Elsevier.

82. A. Belouchrani and A. Cichocki. Robust whitening procedure in blind source separation context. *Electronics Letters*, 36(24):2050–2053, 2000.

83. A. Belouchrani, A. Cichocki, and K. Abed-Meraim. A blind identification and separation technique via multi-layer neural networks. In S. Amari, L. Xu, L.-W. Chan, I. King, and K.-S. Leung, editors, *Progress in Neural Information Processing. Proceedings of the International Conference on Neural Information Processing*, pages 1195–1200 vol.2, Hong Kong, 1996. Springer-Verlag.

84. A. Belouchrani, K. K. Abed-Meraim, J.-F. Cardoso, and É. Moulines. Second-order blind separation of correlated sources. In *Proc. Int. Conference on Digital Sig. Processing*, pages 346–351, Cyprus, 1993.

85. J. Benesty and P. Duhamel. A fast constant modulus adaptive algorithm. In L. Torres, E. Masgrau, and M.A. Lagunas, editors, *Signal Processing V. Theories and Applications. Proceedings of EUSIPCO-90, Fifth European Signal Processing Conference*, pages 241–244 vol.1, Barcelona, Spain, 18-21 Sept. 1990, 1990. Elsevier.

86. J. Benesty and P. Duhamel. Fast constant modulus adaptive algorithm. *IEE Proceedings F (Radar and Signal Processing)*, 138(4):379–387, August 1991.

87. S.E. Bensley and B. Aazhang. Subspace-based channel estimation for code division multiple access communication systems. *IEEE Tran. Communication.*, 44(8):1009–1020, August 1996.

88. A. Benveniste and M. Goursat. Blind equalizers. *IEEE Trans. Communications*, 32(8):871–883, 1984.

89. A. Benveniste, M. Goursat, and G. Ruget. Robust identification of a non-minimum phase system: Blind adjustment of a linear equalizer in data communications. *IEEE Trans. Automatic Contr.*, AC-25, no. 3:385–399, June 1980.

90. N. Benvenuto and T.W. Goeddel. Classification of voiceband data signals using the constellation magnitude. *IEEE Trans. Communications*, 43(11):2759–2770, November 1995.

91. N.J. Bershad and S. Roy. Performance of the 2-2 constant modulus (CM) adaptive algorithm for Rayleigh fading sinusoids in Gaussian noise. In *Proc. IEEE ICASSP*, pages 1675–1678 vol.3, Albuquerque, NM, USA, 3-6 April 1990, 1990.

92. A.G. Bessios. Compound compensation strategies for wireless data communications over the multimodal acoustic ocean waveguide. *IEEE Journal of Oceanic Engineering*, 21(2):167–180, April 1996.

93. A.G. Bessios and F.M. Caimi. Frequency division multiplexing in wireless underwater acoustic LAN's. *Proc. SPIE*, 2556:69–78, 1995.

94. A.G. Bessios and C.L. Nikias. POTEA: the Power Cepstrum and Tricoherence Equalization Algorithm. *IEEE Trans. Communications*, 43(11):2667–2671, November 1995.

95. S. Bhattacharyya, D.H. Szarowski, J.N. Turner, N. O'Connor, and T.J. Holmes. The ML-blind deconvolution algorithm: recent developments. *Proc. SPIE*, 2655:175–186, 1996.

96. P.J. Bickel, C.A.J. Klaassen, Y. Ritov, and J.A. Wellner. *Efficient and Adaptive Estimation for Semiparametric Models.* MD: Johns Hopkins University Press., Baltimore, 1993.

97. P. Binding. Simultaneous diagonalization of several Hermitian matrices. *SIAM Journal Matrix Anal. Appl.*, 4(11):531–536, 1990.

98. E. Bingham and A. Hyvärinen. A fast fixed-point algorithm for independent component analysis of complex-valued signals. *Int. Journal of Neural Systems*, 10(1):1–8, 2000.

99. R.E. Bogner. Blind separation of sources. Technical Report 4559, Defence Research Agency, Malvern, May 1992.

100. P.J. Bones, C.R. Parker, B.L. Satherley, and R.W. Watson. Deconvolution and phase retrieval with use of zero sheets. *Journal of the Optical Society of America A (Optics, Image Science and Vision)*, 12(9):1842–1857, September 1995.

101. S. Bose and B. Friedlander. On the performance output cumulant matching based source separation methods in noise. In *29-th Asilomar Conference Signals, Systems, Comp.*, pages 418–422 vol.1. IEEE, 1996.

102. M. Boumahdi and J-L. Lacoume. Blind identification using the kurtosis: Results of field data processing. In *Proc. of the 1995 IEEE Int. Conference on Acoustics, Speech, and Signal Processing*, volume 3, pages 1980–1983, Detroit, Michigan USA, May 9–12 1995.

103. M. Brandstein and D. Ward (Eds.). *Microphone Arrays: Signal Processing Techniques and Applications.* DSP. Springer, New York, 2001.

104. Brandwood. A complex gradient operator and its application in adaptive theory. *IEE Proceedings*, 130(1):11–16, 1997.

105. R.W. Brockett. Dynamical systems that learn subspaces. In *Mathematical system theory: The influences of R.E. Kalman*, pages 579–592. Springer, Berlin, 1991.

106. R.W. Brockett. Dynamical systems that sort lists, diagonalize matrices, and solve linear programming problems. *Linear Algebra Applications*, 146:79–91, 1991.

107. D.H. Brooks and C.L. Nikias. Cross-bicepstrum and cross-tricepstrum approaches to multichannel deconvolution. In J.L. Lacoume, editor, *Higher Order Statistics. Proceedings of the International Signal Processing Workshop*, pages 141–144, Chamrousse, France, 1992. Elsevier.

108. D.H. Brooks and C.L. Nikias. Multichannel adaptive blind deconvolution using the complex cepstrum of higher order cross-spectra. *IEEE Trans. Signal Processing*, 41(9):2928–2934, September 1993.

109. D.H. Brooks and A.P. Petropulu. Non-iterative blind deconvolution of colored or deterministic signals using higher order cepstra and group delay. In *IEEE SP Workshop on Stat. Signal Array Processing*, pages 132–135, Victoria, BC, 1992.

110. S.R. Brooks, editor. *Mathematics in Remote Sensing*, Danbury, UK, 1989. Clarendon Press.

111. R.G. Brown and P.Y.C. Hwang. *Introduction to Random Signals and Applied Kalman Filtering.* John Wiley & Sons, Inc., 2nd edition, 1992.

112. A. Bunse-Gerstner, R. Byers, and V. Mehrmann. Numerical methods for simultaneous diagonalization. *SIAM Journal Matrix Anal. Appl.*, 4:927–949, 1993.

113. G. Burel. Blind separation of sources - A nonlinear neural algorithm. *Neural Networks*, 5(6):937–947, 1992.

114. C. Byrne. Block-iterative interior point optimization methods for image reconstruction from limited data. *Inverse Problems*, 15:1405–1419, 2000.

115. J.A. Cadzow and X. Li. Blind deconvolution. *Digital Signal Processing*, 5(1):3–20, January 1995.

116. J.A. Cadzow Minimum l_1,l_2 and l_∞ norm approximate solution to na overdetermined system of linear equations. *Digital Signal Processing*, Vol. 12, 524-560, 2002.

117. D. Callaerts, J. Vandewalle, and D. Van Compernolle. OSVD and QSVD in signal separation. In R.J. Vaccaro, editor, *SVD and Signal Processing II*, pages 323–334. Elsevier Science Publishers, 1991.

118. C.N. Canagarajah. *Digital Signal Processing Techniques for Speech Enhancement in Hearing Aids*. PhD thesis, Christ's College, University of Cambridge, 1993.

119. J. Cao, A. Cichocki, and S. Tanaka. Self-scaling and self-adaptive compact time-delay neural network for dynamical nonlinear and nonstationary system identification. *Journal of Signal Processing*, 4(1):37–43, 2000.

120. J. Cao, N. Murata, S. Amari, A. Cichocki, and T. Takeda. MEG data analysis based on ICA approach with pre- & post-processing techniques. In *Proceedings of 1998 International Symposium on Nonlinear Theory and its Applications (NOLTA-98)*, pages 287–290, Switzerland, 1998.

121. J. Cao, N. Murata, S. Amari, A. Cichocki, and T. Takeda. Independent component analysis for single-trial MEG data decomposition and single-dipole source localization. *Neurocomputing*, 2002.

122. J. Cao, N. Murata, and A. Cichocki. Independent component analysis algorithm for online blind separation and blind equalization systems. *Journal of Signal Processing*, 4(2):131–140, March 2000.

123. X.-R. Cao and R.-W. Liu. General approach to blind source separation. *IEEE Trans. Signal Processing*, 44(3):562–571, March 1996.

124. V. Capdevielle, Ch. Serviere, and J. Lacoume. Blind separation of wide-band sources in the frequency domain. In *Proc. of the 1995 IEEE Int. Conference on Acoustics, Speech, and Signal Processing*, volume 3, pages 2080–2083, Detroit, Michigan, USA, May 9–12 1995.

125. J.-F. Cardoso. Blind identification of independent components with higher-order statistics. In *IEEE Workshop on Higher-Order Spectral Analysis*, pages 157–162, Vail, CO, 1989.

126. J.-F. Cardoso. Source separation using higher order moments. In *Proc. IEEE ICASSP*, pages 2109–2112 vol.4, Glasgow, UK, 1989.

127. J.-F. Cardoso. Eigen-structure of the fourth-order cumulant tensor with application to the blind source separation problem. In *Int. Conference on Acoustics Speech and Signal Processing*, pages 2655–2658, Albuquerque, NM, USA, April 3-6 1990.

128. J.-F. Cardoso. Localization and identification with the quadricovariance. *Traitement du Signal*, 7(5):397–406, 1990.

129. J.-F. Cardoso. Super-symmetric decomposition of the fourth-order cumulant tensor. Blind identification of more sources than sensors. In *Proc. IEEE ICASSP*, pages 3109–3112 vol.5, Toronto, 1991.

130. J.-F. Cardoso. Fourth-order cumulant structure forcing: application to blind array processing. In *IEEE SP Workshop Stat. Signal Array Processing*, pages 136–139, Victoria, BC, Canada, 1992.

131. J.-F. Cardoso. Higher-order narrow-band array processing. In J.L. Lacoume, editor, *Proc. Higher Order Statistics*, pages 39–48, Chamrousse, France, 1992. Elsevier.

132. J.-F. Cardoso. Iterative techniques for blind source separation using only fourth order cumulants. In J. Vandewalle e.a., editor, *Signal Processing VI: Theories and Applications (Proc. of the EUSIPCO-92)*, pages 739–742, Brussels, Belgium, 1992. Elsevier.

133. J.-F. Cardoso. Adaptive source separation based on non linear matrix updates. In *Proc. Int. Conference on Digital Sig. Processing*, pages 32–37, Cyprus, 1993.

134. J.-F. Cardoso. How much more DOA information in higher order statistics ? In *Proc. 7th workshop on statistical signal and array processing*, pages 199–202, Quebec City, 1994.

135. J.-F. Cardoso. On the performance of source separation algorithms. In *Signal Processing VII: Theories and Applications (Proc. of the EUSIPCO-94)*, Edinburgh, Scotland, Sept. 13-16 1994. Elsevier.

136. J.-F. Cardoso. The equivariant approach to source separation. In *Proc. NOLTA*, pages 55–60, 1995.

137. J.-F. Cardoso. Infomax and maximum likelihood for blind source separation. *IEEE Signal Processing Letter*, 4:109–111, April 1997.

138. J.-F. Cardoso. Blind signal separation: Statistical principles. *Proceedings. of the IEEE*, 86(10):2009–2025, 1998.

139. J.-F. Cardoso. High-order contrasts for independent component analysis. *Neural Computation*, 11(1):157–192, January 1999.

140. J.-F. Cardoso. On the stability of source separation algorithms. *Journal of VLSI Signal Processing Systems*, 26(1/2):7–14, August 2000.

141. J.-F. Cardoso. On the stability of source separation algorithms. *Journal of VLSI Signal Processing Systems*, 26(1/2):7–14, April 2000. Special issue on Neural networks for signal processing.

142. J.-F. Cardoso and S. Amari. Maximum likelihood source separation: equivariance and adaptivity. In *Proc. of SYSID'97, 11th IFAC symposium on system identification, Fukuoka, Japan.*, pages 1063–1068, 1997.

143. J.-F. Cardoso, A. Belouchrani, and B. Laheld. A new composite criterion for adaptive and iterative blind source separation. In *Int. Conference on Acoustics Speech and Signal Processing*, volume IV, pages 273–276, A.aide, Australia, April 19-22 1994.

144. J.-F. Cardoso, S. Bose, and B. Friedlander. Output cumulant matching for source separation. In *Proc. IEEE SP Workshop on Higher-Order Stat., Aiguablava, Spain*, pages 44–48, 1995.

145. J.-F. Cardoso and P. Comon. Tensor-based independent component analysis. In L. Torres, E. Masgrau, and M. Lagunas, editors, *Signal Processing V: Theories and Applications*, pages 673–676. Elsevier, 1990.

146. J.-F. Cardoso and P. Comon. Independent component analysis, a survey of some algebraic methods. In *Proc. ISCAS Conference*, volume 2, pages 93–96, Atlanta, May 1996.

147. J.-F. Cardoso and D. L. Donoho. Some experiments on independent component analysis of non-Gaussian processes. In *IEEE SP Int. Workshop on High Order Statistics*, pages 74–77, Caeserea, Israel, 1999.

148. J.-F. Cardoso and B.H. Laheld. Equivariant adaptive source separation. *IEEE Trans. Signal Processing*, 44(12):3017–3030, December 1996.

149. J.-F. Cardoso and É. Moulines. A robustness property of DOA estimators based on covariance. *IEEE Tr. on Signal Processing.*, 42(11):3285–3287, November 1994.

150. J.-F. Cardoso and É. Moulines. Asymptotic performance analysis of direction finding algorithms based on fourth-order cumulants. *IEEE Trans. on Signal Processing*, 43(1):214–224, January 1995.

151. J.-F. Cardoso and A. Souloumiac. Blind beamforming for non-Gaussian signals. *IEE Proc. F (Radar and Signal Processing)*, 140(6):362–370, December 1993.

152. J.-F. Cardoso and A. Souloumiac. An efficient technique for the blind separation of complex sources. In *IEEE Signal Processing Workshop on Higher-Order Statistics*, pages 275–279, South Lake Tahoe, 1993.

153. J.-F. Cardoso and A. Souloumiac. Jacobi angles for simultaneous diagonalization. *SIAM Journal Mat. Anal. Appl.*, 17(1):161–164, January 1996.

154. E. De Carvalho and D.T.M. Slock. Maximum-likelihood blind equalization of multiple FIR channels. In *Proc. IEEE ICASSP*, pages 2451–2454 vol. 5, Atlanta, GA, 1996.

155. E. De Carvalho and D.T.M. Slock. Cramér-Rao bounds for semi-blind, blind and training sequence based channel estimation. In *IEEE workshop on Signal Processing Advances in Wireless Communications*, pages 129–132, Paris, April 1997.

156. R.A. Casas, Z. Ding, R.A. Kennedy, C.R. Johnson, Jr., and R. Malamut. Blind adaptation of decision feedback equalizers based on the constant modulus algorithm. In *29-th Asilomar Conference Signals, Systems, Comp.*, pages 698–702 vol.1. IEEE, 1996.

157. L. Castedo, C.-Y. Tseng, A.R. Figueiras-Vidal, and L.J. Griffiths. Behavior of adaptive beamformers based on cyclostationary signal properties in multipath environments. In *27-th Asilomar Conference Signals, Systems Comp.* IEEE, 1993.

158. L. Castedo, C.-Y. Tseng, A.R. Figueiras-Vidal, and L.J. Griffiths. Linearly-constrained adaptive beamforming using cyclostationary signal properties. In *Proc. IEEE ICASSP*, pages 249–252 vol.4, 1994.

159. T.S. Castelein, Y. Bar-Ness, and R. Prasad. Blind linear recursive equalizer with decorrelation algorithm. In *Proc. IEEE ICASSP*, pages 1069–1072 vol.2, Detroit, MI, 1995.

160. R.T. Causey and J.R. Barry. The impact of carrier frequency offset on second-order algorithms for blind channel identification and equalization. In *Proc. IEEE Int. Conference Communic.*, pages 995–999 vol.2, Dallas, TX, 1996.

161. F.R.P. Cavalcanti and J.C.M. Mota. A predictive blind equalizer based in Godard's criterion for 256-QAM digital radio systems. In *Proc. IEEE Int. Conference Communic.*, pages 837–841 vol.2, Dallas, TX, 1996.

162. M. Cedervall, B.C. Ng, and A. Paulraj. Structured methods for blind multi-channel identification. In *Proc. 13th Int. Conference Dig. Sig. Processing*, pages 387–390, Santorini, Greece, July 1997.

163. Y. Censor, D. Gordon, and R. Gordon. Component averaging: An efficient iterative parallel algorithm for large and sparse unstructured problems. *Parallel Computing*, 27:1414–1415, 2001.

164. K.-H. Cha, D.-Y. Jang, J.-W. Hong, and C.-D. Kim. Implementation and evaluation of stereo audio codec using perceptual coding. *Journal of the Korean Institute of Telematics and Electronics*, 33B(4):156–163, April 1996.

165. J. Chambers and S. Lambotharan. Phase inference and error surface analysis of a blind non-minimum phase channel equalizer. In *IEE Colloquium on 'Blind Deconvolution - Algorithms and Applications' (Ref. No.1995/145)*, pages 2/1–6, London, UK, 1995. IEE.

166. J.A. Chambers, D. Mandic, and W. Shirlker "A Normalised gradient algorithm for an adaptive recurrent Perceptron", ICASSP 2000, Istanbul, Turkey

167. C.K. Chan, M.R. Petraglia, and J.J. Shynk. Frequency-domain implementations of the constant modulus algorithm. In *23-th Asilomar Conference Signals, Systems Comp.*, pages 663–669 vol.2, Pacific Grove, CA, USA, 30 Oct.-1 Nov. 1989, 1989. Maple Press.

168. C.K. Chan and J.J. Shynk. Stationary points of the constant modulus algorithm for real Gaussian signals. *IEEE Trans. Acoustics, Speech and Signal Processing*, 38(12):2176–2181, December 1990.

169. D.B. Chan, P.J.W. Rayner, and S.J. Godsill. Multi-channel signal separation. In *Proc. ICASSP*, pages 649–652, May 1996.

170. C. Chang, Z. Ding, S. F. Yau, and F. H. Y. Chan. A matrix-pencil approach to blind separation of colored nonstationary signals. *IEEE Trans. on Signal Processing*, SP-48(3):900–907, March 2000.

171. C. Chatterjee, Z. Kang, and V.P. Roychowdhury. Algorithms for accelerated convergence of adaptive PCA. *IEEE Trans. on Neural Networks*, 11(2):338–355, 2000.

172. C. Chatterjee, V.P. Roychowdhury, and E.K.P. Chong. On relative convergence properties of principal component analysis algorithms. *IEEE Trans. on Neural Network*, 9(2):319–329, March 1998.

173. C. Chatterjee, V.P.Roychowdhury, M.D. Zlotowski, and J. Ramos. Self-organizing and adaptive algorithms for generalized eigen-decomposition. *IEEE Trans. on Neural Network*, 8(6):1518–1530, November 1997.

174. E. Chaumette, P. Comon, and D. Muller. An ICA-based technique for radiating sources estimation; application to airport surveillance. *IEE Proceedings - Part F*, 140(6):395–401, December 1993. Special issue on Applications of High-Order Statistics.

175. R. Chechik, A. Breskin, and A. Gibrekhterman. A proposal for a hadron-blind fast TRD based on secondary electron emission. *Nuclear Physics B, Proceedings Supplements*, 44:364–372, November 1995.

176. L. Chen, H. Kusaka, and M. Kominami. Cumulant-based blind channel equalization. *IEICE Trans. Fundamentals of Electronics, Communications and Computer Sciences*, E79-A(5):727–730, May 1996.

177. L. Chen, H. Kusaka, M. Kominami, and Q. Yin. Blind identification of noncausal AR models based on higher-order statistics. *Signal Processing*, 48(1):27–36, January 1996.

178. L. Chen, H. Kusaka, M. Kominami, and Qingyie Yin. Blind identification using higher order statistics and eigendecomposition. *Trans. of the Institute of Electrical Engineers of Japan, Part C*, 116-C(3):319–324, March 1996.

179. S. Chen and D. Li. A fast minimum error probability blind equalization algorithm. *Chinese Journal of Electronics*, 4(1):59–64, January 1995.

180. S. Chen, S. McLaughlin, P.M. Grant, and B. Mulgrew. Joint channel estimation and data detection using a blind Bayesian decision feedback equalizer. In *IEE Colloquium on 'Blind Deconvolution - Algorithms and Applications' (Ref. No.1995/145)*, pages 4/1–5, London, UK, 1995. IEE.

181. T.-P. Chen. Modified Oja's algorithm for principal subspace and minor subspace extraction. *Neural Processing Letters*, 5(5):105–110, 1997.

182. T.-P. Chen and S. Amari. Unified stabilization approach to principal and minor components. *Neural Networks*, accepted 2001.

183. T.-P. Chen, S. Amari, and Q. Lin. A unified algorithm for principal and minor components extraction. *Neural Networks*, 11:385–390, 1998.

184. T.-P. Chen, S. Amari, and N. Murata. Sequential extraction of minor components. *Neural Processing Letters*, 13:195–201, June 2001.

185. T.-P. Chen and H. Chen. Blind extraction of stochastic and deterministic signals by neural network approach. In *Proc. of 28-th Asilomar Conference on Signals, Systems and Computers*, pages 892–896, 1994.

186. T.-P. Chen, M.A. Fiddy, C.-W. Liao, and D.A. Pommet. Blind deconvolution and phase retrieval from point zeros. *Journal of the Optical Society of America A (Optics, Image Science and Vision)*, 13(7):1524–1531, July 1996.

187. T.-P. Chen, Y. Hua, and W.-Y. Yan. Global convergence of Oja's subspace algorithm for principal component extraction. *IEEE Trans. on Neural Networks*, 9(1):58–67, 1998.

188. Y. Chen and J.C.-I. Chuang. Blind equalization and its application to the PACS system. In C.A.O. Zhigang, editor, *ICCT'96. 1996 International Conference on Communication Technology Proceedings*, pages 841–844 vol.2, Proceedings of International Conference on Communication Technology. ICCT '96, Beijing, China, 5-7 May 1996, 1996. IEEE.

189. Y. Chen and C. Nikias. Fractionally-spaced blind equalization with CRIMNO algorithm. In *Proc. IEEE MILCOM*, pages 221–225 vol.1, San Diego, CA, 1992.

190. Y. Chen, C. Nikias, and J.G. Proakis. CRIMNO: criterion with memory nonlinearity for blind equalization. In *25-th Asilomar Conference Signals, Systems Comp.*, pages 694–698. IEEE, 1991.

191. Y. Chen and C.L. Nikias. Two-dimensional memory nonlinearities and their application to blind deconvolution problems. In *Proc. IEEE SP Workshop on Stat. Signal Array Processing*, pages 210–212, Victoria, BC, 1992. IEEE.

192. Y. Chen, C.L. Nikias, and J.G. Proakis. Blind equalization with criterion with memory nonlinearity. *Optical Engineering*, 31(6):1200–1210, June 1992.

193. Y. Chen, C.L. Nikias, and J.G. Proakis. CRIMNO: criterion with memory nonlinearity for blind equalization. In J.L. Lacoume, editor, *Higher Order Statistics. Proceedings of the International Signal Processing Workshop*, pages 137–140, Chamrousse, France, 1992. Elsevier.

194. Y.-W. Chen, Z. Nakao, and S. Tamura. Blind deconvolution by genetic algorithms. *Proc. SPIE*, 2662:192–196, 1996.

195. K.M. Cheung and S.F. Yau. Blind deconvolution of system with unknown response excited by cyclostationary impulses. In *Proc. IEEE ICASSP*, pages 1984–1987 vol.3, Detroit, MI, 1995.

196. A. Chevreuil and P. Loubaton. On the use of conjugate cyclo-stationarity: a blind second-order multi-user equalization method. In *Proc. IEEE ICASSP*, pages 2439–2442, Atlanta, GA, 1996.

197. A. Chevreuil and Ph. Loubaton. Blind second-order identification of FIR channels: forced cyclostationarity and structured subspace method. *IEEE Signal Processing Letters*, 4(7):204–206, July 1997.

198. C.-Y. Chi and W.-T. Chen. Maximum-likelihood blind deconvolution: non-white Bernoulli-Gaussian case. *IEEE Trans. Geoscience and Remote Sensing*, 29(5):790–795, September 1991.

199. C.-Y. Chi and M.-C. Wu. Inverse filter criteria for blind deconvolution and equalization using two cumulants. *Signal Processing*, 43(1):55–63, April 1995.

200. C.-Y. Chi and M.-C. Wu. A unified class of inverse filter criteria using two cumulants for blind deconvolution and equalization. In *Proc. IEEE ICASSP*, pages 1960–1963 vol.3, Detroit, MI, 1995.

201. H.-H. Chiang and C.L. Nikias. Adaptive deconvolution and identification of nonminimum phase FIR systems based on cumulants. *IEEE Trans. Automatic Control*, 35(1):36–47, January 1990.

202. I. Chiba, W. Chujo, and M. Fujise. Beam space constant modulus algorithm adaptive array antennas. In *Eighth Int. Conference Antennas and Propagation (Conference Publ. No.370)*, pages 975–978 vol.2, Edinburgh, UK, 30 March-2 April 1993, 1993. IEE.

203. I. Chiba, W. Chujo, and M. Fujise. Beam-space CMA adaptive array antennas. *Electronics and Communications in Japan, Part 1 (Communications)*, 78(2):85–95, February 1995.

204. A. Chkeif, K. Abed-Meriam, G. Kawas Kaleh, and Y. Hua. Spatio-temporal blind adaptive multiuser detection. *IEEE Trans. Communications*, 48, May 2000. to be published.

205. S. Choi. Differential Hebbian-type learning algorithms for decorrelation and independent component analysis. *Electronics Letters*, 34(9):900–901, 1998.

206. S. Choi, S. Amari A., Cichocki, and R. Liu. Natural gradient learning with a nonholomonic constraint for blind deconvolution of multiple channels. In *Proc. of the First International Workshop on Independent Component Analysis and Signal Separation - ICA'99*, pages 371–376, Aussois, France, January 11-15 1999.

207. S. Choi, S. Amari, and A. Cichocki. Natural gradient learning algorithms for decorrelation. In N. Kasabov, R. Kozma, K. Ko, R. O'shea, and T. Gedeon, editors, *Progress in Connectionist-Based Information Systems*, volume 1, pages 645–648, 1997.

208. S. Choi, S. Amari, and A. Cichocki. Natural gradient learning for spatio-temporal decorrelation: Recurrent network. *IEICE Trans. Fundamentals*, E-83A(12):2715–2722, Dec. 2000.

209. S. Choi and A. Cichocki. Adaptive blind separation of speech signals: Cocktail party problem. In *International Conference on Speech Processing (ICSP'97)*, pages 617–622, Seoul, Korea, 26-28 Aug. 1997.

210. S. Choi and A. Cichocki. Blind signal deconvolution by spatio-temporal decorrelation and demixing. In J. Principe, L. Gile, N. Morgan, and E. Wilson, editors, *Neural Networks for Signal Processing VII*, pages 426–435. IEEE, 1997.

211. S. Choi and A. Cichocki. A linear feedforward neural network with lateral feedback connections for blind source separation. In *IEEE Signal Processing Workshop on Higher-order Statistics (Banff, Canada)*, pages 349–353, 21-23 July 1997.

212. S. Choi and A. Cichocki. Cascade neural networks for multichannel blind deconvolution. *Electronics Letters*, 34(12):1186–1187, 1998.

213. S. Choi and A. Cichocki. On-line sequential multichannel blind deconvolution: A deflation approach. In *In Proc. 8th IEEE DSP Workshop*, pages 159–162, Utahi, USA, 1998.

214. S. Choi and A. Cichocki. A hybrid learning approach to blind deconvolution of linear MIMO systems. *Electronics Letters*, 35(17):1429–1430, August 19 1999.

215. S. Choi and A. Cichocki. A hybrid learning approach to blind deconvolution of MIMO systems. In *IEEE Signal Processing Workshop on Higher-order Statistics (HOS'99)*, pages 292–295, Ceasarea, Israel, June 14-16 1999.

216. S. Choi and A. Cichocki. An unsupervised hybrid network for blind separation of independent non-Gaussian source signals in multipath environment. *Journal of Communications and Networks*, 1(1):19–25, March 1999.

217. S. Choi and A. Cichocki. Blind separation of nonstationary and temporally correlated sources from noisy mixtures. In *IEEE Workshop on Neural Networks for Signal Processing, NNSP'2000*, pages 405–414, Sydney, Australia, December 11-13 2000.

218. S. Choi and A. Cichocki. Blind separation of nonstationary sources in noisy mixtures. *Electronics Letters*, 36:848–849, April 2000.

219. S. Choi and A. Cichocki. Algebraic differential decorrelation for nonstationary source separation. *Electronics Letters*, 37(23):1414–1415, 2001.

220. S. Choi and A. Cichocki. Blind equalization via approximate maximum likelihood source separation. *Electronics Letters*, 37(1):61–62, Jan. 2001.

221. S. Choi, A. Cichocki, and S. Amari. Adaptive blind deconvolution and equalization with self-adaptive nonlinearities: An information-theoretic approach. In N. Kasabov, R. Kozma, K. Ko, R. O'shea, and T. Gedeon, editors, *Progress in Connectionist-Based Information Systems*, volume 1, pages 641–644, 1997.

222. S. Choi, A. Cichocki, and S. Amari. Blind equalization of SIMO channels via spatio-temporal anti-Hebbian learning rule. In *Proc. of the 1998 IEEE Workshop on NNSP Cambridge*, pages 93–102, UK, 1998. IEEE Press, N.Y.

223. S. Choi, A. Cichocki, and S. Amari. Flexible independent component analysis. In *Proc. of the 1998 IEEE Workshop on NNSP*, pages 83–92, Cambridge, UK, 1998.

224. S. Choi, A. Cichocki, and S. Amari. Fetal electrocardiogram data analysis via flexible independent component analysis. In *The 4th Asia-Pacific Conference on Medical & Biological Engineering (APCMBE'99)*, Seoul, Korea, 1999.

225. S. Choi, A. Cichocki, and S. Amari. Two spatio-temporal decorrelation learning algorithms and their application to multichannel blind deconvolution. In *ICASSP'99*, pages 1085–1088, Phoenix, Arizona, March 15-19 1999.

226. S. Choi, A. Cichocki, and S. Amari. Flexible independent component analysis. *Journal of VLSI Signal Processing*, 26(1/2):25–38, 2000.

227. S. Choi, A. Cichocki, and S. Amari. Local stability analysis of flexible independent component analysis algorithm. In *ICASSP2000*, pages 3426–3429, Istanbul, Turkey, June 5-9 2000.

228. S. Choi, A. Cichocki, and S. Amari. Equivariant nonstationary source separation. *Neural Networks*, 15(1), 2002.

229. S. Choi, A. Cichocki, and A. Belouchrani. Blind separation of second-order nonstationary and temporally colored sources. In *Proceedings of the 11th IEEE Signal Processing Workshop on Statistical Signal Processing*, pages 444–447, Singapore, 2001.

230. S. Choi, A. Cichocki, and A. Belouchrani. Second order nonstationary source separation. *Journal of VLSI Signal Processing*, 2002, to appear.

231. S. Choi, A. Cichocki, and Y. Deville. Differential decorrelation for nonstationary source separation. In *Third International Conference on Independent Component Analysis and Signal Separation (ICA-2001)*, pages 319–322, San Diego, USA, Dec. 9-13 2001.

232. S. Choi, A. Cichocki, L. Zhang, and S. Amari. Approximate maximum likelihood source separation using the natural gradient. In *Proc. IEEE Workshop on Signal Processing Advances in Wireless Communications*, pages 235–238, Taoyuan, Taiwan, 2001.

233. S. Choi, H. Hong, H. Glotin, and F. Berthommier. Multichannel signal separation for cocktail party speech recognition: A dynamic recurrent network. *Neurocomputing*, 2002, to appear.

234. S. Choi, R.-W. Liu, and A. Cichocki. A spurious equilibria-free learning algorithm for the blind separation of non-zero skewness signals. *Neural Processing Letters*, 7(2):61–68, 1998.

235. S. Choi, Y. Lyu, F. Berthommier, H. Glotin, and A. Cichocki. Blind separation of delayed and superimposed acoustic sources: Learning algorithm and experimental study. In *International Conference on Speech Processing (ICSP'99)*, pages 109–114, August 18-20 1999.

236. Y.S. Choi, D.S. Han, and H. Hwang. Joint blind equalization, carrier recovery and timing recovery for HDTV modem. *Proceedings of the SPIE - The International Society for Optical Engineering*, 2094:1357–1363, November 1993.

237. Y.S. Choi, H. Hwang, and D.I. Song. Adaptive blind equalization coupled with carrier recovery for HDTV modem. *IEEE Trans. Consumer Electronics*, 39(3):386–391, 1993.

238. J.C. Christou. Blind deconvolution post-processing of images corrected by adaptive optics. *Proc. SPIE*, 2534:226–234, 1995.

239. M.T. Chu. A continuous Jacobi-like approach to the simultaneous reduction of real matrices. *Lin. Alg. Appl.*, 147:75–96, 1991.

240. K.-I. Chung and C.-T. Lim. Transform methods for PAM signals as asymmetric distribution and performance comparison of bicepstrum blind equalizer using asymmetric distribution. *Journal of the Korean Institute of Telematics and Electronics*, 33B(6):54–63, June 1996.

241. K.I. Chung and C.T. Lim. Asymmetric distribution of PAM signals and blind equalization algorithm using 3rd order statistics. *Journal of the Korean Institute of Telematics and Electronics*, 33A(7):65–75, July 1996.

242. A Cichocki. Neural network for singular value decomposition,. *Electronics Letters*, vol.28, No.8:784–786, 1992.

243. A. Cichocki. Blind separation and extraction of source signals recent results and open problems. In *Proc. of the 4-th Annual Conference of the Institute of Systems, Control and Information Engineers, ISCIIE, Osaka*, pages 43–48, May 21-23 1997.

244. A. Cichocki. Blind identification and separation of noisy source signals - neural networks approaches. *ISCIE Journal*, 42(2):63–73, 1998.

245. A. Cichocki, S. Amari, M. Adachi, and W. Kasprzak. Self–adaptive neural networks for blind separation of sources. In *1996 IEEE International Symposium on Circuits and Systems, ISCAS'96*, volume 2, pages 157–161, Atlanta, USA, May 1996. IEEE.

246. A. Cichocki, S. Amari, and J. Cao. Blind separation of delayed and convolved signals with self-adaptive learning rate. *IEEE International Symposium on Nonlinear Theory and its Applications, NOLTA-96, Kochi Japan*, pages 229–232, Oct 7-9 1996.

247. A. Cichocki, S. Amari, and J. Cao. Neural network models for blind separation of time delayed and convolved signals. *Japanese IEICE Transaction on Fundamentals*, Vol E-82-A No.9:1595–1603, Sept. 1997.

248. A. Cichocki, S. Amari, and R. Thawonmas. Blind signal extraction using self–adaptive non–linear Hebbian learning rule. In *International Symposium on Nonlinear Theory and its*

Applications , NOLTA96, pages 377–380, Research Society on NTA, IEICE,Kochi, Japan, Oct. 1996.

249. A. Cichocki and A. Belouchrani. Sources separation of temporally correlated sources from noisy data using bank of band-pass filters. In *Third International Conference on Independent Component Analysis and Signal Separation (ICA-2001)*, pages 173–178, San Diego, USA, Dec. 9-13 2001.

250. A. Cichocki, R. Bogner, and L. Moszczyński. Improved adaptive algorithms for blind separation of sources. In *Proc. of Conference on Electronic Circuits and Systems, KKTOiUE*, pages 647–652, Zakopane, Poland, 1995.

251. A. Cichocki, R.E. Bogner, L. Moszczyński, and K. Pope. Modified Hérault-Jutten algorithms for blind separation of sources. *Digital Signal Processing*, 7 No.2:80 – 93, April 1997.

252. A. Cichocki and J. Cao. A self-adaptive neural network for on-line blind separation of convolved sources. In B.B. Djordjevic and H.D. Reis, editors, *Proc. of III Int. Workshop-Advances in Signal Processing for NDE of Materials, Topics on Non-destructive Evaluation Series*, pages 207–212, Quebeck, 1998. The American Society for Non-destructive Evaluation Testing, Inc. Quebeck.

253. A. Cichocki, J. Cao, S. Amari, N. Murata, T. Takeda, and H. Endo. Enhancement and blind identification of magnetoencephalographic signals using independent component analysis. In *Proc of th 11th Int. Conference on Biomagentism BIOMAG-98*, pages (169–172, Sendai, Japan, 1999.

254. A. Cichocki, S.C. Douglas, and S. Amari. Robust techniques for independent component analysis (ICA) with noisy data. *Neurocomputing*, 23(1–3):113–129, November 1998.

255. A. Cichocki, S.C. Douglas, S. Amari, and P. Mierzejewski. Independent component analysis for noisy data. In *Proc. of International Workshop on Independence and Artificial Neural Networks*, pages 52–58, Tenerife, 1998.

256. A. Cichocki, R. R. Gharieb, and T. Hoya. Efficient extraction of evoked potentials by combination of Wiener filtering and subspace methods. In *Proc. of IEEE Int. Conf. Acoustics, Speech, Signal Processing, ICASSP-2001*, pages 3117–3120, Utah, USA, May 7-11 2001.

257. A. Cichocki, R.R. Gharieb, and N. Mourad. Extraction of superimposed evoked potentials by combination of independent component analysis and cumulant-based matched filtering. In *Proceedings of the 11th IEEE Signal Processing Workshop on Statistical Signal Processing*, pages 237–240, Singapore, 2001.

258. A. Cichocki, J. Karhunen, W. Kasprzak, and R. Vigário. Neural networks for blind separation with unknown number of sources. *Neurocomputing*, 24(1-3):55–93, February 1999.

259. A. Cichocki and W. Kasprzak. Nonlinear learning algorithms for blind separation of natural images. *Neural Network World*, 6(4):515–523, 1996.

260. A. Cichocki, W. Kasprzak, and S. Amari. Multi-layer neural networks with a local adaptive learning rule for blind separation of source signals. In *to appear in the Proc. of the 1995 Int. Symposium on Nonlinear Theory and its Applications (NOLTA'95)*, volume 1, pages 61–66, Las Vegas, USA, Dec. 10-14 1995.

261. A. Cichocki, W. Kasprzak, and S. Amari. Adaptive approach to blind source separation with cancellation of additive and convolutional noise. In *Third International Conference on Signal Processing, ICSP'96*, volume 1, pages 412–415, Beijing, China, Oct. 1996.

262. A. Cichocki, W. Kasprzak, and S. Amari. Neural network approach to blind separation and enhancement of images. In *Signal Processing VIII. Theories and Applications., EURASIP/LINT Publ., Trieste, Italy*, volume 1, pages 579–582, Sept. 1996.

263. A. Cichocki, W. Kasprzak, and W. Skarbek. Adaptive learning algorithm for principal component analysis with partial data. In R. Trappl, editor, *Cybernetics and Systems '96. Thirteenth European Meeting on Cybernetics and Systems Research*, volume 2, pages 1014–1019. Austrian Society for Cybernetic Studies, Vienna, 1996.

264. A. Cichocki, P. Kostyla, T.Lobos, and Z. Waclawek. Neural networks for real-time estimation of parameters of signals in power systems. *Int Journal of Engineering Intelligent Systems for Electrical Engineering and Communication,*, 6(3):1379–1380, 1998.

265. A. Cichocki and L. Moszczyński. New learning algorithm for blind separation of sources. *Electronics Letters*, 28(21):1986–1987, October 1992.

266. A. Cichocki, L. Moszczyński, and R. Bogner. Improved adaptive algorithms for blind separation of sources. In *Proc. of the Int. Conference on Electronic Circuits and Systems*, pages 647–652, Zakopane, Poland, Oct. 25-28 1995.

267. A. Cichocki, B. Orsier, A.D. Back, and S. Amari. On-line adaptive algorithms in non stationary environments using a modified conjugate gradient approach. In *Proc. of IEEE Workshop on Neural Networks for Signal Processing*, pages 316–325, 1997.

268. A. Cichocki, T. Rutkowski, A. K. Barros, and S.-H. Oh. Blind extraction of temporally correlated but statistically dependent acoustic signals. In *IEEE Workshop on Neural Networks for Signal Processing, NNSP'2000*, pages 455–464, Sydney, Australia, December 11-13 2000.

269. A. Cichocki, I. Sabała, and S. Amari. Intelligent neural networks for blind signal separation with unknown number of sources. In *Proc. of Conference Engineering of Intelligent Systems, ESI-98*, pages 148–154, Tenerife, 1998.

270. A. Cichocki, I. Sabała, S. Choi, B. Orsier, and R. Szupiluk. Self adaptive independent component analysis for sub-Gaussian and super-Gaussian mixtures with unknown number of sources and additive noise. In *Proc. Int. Symposium on Nonlinear Theory and its Applications, NOLTA-97*, pages 731–734, 1997.

271. A. Cichocki, R. Świniarski, and R.E. Bogner. Hierarchical neural network for robust PCA of complex-valued signals. In *World Congress on Neural Networks, WCNN-96*, pages 818–821, San Diego, USA, Sept. 1996. INNS Press, Lawrence Erlbaum Associates Inc. Publ., Mahwah, NJ.

272. A. Cichocki and R. Thawonmas. On-line algorithm for blind signal extraction of arbitrarily distributed, but temporally correlated sources using second order statistics. *Neural Processing Letters*, 12(1):91–98, August 2000.

273. A. Cichocki, R. Thawonmas, and S. Amari. Sequential blind signal extraction in order specified by stochastic properties. *Electronics Letters*, 33(1):64–65, January 1997.

274. A. Cichocki and R. Unbehauen. Neural networks for computing eigenvalues and eigenvectors. *Biological Cybernetics*, 68:155–164, 1992.

275. A. Cichocki and R. Unbehauen. Robust estimation of principal components in real time. *Electronics Letters*, 29(21):1869–1870, 1993.

276. A. Cichocki and R. Unbehauen. *Neural Networks for Optimization and Signal Processing*. John Wiley & Sons, New York, 1994. new revised and improved edition.

277. A. Cichocki and R. Unbehauen. Robust neural networks with on-line learning for blind identification and blind separation of sources. *IEEE Trans. Circuits and Systems I : Fundamentals Theory and Applications*, 43(11):894–906, Nov. 1996.

278. A. Cichocki, R. Unbehauen, L. Moszczyński, and E. Rummert. A new on-line adaptive learning algorithm for blind separation of sources. In *Proc. of the 1994 Int. Symposium on Artificial Neural Networks ISANN-94*, pages 406–411, Tainan, Taiwan,, Dec. 1994.

279. A. Cichocki, R. Unbehauen, and E. Rummert. Robust learning algorithm for blind separation of signals. *Electronics Letters*, 30(17):1386–1387, August 1994.

280. A. Cichocki and S. Vorobyov. Application of ICA for automatic noise and interference cancellation in multisensory biomedical signals. In *Proceedings of the Second International Workshop on ICA and BSS, ICA'2000*, pages 621–626, Helsinki, Finland, 19-22 June 2000.

281. A. Cichocki and L. Zhang. Two-stage blind deconvolution using state-space models (invited). In *Proceedings of the Fifth International Conference on Neural Information Processing(ICONIP'98)*, pages 729–732, Kitakyushu, Japan, Oct. 21-23 1998.

282. A. Cichocki and L. Zhang. Adaptive multichannel blind deconvolution using state-space models. In *Proc of '99 IEEE Workshop on Higher-Order Statistics*, pages 296–299, Caesarea, Israel, June 14-16 1999.

283. A. Cichocki, L. Zhang, and S. Amari. Semi-blind and state-space approaches to nonlinear dynamic independent component analysis. In *Proceedings of 1998 International Symposium on Nonlinear Theory and its Applications (NOLTA-98)*, volume 1, pages 291–294, Crans-Montana, Switzerland, 1998.

284. A. Cichocki, L. Zhang, S. Choi, and S. Amari. Nonlinear dynamic independent component analysis using state-space and neural network models. In *Proc. of the First International Workshop on Independent Component Analysis and Signal Separation - ICA'99*, pages 99–104, Aussois, France, January 11-15 1999.

285. A. Cichocki, L. Zhang, and T. Rutkowski. Blind separation and filtering using state space models. In *The 1999 IEEE International Symposium on Circuits and Systems, (ISCAS'99)*, volume 5, pages 78–81, Orlando, Florida, May 30 - Jun. 2 1999.

286. A. Cichocki, and P. Georgiev. Blind source separation algorithms with matrix constraints. *IEICE Trans. Fundamentals*, volume E86A, March 2003.

287. G. Cirrincione. *A Neural Approach to the Structure from Motion Problem.* PhD thesis, INPG Grenoble, France, 1998.

288. G. Cirrincione, M. Cirrincione, and S. Van Huffel. The GeTLS EXIN neuron for linear regression. In *Proc. of the IEEE International Joint Conference on Neural Networks (IJCNN 2000), Como, Italy*, volume 6, pages 285–289, 2000.

289. M.H. Cohen and A.G. Andreou. Current mode subthreshold MOS implementation of the Hérault-Jutten autoadaptive network. *IEEE Journal Solid-State Circuits*, 42(2):714–727, 1992.

290. M.H. Cohen and A.G. Andreou. Analog CMOS integration and experimentation with an autoadaptive independent component analyzer. *IEEE Trans. on Circuits and Systems-II: Analog and digital signal processing*, 42(2), 1995.

291. P. Comon. Separation of sources using high-order cumulants. In *SPIE Conference on Advanced Algorithms and Architectures for Signal Processing*, volume XII, pages 170–181, San Diego, CA, Aug. 1989.

292. P. Comon. Separation of stochastic processes. In *Workshop on Higher-Order Spectral Analysis*, pages 174–179, Vail, Colorado USA, June 28-30 1989.

293. P. Comon. Independent component analysis and blind identification. *Traitement du Signal*, 7(5):435–450, 1990.

294. P. Comon. Independent component analysis. In *Proc. Int. Sig. Processing Workshop on Higher-Order Statistics*, pages 111–120, Chamrousse, France, July 10-12 1991. Republished in *Higher-Order Statistics*, J.L.Lacoume ed., Elsevier, 1992, pp 29–38.

295. P. Comon. Blind identification in presence of noise. In J. Vandewalle e.a., editor, *Signal Processing VI: Proc. EUSIPCO-90*, pages 835–838 vol.2, Brussels, 1992. Elsevier.

296. P. Comon. Independent component analysis (signal processing). In J.L. Lacoume, editor, *Higher Order Statistics. Proceedings of the International Signal Processing Workshop*, pages 29–38, Chamrousse, France, 1992. Elsevier.

297. P. Comon. MA identification using fourth order cumulants. *Signal Processing*, 26:381–388, 1992.

298. P. Comon. Independent component analysis, and the diagonalization of symmetric tensors. In H. Dedieu, editor, *European Conference on Circuit Theory and Design – ECCTD*, pages 185–190, Davos, Aug 30-Sept 3 1993. Elsevier. invited session.

299. P. Comon. Independent Component Analysis, a new concept ? *Signal Processing, Elsevier*, 36(3):287–314, April 1994. Special issue on Higher-Order Statistics.

300. P. Comon. Tensor diagonalization, a useful tool in signal processing. In M. Blanke and T. Soderstrom, editors, *System Identification (SYSID '94). A Postprint Volume from the IFAC Symposium*, pages 77–82 vol.1, Copenhagen, Denmark, 1995. Pergamon.

301. P. Comon. Contrast functions for blind deconvolution. *IEEE Signal Processing Lett.*, SPL-3, no. 7:209–211, July 1996.

302. P. Comon. *Method and device for the estimation of a signal propagating through a multipath channel*, June 1996. Patent registrated for Thomson-CSF, no 9607858.

303. P. Comon and J.-F. Cardoso. Eigenvalue decomposition of a cumulant tensor with applications. In *Advanced Signal Processing Algorithms, Architectures, and Implementations, SPIE*, volume 1348, pages 361–372, San Diego, California, USA, 10-12 July 1990. SPIE.

304. P. Comon and B. Emile. Estimation of time delays in the blind mixture problem. In *Signal Processing VII: Theories and Applications (Proc. of the EUSIPCO-94)*, pages 482–485, Edinburgh, Scotland, Sept. 13-16 1994. Elsevier.

305. P. Comon and G.H. Golub. Tracking of a few extreme singular values and vectors in signal processing. *Proceedings of the IEEE*, 78(8):1327–1343, August 1990. (published from Stanford report 78NA-89-01, feb 1989).

306. P. Comon, C. Jutten, and J. Hérault. Blind separation of sources, Part II: Problems statement. *Signal Processing*, 24(1):11–20, July 1991.

307. P. Comon and T. Kailath. An array processing technique using the first principal component. In *First International Workshop on SVD and Signal Processing*, September 1987. Extended version published in: SVD and Signal Processing, E.F. Deprettere editor, North Holland, 1988, 301–316.

308. P. Comon and J.L. Lacoume. Noise reduction for an estimated Wiener filter using noise references. *IEEE Trans. on Information Theory*, 32(2):310–313, March 1986.

309. P. Comon and J.L. Lacoume. A robust adaptive filter for noise reduction problems. In *IEEE International Conference on Acoustics, Speech and Signal Processing - ICASSP*, pages 2599–2602, Tokyo, Japan, April 7-11 1986.

310. P. Comon and D.-T. Pham. An error bound for a noise canceller. *IEEE Trans. on ASSP*, 37(10):1513–1517, October 1989.

311. P. Comon and D.-T. Pham. Estimation of the order of a FIR filter for noise cancellation. *IEEE Trans. on Inf. Theory*, 36(2):429–434, March 1990.

312. J.-A. Conchello and Q. Yu. Parametric blind deconvolution of fluorescence microscopy images: preliminary results. *Proc. SPIE*, 2655:164–174, 1996.

313. A.G. Constantinides and E. Fluet. The 'Z-slice': an identification method for noisy MA systems using higher order statistics. In J. Vandewalle e.a., editor, *Signal Processing VI: Proc. EUSIPCO-90*, pages 689–692 vol.2, Brussels, 1992. Elsevier.

314. R. Cristi and M. Kutlu. Blind equalization of encoded sequences by hybrid Markov models. In L.P. Caloba, P.S.R. Diniz, A.C.M. de Querioz, and E.H. Watanabe, editors, *38th Midwest Symposium on Circuits and Systems. Proceedings*, pages 815–818 vol.2, 38th Midwest Symposium on Circuits and Systems. Proceedings, Rio de Janeiro, Brazil, 13-16 Aug. 1995, 1996. IEEE.

315. S. Cruces. *An Unified View of Blind Source Separation Algorithms.* Ph.D. thesis, University of Vigo. Signal Processing Dept., Spain, 1999.

316. S. Cruces and L. Castedo. Blind separation of convolutive mixtures: A Gauss-Newton algorithm. In *Proceedings of the IEEE Signal Processing Workshop on Higher-Order Statistics*, pages 326–330, Banff (Alberta), Canada, July 1997.

317. S. Cruces and L. Castedo. A Gauss-Newton method for blind source separation of convolutive mixtures. In *Proceedings of the ICASSP'98*, volume IV, pages 2093–2096, Seattle, USA, May 1998.

318. S. Cruces and L. Castedo. Stability analysis of adaptive algorithms for blind source separation of convolutive mixtures. *Signal Processing*, 78(3):265–275, 1999.

319. S. Cruces, L. Castedo, and A. Cichocki. An iterative inversion method for blind source separation. In *Proc. of the First International Workshop on Independent Component Analysis and Signal Separation - ICA'99*, pages 307–312, Aussois, France, 1999.

320. S. Cruces, L. Castedo, and A. Cichocki. Novel blind source separation algorithms using cumulants. In *Proceedings of ICASSP'2000*, volume V, pages 3152–3155, Istanbul, Turkey, June 2000.

321. S. Cruces, L. Castedo, and A. Cichocki. Asymptotically equivariant blind source separation using cumulants. *Neurocomputing*, vol 49, pp 87-118, Dec. 2002.

322. S. Cruces, A. Cichocki, and S. Amari. Criteria for the simultaneous blind extraction of arbitrary groups of sources. In *Proceedings of the ICA 2001 Workshop*, San Diego, USA, 2001.

323. S. Cruces, A. Cichocki, and S. Amari. The minimum entropy and cumulants based contrast functions for blind source extraction. In *Bio-Inspired Applications of Connectionism: Proceedings of 6th International Work-Conference on Artificial and Natural Networks*, pages 786–793, Vol. LNCS2085, Granada, Spain, 2001.

324. S. Cruces, A. Cichocki, and L. Castedo. An unified perspective of blind source separation algorithms. In *Proceedings of the Learning'98*, Madrid (Spain), September 1998.

325. S. Cruces, A. Cichocki, and L. Castedo. Blind source extraction in Gaussian noise. In *Proceedings of the Second International Workshop on ICA and BSS, ICA'2000*, pages 63–68, Helsinki, Finland, June 19-22 2000.

326. S. Cruces, A. Cichocki, and L. Castedo. An iterative inversion approach to blind source separation. *IEEE Trans. on Neural Networks*, 11(6):1423–1437, 2000.

327. S. Cruces, A. Cichocki, and S. Amari. On a new blind signal extraction algorithm: Different criteria and stability analysis. *IEEE Signal Processing Letters*, vol. 9 (8), pp. 233 - 236, Aug. 2002.

328. A. Dapena, L. Castedo, and C. Escudero. An unconstrained single stage criterion for blind source separation. In *Proc. IEEE ICASSP*, pages 2706–2709 vol. 5, Atlanta, GA, 1996.

329. B.L.K. Davey, R.G. Lane, and R.H.T. Bates. Blind deconvolution of noisy complex-valued image. *Optics Communications*, 69(5-6):353–356, January 1989.

330. D. Dayton, S. Sandven, and J. Gonglewski. Signal-to-noise and convergence properties of a modified Richardson-Lucy algorithm with Knox-Thompson start point. *Proc. SPIE*, 2827:162–169, 1996.

331. E. de Carvalho and D.T.M. Slock. Maximum-likelihood blind FIR multi-channel estimation with Gaussian prior for the symbols. In *Proc. IEEE ICASSP*, pages 3593–3596, 1997.

332. M. de Courville, P. Duhamel, P. Madec, and J. Palicot. Blind equalization of OFDM systems based on the minimization of a quadratic criterion. In *Proc. IEEE Int. Conference Communic.*, pages 1318–1322 vol.3, Dallas, TX, 1996.

333. J. Declerck, J. Feldmar, F. Betting, and M.L. Goris. Automatic registration and alignment on a template of cardiac stress and rest SPECT images. In *Proceedings of the IEEE Workshop on Mathematical Methods in Biomedical Image Analysis*, pages 212–221, Proceedings of the Workshop on Mathematical Methods in Biomedical Image Analysis, San Francisco, CA, USA, 21-22 June 1996, 1996. IEEE Comput. Society Press.

334. G. Deco and W. Brauer. Nonlinear higher-order statistical decorrelation by volume-conserving neural architectures. *Neural Networks*, 8(4):525–535, 1995.

335. G. Deco and D. Obradovic. *An Information-Theoretic Approach to Neural Computing.* Springer Verlag, New York, February 1996.

336. J.B. Gomez del Moral and E. Biglieri. Blind identification of digital communication channels with correlated noise. *IEEE Trans. Signal Processing*, 44(12):3154–3156, December 1996.

337. L. DeLathauwer, P. Comon, et al. Higher-order power method, application in independent component analysis. In *NOLTA Conference*, volume 1, pages 91–96, Las Vegas, 10–14 Dec 1995.

338. N. Delfosse and P. Loubaton. Adaptive blind separation of independent sources: a deflation approach. *Signal Processing*, 45(1):59–83, July 1995.

339. N. Delfosse and P. Loubaton. Adaptive blind separation of convolutive mixtures. In *Proc. IEEE ICASSP*, pages 2940–2943 vol. 5, Atlanta, GA, 1996.

340. N. Delfosse and P. Loubaton. Adaptive blind separation of convolutive mixtures. In *29-th Asilomar Conference Signals, Systems, Comp.*, pages 341–345 vol.1. IEEE, 1996.

341. F. De la Torre and M. Black. Robust principal component analysis for computer vision. In *Proc. ICCV01*, pages 362–369, 2001.

342. J.-P. Delmas, H. Gazzah, A. P. Liavas, and P. A. Regalia. Statistical analysis of some second-order methods for blind channel identification/equalization with respect to channel undermodeling,. *IEEE Trans. Signal Processing*, 48:1477–1481, 2000.

343. K.W. DeLong, R. Trebino, and W.E. White. Simultaneous recovery of two ultrashort laser pulses from a single spectrogram. *Journal of the Optical Society of America B (Optical Physics)*, 12(12):2463–2466, December 1995.

344. B. Delyon and A. Juditsky. Stochastic approximation with averaging of trajectories. *Stochastic and Stochastic Reports*, 39:107–118, 1992.

345. D. Dembele and G. Favier. A new FIR system identification method based on fourth-order cumulants: application to blind equalization. *Journal of the Franklin Institute*, 334B(1):117–133, January 1997.

346. Y. Deville. A unified stability analysis of the Hérault-Jutten source separation neural network. *Signal Processing*, 51, No.3:229–233, 1996.

347. Y. Deville. Analysis of the convergence properties of a self- normalized source separation neural network. In *Proc. of IEEE Workshop on Signal Processing Advances in Wireless Communications SPAWC-97, Paris*, pages 69–72, April 1997.

348. Y. Deville and L. Andry. Application of blind source separation techniques to multi-tag contactless identification systems. *IEICE Trans. Fundamentals of Electronics, Communications and Computer Sciences*, E79-A(10):1694–1699, October 1996.

349. K.I. Diamantaras and S.Y. Kung. *Principal Component Neural Networks. Theory and Applications.* Adaptive and Learning Systems for Signal Processing, Communications and Control. John Wiley & Sons Inc., New York, 1996.

350. S.N. Diggavi, Y. Cho, and A. Paulraj. Blind estimation of multiple co-channel digital signals in vector FIR channels. In *Proc. IEEE Globecom*, Singapore, November 1995.

351. A. Dinc and Y. Bar-Ness. Bootstrap; a fast blind adaptive signal separator. In *Proc. of the 1992 IEEE Int. Conference on Acoustics, Speech, and Signal Processing*, volume 2, pages 325–328, San Francisco, California, USA, March 23-26 1992. IEEE.

352. A. Dinc and Y. Bar-Ness. Convergence and performance comparison of three different structures of bootstrap blind adaptive algorithm for multisignal co-channel separation. In *Proc. IEEE MILCOM*, pages 913–918 vol.3, San Diego, CA, 1992.

353. Z. Ding. Blind equalization based on joint minimum MSE criterion. *IEEE Trans. Communications*, 42(2-4,):648–654, February 1994.

354. Z. Ding. A blind channel identification algorithm based on matrix outer-product. In *Proc. IEEE Int. Conference Communic.*, pages 852–856 vol.2, Dallas, TX, 1996.

355. Z. Ding. Characteristics of band-limited channels unidentifiable from second-order cyclostationary statistics. *IEEE Signal Processing Letters*, 3(5):150–152, May 1996.

356. Z. Ding. On convergence analysis of fractionally spaced adaptive blind equalizers. In *Proc. IEEE ICASSP*, pages 2431–2434 vol. 5, Atlanta, GA, 1996.

357. Z. Ding. An outer-product decomposition algorithm for multichannel blind identification. In *Proc. IEEE SP Workshop on Stat. Signal Array Processing*, pages 132–135, Corfu, Greece, 1996.

358. Z. Ding. Multipath channel identification based on partial system information. *IEEE Trans. Signal Processing*, 45(1):235–240, January 1997.

359. Z. Ding, C.R. Johnson, Jr., and R.A. Kennedy. Global convergence issues with linear blind adaptive equalizers. In S. Haykin, editor, *Blind Deconvolution*, pages 60–120. Englewood Cliffs, NJ: Prentice-Hall, 1994.

360. Z. Ding and C.R. Johnson, Jr. Existing gap between theory and application of blind equalization. *Proc. SPIE*, 1565:154–165, 1991.

361. Z. Ding and C.R. Johnson, Jr. On the nonvanishing stability of undesirable equilibria for FIR Godard blind equalizers. *IEEE Trans. Signal Processing*, 41(5):1940–1944, May 1993.

362. Z. Ding, C.R. Johnson, Jr., and R.A. Kennedy. Nonglobal convergence of blind recursive identifiers based on gradient descent of continuous cost functions. In *Proc. IEEE Conference on Decision and Control*, pages 225–230 vol.1, Honolulu, HI, 1990.

363. Z. Ding, C.R. Johnson, Jr., and R.A. Kennedy. On the admissibility of blind adaptive equalizers. In *Proc. IEEE ICASSP*, pages 1707–1710 vol.3, Albuquerque, NM, 1990.

364. Z. Ding, C.R. Johnson, Jr., and R.A. Kennedy. Local convergence of 'globally convergent' blind adaptive equalization algorithms. In *Proc. IEEE ICASSP*, pages 1533–1536 vol.3, Toronto, Ont., Canada, 1991.

365. Z. Ding, C.R. Johnson, Jr., and R.A. Kennedy. On the (non)existence of undesirable equilibria of Godard blind equalizers. *IEEE Trans. Signal Processing*, 40(10):2425–2432, October 1992.

366. Z. Ding and R.A. Kennedy. A new adaptive algorithm for joint blind equalization and carrier recovery. In *25-th Asilomar Conference Signals, Systems Comp.*, pages 699–703 vol.2. IEEE, 1991.

367. Z. Ding and R.A. Kennedy. On the whereabouts of local minima for blind adaptive equalizers. *IEEE Trans. Circuits Systems II*, 39(2):119–123, February 1992.

368. Z. Ding, R.A. Kennedy, B.D.O. Anderson, and C.R. Johnson, Jr. Ill-convergence of Godard blind equalizers in data communication systems. *IEEE Trans. Communications*, 39(9):1313–1327, September 1991.

369. Z. Ding, R.A. Kennedy, B.D.O. Anderson, and C.R. Johnson, Jr. Local convergence of the Sato blind equalizer and generalizations under practical constraints. *IEEE Trans. Informat. Th.*, 39(1):129–144, January 1993.

370. Z. Ding and Y. Li. Channel identification using second order cyclic statistics. In *26-nd Asilomar Conference Signals, Systems Comp.*, pages 334–338 vol.1. IEEE, 1992.

371. Z. Ding and Y. Li. *Blind Equalization and Identification.* Marcel Dekker, New York, Basel, 2001.

372. Z. Ding and Z. Mao. Knowledge-based identification of fractionally sampled channels. In *Proc. IEEE ICASSP*, pages 1996–1999, Detroit, May 1995.

373. M.C. Dogan and J.M. Mendel. Blind deconvolution (equalization): Some new results. *Signal Processing*, 53(2-3):109–116, September 1996.

374. K. Dogancay and R.A. Kennedy. A globally admissible off-line modulus restoral algorithm for low-order adaptive channel equalizers. In *Proc. IEEE ICASSP*, pages III/61–64 vol.3, Adelaide, 1994.

375. F. Dominique and J.H. Reed. Despread data rate update multitarget adaptive array for CDMA signals. *Electronics Letters*, 33(2):119–121, January 1997.

376. G. Dong and R.-W. Liu. A neural network for smallest eigenvalue with application to blind equalization. In L.P. Caloba, P.S.R. Diniz, A.C.M. de Querioz, and E.H. Watanabe, editors,

38th Midwest Symposium on Circuits and Systems. Proceedings, pages 811–814 vol.2, 38th Midwest Symposium on Circuits and Systems. Proceedings, Rio de Janeiro, Brazil, 13-16 Aug. 1995, 1996. IEEE.

377. S.C. Douglas. Equivariant algorithms for selective transmission. In *Proc. IEEE International Conference Acoustics, Speech, Signal Processing*, volume II, pages 1133–1136, Seattle, WA, May 1998.
378. S.C. Douglas. Blind source separation: An overview of density-based methods. In *Proc. International Workshop on Acoustic Echo and Noise Control*, pages 8–11, Pocono Manor, PA, September 1999.
379. S.C. Douglas. Equivariant adaptive selective transmission. *IEEE Trans. Signal Processing*, 47(5):1223–1231, May 1999.
380. S.C. Douglas. Prewhitened blind source separation with orthogonality constraints. In *presented at 33rd Annual Asilomar Conference on Signals, Systems, and Computers*, Pacific Grove, CA, October 24-27 1999.
381. S.C. Douglas. Self-stabilized gradient algorithms for blind source separation with orthogonality constraints. *IEEE Trans. Neural Networks*, 1999.
382. S.C. Douglas. Numerically-robust adaptive subspace tracking using householder transformations. In *to be presented at First IEEE Sensor Array and Multichannel Signal Processing Workshop*, Boston, MA, March 16-17 2000.
383. S.C. Douglas. Numerically-stable $o(n^2)$ algorithms for rls adaptive filtering using least-squares prewhitening. In *to be presented at IEEE International Conference Acoustics, Speech, Signal Processing*, Istanbul, Turkey, June 2000.
384. S.C. Douglas. Adaptive algorithms and architectures for blind signal separation and blind deconvolution. In Y.-H. Hu and J.-N. Hwang, editors, *to appear in The Neural Networks for Signal Processing Handbook*, Boca Raton, FL, 2001. CRC/IEEE Press. to be published.
385. S.C. Douglas and S. Amari. Natural gradient adaptation. In S. Haykin, editor, *Unsupervised Adaptive Filtering, Vol. I: Blind Source Separation*, pages 13–61. Wiley, New York, 2000.
386. S.C. Douglas, S. Amari, and S.-Y. Kung. On gradient adaptation with unit-norm constraints. *IEEE Trans. Signal Processing*, 1997.
387. S.C. Douglas, S. Amari, and S.-Y. Kung. Adaptive paraunitary filter banks for spatio-temporal principal and minor subspace analysis. In *Proc. IEEE International Conference Acoustics, Speech, Signal Processing*, volume 2, pages 1089–1092, Phoenix, AZ, March 1999.
388. S.C. Douglas, S. Amari, and S.Y. Kung. Gradient adaptation under unit norm constraints. In *Proc. 9th IEEE Signal Processing Workshop on Statistical Signal and Array Processing*, pages 144–147, Portland, OR, September 1998.
389. S.C. Douglas and A. Cichocki. Convergence analysis of local algorithms for blind decorrelation. In *presented at Neural Information Processing Systems Conference, Workshop on Blind Signal Processing*, Denver, CO, December 2-7 1996.
390. S.C. Douglas and A. Cichocki. Neural networks for blind decorrelation of signals. *IEEE Trans. Signal Processing*, 45(11):2829–2842, November 1997.
391. S.C. Douglas and A. Cichocki. On-line step size selection for training adaptive systems. *IEEE Signal Processing Mag.*, 14(6):45–46, November 1997.
392. S.C. Douglas and A. Cichocki. Adaptive step size techniques for decorrelation and blind source separation. In *Proc. 32nd Asilomar Conference on Signals, Systems, and Computers*, volume 2, pages 1191–1195, Pacific Grove, CA, November 1998.

393. S.C. Douglas, A. Cichocki, and S. Amari. Fast-convergence filtered-regressor algorithms for blind equalisation. *Electronics Letters*, 32(23):2114–2115, 7th November 1996.

394. S.C. Douglas, A. Cichocki, and S. Amari. Multichannel blind separation and deconvolution of sources with arbitrary distributions. In *Proc. IEEE Workshop on Neural Networks for Signal Processing*, pages 436–445, Almelia Island Plantation, FL, September 1997.

395. S.C. Douglas, A. Cichocki, and S. Amari. Quasi-Newton filtered-regressor algorithms for adaptive equalization and deconvolution. In *Proc. IEEE Workshop on Signal Processing Advances in Wireless Communications*, pages 109–112, Paris, France, April 1997.

396. S.C. Douglas, A. Cichocki, and S. Amari. A bias removal technique for blind source separation with noisy measurements. *Electronics Letters*, 34(14):1379–1380, July 1998.

397. S.C. Douglas, A. Cichocki, and S. Amari. Self-whitening algorithms for adaptive equalization and deconvolution. *IEEE Trans. Signal Processing*, 47(4):1161–1165, April 1999.

398. S.C. Douglas and S. Haykin. On the relationship between blind deconvolution and blind source separation. In *Proc. 31st Asilomar Conference on Signals, Systems, and Computers*, volume 2, pages 1591–1595, Pacific Grove, CA, November 1997.

399. S.C. Douglas and S. Haykin. Relationships between blind deconvolution and blind source separation. In S. Haykin, editor, *Unsupervised Adaptive Filtering, Vol II: Blind Deconvolution*, pages 113–145. Wiley, New York, 2000.

400. S.C. Douglas and S.-Y. Kung. Design of estimation/deflation approaches to independent component analysis. In *Proc. 32nd Asilomar Conference on Signals, Systems, and Computers*, volume 1, pages 707–711, Pacific Grove, CA, November 1998.

401. S.C. Douglas and S.-Y. Kung. All-pass vs. unit-norm constraints in contrast-based blind deconvolution. In *Proc. IEEE Workshop on Neural Networks for Signal Processing*, Madison, WI, August 1999.

402. S.C. Douglas and S.-Y. Kung. Gradient adaptive algorithms for contrast-based blind deconvolution. *Journal of VLSI Signal Processing Systems*, 1999.

403. S.C. Douglas and S.-Y. Kung. An ordered-rotation kuicnet algorithm for separating arbitrarily-distributed sources. In *Proc. IEEE International Conference on Independent Component Analysis and Signal Separation*, pages 419–425, Aussois, France, January 1999.

404. S.C. Douglas, S.-Y. Kung, and S. Amari. A self-stabilized minor subspace rule. *IEEE Signal Processing Letters*, 5(12):328–330, December 1998.

405. S.C. Douglas, S.-Y. Kung, and S. Amari. On the numerical stabilities of principal, minor, and independent component analysis algorithms. In *Proc. IEEE International Conference on Independent Component Analysis and Signal Separation*, pages 419–425, Aussois, France, January 1999.

406. S.C. Douglas, S.-Y. Kung, and S. Amari. A self-stabilized minor subspace rule. *Signal Processing Letter*, 5:328–330, Dec. 1998.

407. S.C. Douglas and S.Y. Kung. Kuicnet algorithms for blind deconvolution. In *Proc. IEEE Workshop on Neural Networks for Signal Processing*, pages 3–12, Cambridge, UK, August 1998.

408. M.C. Dovgan and J.M. Mendel. Single sensor detection and classification of multiple sources by higher-order spectra. *IEE Proceedings, Part F*, 140(6):350–355, Dec. 1993.

409. A. Duel-Hallen. A family of multiuser decision-feedback detectors for asynchronous code-division multiple-access channels. *IEEE Tr. Comm.*, 43(2/3/4):421–434, February 1995.

410. A. Edelman, T. Arias, and S. T. Smith. The geometry of algorithms with orthogonality constraints. *SIAM Journal on Matrix Analysis and Applications*, 20(2):303–353, 1998.

411. N.V. Efimov and E.R. Rozendorn. *Linear Algebra and Multi-Dimensional Geometry.* MIR Publishers, Moscow, 1975.

412. S.E. El-Khamy, A.F. Ali, and H.M. EL-Ragal. Fast blind equalization using higher-order-spectra channel-estimation in the presence of severe ISI. In *Proceedings IEEE Symposium on Computers and Communications*, pages 248–254, Alexandria, Egypt, 1995. IEEE Comput. Society Press.

413. H. Elders-Boll, H.D. Schotten, and A. Busboom. Implementation of linear multiuser detectors for asynchronous CDMA systems by linear multi-stage interference cancellation. In *Proc. IEEE ICASSP*, 1998.

414. T.J. Endres, B.D.O. Anderson, C.R. Johnson, Jr., and L. Tong. On the robustness of FIR channel identification from fractionally spaced received signal second-order statistics. *IEEE Signal Processing Letters*, 3(5):153–155, May 1996.

415. A. M. Engebretson. Acoustic signal separation of statistically independent sources using multiple microphones. In *Int. Conference on Acoustics Speech and Signal Processing*, volume II, pages 343–346, Minneapolis, MN, USA, April 27-30 1993.

416. K.-S. Eom, B.-C. Lim, and D.-J. Park. Blind separation algorithm without nonlinear functions. *Electronics Letters*, 32(3):165–166, February 1996.

417. K.-S. Eom and D.-J. Park. Blind separation of sources using decorrelation process. In *Proc. IEEE ISCAS*, pages 434–436 vol.3, Atlanta, GA, 1996.

418. A. Essebbar, J.-M. Brossier, D. Mauuary, and B. Geller. An algorithm for blind equalization and synchronization. *Journal of the Franklin Institute*, 333B(3):339–347, May 1996.

419. B. Farhang-Boroujeny. Pilot-based channel identification: proposal for semi-blind identification of communication channels. *Electronics Letters*, 31(13):1044–1046, June 1995.

420. M. Feder, A.V. Oppenheim, and E. Weinstein. Maximum likelihood noise cancellation using the EM algorithm. *IEEE Trans. on Acoustics, Speech and Signal Processing*, 37(2):204–216, Feb. 1989.

421. M. Feder and E. Weinstein. Parameter-estimation of superimposed signals using the EM algorithm. *IEEE Trans. on Acoustics, Speech and Signal Processing*, 36(4):477–489, April 1988.

422. J.R. Fienup. Reconstruction of an object from the modulus of its Fourier transform. *Optics Letters*, 3(1):27–29, July 1978.

423. A.R. Figueiras-Vidal, A. Artes-Rodriguez, and J.A.A. Hernandez-Mendez. Phase constraining algorithms for data blind equalization. In *Proc. IEEE ICASSP*, pages 1929–1932 vol.3, Toronto, Ont., Canada, 1991.

424. I. Fijalkow. Multichannel equalization lower bound: a function of channel noise and disparity. In *Proc. IEEE SP Workshop on Stat. Signal Array Processing*, pages 344–347, Corfu, Greece, 1996.

425. I. Fijalkow, F. Lopez de Victoria, and C.R. Johnson Jr. Adaptive fractionally spaced blind equalization. In *Proc. 6-th IEEE DSP Workshop*, pages 257–260, Yosemite, 1994. IEEE.

426. I. Fijalkow and P. Loubaton. Identification of rank one rational spectral densities from noisy observations: a stochastic realization approach. *Systems & Control Letters*, 24(3):201–205, February 1995.

427. I. Fijalkow, J.R. Treichler, and C.R. Johnson, Jr. Fractionally spaced blind equalization: loss of channel disparity. In *Proc. IEEE ICASSP*, pages 1988–1991 vol.3, Detroit, MI, 1995.

428. S. Fiori. A theory for learning by weight flow on Stiefel-Grassmann manifolds. *Neural Computation*, 13:1625–1647, July 2001.

429. R.F.H. Fischer, W.H. Gerstacker, and J.B. Huber. Dynamics limited precoding, shaping, and blind equalization for fast digital transmission over twisted pair lines. *IEEE Journal on Selected Areas in Communications*, 13(9):1622–1633, December 1995.

430. D.A. Fish, A.M. Brinicombe, E.R. Pike, and J.G. Walker. Blind deconvolution by means of the Richardson-Lucy algorithm. *Journal of the Optical Society of America A (Optics, Image Science and Vision)*, 12(1):58–65, January 1995.

431. J.D. Fite, S.P. Bruzzone, and B.G. Agee. Blind separation of voice modulated single-side-band using the multi-target variable modulus algorithm. In *Proc. IEEE ICASSP*, pages 2726–2729 vol. 5, Atlanta, GA, 1996. IEEE.

432. S.J. Flockton, D. Yang, and G.J. Scruby. Performance surfaces of blind source separation algorithms. In S. Amari, L. Xu, L.-W. Chan, I. King, and K.-S. Leung, editors, *Progress in Neural Information Processing. Proceedings of the International Conference on Neural Information Processing*, pages 1229–1234 vol.2, Hong Kong, 1996. Springer-Verlag.

433. B.D. Flury and B.E. Neuenschwander. Simultaneous diagonalization algorithms with applications in multivariate statistics. In R.V.M. Zahar, editor, *Approximation and Computation*, pages 179–205. Birkhäuser, Basel, 1995.

434. J. Flusser, T. Suk, and S. Saic. Image features invariant with respect to blur. *Pattern Recognition*, 28(11):1723–1732, November 1995.

435. P. Földiak. Adaptive network for optimal linear feature extraction. In *Proceedings of the IJCNN*, pages 401–405, Washington, DC, 1989.

436. J.R. Fonollosa, J.A.R. Fonollosa, Z. Zvonar, and J. Vidal. Blind multiuser identification and detection in CDMA systems. In *Proc. IEEE ICASSP*, pages 1876–1879 vol.3, Detroit, MI, 1995.

437. G.D. Forney. Maximum-likelihood sequence estimation of digital sequences in the presence of intersymbol interference. *IEEE Trans. Information Theory*, 18(3):363–378, May 1972.

438. J.C. Fort. Stabilite de l'algorithme de separation de sources de Jutten at Hérault. *Traitement du Signal*, no.1:35–42, 1991.

439. M.R. Frater, R.R. Bitmead, and C.R. Johnson, Jr. Escape from stable equilibria in blind adaptive equalizers. In *Proc. IEEE Conference Decision Control*, pages 1756–1761 vol.2, Tucson, AZ, 1992.

440. M.R. Frater and C.R. Johnson, Jr. Local minima escape transients of CMA. In *Proc. IEEE ICASSP*, pages III/37–40 vol.3, Adelaide, 1994.

441. B. Friedlander. A sensitivity analysis of the MUSIC algorithm. *IEEE Trans. Acoustics, Speech., Signal Processing*, 38:1740–1751, October 1990.

442. B. Friedlander and B. Porat. Performance analysis of a null steering algorithm based on direction-of-arrival estimation. *IEEE Trans. Acoustics, Speech., Signal Processing*, 37:461–466, April 1989.

443. B. Friedlander and A. Weiss. Effects of model errors on waveform estimation using the MUSIC algorithm. *IEEE Trans. on Signal Processing*, 42(1):147–155, 1994.

444. M. Fujimoto, N. Kikuma, and N. Inagaki. Behavior of the CMA adaptive array to filtered pi /4-shifted QPSK signals. *Trans. of the Institute of Electronics, Information and Communication Engineers B-II*, J74B-II(9):497–500, September 1991.

445. M. Fujimoto, K. Nishikawa, and K. Sato. A study of adaptive array antenna system for land mobile communications. In *Proceedings of the Intelligent Vehicles '95. Symposium*, pages 36–41, Detroit, MI, 1995. IEEE.

446. K. Fukawa and H. Suzuki. Blind interference cancelling equalizer for mobile radio communications. *IEICE Trans. Communications*, E77-B(5):580–588, May 1994.

447. M. Gaeta and J.L. Lacoume. Source separation versus hypothesis. In J.L. Lacoume, editor, *Higher Order Statistics: Proceedings of the International Signal Processing Workshop on Higher Order Statistics*, pages 271–274, Amsterdam, 1991. Elsevier.

448. W.A. Gardner. A new method of channel identification. *IEEE Trans. Communications*, 39(6):813–817, June 1991.

449. E. Gassiat, F. Monfront, and Y. Goussard. On simultaneous signal estimation and parameter identification using a generalized likelihood approach. *IEEE Trans. Informat. Th.*, 38(1):157–162, January 1992.

450. P. Georgiev and A. Cichocki. Blind source separation via symmetric eigenvalue decomposition. In *Proceedings of Sixth International Symposium on Signal Processing and its Applications, Aug. 2001*, pages 17–20, Kula Lumpur, Malaysia, 2001.

451. P. Georgiev and A. Cichocki. Multichannel blind deconvolution of colored signals via eigenvalue decomposition. In *Proceedings of the 11th IEEE Signal Processing Workshop on Statistical Signal Processing, Aug. 2001*, pages 273–276, Singapore, 2001.

452. P. Georgiev, A. Cichocki, and S. Amari. Nonlinear dynamical system generalizing the natural gradient algorithm. In *Proceedings of the NOLTA 2001*, pages 391–394, Japan, 2001.

453. P. Georgiev, A. Cichocki, and S. Amari. On some extensions of the natural gradient algorithm. In *Proc. Third International Conference on Independent Component Analysis and Blind Signal Separation (ICA 2001)*, pages 581–585, San Diego, USA, 2001.

454. R.W. Gerchberg and W.O. Saxton. A practical algorithm for the determination of phase from image and diffraction plane pictures. *Optik*, 35:237–246, 1972.

455. D. Gerlach and A. Paulraj. Adaptive transmitting antenna arrays with feedback. *IEEE Signal Processing Letters*, 1(10):150–152, October 1994.

456. W.H. Gerstacker, R.F.H. Fischer, and J.B. Huber. Blind equalization for digital cable transmission with Tomlinson-Harashima precoding and shaping. In *Proc. IEEE Int. Conference Communic.*, pages 493–497 vol.1, Seattle, WA, 1995.

457. D. Gesbert and P. Duhamel. Robust blind joint data/channel estimation based on bilinear optimization. In *Proc. IEEE SP Workshop on Stat. Signal Array Processing*, pages 168–171, Corfu, Greece, 1996.

458. D. Gesbert and P. Duhamel. Robust blind channel identification and equalization based on multi-step predictors. In *Proc. IEEE ICASSP*, pages 3621–3624, Munich (Germany), 1997.

459. D. Gesbert, P. Duhamel, and S. Mayrargue. A bias removal technique for the prediction-based blind adaptive multichannel deconvolution. In A. Singh, editor, *Conference Record of The Twenty-Ninth Asilomar Conference on Signals, Systems and Computers*, pages 275–279 vol.1. IEEE Comput. Society Press, 1996.

460. D. Gesbert, P. Duhamel, and S. Mayrargue. Blind multichannel adaptive MMSE equalization with controlled delay. In *Proc. IEEE SP Workshop on Stat. Signal Array Processing*, pages 172–175, Corfu, Greece, 1996.

461. D. Gesbert, P. Duhamel, and S. Mayrargue. On-line blind multichannel equalization based on mutually referenced filters. *IEEE Trans. Signal Processing*, 45(9):2307–2317, September 1997.

462. D. Gesbert and A. Paulraj. Blind multi-user linear detection of CDMA signals in frequency-selective channels. In *Proc. ICC*, 1998.

463. D. Gesbert, J. Sorelius, and A. Paulraj. Blind multi-user MMSE detection of CDMA signals. In *Proc. IEEE ICASSP*, 1998.

464. A. Gharbi and F. Salam. Algorithm for blind signal separation and recovery in static and dynamics environments. In *IEEE Symposium on Circuits and Systems*, pages 713–716, Hong Kong, June 1997.

465. R.R. Gharieb and A. Cichocki. Noise reduction in brain evoked potentials based on third-order correlations. *IEEE Transactions on Biomedical Engineering*, 48:501–512, 2001.

466. R.R. Gharieb and A. Cichocki. Segmentation and tracking of EEG signal using an adaptive recursive bandpass filter. *Medical and Biological Engineering and Computing*, 39:237–248, 2001.

467. D.C. Ghiglia, L.A. Romero, and G.A. Mastin. Systematic approach to two-dimensional blind deconvolution by zero-sheet separation. *Journal of the Optical Society of America A (Optics and Image Science)*, 10(5):1024–1036, May 1993.

468. M. Ghosh and C.L. Weber. Blind deconvolution using a maximum likelihood channel estimator. In *Tenth Annual International Phoenix Conference on Computers and Communications*, pages 448–452, Scottsdale, AZ, 1991. IEEE Comput. Society Press.

469. M. Ghosh and C.L. Weber. Maximum-likelihood blind equalization. *Optical Engineering*, 31(6):1224–1228, June 1992.

470. G.B. Giannakis. Blind equalization of time-varying channels: a deterministic multichannel approach. In *Proc. IEEE SP Workshop on Stat. Signal Array Processing*, pages 180–183, Corfu, Greece, 1996.

471. G.B. Giannakis and S.D. Halford. Blind fractionally-spaced equalization of noisy FIR channels: adaptive and optimal solutions. In *Proc. IEEE ICASSP*, pages 1972–1975 vol.3, Detroit, MI, 1995.

472. G.B Giannakis, Y. Hua, P. Stoica, and L. Tong (eds.). *Signal Processing Advances in Wireless and Mobile Communication*, volume 1 and 2. Prentice Hall, Upper Saddle River, NJ, February 2001.

473. G.B. Giannakis and J.M. Mendel. Identification of nonminimum phase systems using higher order statistics. *IEEE Trans. on Acoustics, Speech, Signal Processing*, 37(3):360–377, March 1989.

474. G.B. Giannakis and E. Serpedin. Blind equalizers of multichannel linear-quadratic FIR Volterra channels. In *Proc. IEEE SP Workshop on Stat. Signal Array Processing*, pages 371–374, Corfu, Greece, 1996.

475. G.B. Giannakis, G. Zhou, and M.K. Tsatanis. On blind channel estimation with periodic misses and equalization of periodically varying channels. In *26-th Asilomar Conference Signals, Systems Comp.*, pages 531–535 vol.1. IEEE, 1992.

476. P.E. Gill, W. Murray, and M.H. Wright. *Practical Optimization.* Academic Press, London, 1981.

477. K. Giridhar, J.J. Shynk, and R.A. Iltis. Bayesian/decision-feedback algorithm for blind adaptive equalization. *Optical Engineering*, 31(6):1211–1223, June 1992.

478. K. Giridhar, J.J. Shynk, and R.A. Iltis. A modified Bayesian algorithm with decision feedback for blind adaptive equalization. In L. Dugard, M. M'Saad, and I.D. Landau, editors, *Adaptive Systems in Control and Signal Processing 1992. Selected Papers from the 4th IFAC Symposium*, pages 511–516, Grenoble, France, 1993. Pergamon Press.

479. K. Giridhar, J.J. Shynk, R.A. Iltis, and A. Mathur. Adaptive MAPSD algorithms for symbol and timing recovery of mobile radio TDMA signals. *IEEE Trans. Communications*, 44(8):976–987, August 1996.

480. M. Girolami. Symmetric adaptive maximum likelihood estimation for noise cancellation and signal separation. *Electronics Letters*, 33, No.17:1437 – 1438, 1997.

481. M. Girolami. Self-organizing neural networks: Independent component analysis and blind signal separation. perspectives in neural computation. In J. Taylor, editor, *Perspectives in Neural Computation.* Springer-Verlag Scientific Publishers, 1999.

482. M. Girolami, A. Cichocki, and S. Amari. A common neural network model for unsupervised exploratory data analysis and independent component analysis,. *IEEE Trans. on Neural Networks*, 9(6):1495–1501, 1998.

483. M Girolami and C Fyfe. Higher order cumulant maximisation using nonlinear Hebbian and anti-Hebbian learning for adaptive blind separation of source signals. In *In Proc. IWSIP-96, IEEE/IEE International Workshop on Signal and Image Processing, Advances in Computational Intelligence*, pages 141 – 144. Elsevier publishing, 1996.

484. M Girolami and C. Fyfe. Extraction of independent signal sources using a deflationary exploratory projection pursuit network with lateral inhibition. In *In Press I.E.E Proceedings on Vision on, Image and Signal Processing Journal.*, 1997.

485. M Girolami and C. Fyfe. Kurtosis extrema and identification of independent components : A neural network approach. In *In Proc. ICASSP-97, I.E.E.E Conference on Acoustics, Speech and Signal Processing*, volume 4, pages 3329 – 3333, 1997.

486. M. Girolami and C. Fyfe. Stochastic ICA contrast maximization using Oja's nonlinear PCA algorithm. *Int. Journal of Neural Systems*, 1997.

487. M. Girolami and C. Fyfe. Temporal model of linear anti-Hebbian learning. *Neural Processing Letters*, 4, No.3:1–10, Jan 1997.

488. M. Girolami and C. Fyfe. Negentropy and kurtosis as projection pursuit indices provide generalized ICA algorithms. In *Advances in Neural Information Processing Systems, NIPS'96 Workshop*, Snowmaas, Dec., 1996.

489. V.P. Godambe. *Estimating Functions.* New York, Oxford University Press., 1991.

490. L.C. Godara. Applications of antenna arrays to mobile communications. part I. *Proc. IEEE*, 85(7):1031–1060, July 1997.

491. D.N. Godard. Self-recovering equalization and carrier tracking in two-dimensional data communication systems. *IEEE Trans. Communications*, 28(11):1867–1875, November 1980.

492. S.H. Goldberg and R.A. Iltis. Joint channel equalization and interference rejection using a modified constant modulus algorithm. In *Proc. IEEE MILCOM*, pages 97–101 vol.1, San Diego, CA, USA, 23-26 Oct. 1988, 1988. IEEE.

493. G. Golub and V. Pereyra. The differentiation of pseudo-inverses and nonlinear least squares problems whose variables separate. *SIAM Journal Num. Anal.*, 10:413–432, 1973.

494. G.H. Golub and C.F. Van-Loan. *Matrix Computations.* Johns Hopkins, third edition, 1996.

495. J. Gomes and V. Barroso. A super-exponential algorithm for blind fractionally spaced equalization. *IEEE Signal Processing Letters*, 3(10):283–285, October 1996.

496. E. Gönen and J.M. Mendel. Applications of cumulants to array processing—part III: Blind beamforming for coherent signals. *IEEE Trans. Signal Processing*, 45(9):2252–2264, September 1997.

497. E. Gönen, J.M. Mendel, and M.C. Dovgan. Applications of cumulants to array processing—part IV: Direction finding in coherent signal case. *IEEE Trans. Signal Processing*, 45(9):2265–2276, September 1997.

498. R. Gooch and J. Lundell. The CM array: An adaptive beamformer for constant modulus signals. In *Proc. IEEE ICASSP*, pages 2523–2526, Tokyo, 1986.

499. R. Gooch, M. Ready, and J. Svoboda. A lattice-based constant modulus adaptive filter. In *20-th Asilomar Conference Signals, Systems Comp.*, pages 282–286. IEEE, 1987.

500. R.P. Gooch and B. Daellenbach. Prevention of interference capture in a blind (CMA-based) adaptive receive filter. In *23-th Asilomar Conference Signals, Systems Comp.*, pages 898–902 vol.2, Pacific Grove, CA, USA, 30 Oct.-1 Nov. 1989, 1989. Maple Press.

501. R.P. Gooch and J.C. Harp. Blind channel identification using the constant modulus adaptive algorithm. In *IEEE Int. Conference Communications '88: Digital Technology - Spanning the Universe*, pages 75–79 vol.1, Philadelphia, PA, USA, 12-15 June 1988, 1988. IEEE.

502. R.P. Gooch and B.J. Sublett. Joint spatial and temporal equalization in a decision-directed adaptive antenna system. In *22-nd Asilomar Conference on Signals, Systems and Computers*, pages 255–259 vol.1. Maple Press, 1988.

503. D.M. Goodman. On the derivatives of the homomorphic transform operator and their application to some practical signal processing problems. *Proc. IEEE*, 78(4):642–651, April 1990.

504. I.F. Gorodnitsky and B. D. Rao. Sparse signal reconstruction from limited data using FOCUSS: A re-weighted minimum norm algorithm. *IEEE Trans. Sig. Proc.*, 45(3):600–616, March 1997.

505. A. Gorokhov and P. Loubaton. Semi-blind second order identification of convolutive channels. In *Proc. IEEE ICASSP*, pages 3905–3908, 1997.

506. A. Gorokhov, P. Loubaton, and É. Moulines. Second order blind equalization in multiple input multiple output fir systems: A weighted least squares approach. In *Proc. IEEE International Conference Acoustics, Speech, Signal Processing, Atlanta, GA*, volume 5, pages 2415–2418, May 1996.

507. Y. Goussard. Blind deconvolution of sparse spike trains using stochastic optimization. In *Proc. IEEE ICASSP*, pages 593–596 vol.4, San Francisco, CA, 1992.

508. D. Graupe. Blind adaptive filtering of unknown speech from unknown noise in a single receiver situation. In *Proc. IEEE ISCAS*, pages 2617–2620 vol.6, San Diego, CA, 1992.

509. W.C. Gray. *Variable Norm Deconvolution.* PhD thesis, Ph.D. Dissertation, Stanford, Univ. Stanford, CA, 1979.

510. M. S. Grewal and A. P. Andrews. *Kalman Filtering: Theory and Practice.* Prentice Hall, 1993.

511. M. Griffiths and M. Hosseini. An application of blind deconvolution for digital speech processing. In *Fifth International Conference on Digital Processing of Signals in Communications (Publ. No.82)*, pages 203–210, Loughborough, UK, 1988. IERE.

512. F. Guglielmi, C. Luschi, and A. Spalvieri. Blind algorithms for joint clock recovery and baseband combining in digital radio. In *Fourth European Conference on 'Radio Relay Systems' (Conference Publ. No.386)*, pages 279–286, Edinburgh, UK, 11-14 Oct. 1993, 1993. IEE.

513. F. Guglielmi, C. Luschi, and A. Spalvieri. Joint clock recovery and baseband combining for the diversity radio channel. *IEEE Trans. Communications*, 44(1):114–117, January 1996.

514. S.F. Gull and J. Skilling. Recent developments at Cambridge (maximum entropy principle). In C.R. Smith and G.J. Erickson, editors, *Maximum-Entropy and Bayesian Analysis and Estimation Problems. Proceedings of the Third Workshop on Maximum Entropy and Bayesian Methods in Applied Statistics*, pages 149–160, Laramie, WY, 1987. Reidel.

515. U. Gummadavelli and J.K. Tugnait. Blind channel estimation and equalization with partial-response input signals. In *Proc. IEEE ICASSP*, pages III/233–236 vol.3, Adelaide, 1994.

516. J. Gunther and A. Swindlehurst. Algorithms for blind equalization with multiple antennas based on frequency domain subspaces. In *Proc. IEEE ICASSP*, pages 2419–2422 vol. 5, Atlanta, GA, 1996.

517. M.I. Gurelli and C.L. Nikias. Evam: An eigenvector-based algorithm for multichannel bind deconvolution of input colored signals. *IEEE Trans. Signal Processing*, SP-43, no. 1:134–149, Jan 1995.

518. F. Gustafsson and B. Wahlberg. Blind equalization by direct examination of the input sequences. In *Proc. IEEE ICASSP*, pages 701–704 vol.4, San Francisco, CA, 1992.

519. F. Gustafsson and B. Wahlberg. On simultaneous system and input sequence estimation. In L. Dugard, M. M'Saad, and I.D. Landau, editors, *Adaptive Systems in Control and Signal Processing 1992. Selected Papers from the 4th IFAC Symposium*, pages 11–16, Grenoble, France, 1993. Pergamon Press.

520. F. Gustafsson and B. Wahlberg. Blind equalization by direct examination of the input sequences. *IEEE Trans. Communications*, 43(7):2213–2222, July 1995.

521. M. Haardt. *Efficient One-, Two-, and Multidimensional High-Resolution Array Signal Processing*. PhD thesis, TU München, Munich, Germany, 1997.

522. B. Halder, B.C. Ng, A. Paulraj, and T. Kailath. Unconditional maximum likelihood approach for blind estimation of digital signals. In *Proc. IEEE ICASSP*, volume 2, pages 1081–1084, Atlanta (GA), May 1996.

523. S.D. Halford and G.B. Giannakis. Optimal blind equalization and symbol error analysis of fractionally-sampled channels. In *29-th Asilomar Conference Signals, Systems, Comp.*, pages 1332–1336 vol.2. IEEE, 1996.

524. L.K. Hansen and G. Xu. Geometric properties of the blind digital co-channel communications problem. In *Proc. IEEE ICASSP*, pages 1085–1088 vol. 2, Atlanta, GA, 1996.

525. L.K. Hansen and G. Xu. A hyperplance-based algorithm for the digital co-channel communications problem. *IEEE Trans. Informat. Th.*, 43(5):1536–1548, September 1997.

526. W.A. Harrison, J.S. Lim, and E. Singer. A new application of adaptive noise cancellation. *IEEE Trans. Acoustics Speech, Signal Processing*, ASSP-34:21–27, February 1986.

527. F. Harroy and J.-L. Lacoume. Maximum likelihood estimators and Cramer-Rao bounds in source separation. *Signal Processing*, 55(2):167–177, December 1996.

528. J. Hartigan. *Clustering Algorithms.* Wiley, 1975.

529. S. Hashimoto and H. Saito. Restoration of shift variant blurred image estimating the parameter distribution of point-spread function. *Systems and Computers in Japan*, 26(1):62–72, January 1995.

530. S. Hashimoto and H. Saito. Restoration of space-variant blurred image using a wavelet transform. *Trans. of the Institute of Electronics, Information and Communication Engineers D-II*, J78D-II(12):1821–1830, December 1995.

531. T. Hastie and W. Stuetzle. Principal curves. *Journal of the American Statistical Association*, 84(406):502–516, 1989.

532. D. Hatzinakos. Stop-and-go sign algorithms for blind equalization. *Proc. SPIE*, 1565:118–129, 1991.

533. D. Hatzinakos. Structures for polyspectra-based blind equalizers. In *Proc. IEEE MILCOM*, pages 1272–1276 vol.3, McLean, VA, 1991.

534. D. Hatzinakos. Blind equalization using stop-and-go adaptation rules. *Optical Engineering*, 31(6):1181–1188, June 1992.

535. D. Hatzinakos. Carrier phase tracking and tricepstrum-based blind equalization. *Computers & Electrical Engineering*, 18(2):109–118, March 1992.

536. D. Hatzinakos. Blind equalization using decision feedback prediction and tricepstrum principles. *Signal Processing*, 36(3):261–276, April 1994.

537. D. Hatzinakos. Blind equalization based on prediction and polycepstra principles. *IEEE Trans. Communications*, 43(2-4,):178–181, February 1995.

538. D. Hatzinakos and C.L. Nikias. Adaptive filtering based on polycepstra. In *Proc. IEEE ICASSP*, pages 1175–1178 vol.2, Glasgow, UK, 1989.

539. D. Hatzinakos and C.L. Nikias. Blind decision feedback equalization structures based on adaptive cumulant techniques. In *Proc. IEEE Int. Conference Communic.*, pages 1278–1282 vol.3, Boston, MA, 1989.

540. D. Hatzinakos and C.L. Nikias. Estimation of multipath channel response in frequency selective channels. *IEEE Trans. Sel. Areas Comm.*, 7(1):12–19, 1989.

541. D. Hatzinakos and C.L. Nikias. Blind equalization based on second and fourth order statistics. In *Proc. IEEE Int. Conference Communic./SUPERCOMM*, pages 1512–1516 vol.4, Atlanta, GA, 1990.

542. D. Hatzinakos and C.L. Nikias. Polycepstra techniques for blind equalization of multilevel PAM schemes. In *Proc. IEEE Int. Conference Communic./SUPERCOMM*, pages 1507–1511 vol.4, Atlanta, GA, 1990.

543. D. Hatzinakos and C.L. Nikias. Blind equalization using a tricepstrum-based algorithm. *IEEE Trans. Communications*, 39(5):669–682, May 1991.

544. S. Haykin. Blind equalization formulated as a self-organized learning process. In *26-th Asilomar Conference Signals, Systems Comp.*, pages 346–350 vol.1. IEEE, 1992.

545. S. Haykin. *Blind Deconvolution.* Prentice Hall, Englewood Cliffs, NJ, 1994.

546. S. Haykin. *Adaptive Filter Theory, Third Edition.* Prentice-Hall, Englewood Cliffs, NJ, 1996.

547. S. Haykin. *Communication Systems, 4th Edition.* John Wiley & Sons, May 15 2000.

548. S. Haykin. *(Ed.) Unsupervised Adaptive Filtering, Volume 1: Blind Source Separation*, volume 1. John Wiley & Sons, March 2000.

549. S. Haykin. *(Ed.) Unsupervised Adaptive Filtering Volume 2 : Blind Deconvolution*, volume 2. John Wiley & Sons, February 2000.

550. S. Haykin and B. Kosko (Eds.). *Intelligent Signal Processing.* IEEE, December 2000.

551. S. Haykin and B. Van Veen. *Signals and Systems.* John Wiley & Sons, September 1998.

552. R. He and J. Reed. A robust co-channel interference rejection technique for current mobile phone system. In *Proc. IEEE VTC*, pages 1195–1199 vol.2, Atlanta, GA, 1996.

553. R. Hecht Nielsen. Replicator neural networks for universal optimal source coding. *Science Volume*, 269:1860–1863, 1995.

554. J. Hérault and C. Jutten. Space or time adaptive signal processing by neural network models. In J. S. Denker, editor, *Neural Networks for Computing. Proceedings of AIP Conference*, pages 206–211, New York, 1986. American Institute of Physics.

555. J. Hérault, C. Jutten, and B. Ans. Détection des grandeurs primitives dans un message composite par une architecture de calcul neuromimétique en apprentissage non supervisé. In *Proc. Xéme colloque GRETSI*, pages 1017–1022, Nice, France, May 20-24 1985.

556. M. Herrmann and H.H. Yang. Perspectives and limitations of self-organizing maps in blind separation of source signals. In S. Amari, L. Xu, L.-W. Chan, I. King, and K.-S. Leung, editors, *Progress in Neural Information Processing. Proceedings of the International Conference on Neural Information Processing*, pages 1211–1216 vol.2, Hong Kong, 1996. Springer-Verlag.

557. J. Hertz, A. Krogh, and R.G. Palmer. *Introduction to the Theory of Neural Computation.* Addison-Wesley, New York, 1991.

558. K. Hilal and P. Duhamel. A convergence study of the constant modulus algorithm leading to a normalized-CMA and a block-normalized-CMA. In J. Vandewalle, R. Boite, M. Moonen, and A. Oosterlinck, editors, *Signal Processing VI - Theories and Applications. Proceedings of EUSIPCO-92, Sixth European Signal Processing Conference*, pages 135–138 vol.1, Brussels, 1992. Elsevier.

559. K. Hilal and P. Duhamel. A general form for recursive adaptive algorithms leading to an exact recursive CMA. In *Proc. IEEE ICASSP*, pages 17–20 vol.4, San Francisco, CA, USA, 23-26 March 1992, 1992.

560. M.J. Hinich. Testing for Gausianity and linearity of a stationary time series. *Journal Time Series Analysis*, 3(3):169–176, 1982.

561. R.S. Holambe, A.K. Ray, and T.K. Basu. Phase-only blind deconvolution using bicepstrum iterative reconstruction algorithm (BIRA). *IEEE Trans. Signal Processing*, 44(9):2356–2359, September 1996.

562. T.J. Holmes. Blind deconvolution of quantum limited incoherent imagery: maximum-likelihood approach. *Journal of the Optical Society of America A (Optics and Image Science)*, 9(7):1052–1061, July 1992.

563. M. Honig, U. Madhow, and S. Verdu. Blind adaptive multiuser detection. *IEEE Trans. Information Th.*, 41(4):944–960, July 1995.

564. G. Hori. A new approach to joint diagonalization. In *Proceedings of the Second International Workshop on ICA and BSS, ICA'2000*, pages 151–155, Helsinki, Finland, 19-22 June 2000.

565. W. Hosseini, R. Lonski, and R. Gooch. A real-time implementation of the Multistage CMA adaptive beamformer. In *27-th Asilomar Conference Signals, Systems Comp.*, pages 643–646. IEEE, 1993.

566. I. Howitt, V. Vemuri, J.H. Reed, and T.C. Hsia. Recent developments in applying neural nets to equalization and interference rejection. In *Virginia Tech's Third Symposium on Wireless Personal Communications Proceedings*, pages 1/1–12, Blacksburg, VA, 1993. Virginia Tech.

567. P.O. Hoyer and A. Hyvärinen. Independent component analysis applied to feature extraction from colour and stereo images. *Network: Computation in Neural Systems*, 11(3):191–210, 2000.

568. P.O. Hoyer. Non-negative sparse coding. Proc. of Workshop on Neural Network for Signal Processing, (NNSP-2002), Martigny, Switzerland, Sept. 4-7, 2002, pp.557-565.

569. Y. Hua. Fast maximum likelihood for blind identification of multiple FIR channels. *IEEE Trans. on Signal Processing*, 44:661–672, March 1996.

570. Y. Hua, K. Abed-Meraim, and M. Wax. Blind system identification using minimum noise subspace. *IEEE Trans. on Signal Processing*, 45(3):770–773, 1997.

571. Y. Hua and W. Liu. Generalized Karhunen-Loeve transform. *IEEE Signal Processing Letters*, 5(6):141–143, June 1998.

572. Y. Hua, Y. Xiang, T.-P. Chen, K. Abed-Meraim, and Y. Miao. A new look at the power method for fast subspace tracking. *Digital Signal Processing*, 9:297–314, 1999.

573. Y. Hua, H.H. Yang, and W. Qiu. Source correlation compensation for blind channel identification based on second-order statistics. *IEEE Signal Processing Letters*, 1(8):119–120, August 1994.

574. Y. Hua, H.H. Yang, and M. Zhou. Blind system identification using multiple sensors. In *Proc. IEEE ICASSP*, pages 3171–3174 vol.5, Detroit, MI, 1995. IEEE.

575. S. Van Huffel. *(Ed.) Recent Advances in Total Least Squares Techniques and Errors in Variables Modeling*, volume 93 of *Siam Proceedings in Applied Mathematics*. SIAM, Philadelphia, USA, 1997.

576. S. Van Huffel and J. Vandewalle (Eds.). *The Total Least Squares Problem : Computational Aspects and Analysis*, volume 9 of *Frontiers in Applied Mathematics*. SIAM, Philadelphia, USA, 1991.

577. R.D. Hughes, E.J. Lawrence, and L.P. Withers, Jr. A robust CMA adaptive array for multiple narrowband sources. In *26-th Asilomar Conference Signals, Systems Comp.*, pages 35–39 vol.1. IEEE, 1992.

578. A.W. Hull and W.K. Jenkins. Performance improvement of blind equalizers using the constant modulus algorithm. In *26-th Asilomar Conference Signals, Systems Comp.*, pages 11–14 vol.1. IEEE, 1992.

579. J. Hurri, A. Hyvärinen, J. Karhunen, and E. Oja. Image feature extraction using independent component analysis. In *Proc. 1996 IEEE Nordic Signal Processing Symposium NORSIG'96*, pages 475–478, Espoo, Finland, Sept. 1996.

580. A. Hyvärinen. Finding cluster directions by non-linear Hebbian learning. In *Proc. Int. Conference on Neural Information Processing*, pages 97–102, Hong Kong, 1996.

581. A. Hyvärinen. Purely local neural principal component and independent component learning. In *Proc. Int. Conference on Artificial Neural Networks*, pages 139–144, Bochum, Germany, 1996.

582. A. Hyvärinen. Simple one-unit neural algorithms for blind source separation and blind deconvolution. In S. Amari, L. Xu, L.-W. Chan, I. King, and K.-S. Leung, editors, *Progress*

in Neural Information Processing. Proceedings of the International Conference on Neural Information Processing, pages 1201–1206 vol.2, Hong Kong, 1996. Springer-Verlag.

583. A. Hyvärinen. A family of fixed-point algorithms for independent component analysis. In *Proc. IEEE Int. Conference on Acoustics, Speech and Signal Processing (ICASSP'97)*, pages 3917–3920, Munich, Germany, 1997.

584. A. Hyvärinen. New approximations of differential entropy for independent component analysis and projection pursuit. In *Proceedings of NIPS'97*, 1997.

585. A. Hyvärinen. One-unit contrast functions for independent component analysis: A statistical analysis. In *Neural Networks for Signal Processing VII (Proc. IEEE Workshop on Neural Networks for Signal Processing)*, pages 388–397, Amelia Island, Florida, 1997.

586. A. Hyvärinen. Fast and robust fixed-point algorithms for independent component analysis. *IEEE Transactions on Neural Networks*, 10(3):626–634, 1999.

587. A. Hyvärinen. Sparse code shrinkage: Denoising of non-Gaussian data by maximum likelihood estimation. *Neural Computation*, 11(7):1739–1768, 1999.

588. A. Hyvärinen and P.O. Hoyer. Emergence of phase and shift invariant features by decomposition of natural images into independent feature subspaces. *Neural Computation*, 12(7):1705–1720, 2000.

589. A. Hyvärinen, J. Karhunen, and E. Oja. *Independent Component Analysis*. John Wiley, New York, 2001.

590. A. Hyvärinen and E. Oja. A neuron that learns to separate one independent component from linear mixtures. In *Proc. IEEE Int. Conference on Neural Networks*, pages 62–67, Washington, D.C., 1996.

591. A. Hyvärinen and E. Oja. Simple neuron models for independent component analysis. *Int. Journal of Neural Systems*, 7(6):671–687, 1996.

592. A. Hyvärinen and E. Oja. A fast fixed-point algorithm for independent component analysis. *Neural Computation*, 9(7):1483–1492, 1997.

593. A. Hyvärinen and E. Oja. One-unit learning rules for independent component analysis. In *Advances in Neural Information Processing System 9 (NIPS*96)*, pages 480–486. MIT Press, 1997.

594. A. Hyvärinen and E. Oja. Independent component analysis by general non-linear Hebbian-like learning rules. *Signal Processing*, 64(3):301–313, 1998.

595. A. Hyvärinen and E. Oja. Independent component analysis by general non-linear Hebbian-like learning rules. *Signal Processing*, 64(3):301–313, 1998.

596. S. Icart and R. Gautier. Blind separation of convolutive mixtures using second and fourth order moments. In *Proc. IEEE ICASSP*, pages 3018–3021 vol. 5, Atlanta, GA, 1996.

597. S. Ikeda. ICA on noisy data: A factor analysis approach. In *Advances in Independent Component Analysis*, pages Ch.11, 201–215, 2000.

598. S. Ikeda and K. Toyama. Independent component analysis for noisy data–MEG data analysis. *Neural Networks*, 13(10):1063–1074, 2001.

599. J. Ilow, D. Hatzinakos, and A.N. Venetsanopoulos. Blind deconvolution based on cumulant fitting and simulated annealing optimization. In *26-th Asilomar Conference Signals, Systems Comp.*, pages 351–355 vol.1. IEEE, 1992.

600. J. Ilow, D. Hatzinakos, and A.N. Venetsanopoulos. Simulated annealing optimization in blind equalization. In *Proc. IEEE MILCOM*, pages 216–220 vol.1, San Diego, CA, 1992.

601. R.A. Iltis and S.H. Goldberg. Joint interference rejection/channel equalization in DS spread-spectrum receivers using the CMA equalizer and maximum-likelihood techniques. In *Proc. IEEE MILCOM*, pages 109–113 vol.1, Washington, DC, USA, 19-22 Oct. 1987, 1987. IEEE.

602. R.A. Iltis, J.J. Shynk, and K. Giridhar. Recursive Bayesian algorithms for blind equalization. In *25-th Asilomar Conference Signals, Systems Comp.*, pages 710–715 vol.2. IEEE, 1991.

603. R.A. Iltis, J.J. Shynk, and K. Giridhar. Bayesian blind equalization for coded waveforms. In *Proc. IEEE MILCOM*, pages 211–215 vol.1, San Diego, CA, 1992.

604. R.A. Iltis, J.J. Shynk, and K. Giridhar. Bayesian algorithms for blind equalization using parallel adaptive filtering. *IEEE Trans. Communications*, 42(2-4,):1017–1032, February 1994.

605. T. Inoue, Y. Fujii, K. Itoh, and Y. Ichioka. Block blind separation of independent spectra in hyperspectral-images. In *1996 International Topical Meeting on Optical Computing. Technical Digest*, pages 40–41 vol.1, Proceedings of International Topical Meeting on Optical Computing - OC 96, Sendai, Japan, 21-25 April 1996, 1996. Japan Society Appl. Phys.

606. Y. Inouye. Blind deconvolution of multichannel linear time-invariant systems of non-minimum phase to appear in statistical methods in control and signal processing. In T. Katayama and S. Sugimoto, editors, *Statistical Methods in Control and Signal Processing.* New York: Marcel Dekker, 1997.

607. Y. Inouye. Criteria for blind deconvolution of multichannel linear time-invariant systems of non-minimum phase. In T. Katayama and S. Sugimoto, editors, *Statistical Methods in Control and Signal Processing*, pages 375–397. Dekker, New York, 1997.

608. Y. Inouye and T. Habe. Blind equalization of multichannel linear time–invariant systems. *The Institute of Electronics Information and Communication Engineers*, 24(24):9–16, 1995.

609. Y. Inouye and T. Habe. Multichannel blind equalization using second- and fourth-order cumulants. In *Proc. IEEE Signal Processing Workshop on Higher Order Statistics*, pages 96–100, 1995.

610. Y. Inouye and K. Hirano. Blind identification of linear multi-input-multi-output systems driven by colored inputs with applications to blind signal separation. In *Proc. IEEE Conference on Decision and Control*, pages 715–720 vol.1, New Orleans, LA, 1995.

611. Y. Inouye and S. Ohno. Adaptive algorithms for implementing the single-stage criterion for multichannel blind deconvolution. In *Proc. of 5th Int. Conference on Neural Information Processing, ICONIP'98*, pages 733–736, Kitakyushu, Japan, October, 21-23 1998.

612. Y. Inouye and T. Ohta. Blind equalization of digital communication channels using fractionally spaced samples. In *Proc. IEEE ISCAS*, pages 165–168 vol.2, Atlanta, GA, 1996.

613. Y. Inouye and T. Sato. Unconstrained optimization criteria for blind deconvolution of multichannel linear systems. In S. Amari, L. Xu, L.-W. Chan, I. King, and K.-S. Leung, editors, *Progress in Neural Information Processing. Proceedings of the International Conference on Neural Information Processing*, pages 1189–1194 vol.2, Hong Kong, 1996. Springer-Verlag.

614. Y. Inouye and T. Sato. Unconstrained optimization criteria for blind equalization of multichannel linear systems. In *Proc. IEEE SP Workshop on Stat. Signal Array Processing*, pages 320–323, Corfu, Greece, 1996.

615. H. Isa and Y. Sato. Performance of implicit blind Viterbi algorithm in QPSK. *Trans. of the Institute of Electronics, Information and Communication Engineers B-II*, J79B-II(4):233–246, April 1996.

616. S.H. Isabelle, A.V. Oppenheim, and G.W. Wornell. Effects of convolution on chaotic signals. In *Proc. IEEE ICASSP*, pages 133–136 vol.4, San Francisco, CA, 1992.

617. K. Itoh. Trend on antenna technologies in the land mobile communication. *Journal of the Institute of Electronics, Information and Communication Engineers*, 73(12):1331–1335, December 1990.

618. K. Itoh, T. Shimamura, and J. Suzuki. Prefiltering for blind equalization. *Trans. of the Institute of Electronics, Information and Communication Engineers A*, J78-A(3):323–331, March 1995.

619. N.K. Jablon. Carrier recovery for blind equalization. In *Proc. IEEE ICASSP*, pages 1211–1214 vol.2, Glasgow, UK, 1989.

620. N.K. Jablon. Joint blind equalization, carrier recovery, and timing recovery for 64-QAM and 128-QAM signal constellations. In *IEEE Int. Conference Communications. BOSTON-ICC/89. World Prosperity Through Communications*, pages 1043–1049 vol.2, Boston, MA, USA, 11-14 June 1989, 1989. IEEE.

621. N.K. Jablon. Joint blind equalization, carrier recovery and timing recovery for high-order QAM signal constellations. *IEEE Trans. Signal Processing*, 40(6):1383–1398, June 1992.

622. N.K. Jablon, C.W. Farrow, and S.-N. Chou. Timing recovery for blind equalization. In *22-nd Asilomar Conference Signals, Systems Comp.*, pages 112–118 vol.1. Maple Press, 1989.

623. J.E. Jackson. *A User Guide to Principal Components*. John Wiley & Sons, New York, 1991.

624. O.L.R. Jacobs. *Introduction to Control Theory*. Oxford University Press, 2nd edition, 1993.

625. O. Jahn, A. Cichocki, A. Ioannides, and S. Amari. Identification and elimination of artifacts from MEG signals using efficient independent components analysis. In *Proc. of th 11th Int. Conference on Biomagentism BIOMAG-98*, pages 224–227, Sendai, Japan, 1999.

626. H. Jamali and S.L. Wood. Error surface analysis for the complex constant modulus adaptive algorithm. In *24-th Asilomar Conference Signals, Systems Comp.*, pages 248–252 vol.1. Maple Press, 1990.

627. H. Jamali, S.L. Wood, and R. Cristi. Experimental validation of the Kronecker product Godard blind adaptive algorithms. In *26-th Asilomar Conference Signals, Systems Comp.*, pages 1–5 vol.1. IEEE, 1992.

628. G.J.M. Janssen and R. Prasad. Propagation measurements in an indoor radio environment at 2.4 GHz, 4.75 GHz and 11.5 GHz. In *42nd VTS Conference*, pages 617–620 vol.2, Denver, 1992. IEEE.

629. S.M. Jefferies and J.C. Christou. Restoration of astronomical images by iterative blind deconvolution. *Astrophysical Journal*, 415(2,):862–874, October 1993.

630. B. Jelonnek and K.D. Kammayer. A blind adaptive equalizer based on a lattice/all-pass configuration. In J. Vandewalle e.a., editor, *Signal Processing VI: Proc. EUSIPCO-92*, pages 1109–1112 vol.2, Brussels, 1992. Elsevier.

631. B. Jelonnek and K.-D. Kammeyer. Improved methods for the blind system identification using higher order statistics. *IEEE Trans. Signal Processing*, 40(12):2947–2960, December 1992.

632. B. Jelonnek and K.D. Kammeyer. Eigenvector algorithm for blind equalization. In *IEEE SP Workshop on Higher-Order Stat.*, pages 19–23, South Lake Tahoe, CA, 1993.

633. B. Jelonnek and K.D. Kammeyer. A closed-form solution to blind equalization. *Signal Processing*, 36(3):251–259, April 1994.

634. R.W. Jenkins and K.W. Moreland. A comparison of the eigenvector weighting and Gram-Schmidt adaptive antenna techniques. *IEEE Trans. Aerospace and Electronic Systems*, 29:568–575, 1993.

635. C.R. Johnson. Admissibility in blind adaptive channel equalization. *IEEE Control Systems Magazine*, 11(1):3–15, January 1991.

636. C.R. Johnson. On the interaction of adaptive filtering, identification and control. *IEEE Signal Processing Magazine*, 12(2):22–37, March 1995.

637. C.R. Johnson, Jr., S. Dasgupta, and W.A. Sethares. Averaging analysis of local stability of a real constant modulus algorithm adaptive filter. *IEEE Trans. Acoustics, Speech and Signal Processing*, 36(6):900–910, April 1988.

638. I.T. Jolliffe. *Principal Component Analysis*. Springer Verlag, New York, 1986.

639. D.L. Jones. A normalized constant-modulus algorithm. In *29-th Asilomar Conference Signals, Systems, Comp.*, pages 694–697 vol.1. IEEE, 1996.

640. V. Jousmäki and R. Hari. Cardiac artifacts in magnetoencephalogram. *Journal of Clinical Neurophysiology*, 13(2):172–176, 1996.

641. B. Juang and L.R. Rabiner. The segmental K-means algorithm for estimating parameters of hidden Markov models. *IEEE Trans. Acoustics, Speech., Signal Processing*, 38(9):1639–1641, September 1990.

642. T.-P. Jung, C. Humphries, T.-W. Lee, S. Makeig, M. McKeown, V. Iragui, and T.J. Sejnowski. Extended ICA removes artifacts from electroencephalographic recordings. *Advances in Neural Information Processing Systems*, 1997.

643. T.-P. Jung, S. Makeig, A.J. Bell, and T.J. Sejnowski. Independent component analysis of electroencephalographic and event-related potential data. In *In: P. Poon, J. Brugge, ed., Auditory Processing and Neural Modeling*. Plenum Press (in press), 1997.

644. T.P Jung, C. Humphries, T-W. Lee, S. Makeig, M. McKeown, V. Iragui, and T. Sejnowski. ICA removes artifacts from electroencephalographic recordings. *Advances in Neural Information Processing Systems*, 10, 1997.

645. T.P Jung, S. Makeig, C. Humphries, T-W. Lee, M. McKeown, V. Iragui, and T.J. Sejnowski. Removing electroencephalographic artifacts by blind source separation. *Psychophysiology*, 37(167-178), March 2000.

646. M. Juntti and B. Aazhang. Finite memory-length linear multiuser detection for asynchronous CDMA communications. *IEEE Trans.. Comm.*, 45(5):611–622, May 1997.

647. C. Jutten and J.-F. Cardoso. Source separation : really blind ? In *Proc. NOLTA*, pages 79–84, 1995.

648. C. Jutten, A. Guerin, and H.-L.-N. Thi. Adaptive optimization of neural algorithms. *Lecture Notes in Computer Science*, 540:54–61, 1991.

649. C. Jutten and J. Hérault. Independent component analysis (INCA) versus principal component analysis. In J.L. Lacoume et al., editor, *Signal Processing IV: Theories and Applications*, pages 643–646. Elsevier, 1988.

650. C. Jutten and J. Hérault. Blind separation of sources I. An adaptive algorithm based on neuromimetic architecture. *Signal Processing*, 24(1):1–10, July 1991.

651. C. Jutten, H.L. N. Thi, E. Dijkstra, E. Vittoz, and J. Caelen. Blind separation of sources, an algorithm for separation of convolutive mixtures. In *Proceedings of Int. Workshop on High Order Statistics*, pages 273–276, Chamrousse (France), 1991.

652. M.F. Kahn and W.A. Gardner. A time-channelized programmable canonical correlation analyzer. In *29-th Asilomar Conference Signals, Systems, Comp.*, pages 346–350 vol.1. IEEE, 1996.

653. T. Kailath. *Linear Systems.* Prentice Hall, Englewood Cliffs, NJ, 1980.

654. K.D. Kammeyer. Constant-modulus algorithms for the adjustment of adaptive receiver filters. *Archiv fur Elektronik und Uebertragungstechnik*, 42(1):25–35, January 1988.

655. K.D. Kammeyer. Closed-form solutions to blind equalization and system identification. *Frequenz*, 49(7-8):138–144, July 1995.

656. K.D. Kammeyer and B. Jelonnek. A cumulant zero-matching method for the blind system identification. In J.L. Lacoume, editor, *Higher Order Statistics. Proceedings of the International Signal Processing Workshop*, pages 153–156, Chamrousse, France, 1992. Elsevier.

657. G. Karam, K. Maalej, V. Paxal, and H. Sari. Variable symbol-rate demodulators for cable and satellite TV broadcasting. *IEEE Trans. Broadcasting*, 42(2):102–109, June 1996.

658. J. Karaoguz and S.H. Ardalan. Use of blind equalization for teletext broadcast systems. *IEEE Trans. Broadcasting*, 37(2):44–54, June 1991.

659. J. Karaoguz and S.H. Ardalan. Blind adaptive channel equalization using the unsupervised cluster formation technique. In *Proc. IEEE ISCAS*, pages 1593–1596 vol.3, San Diego, CA, 1992.

660. J. Karaoguz and S.H. Ardalan. A soft decision-directed blind equalization algorithm applied to equalization of mobile communication channels. In *SUPERCOMM/ICC*, pages 1272–1276 vol.3, Chicago, IL, 1992. IEEE.

661. J. Karhunen. *Recursive Estimation of Eigenvectors of Correlation Type Matrices for Signal Processing Applications.* Dr. Tech. Thesis, Helsinki University of Technology, Finland, 1984.

662. J. Karhunen. Stability of Oja's PCA subspace rule. *Neural Computation*, 6:739–747, 1994.

663. J. Karhunen. Neural approaches to independent component analysis and source separation. In M. Verleysen, editor, *4th European Symposium on Artificial Neural Networks, ESANN '96. Proceedings*, pages 249–266, Proceedings of European Symposium on Artificial Neural Networks, Bruges, Belgium, 24-26 April 1996, 1996. D Facto.

664. J. Karhunen. Nonlinear independent component analysis. In R. Everson and S. Roberts, editors, *ICA: Principles and Practice*, page 25. Cambridge University Press, Cambridge, UK, 2000.

665. J. Karhunen, A. Cichocki, W. Kasprzak, and P. Pajunen. On neural blind separation with noise suppression and redundancy reduction. *Int. Journal of Neural Systems*, 8(2):219–237, April 1997.

666. J. Karhunen, A. Hyvärinen, R. Vigário, J. Hurri, and E. Oja. Applications of neural blind separation to signal and image processing. In *Proc. 1997 Int. Conference on Acoustics, Speech, and Signal Proc. ICASSP-97*, Munich, Germany, April 1997.

667. J. Karhunen and J. Joutsensalo. Learning of robust principal component subspace. In *IJCNN'93. Proceedings of 1993 International Joint Conference on Neural Networks*, volume 3, pages 2409–2412, Nagoya, Japan, 1993.

668. J. Karhunen and J. Joutsensalo. Representation and separation of signals using nonlinear PCA type learning. *Neural Networks*, 7(1):113–127, 1994.

669. J. Karhunen and J. Joutsensalo. Generalization of principal component analysis, optimization problems, and neural networks. *Neural Networks*, 8(4):549–562, 1995.

670. J. Karhunen, E. Oja, L. Wang, R. Vigário, and J. Joutsensalo. A class of neural networks for independent component analysis. *IEEE Trans. on Neural Networks*, 8(3):486–503, May 1997.

671. J. Karhunen and P. Pajunen. Hierarchic nonlinear PCA algorithms for neural blind source separation. In *NORSIG 96 Proceedings – 1996 IEEE Nordic Signal Processing Symposium*, pages 71–74, Espoo, Finland, Sept. 1996.

672. J. Karhunen and P. Pajunen. Blind source separation and tracking using nonlinear PCA criterion: A least-squares approach. In *Proc. 1997 Int. Conference on Neural Networks (ICNN'97)*, volume 4, pages 2147–2152, Houston, Texas, USA, June 1997.

673. J. Karhunen and P. Pajunen. Blind source separation using least-squares type adaptive algorithms. In *Proc. of the Intl. Conference on Acoustics, Speech, and Signal Processing (ICASSP-97)*, volume IV, pages 3361–3364, Munich, Germany, April 1997.

674. J. Karhunen, L. Wang, and J. Joutsensalo. Neural estimation of basis vectors in independent component analysis. In *Proc. of the Int. Conference on Neural Networks*, pages 317–322, Paris, France, Oct. 9–13 1995.

675. J. Karhunen, L. Wang, and R. Vigário. Nonlinear PCA type approaches for source separation and independent component analysis. In *Proc. ICNN'95*, pages 995–1000, Perth, Australia, Nov.-Dec. 1995.

676. P.A. Karjalainen, J.P. Kaipio, A.S. Koistinen, and M. Vauhkonen. Subspace regularization method for the single-trial estimation of evoked potentials. *IEEE Trans. on Biomedical Engineering*, 46:849–860, 1999.

677. J. Karvanen and V. Koivunen. Blind separation methods based on Pearson system and its extensions. *Signal Processing*, Vol. 82, 4: 663-673, 2002.

678. J. Karvanen, J. Eriksson, and V. Koivunen. Maximum likelihood estimation of ICA-model for wide class of source distributions. In *Proceedings of the 2000 IEEE Workshop on Neural Networks for Signal Processing X*, pages 445–454, Sydney, Australia, Dec. 2000.

679. T. Kasami and S. Lin. Coding for a multiple-access channel. *IEEE Trans. Information Theory*, 22(2):129–137, March 1976.

680. M. Kasprzak and A. Cichocki. Hidden image separation from incomplete image mixtures by independent component analysis. In *Proceedings of the 13th International Conference on Pattern Recognition*, pages 394–398 vol.2, Vienna, Austria, 1996. IEEE Comput. Society Press.

681. W. Kasprzak and A. Cichocki. Recurrent least squares learning for quasi-parallel principal component analysis. In *ESANN'96, European Symposium on Artificial Neural Networks*, pages 223–228. D'facto Publications, Brussels, 1996, 1996.

682. M. Kawanabe and N. Murata. Independent component analysis in the presence of Gaussian noise based on estimating functions. In *Proc of the Second Workshop on Independent Component Analysis and Blind Signal Separation*, pages 39–44, 2000.

683. S. M. Kay. *Fundamentals of Statistical Signal Processing.* Prentice Hall, New Jersey, 1993.

684. G. Kechriotis, E. Zervas, and E.S. Manolakos. Using recurrent neural networks for blind equalization of linear and nonlinear communications channels. In *Proc. IEEE MILCOM*, pages 784–788 vol.2, San Diego, CA, USA, 11-14 Oct. 1992, 1992.

685. A.V. Keerthi, A. Mathur, and J.J. Shynk. Direction-finding performance of the multistage CMA array. In *28-th Asilomar Conference Signals, Systems Comp.*, pages 847–852 vol.2. IEEE, 1994.

686. A.V. Keerthi, A. Mathur, and J.J. Shynk. Misadjustment and tracking analysis of the constant modulus array. *IEEE Tr. Signal Processing*, 46(1):51–58, January 1998.

687. C. T. Kelley. Iterative methods for linear and nonlinear equations. *Frontiers in Applied Mathematics 16, SIAM*, 1995.

688. R.A. Kennedy, B.D.O. Anderson, Z. Ding, and C.R. Johnson, Jr. Local stable minima of the Sato recursive identification scheme. In *Proc. IEEE CDC*, pages 3194–3199 vol.6, Honolulu, HI, 1990.

689. R.A. Kennedy and Z. Ding. Blind adaptive equalizers for quadrature amplitude modulated communication systems based on convex cost functions. *Optical Engineering*, 31(6):1189–1199, June 1992.

690. B.H. Khalaj, A. Paulraj, and T. Kailath. 2D RAKE receivers for CDMA cellular systems. In *Proc. Globecom*, volume 1, pages 400–404, San Francisco, CA, November 1994.

691. N. Kikuma, M. Fujimoto, and N. Inagaki. Rapid and stable optimization of CMA adaptive array by Marquardt method. In *Antennas and Propagation Society Symposium 1991 Digest*, pages 102–105 vol.1, London, Ont., Canada, 24-28 June 1991, 1991.

692. N. Kikuma, M. Yamada, and N. Inagaki. Directionally constrained adaptive array using constant modulus algorithm. In *ISAP Japan 1989. Proceedings of the 1989 Int. Conference Antennas and Propagation*, pages 313–316 vol.2, Tokyo, Japan, 22-25 Aug. 1989, 1989. Institute Electron. Inf. Commun. Eng.

693. Y. Kim, S. Kim, and M. Kim. The derivation of a new blind equalization algorithm. *ETRI Journal*, 18(2):53–60, July 1996.

694. A. Klein, G.K. Kaleh, and P.W. Baier. Zero forcing and minimum mean-square-error equalization for multiuser detection in code-division multiple-access channels. *IEEE Trans. Veh. Techn.*, 45(2):276–287, 1996.

695. A. Klein, W. Mohr, R. Thomas, P. Weber, and B. Wirth. Direction of arrival of partial waves in wideband mobile radio channels for intelligent antenna concepts. In *Proc. IEEE Veh. Techn. Conference*, pages 849–853, November 1996.

696. B.U. Koehler, T-W. Lee, and R. Orglmeister. Improving the performance of infomax using statistical signal processing techniques. In *Proceedings International Conference on Artificial Neural Networks (ICANN). Lausanne*, pages 535–540, Oct. 1997.

697. N. Komiya. Ghost reduction by reproduction. *IEEE Trans. Consumer Electronics*, 38(3):195–199, August 1992.

698. K. Kreutz-Delgado and B. D. Rao. Sparse basis selection, ICA and majorization: towards a unified perspective. In *1999 IEEE International Conference on Acoustics, Speech, and Signal Processing*, pages 1081–1084, 1999.

699. H. Krim and M. Viberg. Two decades of array signal processing research: The parametric approach. *IEEE Signal Processing Mag.*, 13(3):67–94, July 1996.

700. V. Krishnamurthi, Y.-H. Liu, S. Bhattacharyya, J.N. Turner, and T.J. Holmes. Blind deconvolution of fluorescence micrographs by maximum-likelihood estimation. *Applied Optics*, 34(29):6633–6647, October 1995.

701. V. Krishnamurthy, S. Dey, and J.P. LeBlanc. Blind equalization of IIR channels using hidden Markov models and extended least squares. *IEEE Trans. Signal Processing*, 43(12):2994–3006, December 1995.

702. M. Krob and M. Benidir. Blind identification of a linear-quadratic mixture: application to quadratic phase coupling estimation. In *IEEE SP Workshop on Higher-Order Stat.*, pages 351–355, South Lake Tahoe, CA, 1993.

703. H. Kubo, K. Murakami, and T. Fujino. Adaptive maximum-likelihood sequence estimation by means of combined equalization and decoding in fading environments. *IEEE Journal on Selected Areas in Communications*, 13(1):102–109, January 1995.

704. D. Kundur and D. Hatzinakos. Blind image restoration via recursive filtering using deterministic constraints. In *Proc. IEEE ICASSP*, pages 2283–2286 vol. 4, Atlanta, GA, 1996.

705. S. Kung and K. Diamantaras. A neural network learning algorithm for adaptive principal component extraction (apex). In *Proc. of the ICASSP-90*, pages 861–864, Albuquerque, NM, 1990.

706. S.Y. Kung, K. Diamantaras, and J. Taur. Neural networks for extracting pure /constrained /oriented principal components. In J.R. Vaccaro, editor, *SVD and Signal Processing*, pages 57–81. Elsevier Science, Amsterdam, 1991.

707. J.L. Lacoume, M. Gaeta, and P.O. Amblard. From order 2 to HOS: new tools and applications. In J. Vandewalle e.a., editor, *Signal Processing VI: Proc. EUSIPCO-92*, pages 91–98 vol.1, Brussels, 1992. Elsevier.

708. J.L. Lacoume, M. Gaeta, and W. Kofman. Source separation using higher order statistics. *Journal of Atmospheric and Terrestrial Physics*, 54(10):1217–1226, 1992.

709. J.L. Lacoume and P. Ruiz. Sources identification: A solution based on cumulants. In *Proc. of the 4th ASSP Workshop on Spectral Estimator and Modeling*, pages 199–203, 1988.

710. J.L. Lacoume and P. Ruiz. Separation of independent sources from correlated inputs. *IEEE Trans. Signal Processing*, 7(1):3074–3078, 1992.

711. J.L. Lacoume and P. Ruiz. Separation of independent sources from correlated inputs. *IEEE Trans. on Signal Processing*, 40(12):3074–3078, Dec. 1992.

712. M. Lagunas, A. Pages-Zamora, and A.Perez-Neira. High order learning in temporal reference array beamforming. In J. Vandevalle, R. Boite, M. Moonen, and A. Oosterlink, editors, *Signal Processing VI: Theories and Applications*, pages 1085–1088. Elsevier, 1992.

713. B. Laheld and J.-F. Cardoso. Adaptive source separation without prewhitening. In *in Proc. EUSIPCO, Edinburgh,1994*, volume SP-43, no. 12, pages 183–186, Dec. 1994.

714. R.H. Lambert. A new method for source separation. In *Int. Conference on Acoustics Speech and Signal Processing*, pages 2116–2119, Detroit, MI, May 9-12 1995.

715. R.H. Lambert. *Multi-Channel Blind Deconvolution: FIR Matrix Algebra and Separation of Multi-Path Mixtures.* PhD thesis, Elec. Eng. Univ. of Southern California, 1996.

716. R.H. Lambert and C.L. Nikias. Optimal equalization cost functions and maximum a posteriori estimation. In *Proc. IEEE Military Comm. Conference Ft. Monmouth NJ*, pages 291–295, Oct. 1994.

717. R.H. Lambert and C.L. Nikias. A sliding cost function algorithm for blind deconvolution. In *29-th Asilomar Conference Signals, Systems, Comp.*, pages 177–181 vol.1. IEEE, 1996.

718. R.G. Lane. Blind deconvolution of speckle images. *Journal of the Optical Society of America A (Optics and Image Science)*, 9(9):1508–1514, September 1992.

719. M.G. Larimore and J.R. Treichler. Convergence behavior of the constant modulus algorithm. In *Proc. IEEE ICASSP*, pages 13–16 vol.1, 1983.

720. L. De Lathauwer. *Signal Processing Based on Multilinear Algebra.* PhD thesis, KU Leuven, Leuven, Belgium, 1997.

721. L. De Lathauwer, P. Comon, B. De Moor, and J. Vandewalle. Higher-order power method – Application in independent component analysis. In *Proc. Int. Symposium Nonlin. Theory and Appl. (NOLTA)*, pages 91–96, Las Vegas, December 1995.

722. L. De Lathauwer, B. De Moor, and J. Vandewalle. Independent component analysis based on higher-order statistics only. In *Proc. IEEE SP Workshop on Stat. Signal Array Processing*, pages 356–359, Corfu, Greece, 1996.

723. L. De Lathauwer, B. De Moor, and J. Vandewalle. A technique for higher-order-only blind source separation. In S. Amari, L. Xu, L.-W. Chan, I. King, and K.-S. Leung, editors, *Progress in Neural Information Processing. Proceedings of the International Conference on Neural Information Processing*, pages 1223–1228 vol.2, Hong Kong, 1996. Springer-Verlag.

724. W.H. Lau, F.L. Hui, S.H. Leung, and A. Luk. Blind separation of impulsive noise from corrupted audio signal. *Electronics Letters*, 32(3):166–168, February 1996.

725. M. Lavielle, É. Moulines, and J.-F. Cardoso. A maximum likelihood solution to DOA estimation for discrete sources. In *Proc. 7th workshop on statistical signal and array processing*, Quebec City, 1994.

726. M. Lavielle, É. Moulines, and J.-F. Cardoso. On a stochastic version approximation of the em algorithm. In *Proc. IEEE SP Workshop on Higher-Order Stat., Aiguablava, Spain*, pages 61–65, 1995.

727. N.F. Law and D.T. Nguyen. Improved convergence of projection based blind deconvolution. *Electronics Letters*, 31(20):1732–1733, September 1995.

728. N.F. Law and D.T. Nguyen. Multiple frame projection based blind deconvolution. *Electronics Letters*, 31(20):1734–1735, September 1995.

729. J.P. LeBlanc, K. Dogancay, R.A. Kennedy, and C.R. Johnson, Jr. Effects of input data correlation on the convergence of blind adaptive equalizers. In *Proc. IEEE ICASSP*, pages III/313–316 vol.3, Adelaide, 1994.

730. J.P. LeBlanc, I. Fijalkow, and C.R. Johnson, Jr. Fractionally-spaced constant modulus algorithm blind equalizer error surface characterization: effects of source distributions. In *Proc. IEEE ICASSP*, pages 2944–2947 vol. 5, Atlanta, GA, 1996.

731. D. D. Lee and H. S. Seung. Learning of the parts of objects by non-negative matrix factorization. Nature, Vol 401, Oct. 1999

732. J. S. Lee, D. D. Lee, S. Choi, K. S. Park, and D. S. Lee. Non-negative matrix factorization of dynamic images in nuclear medicine. IEEE Medical Imaging Conference, San Diego, California, November 4-10, 2001.

733. E. Lee and D. Messerschmitt. *Digital Communication.* Kluwer Publishers, Boston, 1994.

734. G.-H. Lee, R.-H. Park, J.-H. Park, and B.-U. Lee. Shell partition-based constant modulus algorithm. *Journal of the Korean Institute of Telematics and Electronics*, 33B(1):133–143, January 1996.

735. G.-K. Lee, M.P. Fitz, and S.B. Gelfand. Bayesian techniques for equalization of rapidly fading frequency selective channels. *International Journal of Wireless Information Networks*, 2(1):41–54, January 1995.

736. G.-K. Lee, S.B. Gelfand, and M.P. Fitz. Bayesian decision feedback techniques for deconvolution. *IEEE Journal on Selected Areas in Communications*, 13(1):155–166, January 1995.

737. G.-K. Lee, S.B. Gelfand, and M.P. Fitz. Bayesian techniques for blind deconvolution. *IEEE Trans. Communications*, 44(7):826–835, July 1996.

738. T. W. Lee. *Independent Component Analysis: Theory and Applications.* Kluwer Academic Publishers, September 1998.

739. T-W. Lee, A.J. Bell, and R. Lambert. Blind separation of delayed and convolved sources. *Advances in Neural Information Processing Systems*, 9:758–764, 1997.

740. T-W. Lee, A.J. Bell, and R. Orglmeister. Blind source separation of real world signals. In *Proceedings of IEEE International Conference Neural Networks , Houston*, pages 2129–2135, 1997.

741. T-W. Lee, M. Girolami, A.J. Bell, and T.J. Sejnowski. A unifying information-theoretic framework for independent component analysis. *Computers & Mathematics with Applications*, 31(11):1–21, March 2000.

742. T.-W. Lee, M.S. Lewicki, M. Girolami, and T.J. Sejnowski. Blind source separation of more sources using overcomplete representations. *IEEE Signal Processing Letters*, 6(4):87–90, 1999.

743. W. Lee and R.L. Pickholtz. Convergence analysis of a CM array for CDMA systems. *Wireless Personal Communications*, 3(1-2):37–53, 1996.

744. T.K. Leen. Dynamics of learning in feature–discovery networks. *Networks*, 2:85–105, 1991.

745. A. Leshem and A.J. van der Veen. Bounds and algorithm for direction finding of phase modulated signals. In *Proc. IEEE workshop on Stat. Signal Array Processing*, Portland (OR), September 1998.

746. M.S. Lewicki and T.J Sejnowski. Learning overcomplete representation. *Neural Computation*, 12(2):337–365, February 2000.

747. T.R. Lewis and S. Mitra. Application of a blind-deconvolution restoration technique to space imagery. *Proc. SPIE*, 1565:221–226, 1991.

748. D. Li and S. Chen. A fast minimum error probability blind equalization algorithm. *Acta Electronica Sinica*, 23(4):15–19, April 1995.

749. J. Li, B. Halder, P. Stoica, and M. Viberg. Computationally efficient angle estimation for signals with known waveforms. *IEEE Tr. Signal Processing*, 43(9):2154–2163, September 1995.

750. J.-N. Li. Blind equalization with partial response signals. *Frequenz*, 49(11-12):253–258, November 1995.

751. J.-N. Li. Carrier phase recovery and maximum kurtosis deconvolution based blind equalization. *Frequenz*, 49(7-8):151–155, July 1995.

752. J.-N. Li. Convergence analysis of maximum kurtosis deconvolution algorithm with conditional Gaussian model. *Frequenz*, 50(1-2):16–20, January 1996.

753. S. Li and T.J. Sejnowski. Adaptive separation of mixed broad-band sound sources with delays by a beamforming Hérault-Jutten network. *IEEE Journal of Oceanic Engineering*, 20(1):73–79, Jan. 1995.

754. T.-H. Li. Blind identification and deconvolution of linear systems driven by binary random sequences. *IEEE Trans. Informat. Th.*, 38(1):26–38, January 1992.

755. T.-H. Li. Blind deconvolution of linear systems with nonstationary discrete inputs. In *IEEE SP Workshop on Higher-Order Stat.*, pages 160–163, South Lake Tahoe, CA, 1993.

756. T.-H. Li and K. Mbarek. A blind equalizer for nonstationary discrete-valued signals. *IEEE Trans. Signal Processing*, 45(1):247–254, January 1997.

757. Y. Li, A. Cichocki, and L. Zhang. Blind separation of digital sources. In *Proceedings of the 8-th International Conference on Neural Information Processing (ICONIP'2001)*, pages 1018–1022, Shanghai, China, Nov. 14-18 2001.

758. Y. Li and Z. Ding. Global convergence of fractionally spaced Godard equalizers. In *28-th Asilomar Conference Signals, Systems Comp.*, pages 617–621. IEEE, 1994.

759. Y. Li and Z. Ding. Global convergence of fractionally spaced Godard (CMA) adaptive equalizers. *IEEE Trans. Signal Processing*, 44(4):818–826, April 1996.

760. Y. Li and K.J. Ray Liu. Learning characteristics for general class of adaptive blind equalizer. In *Proc. IEEE Int. Conference Communic.*, pages 1020–1024 vol.2, Dallas, TX, 1996.

761. Y. Li and K.J. Ray Liu. On blind equalization of MIMO channels. In *Proc. IEEE Int. Conference Communic.*, pages 1000–1004 vol.2, Dallas, TX, 1996.

762. Y. Li and K.J.R. Liu. Static and dynamic convergence behavior of adaptive blind equalizers. *IEEE Trans. Signal Processing*, 44(11):2736–2745, November 1996.

763. Y. Li and K.J.R. Liu. Blind adaptive spatial-temporal equalization algorithms for wireless communications using antenna arrays. *IEEE Communications Letters*, 1(1):25–27, January 1997.

764. Y. Li, K.J.R. Liu, and Z. Ding. Length- and cost-dependent local minima of unconstrained blind channel equalizers. *IEEE Trans. Signal Processing*, 44(11):2726–2735, November 1996.

765. Y. Li, J. Wang, and J.M. Zurada. Blind extraction of singularly mixed source signals. *IEEE Trans. on Neural Networks*, 11(6):1413–1422, 2000.

766. Q.-L. Liang, C. Zhao, Z. Zhou, and Z. Liu. A convex algorithm for blind equalization based on neural network. In Y. Zhong, Y. Yang, and M. Wang, editors, *Proceedings of International Conference on Neural Information Processing (ICONIP '95)*, pages 237–240 vol.1, Beijing, China, 1995. Publishing House of Electron. Ind.

767. Q.-L. Liang, Z. Zhou, and Z.-M. Lin. A new approach to global minimum and its applications in blind equalization. In *ICNN 96. The 1996 IEEE International Conference on Neural Networks*, pages 2113–2117 vol.4, Proceedings of International Conference on Neural Networks (ICNN'96), Washington, DC, USA, 3-6 June 1996, 1996. IEEE.

768. Q.-L. Liang, Z. Zhou, and Z. Liu. Blind equalization using a hybrid algorithm of multilayer neural network. *High Technology Letters (English Language Edition)*, 2(1):47–50, June 1996.

769. A. P. Liavas, P. A. Regalia, and J.-P. Delmas. Blind channel approximation: Effective channel order determination. *IEEE Trans. Signal Processing*, 47:3336–3344, 1999.

770. J. Lin, F. Ling, and J.G. Proakis. Joint data and channel estimation for TDMA mobile channels. In *IEEE PIMRC '92*, pages 235–239, Boston, MA, 1992.

771. U. Lindgren and A.J. van der Veen. Source separation based on second order statistics — an algebraic approach. In *Proc. IEEE SP Workshop on Stat. Signal and Array Processing*, pages 324–327, Corfu, June 1996.

772. U. Lindgren, T. Wigren, and H. Broman. On local convergence of a class of blind separation algorithms. *IEEE Trans. Signal Processing*, 43:3054–3058, December 1995.

773. B.G. Lindsay. Using empirical partially Bayes inference for increased efficiency. *Ann. Statist.*, pages 914–931, 1985.

774. X. Ling. Blind separation of wide-band signals for the near-field situations. *Acta Electronica Sinica*, 24(7):87–92, July 1996.

775. X.-T. Ling, Y.-F. Huang, and R. Liu. A neural network for blind signal separation. In *Proc. of IEEE Int. Symposium on Circuits and Systems (ISCAS-94)*, pages 69–72, New York, NY, 1994. IEEE Press.

776. R. Linsker. Self-organization in a perceptual network. *Computer*, 21:105–117, 1988.

777. B. Liu, L. Wu, J. Zhang, and Q. Wang. Object recognition using both range and intensity image in machine vision. *Proc. SPIE*, 2665:91–94, 1996.

778. H. Liu and G. Xu. A deterministic approach to blind symbol estimation. *IEEE Signal Processing Letters*, 1(12):205–207, December 1994.

779. H. Liu and G. Xu. Closed-form blind symbol estimation in digital communications. *IEEE Trans. Signal Processing*, 43(11):2714–2723, November 1995.

780. H. Liu and G. Xu. Multiuser blind channel estimation and spatial channel pre-equalization. In *Proc. IEEE ICASSP*, pages 1756–1759 vol.3, Detroit, May 1995.

781. H. Liu and G. Xu. Blind equalization for CDMA systems with aperiodic spreading sequences. In *Proc. IEEE ICASSP*, volume 5, Atlanta (GA), 1996.

782. H. Liu and G. Xu. A subspace method for signature waveform estimation in synchronous CDMA systems. In *29-th Asilomar Conference Signals, Systems, Comp.*, pages 157–161 vol.1. IEEE, 1996.

783. H. Liu and G. Xu. A subspace method for signature waveform estimation in synchronous CDMA systems. *IEEE Trans. Communications*, 44(10):1346–1354, October 1996.

784. H. Liu, G. Xu, and L. Tong. A deterministic approach to blind equalization. In *27-th Asilomar Conference on Signals, Systems and Computers*, pages 751–755 vol.1, 1993.

785. H. Liu, G. Xu, L. Tong, and T. Kailath. Recent developments in blind channel equalization: From cyclostationarity to subspaces. *Signal Processing*, 50(1-2):83–99, April 1996.

786. H. Liu and M.D. Zoltowski. Blind equalization in antenna array CDMA systems. *IEEE Trans. Signal Processing*, 45(1):161–172, January 1997.

787. R.-W. Liu. Blind signal processing: an introduction. In *Proc. IEEE ISCAS*, pages 81–84 vol.2, Atlanta, GA, 1996.

788. X. Liu and J. Hérault. Colour image processing by a neural network model. In *International Neural Network Conference*, pages 3–6, 1990.

789. L. Ljung and T. Söderström. *Theory and Practice of Recursive Identification.* M.I.T. Press, Cambridge, MA, 1983.

790. Y. Long, L. Yanda, and C. Tong. Semi-blind deconvolution using magnitude spectrum or multichannel information. *Acta Electronica Sinica*, 20(2):71–75, February 1992.

791. F. Lopez de Victoria. An adaptive blind equalization algorithm for QAM and QPR modulations: The concentric ordered modulus algorithm, with results for 16 QAM and 9 QPR. In *25-th Asilomar Conference Signals, Systems Comp.*, pages 726–730 vol.2. IEEE, 1991.

792. F. Lopez de Victoria. More on the concentric ordered modulus algorithm for blind equalization of QAM and QPR modulations, with results for 64 QAM, 25 QPR and 49 QPR. In *Proc. IEEE ICASSP*, pages 497–500 vol.4, San Francisco, CA, USA, 23-26 March 1992, 1992.

793. Y. Lou. Adaptive blind equalization criteria and algorithms for QAM data communication systems. In *Ninth Annual International Phoenix Conference on Computers and Communications*, pages 222–229, Scottsdale, AZ, 1990. IEEE Comput. Society Press.

794. Y. Lou. Channel estimation standard and adaptive blind equalization. In *Proc. IEEE ISCAS*, pages 505–508 vol.2, San Diego, CA, 1992.

795. Y. Lou. Channel estimation standard and adaptive blind equalization. *IEEE Trans. Communications*, 43(2-4,):182–186, February 1995.

796. P. Loubaton and P. Regalia. Blind deconvolution of multivariate signals by using adaptive fir lossless filters. In *Signal Processing V: Theories and Applications (Proc. of the EUSIPCO-92)*, pages 1061–1064, Bruxelles, Belgium, 1992. Elsevier.

797. R. W. Lucky. Techniques for adaptive equalization of digital communication systems. *Bell Systems Technical Journal*, 45:255–286, 1966.

798. J. Lundell and B. Widrow. Application of the constant modulus adaptive beamformer to constant and non-constant modulus algorithms. In *21-th Asilomar Conference on Signals, Systems, and Computers*, pages 432–436 vol.1. IEEE, November 1987.

799. D.-S. Luo and A.E. Yagle. Lattice algorithms applied to the blind deconvolution problem. In *Proc. IEEE ICASSP*, pages 2341–2344 vol.4, Glasgow, UK, 1989.

800. H. Luo, R.-W. Liu, and Y. Li. Internal structure identification for layered medium. In *Proc. IEEE ISCAS*, pages 161–164 vol.2, Atlanta, GA, 1996.

801. R. Lupas and S. Verdu. Near-far resistance of multiuser detectors in asynchronous channels. *IEEE Tr. Comm.*, 38(4):496–508, April 1990.

802. S.P. Luttrell. Self-supervised adaptive networks. *IEE Proc. F*, 139(6):371–377, 1992.

803. X. Ma and C.L. Nikias. On blind channel identification for impulsive signal environments. In *Proc. IEEE ICASSP*, pages 1992–1995 vol.3, 1995 International Conference on Acoustics, Speech, and Signal Processing, Detroit, MI, USA, 9-12 May 1995, 1995.

804. X. Ma and C.L. Nikias. Parameter estimation and blind channel identification in impulsive signal environments. *IEEE Trans. Signal Processing*, 43(12):2884–2897, December 1995.

805. O. Macchi and E. Moreau. Self-adaptive source separation part I: convergence analysis of a direct linear network controlled by Hérault-Jutten algorithm. *IEEE Trans. on Signal Processing.*, 1997.

806. F.R. Magee and J.G. Proakis. Adaptive maximum-likelihood sequence estimation for signaling in the presence of intersymbol interference. *IEEE Trans. Information Theory*, 19(1):120–124, Jan 1973.

807. S. Makeig, A. Bell, T.-P. Jung, and T.J. Sejnowski. Independent component analysis in electroencelographic data. In *Advances in Neural Information Processing Systems Cambridge 1996*, pages 145–151. MIT Press, 1996.

808. S. Makeig, T.-P. Jung, A.J. Bell, D. Ghahremani, and T.J. Sejnowski. Blind separation of auditory event-related brain responses into independent components. In *Proceedings of the National Academy of Sciences*, volume 94, pages 10979–10984, 1997.

809. S. Makino, R. Mukai, S. Araki, and S. Katagiri. Separation of speech signal - to realize multiple talker speech recognition (in Japanese). *NTT R and D*, 50:937–944, 2001.

810. R. Makowski. A blind deconvolution method. In L. Torres e.a., editor, *Signal Processing VI: Proc. EUSIPCO-90*, pages 1959–1962 vol.3, Barcelona, Spain, 1990. Elsevier.

811. R. Makowski. A new model-based blind deconvolution method. *Advances in Modelling & Simulation*, 25(1):21–29, 1991.

812. R. Makowski. Blind deconvolution method based on resonance model of wave propagation. In J.L. Lacoume, editor, *Proc. Higher Order Statistics*, pages 149–152, Chamrousse, France, 1992. Elsevier.

813. H.A. Malki and A. Moghaddamjoo. Using the Karhunen-Loeve Transformation in the back-propagation training algorithm. *IEEE Trans. on Neural Networks*, 2:162–165, 1991.

814. R. Mann and K.D. Kammeyer. A pole-zero-tracking constant modulus algorithm. In *Proc. IEEE ICASSP*, pages 1227–1230 vol.2, Glasgow, UK, 23-26 May 1989, 1989.

815. R. Mann, W. Tobergte, and K.D. Kammeyer. Applications of constant modulus algorithms for adaptive equalization of time-varying multipath channels. In V. Cappellini and A.G. Constantinides, editors, *Digital Signal Processing - 87. Proceedings of the International Conference*, pages 421–425, Florence, Italy, 7-10 Sept. 1987, 1987. North-Holland.

816. A. Mansour and C. Jutten. Fourth-order criteria for blind sources separation. *IEEE Trans. on Signal Processing*, 43(8):2022–2025, Aug. 1995.

817. A. Mansour and C. Jutten. A direct solution for blind separation of sources. *IEEE Trans. Signal Processing*, 44(3):746–748, March 1996.

818. S. Marcos, S. Cherif, and M. Jaidane. Blind cancellation of intersymbol interference in decision feedback equalizers. In *Proc. IEEE ICASSP*, pages 1073–1076 vol.2, Detroit, MI, 1995.

819. G. Marques and L.B. Almeida. An objective function for independence. In *ICNN 96. The 1996 IEEE International Conference on Neural Networks*, pages 453–457 vol.1, Proceedings of International Conference on Neural Networks (ICNN'96), Washington, DC, USA, 3-6 June 1996, 1996. IEEE.

820. J.K. Martin and A.K. Nandi. Blind system identification using second, third and fourth order cumulants. *Journal of the Franklin Institute*, 333B(1):1–13, January 1996.

821. M. Martone. Blind multichannel deconvolution in multiple access spread spectrum communications using higher order statistics. In *Proc. IEEE Int. Conference Communic.*, pages 49–53 vol.1, Seattle, WA, 1995.

822. G. Mathew and V. Reddy. Orthogonal eigensubspace estimation using neural networks. *IEEE Trans. on Signal Processing*, 42:1803–1811, 1994.

823. J. Mathews and A. Cichocki. Total least-squares estimation. Technical report, University Utah, USA and Brain Science Institute Riken, Japan, 2000.

824. A. Mathur, A.V. Keerthi, and J.J. Shynk. Cochannel signal recovery using MUSIC algorithm and the constant modulus array. *IEEE Signal Processing Letters*, 2(10):191–194, October 1995.

825. A. Mathur, A.V. Keerthi, and J.J. Shynk. Estimation of correlated cochannel signals using the constant modulus array. In *Proc. IEEE Int. Conference on Communic.*, pages 1525–1529 vol.3, Seattle, WA, 1995.

826. A. Mathur, A.V. Keerthi, J.J. Shynk, and R.P. Gooch. Convergence properties of the multistage constant modulus array for correlated sources. *IEEE Tr. Signal Processing*, 45(1):280–286, January 1997.

827. K. Matsuoka and M. Kawamoto. A neural net for blind separation of nonstationary signal sources. In *IEEE Symposium on Circuits and Systems*, pages 221–226, 1994.

828. K. Matsuoka, M. Ohya, and M. Kawamoto. A neural net for blind separation of nonstationary signals. *Neural Networks*, 8(3):411–419, 1995.

829. M.A. Mayorga and L.C. Ludemann. Shift and rotation invariant texture recognition with neural nets. In *Proceedings of ICNN'94*, pages 4078–4083, Los Alamitos, CA, USA, 1994. IEEE Publ.

830. S. Mayrargue. Spatial equalization of a radio-mobile channel without beamforming using the constant modulus algorithm (CMA). In *Proc. IEEE ICASSP*, volume III, pages 344–347, 1993.

831. S. Mayrargue. A blind spatio-temporal equalizer for a radio-mobile channel using the constant modulus algorithm (CMA). In *Proc. IEEE ICASSP*, pages III/317–320 vol.3, Adelaide, 1994.

832. B.C. McCallum. Blind deconvolution by simulated annealing. *Optics Communications*, 75(2):101–105, February 1990.

833. B.C. McCallum and J.M. Rodenburg. Simultaneous reconstruction of object and aperture functions from multiple far-field intensity measurements. *Journal of the Optical Society of America A (Optics and Image Science)*, 10(2):231–239, February 1993.

834. F. McCarthy. Multiple signal direction-finding and interference reduction techniques. In *Wescon/93 Conference Record*, pages 354–361, San Francisco, CA, September 1993.

835. M. McKeown, T.P. Jung, S. Makeig, G. Brown, S. Kindermann, T-W. Lee, and T. Sejnowski. Transiently task-related human brain activations revealed by independent component analysis. *Proceedings of the National Academic of Sciences*, 1997.

836. M. McKeown, T.P. Jung, S. Makeig, G. Brown, S. Kindermann, T-W. Lee, and T.J. Sejnowski. Spatially independent activity patterns in functional magnetic resonance imaging data during the stroop color-naming task. *Proceedings of the National Academy of Sciences*, 95:803–810, 1998.

837. S. McLaughlin, A. Stogioglou, and J. Fackrell. Introducing higher order statistics (HOS) for detection of nonlinearities. *UK Nonlinear News*, 2(2), Sep. 1995.

838. J.M. Mendel. Tutorial on higher-order statistics (spectra) in signal processing and system theory: theoretical results and some applications. *Proc. IEEE*, 79(3):278–305, March 1991.

839. J.M. Mendel and M.C. Dogan. Higher-order statistics applied to some array signal processing problems. In M. Blanke and T. Soderstrom, editors, *System Identification (SYSID '94). A Postprint Volume from the IFAC Symposium*, pages 101–106 vol.1, Copenhagen, Denmark, 1995. Pergamon.

840. R. Mendoza, J.H. Reed, and T.C. Hsia. Interference rejection using a hybrid of a constant modulus algorithm and the spectral correlation discriminator. In *Proc. IEEE MILCOM*, pages 491–497 vol.2, Boston, 1989. IEEE.

841. R. Mendoza, J.H. Reed, T.C. Hsia, and B.G. Agee. Interference rejection using the time-dependent constant modulus algorithm (CMA) and the hybrid CMA/spectral correlation discriminator. *IEEE Trans. Signal Processing*, 39(9):2108–2111, September 1991.

842. Y. Meng, K.-M. Wong, and K. Lazaris-Brunner. An improved cyclic adaptive beamforming (CAB) algorithm for a multiple agile beam mobile communication system. In *Proc. IEEE SP Workshop on Stat. Signal Array Processing*, pages 148–151, Corfu, Greece, 1996.

843. H. Messer and Y. Bar-Ness. Bootstrapped spatial separation of wideband superimposed signals. In *Signal Processing VI: Theories and Applications*, pages 819–822. Elsevier, 1992.

844. Y. Miao and Y. Hua. Fast subspace tracking and neural network learning by a novel information criterion. *IEEE Trans. on Signal Processing*, 46(7):1967–1979, 1998.

845. T. P. Minka Automatic choice of dimensionality for PCA In *NIPS 13* In Todd K. Leen, Thomas G. Dietterich, and Volker Tresp, editors, Ad- vances in Neural Information Processing Systems *NIPS 13*, pages 598-604, MIT Press, 2001, See alsoTechnical Report 514, MIT Media Laboratory - Perceptual Computing Section, 2000.

846. N. Miura and N. Baba. Extended-object reconstruction with sequential use of the iterative blind deconvolution method. *Optics Communications*, 89(5-6):375–379, May 1992.

847. N. Miura and N. Baba. Segmentation-based multiframe blind deconvolution of solar images. *Journal of the Optical Society of America A (Optics, Image Science and Vision)*, 12(9):1858–1866, September 1995.

848. N. Miura, N. Baba, S. Isobe, M. Noguchi, and Y. Norimoto. Binary star reconstruction with use of the blind deconvolution method. *Journal of Modern Optics*, 39(5):1137–1146, May 1992.

849. C. Mokbel, D. Jouvet, and J. Monne. Deconvolution of telephone line effects for speech recognition. *Speech Communication*, 19(3):185–196, September 1996.

850. L. Molgedey and H.G. Schuster. Separation of a mixture of independent signals using time delayed correlations. *Physical Review Letters*, 72(23):3634–3637, 1994.

851. J. Moody and L.-W. Wu. What is the "true price"?-state space models for high frequency financial data. In S. Amari, L. Xu, L.-W. Chan, I. King, and K.-S. Leung, editors, *Progress in Neural Information Processing. Proceedings of the International Conference on Neural Information Processing*, pages 697–704 vol.2, Hong Kong, 1996. Springer-Verlag.

852. Y.S. Moon, M. Kaveh, and L.B. Nelson. Subspace-based detection for CDMA communications. In *Proc. IEEE ICASSP*, 1998.

853. M. Moonen and E. Deprettere. Parallel filter structures for RLS-type blind equalization algorithms. *Proc. SPIE*, 2563:442–448, 1995.

854. E. Moreau and O. Macchi. Separation auto-adaptive de sources sans blanchment prealable. In *Proc. of the 14th Colloque Gretsi*, pages 325–328, Juan-les-Pins, 13-16 Sept. 1993. In French.

855. E. Moreau and O. Macchi. Two novel architectures for the self adaptive separation of signals. In *Proc. of the IEEE Conference on Communication*, volume 2, pages 1154–1159, Geneva, Switzerland, May 1993. IEEE.

856. E. Moreau and O. Macchi. Complex self-adaptive algorithms for source separation based on high order contrasts. In *Signal Processing VII: Theories and Applications (Proc. of the EUSIPCO-94)*, pages 1157–1160, Edinburgh, Scotland, Sept. 13-16 1994. Elsevier.

857. E. Moreau and O. Macchi. A one stage self-adaptive algorithm for source separation. In *International Conference on Acoustics, Speech and Signal Processing*, volume III, pages 49–52, 1994.

858. E. Moreau and O. Macchi. High order contrasts for self-adaptive source separation. *Int. Journal of Adaptive Control and Signal Processing*, 10 No.1:19–46, Jan 1996.

859. E. Moreau and O. Macchi. New self-adaptive algorithms for source separation based on contrast functions. In *Proc. of IEEE Signal Processing Workshop on Higher Order Statistics*, volume Lake Tahoe, USA, pages 215–219, New York, NY, June, 1993.

860. E. Moreau and N. Thirion. Multichannel blind signal deconvolution using high order statistics. In *Proc. IEEE SP Workshop on Stat. Signal Array Processing*, pages 336–339, Corfu, Greece, 1996.

861. É. Moulines and J.-F. Cardoso. Second-order versus fourth-order MUSIC algorithms. An asymptotical statistical performance analysis. In *Proc. Int. Workshop on Higher-Order Stat., Chamrousse, France*, pages 121–130, 1991.

862. É. Moulines and J.-F. Cardoso. Direction finding algorithms using fourth-order statistics. Asymptotic performance analysis. In *Proc. ICASSP*, pages 437–440, March 1992.

863. É. Moulines, J.-F. Cardoso, A. Gorokhov, and P. Loubaton. Subspace methods for blind identification of SIMO-FIR systems. In *Proc. ICASSP*, volume 5, pages 2449–52, 1996.

864. É. Moulines, K. Choukri, and J.-F. Cardoso. Time-domain procedures for testing that a time-series is Gaussian. In *Proc. 7th workshop on statistical signal and array processing*, Quebec City, 1994.

865. É. Moulines, P. Duhamel, J.-F. Cardoso, and S. Mayrargue. Subspace methods for the blind identification of multichannel FIR filters. *IEEE Trans. Signal Processing*, 43(2):516–525, February 1995.

866. N. Murata, S. Ikeda, and A. Ziehe. *An approach to blind source separation based on temporal structure of speech signals.* BSI RIKEN, Japan, 1998. BSIS Technical Reports No.98-2.

867. N. Murata, S. Ikeda, and A. Ziehe. An approach to blind source separation based on temporal structure of speech signal. *Neurocomputing*, 41(4):1–24, 2001.

868. N. Murata, K. Mueller, A. Ziehe, and S. Amari. Adaptive on-line learning in changing environments. In *NIPS-96*, volume 9. MIT Press, 1997.

869. M.K. Murray and J.W. Rice. *Differential Geometry and Statistics.* New York: Chapman & Hall, 1993.

870. H.L. N.-Thi and C. Jutten. Blind source separation for convolved mixtures. *Signal Processing*, 45 No.2:209–229, 1995.

871. J.-P. Nadal and N. Parga. Nonlinear neurons in the low-noise limit: a factorial code maximizes information transfer. *NETWORK*, 5:565–581, 1994.

872. J.-P. Nadal and N. Parga. Redundancy reduction and independent component analysis: Conditions on cumulants and adaptive approaches. *Neural Computation*, Oct. 1997.

873. J.P. Nadal and N. Parga. ICA: Conditions on cumulants and information theoretic approach. In *Proc.European Symposium on Artificial Neural Networks ESANN'97 Bruges*, pages 285–290, April 16-17 1997.

874. J. Nagy and Z. Strakos. Enforcing nonnegativity in image reconstruction algorithms. In *Mathematical Modeling, Estimation, and Imaging, David C. Wilson, et.al., Eds.*, pages 182–190, 2000.

875. M. Nájar, M.A. Lagunas, and I. Bonet. Blind wideband source separation. In *International Conference on Acoustics, Speech and Signal Processing*, volume IV, pages 65–68, 1994.

876. N. Nakajima. Blind deconvolution of a Hermitian and non-Hermitian function. *Journal of the Optical Society of America A (Optics and Image Science)*, 8(5):808–813, May 1991.

877. N. Nakajima. Blind deconvolution using the maximum likelihood estimation and the iterative algorithm. *Optics Communications*, 100(1-4):59–66, July 1993.

878. A.K. Nandi, D. Mampel, and B. Roscher. Comparative study of deconvolution algorithms with applications in non-destructive testing. In *IEE Colloquium on 'Blind Deconvolution - Algorithms and Applications' (Ref. No.1995/145)*, pages 1/1–6, London, UK, 1995. IEE.

879. A.K. Nandi and V. Zarzoso. Fourth-order cumulant based blind source separation. *IEEE Signal Processing Letters*, 3(12):312–314, December 1996.

880. A. Neri, G. Scarano, and G. Jacovitti. Bayesian iterative method for blind deconvolution. *Proc. SPIE*, 1565:196–208, 1991.

881. B.B. Newman and J. Hildebrandt. Blind image restoration. *Australian Computer Journal*, 19(3):126–133, August 1987.

882. Y. Nishimori. Learning algorithm for ICA by geodesic flows and on orthogonal group. *In Proc. International Joint Conference on Neural Networks (IJCNN'99)*, Washington DC, 10-16 July 1999, Vol.2, pp.933-938.

883. B.-C. Ng, M. Cedervall, and A. Paulraj. A structured channel estimator for maximum-likelihood sequence detection. *IEEE Comm. Letters*, 1(2):52–55, March 1997.

884. T. Nguyen and Z. Ding. Blind CMA beamforming for narrowband signals with multipath arrivals. *Int. Journal Adaptive Control and Signal Processing*, 12(2):157–172, March 1998.

885. H. Niemann and J.-K. Wu. Neural network adaptive image coding. *IEEE Trans. on Neural Networks*, 4:615–627, 1993.

886. S. Niijima and S. Ueno. MEG source estimation using the fourth order MUSIC. *IEICE Trans. on Information and Systems*, E85-D:167–174, 2002.

887. C. Nikias and J. Mendel. Signal processing with higher-order spectra. *IEEE Signal Processing Magazine*, pages 10–37, July 1993.

888. C.L. Nikias. Blind deconvolution using higher-order statistics. In J.L. Lacoume, editor, *Higher Order Statistics. Proceedings of the International Signal Processing Workshop*, pages 49–56, Chamrousse, France, 1992. Elsevier.

889. C.L Nikias and A.P Petropulu. *Higher-Order Spectra Analysis: A Non-Linear Signal Processing Framework.* Prentice-Hall, Englewood Cliffs, NJ, 1993.

890. K. Nishimori, N. Kikuma, and N. Inagaki. The differential CMA adaptive array antenna using an eigen-beamspace system. *IEICE Trans. Communications*, E78-B(11):1480–1488, November 1995.

891. T. Nomura, M. Eguchi, H. Niwamoto, H. Kokubo, and M. Miyamoto. An extension of the Hérault-Jutten network to signals including delays for blind separation. In S. Usui, Y. Tohkura, S. Katagiri, and E. Wilson, editors, *Neural Networks for Signal Processing VI. Proceedings of the 1996 IEEE Signal Processing Society Workshop*, pages 443–452, Neural Networks for Signal Processing VI. Proceedings of the 1996 IEEE Signal Processing Society Workshop, Kyoto, Japan, 4-6 Sept. 1996, 1996. IEEE.

892. S.J. Nowlan and G.E. Hinton. A soft decision-directed LMS algorithm for blind equalization. *IEEE Trans. Communications*, 41(2):275–279, February 1993.

893. R. J. Ober. Balanced canonical forms. In S. Bittanti and G. Picci, editors, *Identification, Adaptation, Learning*, NATO ASI Series, pages 120–183. Springer, 1996.

894. D. Obradovic and G. Deco. Linear feature extraction in non-Gaussian networks. In *Proc. of the 1995 World Congress on Neural Networks (WCNN'95)*, volume 1, pages 523–526, Washington, D.C., USA, July 17-21 1995. INNS.

895. H. Oda and Y. Sato. High speed convergence of blind equalization. *Trans. of the Institute of Systems, Control and Information Engineers*, 6(7):305–318, July 1993.

896. K.-N. Oh and Y.-O. Chin. Modified constant modulus algorithm: blind equalization and carrier phase recovery algorithm. In *Proc. IEEE Int. Conference Communic.*, pages 498–502 vol.1, Seattle, WA, 1995.

897. K.-N. Oh and Y.-O. Chin. A new dual-mode blind equalization algorithm combining carrier phase recovery. *Journal of the Korean Institute of Telematics and Electronics*, 32A(5):14–23, May 1995.

898. K.-N. Oh and Y.-O. Chin. Blind decision feedback equalizer using dual-variance Gaussian clustering algorithm for wireless communications. In *Proc. IEEE VTC*, pages 696–700 vol.2, Atlanta, GA, 1996.

899. T. Ohgane. Characteristics of CMA adaptive array for selective fading compensation in digital land mobile radio communications. *Electronics and Communications in Japan, Part 1 (Communications)*, 74(9):43–53, September 1991.

900. T. Ohgane, N. Matsuzawa, T. Shimura, M. Mizuno, and H. Sasaoka. BER performance of CMA adaptive array for high-speed GMSK mobile communication-a description of measurements in central Tokyo. *IEEE Trans. Vehicular Technology*, 42(4):484–490, November 1993.

901. T. Ohgane, H. Sasaoka, N. Matsuzawa, and T. Shimura. BER performance of CMA adaptive array for a GMSK/TDMA system-a description of measurements in central Tokyo. In *Vehicular Technology Society 42nd VTS Conference. Frontiers of Technology. From Pioneers to the 21st Century*, pages 1012–1017 vol.2, Denver, CO, USA, 10-13 May 1992, 1992. IEEE.

902. T. Ohgane, H. Sasaoka, N. Matsuzawa, K. Takeda, and T. Shimura. A development of GMSK/TDMA system with CMA adaptive array for land mobile communications. In *41st IEEE Vehicular Technology Conference. Gateway to the Future Technology in Motion*, pages 172–177, St. Louis, MO, USA, 19-22 May 1991, 1991. IEEE.

903. T. Ohgane, T. Shimura, N. Matsuzawa, and H. Sasaoka. An implementation of a CMA adaptive array for high speed GMSK transmission in mobile communications. *IEEE Trans. Vehicular Technology*, 42(3):282–288, August 1993.

904. E. Oja. A simplified neuron model as a principal component analyzer. *Journal of Mathematical Biology*, 16:267–273, 1982.

905. E. Oja. *Subspace Methods of Pattern Recognition.* Research Studies Press and J. Wiley, Letchworth, England, 1983.

906. E. Oja. Neural networks, principal components, and subspaces. *Int. Journal on Neural Systems*, 1:61–68, 1989.

907. E. Oja. Principal components, minor components and linear neural networks. *Neural Networks*, 5:927–935, 1992.

908. E. Oja. The nonlinear PCA learning rule and signal separation – mathematical analysis. *Helsinki Univ. of Technology, Lab. of Computer and Information Science, Report A26*, 1995.

909. E. Oja. The nonlinear PCA learning rule in independent component analysis. *Neurocomputing*, 17(1):25–46, 1997.

910. E. Oja and A. Hyvärinenn. Blind signal separation by neural networks. In S. Amari, L. Xu, L.-W. Chan, I. King, and K.-S. Leung, editors, *Progress in Neural Information Processing.*

Proceedings of the International Conference on Neural Information Processing, pages 7–14 vol.1, Hong Kong, 1996. Springer-Verlag.

911. E. Oja and J. Karhunen. On stochastic approximation of the eigenvectors and eigenvalues of the expectation of a random matrix. *Journal of Math. Analysis and Applications*, 106:69–84, 1985.

912. E. Oja and J. Karhunen. *Computational Intelligence*, chapter Signal separation by nonlinear Hebbian learning. IEEE Press, 1995.

913. E. Oja and J. Karhunen. Signal separation by nonlinear Hebbian learning. In M. Palaniswami et al., editors, *Computational Intelligence – A Dynamic System Perspective*, pages 83–97. IEEE Press, 1995.

914. E. Oja, J. Karhunen, and A. Hyvärinen. From neural principal components to neural independent components. In *Proc. Int. Conference on Artificial Neural Networks*, Lausanne, Switzerland, 1997.

915. E. Oja, J. Karhunen, L. Wang, and R. Vigário. Principal and independent components in neural networks – recent developments. In *Proc. of the 7th Italian Workshop on Neural Networks (WIRN-95)*, page 20, Vietri sul Mare, Italy, May 1995.

916. E. Oja, H. Ogawa, and J. Wangviwattana. Principal component analysis by homogeneous neural networks, part ii: Analysis and extensions of the learning algorithms. *IEICE Trans. Inf. & Systems*, E75–D(3):376–382, 1992.

917. E. Oja and L.-Y. Wang. Neural fitting: Robustness by anti-Hebbian learning. *Neurocomputing*, 12:155–170, 1996.

918. E. Oja and L.-Y. Wang. Robust fitting by nonlinear neural units. *Neural Networks*, 9:435–444, 1996.

919. R. Ojeda and J.-F. Cardoso. Non linearity tests for time series using the invariance principle. In *Proc. Int. Symposium on non-linear theory and its applications*, pages 817–820, 1995.

920. D. P. O'Leary. Near-optimal parameters for Tikhonov and other regularization methods. *SIAM J. Sci. Comput.*, 23(4):1161–1171, 2001.

921. B.A. Olshausen and D.J Field. Emergence of simple-cell receptive field properties by learning a sparse code for natural images. *Nature*, 381:607–609, 1996.

922. S. J. Orfanidis. *Optimum Signal Processing: An Introduction.* McGraw-Hill, New York, 2 edition, February 1988.

923. J. Orr and K.-R. Mller. *Neural Networks: Tricks of the Trade.* Springer, Heidelberg, 1998.

924. G.C. Orsak and S.C. Douglas. Code-length-based universal extraction for blind signal separation. In *to be presented at IEEE International Conference Acoustics, Speech, Signal Processing*, Istanbul, Turkey, June 2000.

925. S. Osowski and A. Cichocki. Learning in dynamic neural networks using signal flow graphs. *Int. Journal of Circuit Theory and Applications*, 27:209–228, April 1999.

926. A.J. O'Toole, H. Abdi, K.A. Deffenbacher, and D. Valentin. Low–dimensional representation of faces in higher dimensions of the face space. *Journal of the Optical Society of America A*, 10(3):405–411, 1993.

927. B. Ottersten. Array processing for wireless communications. In *Proc. IEEE workshop on Stat. Signal Array Processing*, pages 466–473, Corfu, June 1996.

928. K. Pahlavan and A.H. Levesque. Wireless data communications. *Proc. IEEE*, 82(9):1398–1430, September 1994.

929. P. Pajunen. Nonlinear independent component analysis by self-organizing maps. In C. von der Malsburg, W. von Seelen, J.C. Vorbruggen, and B. Sendhoff, editors, *Artificial Neural Networks - ICANN 96. 1996 International Conference Proceedings*, pages 815–820, Bochum, Germany, 1996. Springer-Verlag.

930. P. Pajunen. Blind separation of binary sources with less sensors than sources. In *Proc. 1997 Int. Conference on Neural Networks*, volume 3, pages 1994–1997, Houston, Texas, USA, June 1997.

931. P. Pajunen. A competitive learning algorithm for separating binary sources. In *Proc. European Symposium on Artificial Neural Networks (ESANN'97)*, pages 255–260, Bruges, Belgium, April 1997.

932. P. Pajunen, A. Hyvärinen, and J. Karhunen. Nonlinear blind source separation by self-organizing maps. In S. Amari, L. Xu, L.-W. Chan, I. King, and K.-S. Leung, editors, *Progress in Neural Information Processing. Proceedings of the International Conference on Neural Information Processing*, pages 1207–1210 vol.2, Hong Kong, 1996. Springer-Verlag.

933. P. Pajunen and J. Karhunen. A maximum likelihood approach to nonlinear blind source separation. In *Proc. Int. Conference on Artificial Neural Networks (ICANN'97)*, pages 541–546, Lausanne, Switzerland, Oct. 1997.

934. P. Pajunen and J. Karhunen. Self-organizing maps for independent component analysis. In *Proc. of Workshop on Self-Organizing Maps (WSOM'97)*, pages 96–99, Espoo, Finland, June 1997.

935. P. Pajunen and J. Karhunen. Least-squares methods for blind source separation based on nonlinear PCA. *Int. Journal of Neural Systems*, 8(5 and 6):601–612, 1998.

936. D. Pal. Fractionally spaced semi-blind equalization of wireless channels. In *26-th Asilomar Conference Signals, Systems Comp.*, pages 642–645 vol.2. IEEE, 1992.

937. D. Pal. Fractionally spaced equalization of multipath channels: A semi-blind approach. In *Proc. IEEE ICASSP*, pages 9–12 vol.3, 1993.

938. H. Pan, D. Xia, S.C. Douglas, and K.F. Smith. A scalable VLSI architecture for multichannel blind deconvolution and source separation. In *Proc. IEEE Workshop on Signal Processing Systems*, pages 297–306, Boston, MA, October 1998.

939. C.B. Papadias and A. Paulraj. Decision-feedback equalization and identification of linear channels using blind algorithms of the Bussgang type. In *29-th Asilomar Conference Signals, Systems, Comp.*, pages 335–340 vol.1. IEEE, 1996.

940. C.B. Papadias and A. Paulraj. A space-time constant modulus algorithm for SDMA systems. In *Proc. IEEE VTC*, pages 86–90 vol.1, Atlanta, GA, 1996.

941. C.B. Papadias and A. Paulraj. Space-time signal processing for wireless communications : a survey. In *First Signal Processing Workshop on Signal Processing Advances in Wireless Communications SPAWC'97 , Paris, France*, pages 285–288, April 16-18 1997.

942. C.B. Papadias and A.J. Paulraj. A constant modulus algorithm for multi-user signal separation in presence of delay spread using antenna arrays. *IEEE Signal Processing Letters*, 4, No. 6:178–181, June 1997.

943. C.B. Papadias and D.T.M. Slock. New adaptive blind equalization algorithms for constant modulus constellations. In *International Conference on Acoustics, Speech and Signal Processing (ICASSP-94)Adelaide, Australia*, pages III–321–324, April 19-22 1994.

944. C.B. Papadias and D.T.M. Slock. Towards globally convergent blind equalization of constant-modulus signals: a bilinear approach. In *Signal Processing VIII: Proc. EUSIPCO-94*, 1994.

945. C.B. Papadias and D.T.M. Slock. Normalized sliding window constant modulus and decision-directed algorithms: a link between blind equalization and classical adaptive filtering. *IEEE Trans. Signal Processing*, 45(1):231–235, January 1997.

946. A. Papoulis. *Probability, Random Variables, and Stochastic Processes.* McGraw-Hill, NJ, 1991.

947. Y. Park, K.M. Park, I. Song, and H.-M. Kim. Blind channel identification and equalization from second-order statistics and absolute mean. *IEICE Trans. Communications*, E79-B(9):1271–1277, September 1996.

948. L. Parra, K.R. Mueller, C. Spence, A. Ziehe, P. Sajda. Unmixing Hyperspectral Data, Advances in Neural Information Processing Systems 12, MIT Press, pp. 942-948, 2000.

949. T.W. Parsons. Separation of speech from interfering speech by means of harmonic selection. *Journal of the Acoustical Society of America*, 60(4):911–918, Oct. 1976.

950. A.J. Paulraj and T. Kailath. Increasing capacity in wireless broadcast systems using distributed transmission/directional reception (DTDR). U.S. Patent 5345599, September 1994.

951. A.J. Paulraj and C.B. Papadias. Space-time processing for wireless communications. *IEEE Signal Processing Mag.*, 14(6):49–83, November 1997.

952. B.A. Pearlmutter and L.C. Parra. A context-sensitive generalization of ICA. In S. Amari, L. Xu, L.-W. Chan, I. King, and K.-S. Leung, editors, *Progress in Neural Information Processing. Proceedings of the International Conference on Neural Information Processing*, pages 151–157 vol.1, Hong Kong, 1996. Springer-Verlag.

953. P.Z. Peebles. *Probability, Random Variables, and Random Signal Principles.* McGraw-Hill, Singapore, 1993.

954. R.A. Peloso. Adaptive equalization for advanced television. *IEEE Trans. Consumer Electronics*, 38(3):119–126, August 1992.

955. M. Peng, C.L. Nikias, and J.G. Proakis. Adaptive equalization for PAM and QAM signals with neural networks. In *25-th Asilomar Conference Signals, Systems Comp.*, pages 496–500 vol.1. IEEE, 1991.

956. A. Perez-Neira and M. Lagunas. Multiuser array beamforming based on a neural network mapping. In *Proc. of the 1994 IEEE Int. Conference on Acoustics, Speech, and Signal Processing*, pages 9–12, A.aide, Australia, 1994.

957. S. Perreau, L.B. White, and P. Duhamel. A reduced computation multichannel adaptive equalizer based on HMM. In *Proc. IEEE SP Workshop on Stat. Signal Array Processing*, pages 156–159, Corfu, Greece, 1996.

958. R. Peterson and Jr. D.J. Costello. Binary convolutional codes for a multiple-access channel. *IEEE Trans. Information Theory*, 25(1):101–105, Jan 1979.

959. A.P. Petropulu. Blind deconvolution of non-linear random signals. In *IEEE SP Workshop on Higher-Order Stat.*, pages 205–209, South Lake Tahoe, CA, 1993.

960. A.P. Petropulu and C.L. Nikias. Blind deconvolution based on signal reconstruction from partial information using higher-order spectra. In *Proc. IEEE ICASSP*, pages 1757–1760 vol.3, Toronto, Ont., Canada, 1991.

961. A.P. Petropulu and C.L. Nikias. Blind convolution using signal reconstruction from partial higher order cepstral information. *IEEE Trans. Signal Processing*, 41(6):2088–2095, June 1993.

962. A.P. Petropulu and C.L. Nikias. Blind deconvolution of coloured signals based on higher-order cepstra and data fusion. *IEE Proc. F (Radar and Signal Processing)*, 140(6):356–361, December 1993.

963. D.-T. Pham. Blind separation of instantaneous mixture of sources via an independent component analysis. *IEEE Trans. Signal Processing*, 44(11):2768–2779, November 1996.

964. D.-T. Pham. Joint approximate diagonalization of positive definite hermitian matrices. *SIAM Journal on Matrix Analysis and Applications*, 22(4):1136–1152, April 2001.

965. D.-T. Pham and J.-F. Cardoso. Blind separation of instantaneous mixtures of non stationary sources. In *Proceedings of the Second International Workshop on ICA and BSS, ICA'2000*, pages 187–192, Helsinki, Finland, 19-22 June 2000.

966. D.-T. Pham and J.-F. Cardoso. Blind separation of instantaneous mixtures of non stationary sources. *IEEE Trans. on Signal Processing*, 49(9):1837–1848, September 2001.

967. D.-T. Pham, P. Garat, and C. Jutten. Separation of a mixture of independent sources through a maximum likelihood approach. In J. Vandewalle, R. Boite, M. Moonen, and A. Oosterlinck, editors, *Signal Processing VI: Theories and Applications*, pages 771–774, 1992.

968. G. Picchi and G. Prati. Blind equalization and carrier recovery using a 'stop-and-go' decision-directed algorithm. *IEEE Trans. Communications*, 35(9):877–887, September 1987.

969. G. Picchi and G. Prati. A blind Sag-So-DFD-FS equalizer. In *Proc. IEEE Int. Conference Communic.*, pages 957–961 vol.2, Philadelphia, PA, 1988.

970. R. Pickholtz and K. Elbarbary. The recursive constant modulus algorithm: A new approach for real-time array processing. In *27-th Asilomar Conference Signals, Systems Comp.*, pages 627–632. IEEE, 1993.

971. J.C. Platt and F. Faggin. Networks for the separation of sources that are superimposed and delayed. In *Advances in Neural Information Processing Systems 4*, pages 730–737. Morgan Kaufmann, 1992.

972. M. D. Plumbley. Algorithms for non-negative independent component analysis. *to appear in IEEE Trans on Neural Networks*, web page
http://www.eee.kcl.ac.uk/member/staff/m-plumbley/ .

973. M. D. Plumbley. Adaptive lateral inhibition for non-negative ICA. *Proc. ICA-2001, San Diego.* web page
http://www.eee.kcl.ac.uk/member/staff/m-plumbley.

974. B. Polyak. New method of stochastic approximation type. *Automatic Remote Control*, 51:937–946, 1990.

975. H.V. Poor and X. Wang. Adaptive suppression of narrowband digital interferers from spread spectrum signals. In *Proc. IEEE ICASSP*, pages 1061–1064 vol. 2, Atlanta, GA, 1996.

976. K.J. Pope and R.E. Bogner. Blind separation of speech signals. In *Proc. of the Australian Int. Conference on Speech Science and Technology*, Perth, Western Australia, Dec. 6-8 1994.

977. K.J. Pope and R.E. Bogner. Blind signal separation. I. Linear, instantaneous combinations. *Digital Signal Processing*, 6(1):5–16, January 1996.

978. K.J. Pope and R.E. Bogner. Blind signal separation. II. Linear, convolutive combinations. *Digital Signal Processing*, 6(1):17–28, January 1996.

979. B. Porat. *Digital Processing of Random Signals Theory & Methods.* Prentice Hall, NJ, 1993.

980. B. Porat and B. Friedlander. Blind adaptive equalization of digital communication channels using high-order moments. In *Proc. IEEE ICASSP*, pages 1372–1375 vol.2, Glasgow, UK, 1989.

981. B. Porat and B. Friedlander. Blind equalization using fourth order cumulants. In *24-th Asilomar Conference Signals, Systems Comp.*, pages 253–257 vol.1. Maple Press, 1990.

982. B. Porat and B. Friedlander. Blind equalization of digital communication channels using high-order moments. *IEEE Trans. Signal Processing*, 39(2):522–526, February 1991.

983. B. Porat and B. Friedlander. FIR system identification using fourth-order cumulants with application to channel equalization. *IEEE Trans. Automatic Control*, 38(9):1394–1398, September 1993.

984. B. Porat and B. Friedlander. Blind deconvolution of polynomial-phase signals using the high-order ambiguity function. *Signal Processing*, 53(2-3):149–163, September 1996.

985. H. Pozidis and A.P. Petropulu. Cross-correlation based multichannel blind equalization. In *Proc. IEEE SP Workshop on Stat. Signal Array Processing*, pages 360–363, Corfu, Greece, 1996.

986. S. Prakriya and D. Hatzinakos. Blind identification of LTI-ZMNL-LTI nonlinear channel models. *IEEE Trans. Signal Processing*, 43(12):3007–3013, December 1995.

987. S. Prakriya and D. Hatzinakos. Blind identification of nonlinear models using higher order spectral analysis. In *Proc. IEEE ICASSP*, pages 1601–1604 vol.3, Detroit, MI, 1995.

988. C. Prati, F. Rocca, Y. Kost, and E. Damonti. Blind deconvolution for Doppler centroid estimation in high frequency SAR. *IEEE Trans. Geoscience and Remote Sensing*, 29(6):934–941, November 1991.

989. J.C. Principe, B. de Vries, and P. Guedes de Oliveira. The Gamma filter - a new class of adaptive IIR filters with restricted feedback. *i3etsp*, 41:649–656, 1993.

990. J.C. Principe, C. Wang, and H.-C. Wu. Temporal decorrelation using teacher forcing anti-Hebbian learning and its application in adaptive blind source separation. In S. Usui, Y. Tohkura, S. Katagiri, and E. Wilson, editors, *Neural Networks for Signal Processing VI. Proceedings of the 1996 IEEE Signal Processing Society Workshop*, pages 413–422, Neural Networks for Signal Processing VI. Proceedings of the 1996 IEEE Signal Processing Society Workshop, Kyoto, Japan, 4-6 Sept. 1996, 1996. IEEE.

991. J.G. Proakis. *Digital communications.* McGraw-Hill, 2nd edition, 1989.

992. J.G. Proakis and C.L. Nikias. Blind equalization. *Proc. SPIE*, 1565:76–87, 1991.

993. C.G. Puntonet, A. Prieto, and J. Ortega. New geometrical approach for blind separation of sources mapped to a neural network. In *Proceedings International Workshop on Neural Networks for Identification, Control, Robotics, and Signal/Image Processing*, pages 174–182, Proceedings of International Workshop on Neural Networks for Identification, Control, Robotics and Signal/Image Processing, Venice, Italy, 21-23 Aug. 1996, 1996. IEEE Comput. Society Press.

994. V. Radionov and S. Mayrargue. Semi-blind approach to second order identification of SIMO-FIR channel driven by finite alphabet sequence. In *IEEE Int. Conference on Dig. Signal Processing*, volume 1, pages 115–118. IEEE Press, Santorini (Greece), July 1997.

995. R. Raheli and G. Picchi. Synchronous and fractionally-spaced blind equalization in dually-polarized digital radio links. In *Proc. IEEE ICC*, pages 156–161 vol.1, Denver, CO, 1991.

996. M.G. Rahim, B.-H. Juang, W. Chou, and E. Buhrke. Signal conditioning techniques for robust speech recognition. *IEEE Signal Processing Letters*, 3(4):107–109, April 1996.

997. G. Raleigh, S. Diggavi, V. Jones, and A. Paulraj. A blind adaptive transmit antenna algorithm for wireless communication. In *Proc. IEEE ICC*, 1995.

998. J. Ramos and M.D. Zoltowski. Reduced complexity blind 2D RAKE receiver for CDMA. In *Proc. IEEE SP Workshop on Stat. Signal Array Processing*, pages 502–505, Corfu, Greece, 1996.

999. J. Ramos, M.D. Zoltowski, and H. Liu. A low-complexity space-time RAKE receiver for DS-CDMA communications. *IEEE Signal Processing Letters*, 4(9):262–265, September 1997.

1000. A. Ranheim and P. Pelin. Decoupled blind symbol estimation using an antenna array. In *Proc. IEEE SP Workshop on Stat. Signal Array Processing*, pages 136–139, Corfu, Greece, 1996.

1001. B. D. Rao and K. Kreutz-Delgado. Basis selection in the presence of noise. In *Conference Record of the 32rd Asilomar Conference on Signals, Systems and Computers*, pages 752–756, 1998.

1002. B. D. Rao and K. Kreutz-Delgado. An affine scaling methodology for best basis selection. *IEEE Trans. Sig. Proc.*, 47(1):187–200, January 1999.

1003. Y. Rao and J.C. Principe. Time series segmentation using a novel adaptive eigendecomposition algorithm. *Journal of VLSI Signal Processing*, 32(1/2): 7-17, 2002.

1004. T.S. Rappaport. *Wireless Communications: Principles and Practice.* Prentice Hall, Upper Saddle River, NJ, 1996.

1005. M.J. Ready and R.P. Gooch. Blind equalization based on radius directed adaptation. In *Proc. IEEE ICASSP*, pages 1699–1702 vol.3, Albuquerque, NM, USA, 3-6 April 1990, 1990.

1006. V.U. Reddy, C.B. Papadias, and A. Paulraj. Second-order blind identifiability of certain classes of multipath channels using antenna arrays. In *International Conference on Acoustics, Speech, and Signal Processing, Munich, Germany*, pages 3465–3468, April 21-24 1997.

1007. V.U. Reddy, C.B. Papadias, and A.J. Paulraj. Blind identifiability of certain classes of multipath channels for second-order statistics using antenna arrays. *IEEE Signal Processing Letters*, 4, No. 5:138–141, May 1997.

1008. P. A. Regalia and M. Mboup. Properties of some blind equalization criteria in noisy multi-user environments. *IEEE Trans. Signal Processing*, 49:3112–3122, 2001.

1009. E. Ribak. Astronomical imaging by pupil plane interferometry. *Proc. SPIE*, 1038:418–426, 1989.

1010. C. Ringeissen. Combining decision algorithms for matching in the union of disjoint equational theories. *Information and Computation*, 126(2):144–160, May 1996.

1011. C. Riou, T. Chonavel, and P.-Y. Cochet. Adaptive subspace estimation-application to moving sources localization and blind channel identification. In *Proc. IEEE ICASSP*, pages 1648–1651 vol. 3, Atlanta, GA, 1996.

1012. H. Robbins and S. Monro. A stochastic approximation method. *Ann. Math. Stat.*, 22:400–407, 1951.

1013. R.S. Roberts, P.S. Lewis, and O.A. Vela. A pattern recognition algorithm for the blind discrimination of liquid and solid filled munitions. In *29-th Asilomar Conference Signals, Systems, Comp.*, pages 1310–1314 vol.2. IEEE, 1996.

1014. J.A. Rodriguez-Fonollosa and J. Vidal. Adaptive ARMA identification using cumulants. In J.L. Lacoume, editor, *Higher Order Statistics. Proceedings of the International Signal Processing Workshop*, pages 125–128, Chamrousse, France, 1992. Elsevier.

1015. R. Rosipal, M. Girolami, L.J. Trejo, and A. Cichocki. Kernel PCA for feature extraction and de-noising in non-linear regression. *Neural Computing and Applications*, 10:231–243, 2001.

1016. F.J. Ross and D.P. Taylor. An enhancement to blind equalization algorithms. *IEEE Trans. Communications*, 39(5):636–639, May 1991.

1017. P.J. Rousseeuw and D.K. van Driesen. A fast algorithm for the minimum covriance determinant estimator. *Technometrics*, 41: 212–223, 1999.

1018. S. T. Roweis. EM algorithms for PCA and SPCA. In *Advances in Neural Information processing Systems NIPS-98*, volume 10, pages 452–456, 1998.

1019. J. Rubner and P. Tavan. A self organizing network for principal components analysis. *Europhysics Letters*, 10:693–689, 1989.

1020. M.J. Rude and L.J. Griffiths. Incorporation of linear constraints into the constant modulus algorithm. In *Proc. IEEE ICASSP*, pages 968–971 vol.2, Glasgow, UK, 23-26 May 1989, 1989.

1021. M.J. Rude and L.J. Griffiths. A linearly constrained adaptive algorithm for constant modulus signal processing. In L. Torres, E. Masgrau, and M.A. Lagunas, editors, *Signal Processing V. Theories and Applications. Proceedings of EUSIPCO-90, Fifth European Signal Processing Conference*, pages 237–240 vol.1, Barcelona, Spain, 18-21 Sept. 1990, 1990. Elsevier.

1022. M.J. Rude and L.J. Griffiths. An untrained, fractionally-spaced equalizer for co-channel interference environments. In *24-th Asilomar Conference Signals, Systems Comp.*, pages 468–472 vol.1, Pacific Grove, CA, USA, 5-7 Nov. 1990, 1990. Maple Press.

1023. M.J. Rude and L.J. Griffiths. Sensitivity of the linearly-constrained constant-modulus cost function. In *25-th Asilomar Conference Signals, Systems Comp.*, pages 984–988 vol.2. IEEE, 1991.

1024. M. Rupp and S.C. Douglas. A posteriori analysis of adaptive blind equalizers. In *Proc. 32nd Asilomar Conference on Signals, Systems, and Computers*, volume 1, pages 369–373, Pacific Grove, CA, November 1998.

1025. W. Rupprecht. Odd-order Volterra circuits with potential application to blind equalization of linear channels. In *26-th Asilomar Conference Signals, Systems Comp.*, pages 319–323 vol.1. IEEE, 1992.

1026. I. Sabała, A. Cichocki, and S. Amari. Relationships between instantaneous blind source separation and multichannel blind deconvolution. In *Proc. Int. Joint Conference on Neural Networks*, pages 148–152, Alaska USA, 1998.

1027. M.J. Sabin. *Global Convergence and Empirical Consistency of the Generalized Lloyd Algorithm.* PhD thesis, Stanford University, Stanford, CA, 1984.

1028. M. Sabry-Rizk, W. Zgallai, P. Hardiman, and J. O'Riordan. Blind deconvolution homomorphic analysis of abnormalities in ECG signals. In *IEE Colloquium on 'Blind Deconvolution - Algorithms and Applications' (Ref. No.1995/145)*, pages 5/1–9, London, UK, 1995. IEE.

1029. H. Sahlin and U. Lindgren. The asymptotic Cramer-Rao lower bound for blind signal separation. In *Proc. IEEE SP Workshop on Stat. Signal Array Processing*, pages 328–331, Corfu, Greece, 1996.

1030. H. Sakai and K. Shimizu. Two improved algorithms for adaptive subspace filtering. In *Proc. 11th IFAC Symposium System Identification (SYSID'97)*, pages 1689–1694, Kitakyushu, Japan, July 1997.

1031. J. Sala and G. Vazquez. A statistical reference criterion for adaptive filtering. In *Proc. IEEE ICASSP*, pages 1660–1663 vol. 3, Atlanta, GA, 1996.

1032. F. Salam. An adaptive network for blind separation of independent signals. In *Proc. of the 1993 IEEE International Symposium on Circuits and Systems*, volume 1, pages 431–434, Chicago, USA, May 3-6 1993. IEEE.

1033. F. Salam and G. Erten. The state space framework for blind dynamic signal extraction and recovery. In *Proc of '99 Int. Symposium on Circuits and Systems, ISCAS'99*, volume 5, pages 66–69, Orlando, Florida, 1999.

1034. F. Salam and K. Waheed. State space feedforward and feedback structures for blind source recovery. In *Proc. of the Third International Conference on Independent Component Analysis and Signal Separation (ICA-2001)*, pages 248–253, San Diego, USA, Dec. 9-13 2001.

1035. F.S. Samaria and A.C. Harter. Parameterisation of a stochastic model for human face identification. In *Workshop on the Application of Computer Vision*, pages 138–142, Los Alamitos, CA, USA, 1994. IEEE Computer Society Press.

1036. M. Sambur. Adaptive noise canceling for speech signals. *IEEE Trans. Acoustics Speech, Signal Processing*, ASSP-26:419–423, October 1978.

1037. T.D. Sanger. Optimal unsupervised learning in a single–layer linear feedforward neural network. *Neural Networks*, 2:459–473, 1989.

1038. T.D. Sanger. Analysis of the two-dimensional receptive fields learned by the generalized Hebbian algorithm in response to random input. *Biological Cybernetics*, 63:221–228, 1990.

1039. I. Santamaria-Caballero, C. Pantaleon-Prieto, F. Diaz de Maria, and A. Artes-Rodriguez. A new inverse filter criterion for blind deconvolution of spiky signals using Gaussian mixtures. In *Proc. IEEE ICASSP*, pages 1680–1683 vol. 3, Atlanta, GA, 1996.

1040. K. Sasaki and H. Masuhara. Blind-deconvolution analysis of transient curves by the use of a convolved autoregressive model. *Applied Optics*, 35(26):5312–5316, September 1996.

1041. Y. Sato. A method of self-recovering equalization for multilevel amplitude-modulation systems. *IEEE Trans. Communications*, 23:679–682, June 1975.

1042. Y. Sato. Blind equalization and blind sequence estimation. *IEICE Trans. Communications*, E77-B(5):545–556, May 1994.

1043. E.H. Satorius and J.J. Mulligan. Minimum entropy deconvolution and blind equalization. *Electronics Letters*, 28(16):1534–1535, July 1992.

1044. E.H. Satorius and J.J. Mulligan. An alternative methodology for blind equalization. *Digital Signal Processing*, 3(3):199–209, July 1993.

1045. M. Savic, H. Gao, and J.S. Sorenson. Co-channel speaker separation based on maximum-likelihood deconvolution. In *International Conference on Acoustics, Speech and Signal Processing*, volume I, pages 25–28, 1994.

1046. G. Scarano and G. Jacovitti. Sources identification in unknown Gaussian coloured noise with composite HNL statistics. In *Proc. IEEE ICASSP*, pages 3465–3468 vol.5, Toronto, Ont., Canada, 1991.

1047. S.V. Schell. An overview of sensor array processing for cyclostationary signals. In W.A. Gardner, editor, *Cyclostationarity in Communications and Signal Processing*, pages 168–239, New Jersey, USA, 1994. IEEE Press.

1048. S.V. Schell and W.A. Gardner. Maximum likelihood and common factor analysis-based blind adaptive spatial filtering for cyclostationary signals. In *Proc. IEEE ICASSP*, pages IV:292–295. IEEE, 1992.

1049. S.V. Schell and T.E. Shrimpton. Super-exponentially convergent blind fractionally-spaced equalization. In *29-th Asilomar Conference Signals, Systems, Comp.*, pages 703–709 vol.1. IEEE, 1996.

1050. T. Schirtzinger, X. Li, and W.K. Jenkins. A comparison of three algorithms for blind equalization based on the constant modulus error criterion. In *Proc. IEEE ICASSP*, pages 1049–1052 vol.2, Detroit, MI, 1995.

1051. L. Schlicht and G. Ilgenfritz. Simulation of diffusion in 2-D heterogeneous systems: comparison with effective medium and percolation theories. *Physica A*, 227(3-4):239–247, June 1996.

1052. R.O. Schmidt. *A Signal Subspace Approach to Multiple Source Location and Spectral Estimation.* PhD thesis, Stanford University, Stanford, CA, 1981.

1053. R.O. Schmidt. Multiple emitter location and signal parameter estimation. *IEEE Trans. on Antenna and Propagation*, AP-34(3):276–280, March 1986.

1054. J.B. Schodorf and D.B. Williams. A blind adaptive interference cancellation scheme for CDMA systems. In *29-th Asilomar Conference Signals, Systems, Comp.*, pages 270–274 vol.1. IEEE, 1996.

1055. J.B. Schodorf and D.B. Williams. A constrained adaptive diversity combiner for interference suppression in CDMA systems. In *Proc. IEEE ICASSP*, pages 2666–2669 vol. 5, Atlanta, GA, 1996.

1056. J.B. Schodorf and D.B. Williams. Partially adaptive multiuser detection. In *Proc. IEEE VTC*, pages 367–371 vol.1, Atlanta, GA, 1996.

1057. T.J. Schulz. Multiframe blind deconvolution of astronomical images. *Journal of the Optical Society of America A (Optics and Image Science)*, 10(5):1064–1073, May 1993.

1058. M. Segal, E. Weinstein, and B.R. Musicus. Estimate-Maximize algorithms for multichannel time delay and signal estimation. *IEEE Trans. on Signal Processing*, 39(1):1–16, Jan. 1991.

1059. J.H. Seldin and J.R. Fienup. Iterative blind deconvolution algorithm applied to phase retrieval. *Journal of the Optical Society of America A (Optics and Image Science)*, 7(3):428–433, March 1990.

1060. C. Serviere. Blind source separation of convolutive mixtures. In *Proc. IEEE SP Workshop on Stat. Signal Array Processing*, pages 316–319, Corfu, Greece, 1996.

1061. C. Serviere and D. Baudois. Source separation with noisy observations: a noise cancelling application. *Signal Processing*, 42(1):45–57, February 1995.

1062. C. Serviere and V. Capdevielle. Blind adaptive separation of wide-band sources. In *Proc. IEEE ICASSP*, pages 2698–2701 vol. 5, Atlanta, GA, 1996.

1063. N. Seshadri. Joint data and channel estimation using fast blind trellis search techniques. In *IEEE Proc. GLOBECOM*, pages 1659–1663 vol.3, San Diego, CA, 1990. IEEE.

1064. W.. Sethares, G.A. Rey, and C.R. Johnson, Jr. Approaches to blind equalization of signals with multiple modulus. In *Proc. IEEE ICASSP*, pages 972–975 vol.2, Glasgow, UK, 23-26 May 1989, 1989.

1065. W.A. Sethares, R.A. Kennedy, and Z. Gu. An approach to blind equalization of non-minimum phase systems. In *Proc. IEEE ICASSP*, pages 1529–1532 vol.3, Toronto, Ont., Canada, 1991.

1066. B. Shafai and S. Mo. Adaptive deconvolution and identification of nonminimum phase FIR systems using Kalman filter. In *Proc. IEEE ICASSP*, pages 489–492 vol.5, San Francisco, CA, 1992.

1067. A.A. Shah. Fast channel classification. Technical report, Stanford University, 1993.

1068. O. Shalvi and E. Weinstein. New criteria for blind deconvolution of nonminimum phase systems (channels). *IEEE Trans. Informat. Th.*, 36(2):312–321, March 1990.

1069. O. Shalvi and E. Weinstein. Super-exponential methods for blind deconvolution. *IEEE Trans. Informat. Th.*, 39(2):504–519, March 1993.

1070. O. Shalvi and E. Weinstein. Universal method for blind deconvolution. In S. Haykin, editor, *Blind Deconvolution*, pages 121–180. Prentice-Hall Inc., Englewood Cliffs, New Jersey, 1994.

1071. S. Shamsunder and G.B. Giannakis. Modeling of non-Gaussian array data using cumulants: DOA estimation of more sources with less sensors. *Signal Processing*, 30:279–297, 1993.

1072. M. Sharma and R. Mammone. Subword-based text-dependent speaker verification system with user-selectable passwords. In *Proc. IEEE ICASSP*, pages 93–96 vol. 1, Atlanta, GA, 1996.

1073. T. Shimada and S.K. Mitra. Blind adaptive equalization for digital storage systems using timing interpolation. In *Proc. IEEE ICASSP*, pages 1475–1478 vol.3, Albuquerque, NM, 1990.

1074. J. Shin, J.-S. Lee, E.-T. Kim, C.-S. Won, and J.-K. Kim. An improved stop-and-go algorithm for blind equalization. *IEICE Trans. Fundamentals of Electronics, Communications and Computer Sciences*, E79-A(6):784–789, June 1996.

1075. V. Shtrom and H. Fan. Blind equalization: a new convex cost function. In *Proc. IEEE ICASSP*, pages 1779–1782 vol. 3, Atlanta, GA, 1996.

1076. J.J. Shynk and C.K. Chan. A comparative analysis of the stationary points of the constant modulus algorithm based on Gaussian assumptions. In *Proc. IEEE ICASSP*, pages 1249–1252 vol.3, Albuquerque, NM, USA, 3-6 April 1990, 1990.

1077. J.J. Shynk and C.K. Chan. Error surfaces of the constant modulus algorithm. In *1990 IEEE International Symposium on Circuits and Systems*, pages 1335–1338 vol.2, New Orleans, LA, USA, 1-3 May 1990, 1990. IEEE.

1078. J.J. Shynk and C.K. Chan. Performance surfaces of the constant modulus algorithm based on a conditional Gaussian model. *IEEE Trans. Signal Processing*, 41(5):1965–1969, May 1993.

1079. J.J. Shynk, C.K. Chan, and M.R. Petraglia. Blind adaptive filtering in the frequency domain. In *1990 IEEE International Symposium on Circuits and Systems*, pages 275–278 vol.1, New Orleans, LA, USA, 1-3 May 1990, 1990. IEEE.

1080. J.J. Shynk and R.P. Gooch. Convergence properties of the Multistage CMA adaptive beamformer. In *27-th Asilomar Conference Signals, Systems Comp.*, pages 622–626 vol.1. IEEE, 1993.

1081. J.J. Shynk and R.P. Gooch. The constant modulus array for co-channel signal copy and direction finding. *IEEE Trans. on Signal Processing*, 44 No.3:652–660, March 1996.

1082. J.J. Shynk, R.P. Gooch, G. Krishnamurthy, and C.K. Chan. A comparative performance study of several blind equalization algorithms. In *Proc. SPIE Vol. 1565 Adaptive Signal Processing*, pages 102–117. SPIE, 1991.

1083. J.J. Shynk, D.P. Witmer, M.J. Ready, R.P. Gooch, and C.K. Chan. Adaptive equalization using multirate filtering techniques. In *25-th Asilomar Conference Signals, Systems Comp.*, pages 756–762 vol.2. IEEE, 1991.

1084. F.M. Silva and L.B. Almeida. A distributed decorrelation algorithm. In E. Gelenba, editor, *Neural Networks, Advances and Applications*, pages 145–163, Amsterdam, 1991. North–Holland.

1085. W. Skarbek, A. Cichocki, and W. Kasprzak. Principal subspace analysis for incomplete image data in one learning epoch. *Neural Network World*, 6(3):375–382, 1996.

1086. B. Sklar. Rayleigh fading channels in mobile digital communication systems, part I: Characterization. *IEEE Communications Magazine*, 35(7):90–100, July 1997.

1087. B. Sklar. Rayleigh fading channels in mobile digital communication systems, part II: Mitigation. *IEEE Communications Magazine*, 35(7):102–109, July 1997.

1088. D.T.M. Slock. Blind fractionally-spaced equalization, perfect-reconstruction filter banks and multichannel linear prediction. In *Proc. IEEE ICASSP*, pages IV:585–588, 1994.

1089. D.T.M. Slock. Blind joint equalization of multiple synchronous mobile users for spatial division multiple access. In *Proc. 7th Tyrrhenian Int. Workshop on Digital Communications: Signal Processing in telecommunications*, Viareggio, Italy, Sept. 10-14 1995.

1090. D.T.M. Slock. Spatio-temporal training-sequence-based channel equalization and adaptive interference cancellation. In *Proc. ICASSP 96 Conference*, Atlanta, Georgia, May 1996.

1091. D.T.M. Slock and C.B. Papadias. Blind fractionally-spaced equalization based on cyclostationarity. In *Proc. Vehicular Technology Conference*, pages 1286–1290, Stockholm, Sweden, June 1994.

1092. D.T.M. Slock and C.B. Papadias. Further results on blind identification and equalization of multiple FIR channels. In *Proc. Int'l Conference on Acoustics, Speech and Signal Processing, , Detroit, Michigan*, pages 1964–1967, May 8-12 1995.

1093. J. Slotine and W. Li. *Applied Nonlinear Control.* Prentice-Hall, Englewood Cliffs, NJ, 1991.

1094. T. Söderström and P. Stoica. *Instrumental Variable Methods for System Identification.* Springer-Verlag, Berlin, 1983.

1095. H. Sompolinsky, N. Barkai, and H.S. Seung. On–line learning of dichotomies: algorithms and learning curves. In J-H. Oh, C. Kwon, and S. Cho, editors, *Neural Networks: The Statistical Mechanics Perspective*, pages 105–130. World Scientific, Singapore, 1995.

1096. V.C. Soon, L. Tong, Y.F. Huang, and R. Liu. An extended fourth order blind identification algorithm in spatially correlated noise. In *Proc. IEEE ICASSP*, pages 1365–1368 vol.3, Albuquerque, NM, 1990.

1097. V.C. Soon, L. Tong, Y.F. Huang, and R. Liu. A wideband blind identification approach to speech acquisition using a microphone array. In *Proc. IEEE ICASSP*, pages 293–296 vol.1, San Francisco, CA, 1992.

1098. E. Sorouchyari. Blind separation of sources, Part III: Stability analysis. *Signal Processing*, 24(1):21–29, July 1991.

1099. A. Souloumiac and J.-F. Cardoso. Performances en séparation de sources. In *Proc. GRETSI, Juan les Pins,France*, pages 321–324, 1993.

1100. M. Soumekh. Reconnaissance with ultra wideband UHF synthetic aperture radar. *IEEE Signal Processing Magazine*, 12(4):21–40, July 1995.

1101. R. Steele, editor. *Mobile Radio Communications.* IEEE Press, 1994.

1102. B.D. Steinberg. Adaptive beamforming in radio camera imaging. In *24-th Asilomar Conference Signals, Systems Comp.*, pages 2–7 vol.1. Maple Press, 1990.

1103. M. Stewart-Bartlett and T.J. Sejnowski. Independent components of face images: A representation for face recognition. In *Proceedings of the 4th Joint Symposium on Neural Computation*, pages 3–10, 1997.

1104. A.G. Stogioglou and S. McLaughlin. Asymptotic performance analysis for direct and indirect HOS-based deconvolution. In *IEE Colloquium on 'Blind Deconvolution - Algorithms and Applications' (Ref. No.1995/145)*, pages 3/1–7, London, UK, 1995. IEE.

1105. J.V. Stone. Blind source separation using temporal predictability. *Neural Computation*, 13(7):1559–1574, 2001.

1106. J.V. Stone, J. Porrill, and N.R. Porter. Spatiotemporal independent component analysis of fMRI data using skewed probability density functions. *Neuroimage*, pages in–print, 2002.

1107. J. Storck and G. Deco. Factorial learning of multivariate time series. In *Proc. of the Int. Conference on Neural Networks*, pages 149–154, Paris, France, Oct. 9-13 1995.

1108. E. Ström, S. Parkvall, S.L. Miller, and B.E. Ottersten. Propagation delay estimation in asynchronous direct-sequence code-division multiple access systems. *IEEE Tr. Comm.*, 44(1):84–93, January 1996.

1109. A. Stuart and J. Ord. *Kendall's Advanced Theory of Statistics.* Edward Arnold, 1994.

1110. R.J. Stubbs and Q. Summerfield. Evaluation of two voice-separation algorithms using normal and hearing impaired listeners. *Journal of the Acoustical Society of America*, 84(4):1236–1249, Oct. 1988.

1111. R. Sucher. A self-organizing nonlinear filter based on fuzzy clustering. In *Proc. IEEE ISCAS*, pages 101–103 vol.2, Atlanta, GA, 1996.

1112. R. Sucher. A self-organizing nonlinear noise filtering scheme. In *29-th Asilomar Conference Signals, Systems, Comp.*, pages 681–684 vol.1. IEEE, 1996.

1113. X. Sun and S.C. Douglas. Adaptive time delay estimation with allpass constraints. In *presented at 33rd Annual Asilomar Conference on Signals, Systems, and Computers*, Pacific Grove, CA, October 24-27 1999.

1114. H. Suzuki. Adaptive signal processing for optimal transmission in mobile radio communications. *IEICE Trans. Communications*, E77-B(5):535–544, May 1994.

1115. S.C. Swales, M.A. Beach, and D.J. Edwards. Multi-beam adaptive base station antennas for cellular land mobile radio systems. In *Proc. IEEE Veh. Technol. Conference*, pages 341–348, 1989.

1116. S.C. Swales, M.A. Beach, D.J. Edwards, and J.P. McGreehan. The performance enhancement of multibeam adaptive base-station antennas for cellular land mobile radio systems. *IEEE Trans. on Vehicular Technology*, 39(1):56–67, Feb. 1990.

1117. D.L. Swets and J. Weng. Using discriminant eigenfeatures for image retrieval. *IEEE Trans. Pattern Analysis and Machine Intelligence*, 18(8):831–836, 1996.

1118. A. Swindlehurst, S. Daas, and J. Yang. Analysis of a decision directed beamformer. *IEEE Trans. Signal Processing*, 43(12):2920–2927, December 1995.

1119. A. Swindlehurst and J. Yang. Using least squares to improve blind signal copy performance. *IEEE Signal Processing Letters*, 1(5):80–83, May 1994.

1120. A. Swindlehurst, J. Yang, and S. Daas. On the copy of communications signals using antenna arrays. In M. Blanke and T. Soderstrom, editors, *System Identification (SYSID '94). A Postprint Volume from the IFAC Symposium*, pages 119–124 vol.1, Copenhagen, Denmark, 1995. Pergamon.

1121. A.L. Swindlehurst, M.J. Goris, and B. Ottersten. Some experiments with array data collected in actual urban and suburban environments. In *IEEE workshop on Signal Processing Adv. in Wireless Comm.*, pages 301–304, Paris, April 1997.

1122. N. Takagi. A modular inversion hardware algorithm with a redundant binary representation. *IEICE Trans. Information and Systems*, E76-D(8):863–869, August 1993.

1123. K. Takao. Overview of the theory of adaptive antennas. *Electronics and Communications in Japan, Part 1 (Communications)*, 76(7):110–118, July 1993.

1124. K. Takao and H. Matsuda. The choice of the initial condition of CMA adaptive arrays. *IEICE Trans. Communications*, E78-B(11):1474–1479, November 1995.

1125. A. Taleb and G. Cirrincionne. Against the convergence of the minor component analysis neurons. *IEEE trans. Neural Networks*, 10(1):207–210, January 1999.

1126. A. Taleb and C. Jutten. Batch algorithm for source separation in postnonlinear mixtures. In *ICA 99*, pages 155–160, Aussois (France), January 1999.

1127. A. Taleb and C. Jutten. On underdetermined source separation. In *ICASSP 99*, pages 1445–1448, Phoenix (USA), March 1999.

1128. A. Taleb and C. Jutten. Source separation in post nonlinear mixtures. *IEEE Trans. on Signal Processing*, 47(10):2807–2820, 1999.

1129. A. Taleb, J. Solé, and C. Jutten. Blind inversion of Wiener systems. In *IWANN 99*, pages 655–664, Alicante (Spain), June 1999.

1130. S. Talwar and A. Paulraj. Performance analysis of blind digital signal copy algorithms. In *Proc. IEEE MILCOM*, pages 123–127 vol.1, 1994.

1131. S. Talwar, A. Paulraj, and G. Golub. A robust numerical approach for array calibration. In *Proc. IEEE ICASSP*, volume IV, pages 316–319, 1993.

1132. S. Talwar, A. Paulraj, and M. Viberg. Reception of multiple co-channel digital signals using antenna arrays with applications to PCS. In *Proc. ICC*, volume II, pages 790–794, 1994.

1133. S. Talwar, M. Viberg, and A. Paulraj. Blind estimation of multiple co-channel digital signals arriving at an antenna array. In *27-th Asilomar Conference Signals, Systems Comp.*, pages 349–353 vol.1. IEEE, 1993.

1134. S. Talwar, M. Viberg, and A. Paulraj. Blind estimation of multiple co-channel digital signals using an antenna array. *IEEE Signal Processing Letters*, 1(2):29–31, February 1994.

1135. S. Talwar, M. Viberg, and A. Paulraj. Blind estimation of synchronous co-channel digital signals using an antenna array. Part I: Algorithms. *IEEE Trans. Signal Processing*, 44(5):1184–1197, May 1996.

1136. T. Tanaka, R. Miura, I. Chiba, and Y. Karasawa. An ASIC implementation scheme to realize a beam space CMA adaptive array antenna. *IEICE Trans. Communications*, E78-B(11):1467–1473, November 1995.

1137. J. Tang, Z. Li, and L. Li. Blind equalization on time-variant channels. *Acta Electronica Sinica*, 21(7):69–76, July 1993.

1138. S.-H.T. Tang and R.M. Mersereau. Multiscale blind image restoration using a wavelet decomposition. In *Proc. IEEE ICASSP*, pages 2279–2282 vol. 4, Atlanta, GA, 1996.

1139. F. Tarres and J. Fernandez-Rubio. An adaptive array for coherent interference suppression. In M.H. Hamza, editor, *Applied Control, Filtering and Signal Processing. Proceedings of the IASTED International Symposium*, pages 169–170, Geneva, Switzerland, 15-18 June 1987, 1987. ACTA Press.

1140. J.G. Taylor and S. Coombes. Learning of higher order correlations. *Neural Networks*, 6:423–427, 1993.

1141. K. Teramoto and K. Arai. POCS-based blind array processing in incoherent microwave radiometric image reconstruction. In *Proc. IEEE ICASSP*, pages 2702–2705 vol. 5, Atlanta, GA, 1996.

1142. R. Thawonmas and A. Cichocki. Blind signal extraction of arbitrary distributed but temporally correlated signals-neural network approach. *IEICE Transactions, Fundamentals*, E82 A(9):1834–1844, Sept. 1999.

1143. R. Thawonmas and A. Cichocki. Blind extraction of source signals with specified stochastic features. In *Int. Conference on Acoustics Speech and Signal Processing*, volume 4, pages 3353–3357, 97.

1144. R. Thawonmas, A. Cichocki, and S. Amari. A Cascade neural network for blind signal extraction without spurious equilibria. *IEICE Trans. on Fundamentals of Electronics, Communications and Computer Sciences*, E81-A(9):1833–1846, 1998.

1145. H-L.-N. Thi and C. Jutten. Blind source separation for convolutive mixtures. *Signal Processing*, 45(2):209–229, 1995.

1146. H.-L.-N. Thi, C. Jutten, and J. Caelen. Speech enhancement: Analysis and comparison of methods on various real situations. In J. Vandewalle, R. Boite, M. Moonen, and A. Oosterlinck, editors, *Signal Processing VI: Theories and Applications*, pages 303–306, 1992.

1147. E. Thiebaut and J.-M. Conan. Strict a priori constraints for maximum-likelihood blind deconvolution. *Journal of the Optical Society of America A (Optics, Image Science and Vision)*, 12(3):485–492, March 1995.

1148. M.E. Tipping and C.M Bishop. Probabilistic principal component analysis. *Journal of the Royal Statistical Society series*, B(3):600–616, 1999.

1149. B.-E. Toh, D.C. McLernon, and I. Lakkis. Enhancing the super exponential method of blind equalization with the fast RLS Kalman algorithm. *Electronics Letters*, 32(2):92–94, January 1996.

1150. L. Tong. Blind equalization of fading channels. In *26-th Asilomar Conference Signals, Systems Comp.*, pages 324–328 vol.1. IEEE, 1992.

1151. L. Tong. Identifiability of minimal, stable, rational, causal systems using second-order output cyclic spectra. *IEEE Trans. Automatic Control*, 40(5):959–962, May 1995.

1152. L. Tong. Joint blind signal detection and carrier recovery over fading channels. In *Proc. IEEE ICASSP*, pages 1205–1208, Detroit, MI, 1995.

1153. L. Tong. Identification of multivariate FIR systems using higher-order statistics. In *Proc. IEEE ICASSP*, pages 3037–3040 vol. 5, Atlanta, GA, 1996.

1154. L. Tong, Y. Inouye, and R. Liu. Eigenstructure-based blind identification of independent signals. In *International Conference on Acoustics, Speech and Signal Processing*, pages 3329–3332, 1991.

1155. L. Tong, Y. Inouye, and R. Liu. A finite-step global algorithm for the parameter estimation of multichannel MA processes. *IEEE Trans. Signal Proc.*, 40(10):2547–2558, 1992.

1156. L. Tong, Y. Inouye, and R. Liu. Waveform-preserving blind estimation of multiple independent sources. *IEEE Trans. on Signal Processing*, 41(7):2461–2470, July 1993.

1157. L. Tong, R.-W. Liu, V.-C. Soon, and Y.-F. Huang. Indeterminacy and identifiability of blind identification. *IEEE Trans. on Circuits and Systems*, 38(5):499–509, May 1991.

1158. L. Tong, V.C. Soon, Y.F. Huang, and R. Liu. AMUSE: a new blind identification algorithm. In *Proc. IEEE ISCAS*, pages 1784–1787 vol.3, New Orleans, LA, 1990. IEEE.

1159. L. Tong, V.C. Soon, Y. Inouye, Y. Huang, and R. Liu. Waveform-preserving blind estimation of multiple sources. In *Proc. IEEE Conference on Decision and Control*, pages 2388–2393 vol.3, Brighton, UK, 1991.

1160. L. Tong, G. Xu, B. Hassibi, and T. Kailath. Blind channel identification based on second-order statistics: a frequency-domain approach. *IEEE Trans. Information Th.*, 41(1):329–334, January 1995.

1161. L. Tong, G. Xu, and T. Kailath. A new approach to blind identification and equalization of multipath channels. In *25-th Asilomar Conference on Signals, Systems and Computers*, pages 856–860 vol.2, 1991.

1162. L. Tong, G. Xu, and T. Kailath. Blind identification and equalization of multipath channels. In *SUPERCOMM/ICC '92. Discovering a New World of Communications*, pages 1513–1517 vol.3, Chicago, IL, USA, 14-18 June 1992, 1992. IEEE.

1163. L. Tong, G. Xu, and T. Kailath. Fast blind equalization via antenna arrays. In *Proc. IEEE ICASSP*, pages IV:272–274, 1993.

1164. L. Tong, G. Xu, and T. Kailath. Blind identification and equalization based on second-order statistics: A time domain approach. *IEEE Trans. Information Theory*, IT-40, no. 2:340–349, March 1994.

1165. L. Tong, S. Yu, Y. Inouye, and R. Liu. A necessary and sufficient condition of blind identification. In J.L. Lacoume, editor, *Higher Order Statistics. Proceedings of the International Signal Processing Workshop*, pages 263–266, Chamrousse, France, 1992. Elsevier.

1166. L. Tong, S. Yu, and R. Liu. Identifiability of a quadratic equations with unknown coefficients. In *Proc. IEEE ISCAS*, pages 292–295 vol.1, San Diego, CA, 1992.

1167. K. Torkkola. Blind separation of convolved sources based on information maximization. In S. Usui, Y. Tohkura, S. Katagiri, and E. Wilson, editors, *Neural Networks for Signal Processing VI. Proceedings of the 1996 IEEE Signal Processing Society Workshop*, pages 423–432, Neural Networks for Signal Processing VI. Proceedings of the 1996 IEEE Signal Processing Society Workshop, Kyoto, Japan, 4-6 Sept. 1996, 1996. IEEE.

1168. K. Torkkola. Blind separation of delayed sources based on information maximization. In *Proc. Int. Conference Acoustics Speech, Signal Processing (ICASSP)*, volume 4, pages 3509–3513, New York, NY, 1996. IEEE Press.

1169. M. Torlak and G. Xu. Blind multiuser channel estimation in asynchronous CDMA systems. *IEEE Trans. Signal Processing*, 45(1):137–147, January 1997.

1170. J.R. Treichler. Application of blind equalization techniques to voiceband and RF modems. In L. Dugard, M. M'Saad, and I.D. Landau, editors, *Adaptive Systems in Control and Signal Processing 1992. Selected Papers from the 4th IFAC Symposium*, pages 443–451, Grenoble, France, 1993. Pergamon Press.

1171. J.R. Treichler and C.R. Johnson. Blind fractionally-spaced equalization of digital cable TV. In *Proc. IEEE SP Workshop on Stat. Signal Array Processing*, pages 122–130, Corfu, Greece, 1996.

1172. J.R. Treichler and M.G. Larimore. New processing techniques based on constant modulus adaptive algorithms. *IEEE Trans. on Acoustic, Speech and Signal Processing*, ASSP-33, No.2:420–431, April 1985.

1173. J.R. Treichler and M.G. Larimore. The tone-capture properties of CMA-based interference suppressors. *IEEE Trans. Acoustics, Speech, Signal Processing*, 33(4):946–958, August 1985.

1174. J.R. Treichler, V. Wolff, and C.R. Johnson, Jr. Observed misconvergence in the constant modulus adaptive algorithm. In *25-th Asilomar Conference Signals, Systems Comp.*, pages 663–667 vol.2, Pacific Grove, CA, USA, 4-6 Nov. 1991, 1991. IEEE.

1175. J.R. Treichler, S.L. Wood, and M.G. Larimore. Convergence rate limitations in certain frequency-domain adaptive filters. In *Proc. IEEE ICASSP*, pages 960–963 vol.2, Glasgow, UK, 23-26 May 1989, 1989.

1176. T. Trump and B. Ottersten. Estimation of nominal direction of arrival and angular spread using an array of sensors. *Signal Processing*, 50(1-2):57–69, April 1996.

1177. M.K. Tsatsanis and G.B. Giannakis. Time-varying channel equalization using multiresolution analysis. In *Proc. IEEE-SP Int. Symposium Time-Freq. Time-Scale Analysis*, pages 447–450, Victoria, BC, 1992. IEEE.

1178. M.K. Tsatsanis and G.B. Giannakis. Equalization of rapidly fading channels: self-recovering methods. *IEEE Trans. Communications*, 44(5):619–630, May 1996.

1179. M.K. Tsatsanis and G.B. Giannakis. Blind estimation of direct sequence spread spectrum signals in multipath. *IEEE Trans. Signal Processing*, 45(5):1241–1252, May 1997.

1180. M.K. Tsatsanis and G.B. Giannakis. Transmitter induced cyclostationarity for blind channel equalization. *IEEE Trans. Signal Processing*, 45(7):1785–1794, July 1997.

1181. D. Tufts, E. Real, and J. Cooley. Fast approximate subspace tracking (FAST). In *Proc. of the ICASSP'97*, pages 547–550, April 1997.

1182. J.K. Tugnait. Identification of linear stochastic system via second and fourth-order cumulant matching. *IEEE Trans. Information Theory*, 33:393–407, May 1987.

1183. J.K. Tugnait. Adaptive filters and blind equalizers for mixed phase channels. *Proc. SPIE*, 1565:209–220, 1991.

1184. J.K. Tugnait. Blind deconvolution based criteria for parameter estimation with noisy data. In *26-th Asilomar Conference Signals, Systems Comp.*, pages 329–333 vol.1. IEEE, 1992.

1185. J.K. Tugnait. Blind equalization and estimation of digital communication FIR channels using cumulant matching. In *26-th Asilomar Conference Signals, Systems Comp.*, pages 726–730 vol.2. IEEE, 1992.

1186. J.K. Tugnait. A globally convergent adaptive blind equalizer based on second and fourth order statistics. In *SUPERCOMM/ICC*, pages 1508–1512 vol.3, Chicago, IL, 1992. IEEE.

1187. J.K. Tugnait. Blind estimation of digital communication channel impulse response. *IEEE Trans. Communications*, 42(2-4,):1606–1616, February 1994.

1188. J.K. Tugnait. Blind equalization and estimation of digital communication FIR channels using cumulant matching. *IEEE Trans. Communications*, 43(2-4,):1240–1245, February 1995.

1189. J.K. Tugnait. On fractionally-spaced blind adaptive equalization under symbol timing offsets using Godard and related equalizers. In *Proc. IEEE ICASSP*, pages 1976–1979 vol.3, Detroit, MI, 1995.

1190. J.K. Tugnait. Blind equalization and channel estimation for multiple-input multiple- output communications systems. In *Proc. IEEE International Conference Acoustics, Speech, Signal Processing, Atlanta, GA*, volume 5, pages 2443–2446, May 1996.

1191. J.K. Tugnait. Blind equalization and estimation of FIR communications channels using fractional sampling. *IEEE Trans. Communications*, 44(3):324–336, March 1996.

1192. J.K. Tugnait. On blind separation of convolutive mixtures of independent linear signals. In *Proc. IEEE SP Workshop on Stat. Signal Array Processing*, pages 312–315, Corfu, Greece, 1996.

1193. J.K. Tugnait. On fractionally spaced blind adaptive equalization under symbol timing offsets using Godard and related equalizers. *IEEE Trans. Signal Processing*, 44(7):1817–1821, July 1996.

1194. J.K. Tugnait. Spatio-temporal signal processing for blind separation of multichannel signals. *Proc. SPIE*, 2750:88–103, 1996.

1195. J.K. Tugnait. Blind spatio-temporal equalization and impulse response estimation for MIMO channels using a Godard cost function. *IEEE Trans. Signal Processing*, 45(1):268–271, January 1997.

1196. J.K. Tugnait. Identification and deconvolution of multichannel linear non-Gaussian processes using higher order statistics and inverse filter criteria. *IEEE Trans. Signal Processing*, 45:658–672, 1997.

1197. J.K. Tugnait. Adaptive blind separation of convolutive mixtures of independent linear signals. *Signal Processing*, 73(1-2):139–152, February 1999. EURASIP Journal.

1198. J.K. Tugnait and U. Gummadavelli. Blind channel estimation and deconvolution in colored noise using higher-order cumulants. *Journal of the Franklin Institute*, 333B(3):311–337, May 1996.

1199. J.K. Tugnait, O. Shalvi, and E. Weinstein. Comments on 'New criteria for blind deconvolution of nonminimum phase systems (channels)' (and reply). *IEEE Trans. Informat. Th.*, 38(1):210–213, January 1992.

1200. J.K. Tugnait, L. Tong, and Z. Ding. Single-user channel estimation and equalization. *IEEE Signal Processing Magazine*, 17(3):16–28, May 2000. (an invited paper for the special issue on Signal Processing Advances in Wireless and Mobile Communications).

1201. K. Tugnait and B. Huang. Multistep linear predictors-based blind identification and equalization of multiple-input multiple-output channels. *IEEE Trans. Signal Processing*, 48:26–38, January 2000.

1202. M. Turk and A. Pentland. Eigenfaces for recognition. *Journal of Cognitive Neuroscience*, 3:71–86, 1991.

1203. F.-B. Ueng, W.-R. Wu, Y.T. Su, and M.-J. Hsieh. Blind equalization using a self-orthogonalizing algorithm. *Journal of the Chinese Institute of Electrical Engineering*, 3(2):101–114, May 1996.

1204. F.B. Ueng and Y.T. Su. Adaptive blind equalization using second- and higher order statistics. *IEEE Journal on Selected Areas in Communications*, 13(1):132–140, January 1995.

1205. G. Ungerboeck. Adaptive maximum-likelihood receiver for carrier-modulated data transmission systems. *IEEE Trans. Communications*, 22(5):624–636, May 1974.

1206. C. Vaidyanathan, K.M. Buckley, and S. Hosur. A blind adaptive antenna array for CDMA systems. In *29-th Asilomar Conference Signals, Systems, Comp.*, pages 1373–1377 vol.2. IEEE, 1996.

1207. D. Valentin and H. Abdi. Can a linear autoassociator recognize faces from new orientations. *Journal of the Optical Society of America A*, 13, 1996.

1208. D. Valentin, H. Abdi, A.J. O'Toole, and G.W. Cottrell. Connectionist models of face processing: a survey. *Pattern Recognition*, 27:1209–1230, 1994.

1209. R. Vallet. Symbol by symbol MAP detection and the Baum-Welch identification algorithm in digital communications. In J. Vandewalle e.a., editor, *Signal Processing VI: Proc. EUSIPCO-90*, pages 131–134 vol.1, Brussels, 1992. Elsevier.

1210. D. Van Compernolle and S. Van Gerven. Signal separation in a symmetric adaptive noise canceler by output decorrelation. In *International Conference on Acoustics, Speech and Signal Processing*, volume IV, pages 221–224, 1992.

1211. L. van den Steen. Time-averaging circuit with continuous response. *IEE Proc. G.*, 128:111–113, 1981.

1212. A.-J. van der Veen, S. Talvar, and A. Paulraj. A subspace approach to blind space-time signal processing for wireless communication systems. *IEEE trans. on Signal Processing*, 45, No.1:173–190, Jan 1997.

1213. A.J. van der Veen. Resolution limits of blind multi-user multi-channel identification schemes — the band-limited case. In *Proc. IEEE ICASSP*, pages 2722–2725, Atlanta (GA), May 1996.

1214. A.J. van der Veen. Analytical method for blind binary signal separation. *IEEE Trans. Signal Processing*, 45(4):1078–1082, April 1997.

1215. A.J. van der Veen. Algebraic methods for deterministic blind beamforming. *Proc. IEEE*, 86(10), October 1998.

1216. A.J. van der Veen. Joint diagonalization via subspace fitting techniques. In *Proc. of IEEE ICASSP 2001, Salk Lake, USA*, 2001.

1217. A.J. van der Veen, E.F. Deprettere, and A.L. Swindlehurst. Subspace based signal analysis using singular value decomposition. *Proceedings of the IEEE*, 81(9):1277–1308, Sept. 1993.

1218. A.J. van der Veen and A. Paulraj. Analytical solution to the constant modulus factorization problem. In *28-th Asilomar Conference on Signals, Systems, and Computers*, pages 1433–1437. IEEE, October 1994.

1219. A.J. van der Veen and A. Paulraj. A constant modulus factorization technique for smart antenna applications in mobile communications. In F.T. Luk, editor, *Proc. SPIE, "Advanced*

Signal Processing Algorithms, Architectures, and Implementations V", volume 2296, pages 230–241, San Diego, CA, July 1994.

1220. A.J. van der Veen and A. Paulraj. An analytical constant modulus algorithm. *IEEE Trans. Signal Processing*, 44(5):1136–1155, May 1996.

1221. A.J. van der Veen and A. Paulraj. Singular value analysis of space-time equalization in the GSM mobile system. In *Proc. IEEE ICASSP*, pages 1073–1076, Atlanta (GA), May 1996.

1222. A.J. van der Veen, S. Talwar, and A. Paulraj. Blind estimation of multiple digital signals transmitted over FIR channels. *IEEE Signal Processing Letters*, 2(5):99–102, May 1995.

1223. A.J. van der Veen, S. Talwar, and A. Paulraj. Blind estimation of multiple digital signals transmitted over multipath channels. In *Proc. IEEE MILCOM*, volume 2, pages 581–585, San Diego, November 1995.

1224. A.J. van der Veen, S. Talwar, and A. Paulraj. Blind identification of FIR channels carrying multiple finite alphabet signals. In *Proc. IEEE ICASSP*, pages 1213–1216, Detroit, MI, May 1995.

1225. A.J. van der Veen, S. Talwar, and A. Paulraj. A subspace approach to blind space-time signal processing for wireless communication systems. *IEEE Trans. Signal Processing*, 45(1):173–190, January 1997.

1226. A.J. van der Veen and J. Tol. Separation of zero/constant modulus signals. In *Proc. IEEE ICASSP*, pages 3445–3448, Munich (Germany), April 1997.

1227. S. Van Gerven. *Adaptive Noise Cancellation And Signal Separation with Applications to Speech Enhancement.* Dissertation. Katholieke Universiteit Leuven, Department Elektrotechnik, Department Elektrotechnik, Leuven, Belgium, 1996.

1228. S. Van Gerven and D. Van Compernolle. Feedforward and feedback in a symmetric adaptive noise canceler: "stability analysis in a simplified case". In J. Vandevalle, R. Boite, M. Moonen, and A. Oosterlink, editors, *Signal Processing VI: Theories and Applications*, pages 1081–1084. Elsevier, 1992.

1229. S. Van Gerven and D. Van Compernolle. On the use of decorrelation in scalar signal separation. In *Proc. IEEE ICASSP*, pages III.57–60, 1994.

1230. S. Van Gerven and D. Van Compernolle. On the use of decorrelation in scalar signal separation. In *International Conference on Acoustics, Speech and Signal Processing*, volume III, pages 57–60, 1994.

1231. S. Van Gerven, D. Van Compernolle, H.L. N.-Thi, and C. Jutten. Blind separation of sources: A comparative study of a 2–nd and a 4–th order solution. In *Signal Processing VII.*, Proceedings of EUSIPCO–94, pages 1153–1156, Lausanne, CH., 1994. EUSIP Association.

1232. H.L. Van Trees. *Detection, Estimation and Modulation Theory*, volume I. Wiley, New York, 1968.

1233. J. Vanderschoot, D. Callaerts, W. Sansen, J. Vandewalle, G. Vantrappen, and J. Janssens. Two methods for optimal MCEG elimination and FECG detection from skin electrode signals. *IEEE Trans. on Biomedical Engineering*, BME-34(3):233–243, March 1987.

1234. M.C. Vanderveen, C.B. Papadias, and A. Paulraj. Joint angle and delay estimation (jade) for multipath signals arriving at an antenna array. *IEEE Communications Letters*, 1, No.1:12–14, Jan. 1997.

1235. V. Vapnik. *Statistical Learning Theory.* John Wiley & Sons, 1998.

1236. M. Varanasi and B. Aazhang. Multistage detection in asynchronous code-division multiple-access communications. *IEEE Tran.. Communication.*, 38(4):509–519, April 1990.

1237. S. Verdu. Minimum probability of error for asynchronous Gaussian multiple-access channels. *IEEE Trans. on Information Theory*, 32:pp. 2461–2470, 1986.

1238. S. Verdu. *Multiuser Detection.* JAI Press, Greenwich, CT, 1993.

1239. S. Verdu, B.D.O. Anderson, and R.A. Kennedy. Blind equalization without gain identification. *IEEE Trans. Informat. Th.*, 39(1):292–297, January 1993.

1240. J. Vidal and J.A. Rodriguez Fonollosa. Impulse response recovery of linear systems through weighted cumulant slices. *IEEE Trans. Signal Processing*, 44(10):2626–2631, October 1996.

1241. M. Vidyasagar. *Nonlinear Systems Analysis.* Prentice-Hall, Englewood Cliffs, NJ, 1993.

1242. R. Vigário, A. Hyvärinen, and E. Oja. ICA fixed-point algorithm in extraction of artifacts from EEG. In *Proc. 1996 IEEE Nordic Signal Processing Symposium NORSIG'96*, pages 383–386, Espoo, Finland, Sept. 1996.

1243. R. Vigário and E. Oja. Signal separation and feature extraction by nonlinear PCA network. In *Proc. 9th Scand. Conference on Image Analysis*, pages 811–818, Uppsala, Sweden, June 1995.

1244. R. N. Vigário. Extraction of ocular artifacts from EEG using independent component analysis. *Electroencephalography and Clinical Neurophysiology*, 103:395–404, 1997.

1245. R. N. Vigário, M. Jousmäki, R. Hämäläinen, E. Hari, and E. Oja. Independent component analysis for identification of artifacts in magnetoencephalographic recordings. In *Advances in Neural Information Processing System 10 (Proceedings NIPS*97)*, pages 229 – 235, 1997.

1246. E.A. Vittoz and X. Arreguit. CMOS integration of Hérault-Jutten cells for separation of sources. In C. Mead and M. Ismail, editors, *Analog VLSI Implementation of Neural Systems*, pages 57–84. Kluwer, Boston, 1989.

1247. S. Vorobyov and A. Cichocki. Hyper radial basis functions neural networks for interference cancellation with nonlinear processing of reference signal. *Digital Signal Processing*, 11:204–221, 2001.

1248. S. Vorobyov and A. Cichocki. Blind noise reduction for multisensory signals using ICA and subspace filtering, with applications to EEG analysis. *Biological Cybernetics*, vol.86, No. 4: pp.293-303, 2002.

1249. S. Vorobyov, A. Cichocki, and Y. V. Bodyanskiy. Adaptive noise cancellation for multi-sensory signals. *Fluctuation and Noise Letters*, 1:R13–R24, 2001.

1250. B.W. Wah, A. Ieumwananonthachai, S. Yao, and T. Yu. Statistical generalization: theory and applications. In *Proceedings. International Conference on Computer Design: VLSI in Computers and Processors*, pages 4–10, Austin, TX, 1995. IEEE Comput. Society Press.

1251. K. Waheed and F. M. Salam. State-space blind source recovery for mixtures of multiple source distributions. In *IEEE Int. Symposium on Circuits and Systems (ISCAS-2002)*, page in print, Scottsdale, Arizona, USA, May 26-29 2002.

1252. C. Wang, H.-C. Wu, and J.C. Principe. Crosscorrelation estimation using teacher forcing Hebbian learning and its application. In *ICNN 96. The 1996 IEEE International Conference on Neural Networks*, pages 282–287 vol.1, Proceedings of International Conference on Neural Networks (ICNN'96), Washington, DC, USA, 3-6 June 1996, 1996. IEEE.

1253. J. Wang and Z. He. Criteria and algorithms for blind source separation based on cumulants. *International Journal of Electronics*, 81(1):1–14, July 1996.

1254. L. Wang and J. Karhunen. A unified neural bigradient algorithm for robust PCA and MCA. *International Journal of Neural Systems*, 7(1):53–67, March 1996.

1255. L. Wang, J. Karhunen, E. Oja, and R. Vigário. Blind separation of sources using nonlinear PCA type learning algorithms. In *Proc. of the 1995 Int. Conference on Neural Networks and Signal Processing (ICNNSP95)*, Nanjing, P.R. China, Dec. 10–13 1995.

1256. X. Wang and H.V. Poor. Blind equalization and multiuser detection in dispersive CDMA channels. *IEEE Tr. Comm.*, 46(1):91–103, 1998.

1257. X.F. Wang, W.S. Lu, and A. Antoniou. Efficient implementation of linear multiuser detectors. In *Proc. IEEE ICASSP*, 1998.

1258. Y. Wang, Y.C. Pati, Y.M. Cho, A. Paulraj, and T. Kailath. A matrix factorization approach to signal copy of constant modulus signals arriving at an antenna array. In *Proc. 28-th Conference on Informat. Sciences and Systems*, Princeton, NJ, March 1994.

1259. Z. Wang and E.M. Dowling. Stochastic conjugate gradient constant modulus blind equalizer for wireless communications. In *Proc. IEEE Int. Conference Communic.*, pages 832–836 vol.2, Dallas, TX, 1996.

1260. M. Wax and T. Kailath. Detection of signals by information theoretic criteria. *IEEE Trans. Acoustics, Speech., Signal Processing*, 33:387–392, April 1985.

1261. M. Wax and J. Sheinvald. A least-squares approach to joint diagonalization. *IEEE Signal Processing Letters*, 4(2):52–53, February 1997.

1262. M. Wax and I. Ziskind. On unique localization of multiple sources by passive sensor arrays. *IEEE Trans. Acoustics, Speech., Signal Processing*, 37:996–1000, July 1989.

1263. V. Weerackody and S.A. Kassam. Blind equalization using lattice filters. In *Proc. IEEE Int. Conference Communic.*, pages 376–379 vol.1, Philadelphia, PA, 1988.

1264. V. Weerackody and S.A. Kassam. Blind adaptive equalization using dual-mode algorithms. In *24-th Asilomar Conference Signals, Systems Comp.*, pages 263–267 vol.1. Maple Press, 1990.

1265. V. Weerackody and S.A. Kassam. Variable step-size blind adaptive equalization algorithms. In *1991 IEEE International Symposium on Circuits and Systems*, pages 718–721 vol.1, Singapore, 1991. IEEE.

1266. V. Weerackody and S.A. Kassam. Dual-mode type algorithms for blind equalization. *IEEE Trans. Communications*, 42(1):22–28, January 1994.

1267. V. Weerackody, S.A. Kassam, and K.R. Laker. A convergence model for the analysis of some blind equalization algorithms. In *Proc. IEEE ISCAS*, pages 2136–2139 vol.3, Portland, OR, 1989.

1268. V. Weerackody, S.A. Kassam, and K.R. Laker. Convergence analysis of an algorithm for blind equalization. *IEEE Trans. Communications*, 39(6):856–865, June 1991.

1269. V. Weerackody, S.A. Kassam, and K.R. Laker. Sign algorithms for blind equalization and their convergence analysis. *Circuits, Systems, and Signal Processing*, 10(4):393–431, 1991.

1270. V. Weerackody, S.A. Kassam, and K.R. Laker. A simple hard-limited adaptive algorithm for blind equalization. *IEEE Trans. Circuits Systems II*, 39(7):482–487, July 1992.

1271. E. Weinstein, M. Feder, and A.V. Oppenheim. Multi-channel signal separation by decorrelation. *IEEE Trans. on Speech and Audio Processing*, 1(4):405–413, Oct. 1993.

1272. E. Weinstein, A.V. Oppenheim, M. Feder, and J.R. Buck. Iterative and sequential algorithms for multisensor signal enhancement. *IEEE Trans. on Signal Processing*, 42(4):846–859, April 1994.

1273. A.J. Weiss and B. Friedlander. "Almost blind" steering vector estimation using second-order moments. *IEEE Trans. Signal Processing*, 44(4):1024–1027, April 1996.

1274. K. Wesolowski. Analysis and properties of the modified constant modulus algorithm for blind equalization. *European Trans. on Telecommunications and Related Technologies*, 3(3):225–230, May 1992.

1275. K. Wesolowski. On acceleration of adaptive blind equalization algorithms. *Archiv fur Elektronik und Uebertragungstechnik*, 46(6):392–399, November 1992.

1276. L.B. White. Blind equalization of constant modulus signals using an adaptive observer approach. *IEEE Trans. Communications*, 44(2):134–136, February 1996.

1277. L.B. White, S. Perreau, and P. Duhamel. Reduced computation blind equalization for FIR channel input Markov models. In *Proc. IEEE Int. Conference Communic.*, pages 993–997 vol.2, Seattle, WA, 1995.

1278. B. Widrow. Adaptive noise cancelling : Principles and applications. *Proc. IEEE*, 63:1692–1716, December 1975.

1279. B. Widrow and S.D. Stearns. *Adaptive Signal Processing.* Prentice–Hall, Englewood Cliffs, NJ, 1985.

1280. B. Widrow and E. Walach. *Adaptive Inverse Control.* Upper Saddle River, NJ: Prentice-Hall PTR, 1996.

1281. T. Wigren. Convergence analysis of recursive identification algorithms based on the nonlinear Wiener model. *IEEE Trans. Automatic Contr.*, 39:2191–2206, November 1994.

1282. R.P. Wildes and et al. A system for automated iris recognition. In *Workshop on the Application of Computer Vision*, pages 121–128, Los Alamitos, CA, USA, 1994. IEEE Computer Society Press.

1283. M.E. Wilhoyte and T. Ogunfunmi. A decision-directed Bayesian equalizer. In *29-th Asilomar Conference Signals, Systems, Comp.*, pages 325–329 vol.1. IEEE, 1996.

1284. R. Williams and D. Zipser. A learning algorithm for continually running fully recurrent neural networks. *Neural Computation*, 1(2):270–280, 1989.

1285. D. Williamson and I. Mareels. Quantization issues in blind equalization: a case study. In *IEEE Pacific Rim Conference on Communications, Computers and Signal Processing*, pages 657–660 vol.2, Victoria, BC, 1991. IEEE.

1286. J.H. Winters. Optimum combining in digital mobile radio with co-channel interference. *IEEE Jour. on Selected Areas in Comm.*, 2(4):528–539, 1984.

1287. C.-F. Wong and T.L. Fine. Adaptive blind equalization using artificial neural networks. In *ICNN 96. The 1996 IEEE International Conference on Neural Networks*, pages 1974–1979 vol.4, Proceedings of International Conference on Neural Networks (ICNN'96), Washington, DC, USA, 3-6 June 1996, 1996. IEEE.

1288. C.-F. Wong and T.L. Fine. Blind signal processing: adaptive blind equalization using artificial neural networks. In S. Amari, L. Xu, L.-W. Chan, I. King, and K.-S. Leung, editors, *Progress in Neural Information Processing. Proceedings of the International Conference on Neural Information Processing*, pages 1217–1222 vol.2, Hong Kong, 1996. Springer-Verlag.

1289. H.E. Wong and J.A. Chambers. Two-stage interference immune blind equalizer which exploits cyclostationary statistics. *Electronics Letters*, 32(19):1763–1764, September 1996.

1290. G. Woodward and B. Vucetic. Adaptive detection for DS-CDMA. *Proceedings IEEE*, 86(7):1413–1434, July 1998.

1291. H.-S. Wu. Minimum entropy deconvolution for restoration of blurred two-tone images. *Electronics Letters*, 26(15):1183–1184, July 1990.

1292. J. Wu and Z.-M. Liu. A new scheme of decision-feedback blind equalization based on strictly convex cost function. *Acta Electronica Sinica*, 23(1):21–27, January 1995.

1293. J.-H. Wu and Z.-M. Liu. Blind equalization using a Kalman filter. In *ICCT '92. Proceedings of 1992 International Conference on Communication Technology*, pages 13.05/1–4 vol.1, Beijing, China, 1992. Int. Acad. Publishers.

1294. J.-H. Wu and Z.-M. Liu. A novel series of blind equalization criteria. In *ICCT '92. Proceedings of 1992 International Conference on Communication Technology*, pages 03.07/1–4 vol.1, Beijing, China, 1992. Int. Acad. Publishers.

1295. W.-R. Wu and M.J. Hsieh. Blind equalization using a fast self-orthogonalizing algorithm. In J. Vandewalle e.a., editor, *Signal Processing VI: Proc. EUSIPCO-92*, pages 1625–1628 vol. 3, Brussels, 1992. Elsevier.

1296. Z. Xie, R.T. Short, and C.K. Rushforth. A family of suboptimum detectors for coherent multiuser communications. *IEEE J. Sel. Areas Comm.*, 8(4):683–690, May 1990.

1297. D. Xu, J.C. Principe, and C.H. Wu. Generalized eigendecomposition with an on-line local algorithm. *IEEE Signal Processing Letters*, 5(11), November 1998.

1298. G. Xu, H. Liu, L. Tong, and T. Kailath. A least-squares approach to blind channel identification. *IEEE Trans. Signal Processing*, SP-43, no. 12:2982–2993, Dec. 1995.

1299. G. Xu, L. Tong, and H. Liu. A new algorithm for fast blind equalization of wireless communication channels. In *Proc. IEEE ICASSP*, pages IV:589–592, 1994.

1300. G. Xu, H. Zha, G. Golub, and T. Kailath. Fast algorithms for updating signal subspaces. *IEEE Trans. on Circuits and Systems*, 41:537–549, 1994.

1301. L. Xu. Least mean square error reconstruction principle for self-organizing neural-nets. *Neural Networks*, 6:627–648, 1993.

1302. L. Xu, C.-C. Cheung, J. Ruan, and S. Amari. Nonlinearity and separation capability: Further justification for the ICA algorithm with a learned mixture of parametric densities. In *Proc.European Symposium on Artificial Neural Networks, ESANN'97, Bruges*, pages 291–296, April ,16-17 1997.

1303. L. Xu, E. Oja, and C.Y. Suen. Modified Hebbian learning for curve and surface fitting. *Neural Networks*, 5:393–407, 1992.

1304. L. Xu and A. Yuille. Self organizing rules for robust principal component analysis. In *Advances in Neural Information Processing Systems, NIPS 5 (Eds. J. Hanson, J.D. Cowan and C.L. Giles*, pages 467–474, 1993.

1305. G.-Q. Xue, X.-H. Yu, and S.-X. Cheng. Blind sequence detection algorithm for uplink CDMA RAKE receiver. In *Proc. IEEE Int. Conference Communic.*, pages 842–846 vol.2, Dallas, TX, 1996.

1306. T. Yahagi and M.K. Hasan. Estimation of noise variance from noisy measurements of AR and ARMA systems: application to blind identification of linear time-invariant systems.

IEICE Trans. Fundamentals of Electronics, Communications and Computer Sciences, E77-A(5):847–855, May 1994.

1307. Y. Yamashita and H. Ogawa. Relative Karhunen-Loeve transform. *IEEE Trans. on Signal Processing*, 44:371–378, Feb. 1996.

1308. K. Yamazaki and R.A. Kennedy. On globally convergent blind equalization for QAM systems. In *Proc. IEEE ICASSP*, pages III/325–328 vol.3, Adelaide, 1994.

1309. K. Yamazaki and R.A. Kennedy. Blind equalization for QAM systems based on general linearly constrained convex cost functions. *International Journal of Adaptive Control and Signal Processing*, 10(6):707–744, November 1996.

1310. K. Yamazaki, R.A. Kennedy, and Z. Ding. Candidate admissible blind equalization algorithms for QAM communication systems. In *SUPERCOMM/ICC*, pages 1518–1522 vol.3, Chicago, IL, 1992. IEEE.

1311. K. Yamazaki, R.A. Kennedy, and Z. Ding. Globally convergent blind equalization algorithms for complex data systems. In *Proc. IEEE ICASSP*, pages 553–556 vol.4, San Francisco, CA, 1992.

1312. B. Yang. Projection approximation subspace tracking. *IEEE Trans. on Signal Processing*, 43(1):95–107, 1995.

1313. H.H. Yang and S. Amari. A stochastic natural gradient descent algorithm for blind signal separation. In S. Usui, Y. Tohkura, S. Katagiri, and E. Wilson, editors, *Neural Networks for Signal Processing VI. Proceedings of the 1996 IEEE Signal Processing Society Workshop*, pages 433–442, Neural Networks for Signal Processing VI. Proceedings of the 1996 IEEE Signal Processing Society Workshop, Kyoto, Japan, 4-6 Sept. 1996, 1996. IEEE.

1314. H.H. Yang and S. Amari. Two gradient descent algorithms for blind signal separation. In C. von der Malsburg, W. von Seelen, J.C. Vorbruggen, and B. Sendhoff, editors, *Artificial Neural Networks - ICANN 96. 1996 International Conference Proceedings*, pages 287–292, Bochum, Germany, 1996. Springer-Verlag.

1315. H.H. Yang and S. Amari. Adaptive on-line learning algorithms for blind separation: Maximum entropy and minimal mutual information. *Neural Comput.*, 9:1457–1482, 1997.

1316. H.H. Yang and S. Amari. A stochastic natural gradient descent algorithm for blind signal separation. In *Proc. of the IEEE Workshop on Neural Networks and Signal Processing, Japan.*, 1997.

1317. H.H. Yang, S. Amari, and A. Cichocki. Information back propagation for blind separation of sources from nonlinear mixture. In *Proc. ICNN-97, Huston, USA*, June 1997.

1318. H.H. Yang, S. Amari, and A. Cichocki. Information-theoretic approach to blind separation of sources in non-linear mixture. *Signal Processing*, 64(3):291–300, 1998.

1319. H.H. Yang and E.-S. Chng. An on-line learning algorithm for blind equalization. In S. Amari, L. Xu, L.-W. Chan, I. King, and K.-S. Leung, editors, *Progress in Neural Information Processing. Proceedings of the International Conference on Neural Information Processing*, pages 317–321 vol.1, Hong Kong, 1996. Springer-Verlag.

1320. T. Yang. Blind signal separation using cellular neural networks. *International Journal of Circuit Theory and Applications*, 22(17):399–408, 1994.

1321. X. Yang, T.K. Sarkar, and E. Arvas. A survey of conjugate gradient algorithms for solution of extreme eigen-problems of a symmetric matrix. *IEEE Trans. on Acoustics, Speech and Signal Processing*, 37(10):1550–1556, 1989.

1322. Y.-G. Yang, N.-I. Cho, and S.-U. Lee. Fast blind equalization by using frequency domain block constant modulus algorithm. In L.P. Caloba, P.S.R. Diniz, A.C.M. de Querioz, and E.H. Watanabe, editors, *38th Midwest Symposium on Circuits and Systems. Proceedings*, pages 1003–1006 vol.2, 38th Midwest Symposium on Circuits and Systems. Proceedings, Rio de Janeiro, Brazil, 13-16 Aug. 1995, 1996. IEEE.

1323. D. Yellin and B. Friedlander. Blind multi-channel system identification and deconvolution: performance bounds. In *Proc. IEEE SP Workshop on Stat. Signal Array Processing*, pages 582–585, Corfu, Greece, 1996.

1324. D. Yellin and B. Porat. Blind identification of FIR systems excited by discrete-alphabet inputs. *IEEE Trans. Signal Processing*, 41(3):1331–1339, March 1993.

1325. D. Yellin and E. Weinstein. Multi-channel signal separation based on cross-bispectra. In *IEEE Signal Processing Workshop on Higher-Order Statistics*, pages 270–274, 1993.

1326. D. Yellin and E. Weinstein. Criteria for multichannel signal separation. *IEEE Trans. on Signal Processing*, 42(8):2158–2168, Aug. 1994.

1327. D. Yellin and E. Weinstein. Multichannel signal separation: methods and analysis. *IEEE Trans. Signal Processing*, 44:106–118, Jan 1996.

1328. A. Yeredor. Blind source separation via the second characteristic function. *Signal Processing*, 80(5):897–902, 2000.

1329. Y.-L. You and M. Kaveh. A regularization approach to blind restoration of images degraded by shift-variant blurs. In *Proc. IEEE ICASSP*, pages 2607–2610 vol.4, Detroit, MI, 1995.

1330. A. Ypma and A. Leshem. Blind separation of machine vibration with bilinear forms. In *Proceedings of the Second International Workshop on ICA and BSS, ICA'2000*, pages 405–410, 19-22, June 2000.

1331. A. Ypma, A. Leshem, and R.P.W. Duin. Blind separation of rotating machine sources: bilinear forms and convolutive mixture. *Neurocomputing*, 2002.

1332. A. Ypma and P. Pajunen. Rotating machine vibration analysis with second-order independent component analysis. In *Proceedings of First International Workshop on Independent Component Analysis and Signal Separation (ICA'99)*, pages 37–42, Aussois, France, January 1999.

1333. X.-H. Yu and Z.-Y. Wu. A novel blind adaptive equalization technique. In *Proc. IEEE ICASSP*, pages 1445–1448, vol.3, Albuquerque, NM, 1990.

1334. N. Yuen and B. Friedlander. Asymptotic performance analysis of blind signal copy using fourth-order cumulants. *International Journal of Adaptive Control and Signal Processing*, 10(2-3):239–265, March 1996.

1335. A.L. Yuille, D.M. Kammen, and D.S. Cohen. Quadrature and the development of orientation selective cortical cells by Hebb rules. *Biological Cybernetics*, 61:183–194, 1989.

1336. W.I. Zangwill. *Nonlinear Programming: A Unified Approach.* Prentice Hall Inc., New Jersey, 1969.

1337. S. Zazo, J.M. Paez-Borrallo, and I.A.P. Alvarez. A linearly constrained blind equalization scheme based on Bussgang type algorithms. In *Proc. IEEE ICASSP*, pages 1037–1040 vol.2, Detroit, MI, 1995.

1338. H.H. Zeng and L. Tong. Connections between the least-squares and the subspace approaches to blind channel estimation. *IEEE Trans. Signal Processing*, 44:1593–1596, June 1996.

1339. H.H. Zeng, S. Zeng, and L. Tong. On the performance of blind equalization using the second-order statistics. In *Proc. IEEE ICASSP*, pages 2427–2430 vol. 5, Atlanta, GA, 1996.

1340. S. Zeng, H.H. Zeng, and L. Tong. Blind channel equalization via multiobjective optimization. In *Proc. IEEE SP Workshop on Stat. Signal Array Processing*, pages 160–163, Corfu, Greece, 1996.

1341. S. Zeng, H.H. Zeng, and L. Tong. Blind channel estimation by constrained optimization. In *Proc. IEEE ISCAS*, pages 89–92 vol.2, Atlanta, GA, 1996.

1342. S. Zeng, H.H. Zeng, and L. Tong. Blind equalization using CMA: performance analysis and a new algorithm. In *Proc. IEEE Int. Conference Communic.*, pages 847–851 vol.2, Dallas, TX, 1996.

1343. E. Zervas and J. Proakis. A sequential algorithm for blind equalization. In *Proc. IEEE MILCOM*, pages 231–235 vol.1, San Diego, CA, 1992.

1344. E. Zervas, J. Proakis, and V. Eyuboglu. Effects of constellation shaping on blind equalization. *Proc. SPIE*, 1565:178–187, 1991.

1345. P. Zetterberg. *Mobile Cellular Communications with Base Station Antenna Arrays: Spectrum Efficiency, Algorithms and Propagation Models.* PhD thesis, Royal Institute Technology, Stockholm, Sweden, June 1997.

1346. P. Zetterberg and B. Ottersten. The spectrum efficiency of a base station antenna array system for spatially selective transmission. *IEEE Trans. Vehicular Technology*, 44(3):651–660, August 1995.

1347. B. Zhang, M.N. Shirazi, and H. Noda. Blind restoration of degraded binary Markov random field images. *Graphical Models and Image Processing*, 58(1):90–98, January 1996.

1348. L. Zhang, S. Amari, and A. Cichocki. Natural gradient approach to blind separation of over- and under-complete mixtures. In *Proc. of the First International Workshop on Independent Component Analysis and Signal Separation - ICA'99*, pages 455–460, Aussois, France, January 11-15 1999.

1349. L. Zhang, S. Amari, and A. Cichocki. Semiparametric approach to multichannel blind deconvolution of nonminimum phase systems. In S.A. Solla, T.K. Leen, and K.-R. Müller, editors, *Advances in Neural Information Processing Systems*, volume 12, pages 363–369. MIT Press, Cambridge, MA, 2000.

1350. L. Zhang, S. Amari, and A. Cichocki. Equi-convergence algorithm for blind separation of sources with arbitrary distributions. In J.Mira and A. Prieto, editors, *Bio-Inspired Applications of Connectionism*, volume LNCS 2085, pages 626–833. Springer, Granada, Spain, 2001.

1351. L. Zhang, S. Amari, and A. Cichocki. Semiparametric model and superefficiency in blind deconvolution. *Signal Processing*, 81:2535–2553, Dec. 2001.

1352. L. Zhang and A. Cichocki. Blind deconvolution/equalization using state-space models. In *Proceedings of the 1998 IEEE Workshop on Neural Networks for Signal Processing (NNSP'98)*, pages 123–131, Cambridge, UK, August 31 - September 2 1998.

1353. L. Zhang and A. Cichocki. Blind separation of filtered source using state-space approach. In *Neural Information Processing Systems, NIPS'98*, pages 648–654, Denver, USA, 1998.

1354. L. Zhang and A. Cichocki. Blind separation/deconvolution of sources using canonical stable state-space models. In *Proceeding of the 1998 international Symposium on Nonlinear Theory and its Applications (NOLTA'98)*, pages 927–930, Crans-Montana, Switzerland, Sept, 14-17 1998.

1355. L. Zhang and A. Cichocki. Information backpropagation learning algorithm for blind dynamic separation. In *Proceeding of IASTED International Conference Signal and Image Processing (SIP'98)*, pages 1–5, Las Vegas, October. 28-31 1998.

1356. L. Zhang and A. Cichocki. Blind separation of filtered source using state-space approach. In M.S. Kearns, S.A. Solla, and D.A. Cohn, editors, *Advances in Neural Information Processing Systems, NIPS-98*, volume 11, pages 648–654. MIT press, Cambridge, MA, 1999.

1357. L. Zhang and A. Cichocki. Adaptive blind source separation for tracking active sources of biomedical data. In *Proc of Workshop on Signal Processing and Applications*, page Paper No. 45 at CDROM, Brisbane, Australia, 2000.

1358. L. Zhang and A. Cichocki. Blind deconvolution of dynamical systems: A state space approach (invited paper). *Japanese Journal of Signal Processing*, 4(2):111–130, March 2000.

1359. L. Zhang and A. Cichocki. Natural gradient approach to blind deconvolution of dynamical systems. In *Proceedings of the Second International Workshop on ICA and BSS, ICA'2000*, pages 27–32, Helsinki, Finland, 19-22 June 2000.

1360. L. Zhang and A. Cichocki. Feature extraction and blind separation of convolutive signals. In *Proceedings of the 8-th International Conference on Neural Information Processing (ICONIP'2001)*, pages 789–794, Shanghai, China, Nov. 14-18 2001.

1361. L. Zhang, A. Cichocki, and S. Amari. Geometrical structures of FIR manifolds and their application to multichannel blind deconvolution. In *Proceeding of Int'l IEEE Workshop on Neural Networks for Signal Processing (NNSP'99)*, pages 303–312, Madison, Wisconsin, USA, August 23-25 1999.

1362. L. Zhang, A. Cichocki, and S. Amari. Multichannel blind deconvolution of nonminimum phase systems using information backpropagation. In *Proc. of the Fifth International Conference on Neural Information Processing (ICONIP'99)*, pages 210–216, Perth, Australia, Nov. 16-20 1999.

1363. L. Zhang, A. Cichocki, and S. Amari. Natural gradient algorithm to blind separation of overdetermined mixture with additive noises. *IEEE Signal Processing Letters*, 6(11):293–295, 1999.

1364. L. Zhang, A. Cichocki, and S. Amari. Semiparametric approach to blind separation of dynamical systems. In *Proceedings of NOLTA'99*, pages 707–710, Hawaii, USA, Nov.28-Dec. 2 1999.

1365. L. Zhang, A. Cichocki, and S. Amari. Estimating function approach to multichannel blind deconvolution. In *IEEE APCCAS 2000*, pages 587–590, Tianjin, China, 2000.

1366. L. Zhang, A. Cichocki, and S. Amari. Kalman filter and state-space approach to multichannel blind deconvolution. In *Neural Network for Signal Processing X, Proc. of the IEEE Workshop on Neural Networks for Signal Processing, NNSP'2000*, pages 425–434, Sydney, Australia, December 11-13 2000. IEEE.

1367. L. Zhang, A. Cichocki, and S. Amari. Geometrical structures of FIR manifold and multichannel blind deconvolution. *Journal of VLSI Signal Processing*, pages 31–44, 2002.

1368. L. Zhang, A. Cichocki, and S. Amari. Multichannel blind deconvolution of nonminimum phase systems using filter decomposition. *submitted*, (April 5, 1999).

1369. L. Zhang, A. Cichocki, J. Cao, and S. Amari. Semiparametric approach to blind deconvolution. Technical report, The Institute of Electronics, Information & Communication Engineers of Japan (IEICE), 1999.

1370. Q. Zhang and Z. Bao. Dynamical system for computing the eigenvalue of positive definite matrix. *IEEE Trans. on Neural Network*, 6:790–794, 1995.

1371. C.-L. Zhao, Z.-M. Liu, and Z. Zhou. Blind equalization and parameters estimation of nonminimum phase channels using fourth order cumulants. In C.A.O. Zhigang, editor, *ICCT'96. 1996 International Conference on Communication Technology Proceedings*, pages 528–531 vol.1, Proceedings of International Conference on Communication Technology. ICCT '96, Beijing, China, 5-7 May 1996, 1996. IEEE.

1372. F. Zheng, S. Laughlin, and B. Mulgrew. Robust blind deconvolution algorithm: variance approximation and series decoupling. *Electronics Letters*, 26(13):921–923, June 1990.

1373. F.-C. Zheng, S. McLaughlin, and B. Mulgrew. Blind equalization of nonminimum phase channels: higher order cumulant based algorithm. *IEEE Trans. Signal Processing*, 41(2):681–691, February 1993.

1374. J. Zhu, X.-R. Cao, and M.L. Liou. A unified algorithm for blind separation of independent sources. In *Proc. IEEE ISCAS*, pages 153–156 vol.2, Atlanta, GA, 1996.

1375. J. Zhu, X.-R. Cao, and R.-W. Liu. Blind source separation based on output independence - theory and implementation. In *Proc. of Int. Symposium on Nonlinear Theory and its Applications, NOLTA-95*, volume 1, pages 97–102, Las Vegas, USA, 1995.

1376. J. Zhuang, X. Zhu, and D. Lu. A blind equalization algorithm for multilevel digital communication system. *Acta Electronica Sinica*, 20(7):28–35, July 1992.

1377. J. Zhuang, X.-L. Zhu, and D. Lu. On the control function of blind equalization algorithms (digital radio). *Modelling, Simulation & Control A*, 40(2):57–64, 1991.

1378. X. Zhuang and A. Lee Swindlehurst. Blind equalization via blind source separation techniques. *Signal Processing*, 2002.

1379. M. Zibulevsky, P. Kisilev, Y. Y. Zeevi and B. A. Pearlmutter. Blind source separation via multinode sparse representation. *In: Advances in Neural Information Processing Systems* 14, Morgan Kaufmann, 1049-1056, 2002.

1380. A. Ziehe, K.-R. Müller, G. Nolte, B.-M. Mackert, and G. Curio. Artifact reduction in biomagnetic recordings based on time-delayed second order correlations. *IEEE Trans. on Biomedical Engineering*, 47:75–87, 2000.

1381. I. Ziskind and Y. Bar-Ness. Localization of narrow-band autoregressive sources by passive sensor arrays. *IEEE Trans. on Signal Processing*, 40(2):484–487, Feb. 1992.

1382. M.-Y. Zou and R. Unbehauen. New algorithms of two-dimensional blind deconvolution. *Optical Engineering*, 34(10):2945–2956, October 1995.

13

Appendix – Mathematical Preliminaries

He who would perfect his work must first sharpen his tools.

—(Confucius)

13.1 MATRIX ANALYSIS

13.1.1 Matrix inverse update rules

Neumann expansion of the inverse matrix:

$$(\mathbf{I}+\mathbf{A})^{-1}=\mathbf{I}-\mathbf{A}+\mathbf{A}^2-\ldots, \tag{13.1}$$

if the eigenvalues of $\mathbf{A}$ have absolute values less than one.
Matrix inversion lemma:

$$(\mathbf{A}+\mathbf{BCD})^{-1}=\mathbf{A}^{-1}-\mathbf{A}^{-1}\mathbf{B}\left(\mathbf{C}^{-1}+\mathbf{DA}^{-1}\mathbf{B}\right)^{-1}\mathbf{DA}^{-1}. \tag{13.2}$$

Here, it is assumed that $\mathbf{A}\in\mathbb{R}^{n\times n}$, $\mathbf{B}\in\mathbb{R}^{n\times m}$, $\mathbf{C}\in\mathbb{R}^{m\times m}$ and $\mathbf{D}\in\mathbb{R}^{m\times n}$ and that indicated inverse exists. In the special case when $\mathbf{D}=\mathbf{B}^T$ we obtain

$$(\mathbf{A}+\mathbf{BCB}^T)^{-1}=\mathbf{A}^{-1}-\mathbf{A}^{-1}\mathbf{B}(\mathbf{C}^{-1}+\mathbf{B}^T\mathbf{A}^{-1}\mathbf{B})^{-1}\mathbf{B}^T\mathbf{A}^{-1}. \tag{13.3}$$

A special case known as Woodbury identity results for $\mathbf{B}=\mathbf{u}\in\mathbb{R}^n$ and $\mathbf{D}=\mathbf{v}^T\in\mathbb{R}^{n\times n},\mathbf{C}=\mathbf{I}$

$$\left(\mathbf{A}+\mathbf{uv}^T\right)^{-1}=\mathbf{A}^{-1}-\frac{\mathbf{A}^{-1}\mathbf{uv}^T\mathbf{A}^{-1}}{1+\mathbf{v}^T\mathbf{A}^{-1}\mathbf{u}}. \tag{13.4}$$

$$\mathbf{A}^{-1} = \begin{pmatrix} \mathbf{B}^{-1} + \mathbf{B}^{-1}\mathbf{D}\mathbf{F}^{-1}\mathbf{C}\mathbf{B}^{-1} & -\mathbf{B}^{-1}\mathbf{D}\mathbf{F}^{-1} \\ -\mathbf{F}^{-1}\mathbf{C}\mathbf{B}^{-1} & \mathbf{F}^{-1} \end{pmatrix}, \tag{13.5}$$

where matrices $\mathbf{A} = \begin{pmatrix} \mathbf{B} & \mathbf{D} \\ \mathbf{C} & \mathbf{Q} \end{pmatrix}$ and $\mathbf{F} = \mathbf{Q} - \mathbf{C}\mathbf{B}^{-1}\mathbf{D}$ are nonsingular.

13.1.2 Some properties of determinant

$$\det \begin{pmatrix} \mathbf{A} & \mathbf{B} \\ \mathbf{D} & \mathbf{C} \end{pmatrix} = \det(\mathbf{C}) \det(\mathbf{A} - \mathbf{B}\mathbf{C}^{-1}\mathbf{D}) = \det(\mathbf{A}) \det(\mathbf{C} - \mathbf{D}\mathbf{A}^{-1}\mathbf{B}), \tag{13.6}$$

where $\mathbf{A} \in \mathbb{R}^{s \times s}$, $\mathbf{C} \in \mathbb{R}^{r \times r}$ are nonsingular, $\mathbf{B} \in \mathbb{R}^{s \times r}$, $\mathbf{D} \in \mathbb{R}^{r \times s}$.
In particular,

$$\det(\mathbf{I}_r + \mathbf{A}\mathbf{B}) = \det(\mathbf{I}_s + \mathbf{B}\mathbf{A}), \tag{13.7}$$

where $\mathbf{I}_r \in \mathbb{R}^{r \times r}$ and $\mathbf{I}_s \in \mathbb{R}^{s \times s}$ are identity matrices, $\mathbf{A} \in \mathbb{R}^{r \times s}$, $\mathbf{B} \in \mathbb{R}^{s \times r}$.

$$\det(\mathbf{I} + \mathbf{X}\mathbf{X}^T) = 1 + \|\mathbf{X}\|^2. \tag{13.8}$$

13.1.3 Some properties of the Moore-Penrose pseudo-inverse

$$\mathbf{A}^+ = \mathbf{A}^{-1} \text{ for non-singular } \mathbf{A} \tag{13.9}$$

$$(\mathbf{A}^+)^+ = \mathbf{A} \tag{13.10}$$

$$(\mathbf{A}^T)^+ = (\mathbf{A}^+)^T \tag{13.11}$$

$$\mathbf{A}^+ = \mathbf{A} \text{ if } \mathbf{A} \text{ is symmetric and idempotent} \tag{13.12}$$

$$\mathbf{A}\mathbf{A}^+ \text{ and } \mathbf{A}^+\mathbf{A} \text{ are idempotent} \tag{13.13}$$

$$\mathbf{A}, \quad \mathbf{A}^+, \quad \mathbf{A}\mathbf{A}^+ \text{ and } \mathbf{A}^+\mathbf{A} \text{ have the same rank} \tag{13.14}$$

$$\mathbf{A}^T\mathbf{A}\mathbf{A}^+ = \mathbf{A}^T = \mathbf{A}^+\mathbf{A}\mathbf{A}^T \tag{13.15}$$

$$\mathbf{A}^T(\mathbf{A}^+)^T\mathbf{A}^+ = \mathbf{A}^+ = \mathbf{A}^+(\mathbf{A}^+)^T\mathbf{A}^T \tag{13.16}$$

$$(\mathbf{A}^T\mathbf{A})^+ = \mathbf{A}^+(\mathbf{A}^+)^T \tag{13.17}$$

$$(\mathbf{A}\mathbf{A}^T)^+ = (\mathbf{A}^+)^T\mathbf{A}^+\mathbf{A}(\mathbf{A}^T\mathbf{A})^+\mathbf{A}^T\mathbf{A} = \mathbf{A} = \mathbf{A}\mathbf{A}^T(\mathbf{A}\mathbf{A}^T)^+\mathbf{A} \tag{13.18}$$

$$\mathbf{A}^+ = (\mathbf{A}^T\mathbf{A})^+\mathbf{A}^T = \mathbf{A}^T(\mathbf{A}\mathbf{A}^T)^+ \tag{13.19}$$

$$\mathbf{A}^+ = (\mathbf{A}^T\mathbf{A})^{-1}\mathbf{A}^T \text{ if } \mathbf{A} \text{ has full column rank} \tag{13.20}$$

$$\mathbf{A}^+ = \mathbf{A}^T(\mathbf{A}\mathbf{A}^T)^{-1} \text{ if } \mathbf{A} \text{ has full row rank} \tag{13.21}$$

$$\mathbf{A} = \mathbf{0} \leftrightarrow \mathbf{A}^+ = \mathbf{0} \tag{13.22}$$

$$\mathbf{AB} = \mathbf{0} \leftrightarrow \mathbf{B}^+\mathbf{A}^+ = \mathbf{0} \tag{13.23}$$

$$\mathbf{A}^+\mathbf{B} = \mathbf{0} \leftrightarrow \mathbf{A}^T\mathbf{B} = \mathbf{0} \tag{13.24}$$

$$(\mathbf{A} \otimes \mathbf{B})^+ = \mathbf{A}^+ \otimes \mathbf{B}^+, \text{ where } \otimes \text{ is the Kronecker product symbol} \tag{13.25}$$

13.1.4 Matrix Expectations

$$E\{\mathbf{x}\} = \mathbf{m_x} \tag{13.26}$$

$$E\{(\mathbf{x} - \mathbf{m_x})(\mathbf{x} - \mathbf{m_x})^T\} = \mathbf{R_{xx}} \tag{13.27}$$

$$E\{\mathrm{tr}(\mathbf{A})\} = \mathrm{tr}(E\{\mathbf{A}\}) \tag{13.28}$$

$$E\{\mathbf{Ax} + \mathbf{b}\} = \mathbf{Am_x} + \mathbf{b} \tag{13.29}$$

$$E\{(\mathbf{Ax} + \mathbf{a})(\mathbf{Bx} + \mathbf{b})^T\} = \mathbf{AR_{xx}B}^T + (\mathbf{Am_x} + \mathbf{a})(\mathbf{Bm_x} + \mathbf{b})^T \tag{13.30}$$

$$E\{\mathbf{xx}^T\} = \mathbf{R_{xx}} + \mathbf{m_x m_x}^T \tag{13.31}$$

$$E\{\mathbf{xa}^T\mathbf{x}\} = (\mathbf{R_{xx}} + \mathbf{m_x m_x}^T)\mathbf{a} \tag{13.32}$$

$$E\{(\mathbf{Ax})(\mathbf{Ax})^T\} = \mathbf{A}(\mathbf{R_{xx}} + \mathbf{m_x m_x}^T)\mathbf{A}^T \tag{13.33}$$

$$E\{(\mathbf{x} + \mathbf{a})(\mathbf{x} + \mathbf{a})^T\} = \mathbf{R_{xx}} + (\mathbf{m_x} + \mathbf{a})(\mathbf{m_x} + \mathbf{a})^T \tag{13.34}$$

$$E\{(\mathbf{Ax} + \mathbf{a})^T(\mathbf{Bx} + \mathbf{b})\} = \mathrm{tr}(\mathbf{AR_{xx}B}^T) + (\mathbf{Am_x} + \mathbf{a})^T(\mathbf{Bm_x} + \mathbf{b}) \tag{13.35}$$

$$E\{\mathbf{x}^T\mathbf{x}\} = \mathrm{tr}(\mathbf{R_{xx}}) + \mathbf{m_x}^T\mathbf{m_x} = \mathrm{tr}(\mathbf{R_{xx}} + \mathbf{m_x m_x}^T) \tag{13.36}$$

$$E\{\mathbf{x}^T\mathbf{ax}^T\} = \mathbf{a}^T(\mathbf{R_{xx}} + \mathbf{m_x m_x}^T) \tag{13.37}$$

$$E\{\mathbf{x}^T\mathbf{Ax}\} = \mathrm{tr}(\mathbf{AR_{xx}}) + \mathbf{m_x}^T\mathbf{Am_x} = \mathrm{tr}(\mathbf{A}(\mathbf{R_{xx}} + \mathbf{m_x m_x}^T)) \tag{13.38}$$

$$E\{(\mathbf{Ax})^T(\mathbf{Ax})\} = \mathrm{tr}(\mathbf{AR_{xx}A}^T) + \mathbf{m_x}^T\mathbf{A}^T\mathbf{Am_x} = \mathrm{tr}(\mathbf{A}(\mathbf{R_{xx}} + \mathbf{m_x m_x}^T)\mathbf{A}^T) \tag{13.39}$$

$$E\{(\mathbf{x}+\mathbf{a})^T(\mathbf{x}+\mathbf{a})\} = \mathrm{tr}(\mathbf{R}_{\mathbf{x}\mathbf{x}}) + (\mathbf{m}_\mathbf{x}+\mathbf{a})^T(\mathbf{m}_\mathbf{x}+\mathbf{a}) = \mathrm{tr}(\mathbf{R}_{\mathbf{x}\mathbf{x}}) + \|\mathbf{m}_\mathbf{x}+\mathbf{a}\|^2 \quad (13.40)$$

$$\begin{aligned} E\{(\mathbf{A}\mathbf{x}+\mathbf{a})(\mathbf{B}\mathbf{x}+\mathbf{b})^T(\mathbf{C}\mathbf{x}+\mathbf{c})\} &= \mathbf{A}\mathbf{R}_{\mathbf{x}\mathbf{x}}\mathbf{B}^T(\mathbf{C}\mathbf{m}_\mathbf{x}+\mathbf{c}) \\ &+ \mathbf{A}\mathbf{R}_{\mathbf{x}\mathbf{x}}\mathbf{C}^T(\mathbf{B}\mathbf{m}_\mathbf{x}+\mathbf{b}) \\ &+ \mathrm{tr}(\mathbf{B}\mathbf{R}_{\mathbf{x}\mathbf{x}}\mathbf{C}^T)(\mathbf{A}\mathbf{m}_\mathbf{x}+\mathbf{a}) \\ &+ (\mathbf{A}\mathbf{m}_\mathbf{x}+\mathbf{a})(\mathbf{B}\mathbf{m}_\mathbf{x}+\mathbf{b})^T(\mathbf{C}\mathbf{m}_\mathbf{x}+\mathbf{c}) \end{aligned} \quad (13.41)$$

$$E\{\mathbf{x}\mathbf{x}^T\mathbf{x}\} = 2\mathbf{R}_{\mathbf{x}\mathbf{x}}\mathbf{m}_\mathbf{x} + (\mathrm{tr}(\mathbf{R}_{\mathbf{x}\mathbf{x}}) + \mathbf{m}_\mathbf{x}\mathbf{m}_\mathbf{x}^T)\mathbf{m}_\mathbf{x} \quad (13.42)$$

13.1.5 Differentiation of a scalar function with respect to a vector

$$\frac{\partial}{\partial \mathbf{x}}(\mathbf{y}^T\mathbf{x}) = \frac{\partial}{\partial \mathbf{x}}(\mathbf{x}^T\mathbf{y}) = \mathbf{y} \quad (13.43)$$

$$\frac{\partial}{\partial \mathbf{x}}(\mathbf{x}^T\mathbf{A}) = \mathbf{A} \quad (13.44)$$

$$\frac{\partial}{\partial \mathbf{x}}(\mathbf{x}^T) = \mathbf{I} \quad (13.45)$$

$$\frac{\partial}{\partial \mathbf{x}}(\mathbf{x}^T\mathbf{x}) = 2\mathbf{x} \quad (13.46)$$

$$\frac{\partial}{\partial \mathbf{x}}(\mathbf{x}^T\mathbf{A}\mathbf{y}) = \mathbf{A}\mathbf{y} \quad (13.47)$$

$$\frac{\partial}{\partial \mathbf{x}}(\mathbf{y}^T\mathbf{A}\mathbf{x}) = \mathbf{A}^T\mathbf{y} \quad (13.48)$$

$$\frac{\partial}{\partial \mathbf{x}}(\mathbf{x}^T\mathbf{A}\mathbf{x}) = (\mathbf{A}+\mathbf{A}^T)\,\mathbf{x} \quad (13.49)$$

$$\frac{\partial}{\partial \mathbf{x}}(\mathbf{x}^T\mathbf{A}\mathbf{x}) = 2\mathbf{A}\mathbf{x} \text{ if } \mathbf{A} \text{ is symmetric} \quad (13.50)$$

$$\frac{\partial}{\partial \mathbf{x}}(\mathbf{x}^T\mathbf{A}\mathbf{x}\mathbf{x}^T) = (\mathbf{A}+\mathbf{A}^T)\mathbf{x}\mathbf{x}^T + \mathbf{x}^T\mathbf{A}\mathbf{x}\mathbf{I} \quad (13.51)$$

$$\frac{\partial}{\partial \mathbf{x}}\left[\mathbf{a}^T(\mathbf{x})\,\mathbf{Q}\mathbf{a}(\mathbf{x})\right] = \nabla_\mathbf{x}\mathbf{a}^T(\mathbf{x})\,(\mathbf{Q}+\mathbf{Q}^T)\mathbf{a}(\mathbf{x}) \quad (13.52)$$

$$\frac{\partial}{\partial \mathbf{x}}\left[\mathbf{a}^T(\mathbf{x})\,\mathbf{b}(\mathbf{x})\right] = \left[\frac{\partial \mathbf{a}(\mathbf{x})}{\partial \mathbf{x}}\right]^T \mathbf{b}(\mathbf{x}) + \left[\frac{\partial \mathbf{b}(\mathbf{x})}{\partial \mathbf{x}}\right]^T \mathbf{a}(\mathbf{x}) \quad (13.53)$$

13.1.6 Matrix differentiation

$$\frac{d}{dt}(\mathbf{A}+\mathbf{B}) = \frac{d\mathbf{A}}{dt} + \frac{d\mathbf{B}}{dt}. \tag{13.54}$$

$$\frac{d}{dt}(\mathbf{ABC}) = \frac{d\mathbf{A}}{dt}\mathbf{BC} + \mathbf{A}\frac{d\mathbf{B}}{dt}\mathbf{C} + \mathbf{AB}\frac{d\mathbf{C}}{dt}. \tag{13.55}$$

$$\frac{d}{dt}\mathbf{A}^n = \frac{d\mathbf{A}}{dt}\mathbf{A}^{n-1} + \mathbf{A}\frac{d\mathbf{A}}{dt}\mathbf{A}^{n-2} + \ldots + \mathbf{A}^{n-1}\frac{d\mathbf{A}}{dt}. \tag{13.56}$$

$$\frac{d}{dt}\mathbf{A}^{-1} = -\mathbf{A}^{-1}\frac{d\mathbf{A}}{dt}\mathbf{A}^{-1} \quad for \quad \det(\mathbf{A}) \neq \mathbf{0}. \tag{13.57}$$

$$\frac{d}{dt}(\det(\mathbf{A})) = \operatorname{tr}\left(\frac{d\mathbf{A}}{dt}\mathbf{A}\right). \tag{13.58}$$

$$\frac{d}{dt}\log\det\mathbf{A} = \operatorname{tr}[\mathbf{A}^{-1}\frac{d}{dt}\mathbf{A}]. \tag{13.59}$$

$$\frac{\partial\det(\mathbf{X})}{\partial\mathbf{X}} = \frac{\partial\det(\mathbf{X}^T)}{\partial\mathbf{X}} = \det(\mathbf{X})\left(\mathbf{X}^T\right)^{-1} \quad \text{if } \det(\mathbf{X}) \neq \mathbf{0}. \tag{13.60}$$

$$\frac{\partial\det(\mathbf{X}^k)}{\partial\mathbf{X}} = k\det(\mathbf{X}^k)\left(\mathbf{X}^T\right)^{-1}. \tag{13.61}$$

$$\frac{\partial\log(\det(\mathbf{X}))}{\partial\mathbf{X}} = \left(\mathbf{X}^T\right)^{-1} \quad \text{if } \det(\mathbf{X}) \neq \mathbf{0}. \tag{13.62}$$

$$\frac{\partial\det(\mathbf{AXB})}{\partial\mathbf{X}} = \det(\mathbf{AXB})\mathbf{A}^T(\mathbf{B}^T\mathbf{X}^T\mathbf{A}^T)^{-1}\mathbf{B}^T \quad \text{if} \quad \det\left(\mathbf{B}^T\mathbf{X}\mathbf{A}^T\right)^{-1} \neq \mathbf{0}. \tag{13.63}$$

$$\frac{\partial\det(\mathbf{AXB})}{\partial\mathbf{X}} = \det(\mathbf{AXB})(\mathbf{X}^{-1})^T\det(\mathbf{X}) \neq \mathbf{0} \quad \text{if } \det\mathbf{A} \neq 0, \det\mathbf{B} \neq 0, \det\mathbf{X} \neq 0. \tag{13.64}$$

$$\frac{\partial\log(\det(\mathbf{XX}^T))}{\partial\mathbf{X}} = 2(\mathbf{XX}^T)^{-1}\mathbf{X} \quad \text{if } \det(\mathbf{XX}^T) \neq 0. \tag{13.65}$$

$$\frac{\partial\det(\mathbf{X}^T\mathbf{CX})}{\partial\mathbf{X}} = \det(\mathbf{X}^T\mathbf{CX})(\mathbf{C}+\mathbf{C}^T)\mathbf{X}(\mathbf{X}^T\mathbf{CX})^{-1} \tag{13.66}$$

$$\text{for real matrices, if } \det\mathbf{X}^T\mathbf{CX} \neq 0. \tag{13.67}$$

$$\frac{\partial\det(\mathbf{X}^T\mathbf{CX})}{\partial\mathbf{X}} = 2\det(\mathbf{X}^T\mathbf{CX})\mathbf{CX}(\mathbf{X}^T\mathbf{CX})^{-1}, \tag{13.68}$$

$$\text{if } \mathbf{C} \text{ is real and symmetric, and if } \det\mathbf{X}^T\mathbf{CX} \neq 0. \tag{13.69}$$

$$\frac{\partial \log(\det(\mathbf{X}^T\mathbf{C}\mathbf{X}))}{\partial \mathbf{X}} = 2\mathbf{C}\mathbf{X}(\mathbf{X}^T\mathbf{C}\mathbf{X})^{-1} \text{ if } \mathbf{C}, \tag{13.70}$$

$$\text{is real and symmetric, and if } \det \mathbf{X}^T\mathbf{C}\mathbf{X} \neq 0. \tag{13.71}$$

$$D_{\mathbf{X}}(\mathbf{A}\mathbf{X}\mathbf{B}) = \mathbf{B}^T \otimes \mathbf{A}, \tag{13.72}$$

where $D_{\mathbf{X}}\mathbf{F}(\mathbf{X})$ means the Jacobian matrix of the matrix function $\mathbf{F}$ at $\mathbf{X}$:

$$D_{\mathbf{X}}\mathbf{F}(\mathbf{X}) = \frac{\partial \text{vec} F(\mathbf{X})}{\partial \text{vec}\mathbf{X}^T}.$$

$$D_{\mathbf{X}}\mathbf{A}\mathbf{X}^{-1}\mathbf{B} = -(\mathbf{X}^{-1}\mathbf{B})^T \otimes (\mathbf{A}^T\mathbf{X}^{-1}). \tag{13.73}$$

$$\frac{\partial(\mathbf{a}^T\mathbf{X}\mathbf{b})}{\partial \mathbf{X}} = \mathbf{a}\mathbf{b}^T. \tag{13.74}$$

$$\frac{\partial(\mathbf{a}^T\mathbf{X}^T\mathbf{b})}{\partial \mathbf{X}} = \mathbf{b}\mathbf{a}^T. \tag{13.75}$$

$$\frac{\partial(\mathbf{a}^T\mathbf{X}^T\mathbf{X}\mathbf{b})}{\partial \mathbf{X}} = \mathbf{X}(\mathbf{a}\mathbf{b}^T + \mathbf{b}\mathbf{a}^T). \tag{13.76}$$

$$\frac{\partial(\mathbf{a}^T\mathbf{X}^T\mathbf{X}\mathbf{a})}{\partial \mathbf{X}} = 2\mathbf{X}\mathbf{a}\mathbf{a}^T. \tag{13.77}$$

$$\frac{\partial(\mathbf{a}^T\mathbf{X}^T\mathbf{C}\mathbf{X}\mathbf{b})}{\partial \mathbf{X}} = \mathbf{C}^T\mathbf{X}\mathbf{a}\mathbf{b}^T + \mathbf{C}\mathbf{X}\mathbf{b}\mathbf{a}^T. \tag{13.78}$$

$$\frac{\partial(\mathbf{a}^T\mathbf{X}^T\mathbf{C}\mathbf{X}\mathbf{a})}{\partial \mathbf{X}} = (\mathbf{C} + \mathbf{C}^T)\mathbf{X}\mathbf{a}\mathbf{a}^T. \tag{13.79}$$

$$\frac{\partial(\mathbf{a}^T\mathbf{X}^T\mathbf{C}\mathbf{X}\mathbf{a})}{\partial \mathbf{X}} = 2\mathbf{C}\mathbf{X}\mathbf{a}\mathbf{a}^T \text{ if } \mathbf{C} \text{ is symmetric.} \tag{13.80}$$

$$\frac{\partial((\mathbf{X}\mathbf{a} + \mathbf{b})^T\mathbf{C}(\mathbf{X}\mathbf{a} + \mathbf{b}))}{\partial \mathbf{X}} = (\mathbf{C} + \mathbf{C}^T)(\mathbf{X}\mathbf{a} + \mathbf{b})\mathbf{a}^T. \tag{13.81}$$

13.1.7 Trace

1. The trace of matrix $\mathbf{A}$, denoted by $\text{tr}(\mathbf{A})$ is the sum of the diagonal elements of $\mathbf{A}$. The trace is invariant under circular permutation of its argument, consequently:

$$\text{tr}\,(\mathbf{A}\mathbf{B}\mathbf{C}) = \text{tr}\,(\mathbf{B}\mathbf{C}\mathbf{A}) = \text{tr}\,(\mathbf{C}\mathbf{A}\mathbf{B}) \tag{13.82}$$

2. The trace is a linear operator:

$$\text{tr}\,(\mathbf{A} + \mathbf{B}) = \text{tr}\,(\mathbf{A}) + \text{tr}\,(\mathbf{B}) \text{ and } \text{tr}\,(\alpha\mathbf{A}) = \alpha\, \text{tr}\,(\mathbf{A}) \tag{13.83}$$

3. The following expectation of the quadratic Hermitian form holds:

$$\begin{aligned}
E\left\{\mathbf{x}^T\mathbf{A}\mathbf{x}\right\} &= E\left\{\operatorname{tr}\left(\mathbf{x}^T\mathbf{A}\mathbf{x}\right)\right\} = E\left\{\operatorname{tr}\left(\mathbf{A}\mathbf{x}\mathbf{x}^T\right)\right\} \\
&= \operatorname{tr}\left(\mathbf{A}E\left\{\mathbf{x}\mathbf{x}^T\right\}\right) = \operatorname{tr}\mathbf{A}\mathbf{R}_{\mathbf{x}\mathbf{x}} \qquad (13.84) \\
\operatorname{var}\left\{\mathbf{x}^T\mathbf{A}\mathbf{x}\right\} &= E\left\{\left(\mathbf{x}^T\mathbf{A}\mathbf{x}\right)^2\right\} = 2\operatorname{tr}(\mathbf{A}\mathbf{R}_{\mathbf{x}\mathbf{x}})^2 + 4\mathbf{m}_{\mathbf{x}}\mathbf{A}\mathbf{R}_{\mathbf{x}\mathbf{x}}\mathbf{A}\mathbf{m}_{\mathbf{x}}, \qquad (13.85)
\end{aligned}$$

if the random $n \times 1$ vector $\mathbf{x}$ has normal distribution.

13.1.8 Matrix differentiation of trace of matrices

$$\frac{\partial \operatorname{tr}(\mathbf{X})}{\partial \mathbf{X}} = \mathbf{I}. \qquad (13.86)$$

$$\frac{\partial \operatorname{tr}(\mathbf{A}\mathbf{X})}{\partial \mathbf{X}} = \mathbf{A}^T. \qquad (13.87)$$

$$\frac{\partial \operatorname{tr}\left(\mathbf{A}\mathbf{X}^T\right)}{\partial \mathbf{X}} = \mathbf{A}. \qquad (13.88)$$

$$\frac{\partial \operatorname{tr}\left(\mathbf{X}^k\right)}{\partial \mathbf{X}} = k\left(\mathbf{X}^{k-1}\right)^T. \qquad (13.89)$$

$$\frac{\partial \operatorname{tr}\left(\mathbf{A}\mathbf{X}^k\right)}{\partial \mathbf{X}} = \left(\sum_{i=0}^{k-1} \mathbf{X}^i\mathbf{A}\mathbf{X}^{k-i-1}\right)^T. \qquad (13.90)$$

$$\frac{\partial \operatorname{tr}\left(\mathbf{X}^T\mathbf{A}\mathbf{X}\right)}{\partial \mathbf{X}} = \left(\mathbf{A} + \mathbf{A}^T\right)\mathbf{X}. \qquad (13.91)$$

$$\frac{\partial \operatorname{tr}\left(\mathbf{X}\mathbf{A}\mathbf{X}^T\right)}{\partial \mathbf{X}} = \mathbf{X}\left(\mathbf{A} + \mathbf{A}^T\right). \qquad (13.92)$$

$$\frac{\partial \operatorname{tr}\left(\mathbf{A}\mathbf{X}^T\mathbf{B}\right)}{\partial \mathbf{X}} = \mathbf{B}\mathbf{A}. \qquad (13.93)$$

$$\frac{\partial \operatorname{tr}(\mathbf{A}\mathbf{X}\mathbf{B})}{\partial \mathbf{X}} = \mathbf{A}^T\mathbf{B}^T. \qquad (13.94)$$

$$\frac{\partial \operatorname{tr}(\mathbf{A}\mathbf{X}\mathbf{B}\mathbf{X})}{\partial \mathbf{X}} = \mathbf{A}^T\mathbf{X}^T\mathbf{B}^T + \mathbf{B}^T\mathbf{X}^T\mathbf{A}^T. \qquad (13.95)$$

$$\frac{\partial \operatorname{tr}\left(\mathbf{A}\mathbf{X}\mathbf{B}\mathbf{X}^T\right)}{\partial \mathbf{X}} = \mathbf{A}\mathbf{X}\mathbf{B} + \mathbf{A}^T\mathbf{X}\mathbf{B}^T. \qquad (13.96)$$

$$\frac{\partial \operatorname{tr}\left(\mathbf{A}\mathbf{X}\mathbf{X}^T\mathbf{B}\right)}{\partial \mathbf{X}} = \left(\mathbf{A}^T\mathbf{B}^T + \mathbf{B}\mathbf{A}\right)\mathbf{X}. \qquad (13.97)$$

$$\frac{\partial \operatorname{tr}\left(\mathbf{X}^{-1}\right)}{\partial \mathbf{X}} = -\left[\mathbf{X}^{-1}\mathbf{X}^{-1}\right]^T. \qquad (13.98)$$

$$\frac{\partial \mathrm{tr}\left(\mathbf{A}\mathbf{X}^{-1}\mathbf{B}\right)}{\partial \mathbf{X}} = -\left(\mathbf{X}^{-1}\mathbf{B}\mathbf{A}\mathbf{X}^{-1}\right)^T. \tag{13.99}$$

$$\frac{\partial \mathrm{tr}\left(\det\left(\mathbf{A}\mathbf{X}\mathbf{B}\right)\right)}{\partial \mathbf{X}} = \det\left(\mathbf{A}\mathbf{X}\mathbf{B}\right)\left(\mathbf{X}^{-1}\right)^T, \tag{13.100}$$

if $\mathbf{A}, \mathbf{B}, \mathbf{X}$ are nonsingular.

$$\frac{\partial \mathrm{tr}\left(e^{\mathbf{X}}\right)}{\partial \mathbf{X}} = e^{\mathbf{X}^T}. \tag{13.101}$$

$$\frac{\partial \mathrm{tr}\left(\log(\mathbf{I}_n - \mathbf{X})\right)}{\partial \mathbf{X}} = -((\mathbf{I}_n - \mathbf{X})^{-1})^T, \tag{13.102}$$

where $\mathbf{X}$ is $n \times n$ matrix, whose eigenvalues are smaller than one in absolute value. Here by definition,

$$\log(\mathbf{I}_n - \mathbf{X}) = -\sum_{k=1}^{\infty} \frac{1}{k}\mathbf{X}^k. \tag{13.103}$$

13.1.9 Important Inequalities

The Cauchy-Schwartz inequality

$$(\mathbf{x}^T\mathbf{y})^2 \leq (\mathbf{x}^T\mathbf{x})(\mathbf{y}^T\mathbf{y}) \tag{13.104}$$

with equality if and only if $\mathbf{x}$ and $\mathbf{y}$ are linearly dependent.

$$\left(\sum_i x_i y_i\right)^2 \leq \left(\sum_i x_i^2\right)\left(\sum_i y_i^2\right). \tag{13.105}$$

$$(\mathbf{x}^T\mathbf{A}\mathbf{y})^2 \leq (\mathbf{x}^T\mathbf{A}\mathbf{x})(\mathbf{y}^T\mathbf{A}\mathbf{y}), \tag{13.106}$$

if $\mathbf{A}$ is positive semi-definite, with equality if and only if $\mathbf{A}\mathbf{x}$ and $\mathbf{A}\mathbf{y}$ are linearly dependent.

$$(\mathbf{x}^T\mathbf{y})^2 \leq (\mathbf{x}^T\mathbf{A}\mathbf{x})(\mathbf{y}^T\mathbf{A}^{-1}\mathbf{y}), \tag{13.107}$$

if $\mathbf{A}$ is positive definite, with equality if and only if $\mathbf{x}$ and $\mathbf{A}^{-1}\mathbf{y}$ are linearly dependent.

$$\left(\int_\Omega fg\,dv\right)^2 \leq \int_\Omega f^2\,dv \int_\Omega g^2\,dv \tag{13.108}$$

with equality if and only if f and g are linearly dependent. The Lagrange inequality

$$\left(\sum_i x_i^2\right)\left(\sum_i y_i^2\right) - \left(\sum_i x_i y_i\right)^2 \leq \sum_{i<j}(x_i y_j - x_j y_i)^2. \tag{13.109}$$

Holder's inequality

if $x_i, y_i \geq 0$, $i = 1, 2, \ldots, n$ and $\left(\frac{1}{p}\right) + \left(\frac{1}{q}\right) = 1$, $p > 0$

$$\sum_i x_i y_i \leq \left(\sum_i x_i^p\right)^{\frac{1}{p}} \left(\sum_i y_i^q\right)^{\frac{1}{q}}, \tag{13.110}$$

if $\frac{1}{p} + \frac{1}{q} + \frac{1}{r} + \ldots \leq 1$

$$\sum_i x_i y_i z_i \ldots \leq \left(\sum_i x_i^p\right)^{\frac{1}{p}} \left(\sum_i y_i^q\right)^{\frac{1}{q}} \left(\sum_i z_i^r\right)^{\frac{1}{r}} \ldots \tag{13.111}$$

Minkowski's inequality

$$\left[\sum_i (x_i + y_i)^k\right]^{\frac{1}{k}} \leq \left(\sum_i x_i^k\right)^{\frac{1}{k}} + \left(\sum_i y_i^k\right)^{\frac{1}{k}}. \tag{13.112}$$

Hadamard's inequality for complex matrix $\mathbf{B}$

$$\det(\mathbf{B})^2 \leq \prod_{j=1}^{n} \left(\sum_{k=1}^{n} b_{ik}^2\right) \tag{13.113}$$

$$\det(\mathbf{B}) \leq 1 \text{ if } \sum_{k=1}^{n} b_{ik}^2 = 1, \quad i = 1, 2, \ldots, n \tag{13.114}$$

$$\det(\mathbf{B}) \leq M^n n^{\frac{n}{2}} \text{ if } |b_{ij}| \leq M. \tag{13.115}$$

If $\mathbf{B}$ is positive semi-definite, then

$$\det \mathbf{B} \leq \prod_{i=1}^{n} b_{ii}. \tag{13.116}$$

Jensen's inequality for convex function

$$E\{f(x)\} \geq f\left[E\{x\}\right]. \tag{13.117}$$

13.1.10 Inequalities in Information Theory

Let $\sum a_i$ and $\sum b_i$ be convergent series of positive numbers such that $\sum a_i \geq \sum b_i$ then

$$\sum a_i \log \frac{b_i}{a_i} \leq 0. \tag{13.118}$$

Further, if $a_i \leq 1$ and $b_i \leq 1$ for all i

$$2 \sum a_i \log \frac{a_i}{b_i} \geq \sum a_i (a_i - b_i)^2. \tag{13.119}$$

13.2 DISTANCE MEASURES

13.2.1 Geometric distance measures

The distance measures between two n-dimensional vectors $\mathbf{x} = [x_1, x_2, \ldots, x_n]^T$ and $\mathbf{y} = [y_1, y_2, \ldots, y_n]^T$ can be defined as follows:

Minkowski L_p metric, $p \geq 1$

$$d(\mathbf{x}, \mathbf{y}) = \left(\sum_{i=1}^{n} |x_i - y_i|^p \right)^{\frac{1}{p}} \tag{13.120}$$

L_2 Euclidean metric

$$d(\mathbf{x}, \mathbf{y}) = \sqrt{\sum_{i=1}^{n} (x_i - y_i)^2} \tag{13.121}$$

L_1 metric

$$d(\mathbf{x}, \mathbf{y}) = \sum_{i=1}^{n} |x_i - y_i| \tag{13.122}$$

Camberra metric

$$d(\mathbf{x}, \mathbf{y}) = \sum_{i=1}^{n} \frac{|x_i - y_i|}{x_i + y_i} \tag{13.123}$$

Chebyshev metric

$$d(\mathbf{x}, \mathbf{y}) = \max_{i=1,2,\ldots,n} |x_i - y_i| \tag{13.124}$$

City block metric, with given weights $a_i \geq 0$

$$d(\mathbf{x}, \mathbf{y}) = \sum_{i=1}^{n} a_i |x_i - y_i| \tag{13.125}$$

Quadratic metric, with given weighting symmetric positive definite matrix $\mathbf{Q}$

$$d(\mathbf{x}, \mathbf{y}) = (\mathbf{x} - \mathbf{y})^T \mathbf{Q} (\mathbf{x} - \mathbf{y}) \tag{13.126}$$

13.2.2 Distances between sets

Let X, Y be two sets in $\mathbb{R}^n$, and let $d(x, y)$ be a metric on $\mathbb{R}^n$. Then we have the following set of distances between X and Y:

Nearest neighbor distance

$$D(X,Y) = \min_{x \in X, y \in Y} d(x,y). \tag{13.127}$$

Farthest neighbor distance

$$D(X,Y) = \max_{x \in X, y \in Y} d(x,y). \tag{13.128}$$

Hausdorff metric, depending on a given norm $\|.\|$ in $\mathbb{R}^n$

$$D(X,Y) = \inf\{c \in \mathbb{R} : \; X \subseteq (Y + cN_{\|.\|}) \text{ and } Y \subseteq (X + cN_{\|.\|})\}, \tag{13.129}$$

where $N_{\|.\|}$ is the open unit ball in $\mathbb{R}^n$ (with respect to the norm $\|.\|$).

13.2.3 Discrimination measures

The information measures $S(\Omega/z)$, where $\Omega = \{X, Y\}$ is the set of classes and z a point in the space, express a measure of separation or discrimination between the classes X and Y at point z. These information measures also depend on:

1. The *a priori* probabilities $\Pr(X)$, $\Pr(Y)$ of the classes X and Y
2. The conditional probability density functions $p(z/X)$, $p(z/Y)$
3. The *a posteriori* probabilities $p(X/z)$ and $p(Y/z)$ of the classes, conditional on z
4. The combined probability density $p(z) = \Pr(X)p(z/X) + \Pr(Y)p(z/Y)$

The following is a list of major separation measures:

Shannon entropy

$$S(\Omega/z) = E\{-\Pr(X/z)\log\Pr(X/z) - \Pr(Y/z)\log\Pr(Y/z)\} \tag{13.130}$$

Average quadratic entropy

$$S(\Omega/z) = E\{\Pr(X/z)[1 - \Pr(X/z)] + \Pr(Y/z)[1 - \Pr(Y/z)]\} \tag{13.131}$$

Bayesian distance

$$S(\Omega/z) = E\{\Pr(X/z)^2 + \Pr(Y/z)^2\} \tag{13.132}$$

Kullback-Leibler divergence

$$S(\Omega/z) = E\left\{\left[\frac{\Pr(X/z)}{\Pr(X)} - \frac{\Pr(Y/z)}{\Pr(Y)}\right]\log\left[\frac{\Pr(Y)\Pr(X/z)}{\Pr(X)\Pr(Y/z)}\right]\right\} \tag{13.133}$$

Bhattacharyya coefficient

$$S(\Omega/z) = -\log(\rho), \tag{13.134}$$

where $\rho = \int [p(X/z)p(Y/z)]^{\frac{1}{2}}\, dx$, $0 < \rho < 1$.

Chernov bound for $0 \leq s \leq 1$

$$S(\Omega/z) = E\{\Pr(X/z)^s \Pr(Y/z)^{1-s}\} \tag{13.135}$$

Kolmogorov variational distance for $s > 0$

$$S(\Omega/z) = \frac{1}{2} E\{|\Pr(X/z) - \Pr(Y/z)|^s\} \tag{13.136}$$

Generalized Traubert distance measure

$$S(\Omega/z) = E\left\{\frac{\Pr(X/z)^n + \Pr(Y/z)^n}{\Pr(X/z)^{n-1} + \Pr(Y/z)^{n-1}}\right\}. \tag{13.137}$$

14

Glossary of Symbols and Abbreviations

Principal Symbols

$\mathbf{A} = [a_{ij}]$	matrix (mixing or state-space matrix)
a_{ij}	ij-th element of matrix $\mathbf{A}$
$\arg\max_{\theta} J(\theta)$	denotes the value of θ that maximizes $J(\theta)$
b_i	i-th element of vector $\mathbf{b}$
$\mathbf{D}$	diagonal scaling matrix
$\det(\mathbf{A})$	determinant of matrix $\mathbf{A}$
$\operatorname{diag}(d_1, d_2, \ldots, d_n)$	diagonal matrix with elements $d_1, d_2, \ldots, d_n$ on main diagonal
$d(n)$	desired response
$\mathbf{e}_i$	natural unit vector in ith direction
exp	exponential
$E\{\cdot\}$	expectation operator
E_x	expected value with respect to p.d.f. of $\mathbf{x}$
$\mathbf{f}(\mathbf{y}) = [f_1(y_1), \ldots, f_n(y_n)]^T$	nonlinear transformation of vector $\mathbf{y}$
$\mathbf{g}(\mathbf{y}) = [g_1(y_1), \ldots, g_n(y_n)]^T$	nonlinear transformation of vector $\mathbf{y}$

$\mathbf{H}$	mixing matrix
$\mathbf{H}^{-1}$	inverse of a nonsingular matrix $\mathbf{H}$
$\mathbf{H}^{+}$	pseudo-inverse of a matrix $\mathbf{H}$
$H(z)$	transfer function of discrete-time linear filter
$\mathbf{H}(z)$	matrix transfer function of discrete-time filter
$H(\mathbf{y}) = \langle \log p_y(\mathbf{y}) \rangle$	entropy
$\mathbf{I}$ or $\mathbf{I}_n$	identity matrix or identity matrix of dimension $n \times n$
Im( )	imaginary part of
j	$\sqrt{-1}$
$J(\mathbf{w})$	cost function
$KL(\mathbf{y}\|\|\mathbf{s}) = \left\langle \log \frac{p_y(\mathbf{y})}{p_s(\mathbf{y})} \right\rangle$	Kullback-Leibler divergence, relative entropy
$KL(p_y(\mathbf{y})\|\|p_s(\mathbf{s}))$	as above
log	natural logarithm
k	discrete-time or number of iterations applied to a recursive algorithm
n	number of inputs and outputs
N	data length
m	number of sensors
$p(x)$ or $p_x(x)$	probability density function (p.d.f.) of $\mathbf{x}$
$p_y(\mathbf{y})$	probability density function (p.d.f.) of $\mathbf{y}(k)$
$\mathbf{P}$	permutation matrix
$r_{xy}[\tau]$	cross-correlation function of discrete-time processes $x[n]$ and $y[n]$
$r_{xy}(\tau)$	cross-correlation function of continuous-time processes $x(t)$ and $y(t)$
$\mathbf{R_x}$ or $\mathbf{R_{xx}}$	covariance matrix of $\mathbf{x}$
$\mathbf{R_{xy}}$	covariance matrix of $\mathbf{x}$ and $\mathbf{y}$
$\mathbf{R_{fg}}$	correlation matrix between $\mathbf{f}(\mathbf{y})$ and $\mathbf{g}(\mathbf{y})$
$\mathbb{R}^M$	real M-dimensional parameter space
Re( )	real part

$\mathbf{s}$	vector of source signals
$s(t)$	continuous-time signal
$\mathbf{s}(k) = [s_1(k), \ldots, s_n(k)]^T$	vector of (input) source signals at k-th sample
$\mathbf{S}(z)$	Z-transform of source signal vector $\mathbf{s}(k)$
$\text{sign}(x)$	sign function ($= 1$ for $x > 0$ and $= -1$ for $x < 0$)
t	continuous time
$\text{tr}(\mathbf{A})$	trace of matrix $\mathbf{A}$
$\mathbf{W} = [w_{ij}]$	separating (demixing) matrix
$\mathbf{W}^H = (\mathbf{W}^*)^T$	transposed and complex conjugated (Hermitian) of $\mathbf{W}$
$\mathbf{W}(z)$	matrix transfer function of deconvoluting filter
$\mathbf{x}(k)$	observed (sensor or mixed) discrete-time data
$\|x\|$	absolute value (magnitude) of x
$\|\mathbf{x}\|$	norm (length) of vector $\mathbf{x}$
$\mathbf{y}$	vector of separated (output) signals
z^{-1}	unit-sample (delay) operator
$\mathcal{Z}$	z transform
$\mathcal{Z}^{-1}$	inverse z transform
δ_{ij}	Kronecker delta
η	learning rate for discrete-time algorithms
$\Gamma(x)$	Gamma function
λ_{max}	maximum eigenvalue of correlation matrix $\mathbf{R}$
λ_{min}	minimum eigenvalue of correlation matrix $\mathbf{R}$
$\mathbf{\Lambda}$	diagonal matrix
$\kappa_4(y)$	kurtosis of random variable y
$\kappa_p(y)$	p-th cumulant
μ	learning rate for continuous-time algorithms
$\mathbf{\Phi}$	cost function
$\varphi(y)$	activation function of a neuron, expressed as a function of input y
$\varphi_i(\cdot)$	nonlinear activation function of neuron i

σ^2	variance
$\mathbf{\Theta}$	unknown parameter (vector)
$\hat{\theta}$	estimator of θ
ω	normalized angular frequency; $0 < \omega \leq 2\pi$
$\triangle w_i$	small change applied to weight w_i
∇	gradient operator
$\nabla_{w_i} J$	gradient of J with respect to variable w_i
$\nabla_{\mathbf{W}} J$	gradient of cost function J with respect to matrix $\mathbf{W}$
$[\cdot]^+$	superscript symbol for pseudo-inversive of a matrix
$[\cdot]^T$	transpose
$[\cdot]^*$	complex conjugate
$[\cdot]^H$	complex conjugate, transpose
$\langle \cdot \rangle$	average operator
$\star$	convolution
$\widehat{}$	denotes estimator
$\otimes$	Kronecker product

Abbreviations

i.i.d.	independent identical distribution
cdf	cumulative density function
pdf	probability density function
BSE	Blind Signal Extraction
BSS	Blind Signal Separation
BSD	Blind Signal Deconvolution
CMA	Constant Modulus Algorithm
CLT	Central Limit Theorem
FIR	Finite Impulse Response
ICA	Independent Component Analysis
IIR	Infinite Impulse Response

ISI	Intersymbol Interference
LMS	Least Mean Squares
MCA	Minor Component Analysis
MBD	Multichannel Blind Deconvolution
MED	Maximum Entropy Distribution
MIMO	Multiple-Input, Multiple-Output
PAM	Pulse-Amplitude Modulation
PCA	Principal Component Analysis
QAM	Quadrature Amplitude Modulation
RLS	Recursive Least Squares
SIMO	Single-Input, Multiple-Output
SISO	Single Input, Single Output
SVD	Singular Value Decomposition
TLS	Total Least Squares
ETLS	Extended Total Least Squares
GTLS	Generalized Total Least Squares

Index